3rd Annual International Conference on Cloud Technology and Communication Engineering (CTCE 2019)

IOP Conference Series: Materials Science and Engineering
Volume 719

Wuhan, China
15 - 16 November 2019

Editor:

Simon K.S. Cheung

ISBN: 978-1-7138-2578-4
ISSN: 1757-8981

Printed from e-media with permission by:

Curran Associates, Inc.
57 Morehouse Lane
Red Hook, NY 12571

proceedings.com

Some format issues inherent in the e-media version may also appear in this print version.

Institute of Physics
Dirac House, Temple Back
Bristol BS1 6BE UK

Phone: 44 1 17 929 7481
Fax: 44 1 17 920 0979

techtracking@iop.org

Additional copies of this publication are available from:

Curran Associates, Inc.
57 Morehouse Lane
Red Hook, NY 12571 USA
Phone: 845-758-0400
Fax: 845-758-2633
Email: curran@proceedings.com
Web: www.proceedings.com

TABLE OF CONTENTS

CHAPTER 1: CLOUD COMPUTING

CHAPTER 2: BIG DATA

CHAPTER 4: WIRELESS COMMUNICATION

CHAPTER 5: ARTIFICIAL INTELLIGENCE

IOP Conf. Series: Materials Science and Engineering **719** (2020) 011001 doi:10.1088/1757-899X/719/1/011001

Preface

2019 the 3rd annual International Conference on Cloud Technology and Communication Engineering[CTCE 2019] has been held in Wuhan, China during November 15-16, 2019. Aiming at providing an excellent international academic forum for all the researchers and practitioners, the conference enjoys a wide spread participation among all over the universities and research institutes. Any paper and topic concentrates on related fields of Cloud Computing, Big data, Signal Processing, Wireless Communication and Artificial Intelligence are broadly welcomed. Besides that we also look forward to making the conference perfect circumstances to share resources, exchange opinions and inspire studying.

CTCE2019 proceeding tends to collect the most up-to-date, comprehensive, and worldwide state-of-art knowledge on cloud technology and communication engineering. All the accepted papers have been submitted to strict peer-review by 2-4 expert referees, and selected based on originality, significance and clarity for the purpose of the conference. The conference program is extremely rich, profound and featuring high-impact presentations of selected papers and additional late-breaking contributions. We sincerely hope that the conference would not only show the participants a broad overview of the latest research results on related fields, but also provide them with a significant platform for academic connection and exchange.

The Technical Program Committee members have been working very hard to meet the deadline of review. To conclude, we would like to express our sincere gratitude to all the TPC members and organizers for their hard work, precious time and endeavor preparing for the conference. Our deepest thanks also go to the volunteers and staffs for their long-hours work and generosity they've given to the conference. The last but not least, we would like to thank each and every of the authors, speakers and participants for their great contributions to the success of CTCE 2019.

CTCE2019 Organizing Committee
Wuhan, China

IOP Conf. Series: Materials Science and Engineering **719** (2020) 011001 doi:10.1088/1757-899X/719/1/011001

CTCE 2019 Committee

Editors

Prof. Simon K.S. Cheung, The Open University of Hong Kong, Hong Kong

Technical Program Committee

Prof. Chi-Wai Chow, National Chiao Tung University, Taiwan

Prof. Chuan-Ming Liu, National Taipei University of Technology, Taiwan

Dr. Xiao-feng Qiu, Beijing University of Posts and Telecommunications, Beijing, China

Prof. Qiao Guo, Beijing Institute of Technology, Beijing, China

Dr. Mohd Helmy Abd Wahab, Universiti Tun Hussein Onn Malaysia, Malaysia

Dr. Zhi-Xiao Wang, Xi'an University of Technology, Xi'an, China

Dr. Hua-Yi Lin, China University of Technology, Taiwan

Dr. Michel Albonico, Federal University of Technology, Paraná, Brazil

Dr. Ying Ma, Xiamen University of Technology, Xiamen, China

Dr. Yu-Lin Song, Asia University Taiwan

Prof. Samir Ladaci, National Polytechnic School of Constantine, Algeria

Dr. Omer Mohsin Mubarak, Jouf University, Kingdom of Saudi Arabia, Australia

Dr. Yong Ma, Wuhan University of Technology, Wuhan, China Dr. Daniele Codetta-Raiteri, Università di Torino, Italy

Dr. Zainab Ali Khalaf, Universiti Sains Malayisa, Basra, Iraq

Dr. Cheng-Yuan Chang, National United University, Taiwan

Prof. Loc Nguyen, Loc Nguyen's Academic Network, Vietnam

Prof. Bachir ACHOUR, University of Biskra, Algeria

Dr. Rocío Pérez de Prado, University of Jaen, Spain

Dr. Tomasz RAK, Rzeszow University of Technology, Poland

Prof. Farid ABED-MERAIM, Arts et Métiers ParisTech, France

Prof. Samir LADACI, National Polytechnic School of Constantine, Algeria

Dr. Fateh Mebarek-Oudina, Skikda University, Algeria

Dr. Jianwei Liu, China University of Petroleum, China

Dr. Jose Anand, KCG College of Technology, India

Dr. Rafal Drezewski, AGH University of Science and Technology, Poland

IOP Conf. Series: Materials Science and Engineering **719** (2020) 011001 doi:10.1088/1757-899X/719/1/011001

Peer review statement

All papers published in this volume of *IOP Conference Series: Materials Science and Engineering* have been peer reviewed through processes administered by the proceedings Editors. Reviews were conducted by expert referees to the professional and scientific standards expected of a proceedings journal published by IOP Publishing.

Research and Design of Road Intelligent Parking Based on IoT

Lei Wang[1,2,3], Kunqin Li[1,2], and Yong Luo[1,2]

[1]Computer and Information Engineering College, Guizhou University of Commerce, Guiyang, 550014, China

[2]Supported by Youth Science and Technology Talent Growth Project of Education Department of Guizhou Province: Qianjiaohe KY Zi [2018]271

[3]Email:460576177@qq.com

Abstract. Development of auto industry leads to parking difficulty and traffic jam in the city, which decreases operating efficiency of the city and happiness of people. Under the circumstances, parking space is planned on both sides of roads, by which the traffic jam is relieved to a certain extent. But some problem exists in management of the parking space as follows: Firstly, the management cost is high such as Human Resources investment. Secondly, rate of received fees (about 30 percent) is low because many car owners don't pay parking fees consciously. For the above problems, in this paper, the road intelligent parking request is firstly analyzed, and then an intelligent management and fee collecting solution for the road parking based on IoT is put forward. Besides, a management and control platform of road intelligent parking is designed. Finally key technologies for this system are analyzed. The purpose is to realize intelligent management, lower management cost, and higher fee collecting rate for the road parking.

1. Introduction

As rapid development of auto industry, population of private cars is larger and larger. So the parking difficulty is increasingly prominent. Because of lack of advanced managing method, traffic jam is becoming more and more serious, which decreases operating efficiency of the city and happiness of people [1][2][3][4][5][6]. Under the circumstances, parking space is planned on both sides of municipal roads, by which the traffic jam is relieved to a certain extent. But some problem exists during management for the parking space of municipal roads as follows: Firstly, the management cost is high such as Human Resources investment. Secondly, rate of received fees (about 30 percent) is low because many car owners don't pay parking fees consciously. Therefore, how to solve the above problems and improve management intelligence level is a problem faced in the management of municipal road parking space.

Fortunately, at present, the rapid development of Internet of Things (IoT) technology has become the technical support of intelligent parking in cities, which is the best way to solve the "parking difficulty". IoT is the third wave of information industry after computer and internet. IoT is a technology that connects all kinds of objects with the internet by information sensing device (such as RFID devices, sensors, cameras, etc.), various communication means (such as wireless, cable), and the agreed agreement. The purpose is to realize intelligent perception, identification, location, tracking and monitoring and management for the objects. The ultimate goal is to realize the communication between humans, humans and things, things and things [3]. In terms of intelligent parking, the technical architecture of the IoT is shown in Figure 1.

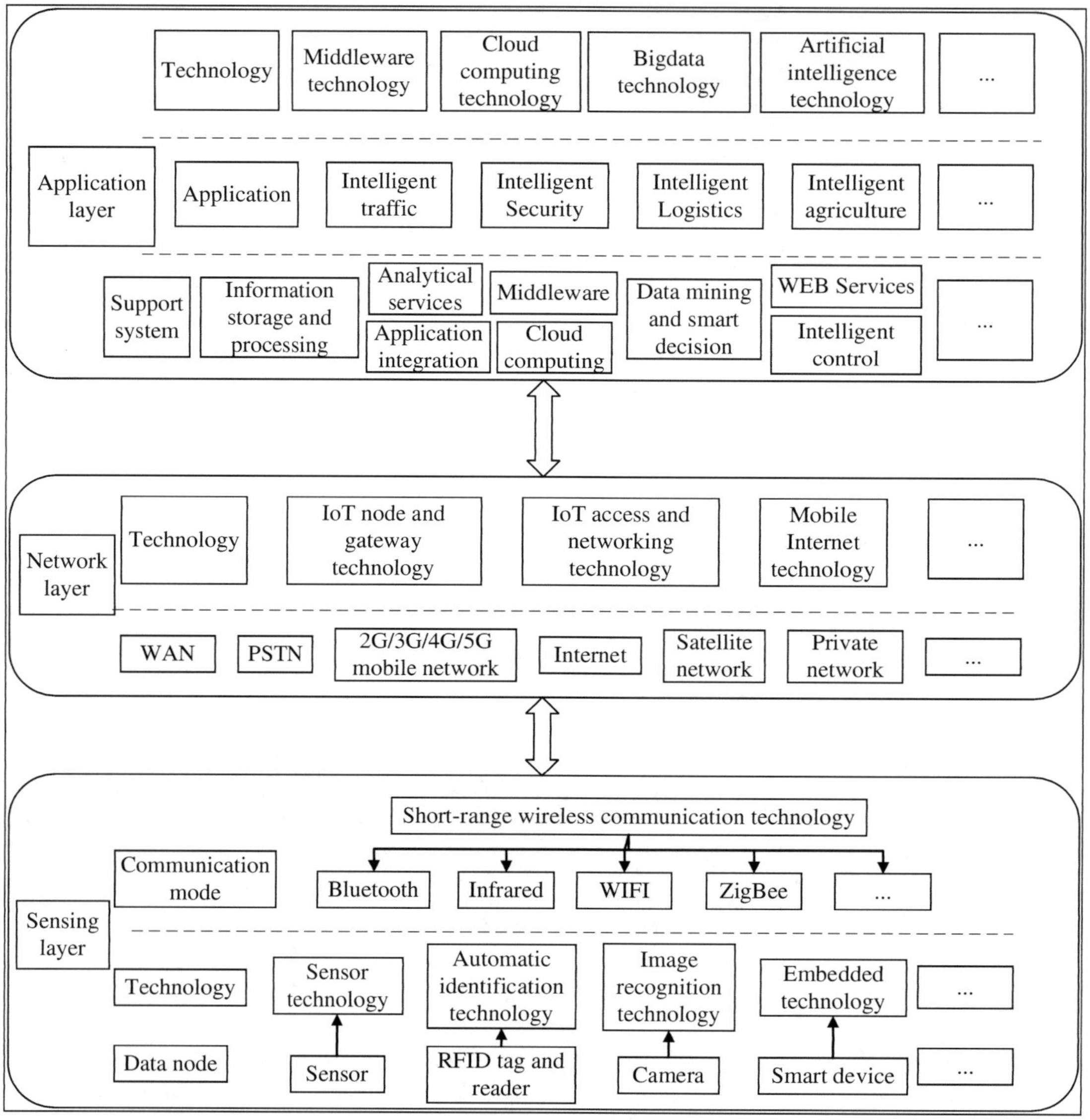

Figure 1. Technical architecture of the IoT

As shown in Figure 1, the function of the sensing layer is to comprehensively perceive and recognize cars and obtain the license plate information. The function of the network layer is to transfer information acquired by the sensing layer reliably to the application layer. Application layer is the deep integration of the IoT and industry professional technology, which can realize an intelligent parking industry.

2. Request Analysis of Road Intelligent Parking System

2.1. Market Size of Intelligent Parking

At present, there are more than 30 private cars for every 100 families, which accounts for about 92 percent of small passenger cars. In china, 49 cities had more than one million cars, 18 cities more than two million, and 6 cities more than three million by 2016. In Guizhou province, the car population is about 3.5 million by 2016. The above data is shown in Figure 2.

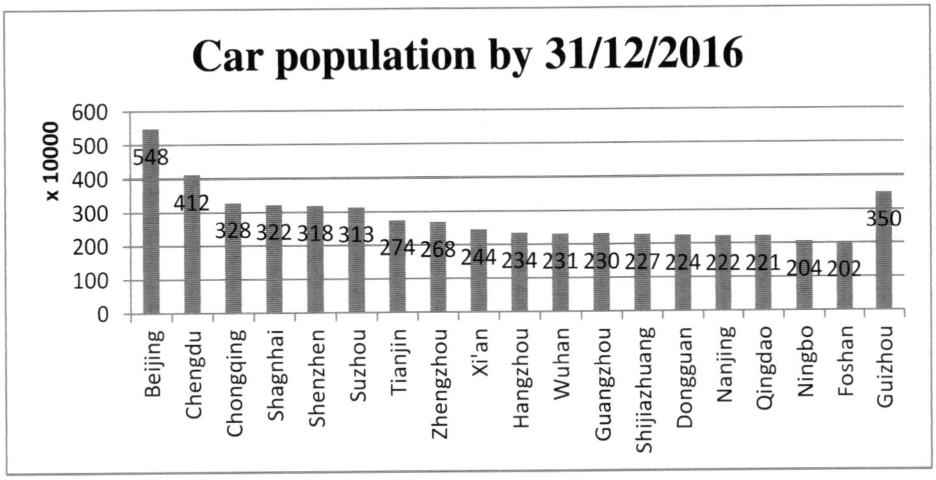

Figure 2. Car population by 2016

As shown in Figure 2, such a large car population means that a huge market size exists[4], as shown in Figure 3.

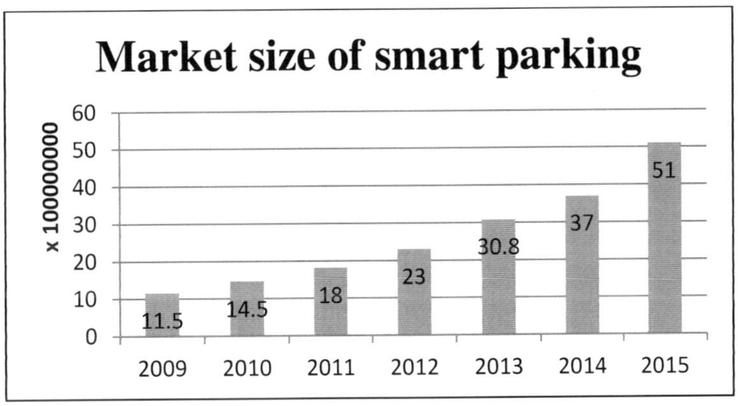

Figure 3. Market size of intelligent parking

As shown in Figure 3, with the development of economy and the improvement of people's living standard, the market size and request of intelligent parking increase year by year. So there will be a broad market prospect at the intelligent parking area.

On the other hand, in a city, new road infrastructure induces new traffic, but request tends to exceed supply. And parking lots are always in short supply in cities, especially in big cities[2][5][6]. So road parking space is planned in most streets of the city. Therefore, intelligent parking system for road parking space will have a broad market prospect.

2.2. Current Situation of Road Parking Management

At present, "intelligent parking" solution developed by "Songli group co. LTD" is widely used in municipal road parking management. The hardware is mainly composed of front-end video pile and mobile inspection end. Specific distribution of the front-end video piles is shown in Figure 4 [7].

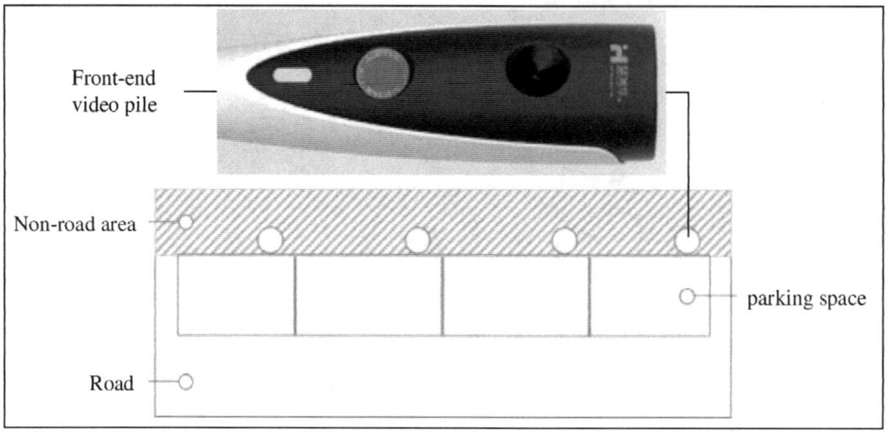

Figure 4. Distribution of the front-end video piles

As shown in Figure 4, the function of the front-end video pile is to take photos of license plate after a car entering parking space. Each video pile contains two cameras distributed at a certain angle, which can scan the license plates of two adjacent parking spaces through a certain layout method, so as to scan and identify all cars entering the parking spaces. The scanning mode is shown in Figure 5 [7].

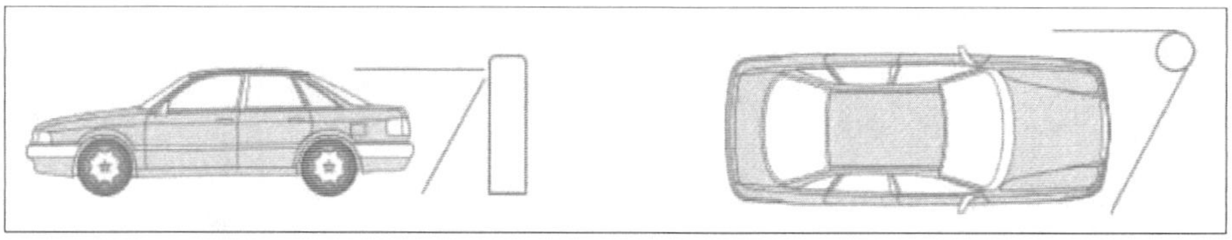

Figure 5. Scanning mode of the front-end video pile

As can be seen from Figure 5, in this solution, the lateral license plate recognition method is adopted, which can be used to recognize the license plate whether the car stops in the right direction or in the wrong direction [7].

Since there is no access restricting measures for the cars in this solution, manual management and charging are adopted at present. The specific implementation process is as follows: (1) each person is responsible for about 50 parking spaces; (2) the car enters freely. When the car leaves, the management personnel will hold the mobile inspection end and input the license plate for billing display. The owner can leave after paying the fee.

In this solution, there are the following problems: (1) high management cost. The annual management fee for each 50 parking spaces is about 30,000 yuan, so when the rate is low, the enterprise parking income is low; (2) when the owner leaves, if the manager does not find out, the charge will be delayed or missed, resulting in a low charging rate; (3) some car owners have a low consciousness of paying the fees. When managers collect fees, they refuse to pay for it, which will have a negative social impact and lead to a low charging rate.

3. Design of Intelligent Road Parking System

3.1. System Function

In view of the existing problems in the current municipal road parking management, a new solution is proposed in this paper, including some functions such as parking monitoring, parking guidance, cars identification, cars restriction, billing and payment and so on.

(1) Parking monitoring: Through the front video pile, the use of each parking space can be monitored and collected in real time in this system. It is convenient for parking space managers to know the details of parking space use in real time, and meanwhile provide the real-time parking space

remaining number for car owners to guide them to park smoothly.

(2) Parking guidance: On the basis of parking monitoring, statistical analysis of the remaining parking spaces will be made in real then released to the public. Before entering the road, the car owners will be informed of the number of remaining parking spaces in this section. They can independently choose appropriate parking space according to the prompt information of the system.

(3) Cars identification: After the car enters the parking space, the license plate can be imaged by the front-end video pile, and image data is transferred to the background computer through the network for calculation and processing. Finally license plate can be recognized by using a certain algorithm and character segmentation technology.

(4) Cars restriction: When the car enters the parking space, its license plate can be recognized by the system and then timed. After a certain period of time (free period stipulated by the system), the car will be restricted by controlling the actuator. When the owner needs to drive away, the actuator will automatically reset and release the car after paying the parking fee online.

(5) Billing and payment: When the car enters the parking space, the license plate is recognized, and the time and parking fee is automatically calculated. When the car owner needs to drive away, she can scan the QR code on the front-end video pile, input the license plate, or download the designated APP, bind the phone number and the license plate, after which the fee is billed. Car owners can pay and leave according to consumption.

In conclusion, this system aims to realize unattended intelligent municipal road parking management, standardize parking order, reduce enterprise management costs and increase the collecting rate.

3.2. Technical Solution of the System

STM32 is planned to be used as the control core to complete the automatic control and intelligent processing of the actuator motor in this system. The solution framework of the system is shown in Figure 6.

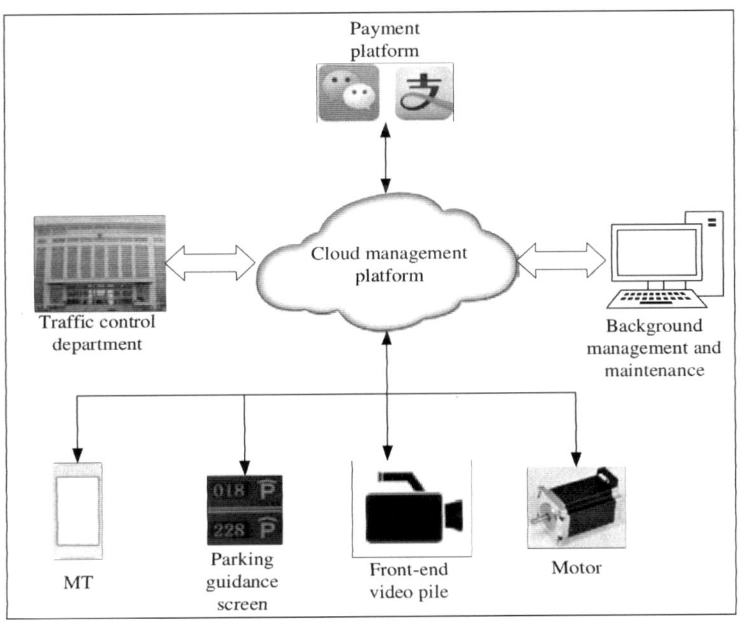

Figure 6. Solution framework of the system

As shown in Figure 6, the workflow of this system is as follows: (1) A parking guidance screen is placed near the road entrance to display the real-time remaining number of parking spaces; (2) When the car stops well, the license plate is imaged by the front-end video pile, and then transferred to data center for calculation and recognition, and start timing;(3) After a certain period of time (free time set), the background computer sends control command to STM32 through the gateway to control the actuator to restrict the car;(4) When the car leaves, on one hand, the owners can download and install the corresponding intelligent parking management APP, bind mobile phone and license plate, then she can query the parking fee and pay for it in the APP. On the other hand, she can scan the QR code (WeChat or Alipay) on the front-end video pile, and then enter the license plate number to complete payment. After the payment finished, the background sends command to the STM32 through gateway to control the actuator are to reset and release the car.

3.3. Intelligent Parking Management Platform Design

According to the actual situation of intelligent parking on municipal roads designed in this paper, the following objectives should be satisfied in the design of the management platform :(1) the interface of the system should be beautiful and easy to be operated;(2) query function based on license plate and parking space should be supported;(3) the function of intelligent charging should be realized; (4) the platform should be simple to be maintained and later updating service should be supported. According to the system function in section 3.1 and solution framework in section 3.2, function and structure of the management platform is shown in Figure 7.

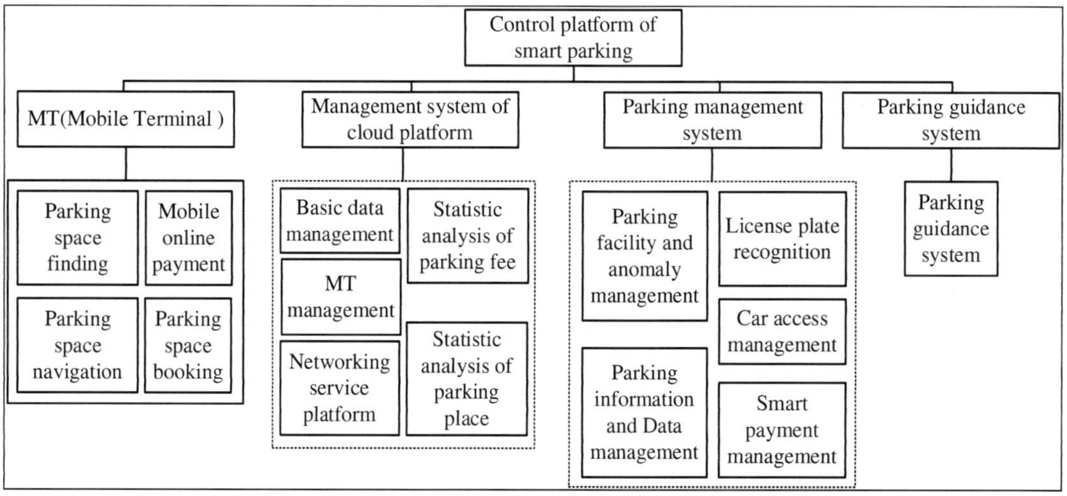

Figure 7. Function and structure of the management platform

4. Key Technologies of Road Intelligent Parking

According to the function and structure of the management platform, the required key technologies are introduced in the following section.

4.1. License Plate Recognition Technology

License plate recognition technology mainly consists of license plate location, character segmentation and character recognition. The recognition effect depend on the above three parts. The identification flow is shown in Figure 8.

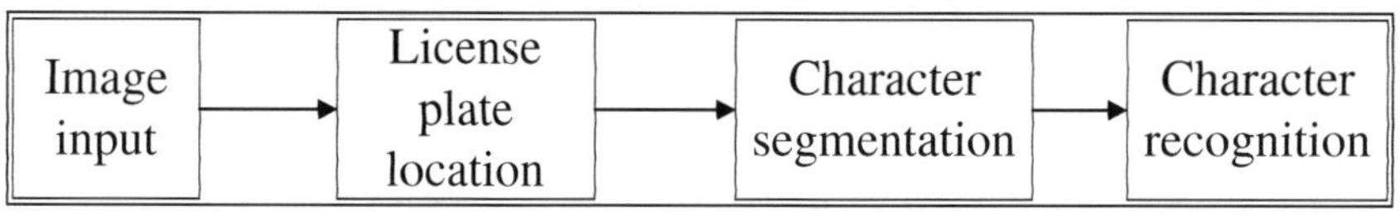

Figure 8. License plate identification flow

In Figure 8, the methods of license plate location mainly include the method based on the variation feature of horizontal gray scale, the method based on edge detection, the method based on the color feature of license plate, and the method based on mathematical morphology. Character segmentation algorithms mainly include recognition based segmentation algorithm, direct segmentation algorithm and adaptive segmentation and clustering algorithm. The character recognition algorithms mainly include template matching, neural network, support vector machine, Bayes classifier and feature matching.

4.2. Information Transmission Technology

In intelligent parking, data of the sensing layer, such as the license plate number, entry time and image, are transmitted to the application layer through the network layer. The information transmission can be divided into wire transmission and wireless transmission (such as ZigBee, Bluetooth, Wi-Fi, etc.). According to the technical implementation, it can be divided into: fiber optic communication technology, Wi-Fi technology, WIMAX technology, 2G/3G/4G/5G communication technology, etc.

Among them, faced with more and larger data transmission volume in the future, 5G technology has great advantages due to its higher transmission rate and lower delay. The characteristics of 5G technology are as follows.

(1) Millimeter wave: with the millimeter wave, ultra-high speed transmission rate, narrow beam can be supported, and a large number of devices can be connected. Besides it can be controlled flexibly. (2) Micro base station: due to the small coverage of millimeter-wave, more base stations are needed, and the use of micro base station can reduce the cost. (3) Massive MIMO: the more antennas there are, the higher the spectrum efficiency and transmission reliability. (4) Beam forming: an antenna array is arranged on the base station. By controlling the phase of the radio frequency signal, the wave lobe of the electromagnetic wave after interaction becomes very narrow and points to the serviced mobile phone, and also the direction of it can change according to the mobile phone's movement. (5) D2D: in 5G era, if two users in the same base station communicate with each other, the data will no longer be forwarded through the base station, but directly from mobile phone to mobile phone.

4.3. Self-service Payment Technology

On one hand, the self-service payment of parking fee helps saving the time of entering and leaving the parking space and greatly alleviates the parking queue problem. On the other hand, it can help realizing the unattended parking, reducing the human cost, and realizing the intelligent management and operation of parking.

According to the system functions and workflow described in sections 3.1 and 3.2, the principle of self-service payment in this system is shown in Figure 9.

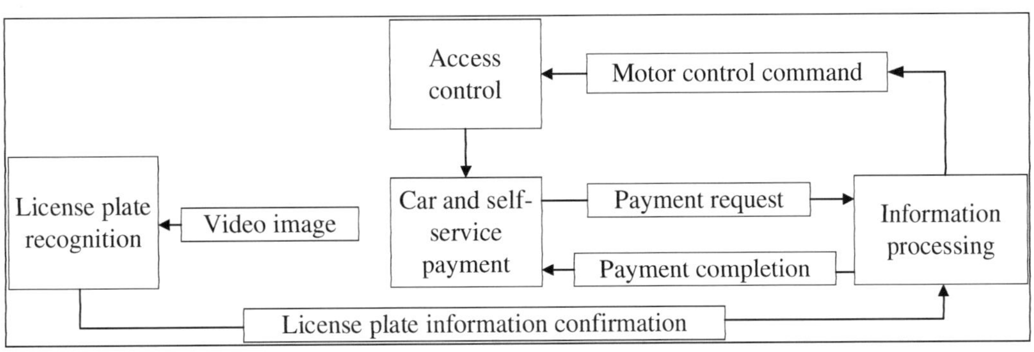

Figure 9. Self-service payment principle

5. Conclusion

In this paper, the existing problems of municipal road parking management and intelligent parking system architecture based on the IOT are firstly analyzed. Secondly, the market request of intelligent parking industry in China is analyzed, and also the current municipal road parking management system solution and the existing problems are introduced. Thirdly, a new solution is proposed, and its workflow and management platform are designed.

With the proposed solution, the following goals can be achieved :(1) to achieve intelligent management of municipal road parking space;(2) to save human resources and reduce management costs and Generate revenue for the management company;(3) to restrain car owners from not paying parking fees and increase the collecting rate;(4) to regulate the road parking order and alleviate urban traffic jam.

6. References

[1] Zhizheng Chen. Research of Parking Lot System[D]. Dalian Maritime University, 2013.
[2] Songke Yu. Investigation of Intelligent Self-Service Parking System[D]. Chengdu University of Technology, 2016.
[3] Suhong Zhang. Research on Intelligent Parking System Based on Internet of Things[D]. Yanshan University, 2016.
[4] Hang Li. Research on the Key Technologies for Intelligent Park Based on Wireless Network[D]. Xinhua University, 2014.
[5] Haijun Meng. Design and Realization of Intelligent Parking Solution[D]. Dalian University of Technology, 2013.
[6] Min Zhang. Research on Vehicle Localization and Parking Guidance in Large Scale Parking Lot[D]. Tianjin University, 2014.
[7] Songli holding group co. LTD. http://www.wit-parking.com.

How to Implement the Commodity Publishing System in Ecommerce Platform

Lijun Li and Hong He*

School of Mechanical, Electrical & Information Engineering, Shandong University, Weihai, China
*Corresponding author
Email: hehong@sdu.edu.cn

Abstract. The Commodity publishing system is one of the most important software component of the ecommerce platform system. A simple and practical publishing system is necessary for the ecommerce platform system blooming everywhere. We describe how to design a commodity publishing system, including the system design features, the database design and software implementation. The method is helpful for software engineers.

1. Introduction

An ecommerce platform includes two core systems, namely, the mall system and the commodity management system. Consumers can choose and buy commodities in the mall system. The foundation of the mall system is the commodity management system [1]. The commodity publishing system is of great importance in the commodity management system, which belongs to a module of modular design of ecommerce platform system [2]. This article describes how to design a commodity publishing system, including the design features of the commodity publishing system, the database design and the implementation of the commodity publishing system [3-4]. The publishing system we designed is simple and practical. It could be used in general development of ecommerce platform systems.

2. Design Features of the Commodity Publishing System

The commodity publishing system described following has four distinctive designs.

(1) This system can not only release domestically produced goods, but also publish imported goods. When the goods are released, you need to choose the "Is import" attribute. If you click Yes, you need to fill in the additional import attributes, including the presence or absence of the country of origin certificate, the customs clearance status, the import port, the customs declaration number, the delivery time and the import of the import documents.

(2) A drop-down list or a multi-select box. If the seller user feels that the attribute selection value provided by the system cannot meet its needs, you can customize the added attribute. For the drop-down list, if the existing value of the system is insufficient to meet the user's needs, the user can select "Other". After the selection, an input box will appear next to the drop-down list, and the user can input the attribute value that he wants to select in the input box. The same is true for the multi-box function. Click the "Add Attribute" button after the checkbox to allow the user to customize the attributes that you want to select.

(3) The system has designed two kinds of quotation methods, one is to quote according to specifications, and the other is to quote according to quantity to meet the different needs of different types of goods. For example, the mobile phone may be quoted according to the specifications, and the

mobile phone price of the 4G memory mobile phone and the 8G memory mobile phone are different. Although the different sizes of clothes are the same price, purchasing 50 pieces may get more discounts than buying 10 pieces of clothing. At this time, you need to quote according to the quantity.

- Specification quotation. The user needs to fill in the batch, that is, this product can be purchased from a few pieces, for example, one piece, and need to fill in the price corresponding to different specifications, as well as the stock and weight of different specifications of the goods.
- Quantity quote. The user needs to fill in the price range, the price of each price range, and the item number, inventory, and weight of each specification.

(4)The system not only supports user-defined values of added attributes, but also supports user-defined addition of up to three attributes, custom attribute names and attribute values.

3. Commodity Publishing System Database Design

The database logic design of the commodity publishing system consists of six tables, including cate_id, cate_attr, attr_id, attr_value, goods, goods_attr. The cate_id table stores the classification of goods. The categories of goods in this system use three-level classification. The first level is a large category, such as daily necessities, office supplies, etc., and the second category is a small category subdivision under a large category, for example, there are secondary classifications such as clocks and cups under the department store category. The third-level category is specific to a certain commodity. For example, there are specific categories of cups, such as thermos cups and coffee cups. The cate_attr table stores the class attributes of each specific three-level class, that is, information about which attributes need to be added when the item is published. The attr_id table stores all attribute names and attribute ids. The attr_value table stores the values corresponding to the attributes under each category. Here, the pre-selected values of the attributes are stored, and which value is selected when the user adds the items is stored in the goods_attr table. After the user adds the commodity information in the system, the data is mainly stored in the goods table and the goods_attr table. The goods table stores the public information of each commodity, including the category number, product number, item number, inventory, price, etc., and the goods_attr table stores the value of the unique attribute of each item, including the product number, attribute id, and value. The conceptual design of the released commodity system is shown in Figure 3-1 and Figure 3-2.

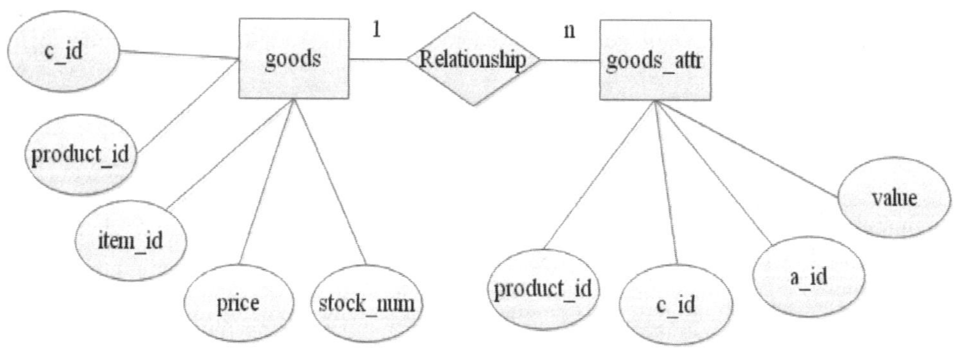

Figure 3-1. Commodity storage ER diagram

IOP Conf. Series: Materials Science and Engineering **719** (2020) 012002 doi:10.1088/1757-899X/719/1/012002

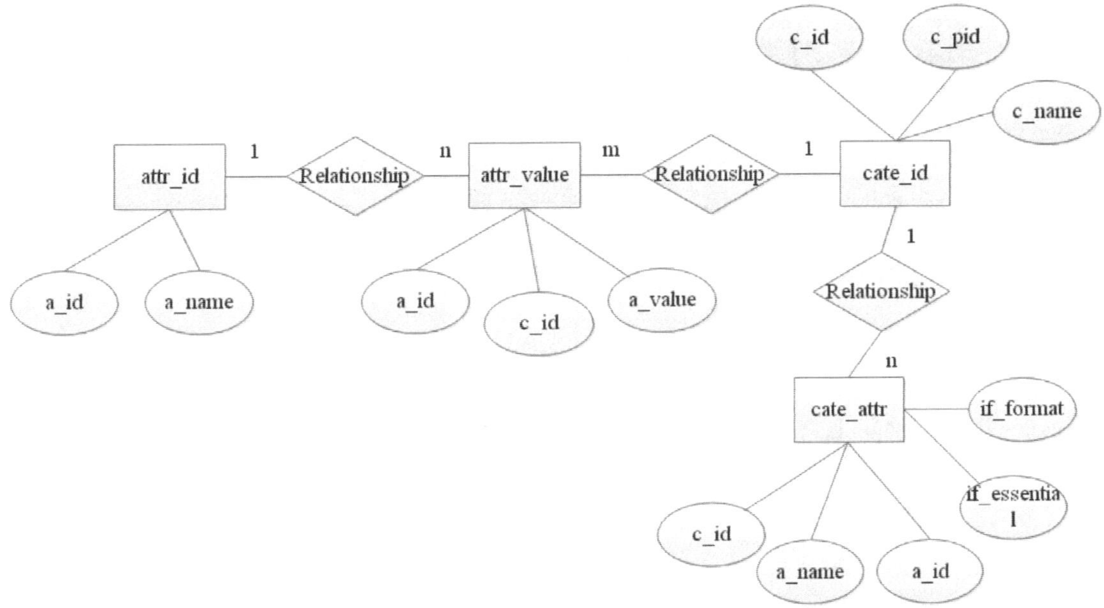

Figure 3-2. Commodity publishing ER diagram

4. Commodity Publishing System Implementation

The foreground of the system adopts the Bootstrap framework, the background uses Java language, the data storage uses MySQL database, and the development framework used is Spring. Spring MVC and MyBatis framework. The Spring architecture can be divided into Dao layer, Service layer and Web layer. The database and business logic and presentation layer are separated, high cohesion and low coupling, which increases the flexibility of design, improves the reuse rate of code, and enables the system to open to the expansion [5-6]. The MyBatis framework encapsulates the code for database connections, enabling users to interact with the database more efficiently. The system page uses the FreeMaker template engine to speed up page rendering. System login and privilege management use the Spring Security framework, which can sets different system roles, such as buyer users, seller users, administrators, etc. Different roles have different access rights to the system [7-8]. The specific implementation is as follows.

4.1. Select the Category of the Item You Want to Publish

The first choice is to choose the first category, and then the second category. After the first two categories are selected, the third category is specific to a small commodity category. After selecting the three-level category, you can click the button with Next to fill in the details. If you don't know which category of the commodity you want to publish, you can enter the keyword of the commodity category to be published in the search field, click the search button, you can directly search for the third-level category, and after searching for the result, select the corresponding class. For example, select the first category of office stationery and the secondary category of writing stationery and the third category of pens, as shown in Figure 4-1.

IOP Conf. Series: Materials Science and Engineering **719** (2020) 012002 doi:10.1088/1757-899X/719/1/012002

Figure 4-1. Select the commodity category page

4.2. Fill in the Commodity Details

After selecting the category, fill in the details of the commodity of a certain category, and the attribute marked with an asterisk is a required attribute. The following is an example of publishing a pen to introduce the implementation of the system, as shown in Figure 4-2.

Figure 4-2. Fill in the commodity details page

For the four design features given earlier, we forward a detailed implementation here.

IOP Conf. Series: Materials Science and Engineering **719** (2020) 012002 doi:10.1088/1757-899X/719/1/012002

- Fill in import attributes for cross-border goods. If you need to publish imported goods, if you choose “Yes” in the import attribute, there will be import attributes that need to be filled out, as shown in Figure 4-3.

Figure 4-3. Fill in the import properties page

- Customize the drop-down list property and the value of the checkbox. For example, for the material property, if the user feels that the material provided by the system is not the material he wants, he can select other first. After selecting other, an input box will appear next to the drop-down box of the material, and the user can fill in the actual material of the product in this input box. The multi-select box can also be customized to add attribute values. Click the Add button to customize the value of the added attribute
- Two quotations are implemented. If you choose the specification quotation, you need to fill in the price under different specifications.

If you choose to quote by quantity, you need to fill in the correspondence between the price and the quantity range. Figure 4-4 shows that at least three pens can be purchased, the unit price of more than or equal to three pens less than 10 is 35 yuan, the unit price of more than or equal to 10 pens less than 50 is 33 yuan, and the unit price of more than or equal to 50 pens is 30 yuan.

Figure 4-4. Quantity quote realization

• To customize the attribute name and attribute value of the commodity. Click the "Custom Add Product Attributes" button to customize the added attribute name and attribute value.

In addition, the system provides a rich text box that allows the user to customize the details of the commodity. At the bottom of the page, the user can also choose which of the custom store categories. Finally, click on the "Add Item" button to enter the published commodity information into the system.

5. Conclusions

The methods of implement the commodity publishing system in ecommerce platform are from the real software engineering. Maybe there are better methods. We hope the article will be helpful for the new engineers who are trying basic development.

6. References

[1] Roberto Aringhieri, Davide Duma, Vito Fragnelli. Modeling the rational behavior of individuals on an ecommence system. Operations Research Perspectives, Volume 5, 2018, Pages 22-31.

[2] Emmanuel Francalanza, Mark Mercieca, Alec Fenech. Modular System Design Approach for Cyber Physical Production Systems. Procedia CIRP, Volume 72, 2018, Pages 486-491.

[3] Umut Tosun. Distributed Database Design: A Case Study [J]. Procedia Computer Science, Volume 37, 2014, Pages 447-450.

[4] Sahithi Tummalapalli, Venkata rao Machavarapu. Managing Mysql Cluster Data Using Cloudera Impala [J]. Procedia Computer Science, Volume 85, 2016, Pages 463-474.

[5] Enrico Fregnan, Tobias Baum, Fabio Palomba, AlbertoBacchelli. A survey on software coupling relations and tools [J]. Information and Software Technology, Volume 107, March 2019, Pages 159-178.

[6] Dragos-Paul Pop, Adam Altar. Designing an MVC Model for Rapid Web Application Development. Procedia Engineering, Volume 69, 2014, Pages 1172-1179.

[7] Andrii Boiko, Vira Shendryk. System Integration and Security of Information Systems. Procedia Computer Science, Volume 104, 2017, Pages 35-42.

[8] Zoran Vučković, Dragan Vukmirović, Marina Jovanović Milenković, Slobodan Ristić, Katarina Prljić. Analyzing of ecommence user behavior to detect identity theft. Physica A: Statistical Mechanics and its Applications, Volume 511, 1 December 2018, Pages 331-335.

Design of Positioning System for Glass Plate Printer

Ping Xu [1,*] **and Wenxuan Lai**[1]

[1] School of Electromechanical and Architectural Engineering, Jianghan University, Wuhan 430056, China
1579356280@qq.com

Abstract. Glass plate printing has been widely used in production and life. However, for fragile glass plate, high precision positioning system is the key in the printing process. In this paper, a motor driving and positioning system consisting of operating mechanism, positioning mechanism and lifting mechanism is proposed. The structure and design scheme of each mechanism are given. Finally, the finite element analysis and stress checking of key parts are carried out. The results provide a new idea for the design of similar systems and have a good reference value

1. Introduction

Printing press is a machine for printing words and images. China was the first country to invent printing press. The invention and development of printing press play an important role in the spread of civilization and human culture. For glass plate printing machine, before glass plate printing, it is necessary to carry out the process of feeding, positioning and dust removal. Because the glass plate is light and fragile, the traditional mechanical positioning clamping device cannot meet the requirements. Therefore, this paper proposes a positioning system for glass plate printing press, which makes the positioning process more convenient, fast and efficient.

2. Scheme Design

For any workpiece, whether in the formulation of its processing technology or in the design of its fixture, the positioning problem is very important. It involves what positioning elements should be used for positioning, and how to analyze and calculate positioning errors, so as to take measures to control the magnitude of the control errors and meet the requirements of workpiece processing technology. In the design process of glass plate printing press, the planning of positioning scheme is an important link. The quality of product processing and the success or failure of fixture design are directly affected by positioning scheme planning.

According to the requirement, the glass plate with large area and thin thickness should be positioned in the design. The requirements for positioning system are:

(1) Sufficient accuracy.
(2) Good wear resistance.
(3) Adequate strength and stiffness.
(4) Good Technology.

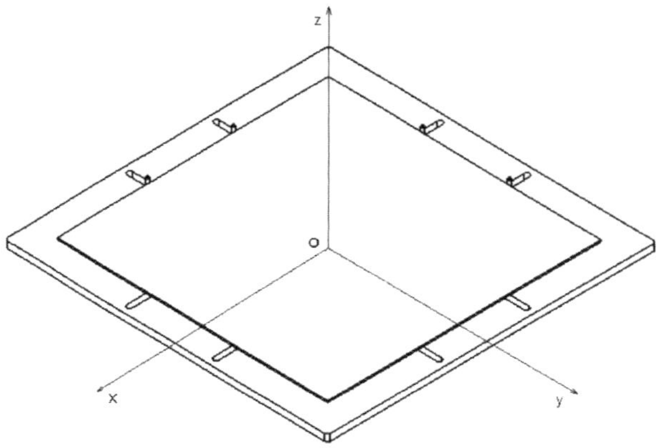

Figure 1. Position platform

Any workpiece in space position is uncertain. If there is no restriction on it, it can move in any direction or rotate along any axis. The possibility of motion of this kind of workpiece is called the degree of freedom of the workpiece. Placing the workpiece in the space rectangular coordinate system, the workpiece has six degrees of freedom, namely, the degree of freedom of movement along three coordinate axes (x axis, y axis, Z axis) and the degree of freedom of rotation around three coordinate axes. Taking the centre of the position platform as the origin, the space rectangular coordinate system shown in Fig. 1 is established. The positioning process of glass plate should limit all six degrees of freedom of glass plate. Among them, the initial positioning bottom plate as a plane can hinder the glass plate moving along Z axis and rotating around X axis and Y axis, limiting three degrees of freedom. The rest of the degrees of freedom can be accomplished in combination with the initial positioning of the bottom plate, the column, the bottom plate and other components.

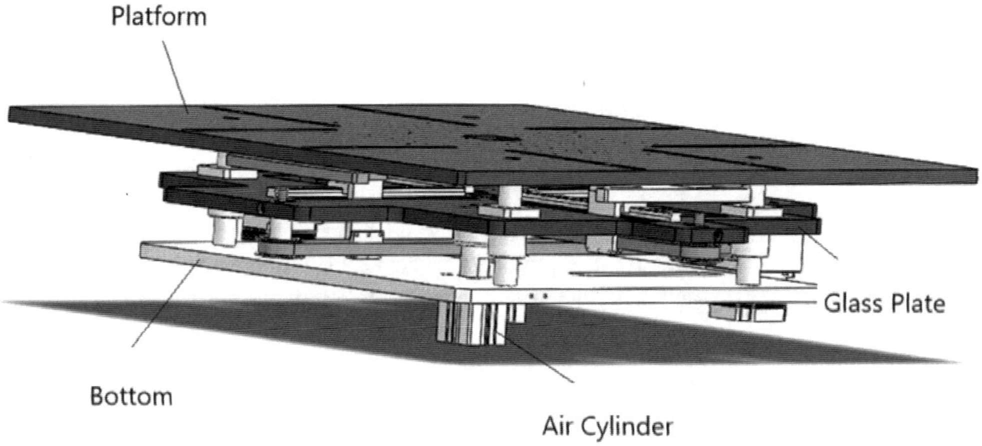

Figure 2. Position System

In the printing process, the feeder places the glass plate on the position platform, the glass plate triggers the photoelectric switch in the platform centre, the PLC commands the motor driver to make the motor move, and the positioning column in two directions (X direction and Y direction) moves toward the glass plate through the synchronous toothed belt in the motion mechanism, so that the glass plate is located in the platform centre location. Subsequently, the vacuum generator acts by adsorbing the glass plate on the platform through the air holes on the platform to complete the initial positioning function; at the same time, the cylinder fixed on the bottom plate acts, and the piston rod connecting the initial positioning bottom plate drops, so that the positioning rod falls below the platform, and the positioning column returns to its original position under the action of synchronous belt. The glass plate is in the process of dust removal. After dust removal, the manipulator takes the glass plate off and puts it on the printing platform, and the positioning device returns to its original position, waiting for the next positioning work. As shown in Figure 2.

The function of vertical lifting mechanism is lifting tray. After the manipulator grabs the workpiece on the flat belt conveyor, the feature points of the workpiece are identified by the camera, and then the workpiece is put into the pallet. For each workpiece, the motor drives the ball screw to drop one unit vertically, waiting for another workpiece until the tray is full. The tray with full workpiece is sent out by the transverse discharging mechanism. The linear module of ball screw is driven by stepping motor, which can achieve better servo control. Because of the need to achieve multi-point residence and infinite speed regulation, the ball screw nut actuator driven by stepping motor is selected.

3. Mechanism Composition

The positioning system is mainly composed of operating mechanism, positioning mechanism and lifting mechanism. The function of the operating mechanism is to complete the opposite movement of the positioning column in the X direction and the Y direction. The operation mechanism is mainly composed of synchronous toothed belt and two positioning columns. The synchronous belt and the guide rail are mounted on the initial positioning base plate, and the two positioning columns are mounted on both sides of the belt and are located in the centrosymmetrical position. When the motor rotates, the synchronous drive drives the positioning column to move in opposite direction on the guide rail. There are three photoelectric sensors installed on the initial positioning base plate, two of which play a limiting role and one is used as the origin of calculation. The induction plate is mounted on one side of the positioning column, as shown in Figure 3.

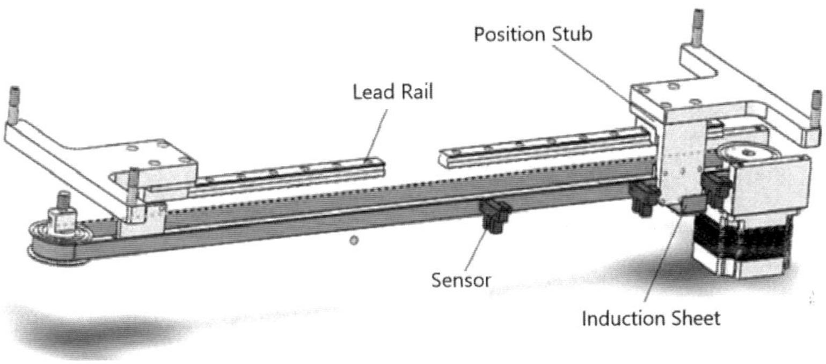

Figure 3. Operating Mechanism

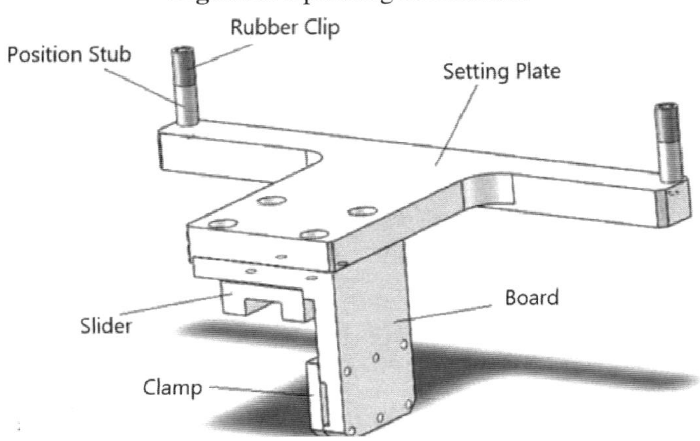

Figure 4. Location mechanism

The positioning mechanism consists of a splint, a press plate, a slider, a fixing plate of a positioning rod and two positioning rods. Each part is connected by a screw. According to the minimum specification of glass plate, the distance between two positioning rods can be determined. In order to protect the glass plate from collision, the positioning rod is equipped with rubber chucks. The positioning column is fixed on the synchronous belt by splint and pressure plate. The positioning column slides on the guide rail under the synchronous drive to complete the positioning process, as shown in Figure 4.

The function of the lifting mechanism is to complete the lifting and lifting of the operating mechanism so as to ensure that the positioning process does not interfere with other processes. Before discharging, the lifting mechanism is in the lifting state, and the positioning rod exposes the positioning platform to prevent the glass plate from being outside the positioning range of the positioning rod. After the positioning is completed, the cylinder action reduces the operating mechanism to prevent interference with the following dust removal process. After the dust removal is completed, the operating mechanism is lifted up to prepare for the next positioning. The lifting mechanism consists of bottom plate, initial positioning bottom plate, two cylinders and equal height supporting flange, four columns and connecting flange. Among them, the cylinder is installed on the bottom plate, and the flange of equal height support is fixed on the initial positioning bottom plate. It cooperates with the piston rod of the cylinder to support the initial positioning bottom plate and rectify the operation mechanism. Connecting flanges are installed on the initial positioning base plate to enable it to slide on the column installed on the base plate, so as to achieve a smooth lifting motion, as shown in Figure 5.

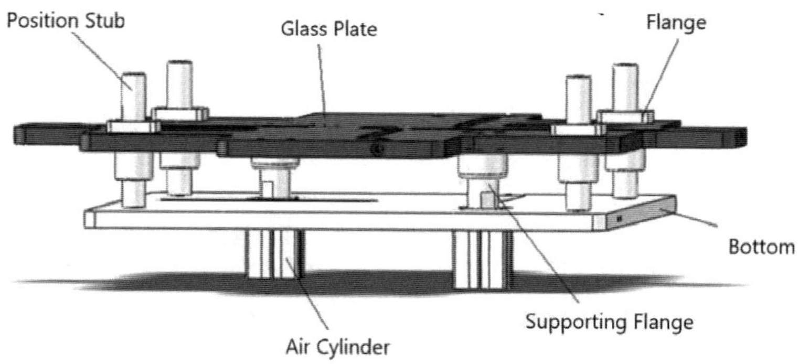

Figure 5. Lifting Mechanism

4. Stress Analysis

During the operation of positioning system of glass plate printing press, two flanges of equal height support the whole operating mechanism to move frequently, so it is necessary to calculate and analyze the strength of supporting flange. Finite element analysis is an important tool in mechanical design. The basic idea is to divide the continuous physical model into a finite number of elements. Then the physical quantities of each element are simulated to approximate the physical quantities of the whole body. Finite element method (FEM) not only has the advantages of high accuracy, but also can be applied to all kinds of complex shapes, so it has become one of the commonly used engineering analysis methods. As shown in Fig. 6, the finite element method is used to establish the model of supporting flange and mesh it.

IOP Conf. Series: Materials Science and Engineering **719** (2020) 012003 doi:10.1088/1757-899X/719/1/012003

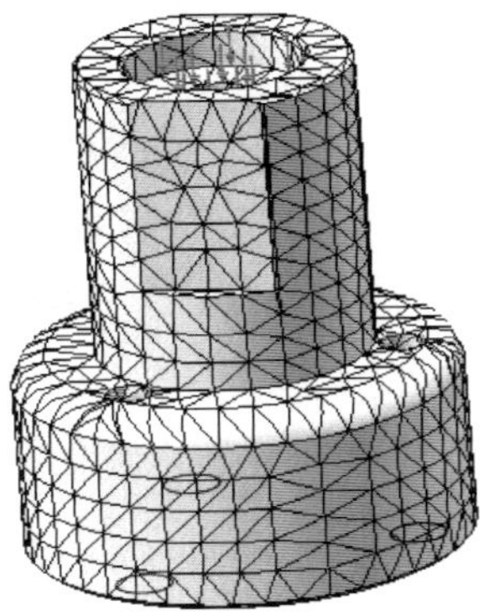

Figure 6. Finite Element Model of Supporting Flange

According to the designed parameters, the stress distribution of parts can be simulated by software program calculation, as shown in Figure 7. From Figure 7, it can be found that the position where the maximum stress occurs on the flange is at the junction of the flange with the cylinder. Referring to the material parameters, the value is less than the allowable stress value, so the strength meets the requirements.

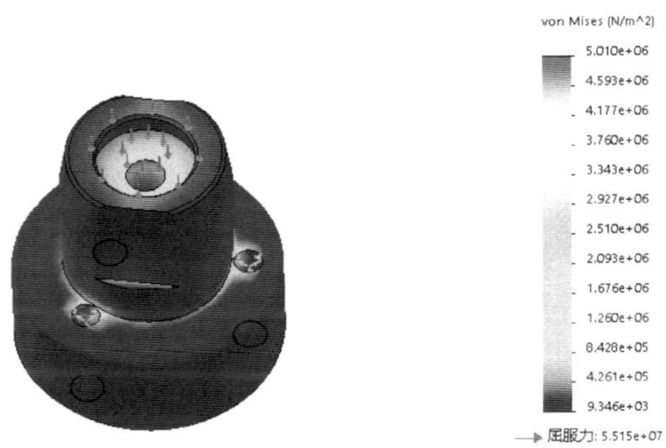

Figure 7. Finite Element Model of Supporting Flange

5. Conclusions

In this paper, a more efficient and automatic positioning system for glass printing press is designed, which includes operating mechanism, positioning mechanism and lifting mechanism. Synchronized belt drive is used in the operation mechanism to ensure the accuracy and efficiency of positioning. The pneumatic transmission is used in the lifting mechanism, which makes the platform have better bearing capacity and quick response. The travel control method is used in the positioning mechanism to ensure that the positioning elements do not interfere. Finally, finite element modelling and analysis of flanges are carried out to ensure that their strength meets the requirements.

6. References

[1] Aoyama T, Takahata N. Development of the automatic positioning system of microtool edge in micromachining of glass plate for microfluidic chips[J]. CIRP Annals - Manufacturing Technology, 2010, 59(1):551-554.

[2] Li Y, Wu H. Design and Implementation of the License Plate Positioning System Based on the DSP[C]// International Conference on Computer Sciences & Applications. 2014.

[3] Xu, Qingsong. Design and testing of a novel multi-stroke micropositioning system with variable resolutions[J]. Review of Scientific Instruments, 2014, 85(2):1-11.

[4] Sandoz P, Jacquot M. Lensless vision system for in-plane positioning of a patterned plate with subpixel resolution.[J]. Journal of the Optical Society of America A Optics Image Science & Vision, 2011, 28(12):2494-500.
[5] Lu J L, Lin M X. A Cutting Line Positioning and Defects Detection System for Polyurethane Plate Based on Computer Vision[J]. Advanced Materials Research, 2012, 562-564:119-122.

IOP Conf. Series: Materials Science and Engineering **719** (2020) 012004 doi:10.1088/1757-899X/719/1/012004

Cryptanalysis of Security Analysis and Enhancements of a Remote User Authentication Scheme

Min-Shiang Hwang[1,2], Hung-Wei Yang[1] and Cheng-Ying Yang[3,*]

[1]Department of Computer Science and Information Engineering, Asia University Taichung, Taiwan 41354 (Email: mshwang@asia.edu.tw)
[2]Department of Medical Research, China Medical University Hospital, China Medical University, Taichung, Taiwan 40402
[3]Department of Computer Science, University of Taipei, Taipei, Taiwan
(*Email: cyang@utaipei.edu.tw)

Abstract. The main purpose of user authentication schemes is to verify the authorized user using a server via an insecure channel. With the authentication, a server and a user could have a mutual authentication. In 2019, Cao proposed an improvement of a user authentication scheme. The scheme was postulated in that it could protect from the several possible attacks and have the following advantages: Identity preservation, not only to resist the slow wrong password detection, to resist the user masquerading, the password guessing, and the sever masquerading attacks, but also to have a mutual authentication between servers and users. However, we will show his scheme is not with the capacity to against the denial of service and on-line password guessing attacks in this article. In order to improve that authentication scheme, this work proposes an enhanced remote authentication scheme with the capacity to resist those vulnerabilities as shown in Cao-Sun-Cao's scheme.

1. Introduction

As cloud-computing technologies matures, more and more data owners are beginning to make heavy use of cloud system services to reduce the cost of building systems and devices [1-3]. However, how to ensure the security of cloud services is also an important issue [4-5]. A remote authentication scheme provides to verify an authorized user in a cloud-computing server via an insecure channel. Both of server and user could have a mutual authentication with a remote user authentication scheme [6-8]. Many remote user authentication schemes had been proposed [9-22]. A secure and simple authentication could be applied to practical [23-24].

Based on using a smart card, Chang and Lee proposed an easy to implement and practical remote user authentication scheme in 2013 [25]. Without withstanding the denial of service attack and the on-line guessing identity attack, Chiou et al. showed the drawbacks in Chang-Lee's scheme. In order to overcome the disadvantage in their scheme, Chang and Lee proposed an improved scheme to withstand these vulnerabilities [26].

Based on Hash function, Hsieh and Leu presented a simple and practical remote user authentication [27]. However, Cao, Sun, and Cao showed that Hsieh-Leu's scheme have not user anonymity in 2019 [28]. Besides, the scheme is vulnerable to the following attacks: the slow wrong password detection, the password guessing, and the masquerading attacks. To resist those possible attacks, Cao, Sun, and Cao proposed another authentication scheme for the remote users [28]. Cap et al. claimed the proposed scheme could against several possible types of attacks and have the following advantages including Identity preservation, to resist slow wrong password detection, to resist user masquerading attack, to resist sever masquerading attack, to resist password guessing attack, and to provide mutual authentications. Even though that proposed scheme is with the advantages above, with user's smart card, that scheme could not withstand the attacks including denial of service and on-line password guessing attacks. In this work, an improved remote user authentication scheme is proposed to withstand these vulnerabilities as those in Cao-Sun-Cao's scheme.

2. Review of Cao-Sun-Cao's Remote User Authentication Scheme [28]

According to Cao-Sun-Cao's scheme, there are two entities, server S and user U_i [28]. There are four phases in their authentication scheme: Registration phase, login phase, authentication phase, and password change phase. The following gives the description individually.

2.1. Registration Phase

Two phases, Initial registration and Re-registration, are included in the registration. For a new user U_i, registration is the initial phase to use the system and the server authorizes user U_i to be legal. After this initial registration phase, the server S issues a valid smart card to user U_i. These messages, $\{f_u, R, h(\cdot), b, n\}$, are embedded in the issued smart card, where

$$R = P \oplus h(b \oplus PW_i),$$
$$P = h(EID_i \oplus x), \text{ and}$$
$$EID_i = h(h(ID_i)||n)$$

Here, $h(\cdot)$ denotes a hash function; b denotes a random number according to user decision. ID_i and PW_i are the user's identity and password, respectively. Secret key x is for the server, and n is an entry of U_i in the server's database. There are three attributes in the server's database. They are n, EID_i, and $V = h(h(PW_i) \oplus h(x))$.

User could invalid his smart card if the issued smart card is lost. If the user wants to have a valid smart card, he should begin with the registration phase.

2.2. Login Phase

Once if user U_i wants to access the resource, with the smart card, user U_i has to give the identity ID_i, the password PW_i to the terminal device to connect the remote system with the following steps.

1) The smart card S calculates $fu` = h(ID_i \oplus h((PW_i))$ and, then, it compares the difference between $f_u`$ and f_u stored in the issued card. If $f_u`$ is not equal to f_u, the processor terminates the connection to service.
2) The end of user sends a login request with the message $\{EID_i, M_2, M_3, T_1\}$ to the server, where $M_2 = M_1 \oplus R_c$, $M_1 = R \oplus h(b \oplus PW_i)$, $M_3 = h(M_1||R_c||T_1)$, and $EID_i = h(h(ID_i)||n)$. Also, R, b, and n are the stored number in the smart card, R_c is a random number, and T_1 is a time stamp in the terminal device.

2.3. Authentication Phase

User U_i and server S authenticates each other with the following steps.

1) Initially, the time stamp T_1 is checked if it is valid or not by server S. If T_s-$T_1 > \varepsilon$, the server stops the connection to the user, where, T_s is the time stamp in the server. Time difference ε could be the threshold for the communication delay or the tolerance for synchronization between the client device and the server.
2) Server S queries EID_i from the server's database. If EID_i is not a valid, the server stops the connection.
3) Server S computes $M_3`$ and checks if $M_3`$ is equal or not to M_3, where $M3` = h(h(EIDi \oplus x)||M2 \oplus h(EIDi \oplus x)||T_1)$ If it is not, the server stops the connection.
4) The server sends $\{EIDi, M_4, M_5, T_2\}$ to the user, where $M_4 = h(EID_i \oplus x) \oplus Rs$, $M_5 = h(h(EID_i \oplus x)||Rs||T_2)$, R_s is a random number generated in the server, and T_2 is the time stamp of the server.

5) The user computes M_5\` and checks weather $M_5$\` is equal to M_5 or not, where $M_5\text{`} = h(M_1||M_4 \oplus M_1)||T_2)$. If it is not, the user stops the connection.
6) The user computes the session key s_k and M_6 according to $s_k = h(M_1||R_c||M_4 \oplus M_1||T_2||T_3)$, $M_6 = h(M_1||R_c||M_4 \oplus M_1||T_3)$, where, T_3 is the valid time stamp of the user. Then, the user sends message $\{M_6, T_3\}$ to the server.
7) The server computes M_6\` and checks weather $M_6$\` is equal to M_6 or not, where $M_6\text{`} = h(h(EID_i \oplus x)||M_2 \oplus h(EID_i \oplus x)||R_s||T_3)$. If it is not, the user stops the connection.
8) S calculates and obtains the session key s_k and M_7 according to $s_k = h(h(EID_i \oplus x)||M_2 \oplus h(EID_i \oplus x)||R_s||T_2||T_3)$ and $M_7 = h(h(EID_i \oplus x)||M_2 \oplus h(EID_i \oplus x)||R_s||T_4)$, where, T_4 is a valid server's timestamp. Then, S sends $\{M_7, T_4\}$ to the user.
9) The user computes M_7\` and checks weather $M_7$\` is equal to M_7 or not. If it is, the user checks the legal server, and confirms the sharing session key s_k.

3. Weakness in Cao-Sun-Cao's Remote User Authentication

This section shows the weakness in Cao-Sun-Cao's user authentication scheme [28]. The authentication could not resist the attacks from the denial of service attacks and the on-line password guessing with user's smart card attacks.

3.1. On-Line Password Guessing with User'S Smart Card Attacks

The adversary might use password guessing attack if he/she has a chance to hold the user's smart card. The following event might be hold at the login steps in Cao-Sun-Cao's authentication.

1) With guessing user's password, the adversary inputs ID_i and PW_i\` with the user's smart card.
2) Smart card S calculates $f_u\text{`} = h(ID_i \oplus h((PW_i\text{`}))$ and compares the difference between f_u\` and $f_u$ stored in smart card. If $f_u$\` is not equal to f_u, the smart card terminates the service. Otherwise, the smart card begins to send the login request messages: $\{EID_i, M_2, M_3, T_1\}$ to the server.
3) The adversary monitors the login request message. If there are messages send to the server, this implies the password was guessed. Otherwise, the adversary repeatly guesses a password and performs Steps 1-3.

3.2. Denial of Service Attack

The login request message $\{EID_i, M_2, M_3, T_1\}$ might be interrupted by the adversary in Step 2 of the login phase. Then, the adversary uses a fake message $\{EID_i, M_2, M_3, T_1\text{`}\}$ to send to the server, where $T_1$\` is updated time stamp. The server will process the following steps in the authentication phase.

1) 1) The server checks whether the time stamp T_1 is valid or not. If the difference $(T_s\text{-}T_1)$ is larger than the time difference ε, the server stops the connection to the user, where T_s is the time stamp of the server. ε is a threshold for the communication delay or the tolerance for synchronization between the client device and the server. In this step, the server will pass the verification because the T_1\`is a new current time stamp.
2) Server S queries EID_i from the database. If EID_i not exists, the server stops the connection. In this step, the server will pass the verification because the EID_i is a legal ID_i.
3) The server S computes M_3\` and checks weather $M_3$\` is equal to M_3 or not, where $M3\text{`} = h(h(EID_i \oplus x)||M_2 \oplus h(EID_i \oplus x)||T_1)$. If it is not, the server stops the connection. In this step, the server has an ability to check the illegal login request and will stop the connection to the user. However, the server needs to compute 2 X-ORs, 2 hash functions, and one comparison computations.

In summary, for an illegal login request message, the server needs to compute 2 X-ORs, 2 hash functions, 2 comparisons computations, and a query to a database. These computations will slow down the server's performance and unable to provide normal services.

4. The Proposed Authentication Scheme

To improve those weaknesses in Cao-Sun-Cao's scheme, this work proposes an authentication scheme. Compared with those phases in Cao-Sun-Cao's scheme, this proposed scheme modifies the registration in login phase and the password change phases in authentication phase.

4.1. Login Phase
For the remote access, user U_i inputs his/her identity ID_i and password PW_i to the end device with the smart card according to the following steps.

1) Smart card S calculates $f_u` = h(ID_i \oplus h((PW_i))$ and compares the difference between $f_u`$ and f_u stored in the smart card. If $f_u`$ is not equal to f_u, smart card S cumulates the number of wrong login requests and terminates the service to the user. If the number of the wrong login requests is larger than 3, the smart card automatically idles for a period (i.e. one day).
2) If $f_u`$ is equal to f_u, smart card S resets the number of the wrong login requests.
3) The user begins to send the login request message $\{EID_i, M_2, M_3, N_1\}$ to the server, where $M_2 = M_1 \oplus R_c$, $M_1 = R \oplus h(b \oplus PW_i)$, $M_3 = h(M_1||R_c||T_1)$, $EID_i = h(h(ID_i)||n)$, $N_1 = T_1 \oplus n$. Also, R, b, and n are retrieved from the smart card, R_c is a random number and T_1 is a current time stamp of the terminal device.

4.2. Authentication Phase
For authentication in the proposed scheme, Step 2 to Step 9 are kept the same as those authentication phase in Cao-Sun-Cao's remote user scheme. The modified Step 1 in the proposed scheme is the following.

1) Server S queries EID_i from the database. If EID_i does not exist, the server disconnects to the user. Otherwise, the server retrieves n from the database. Next the server obtains the time stamp T_1 by $T_1=N_1 \oplus n$.

5. Security Analysis of the Proposed Scheme
This section shows that the proposed scheme could resist the attacks from on-line password guessing with user's smart card and the denial of service in Cao-Sun-Cao's remote user authentication.

5.1. Resist The Attack from On-Line Password Guessing with User's Smart Card
In login phase, the adversary might use the attack from on-line password guessing when he/she hold the user's smart card.

1) The adversary begins to conjectures the password $PW_i`$ and inputs $ID_i$ and $PW_i`$ with the user's smart card.
2) Smart card S calculates $f_u` = h(ID_i \oplus h((PW_i`))$ and compares $f_u`$ and $f_u$ stored in smart card. If $f_u`$ is not equal to f_u, the smart card cumulates the number of wrong login requests and terminates the service to the user. Once if the number of the wrong login requests is larger than 3, the card automatically idles for a period (i.e. one day). Otherwise, the login request message $\{EID_i, M_2, M_3, T_1\}$ is sent to the server by the card.
3) The adversary monitors the login request message. If there are messages send to the server, this implies the password was guessed. Otherwise, the adversary repeatly guesses a password and performs Steps 1-3.

If the number of the wrong login requests is larger than 3, the smart card automatically idles for a period (i.e. one day). With this proposed authentication, it could resist the attack from on-line password guessing with user's smart card.

5.2. Resist Denial of Service Attack
The login request messages, $\{EID_i, M_2, M_3, N_1\}$, could be intercepted by the adversary in Step 3 of login phase in our proposed user authentication scheme. Since the adversary did not know the n, he/she is thus unable to masquerade $N_1`$ with the valid time stamp. The server authenticates these illegal login request messages in Step 1 of the authentication phase in our proposed scheme. In this step, the server has an ability to check the illegal login request and will stop the connection to the user. The server only needs one X-OR, one comparison, and a query to a database. These computations are less than that of Cao-Sun-Cao's scheme.

6. Conclusion

In this article, we have shown the weaknesses in Cao-Sun-Cao's scheme for remote users. Their scheme could not against the denial of service attack and the on-line guessing password attack with user's smart card. For the purpose of improvement, this work proposes a user authentication scheme for remote users. With security analysis, this proposed scheme has an ability to withstand these vulnerabilities as those in Cao-Sun-Cao's scheme.

7. Acknowledgments

We gratefully acknowledge the anonymous reviewers for their valuable comments. This study was supported by the Ministry of Science and Technology (Taiwan), under grant MOST 108-2410-H-468-023.

8. References

[1] Yang C, Chen Q, and Liu Y. Fine-grained outsourced data deletion scheme in cloud computing [J]. *International Journal of Electronics and Information Engineering*, 2019, 11(2): 81--98.
[2] Liu L, Cao Z, Mao C. A note on one outsourcing scheme for big data access control in cloud [J]. International Journal of Electronics and Information Engineering, 2018, 9(1): 29-35.
[3] Rezaei S, Doostari M A, and Bayat M. A lightweight and efficient data sharing scheme for cloud computing [J]. International Journal of Electronics and Information Engineering, 2018, 9(2): 115-131.
[4] AbdElminaam D S. Improving the security of cloud computing by building new hybrid cryptography algorithms [J]. *International Journal of Electronics and Information Engineering*, 2018, 8(1): 40-48.
[5] Al-Shaikhly M H R, El-Bakry H M, and Saleh A A. Cloud security using Markov chain and genetic algorithm [J]. *International Journal of Electronics and Information Engineering*, 2018, 8(2): 96-106.
[6] Bayat M, Atashgah M B, Barari M, and Aref M R. Cryptanalysis and improvement of a user authentication scheme for internet of things using elliptic curve cryptography [J]. *International Journal of Network Security*, 2019, 21(6): 897-911.
[7] Hou G and Wang Z. A robust and efficient remote authentication scheme from elliptic curve cryptosystem [J]. *International Journal of Network Security*, 2017, 19(6): 904-911.
[8] Hwang M S and Li L H. A new remote user authentication scheme using smart cards [J]. *IEEE Transactions on Consumer Electronics*, 2000, 46(1): 28-30.
[9] Wan T, Liu X, Liao W, and Jiang N. Cryptanalysis and improvement of a smart card based authentication scheme for multi-server architecture using ECC. *International Journal of Network Security*, 2019, 21(6): 993-1002.
[10] Annamalai P, Raju K, and Ranganayakulu D. Soft biometrics traits for continuous authentication in online exam using ICA based facial recognition [J]. *International Journal of Network Security*, 2018, 20(3): 423-432.
[11] Chiou S Y. An efficient RFID authentication protocol using dynamic identity [J]. *International Journal of Network Security*, 2019, 21(5): 728-734.
[12] Cao S Q, Sun Q, and Cao L L. Security analysis and enhancements of a remote user authentication scheme [J]. *International Journal of Network Security*, 2019, 21(4): 661-669.
[13] Chen TY, Lee CC, Hwang MS, and Jan JK. Towards secure and efficient user authentication scheme using smart card for multi-server environments. *The Journal of Supercomputing*, 2013, 66(2): 1008-1032.
[14] Chiou SF, Pan HT, Cahyadi EF, and Hwang MS. Cryptanalysis of the mutual authentication and key agreement protocol with smart cards for wireless communications. *International Journal of Network Security*, 2019, 21(1): 100-104.
[15] Liu Y, Chang C C, and Chang S C. An efficient and secure smart card based password authentication scheme [J]. *International Journal of Network Security*, 2017, 19(1): 1-10.

[16] Hong S. Authentication techniques in the internet of things environment: A survey [J]. *International Journal of Network Security*, 2019, 21(3): 462-470.
[17] Zhang X, Wang B, and Zhang W. A robust authentication protocol for multi-server architecture using elliptic curve cryptography [J]. *International Journal of Network Security*, 2019, 21(2): 191-198.
[18] Agidi R C. Biometrics: The future of banking and financial service industry in Nigeria [J]. International Journal of Electronics and Information Engineering, 2018, 9(2): 91-105.
[19] Liu Y and Chang C C. A cloud-assisted passenger authentication scheme for Japan rail pass based on image morphing [J]. *International Journal of Network Security*, 2019, 21(2): 211-220.
[20] Tian X, Tian F, Zhang A, and Chen X. Privacy-preserving and dynamic authentication scheme for smart metering [J]. *International Journal of Network Security*, 2019, 21(1): 62-70.
[21] Tarek E, Ouda O, and Atwan A. Image-based multimodal biometric authentication using double random phase encoding [J]. *International Journal of Network Security*, 2018, 20(6): 1163-1174.
[22] Wei CH, Hwang MS, and Chin AYH. A mutual authentication protocol for RFID. *IEEE IT Professional*, 2011, 13(2): 20-24.
[23] Chiou SY, Ko WT, and Lu EH. A secure ECC-based mobile RFID mutual authentication protocol and its application. *International Journal of Network Security*, 2018, 20(2): 396-402.
[24] Guo C, Chang CC, and Chang SC. A secure and efficient mutual authentication and key agreement protocol with smart cards for wireless communications. *International Journal of Network Security*, 2018, 20(2): 323-331.
[25] Chang CC and Lee CY. A smart card-based authentication scheme using user identity cryptography. *International Journal of Network Security*, 2013, 16(1): 139-147.
[26] Chiou S F, Cahyadi E F, Yang C Y, and Hwang M S, An improved Chang-Lee's smart card-based authentication scheme [C]. *Journal of Physics: Conference Series*, ICSP 2019, Mar. 29-31, 2019.
[27] Hsieh W B and Leu J S, Exploiting hash functions to intensify the remote user authentication scheme [J]. *Elsevier Advanced Technology Publications*, 2012, 31(6): 791-798.
[28] Cao S Q, Sun Q, and Cao L L. Security analysis and enhancements of a remote user authentication scheme [J]. *International Journal of Network Security*, 2019, 21(4): 661-669.

A Novel Weight-based Leader Election Approach for Split Brain in Distributed System

Feng Jiang, Yongyang Cheng, Changkun Dong and Erdong Yu

China Telecom Cloud Computing Corporation, Beijing, China.
Email: {jiangfeng, chengyy2, dongck and yued}@chinatelecom.cn}

Abstract. The rapid development of High Availability (HA) system has attracted growing attention from both industry and academia. Due to the network and hardware failure, a complete system would split into two or more separate partitions, which begin to compete for shared resources, resulting in system chaos and data corruption. In this paper, we propose a novel weight-based leader election approach for split brain in a distributed system. The leader of separate partitions would be elected by embedding an arbitration program in the client. Unlike the traditional mode, the leader election in our proposed approach is lightweight and fully automatic, meaning that no additional hardware resources or manual intervention are required. The approach presented in this paper has been validated to be valid through qualitative and quantitative experiments.

1. Introduction

With the development of information technology, the business scenarios become more complex and the amount of data to be processed becomes larger. In order to improve the performance in scalability, availability and reliability, the distributed storage system uses multiple servers to share storage load. As we know, the distributed system has advantages of low cost, flexible expansion capability and unified resource pool management by nature [1]. Thus, it has gradually replaced the traditional storage and becomes a powerful way for processing massive data effectively. However, due to the network and hardware failure, a complete system would split into two or more separate partitions. A split brain condition is the result of the partition, where each side believes the other is dead, and then proceeds to take over shared resources as though the other side no longer owns any resources. In this case, if each side continues to respond to reads and writes while they are unable to communicate with each other, the whole distributed system would diverge and no longer be consistent. Thus, the key to solve the problem of split brain is to select a leader from given candidates based on some strategies. In general, there are several common approaches of leader election to avoid split brain. In work [2], an approach based on dedicated access is proposed, which enforces that the separate partitions could continue to communicate to select the leader by adding a heartbeat line. This is the easiest way to avoid split brain from the perspective of engineering practice. Work [3] and [4] use additional detection nodes, like database, IP address or USB, as additional detection nodes, to form a new voting group with original partitions to conduct the leader election. In this approach, only the partition that could normally connect to the additional detection node has the right to vote as a leader. In work [5] and [6], a logical concept "region leader" is proposed to avoid the stale read, which requires that all reading and writing requests from clients must go through the "region leader" before they are processed, which guarantees the consistency of data reading and writing at the cost of client's availability.

Although these mentioned approaches could reduce the probability of split brain occurring in some scenarios, challenges still remain to be solved. First of all, there is also the possibility of disconnection of the heartbeat line. In this case, the problem of split brain remains unsolved. Furthermore, the detection node-based approach works on the server layer, which not only leads to additional resource costs, but also requires manual completion. Finally, the region leader-based approach only could solve the split brain when a whole system is divided into three or more separate partitions, but no longer works when there are only two separate partitions.

The main contributions of our work contain the following.

- The arbitration process is lightweight. User only needs to install a program in the client instead of any additional hardware resources or manual interventions.
- We assign weights to voters according to their service levels. In this way, we could get the minimum weights to guarantee the whole system high available.
- We have improved the universality of the leader election approach to solve split brain so that it could keep working when there are only two separate partitions.

2. Our proposed Leader Election Approach

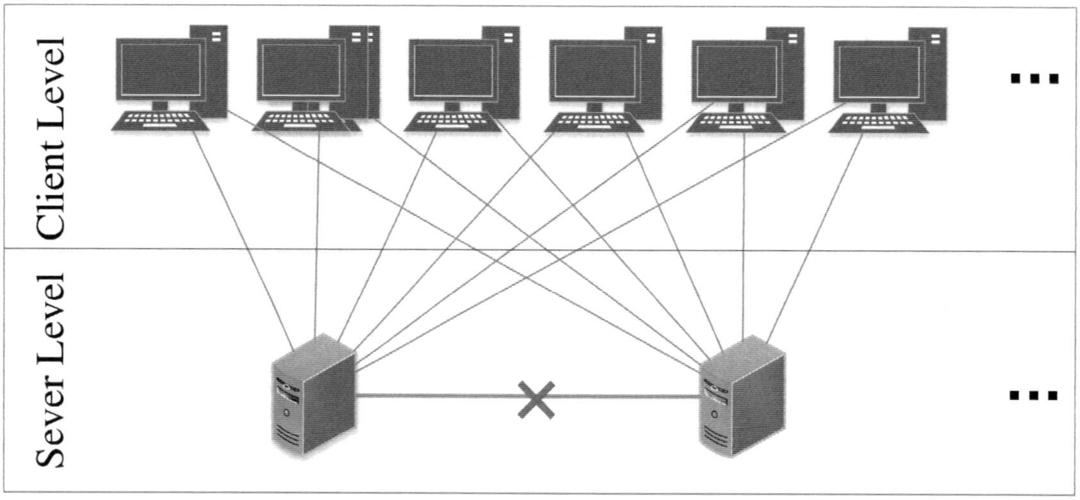

Figure 1. The scenario of split brain in a distributed system.

In order to solve the above mentioned problem, we propose a novel weight-based leader election approach for split brain in a distributed system. Figure 1 illustrates the scenario of split brain in a distributed system. In the traditional approaches, all voters have the same voting rights, meaning that all servers and additional detection nodes would immediately take part in the leader election when split brain occurs. However, only the existing server partitions could not elect a leader would the additional detection nodes be required for arbitration. Thus, in our proposed approach, voters are first assigned different roles according to their levels when split brain occurs. The server has three kinds of roles: speaker, senator and leader. On the contrary, the arbitration program in the client only has one role: senator. When a leader server could not receive the heartbeat signal in a given time threshold, the leader election begins and the role of servers becomes a speaker. In this case, speakers independently vote according to specific rules. Once a server receives votes from majority of speakers, its role becomes a leader and the election ends. To avoid an infinite loop, we set a threshold for the leader election. The role of servers becomes a senator and the arbitration program in the client is required if the threshold times out. Similarly, once a server receives votes from majority of senators, its role becomes a leader and the election ends. We also set a threshold for senators to start a new round of voting. Figure 2 illustrates the role change of voters in a leader election process.

In a large-scale distributed system, business process applications and databases are usually deployed across regions, which become quite sensitive to access delays. If the leader node randomly switches to a server in another area, it might lead to massive accesses to remote resources, greatly increasing response time. Thus, we assign different weights to corresponding voters, which could be flexibly modified according to resource deployment requirements. We design a two-stage process for leader election. The first stage is the unified lease time in the distributed system, while the second stage is decided by the weight, which means that the senator whose weight is larger would earlier conduct the self-election operation. To further quantitatively explain our proposed approach, we assume that there are m servers and n clients, where m>1. We define the variable $Sweight_i$ to represent the weight of server i and $Cweight_j$ to represent the weight of client j. Thus, the minimum weights to keep the system high available could be expressed as Equation (1).

$$Weightmin > \frac{1}{2}(\sum_{i=1}^{m} Sweighti + \sum_{j=1}^{n} Cweightj) \quad (1)$$

In addition, there is a weight detection mechanism to ensure the priority of voter weights. The leader node broadcasts and detects all the voter nodes that could be connected. If the voter node with higher weight is found, the new node would actively initiate a role transfer to take over the leader role.

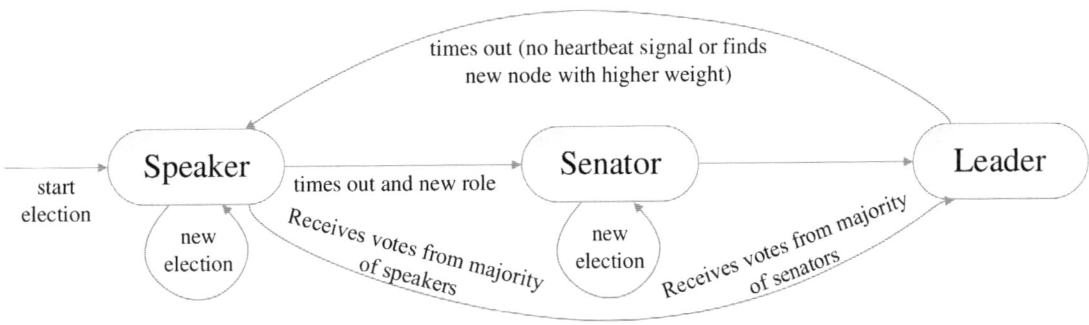

Figure 2. The role change of voters in a leader election process.

3. Performance Evaluation

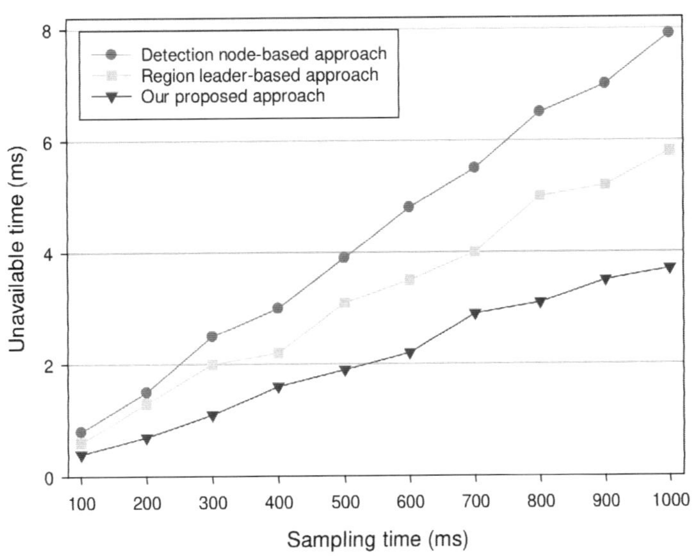

Figure 3. The comparison of unavailable time among different approaches.

To further quantitatively evaluate our proposed approach, we perform a series of quantitative experiments, which are performed on six PCs, each with 6GB of memory, 3.07 GHz CPUs. Furthermore, four PCs run as servers and two PCs run as clients. These servers form a distributed system and communicate with each other through heartbeat lines. In order to simulate the scenario of split brain, we randomly assign weights to PCs and cut off heartbeat lines among servers every 100 milliseconds. The sampling time is 1000 milliseconds and final results are illustrated as follows. Figure 3 illustrates that the unavailable time in our proposed approach is significantly lower than that in other two approaches. Figure 4 shows the average CPU utilization of arbitration program in the clients, which only takes about 2 percent of the whole CPU.

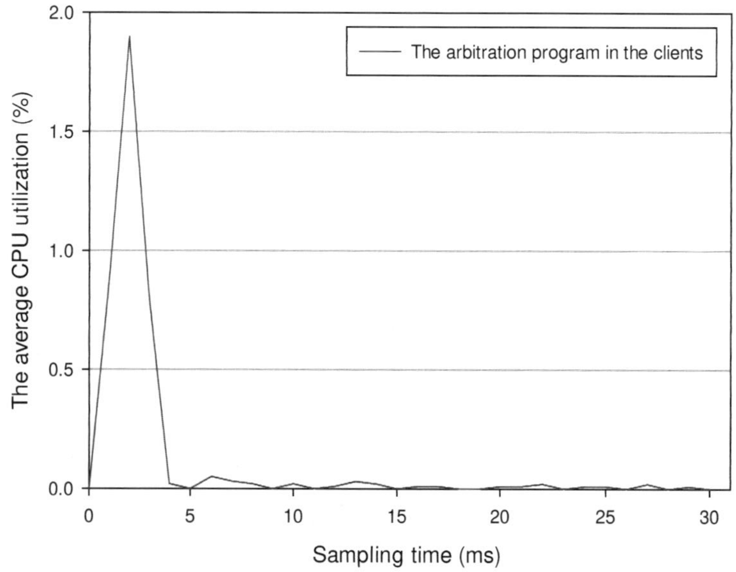

Figure 4. The CPU utilization in a leader election process.

4. Conclusions

In this paper, we present a novel weight-based leader election approach for split brain. To reach the goal, we initially install an arbitration program in the clients. Furthermore, we assign weights to voters and make them automatically perform the leader election. Finally, we validate our proposed approach to be valid through a series of quantitative experiments.

5. References

[1] C. A. Vissers, G. Scollo, M. Sinderen and E. Brinksma, "Specification styles in distributed systems design and verification", Theoretical Computer Science, vol. 89, pp. 179-206, 2017.

[2] M. Gerami, M. Xiao and J. Li, "Repair for Distributed Storage Systems with Packet Erasure Channels and Dedicated Nodes for Repair", IEEE Transactions on Communications, vol. 64, pp. 1367-1383, 2016.

[3] R. Zhang, P. Isola and A. A. Efros, "Split-Brain Autoencoders: Unsupervised Learning by Cross-Channel Prediction", In: IEEE Conference on Computer Vision and Pattern Recognition, vol. 1, pp. 645-654, 2017.

[4] D. U. S. Rajkumar and R. Vayanaperumal, "A leader based intrusion detection system for preventing intruder in heterogeneous wireless sensor network", IEEE Bombay Section Symposium, 2015.

[5] C. E. Ren, Z. P. Shi and T. Du, "Distributed Observer-Based Leader-Following Consensus Control for Second-Order Stochastic Multi-Agent Systems", IEEE Access, vol. 6, pp. 20077-20084, 2018.
[6] A. Das, C. Zhou and A. Sen, "Region based fault-tolerant distributed file storage system design under budget constraint", in 6th International Workshop on Reliable Networks Design and Modeling, 2014.

An Improved Hydraulic Driven Feeding and Loading Wagon

Yinglong Liao[1], Jiangang Yi[1,*], Bangxu Fang [1] and Changjun Li[1]

[1] School of Electromechanical and Architectural Engineering, Jianghan University, Wuhan 430056, China
469883553@qq.com

Abstract. In the process of rapid development manufacturing industry, feeding and loading wagon plays an important role in the connection and switching of each industrial station. In this paper, a hydraulic-driven feeding and loading system is proposed, and the system scheme and structure are given. On the basis of it, the finite element analysis and verification of the support are carried out. The designed wagon not only saves manpower and material resources, but also improves the work efficiency to a great extent.

1. Introduction

With the development of manufacturing industry, how to further improve production efficiency is the key problem to be solved. Feeding and loading system plays an important role in the connection and switching of each station in the production line. In the process of material transportation, the loading system not only reduces the consumption of manpower and material resources, but also greatly improves the work efficiency. Traditional hopper vibration loading system can realize automatic positioning, but it is only suitable for small size and simple shape of workpiece. In this paper, an improved hydraulic loading system is proposed, which can be used in the case of complex workpiece shape and frequent start-up.

2. Scheme Design

Common loading systems are shown in Figure 1. The electromagnetic field between the core coil and armature is generated by electrification or power failure, which changes the magnetic force to produce vibration between the base and the groove. However, in the process of frequent vibration feeding, it is easy to cause fracture between grooves. When the workpiece size is large, the hopper vibration loading system is difficult to complete the feeding task. In addition, when choosing the scheme of automatic feeding system, the following problems should also be considered:

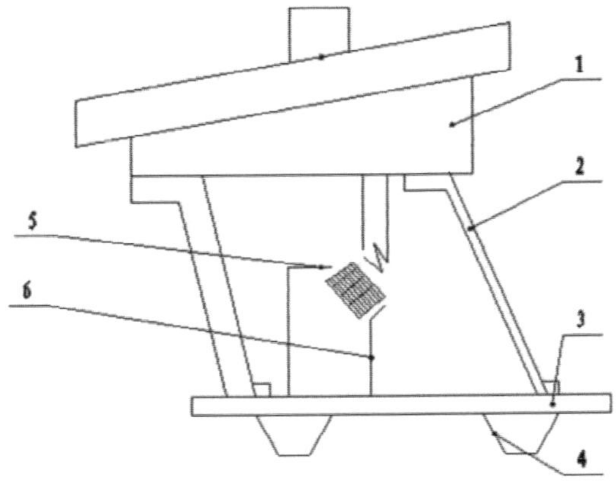

1. Groove body; 2. Leaf spring; 3. Base; 4. Shock-absorbing rubber frame; 5. Iron core coil; 6. armature; 7. Workpiece

Figure 1. Hopper Vibration Loading System

(1) The connection between automatic feeding mechanism and feeding system.
(2) The length of pipe and rod workpiece.
(3) Composition of brake mechanism and feed mechanism.

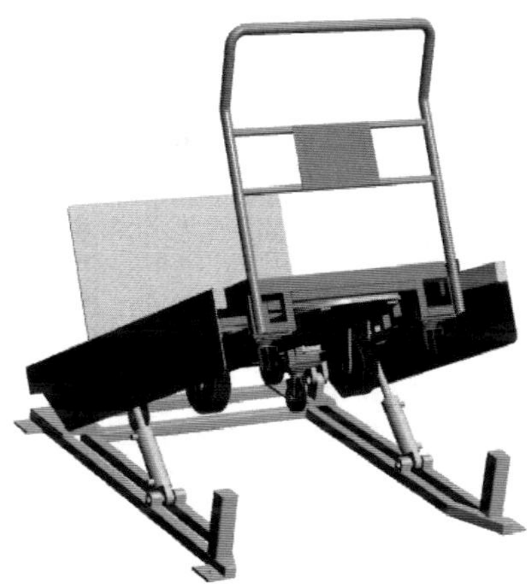

Figure 2. Hydraulic drive feeding system

According to the requirements, a hydraulic drive feeding system is designed as shown in Fig. 2. Among them, the hydraulic components are located in the overturning mechanism. In order to avoid the radial shear force generated in the processing or assembly process, the bottom cross beam is removed and the flat panel used to support the material box is replaced by a horizontal sliding plate. In this way, when the workpiece material box is lifted, the sliding plate can be directly inserted into the bottom of the car, the car can be lifted, and then the material can be reversed. After feeding, reset through hydraulic cylinder. The next feeding process is then carried out.

3. System Composition

The performance of the machine is mainly reflected in its efficiency. For the feeding process, the efficiency is reflected in the speed of feeding. When the material completes a process, it will be placed in the material box and transported to the next process area by trolley or forklift. After reaching the next process, the first step is to feed. In the actual environment, accidental operations may occur when picking up materials. This also involves the safety protection of the machine. For the process of human-computer interaction, it is necessary to prevent accidental operation and take targeted measures to ensure the safety of personnel. Misoperation is usually caused by internal factors or by the operator. Therefore, safety device, automatic locking device and switch control should be considered in the design of feeding system.

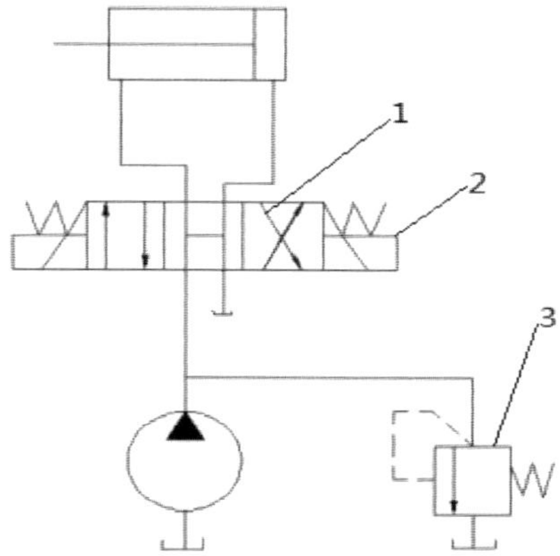

1. Three-position four-way commutation valve; 2. Electromagnet; 3. Overflow valve

Figure 3. Hydraulic Commutation Circuit

Because there is inclined ascent and descent process in the design of hydraulic system, considering the two cases of reversal and return, commutation circuit is needed. In the commutation circuit, the commutation valve is usually used to play the role of commutation. The selection of commutation valve needs to consider the rationality of the system. Commutation loops generally adopt the commutation mode of different executable components such as two-way four-way, three-way four-way or five-way commutation valves. Among them, the use of commutation valve is the most convenient one. The advantage of using solenoid valve for commutation is its fast action, but the disadvantage is that commutation has impact. The use of alternating current solenoid valve is another way of commutation, but this way is not suitable for frequent commutation operation. As shown in Figure 3, the circuit uses a three-way four-way electromagnetic commutation valve for commutation. Specific operation is: when the electro-hydraulic valve moves to the middle position, the piston is stationary and the hydraulic pump unloads; when the reversing valve is on the right or left side, the piston of the hydraulic cylinder will move to the left or right accordingly.

In the commutation loop, the bidirectional variable hydraulic pump can also be used to realize the commutation operation. The specific operation is: when the oil outlet on the left side of the bidirectional variable hydraulic pump is discharged, the piston of the hydraulic cylinder moves to the right; by adjusting the variable mechanism of the bidirectional variable hydraulic pump, the oil outlet on the right side of the bidirectional variable hydraulic pump is discharged, and the piston of the hydraulic cylinder moves to the left. Because of the advantages of three-way four-way reversing valve in reversing circuit, such as small leakage between oil outlets, reliable reversing, fast reversing speed and smooth impact-free reversing, the hydraulic commutation circuit shown in Figure 3 is adopted.

4. Support Analysis

Because the "L" sliding fork plate is very convenient to handle the workpiece and can be directly inserted into the bottom of the workpiece box, the "L" sliding fork plate is used as the support of the workpiece box in the design of the feeding rack. The "L" sliding fork plate can be divided into horizontal and vertical parts. The horizontal part is used to carry the material, and the vertical part is used to connect with the main body of the overturning feeder. For the vertical part of the connection between fork plate and the main body of the overturning feeding rack, there are two types: hook type and articulated type. The hook type is directly attached to the connected fork rack. Because the feeding rack

IOP Conf. Series: Materials Science and Engineering **719** (2020) 012006 doi:10.1088/1757-899X/719/1/012006

can realize the overturning motion, the hook type "L" sliding fork plate is not suitable. The articulated fork plate is provided with a pin hole in its vertical part. The pin hole can be connected with the supporting optical axis of the overturning feeding frame to realize the rotation around the axis, as shown in Figure 4.

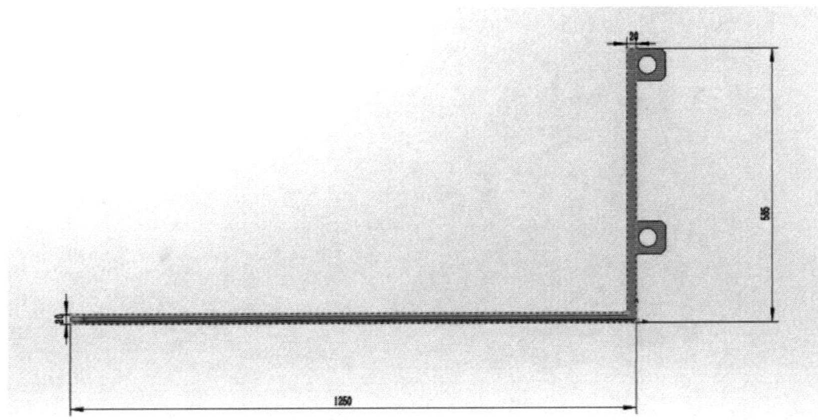

Figure 4. The "L" Sliding Fork Plate

The strength and stiffness of single sliding fork plate are not up to requirements. It can increase the thickness of the sliding fork plate or replace the material, but these two methods are not suitable from the cost point of view. Therefore, the strength and stiffness of the L-shaped sliding fork plate can be effectively improved by adding stiffeners to the bottom of the dangerous section of the sliding fork plate. The improved sliding fork plate is shown in Figure 5. Four stiffeners are added to the bottom of the horizontal part of the sliding fork plate. The strength and stiffness of the structure are calculated to meet the requirements.

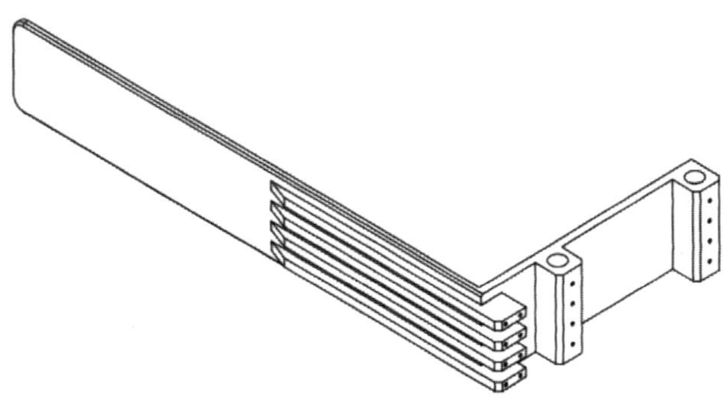

Figure 5. Sliding Fork with Stiffeners

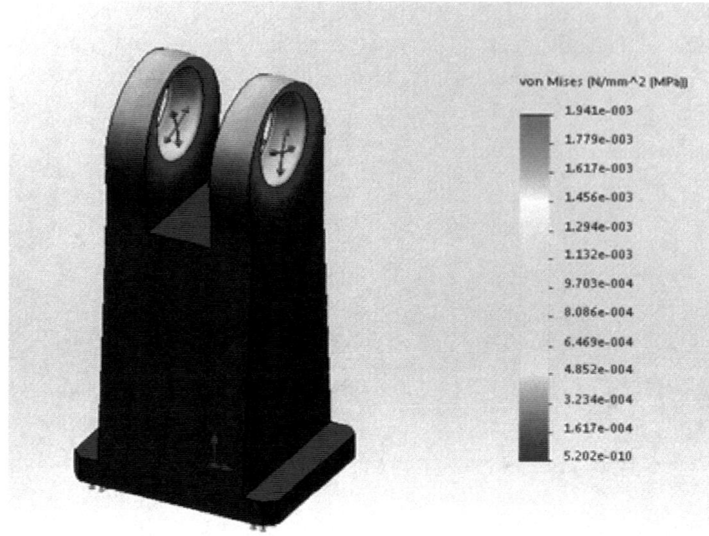

Figure 6. Analysis of Support Pedestal

The bearing connected with sliding fork plate should be analyzed because the support base connects the sliding fork plate with the loading system body. In the process of working, the support is always in the state of force, and the transmission of moment must be completed. At the hole-axle connection of the support seat, the stress distribution should be paid special attention. For this reason, a three-dimensional model as shown in Figure 6 is established by using the finite element analysis method. Taking into account the actual value, it can be found that the hole-axle connection of the support seat is the most concentrated position of the working load, and its value is less than the allowable stress value of the material, so it meets the requirements.

5. Conclusions

The main purpose of designing hydraulic loading system is to solve the problem of frequent manual operation of workpiece box in loading operation. In order to achieve the main functions required by the site, the driving mode and structure of the loading system were improved, including the determination of the overturning part of the feeding frame, the transmission part of the hydraulic mechanism and the support part. On this basis, the finite element analysis of key stress components is carried out. The results of calculation and practical application show that the hydraulic loading system can meet the requirements of production line and greatly improve the working efficiency.

6. References

[1] Chen C , Tang H , Qi S , et al. An electro-hydraulic servo control system research for CFETR blanket RH[J]. Fusion Engineering and Design, 2014, 89(11):2806-2813.

[2] Quan L X, Tian, De Zhi, Yu, Bin. Simulation Research on Load-Imitating Test Equipment for Hydraulic Drive Unit of a Quadruped Robot Based on Feed-Forward Decoupling Control[J]. Applied Mechanics & Materials, 2013, 365-366:839-846.

[3] Li-Fa F , Xin-Yu Y , Yu Z . An Overall Evaluation of the Status of a Steam Turbine DEH(Digital Electro-hydraulic) Control System Based on Entropy Weights and a Multistage Physical Element Analysis[J]. Journal of Engineering for Thermal Energy and Power, 2009.

[4] Xu, Li. Study and Development on an Electro-Hydraulic Pressure Control System of the Riveting Machine[J]. Applied Mechanics and Materials, 2013, 321-324:1641-1644.

[5] Ba D X , Ahn K K . Adaptive output feedback control of an electro-hydraulic system[C]// 2015 15th International Conference on Control, Automation and Systems (ICCAS). IEEE, 2015.

[6] Yang J L, Ruan, J, Yuan, Ju Long, et al. An Electro-Hydraulic Grinding Feed Drive Digital Control System[J]. Key Engineering Materials, 2001, 202-203(202):6.

IOP Conf. Series: Materials Science and Engineering **719** (2020) 012007 doi:10.1088/1757-899X/719/1/012007

Evaluation of Rock Slope Stability based on Multi-Dimension Cloud Model

Shuyuan Ran

China Energy Engineering Group Jiangsu Power Design Institute Co. Ltd, 13655172277

Email: ranshuyuan@jspdi.com.cn

Abstract. In this paper, slope stability is classified into five grades in terms of very steady, steady, basically steady, unsteady, and very unsteady. Ten indicators of slope height, tendency difference between structural plane and slope, structural plane inclination, etc., were selected. To solve the uncertainty and randomness, multi-dimension cloud model was applied with weights determined by projection pursuit method. The grade of steadiness was determined by maximum subject degree. The method was applied to 34 rock slopes to assess stability. The accuracy of the method reaches 50%, which indicates that the method has higher accuracy, and can be applied to other fields.

1. Introduction

The stability of rock slope is of great significance to the safety of people's lives and property. Since various evaluation indexes affected the rock slope stability, it is essential to determine the index system of rock slope stability scientifically and reasonably. Chen established an evaluation system with 10 indexes including slope height, the difference between structural plane dip and slope tendency, structural plane dip, the difference between structural plane dip and slope dip, rock uniaxial compressive strength, rock quality index, discontinuous plane spacing, structural plane characteristic value, groundwater level condition and slope excavation coefficient [1]. Affected by various fuzzy and random uncertain parameters, Zhu set up five indexes including rock quality, controlled structural plane, structural surface condition coefficient, slope height correction coefficient, and slope excavation method, and evaluated the stability of rock slope by grey correlation analysis method [2]. Wen adopts SAPSO-ELM based on extreme learning machine to select pore water pressure, cohesion, slope height, internal friction angle, weight of rock and slope angle as main indexes to effectively predict slope stability [3]. Based on the weighted least square method and matter element concept, Zhou adopted the evaluation index system for stability evaluation for high and steep rock slope, considering height, slope, rock quality, bulk density, cohesion, internal friction angle, slope aspect and strata occurrence, rock weathering degree, annual maximum rainfall and seismic intensity [4]. Wang established probability density function and membership function of slope safety reserve based on mathematical statistics and fuzzy mathematics, and applied to stability evaluation of bedding rock slope [5]. Li improved the slope stability probability classification (SSPC) method, which can quickly evaluate the stability of hydroelectric rock slope [6].

Since the stability of rock slope is affected by various factors, and uncertainties and randomness exist among the indicators, in this paper multi-dimensional cloud model is applied, which consider the uncertainty and randomness of the evaluation indicators synthetically, and the weights of the evaluation indicators were determined by projection pursuit method. The method proposed can overcome the subjectivity during the process of the weight determination. The method was applied to

34 rock slope or stability evaluation, and compared with the actual stability of the slope These guidelines, written in the style of a submission to Material Science and Engineering: Conf. Ser., show the best layout for your paper using Microsoft Word. If you don't wish to use the Word template provided, please use the following page setup measurements.

2. Stability Evaluation of Rock Slope Based on Multidimensional Cloud Model

2.1. Evaluation Indicators and Classification Criteria

Based on <Technical Regulations for Engineering Geological Survey of Slopes in Hydropower and Hydraulic Engineering> (DL/T5337-2006), the stability of rock slopes is divided into five grades: very stable, stable, basically stable, unstable and very unstable. According to the revised rock quality classification criterion (CSMR) recommended by the regulations, eight indexes are selected including slope height h (m), difference between structural plane and slope inclination $\Delta\beta$ (°), difference between structural plane inclination and slope angle $\Delta\alpha$ (°), rock uniaxial compressive strength σ_c (MPa), discontinuous plane spacing J_s (cm), structural plane characteristic value η, groundwater level condition C_w, slope excavation method coefficient K_c.The correlation analysis among the indexes indicated that the correlation is small, which proved that the eight indicators van be used to establish the index system to evaluate slope stability. The classification of each evaluation index is shown in Table 1.

Table 1. Grading threshold of rock slope stability evaluation index

Grade	h	$\Delta\beta$	$\Delta\alpha$	σ_c	J_s	η	C_w	K_c
	(m)	(°)	(°)	(MPa)	(cm)			
very stable	0~10	30~18	10~90	100~2	100~2	25~3	12~1	12~1
stable	10~30	20~30	3~10	60-100	50~10	20~2	9~12	8~12
basically stable	30~60	10~20	-3~3	30~60	30~50	10~2	6~9	4~8
unstable	60~100	5~10	-10~-	15~30	5~30	5~10	3~6	0~4
very unstable	100~100	0~5	-90~-	0~15	0~5	0~5	0~3	-3~0

2.2. Weight Determined by Projection Pursuit Method

Projection pursuit method can avoid the subjective weighting problem in the evaluation process, so projection pursuit method is used to determine the weight of evaluation index. The weighting steps of projection pursuit method are as follows:

First of all, normalize the grading evaluation indicators to obtain $\{x_{ij}^*|i=1,2,...,n; j=1,2,...,m\}$, where n is the number of samples, m is the number of indicators. For positive indicators,

$$x_{ij}^* = \frac{x_{ij} - x_{j\min}}{x_{j\max} - x_{j\min}} \tag{1}$$

For negative indicators,

$$x_{ij}^* = \frac{x_{j\max} - x_{ij}}{x_{j\max} - x_{j\min}} \tag{2}$$

where x_{ij} is the value of the jth indicator for the ith sample.

Construct projection index function

Assume P is m-dimension unit vector $P=\{P_1,P_2,...,P_m\}$, and $\sum_{j=1}^{m} P_j^2 = 1$. Then One-dimensional projection pursuit value is

$$z_i = \sum_{j=1}^{m} z_{ij} = \sum_{j=1}^{m} P_j x_{ij}^* \tag{3}$$

Where z_{ij} is the projection component value for the ith sample and the jth indicator. The aim of projection pursuit method is to find the best projection direction to maximize the projection index function $H(P)$.

$$\max H(P) = S_z D_z \tag{4}$$

Where S_z is the standard deviation for projection value z_i, D_z is the local density of the projection value z_i.

$$S_z = \sqrt{\frac{\sum_{i=1}^{n}(z_i - E(z))^2}{n-1}} \tag{5}$$

$$D_z = \sum_{i=1}^{n}\sum_{j=1}^{n}(R - r_{ij}) \times f(R - r_{ij}) \tag{6}$$

Where $E(z)$ is the mathematical expectations of z_i, R is the window width for local density, i.e., density window width, r_{ij} is the distance between samples, $r_{ij} = |z_i - z_j|$, $f(t)$ is step function, When $t \geq 0$, the value is 1.0, and when $t < 0$, the value is 0.

Determine the weights

The weights were determined by

$$w_j = \frac{P_j^2}{\sum_{j=1}^{m} P_j^2} \tag{7}$$

2.3. Multi-Dimension Cloud Model

Cloud model is put forward by Deyi Li, which can transform stochastic and fuzzy problems from qualitative description to quantitative expression. It is widely applied in rock burst prediction, water quality evaluation, slope stability evaluation and other fields [1, 7, 8, 10]. Cloud model uses three parameters to characterize its digital characteristics: mathematical expectation, entropy and hyper-entropy. Among them, E_x is the mathematical expectation of cloud droplets, which refers to the central value of cloud droplets, E_n represents the uncertainty of qualitative concepts, which is determined by the randomness and fuzziness of qualitative concepts, and H_e is the uncertainty measure of entropy, i.e. the entropy of entropy, which is determined by the randomness and fuzziness of entropy. The calculation of three parameters is follows:

$$\begin{cases} E_x = \dfrac{B_{\max} + B_{\min}}{2} \\ E_n = \dfrac{B_{\max} + B_{\min}}{6} \\ He = k \end{cases} \tag{8}$$

Where $B_{\max}$ and $B_{\min}$ are upper and lower thresholds of each evaluation index grade, respectively, k is usually assumed to be 0.1.

Assume the dimensions of cloud model are irrelative with expectations of $E_{x1}, E_{x2}, ..., E_{xm}$, entropy is expressed by $E_{n1}, E_{n2}, ..., E_{nm}$, hyper-entropy is expressed by $H_{e1}, H_{e2}, ..., H_{em}$, then

$$\mu_i = e^{-w_1 \frac{(x_i - E_{x1})^2}{2E_{n1}^2} - w_2 \frac{(x_i - E_{x2})^2}{2E_{n2}^2} - \ldots - w_m \frac{(x_i - E_{xm})^2}{2E_{nm}^2}} \tag{9}$$

Where $w_1, w_2, \ldots, w_m$ are the weights of m-dimension parameters, μ_i is the complex membership degree of samples.

The rock slope stability grade is determined according to the principle of maximum degree of membership.

3. Application

The multi-dimensional cloud model is used to evaluate the stability of 34 typical slopes [1]. Firstly, the weights of indicators were determined by projection pursuit method to be 0.08, 0.127, 0.126, 0.133, 0.134, 0.138, 0.131, and 0.131, respectively. The membership degree were determined by multi-dimension cloud model, which are shown in Table 2. The slope stability level is obtained according to the principle of maximum membership degree. Since there are only three statuses of unstable, locally unstable, and stable, the status of very unstable and unstable are taken as unstable status, basically stable is taken as locally unstable status, and very stable is taken as stable status.

From Table 2, it can be seen that stability results of 17 slopes in 34 actual slopes are in accordance with the actual stability state. In addition, 11 actual slopes using multi-dimensional cloud model are basically in agreement with the actual status with only one grade difference. Only 6 slopes have a large difference from the actual evaluation results, accounting for 17.6%. However, with CSMR method only 14 evaluation results are consistent with the actual situation, therefore the multi-dimensional cloud model based on projection pursuit can be applied to evaluate the stability of rock slope more accurately, and this method can be applied to other fields, such as water eutrophication assessment, river health assessment, and so on.

4. Conclusion

In this paper, the index system of rock slope stability evaluation is established by correlation analysis. The evaluation grade is divided into five grades: very stable, stable, basically stable, unstable and very unstable. The multi-dimensional cloud model is established by using eight parameters, such as slope height, difference between structural plane, and slope inclination, etc. The weight of the index is determined by projection pursuit method objectively. The proposed method is applied to the stability evaluation of 34 actual slopes. The evaluation results are close to the actual stability state. It indicated that the method can be applied to evaluate the stability of rock slopes and other fields.

Table 2. Comparison of the evaluation results between proposed method and other methods

ope cases	Results					Metho in t
	Very	Stable	Basically stable	Unstable	Very	
of Zhexi Hydropower Station	0.107	0.718	**0.733**	0.525	0.153	
at Lijiaxia Dam	0.064	**0.543**	0.541	0.377	0.153	
at Lijiaxia Dam	0.111	**0.505**	0.451	0.288	0.105	
ngqiao II Hydropower Station	0.035	0.497	**0.685**	0.651	0.410	
gfen Hydropower Station	0.058	0.264	**0.327**	0.302	0.201	
e of Miaojiaba Hydropower Station	0.128	**0.565**	0.557	0.404	0.120	
oling Reservoir Area	0.086	**0.630**	0.605	0.426	0.193	
oshan Hydropower Station	0.094	0.640	**0.658**	0.485	0.171	
ank abutment slope	0.093	0.649	**0.671**	0.501	0.214	
ouse of Nanyi Reservoir	0.086	**0.573**	0.524	0.351	0.190	
hushan Hydropower Station	0.044	0.527	0.649	0.564	0.360	
of Sanbanxi Hydropower Station	0.111	0.539	**0.575**	0.438	0.126	
sion tunnel of Taipingyi Hydropower	0.073	0.508	0.722	**0.745**	0.341	
rsion tunnel of Lijiaxia Hydropower	0.077	0.657	**0.820**	0.725	0.330	
r Station Taishui Canal Slope	0.205	0.811	**0.845**	0.652	0.200	
ingyi Hydropower Station	0.095	0.636	**0.809**	0.751	0.261	
of Manwan Hydropower Station	0.091	0.641	**0.827**	0.775	0.316	
nal of Ertan Hydropower Station	0.075	0.509	0.718	**0.747**	0.357	
tation Hongdong inlet slope	0.065	0.513	**0.653**	0.616	0.308	
iaojiaba Hydropower Station	0.100	**0.320**	0.280	0.185	0.100	
er Station Dahuang Yai slope	0.103	**0.658**	0.650	0.459	0.180	
Gongboxia Hydropower Station	0.111	**0.611**	0.594	0.416	0.153	
se of Tianshengqiao I Hydropower	0.105	0.768	**0.905**	0.749	0.230	
hengqiao I Hydropower Station	0.128	0.746	**0.757**	0.570	0.263	
Xiaoxia Shi Hydropower Station	0.143	**0.438**	0.421	0.292	0.087	
shixia Hydropower Station	0.180	**0.654**	0.574	0.365	0.079	
Xiangwan Hydropower Station	0.076	**0.613**	0.542	0.345	0.169	
shixia Hydropower Station	0.077	0.597	**0.672**	0.538	0.219	
qiao second stage hydropower station	0.069	**0.592**	0.549	0.357	0.164	
Tianshengqiao II Hydropower Station	0.064	0.569	**0.657**	0.531	0.201	
Miaojiaba Hydropower Station		**0.608**	0.508	0.318	0.074	
he Three Gorges of the Yangtze River		**0.326**	0.313	0.214	0.115	
f Pubugou Hydropower Station	0.266	**0.650**	0.518	0.301	0.050	
Tianshengqiao II Hydropower Station	0.113	0.283	**0.306**	0.237	0.065	

table, U=Unstable, and LS=Locally stable

IOP Conf. Series: Materials Science and Engineering **719** (2020) 012007 doi:10.1088/1757-899X/719/1/012007

5. Reference

[1] Chen Zhongyuan, Dai Zihang. Evaluation method of rock slope stability based on cloud model. *Journal of Water Resources and Architectural Engineering*, 2018, 16 (6): 62-66

[2] Zhu Yuping, Mo Haihong. Application of gray correlation analysis to rock slope stability estimation. *Chinese Journal of Rock Mechanics and Engineering*, 2004, 23 (6): 915-919.

[3] Wen Tingxi, Zhu Jing. On the prediction for the slope stability based on the SAPSO-ELM. *Journal of Safety and Environment*, 2018 (6): 2146-2150.

[4] Zhou Jing, Li Fuyuan, He Zhimin, et al. Stability evaluation of high steep rock slope based on the weighted least-square method and extencies. *Subgrade Engineering*, 2018, 4: 1-6.

[5] Wang Shaoyang, Zhang Hui. Fuzzy random reliability analysis of the stability of bedding rock slope. *Water Power*, 2018, 44 (11): 31-34.

[6] Li Xunzhen, Li Jiming, Li Shengwei. Stability probability classification method for rocj slope in hydropower engineering regions. *Chinese Journal of Rock Mechanics and Engineering*, 2011, 30: 4025-4032.

[7] Guo Jiang, Zhang Weixing, Zhao Yan. A multidimensional cloud model for rockburst prediction. *Chinese Journal of Rock Mechanics and Engineering*, 2018, 37 (5): 1199-1206.

[8] Zeng Debiua. Entrophication assessment thod based on multidimensional cloud model. *Dissertationof Nanjing University*[M], 2015.

[9] Li Xiuzhe, Kong Jiming, Wang Huacheng. Modification of rock slope stability classifications system by continuous functions and its applications. *Chinese Journal of Rock Mechanics and Engineering*, 2010, 29: 3439-3446.

[10] Guo Rongxiao, Xia Jingbo, Dong Shufu, et al. Multiple attribute evaluation method based on multidimensional cloud model. *Computer Science*, 2010, 37, 11: 75-77.

IOP Conf. Series: Materials Science and Engineering **719** (2020) 012008 doi:10.1088/1757-899X/719/1/012008

Research on Security Risk Assessment Based on the Improved FAHP

Wenmin Li[1,*], Ye Liang[2], Wenyu Wang[1], Xueqi Jin[3], Piaohong Kong[3], Zhengwei Jiang[3] and Lisong Shao[2]

[1]State Key Laboratory of Networking and Switching Technology, Beijing University of Posts and Telecommunications, Beijing, 100876, China
[2]NARI Group Corporation, Nanjing 211106, China
[3]State Grid Zhejiang Electric Power Co. Ltd., Hangzhou 310007, China
*liwenmin02@outlook.com

Abstract. In order to reduce the subjectivity of information security risk assessment process and improve assessment efficiency, we propose a new method of information security risk assessment based on improved FAHP (Fuzzy Analytic Hierarchy Process) to analyse the information security-related standards for domestic and international risk assessment. We establish a Hierarchical Security Assessment Model and introduce refinement indicators and Intuitionistic Fuzzy Sets to reduce subjective judgment factors in the assessment of traditional risk. We then applied an e-commerce company in case analyse the security risk and the results are satisfactory and in line with the actual situation of the company. The indicator system of this method is more objective and comprehensive and the evaluation process is more efficient, which provide new ideas for risk assessment of existing information security companies.

1. Introduction

Network security threats such as APT and account fraud are increasing unabated and the scope of which is extending and threat every internet user. The threat posed by new technologies such as cloud computing and big data, which need to be solved, are making urgent security issues. In December 2017, Gartner predicted[1] that the total security expenditure of global enterprises will reach 96.3 billion dollars next year, an increase of 8% compared with 2017. In order to ensure the safety of the organization business, safety investment will be increased within five years. Besides, it is an inevitable trend to ensure the rationality, compliance and correctness of business security with the help of internal and external forces.

Information Security Risk Assessment is an assessment of the threat, vulnerability and hidden risk in an organization's information security management system, which runs through the organization, strategy, operation, technology, evaluation and other stages of the information security system [2].

Laarhoven et al. [3] first proposed the combination of AHP and Fuzzy theory in 1983. Bian et al. [4] combined AHP with fuzzy comprehensive evaluation by using AHP_FCE model, taking into account the complexity of relationship within the system and the fuzziness of value system. Gu et al. [5] combined with three evaluation technologies AHP-FAHP-FCE to build a model and solve the security situation evaluation from different perspectives. Zhao et al. [6] used AHP and Fuzzy Logic method to evaluate the risk of network security and wireless network security respectively. Yu et al. [7] analyzed the risk factors involved in the information system with the help of the Fuzzy Sets theory and gave the level description. However, these methods do not completely get rid of the subjective influence of experts' evaluation of the judgment matrix. At the same time, due to the high complexity of the evaluation method and the high industry barriers, the theoretical research is far from the practical application, and there is no real practical risk evaluation scheme.

In this paper, a Fuzzy Analytic Hierarchy Process (FAHP) based information security risk assessment scheme is proposed. The objective and perfect general index system is used to replace the subjective evaluation of experts in the traditional assessment process. A scientific hierarchical security assessment model is established from the five dimensions of risk analysis, risk prevention, risk control, risk treatment, monitoring and improvement, and the detailed index system and Intuitionistic Fuzzy Sets are combined, which makes complex theoretical research more practical.

2. Construction of Information Security Risk Assessment Scheme Index System

2.1. Risk Assessment Related Definitions

According to GB / T 20498-2007 [8], the function of assessing information security risk is as follows:

$$r=\varphi(w,v,t) \tag{1}$$

Where r, w, v and t respectively represent information security risk value, asset scope, vulnerability impact and consequences caused by threats. Security risk needs to consider the possibility and impact of risks. Vulnerability comes from organizational assets, and threats use vulnerability to cause risk impact on the organization. Therefore, a scientific and comprehensive risk assessment index system should cover w, v, t and related aspects, so as to make the most objective and real assessment of the complex information security management system.

2.2. Index System Construction

This paper divides the information security index system into target level, criterion level (primary indicator) and indicator level (secondary indicator) according to the hierarchical concept of AHP method, covering five aspects: strategy, organization, operation, technology and evaluation. The criterion level consists of five dimensions including risk analysis, risk prevention, risk control, risk treatment and monitoring and improvement, as shown in Table 1:

Table 1. Information security risk assessment index system

target level	criterion level	indicator level
Risk Assessment(A)	Risk Analysis(B_1)	Business Concept (C_1) , Risk Identification (C_2) Risk Classification (C_3)
	Risk Prevention(B_2)	Security Organizations (C_4), Training And Drill (C_5) Business Continuity (C_6)
	Risk Control(B_3)	Security Strategy And Procedures(C_7), Personnel Safety (C_8), Physical Environment Security (C_9), Network Partition (C_{10}) Access Control (C_{11})
	Risk Treatment (B_4)	Event Response (C_{12}), System Maintenance (C_{13}) Information And Document Management (C_{14})
	Monitoring and Improvement (B_5)	Consistency (C_{15}) Review of Improvement And Maintenance (C_{16})

Risk analysis consists of three secondary indicators: business concept, risk identification and risk classification. Risk prevention includes three secondary indicators: safety organization, training and drill, and business continuity. Risk control includes five secondary indicators: security strategies and regulations, personnel security, physical environment security, network partition and access control. Risk management consists of three secondary indicators: event response, system maintenance and information and document management. Monitoring and improvement are composed of two secondary indicators: consistency, review improvement and maintenance.

3. Information Security Risk Assessment based on FAHP

3.1. Overview of Intuitionistic Fuzzy Sets

The Fuzzy Analytic Hierarchy Process (FAHP) overcomes the limitations of the traditional analytic hierarchy process, such as the difficulty of matrix consistency, the complexity of adjusting the judgment matrix to meet the consistency [9], and increases the objectivity of the evaluation results.

Definition 1 [10-11] if x is a fixed non-empty set, $A = \{\langle x, \mu_A(x), v_A(x)\rangle | x \in X\}$, such triples are Intuitionistic Fuzzy Set, where $\mu_A(x), v_A(x)$ represent the membership and non-membership degree of the element x in X, and satisfy $\mu_A\colon X \longrightarrow [0,1], v_A\colon X \longrightarrow [0,1], 0 \leq \mu_A(x) + v_A(x) \leq 1, x \in X$.

Definition 2 [10-11] In every Fuzzy Set of X,$\pi_A(x) = 1 - \mu_A(x) - v_A(x)$ is the degree of hesitation that element x belongs to X, and $0 \leq \pi_A(x) \leq 1, x \in X$.

Intuitionistic Fuzzy Set in this paper is $\alpha = (\mu_\alpha, v_\alpha, \pi_\alpha)$, We describe the corresponding risk assessment index questionnaire. For example,$(\mu_\alpha, v_\alpha, \pi_\alpha) = (0.6,0.3,0.1)$, or$\mu_\alpha = 0.6, v_\alpha = 0.3, \pi_\alpha = 0.1$. The meaning is: suppose there are ten detailed questions in a secondary index of the index questionnaire, among which 6 questions are "Yes", 3 questions are "No", and 1 question is "Not Sure ".

3.2. Information Security Risk Assessment Process

3.2.1. Construction of information security risk assessment index system

According to the hierarchical concept of AHP method, the information security index system is divided into target layer, criterion layer and indicator layer. The criterion layer is composed of five dimensions, which are risk analysis, risk prevention, risk control, risk treatment, monitoring and improvement, as shown in table 1. An instance of the detailed questionnaire part of the indicator system is shown in Figure 1.

1. Does the enterprise set the recovery goal of the system in advance for the establishment of business continuity plan?
2. Does the recovery objective of the system include the recovery of all communication links, the functions of industrial automation and control system, and the recovery time objective or time to recover these links and functions?
1. Does the enterprise have an early assessment of the impact and consequences of each system?
2. Is there a corresponding assessment of the impact of simultaneous failures of multiple systems?
3. Do you ensure that the backup system does not affect normal plant operations?

Figure 1. Business continuity indicator refinement questionnaire

3.2.2. Construction of judgment matrix of Intuitionistic Fuzzy Sets

According to the importance preference degree of information security indicators, this paper defines the quantitative scale of importance evaluation [26], as shown in Table 2.

Table 2. Quantitative scale of importance

Importance evaluation	Intuitionistic Fuzzy number
Factor i is much more important than factor j	(0.90,0.10,0.00)
Factor i is much more important than factor j	(0.80,0.15,0.05)
Factor i is more important than factor j	(0.70,0.20,0.10)
Factor i is more important than factor j	(0.60,0.25,0.15)
Factor i is as important as factor j	(0.50,0.30,0.20)
Factor j is more important than factor i	(0.40,0.45,0.15)
Factor j is more important than factor i	(0.30,0.60,0.10)
Factor j is much more important than factor i	(0.20,0.75,0.05)
Factor j is much more important than factor i	(0.10,0.90,0.00)

We construct the judgment matrix of Intuitionistic Fuzzy Sets. The judgment matrix of Intuitionistic Fuzzy Sets is $R = (r_{ij})_{n\times n}(i,j = 1,2 \ldots n)$, i and j respectively represent the rows and columns of the matrix, where $r_{ij} = (\mu_{ij}, v_{ij})$, μ_{ij} indicates the importance of index i relative to index j, γ_{ij} indicates the importance of index j relative to index I, and $0 \leq \mu_{ij} + v_{ij} \leq 1$. In order to convert the scoring results of the index questionnaire into the relative importance elements of the judgment matrix of Intuitionistic Fuzzy Sets, this paper considers the universality of the safety index and the rationality of the limit case, and defines the fuzzy mapping matrix $M = (m_{ij})_{n\times n}$ by combining the fitting function with the quantitative scale, where i and j respectively represent the rows and columns of the matrix, and m_{ij} represents two comparative indexes. The importance of index with score i compared with index with score j is m_{ij}. A complete judgment matrix of Intuitionistic Fuzzy Sets can be obtained by normalizing the reciprocal distance of the four vertices mapped to the matrix region as the weight. The pseudo code is given in algorithm 1.

algorithm **1:** Inverse distance normalized weighting

Input: Fuzzy mapping matrix M , 0-1 cut distance, list A of Fuzzy Membership Degree of each index;

Output: Judgment matrix of Intuitionistic Fuzzy Sets $R = (r_{ij})_{n\times n}$;

1. **for** $index = i$ in A **do**
2. **for** $index = j$ in A **do**
3. Solve the minimum rectangle position of $(A[i], A[j])$ in M, and get the four vertex coordinates LU, LB, RU, RB *of the rectangle.*
4. Calculate the distance between $(A[i], A[j])$ and the vertex of the rectangle, and get the values of *D_LU,D_LB,D_RU,D_R.*
5. Normalized distance: *Norm(D_LU, D_LB, D_RU, D_RB)*;
6. Weighted weight: $M[i][j] = S[i][j] \times D\_LU + S[i+1][j] \times D\_RU + S[i][j+1] \times D\_LB + S[i+1][j+1] \times D\_RB$
7. **end** for
8. **end** for

3.2.3. Construction consistency judgment matrix of Intuitionistic Fuzzy Sets

In this paper, $\bar{R} = (\bar{r}_{ij})_{n\times n}$, $\bar{r}_{ij} = (\bar{\mu}_{ij}, \bar{v}_{ij})$ is used to express the consistency judgment matrix of Intuitionistic Fuzzy Sets, which is based on the judgment matrix $R = (r_{ij})_{n\times n}$, $r_{ij} = (\mu_{ij}, v_{ij})$, The process is as follows [25]:

(1) When $j > i + 1$:

$$\mu_1 = \sqrt[j-i-1]{\prod_{t=i+1}^{j-1} \mu_{it}\mu_{tj}} \,, \mu_2 = \sqrt[j-i-1]{\prod_{t=i+1}^{j-1} (1-\mu_{it})(1-\mu_{tj})} \,, \bar{\mu}_{ij} = \frac{\mu_1}{\mu_1+\mu_2} \tag{2}$$

$$v_1 = \sqrt[j-i-1]{\prod_{t=i+1}^{j-1} v_{it}v_{tj}} \,, v_2 = \sqrt[j-i-1]{\prod_{t=i+1}^{j-1} (1-v_{it})(1-v_{tj})} \,, \bar{v}_{ij} = \frac{v_1}{v_1+v_2} \tag{3}$$

(2)When $j < i, \bar{r}_{ij} = (\bar{v}_{ji}, \bar{\mu}_{ji})$;

(3)When $j = i + 1, \bar{r}_{ij} = (\mu_{ij}, v_{ij})$.

3.2.4. Consistency test

According to the distance measure between Fuzzy Sets [13][14], the consistency test process of the judgment matrix of Fuzzy Sets is as follows [13]:

$$\Delta_\mu = \sum_{i=1}^{n}\sum_{j=1}^{n}(|\bar{\mu}_{ij} - \mu_{ij}|) \,, \Delta_v = \sum_{i=1}^{n}\sum_{j=1}^{n}(|\bar{v}_{ij} - v_{ij}|) \,, \Delta_\pi = \sum_{i=1}^{n}\sum_{j=1}^{n}(|\bar{\pi}_{ij} - \pi_{ij}|),$$

$$d(\bar{R}, R) = (\Delta_\mu + \Delta_v + \Delta_\pi)/2(n-1)(n-2) \tag{4}$$

Where R is the judgment matrix of Intuitionistic Fuzzy Sets, $\bar{R}$ is the consistency judgment matrix of Intuitionistic Fuzzy Sets, $i, j = 1,2 \dots n$. If $d(\bar{R}, R) < 0.1$, then $\bar{R}$ passes the consistency test. If $d(\bar{R}, R) \geq 0.1$, it needs to adjust the consistency.

3.2.5. Consistency adjustment
When $\bar{R}$does not meet the consistency, the iterative parameter σ needs to be introduced, and the value of σ needs to be reduced in a certain step to change the consistency judgment matrix of Intuitionistic Fuzzy Sets, so that it finally meets the consistency condition. In this paper, $\sigma = 1 (\sigma > 0)$, and the step $- 0.01$ is used for iteration. The process is as follows [13]:

$$\tilde{\mu}_{ij} = \frac{(\mu_{ij})^{1-\sigma}(\bar{\mu}_{ij})^{\sigma}}{(\mu_{ij})^{1-\sigma}(\bar{\mu}_{ij})^{\sigma}+(1-\mu_{ij})^{1-\sigma}(1-\bar{\mu}_{ij})^{\sigma}} \quad \tilde{v}_{ij} = \frac{(v_{ij})^{1-\sigma}(\bar{v}_{ij})^{\sigma}}{(v_{ij})^{1-\sigma}(\bar{v}_{ij})^{\sigma}+(1-v_{ij})^{1-\sigma}(1-\bar{v}_{ij})^{\sigma}} \tag{5}$$

Where $i, j = 1,2 \dots n$. Through the above formula, the consistency judgment matrix of the adjusted Intuitionistic Fuzzy Sets $\tilde{R} = (\widetilde{r_{ij}})_{n \times n}, \widetilde{r_{ij}} = (\tilde{\mu}_{ij}, \tilde{v}_{ij})$ is obtained, and the consistency is checked again by publicity (6). If it is satisfied, the consistency is checked. If it is not satisfied, step (5) is repeated until it passes the consistency test.

$$\Delta_{\mu} = \sum_{i=1}^{n}\sum_{j=1}^{n}(|\tilde{\mu}_{ij} - \mu_{ij}|) \text{ , } \Delta_{v} = \sum_{i=1}^{n}\sum_{j=1}^{n}(|\tilde{v}_{ij} - v_{ij}|) \text{ , } \Delta_{\pi} = \sum_{i=1}^{n}\sum_{j=1}^{n}(|\tilde{\pi}_{ij} - \pi_{ij}|),$$

$$d(\tilde{R}, R) = (\Delta_{\mu} + \Delta_{v} + \Delta_{\pi})/2(n-1)(n-2) \tag{6}$$

3.2.6. Determine the weight
Finally, the weight of each index is calculated by the consistency judgment matrix of Intuitionistic Fuzzy Sets through consistency test, as formula (7), where $i, j = 1,2 \dots n$.

$$\omega_i = \left(\frac{\sum_{j=1}^{n}\bar{\mu}_{ij}}{\sum_{i=1}^{n}\sum_{j=1}^{n}(1-\bar{v}_{ij})}, 1 - \frac{\sum_{j=1}^{n}(1-\bar{v}_{ij})}{\sum_{i=1}^{n}\sum_{j=1}^{n}\bar{\mu}_{ij}}\right) \tag{7}$$

3.2.7. Final evaluation results
According to the operation rule of Intuitionistic Fuzzy Sets[15], the weights of each secondary index and its corresponding primary index are weighted and carried out set operation, and the process is as follows [13]:

$$\omega_1 \otimes \omega_2 = (\mu_{\omega1}\mu_{\omega2}, v_{\omega1} + v_{\omega2} - v_{\omega1}v_{\omega2}), \ \omega_1 \oplus \omega_2 = (\mu_{\omega1} + \mu_{\omega2} - \mu_{\omega1}\mu_{\omega2}, v_{\omega1}v_{\omega2}) \tag{8}$$

Calculate the weight of indicators at all levels as formula (9), refer to the index system in this paper, $n = 16, j = 5$. Calculate total weight as formula (10). In order to facilitate the rating of the organization's security status, this paper defines the scoring function of information security risk assessment according to the actual situation as formula (11).

$$\omega(C_i) = \omega_{B_j} \otimes \omega_{C_i}, i = 1,2 \dots n, j = 1,2 \dots m \tag{9}$$

$$W = \oplus_{i=1}^{n}\omega(C_i) \ = 1,2 \dots n \tag{10}$$

$$h(W) = 1 - 0.5(1 - \mu_W)(1 + \pi_W) \tag{11}$$

3.2.8. Maturity level division
In this paper, CMMI [16] is used as the basis for the classification of information security level. Combined with the final evaluation results, the model is divided into five levels (initial level, spontaneous management level, process definition level, quantitative control level and optimal level) by using interval division method, which marks the five levels of maturity of enterprise information security capability, as shown in Table 3.

Table 3. Enterprise information security maturity classification

Maturity level	Grade standards	Interval
Initial Level 1	The importance and necessity of this safety control measure have been realized, but no standardized process or mode has been formed, and the implementation and management of the control measure are unorganized.	(0,0.2)
Spontaneous Management Level 2	It has started to follow some conventional process or mode, but it lacks of documented and standardized system support, and there is no education, training and responsibility definition related to it.	(0.2,0.4)
Process Definition Level 3	It has been supported by standardized and documented systems, as well as related education, training and responsibility definitions, but it is difficult to find problems in the implementation process due to the lack of monitoring and quantitative management.	(0.4,0.6)
Quantitative Control Level 4	Comprehensive monitoring and quantitative management have been implemented to ensure that the implementation process is carried out in strict accordance with the system requirements, and automation tools are also applied to this control measure.	(0.6,0.8)
Optimal Level 5	It has reached or exceeded the level of best practice, and made continuous improvement according to the implementation effect or reference benchmark level.	(0.8,1)

4. Case Analysis-an E-Commerce Company

4.1. Information Security Risk Assessment of

Taking an e-commerce company A as an example, this paper makes a risk assessment on the current situation of information security of the e-commerce company by using the improved FAHP scheme. The company has 11 departments, including management, business travel business department and credit business department, and 11 people participate in filling in the information security index questionnaire. Due to the length, this paper only shows the process of evaluating the results of one of the questionnaires.

4.2. Construction of judgment matrix of Intuitionistic Fuzzy Sets

5 primary indicators and 16 Secondary indicators of the questionnaire are as S_B, S_C : After normalization and algorithm 1 processing, the Intuitionistic Fuzzy Sets judgment matrix R_B of the primary indicators is obtained:

$$S_B=\begin{Bmatrix} & (33,10,9) & (39,12,5) & \\ (102,13,23) & (52,9,12) & (26,4,8) & \end{Bmatrix} S_C=\begin{pmatrix} (2,0,3) & (12,5,3) & (19,5,3) & (9,5,0) \\ (18,4,1) & (12,3,4) & (12,1,4) & (14,3,3) \\ (20,1,5) & (6,0,3) & (50,8,8) & (20,1,5) \\ (21,4,4) & (11,4,3) & (11,2,4) & (15,2,4) \end{pmatrix}$$

$$R_B=\begin{bmatrix} (0.50,0.30) & (0.45,0.30) & (0.45,0.27) \\ (0.60,0.30) & (0.50,0.30) & (0.50,0.27) \\ (0.60,0.35) & (0.50,0.35) & (0.50,0.30) \\ (0.55,0.35) & (0.50,0.35) & (0.70,0.30) \\ (0.60,0.35) & (0.50,0.35) & (0.60,0.30) \\ & (0.45,0.27) & (0.45,0.27) \\ & (0.50,0.27) & (0.50,0.27) \\ & (0.55,0.30) & (0.60,0.30) \\ & (0.50,0.30) & (0.70,0.30) \\ & (0.60,0.35) & (0.50,0.30) \end{bmatrix}$$

4.3. Consistency Test

First, the consistency judgment matrix $\bar{R}_B$ of Intuitionistic Fuzzy Sets of the primary indicators is obtained by formula (2) and (3) , and then $d(\bar{R}_B,R_B)\approx 0.4335>0.1$ is obtained by formula (4), which does not meet the consistency condition. Using the same method, the consistency judgment matrix of Intuitionistic Fuzzy Sets of secondary indicators, $\bar{R}_{C_1},\bar{R}_{C_2},\bar{R}_{C_3},\bar{R}_{C_4},\bar{R}_{C_5}$, is obtained.

$$\bar{R}_B=\begin{bmatrix}(0.50,0.50) & (0.45,0.30) & (0.45,0.14)\\ (0.30,0.45) & (0.50,0.50) & (0.50,0.27)\\ (0.14,0.45) & (0.27,0.50) & (0.50,0.50)\\ (0.14,0.45) & (0.14,0.50) & (0.30,0.50)\\ (0.14,0.45) & (0.14,0.50) & (0.16,0.50)\\ & (0.45,0.14) & (0.45,0.14)\\ & (0.50,0.14) & (0.50,0.14)\\ & (0.55,0.30) & (0.50,0.16)\\ & (0.50,0.50) & (0.70,0.30)\\ & (0.30,0.50) & (0.50,0.50)\end{bmatrix}\quad \bar{R}_{C_3}=\begin{bmatrix}(0.50,0.50) & (0.50,0.30) & (0.60,0.16)\\ (0.30,0.50) & (0.50,0.50) & (0.60,0.30)\\ (0.16,0.65) & (0.30,0.60) & (0.50,0.50)\\ (0.16,0.65) & (0.16,0.78) & (0.30,0.70)\\ (0.16,0.60) & (0.16,0.60) & (0.16,0.78)\\ & (0.65,0.16) & (0.60,0.16)\\ & (0.78,0.16) & (0.60,0.16)\\ & (0.70,0.30) & (0.78,0.16)\\ & (0.50,0.50) & (0.60,0.30)\\ & (0.30,0.60) & (0.50,0.50)\end{bmatrix}$$

$$\bar{R}_{C_1}=\begin{bmatrix}(0.50,0.50) & (0.40,0.45) & (0.35,0.23)\\ (0.45,0.40) & (0.50,0.50) & (0.45,0.27)\\ (0.33,0.28) & (0.56,0.36) & (0.50,0.50)\end{bmatrix}\quad \bar{R}_{C_4}=\begin{bmatrix}(0.50,0.50) & (0.70,0.30) & (0.78,0.19)\\ (0.30,0.70) & (0.50,0.50) & (0.60,0.35)\\ (0.19,0.78) & (0.35,0.60) & (0.50,0.50)\end{bmatrix}$$

$$\bar{R}_{C_2}=\begin{bmatrix}(0.50,0.50) & (0.55,0.25) & (0.60,0.13)\\ (0.25,0.55) & (0.50,0.50) & (0.55,0.30)\\ (0.13,0.60) & (0.30,0.55) & (0.50,0.50)\end{bmatrix}\quad \bar{R}_{C_5}=\begin{bmatrix}(0.50,0.50) & (0.45,0.30)\\ (0.30,0.45) & (0.50,0.50)\end{bmatrix}$$

Then, by formula (5), we take σ=1 and iterate with step size − 0.01, so that it finally meets the consistency constraint condition. Finally, when σ=0.19, we get the adjusted consistency judgment matrix $\tilde{R}_B$ of Intuitionistic Fuzzy Sets. After the test by formula (8), $d(\tilde{R}_B,R_B)\approx 0.0977<0.1$, the consistency test is passed. The same process is used to deal with the secondary index layer. After iteration, $d(\tilde{R}_{C_1},R_{C_1})\approx 0.0958$, $d(\tilde{R}_{C_2},R_{C_2})\approx 0.0972$, $d(\tilde{R}_{C_2},R_{C_2})\approx 0.0999$, $d(\tilde{R}_{C_2},R_{C_2})\approx 0.0959$, σ are 0.13, 0.11, 0.18, 0.10 respectively, and the adjusted judgment matrices of consistency of Intuitionistic Fuzzy Sets are $\tilde{R}_{C_1},\tilde{R}_{C_2},\tilde{R}_{C_3},\tilde{R}_{C_4},\tilde{R}_{C_5}$, as follows:

$$\tilde{R}_{C_2}=\begin{bmatrix}(0.50,0.32) & (0.55,0.25) & (0.51,0.23)\\ (0.51,0.46) & (0.50,0.32) & (0.55,0.30)\\ (0.45,0.47) & (0.52,0.32) & (0.50,0.32)\end{bmatrix}\quad \tilde{R}_{C_5}=\begin{bmatrix}(0.50,0.50) & (0.45,0.30)\\ (0.30,0.45) & (0.50,0.50)\end{bmatrix}$$

$$\tilde{R}_{C_1}=\begin{bmatrix}(0.32,0.32) & (0.40,0.45) & (0.35,0.33)\\ (0.58,0.27) & (0.50,0.32) & (0.45,0.27)\\ (0.33,0.28) & (0.56,0.36) & (0.50,0.32)\end{bmatrix}\quad \tilde{R}_{C_4}=\begin{bmatrix}(0.50,0.31) & (0.70,0.30) & (0.58,0.33)\\ (0.57,0.37) & (0.50,0.32) & (0.60,0.35)\\ (0.50,0.32) & (0.44,0.30) & (0.50,0.32)\end{bmatrix}$$

$$\tilde{R}_B=\begin{bmatrix}(0.50,0.33) & (0.45,0.30) & (0.45,0.24)\\ (0.54,0.33) & (0.50,0.34) & (0.50,0.27)\\ (0.49,0.37) & (0.45,0.38) & (0.50,0.34)\\ (0.49,0.37) & (0.41,0.38) & (0.46,0.34)\\ (0.49,0.37) & (0.41,0.38) & (0.42,0.34)\\ & (0.45,0.24) & (0.45,0.24)\\ & (0.50,0.24) & (0.50,0.24)\\ & (0.50,0.30) & (0.50,0.27)\\ & (0.50,0.34) & (0.50,0.30)\\ & (0.46,0.34) & (0.50,0.34)\end{bmatrix}\quad \tilde{R}_{C_3}=\begin{bmatrix}(0.50,0.33) & (0.46,0.33) & (0.60,0.35)\\ (0.50,0.30) & (0.50,0.33) & (0.63,0.35)\\ (0.60,0.27) & (0.60,0.30) & (0.50,0.33)\\ (0.53,0.27) & (0.56,0.27) & (0.70,0.30)\\ (0.60,0.27) & (0.60,0.27) & (0.56,0.27)\\ & (0.42,0.36) & (0.59,0.5)\\ & (0.42,0.39) & (0.60,0.35)\\ & (0.55,0.37) & (0.42,0.38)\\ & (0.50,0.33) & (0.63,0.35)\\ & (0.56,0.27) & (0.50,0.33)\end{bmatrix}$$

4.4. Evaluation Results

Take the $\tilde{R}_B$ passing the consistency test into formula (7) to get the weight of each primary indicator, and use the same method to calculate the secondary indicators $\tilde{R}_{C_1}, \tilde{R}_{C_2}, \tilde{R}_{C_3}, \tilde{R}_{C_4}, \tilde{R}_{C_5}$ of consistency test to get the final weight information, as shown in Table 4:

Table 4. Final weight information of all levels of indicators

Primary indicators (ω_{B_i})	Secondary indicators(ω_{C_i})
(0.1344, 0.6935)	(0.1770, 0.5260), (0.2524, 0.4644), (0.2289, 0.4907)
(0.1484, 0.6985)	(0.2599, 0.5216), (0.2605, 0.5822),(0.2445, 0.5888)
(0.1426, 0.7186)	(0.1610, 0.7396), (0.1603, 0.7443), (0.1761, 0.7514) (0.1471, 0.7619), (0.1622, 0.7635)
(0.1377, 0.7246)	(0.2905, 0.5809), (0.2731, 0.5923), (0.2369, 0.5776)
(0.1328, 0.7277)	(0.4222, 0.3143), (0.3556, 0.4000)

Finally, weight aggregation is carried out by formula (9)~(10), the total weight $W=(0.4174, 0.1299)$ can be obtained, and the final risk assessment score $h(W)=1\text{-}0.5(1\text{-}0.4174)(1+0.1299)=0.5989$ can be obtained by formula (11). According to the information security maturity rating table, this questionnaire shows that company a is at Level 3 (process definition level).

The average of 11 results was taken, and the final total score was 0.5867, which was at the level of process definition. Last year, the company obtained 0.54 points through other risk assessment schemes, indicating that the company is at the third level of maturity. The scheme proposed in this paper shows that the company's score is higher, because the company has made supplementary improvements in safety planning and other aspects after last year's assessment. Although it has been supported by standardized and documented systems, it is difficult to find problems in the implementation process due to the lack of monitoring and quantitative management, so it is still at the third level. This scheme is more efficient and less time-consuming, and because the index system is designed according to the five aspects of strategy, organization, operation, technology and evaluation, the evaluation results can clearly reflect the gap between the company and the target level. In addition, compared with the scheme of [25], the scheme proposed in this paper is better in dealing with the limit case (the questionnaire scores are all ' Yes ' or ' No ').

5. Conclusion

In order to solve the problems in traditional information security risk assessment, this paper proposes an information security risk assessment scheme based on improved FAHP, and establishes a five-dimensional detailed index system. Taking an e-commerce company as an example, this scheme is used to evaluate the information security risk of it. The results show that the evaluation results of the scheme are consistent with the actual situation, and the evaluation process is efficient.

6. Acknowledgement

This research was funded by the Science and Technology Project Funding of State Grid Corporation of China (Research and Application of Network Security Situation Awareness Technology in Power Monitoring System in 2018-2019, Contract no.: 521104180028).

7. References

[1] *Forecast: Information Security,* Worldwide, 2015-2021, 3Q17 Update
[2] Li Z, Jianfen P, Yuge D U, et al. Information security risk assessment survey[J]. *Journal of Tsinghua University(Science and Technology)*, 2012, 25(1): 1578-1584.
[3] Laarhoven P J M V, Pedrycz W. A fuzzy extension of Saaty's priority theory[J]. *Fuzzy Sets & Systems*, 1983, 11(1): 199-227.
[4] Bian N, Wang X, Li M. Network security situational assessment model based on improved

AHP_FCE[C]//*International Conference on Advanced Computational Intelligence*. 2013.

[5] Zhao-Jun G U, Rui-Li W. A security situation assessment model of information system based on improved fuzzy analytical hierarchy process[J]. *Computer Engineering & Science*, 2016.

[6] Zhao D M, Wang J H, Ma J F. Fuzzy Risk Assessment of the Network Security[C]// *International Conference on Machine Learning & Cybernetics*. 2009.

[7] Yu Fu, Xiaoping Wu, Qing Ye, et al. Study on security risk assessment of information system based on Fuzzy Set and Entropy Weight Theory [J]. Journal of Electronics, 2010, 38(7): 1489-1494.

[8] GB/T 20984-2007, Code for risk assessment of information security [S]. Beijing: China Standards Press.

[9] Jijun Zhang. Fuzzy Analytic Hierarchy Process (FAHP)[J]. Fuzzy system and Mathematics, 2000, 14(2): 80-88.

[10] Atanassov K, Pasi G, Yager R. Intuitionistic fuzzy interpretations of multi-person multi-criteria decision making[C]// *Intelligent Systems, First International IEEE Symposium*. 2002.

[11] Atanassov K T. *On Intuitionistic Fuzzy Sets Theory*[M]. 2012.

[12] Xu Z, Liao H. Intuitionistic Fuzzy Analytic Hierarchy Process[J]. *IEEE Transactions on Fuzzy Systems*, 2014, 22(4): 749-761.

[13] Fahmi A, Derakhshan A, Kahraman C . Human resources management using interval valued intuitionistic fuzzy analytic hierarchy process[C]// *IEEE International Conference on Fuzzy Systems. IEEE*, 2015.

[14] Szmidt E, Kacprzyk J. Amount of Information and Its Reliability in the Ranking of Atanassov's Intuitionistic Fuzzy Alternatives[M]// Recent Advances in Decision Making. *Springer Berlin Heidelberg*, 2009.

[15] Xia M, Xu Z, Zhu B. Some issues on intuitionistic fuzzy aggregation operators based on Archimedean t-conorm and t-norm[J]. *Knowledge-Based Systems*, 2012, 31(none): 78-88.

[16] Chrissis M B, Konrad M, Shrum S. CMMI Guidlines for Process Integration and Product Improvement[M]. 2003.

Study of Internet of Things Technology Applied in Intelligent Substation Auxiliary System

Chuankai Yang[1], Jingfeng Wu[1], Bingquan Tao[2*], Xiaolan Zhang[1], Liangshu Li[1] and Haoyue Qu[2]

[1] State Grid Shaanxi Electric Power Research Institute, Xi'an, Shaanxi Province, China
[2] School of Electrical Engineering and Automation, Wuhan University, Wuhan, Hubei Province, China
Email: 1292710292@qq.com

Abstract. In order to improve the linkage of related auxiliary systems in intelligent substation and realize the complete monitoring and control functions of the equipment in the station, this paper integrates the technologies of Internet of Things, edge computing and big data to establish an auxiliary system of IoT intelligent substation. The overall structure of the system is introduced in detail. The effective calculation and reduction of the computing task of a single edge node is achieved by the application function and the edge calculation algorithm of task balancing allocation and scheduling, effectively improving the immediacy and security of the intelligent substation auxiliary system of the Internet of Things. The construction of transportation inspection has certain guiding significance.

1. Introduction

With the expansion of the scale of the power grid, the number of substations has increased dramatically. In order to improve labor efficiency and rationally allocate human resources, the substation is generally implemented with unattended and centralized monitoring modes. In this mode, the safety issues in the substation are particularly prominent. In recent years, substation equipment has been destroyed, and theft has occurred from time to time, causing some power outages. Therefore, it is an urgent need to implement practical and feasible technical means to monitor the relevant auxiliary systems in the substation, so that the safety control and intelligent control of the unattended substation becomes an urgent need. With the development of intelligent technology, intelligent robot inspection has become a possibility to replace manual inspection, which greatly reduces manpower and provides assistance for the rational deployment of personnel. However, since the patrol route and the patrol time of the robot are relatively fixed, the patrol content and the corresponding control functions are relatively independent, and they are not combined with the related auxiliary control system, the complete monitoring and control functions in the station cannot be realized[1]-[3].

At present, there are many kinds of production auxiliary systems such as video monitoring system, robot inspection system, conventional auxiliary control system and security perimeter system in substation. Each system only realizes some basic functions such as data acquisition, equipment monitoring and auxiliary inspection. Various systems are relatively independent, and the application scenarios are incomplete. There is still room for improvement in advanced applications such as data sharing, information interaction, intelligent decision-making, and intelligent linkage. Therefore, series of substation intelligent IoT integrated application systems are urgently needed, and the advanced application and data analysis functions of “substation operation task-oriented” are realized by collecting multi-dimensional big data.

Therefore, this paper applies the Internet of Things technology to the substation auxiliary system, and builds a set of IoT intelligent substation auxiliary system with its own characteristics. Combined with related equipment or system on the measurement and control network platform, it realizes security fire monitoring, environmental monitoring, power monitoring and operation assistance, maintenance assistance, operation status monitoring, pre-alarm and other system integration application, linkage control, asset management and other functions, achieving intelligent substation safe operation and safe operation management, improving information level of the intelligent substation full station, and providing decision-making technical support for the grid safe operation.

2. System Architecture

The Internet of Things (IoT) is an intelligent sensing network that collects data information through intelligent sensing devices of objects and transmits them via the network. The data is transmitted to a designated information processing center for processing to realize information interaction between people and objects, objects and objects. The academic community defines the Internet of Things as a network that intelligently identifies, locates, tracks, monitors, and manages through information sensing devices such as radio frequency identification, infrared sensors, global positioning systems, laser scanners, etc., connectting any item to the Internet for information exchange and communication according to the agreed protocol.

The whole operation process of the Internet of Things can be called IoT technology. In this paper, the technology is applied to the intelligent substation auxiliary system, and an IoT intelligent substation auxiliary system is constructed by using the characteristics of the Internet of Things. The system integrates various Internet of Things applications such as video surveillance, online monitoring, security perimeter, fire monitoring, environmental monitoring, infrared imaging, inspection robots, etc. It also realizes vertical integration of information between various IoT applications, horizontal data penetration, and highly integrated services, achieving information sharing and data integration between business applications flexibly. Moreover, it uses big data analysis, intelligent algorithm, image recognition and other technologies to realize automatic reading of equipment meter, automatic identification of circuit breaker and disconnector switching status, intelligent analysis of abnormal alarm, automatic development of inspection task, etc. In the development of the substation auxiliary system, the B/S architecture is adopted to realize the remote transmission and access of data through the power private network, which satisfies the system requirements of the sub-plant level deployment, operation and maintenance master station and the company multi-level application.

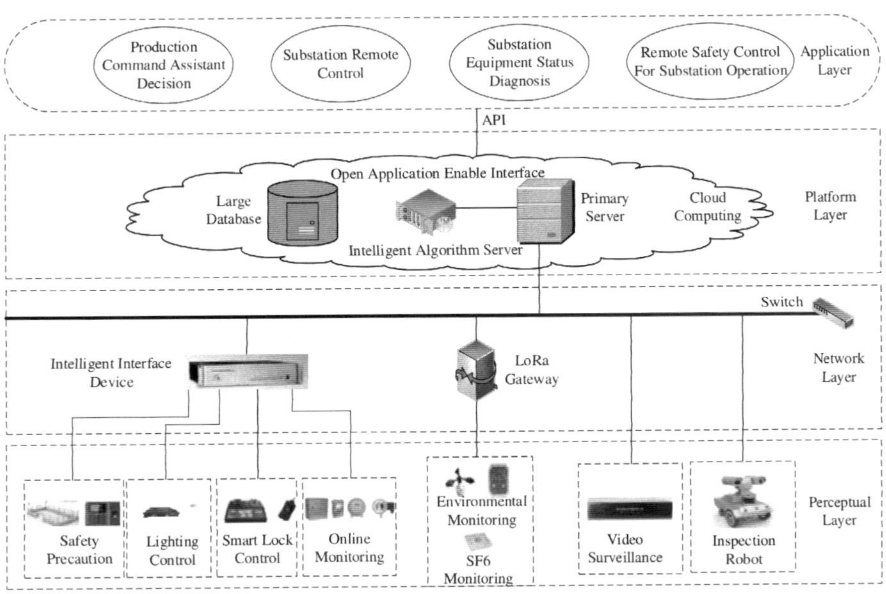

Figure 1. System overall structure

The overall structure of the IoT intelligent substation auxiliary system is shown in Figure 1. Layered and distributed structure design is used to divide the system into sensing layer, network layer, platform layer and application layer. The subsystems at the bottom layer adopt independent networking and independent computing. The back-end system and the sensing layer device transmit data through the wired power private network, 1.8G wireless private network and LoRa wireless network. The system has terminal ubiquitous technical features such as access, platform open sharing, computing cloud edge collaboration, data-driven services, and application-on-demand customization. The whole system adopts the combination of central computing and edge computing to improve data processing efficiency and reduce the load of the system center server. Each subsystem and sensing device completes the front-end processing of various data collection, calculation, analysis, etc., to realize the edge of the data. The system background center server completes the integration, analysis and processing of panoramic big data, and realizes the data analysis at the center.

The communication protocol architecture of the intelligent substation auxiliary system of the Internet of Things is shown in Figure 2. The traditional substation auxiliary equipment such as fire protection, security, and light control in the Internet of Things system communicates with the Internet of Things system through the intelligent interface equipment through the DL/T 860 standard protocol. The Video surveillance equipment communicates with the Internet of Things system using standard national network B interface. The data of the inspection robot is transmitted through a dedicated SDK (Software Development Kit) protocol. The main transformer vibration and oil chromatographic monitoring data are processed by a dedicated server and communicated with the Internet of Things system using the IEC61850 protocol. The IoT system receives various types of telemetry, remote control, and remote signaling data of the substation station control layer SCADA (monitoring and data acquisition system)/human machine operator station through the forward isolation device to realize the real-time linkage and data analysis functions between the main devices and auxiliary devices, and provides the double criterion of circuit breaker and isolation switch position recognition for one-button remote sequence control.

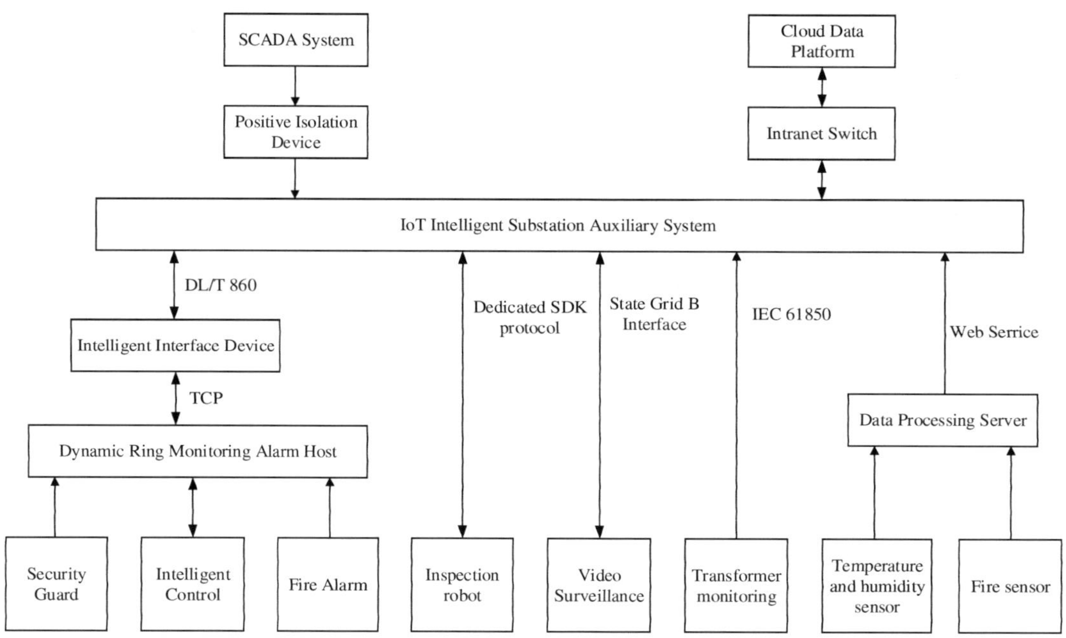

Figure 2. System communication protocol architecture

3. Edge Computing Application

With the rapid development of intelligent substations, the number of process layer devices and online monitoring systems are increasing day by day, and higher requirements are placed on the centralized data processing mode of intelligent substations. The previous data processing of the substation is based on the network cloud computing mode of the centralized core node. However, the cloud computing has the problem of real-time acquisition and analysis of the large number of inspection equipment data of the substation. For economic reasons, in order to prevent the centralized node from becoming a potential risk point and a factor to inhibit development, the system adopts the edge computing mode, and uses the edge node to acquire the sensing node at the edge of the network. The response of the data not only improves the response speed of the system, but also reduces the requirements of system transmission [4]. The specific design structure is shown in Figure 3.

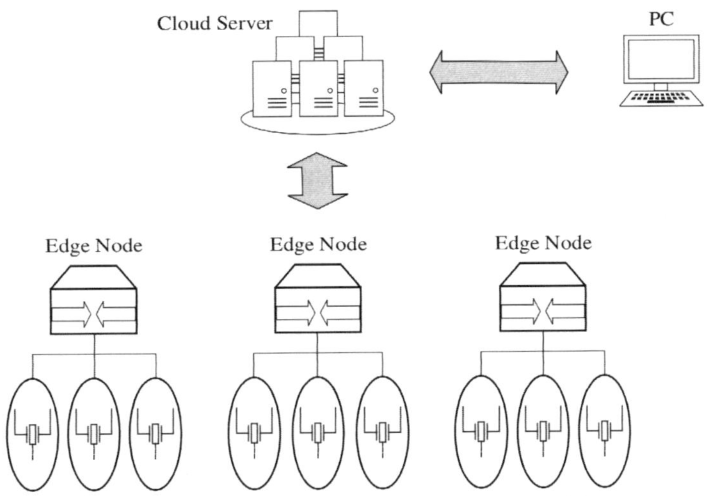

Figure 3. Edge computing architecture in the IoT environment

The edge calculation algorithm is oriented to the linear network structure of the IoT intelligent substation auxiliary system composed of multiple edge nodes. Each edge node simultaneously performs edge technology tasks and exchanges data with external devices, and then transmits the calculation to the cloud server. Therefore, in order to improve the overall network efficiency, it is necessary to balance the computational workload of each edge node with the network transmission calculation work to ensure collaborative processing between the edge nodes. In this paper, the edge calculation algorithm of task equilibrium allocation and scheduling is used to effectively control to reduce the computational task of a single edge node, and improve the real-time and security of the intelligent substation auxiliary system of the Internet of Things. The performance parameters between the cloud server and the edge nodes mainly include network traffic, delay, and response speed. During the execution of the edge computing task, the total delay of the transmission can be used to characterize the workload and working state of the edge node [5].

Assuming that the transmission rate between the edge node is expressed as S_i, the cloud server is u_i and the data volume of the cloud server transmission task is D_i, the transmission task transmission delay of the cloud server can be expressed as:

$$Tu_i = \frac{D_i}{u_i} \tag{1}$$

After the edge node S_i receives the task and the processing task rate p_i, the time for processing the assigned task can be expressed as:

$$Tp_i = \frac{D_i}{p_i} \tag{2}$$

Assuming that the processing data of the j-th sensing node connected by the edge node S_i is D_{ij}, and the transmission rate is n_{ij}, the data time calculated by the edge node and uploaded by the sensing node can be described as:

$$Tn_{ij} = \frac{D_{ij}}{n_{ij}} \tag{3}$$

Therefore, the total delay of the edge computing task connected by edge node S_i to complete a cloud server with a data volume of D_i is:

$$T_{\text{delay}} = Tu_i + Tp_i + 2\max(Tn_{ij}) + \sum_j \frac{D_{ij}}{u_i} \tag{4}$$

The total delay T_{delay} can characterize the working delay of the edge node computing task load. Based on T_{delay}, the working state of the edge node is judged by the formula (5), and the corresponding adjustment is made.

$$\begin{cases} 0 < T_{\text{delay}} \le T_{\text{normal}} & \text{Normal working condition} \\ T_{\text{normal}} < T_{\text{delay}} \le T_{\text{overwork}} & \text{Busy working condition} \\ T_{\text{delay}} \ge T_{\text{overwork}} & \text{Fault working condition} \end{cases} \tag{5}$$

Among them, T_{normal} is the maximum normal working delay, and T_{overwork} is the maximum busy working delay. The corresponding state transition diagram is shown in Figure 4. According to this state transition diagram, the system performs edge calculation algorithm for edge node to perform task balancing allocation and scheduling. The corresponding edge calculation algorithm for different working states can make the edge node work more flexible and stable.

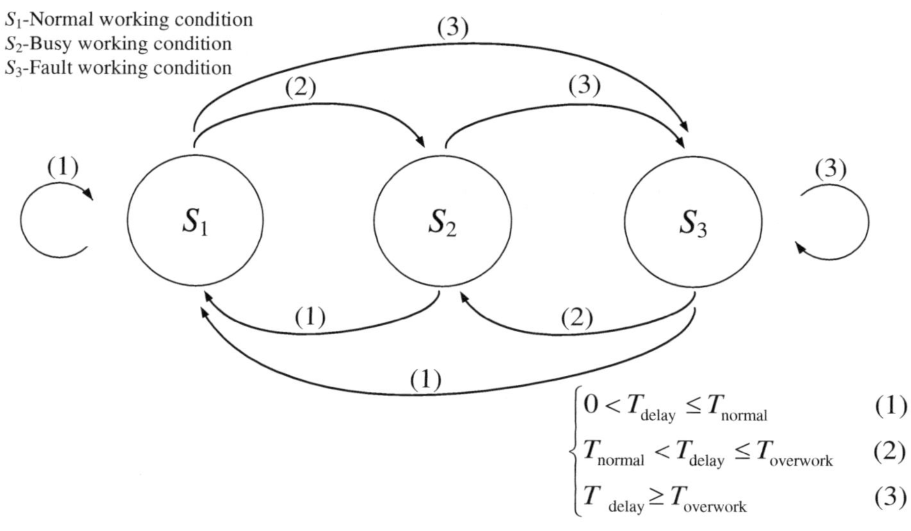

Figure 4. Edge calculation work state transition diagram

4. Application Function Design

The IoT intelligent substation auxiliary system based on edge computing is a basic management and information platform for people, machines and objects. It is the intermediate link of intelligent substation auxiliary system, which realizes the aggregation, calculation and distribution of energy information, as well as the control ability of equipment. Through unified technical standards, the system is oriented to the actual needs of operation and maintenance on energy flow, data flow, business flow and other elements. Other functions such as sharing data and services, supporting standardized access to all equipment status monitoring data of substations, and opening and closing functions for internal and external functions can be realized simultaneously [6]. The main functions are as follows:

(1) The intelligent inspection function.

The system mainly carries out daily inspections such as visual inspection, meter transcript, oil level monitoring, and location identification of circuit breakers, isolating switches, transformers, reactors, capacitors, etc., through video surveillance and data processes by substation monitoring equipments, inspection robots and so on. The system can automatically generate operation and maintenance inspection reports according to the requirements of substation operation and maintenance, for remote operation and judgment of operation and maintenance personnel.

(2) Comprehensive sensing of equipment status.

The system uses the main transformer online monitoring sensor, video monitoring, temperature and humidity sensor and other equipments to realize the full-scale self-real-time sensing of the substation power equipment status, and judge whether the data is normal according to the threshold and other criteria. When the device data is abnormal, all information of the device can be obtained in real time through the system. Moreover, the longitudinal analysis of the historical data, the phase comparison between the phase devices and the same type devices are performed, and the device state is initially diagnosed, thereby realizing the autonomous rapid sensing and early warning of the device state. For abnormal equipments, warning information is timely pushed to the operation personnel, the status monitoring strategy is timely adjusted, and data is uploaded to the platform layer for more accurate diagnosis and analysis.

(3) The equipment operation linkage.

The system and the substation station control layer SCADA system communicate through the forward isolation device. The SCADA system transmits the operation and alarm information of the primary device such as remote control, remote communication, and telemetry to the Internet of Things system through UDP (User Datagram Protocol) message format. When the device is operated once in the SCADA system, the IoT system server receives the corresponding UDP message and then intelligently links the video camera to access the corresponding video picture. When the SCADA system has important information such as SF6 alarm and protection trip, the IoT system automatically starts the corresponding inspection task to provide detailed inspection results for the operation and maintenance personnel.

5. Conclusions

This paper integrates Internet of Things, edge computing, big data and other technologies to establish an intelligent substation auxiliary system for the Internet of Things. The status of substation power equipment are comprehensively perceived and analyzed to increase mutual verification between data and reduce false alarms, and make the power equipment state perception accuracy higher. By applying the Internet of Things technology to the substation auxiliary system, intelligence and informatization of the substation operation management are realized. This method can ensure the stability of the substation management operation, improve the operation efficiency of the power grid to a certain extent, enhance the strength of the power grid, and lay a foundation for the strong operation of the power grid.

6. Acknowledgments

This work is supported by the Science and Technology Project of State Grid Corporation of China under Project of "Research on the third generation substation auxiliary system" with NO. 520530180015.

7. References

[1] Jia Hao, Wang Zhe, JiaHao, et al. *Application of Internet of Things Technology in Substation Assist System*. Electric Power Science & Engineering, 2011, 27(4): 53-57.

[2] C. Cai, J. Wang, R. Liu, Z. Fang, P. Zhang, M. Long, M. Hu, and Z. Lin, "*Resonant Wireless Charging System Design for 110-kV High-Voltage Transmission LineMonitoring Equipmen*t," IEEE Transactions on Industrial Electronics, vol. 66, no. 5, pp. 4118–4129, 2019.
[3] C. Cai, J.Wang, Z. Fang, P. Zhang, M. Hu, J. Zhang, L. Li, and Z. Lin, "*Design and Optimization of Load-Independent Magnetic Resonant Wireless Charging System for Electric Vehicles*," IEEE Access, vol. 6, pp. 17264–17274, 2018.
[4] Gong Gangjun, Luo Anqin, Chen Zhimin, et al. *Active Grid Information Physics System Based on Edge Computing*. Power System Technology, 2018, 42(10).
[5] Gou Ying, Li Jimin, Wei Xing. *IoT deep learning and task offload scheduling strategy for edge computing*. Computer Applications and Software, 2019(8): 125-129.
[6] Wang Yanru, Liu Haifeng, Li Lin, et al. *Application of Image Recognition Technology Based on Edge Intelligent Analysis in Online Monitoring of Transmission Lines*. Electric Power Information and Communication Technology, 2019(7).

IOP Conf. Series: Materials Science and Engineering **719** (2020) 012010 doi:10.1088/1757-899X/719/1/012010

Research on Computing Resources Sharing of University Laboratories in Education Cloud Environment

Nian Xiong[1], Shan Zhou[1], Zujian Wu[2], Zhenyu He[1] and Yujuan Quan[1,*]

1 College of Information Science and Technology, Jinan University, 601 West Huangpu Ave, Guangzhou, China

2 College of Birmingham Joint Institute, Jinan University, 601 West Huangpu Ave, Guangzhou, China

Email: nancy0122@163.com

Abstract. In view of the common problems of computing resources sharing existing in university laboratories, and combined with the advantages of cloud computing technology in the current education cloud environment, this paper explores the computing resources sharing in colleges and universities. The integration scheme of laboratory computing resources is designed, the sharing mode and operation management mechanism of computing resources are proposed, and the application domain model of computing resources is studied, which is helpful to promote the standardized development of resource sharing, and provides a reference for resource sharing implementation in the education industry.

1. Introduction

In recent years, education informatization has risen to the height of national education strategy. Along with the rapid development of cloud computing technology, "Internet plus Cloud" has penetrated into various industries, and " Education Cloud " came into being[1][2].Generally speaking, "Education Cloud" provides a new type of shared computing model, which integrates the entire IT resources to form a huge resource pool and realize the real resource sharing[3]. It is the migration application of cloud computing in the field of education. In the era of "Education Cloud", resource sharing has ushered in an unprecedented opportunity. This paper studies the problem of computing resources sharing from the perspective of diversification, designs the integration scheme of laboratory computing resources, proposes the sharing mode and operation management mechanism of computing resources, and explores the application fields of computing resources.

2. Analysis of the Development and Current Situation of Computing Resources

Computing resources sharing which has been included in national development strategy in the United States, Japan, South Korea and some developed countries, is a topic of global education technology[1][4].The concept of cloud computing was originally proposed by IBM in the United States. IBM has customized the Virtual Computing Laboratory(VCL) to serve colleges and universities. Then, many well-known universities have successively carried out research on cloud resources sharing cooperation. For example, Stanford University, Massachusetts Institute of Technology and University of California at Berkeley jointly develop a cloud resources cross-university program. University of California at Berkeley, Cornell University and University of Massachusetts Amherst jointly conduct research on cloud computing resources sharing. High-quality resources among universities in developed countries can be highly shared across universities and states, creating favorable conditions for enhancing the innovation competitiveness of universities and countries. In recent years, with the active promotion and guidance of our government's policies, cloud computing technology has developed rapidly. At present, domestic colleges and universities rely on cloud computing to actively build platforms for teaching and scientific research, however, the platform is mainly open to teachers and students in a limited range. At present, domestic universities rely on cloud computing to actively build platforms for teaching and research[3]. However, the platform is mainly open to a limited range of teachers and students. According to a statistical study in Zhejiang Province, the proportion of open laboratory resources in colleges and universities in the province is relatively low. Universities with an open rate below 10% account for 64.29%, and 10% to 30% account for 19.05%.In recent years, it was found that the following problems about the current status of resources sharing platforms are common.

2.1. Decentralized Management and Redundant Construction

Many colleges affiliated to universities generally build their own computer labs for experimental teaching. In addition to computer laboratories for experimental teaching, some science and technology colleges will also build research laboratories for high-performance computing, which makes computing resources are relatively scattered. It is not conducive to unified management and construction, and there are problems of redundant construction.

2.2. Insufficient Depth and Scope of Resources Sharing

Some universities have built a resource sharing platform based on cloud computing, but it is only open to teachers and students within the campus. There are many barriers among campuses. Resource sharing should not be involved in the school alone, but also be extended to a certain region, especially the whole province and most schools across the country. In addition, the functions of the existing resources sharing platform are relatively simple, some of which are for experimental teaching and some for scientific research. We should expand the functions, integrate teaching and research, and enable resource sharing to communicate in a deeper level.

2.3. Lack of Standardized Resources Sharing Architecture

In terms of the current situation of laboratory resources sharing in colleges and universities, most of them do not have a perfect sharing architecture and mechanism, and there are no overall sharing plans. The content and sharing mode of shared resources show great randomness.

3. Integration Scheme of Computing Resources in University Laboratories

The resource integration and sharing in university laboratories are not isolated, but interactive and mutually promoting. With the continuous fusion development of modern disciplines, we should continue to improve and optimize the laboratory's own functions, expand the scope of laboratory's computing resources to meet the needs of the development of teaching and research[5]. It is necessary to implement the scientific integration program of computing resources, to achieve the mutual openness,, mutual cooperation, resources sharing and collaborative innovation among laboratories, colleges, schools, regions, and school enterprises, so that the laboratories mainly based on computing resources can transform into the integrated service of teaching, scientific research and management. The integration scheme of computing resources in colleges and universities is carried out according to different types and levels, as shown in Figure 1.

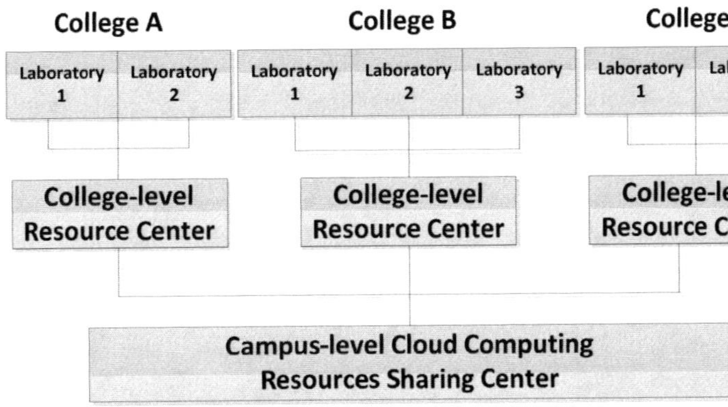

Figure 1. Integration scheme of computing resources in university laboratories

3.1. Integration of Computing Resources in Different Functional Laboratories within College

In the early stage of laboratories development, the traditional laboratories mode were decentralized, closed and inefficient[6]. Most laboratories were set up according to the curriculum, and belonged to the teaching and research sections. For example, in our university, three departments affiliated to Information Science and Technology college have their own image processing laboratory, computer application laboratory, data structure laboratory, information security laboratory, etc. From the actual situation, it is possible to centrally integrate and manage large-scale computing resources which contain mainly computers and servers, to establish a college-level data resources experiment center, and to carry out comprehensive and innovative interdisciplinary experimental projects and scientific research[7]. It will fully reflect the continuity and integrity of the entire laboratories of college.

3.2. Integration of Computing Resources of in Same Functional Laboratories Across Colleges

After adjusting the computing resources of each college laboratories, due to the basic supporting role of computing resources[8], there are still computer laboratories with similar functions in different colleges. It is necessary to integrate the computing resources across colleges, establish a campus-level cloud computing resources sharing center, and provide computing resource data services for the whole campus. The resources integration work of the laboratories has fully utilized the cross-disciplinary advantages of computers and achieved the optimization of resource sharing.

4. Sharing Mode and Operation Management Mechanism of Computing Resources

4.1. Open-sharing Mode

Through the integration of superior resources and teams, a multi-dimensional and integrated management and service framework of teaching and research is formed, which is based on the first level teaching and research platform, linked by the second level teaching and research platform, and centered on the third level teaching experiment centers and various laboratories. It is open for undergraduates, graduate students, teachers and functional laboratories and connects with regional universities, scientific research institutions, enterprises, other social organizations and public welfare, so that all kinds of platform resources can be organically combined and connected with each other. Laboratories resources have changed from mutual non-communication to information resources sharing and complementary advantages. The resources open-sharing mode based on cloud computing is designed, which proposes an open-sharing service mode of four-dimensional linkage, complementary advantages and various levels, as shown in Figure 2.

IOP Conf. Series: Materials Science and Engineering **719** (2020) 012010 doi:10.1088/1757-899X/719/1/012010

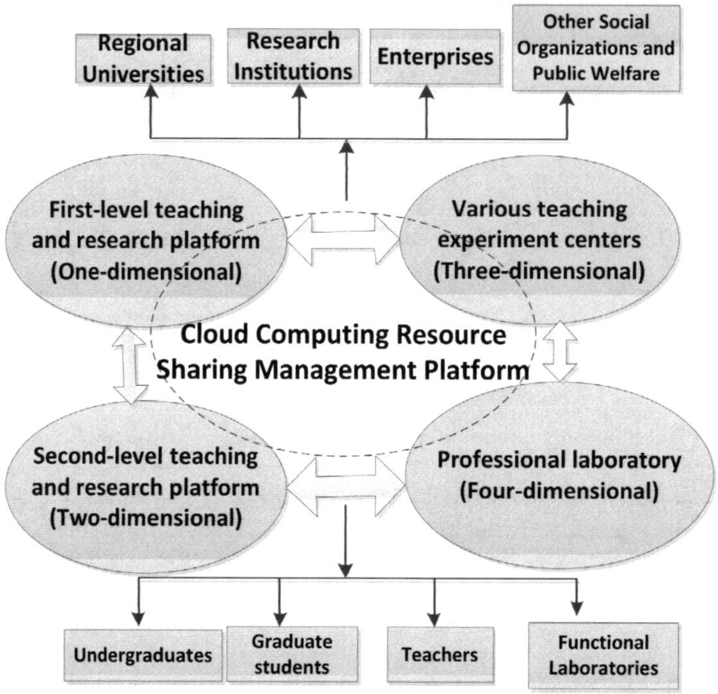

Figure 2. Resources open-sharing mode based on cloud computing

4.2. Sharing Management Mechanism

As shown in Figure 3, the shared laboratory management mechanism improves the traditional manual reservation mode in the management mode to realize the information reservation service management. Through close communication within and among universities, information feedback mechanism is established to evaluate the use efficiency of resources. At the same time, through platform real-time monitoring, data indicators are collected and data such as access time, response speed, connection time, peak valley is mined for analysis, in order to adjust the background allocation strategy and online scheduling in real time for performance optimization. Background allocation strategy and online scheduling are adjusted in real time to optimize performance. The platform has stepped onto the scientific and standardized management track, and the overall operation management mechanism has realized the virtuous circle of the closed system.

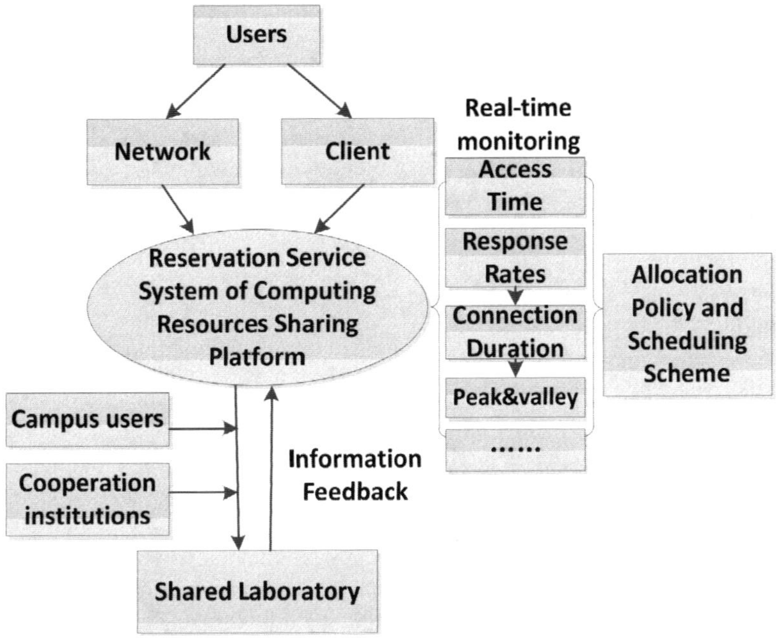

Figure 3. Management mechanism of shared laboratory

5. Application Model of Computing Resources

The sharing and management of computing resources in colleges and universities can promote the development of education in universities and provide a strong technical platform for the development of teaching, research and management[9][10]. From the perspective of the application field of computing resources, the application model is designed as shown in Figure 4. First, the teaching resources sharing can absorb advanced education concepts such as micro-curriculum and MOOC, then integrate high-quality information resources to establish a shared library of teaching resources, which is shared and exchanged among universities. At the same time, an open question bank system is established for the practical courses in need, which can provide mock examinations and online exercises, so that the experimental habits of teachers and students are not limited by time and place, and meet the diverse learning needs. Second, scientific research resource sharing can provide high-performance computing resources for colleges and universities and scientific research institutions to carry out information security such as information security, big data, deep learning, etc, as well as provide technical support for application of medical information and data analysis of economic management research[11].In addition, it can also provide platform support for the construction of national or provincial engineering technology centers, college students' innovation practice bases and other innovative laboratories, and will actively promote the development of national, provincial and ministerial scientific research projects. Third, management resource sharing is mainly used in information management systems and digital libraries university[5], which provide unified resource allocation and management to serve teachers and students.

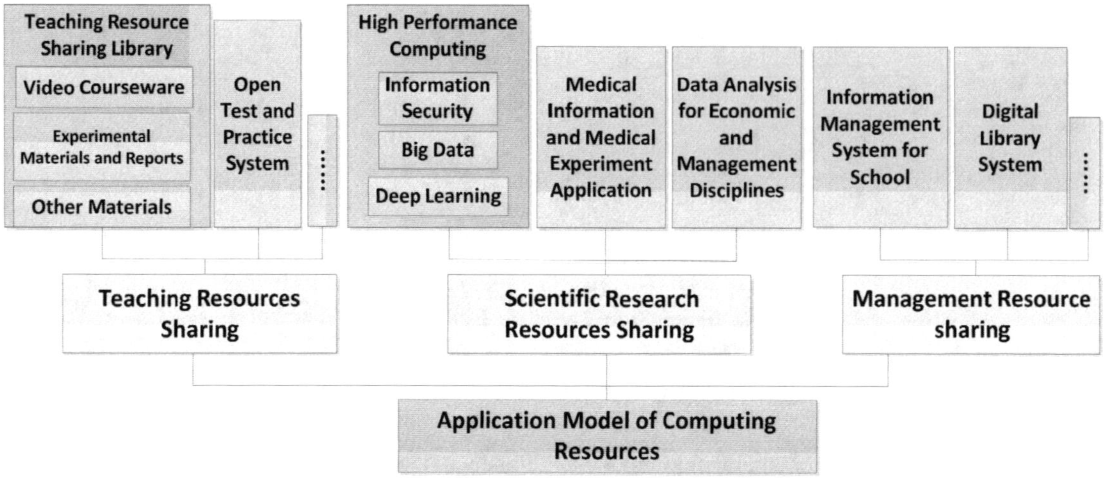

Figure 4. Application model of computing resources

6. Conclusion

The resources open-sharing based on cloud environment which is a major trend in the reform and development of university laboratories in China, has great significance. The Chinese government fully recognizes the importance of education informatization and resources sharing, and has successively issued a series of strategic development documents. The "National Medium and LongTerm Science and Technology Development Plan(2006-2020)" requires that the scientific and technological resources sharing be listed as an important part of the national construction for science and technology innovation system[12].In early 2019, "Outline Development Plan for the Guangdong-Hong Kong-Macao Greater Bay Area" advocated innovation and integration of high-quality education resources in the Guangdong-Hong Kong-Macao Greater Bay Area[13]. The development of open-sharing computing resources is conducive to optimize laboratories resources allocation in colleges and universities, greatly saving the investment in the overall laboratory construction costs. At the same time, it is conducive to improve the management efficiency of the laboratories and the experimental platforms, greatly increasing the utilization rate of laboratories resources. It is also conducive to realize the balance and fair distribution of education resources, and promote the cooperation development of the overall education level in colleges and universities. This paper first analyzes the problems existing in the current university laboratories computing resources sharing, combines computing resources sharing and cloud computing technology with its advantages in the current educational cloud environment. Then the integration scheme of computing resources between college-level and campus-level is proposed, and the sharing mode of computing resources and the operation management mechanism of shared labs are also proposed. Finally, application model of computing resources is studied. The realization of computing resources open-sharing which can break the barriers among laboratories, universities and regions, provide a reference for the co-construction and sharing of laboratories resources in higher education institutions in China, accelerate the pace of co-construction and sharing of high-quality resources in higher education institutions, and promote the development of resources sharing in higher education institutions.

7. Acknowledgment

This work is supported in part by the following funding: National Key R&D Program of China(2018YFC2002500), Funding of National Natural Science Foundation of China(61602209)and Guangdong Provincial Higher Education Teaching Research and Reform Projects ("Reform and Practice of College Computer Fundamentals Using Data as the Mainline,2017"). Yujuan Quan is the corresponding author.

8. References

[1] Haiqing Xu. *Application of Cloud Computing Information Processing System in Network Education*. ATCI 2019 Proc.pp1809-15.

[2] Boqin Liu, Hanrong Chen, He Junmei. *Research on the Platform of Online Education Platform Based on Cloud Computing*. ICCSE 2019 Proc.pp22-25.

[3] Nian Xiong, Xiaoli Liu, Shan Zhou, Zhenyu He, Longxin Lin and Yaosheng Lu. *Research on the Construction and Application of A Cloud Computing Experiment Platform for Computer Science General Education Courses*.ICNISC2019 Proc.pp108-15.

[4] Florin Anton, Theodor Borangiu, Silviu Răileanu, Silvia Anton, Nick Ivănescu, Iulia Iacob. *Secure Sharing of Robot and Manufacturing Resources in the Cloud for Research and Development*. RAAD 2019:Advances in Service and Industrial RoboticsProc.pp535-543.

[5] Ningning Sun, Yong Li, Lihua Ma, Wenyong Chen, Dunn Cynthia. *Research on Cloud Computing in the Resource Sharing System of University Library Services*. Evolutionary Intelligence, v 12, n 3, p 377-384, September 1, 2019.DOI: 10.1007/s12065-018-0195-8

[6] Cheng Qian, Yingfeng Zhang, Yang Liu, Zhe Wang. *A Cloud Service Platform Integrating Additive and Subtractive Manufacturing with High Resource Efficiency*. Journal of Cleaner Production.DOI:10.1016/j.jclepro.2019.118379.vol241.pp1-16.

[7] Yuanyuan Wei, Ping Zhu, Wanping Wu and Xinrong Hu. *Teaching Resources Construction of Fundamentals of Computers in the Cloud Platform*. ICCSE2018 Proc.pp423-27.

[8] Qingxian An, Yao Wen, Junfei Chu and Xiaohong Chen. *Profit Inefficiency Decomposition in a Serial-structure System with Resource Sharing*. Journal of the Operational Research Society. DOI:10.1080/01605682.2018.1510810.vol 70, 2019.pp2112–26.

[9] Lina Zhao, Guangrning Li. *Research on Higher Education Resources Sharing System Based on SOA*.CSIP 2012.pp1004-1007.

[10] Gianluca Cornetta, Javier Mateos, Abdellah Touhafi and Gabriel-Miro Muntean. *Design Simulation and Testing of a Cloud Platform for Sharing Digital Fabrication Resources for Education*.JournalofCloudComputing.vol8,2019.DOI:10.1186/s13677-019-0135-x.pp2-22.

[11] Di Lu, Yuping Qiu, Chun Qian, Xi Wang and Wei Tan. *Design of Campus Resource Sharing Platform based on SSM Framework*. SAMSE 2018 Proc.pp490-97.

[12] *National Medium and Long Term Science and Technology Development Plan (2006-2020)*.[2008-02-05].http://www.gov.cn/ztzl/kjfzgh/content_883887.htm.

[13] *Outline Development Plan for the Guangdong-Hong Kong-Macao Greater Bay Area*.[2019-02-18]. http://www.gov.cn/zhengce/2019-02/18/content_5366593.htm.

IOP Conf. Series: Materials Science and Engineering **719** (2020) 012011 doi:10.1088/1757-899X/719/1/012011

A Game-based Method for User Privacy Protection in Big Data

Xiaohui Li[1], Chaoyang Chen[1], Feng Liu[2], Bo Li[1] and Yajun Wang[1]

1,College of Electrical and Information Engineering ,Liaoning University of Technology, Jinzhou, Liaoning, 121001, China
2,Shenyang institute of computing technology, Chinese academy of sciences, Shenyang, Liaoning,110168, China
Email: lhxlxh@163.com

Abstract. In order to protect the user privacy in the big data environment, We propose a security method based on game theory through trust management mechanism. The main contribution of this paper include:(1)analysis service game relationship in the big data environment ,(2)combine trust evaluation with decision management, and apply game theory to design a user-centric scheme to protect privacy,(3)proposing a service selection method. By using game theory to restrict the service interaction between the users and services in big data, we introduce some factors corresponding game model. we then apply the trust evaluation and decision management to describe the relationship and trust quantify in which we propose a method for service selection to implement user privacy protection. Simulation results show that the proposed method is effective and can apply the dynamic game of incomplete information relationship between user and service to make decisions on the service selection, thus providing customized privacy protection .

1. Introduction

In the big data environment, the need for data protection (especially in a shared multi-tenant environment) is the main problem of users to consider, which is the main fetters to carry out big data, user data protection big data environment has become a well studied hot spots. How shared storage, computing and network resources and the environment effectively isolate and protect workloads and data need to rely on some technical, many researchers have introduced trusted tools , to try to solve the user data in the big data environment protection issues, and made some research . Studies have shown that trust in the big data environment should be trusted service centers, the trust is a two-way behavioral data and storage services have side trust between the parties, by means of both credibility and dual constraints together contributed data reasonable and lawful use without being abused. For users, the service can select trusted party, agreed a mutually satisfactory security mechanisms in order to achieve maximum security. On the service side, once integrity will have no place to lose profits than the integrity of an important enterprise. Therefore, big data environments can use data protection and service end user behavior trust bidirectional way to resolve it.

2. Service Game Analysis in Big Data

2.1. Related Theory

The basic elements of the game are:

(1)participants: that is decision-making body of the game selecting actions to maximize the effectiveness ,who can be individuals or organizations.

(2)strategies: that person's participation in decision-making variables and the selective rules of decision variables, which tells what kind of action involved under what conditions.

(3)orders: game party action priorities order ,game can be divided into static game and dynamic game.

(4)payoffs: Results benefit or payment, corresponding to the decision-making game party, the proceeds or the failure of the game to participate in what people are most concerned about.

Divided from different angles, the game has a variety of classification game: game action according to the order, the game can be divided into static and dynamic. If all the players actions while selecting called static game, players' actions have successively divided, is called dynamic game; according to people involved in the game to other participants understand the extent of information that can be divided into complete information game and non-complete information game. If each participant to have an accurate knowledge of the characteristics of all the other participants, strategies is called complete information game, otherwise it's not complete information game; according to the demand side of the interests of the game can be divided into cooperative game and non-cooperative game, organizations seeking to maximize the interests of cooperative game, participants get the most out of the first collaborative community interest groups and then assign interests rather than the interests of individuals seeking to maximize cooperative game, not sign the agreement or the existence of collaboration between participants.

2.2. Mechanism Design

Mechanism design theory is one of the core problems of modern economics, it is for any discussion of a given economic or social goals, in a free choice of decentralized decision-making conditions, whether and how to design an economic mechanism that allows economic activity participants consistent with personal interests and goals. Designers can be as large as the entire economic and social system designers, small to have only two participants (in their own best interests as the goal) of the principal economic organization and management.

People face with an incomplete information society, all personal information cannot be completely one person to master, so the use of decentralized decision-making, with incentives to motivate people to do the designers want to do, or to realize the designers want to achieve goal, it is a matter of economic mechanism design theory to the study. Incentive mechanisms, that is, under the mechanism designed so that each participant while in pursuit of personal interests to achieve goals set by the designer. To achieve a certain goal, the first to make this goal within the scope of the technical feasibility; secondly, to make it meet the individual rationality, namely participation, if you do not participate in the game provided, because there is a better choice, then the mechanism design is invalid; third, to meet incentive compatible constraint, to make voluntary self-interested behavior to achieve personal goals.

For a variety of technical integration of big data environment, the user's choice to face multiple problems, mechanism design theory this has important practical significance. services and users as a resource requester is also the provider of resources, are the dual roles, but the identity is not fixed, access control permissions granted can not fully guarantee safety. From the user's privacy protection, we need a variety of services to compare and study the merits of the degree of basic standards and a unified framework, the need for a general theory can research and compare various service providers to consider the service selection problem, and mechanism design theory is undoubtedly a good choice in this regard.

2.3. Dynamic Service Analysis

The service interaction big data environment is described as the game process between users and service: the service both quantify and make service selection based on trust, because the trust itself is a psychological expectation, both services and interactive process cannot fully grasp the information with each other, so the process is non-complete information game.

The two sides were based on trust game, if the user trust services, the service is accepted, otherwise rejected; services provide users may take integrity or fraud. Four factors corresponding game model are as follows:

(1) Participants: big data environment service parties, collectively referred to users $P-c$ and services $P-s$;

(2) Strategy: In the service interaction process, the strategy space users $S\_c=\{accept, reject\}$; strategy space for services $S\_s=\{honest, dishonest\}$;

(3) the order of the game: the interactive process based on service, services are the first actors, users are after actors;

(4) payoffs: the service side of honest or dishonest policy, the service participants benefits or losses obtained.

In the big data environment to benefit form the game to make the following assumptions, when the user chooses to accept strategy, users benefit $C-Income$, services benefit $S-Income$; users are dishonest to provide services, user loss $C-Loss$, services proceeds $S_d-Income$; users to opt out strategy, service benefits both sides are 0. Payoffs matrix shown in Table 1.

Table 1. Payoffs matrix table

	accept	refuse
honest	S_Income, C_Income	0,0
dishonest	S_d_Income, C_Loss	0,0

3. Method Design

3.1. Design Idea

The first is the user's information release and services to provide a conflict of interest exists honesty, integrity and provide services need to pay the cost, from the game theory analysis, for each participant, it will choose strategies to maximize their effectiveness. services in order to maximize revenue, will choose the acts of bad faith, that the participants did not have any normative constraints under service commitment is not credible; In a big data environment, trust between users and services is unequal, service users' trust in the is a service that users can choose the premise, the purpose is to protect the environment of services are not affected by the user, and the users trust services demand security services. In other words, although users and services while both the dual role of resource requests and resources provided, but the service selection cooperation is to establish a good reputation, there is no users to establish a good reputation problems. Game theory analysis between users and service information is asymmetric. No matter how the user incentive services, services would choose to maximize the effectiveness of their strategies. When users to maximize the effectiveness of the policy will be made to receive services, but services in order to maximize revenue, will choose non-faith conduct, even if such an impact on reputation.

Hence the need to make services through incentives to make the integrity of the service provider, an increase of user data release granular, increasing the income of services, users with the release of more data, the more we need to protect the integrity of the service. Using trust to balance the user information release and conflicts of interest between services, services and its service providers trust quantify historical behavior are closely related.

In the dynamic game of incomplete information, the involvement of people with the revaluation of action sequences and correction of information is an important factor in the decision game results. Trust in the -based service acquisition game, services are the first actors to provide the type of service, service users to take action by observing the catalog, which is determined by the order of the game. So this game have information on the results of the game is important. Before the game, from the observation, the accumulation of pre-game or subjective inference, this information is called a priori information (prior probability) is also known as a priori probabilities. In the dynamic game, the game is the process involved in the process who initially released and reveal private information, and is the process by participants observed results of prior information be corrected, and a priori information corrected for a posteriori information (posterior probability), also known as the posterior probability.

Game information correction process includes two dynamic decision-making processes and their integration interact and influence each other. Trust in the -based service acquisition for analysis, used to calculate the credibility and trust services directly belong to the history of trust and other information prior information on services user feedback evaluation belong posteriori information.

3.2. Users Data Protection Method Based on Game

The program of the big data environment between the user and the service acquisition process deem incomplete information game, on the one hand by the rules of the game to build services offer incentives security; on the other hand, users of the trust relationship management services to enhance user efforts to protect its own resources. Specific user data big data environment protection method designed with the following objectives:

(1) The behavior of a big data environment service interactive mode, seen as services and between users of the game, trust is important information for both games, the application rules of the game, encouraging both behavior is credible.

(2) Game organization and management information including quantitative methods and trust service selection methods.

(3) Big data environment, services both responsible for their behavior, if there is fraud or malicious attacks, service interactions will affect future that pay the price, so its services according to certain criteria to obtain, make rational decisions.

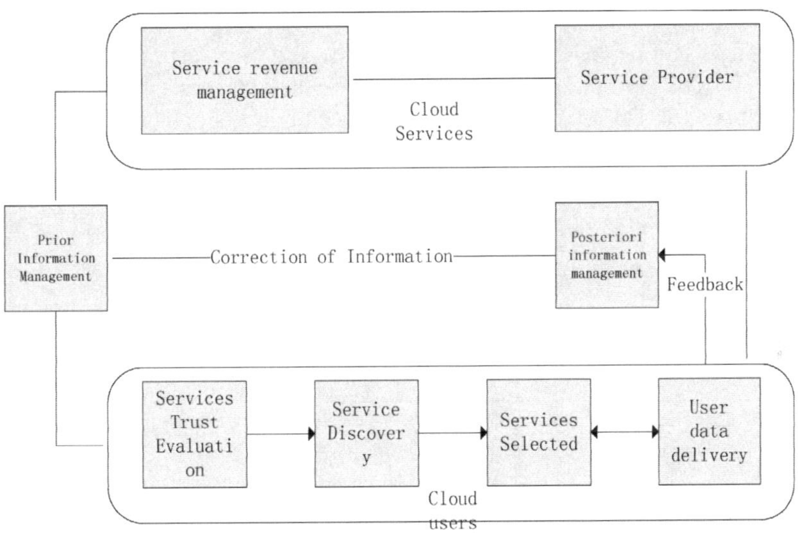

Figure 1. Work flow chart

3.3. Service Selection Method

Under the trust game as a service information interaction between the two sides, and how the trust the same or similar service resources, select one of the most appropriate services to meet the security conditions in the premise of the program, to demand personalized service is controlled one important issue of user data protection solution implemented, this section will give service selection module approach instance programs.

Services option is to evaluate the choice of alternative service problems, the number is limited classified as attribute decision making, you can choose to study theory. Known for service attribute weights, weights based on decision analysis and decision-making matrix, sort, the more common methods are: linear distribution method. Because individual needs and preferences trust users, using the linear assignment.

Linear distribution method is an attribute decision making problems to sort methods. The law does not require the given decision matrix, only need to know the various services of the priorities for the election of each property, use this order to solve the corresponding binary linear programming, so that the corresponding right and reaches the maximum, in order to determine the most likely location solutions. The need to construct a weight matrix method to estimate the location of each program aligned with the n and m for selected service attributes, the general steps are as follows:

Step 1: single objective sorting special case. The program may appear side by side, the virtual target method, which set the goal g_k for the corresponding weight w_k for the g_k decomposition h virtual target $g_{k1}, g_{k2}, ..., g_{kh}$ and empower heavy w_k / h, do the equivalent processing;

Step 2: The program sort by a single goal;

Step 3: Given weights and program line up for each target, construct weight matrix;

Step 4: Construct and solve linear programming problems, the definition of binary permutation matrix P, whose elements P_{ik} is 0 or 1. According to the principle of linear distribution method, linear structure following questions:

$$\max \sum_{i=1}^{n} \sum_{k=1}^{n} w_{ik} P_{ik} \quad s.t. \sum_{k=1}^{n} P_{ik} = 1, i = 1, ..., n \tag{1}$$

Solving this problem is to find a set of values $P_{ik} = 1$ is the right and the corresponding maximum.

In a big data environment, due to the demand for users personalized service, as the service selection policy makers consider the preference structure to describe the need for trust in services, from spatial and temporal properties and behavior of the user feedback and other aspects of the trust describe the service trust attribute (the attribute weights and trust services are described in detail in the article quantified). When users access to services attribute specific weight, based on the right lineament analysis matrix, selection of service plans will be sorted. Here is an examples service selected linear assignment:

Assume elected service $(X_1, X_2 X_3, X_4)$, known weight of each attribute $W^T = (0.232, 0.142, 0.51, 0.116)$, according to the preference of service users, as shown in Table 2 single attribute sort.

Table 2. Service single attribute sorting

Attributes Order	Time	Space	Users trust	Historical behavior
1	X_3	X_1	X_2	X_3
2	X_4	X_2	X_1	X_4
3	X_2	X_4	X_4	X_1
4	X_1	X_3	X_3	X_2

Weight matrix structure $\overline{W}$:

$$\overline{W} = \begin{bmatrix} 0.142 & 0.51 & 0.116 & 0.232 \\ 0.51 & 0.142 & 0.232 & 0.116 \\ 0.348 & 0 & 0 & 0.652 \\ 0 & 0.348 & 0.652 & 0 \end{bmatrix}$$

Weight matrix for service by sorting, sorting method is as follows:

If $\overline{W}$ all elements of the ith row of the first one in the maximum value, the service X_i in the first row, and so on, should have multiple services on the same ranking, you need to consider the following order of precedence values. The priority of the following four services: $X_2 > X_1 > X_4 > X_3$, corresponding rights $w_{21}+w_{12}+w_{43}+w_{34}=0.51+0.51+0.652+0.652=2.324$ .The results achieve $\overline{W}$ the corresponding right to maximum.

4. Conclusion

In this paper, we proposed a user privacy protection method in the big data environment. Firstly, from the perspective of game theory, We describe the service interaction in the big data environmentas the behavioral game between the service and the user, the trust and decision (service selection) as the game Information. Applying game constraints, the method motivated both parties to implement credible behavior, the user's service selection makes decisions through a trust mechanism to provide customized protection for user privacy data.

5. Acknowledgments

This work is supported by the National Natural Science Foundation of China (under Grant 61802161, 51679116),Science and Technology Fund Project of Liaoning Province (under Grant 20170540448), the Innovative Talents Support Plan of Colleges and Universities of Liaoning Province (under Grant LR2017068).

6. References

[1] Privacy-preserving publishing of opinion polls[J] . Ahmed Abdalaal, Mehmet Ercan Nergiz, Yucel Saygin. Computers & Security. 2013
[2] Parallel search over encrypted data under attribute based encryption on the Cloud Computing[J] . Thouraya Bouabana-Tebibel, Abdellah Kaci. Computers & Security. 2015
[3] PPTD: Preserving personalized privacy in trajectory data publishing by sensitive attribute generalization and trajectory local suppression[J] . Elahe Ghasemi Komishani, Mahdi Abadi, Fatemeh Deldar. Knowledge-Based Systems . 2015
[4] The new casper: query processing for location services without compromising privacy. M.F. Mokbel, C.Y. Chow, W. G. Aref. Proc of the 32nd international conference on Very Large Data Bases (VLDB) . 2006
[5] Building access control policy model for privacy preserving and testing policy conflicting problems[J] . Hua Wang, Lili Sun, Elisa Bertino. Journal of Computer and System Sciences. 2014
[6] Priv Bayes [J] . Jun Zhang, Graham Cormode, Cecilia M. Procopiuc, Divesh Srivastava, Xiaokui Xiao. ACM Transactions on Database Systems (TODS) . 2017 (4)
[7] Tools for privacy preserving distributed data mining[J] . Chris Clifton, Murat Kantarcioglu, Jaideep Vaidya, Xiaodong Lin, Michael Y. Zhu. ACM SIGKDD Explorations Newsletter . 2002 (2)
[8] PHOABE: Securely outsourcing multi-authority attribute based encryption with policy hidden for cloud assisted IoT[J] . Sana Belguith, Nesrine Kaaniche, Maryline Laurent, Abderrazak Jemai, Rabah Attia. Computer Networks . 2018

IOP Conf. Series: Materials Science and Engineering **719** (2020) 012012 doi:10.1088/1757-899X/719/1/012012

Investigation and Thinking on the Information Construction of South China University of Technology

Fang Fang and XingE Luo*

Network Engineering and Research Center, South China University of Technology, Guangzhou, China, 510641
Email: ffang@scut.edu.cn; Xluo@scut.edu.cn

Abstract. In order to better understand and improve the information construction of our university, further strengthen the network security and informationization work, and improve the quality of service, the network center conducted investigation on information construction and satisfaction of the university. Based on the statistics of the survey results and other related data support, we try to investigate and research the status quo of information construction in the university, explore the application of information technology in the whole university teaching, research, management and campus life, and analyse its practical application effects. Finally, it will provide reference and direction for the future work of the network center.

1. Introduction

Universities use information technology to change the information transfer process between departments, and promote the transformation of organizational models, management structures and operational methods, so as to achieve the goal of optimizing management processes, improving work efficiency and promoting teaching and research. Exploring the new application methods and modes of information technology in teaching, scientific research, management and campus life, and analysing the actual use effects, it is not only important for solving the practical problems of campus informationization, but also important for promoting the construction of "Double First-rate" university. [1]

In recent years, the network center has been committed to information construction, dealing with informationization issues in a consultative manner, and listening to ideas from teachers and students. In order to strengthen the guidance of planning, we conducted full research and communication, and learned from the experience of other universities, paying attention to user experience and improving satisfaction.

On the one hand, we want to know what problems have been improved through previous efforts, and what problems still exist at present, and provide support for the next stage of information construction. On the other hand, we hope to improve the participation of teachers and students through the investigation, create a good application environment, promote effective application in teaching and daily life, and strive to realize the transformation of information construction from management to service.

2. Form of the Investigation

Our research forms include questionnaires and interviews, using statistical analysis software to analyse the collected data, interpreting the data results, summarizing conclusions and making recommendations.

The questionnaires are aimed at teachers and students. The interviews targeted approximately 80% of the departments in the university, including management departments and colleges. We conducted face-to-face discussions to ensure that the survey truly reflects the needs. Many colleges have collected opinions and suggestions from teachers and students. The leaders personally participated in the interview and discussed issues of mutual concern.

3. Content of Investigation

3.1. Questionnaires

3.1.1. Application system usage

This part is the frequency and satisfaction of teachers and students on the use of information technology applications. The systems involved are: homepage, electronic map, portal system, unified authentication account, scientific research management system, student management system, personnel system, teaching administration system, financial system, mail system, Office Automation system, virtual private network, enrol system and departure system, WeChat Enterprise account, etc.

3.1.2. Attention to information construction events

This part is the awareness of teachers and students about the main events of university information construction in the past two years. These events include: opening E-card WeChat recharge function; software and information service purchase approval platform, school new version information portal, welcome new, school leaving, unified messaging platform and other systems start running; Cloud platform service; expansion of mailbox space; student dormitory wireless network installation project; opening IPv6 network to access foreign academic resources; expanding campus network export broadband, etc.

3.2. Interview

3.2.1. The Construction of the application system

Learn about the teacher's opinions, suggestions and needs for the current application system.

3.2.2. Service situation

Learn about the opinions, suggestions and needs of teachers on the information service.

3.2.3. Guarantee system

Learn about teachers' opinions, suggestions and needs on information system management and network security.

3.2.4. Other related issues

Public databases, data integration and sharing, wired, wireless network quality, etc.

3.3. Statistical Analysis

We received a total of 1159 valid questionnaires from students and 364 valid questionnaires from teachers. The user categories of teachers are shown in the Table 1.

Table 1. Type of users participating in the survey

	Type	Number	Proportion(%)
Teachers	Teaching and research staff	216	59.3%
	Administrative staff	100	27.5%
	Engineering staff	44	12.1%
	Support staff	4	1.1%
Students	Bachelor	697	60.1%
	Mater/Doctor	462	39.9%

The three information systems with the highest frequency of student use in 2017 are the unified authentication account system, homepage and educational system. The overall usage frequency is shown in Table 2. (1 point = "not used", 2 points = "occasional use", 3 points = "use once a few weeks", 4 points = "use once a week", 5 points = "use multiple times a week", 6 points = “Use once a day”, 7 points = “Used multiple times per day”, the higher the mean value, the higher the frequency of use) [2]

Table 2. The usage frequency of the information systems

	System	Usage frequency
1	Unified Authentication Account	4.18
2	Homepage	3.72
3	Teaching Administration System	3.61
4	Portal System	3.58
5	Student Management System	3.47
6	WeChat Enterprise Account	3.01
7	Mail system	2.49
8	Virtual Private Network	2.27
9	Financial System	2.1
10	Electronic Map	1.44

This statistic shows that the frequency of application system use by teachers and students has improved overall, the overall satisfaction of the application system has increased, and the awareness of information construction events has yet to be improved. The overall awareness of teachers and students on information construction events still needs to be improved. The enroll system and departure system were well received by the teachers. [2]

Teachers and students are satisfied with the overall network center service, including the satisfaction of hotline telephone service personnel, fault handling personnel, service hall staff, telephone service processing, E-card processing procedures, and fault handling speed. [2]

4. Thinking and Countermeasures

In April 2018, the Ministry of Education issued the "Education Informatization 2.0 Action Plan", proposing to achieve the development goal of "three full, two high and one big" by 2022. That is, the teaching application covers all teachers, and the learning application covers all school-age students. Digital campus construction covers all schools; improve the application level of information technology, improve the information literacy of teachers and students; build an "Internet + education" platform. Promote the transformation from education-specific resources to educational resources, from improving teachers' and students' information technology application capabilities to comprehensively improving their information literacy, from fusion applications to innovation development, striving to build a new model of talent training under the "Internet +" condition developing a new model of Internet-based education services and exploring new models of education governance in the information age. [3]

Serving the “Double First-rate” construction is the core task of informationization. Informatization needs to be deeply integrated with teaching and management. Informatization work is not only a single service guarantee or supplementary work, but also a deep integration and innovation integration with various business fields of the university to promote its business transformation. The problem of informatization involves the management and service concept of universities, institutional issues and human problems.

We need to consider what kind of infrastructure and information services can be provided to university from a university perspective, and understand the university development strategy and phased requirements. "Application-driven, People-oriented." The process of informationization is not only technology, but also the embodiment of management and service concepts. [4]

4.1. Multi-participation
Informatization work aims to improve the comprehensive management ability of university. It is necessary to attract more teachers and students to participate in it. Only in this way can informationization play a role in supporting university reform and development and providing better services for teachers and students. At the same time, we need to gradually expand the participants of education informatization. In addition to university teachers and students, we also need to learn to cooperate with other partners to introduce participants in the Internet.

4.2. Improve the Information Literacy of Teachers and Students
Improving the application of information technology is a technical measure. It is more fundamental to improve the information literacy of teachers and students. It is to ensure that teachers and students not only use the information technology they need, but more importantly, they have the thinking and actions of the information society.

Informatization literacy is reflected both in the knowledge level and in the spiritual level. We need to improve information literacy as a long-term job and integrate into the daily life of the university. It is necessary not only to educate students to correctly understand the cultural, ethical and social issues related to information technology, but also to lay the necessary foundation for adapting to the learning, work and life of the information society; it is also necessary to increase the training of teachers' information literacy and establish an assessment system to enable teachers adapt to the requirements of the information society for education.

4.3. Infrastructure
The high-speed campus network with strong stability is still the most basic appeal of teachers and students. Wireless coverage needs to be improved. Find ways to reduce the failure rate of the office network due to internal problems in the building.

4.4. Mobile Information Service
Teachers and students are very interested in new media and mobile, it is necessary to improve the mobile Internet application rate. Apply the "WeChat Enterprise account" mobile application as an entry point and innovate the management service model. The information portal is built on mobile application support, and is fully compatible with mainstream terminals such as mobile phones, tablets, and personal computers. The mobile terminal can obtain information content, service functions, and comfortable browsing experience consistent with the PC.

4.5. Data Sharing
Based on the public database platform, many information systems covering the fields of teaching affairs, scientific research, management, and services continue to expand and update, and comprehensively serve the various undertakings of university reform and development. Data sharing is an informational event that teachers and students are most concerned about. The data sharing service platform should be enabled as soon as possible to gradually solve the problem of repeated filling.

4.6. Information Security
The spirit of the "National Network Security and Informatization Work Conference" held in April 2018: improve the leadership system, strengthen overall planning, focus on key tasks, encourage multi-participation, focus on training publicity, and implement safety responsibilities. Building a safe and orderly educational information environment, resisting the intrusion of bad information, and ensuring network and data security are the concerns of teachers. We need to implement a full lifecycle security management of the website, including: website registration and filing, website security access check, website operation security monitoring and so on. We will provide information security training and provide regular training for different types of people, hoping to be included in the assessment content of each department.

4.7. Expand Publicity

Give full play to the role of information officers and network administrators of various departments. When new business, new systems, and convenience measures are activated, information officers and network administrators can push messages within their respective departments. We need to integrate existing channels for access to information services, identify which channels can actually obtain effective information, and then use effective communication methods to enhance publicity and promotion, and optimize the promotion of campus information services.

5. Summary

From 2000 to 2017, our university's informatization has gone through three stages: infrastructure construction, application promotion and service promotion. Education informatization has shifted from the application of information technology itself to how to use information technology to promote university teaching, research and management changes. At the deeper level, it affects the core issues of the overall transformation and development of the university.

In the process of “Double First-rate” construction, the university is indispensable for information construction. We need to change our philosophy, participate in the whole process, multi-participation, find the direction to solve the problem, and better meet the teachers and students' good yearning for the smart campus.

6. References

[1] Zhao guodong, Wang tingting, Yan yan, Li zhigang, Journal of Distance Education, Analysis of Campus Information Application and Development Status, 2011(6),pp31-38.

[2] Hu bing, Mao yanhua, Yao qiong, Li zhen, He dejun, Survey report on the use and satisfaction of the campus network of South China University of Technology in 2017, 2017.12 pp2-4,pp7-10, pp54,pp64.

[3] "Education Informatization 2.0 Action Plan", 2018.4.

[4] Shen fuke, Informatization is light on "information" and focuses on "informatization", China Education Network, 2018.4.

IOP Conf. Series: Materials Science and Engineering **719** (2020) 012013 doi:10.1088/1757-899X/719/1/012013

The Distribution Network Operation Online Monitoring and Analysis System of State Grid Corporation (Phase 5)

Jian Zhang[1], Xin Xuan[1], Zheng Xu[1] and Xiufeng Shao[2]

1) Beijing China Power Information Technology Co., Ltd., State Grid Corporation, Beijing, 100085
2) Department of Information, BeiJing City University, Beijing, 100083

Abstract. Under the organization of Information & Communication Department and the Operation & Monitor Center in State Gird Corporation's headquarter, and after two years of construction from 2015 to 2016, the operational efficiency research and monitoring work was carried out at the three-level operation and monitor centers including headquarters, provincial and prefecture level. This provided quantitative evaluation and decision-making basis for the state grid planning and production operations. The leaders of Corporation highly affirmed the results of the special analysis of the operational efficiency of the distribution network assets which is actively explored by the operation & Monitor center. The leaders also put forward the higher requirement on this. To this end, improving the online monitoring and analysis system of the State Gird Corporation's network distribution operation has stronger practicality.

1. Introduction

Under the organization of Information & Communication Department and the Operation & Monitor Center in State Gird Corporation's headquarter, and after two years of construction from 2015 to 2016, monitoring and other functions of the operational efficiency of the 110kV and below power grids has been realized from three dimensions[1]: single equipment, substation and administrative district. It covers five major functional modules: load monitoring, efficiency monitoring, coordination degree [2] monitoring, decision analysis, and simulation prediction; the implementation scope is company headquarters and 27 provincial (city) companies.

The online monitoring and analysis system of the State Gird distribution network operation integrates key information such as equipment, load and assets from the PMS, EMS, ERP systems, which is realized the cross-platform data fusion of distribution network data. It lays a solid foundation for the on-line monitoring analysis and follow-up business development. The system also realized the online monitoring and analysis function of distribution network operation efficiency, completed the distribution network efficiency from offline qualitative description to online quantitative monitoring. Also, from the perspective of efficiency, this system provides the auxiliary decision support for the distribution network's planning, construction and operation[3].

Through the operational efficiency research and monitoring work carried out at the headquarters as well as provincial & prefecture-level operation and monitor centers, quantitative evaluation and decision-making basis were provided for the company's power grid planning and production operations. The leaders of Corporation highly affirmed the results of the special analysis of the operational efficiency of the distribution network assets which was actively explored by the operation & Monitor center, also put forward the higher requirement as: 1) further expanding the scope of power grid operation's monitoring, carrying out monitoring of equipment operation which the efficiency are 220kV or above, 2) deepen the monitoring capacity of power grid operation, for example, carrying out

IOP Conf. Series: Materials Science and Engineering **719** (2020) 012013 doi:10.1088/1757-899X/719/1/012013

the power supply capacity monitoring, courts voltage monitoring, load Monitoring and courts power outage monitoring, thus realizing the value for the application of power grid operation monitoring in 3 aspects that are precision investment, lean operation and maintenance, and quality service.

2. Overall Structure

The design of the online monitoring and analysis system of the State Grid Corporation's distribution network operation follows the idea of "business-driven, application-oriented, inheritance development, and continuous improvement". The overall business structure [4] was formed through the analysis of business structure and experience of the operational monitoring (control) center, also the overall system application structure design and data structure design were deducted. Then, the technical structure was designed as the requirement of the application Structure and data structure, and based on all above, the special technical solutions' design was done, accordingly, it formed the complete structure of distribution network operation online monitoring and analysis system to ensure effective support of the system.

2.1. Business Structure

The construction of distribution network online monitoring and analysis system was carried out to enhance information level and implement the intensification and specialization management of power gird operation and monitoring work. The construction work was in accordance with the plan for the company's grid operation information construction, combined with the grid operation monitoring business needs, as well as based on fully investigating the status quo of grid operation and extensively collecting opinions and suggestions from various units of the company[5]. It mainly includes five major businesses: operational efficiency, power supply capability, decision support, platform monitoring and data quality as Figure 1.

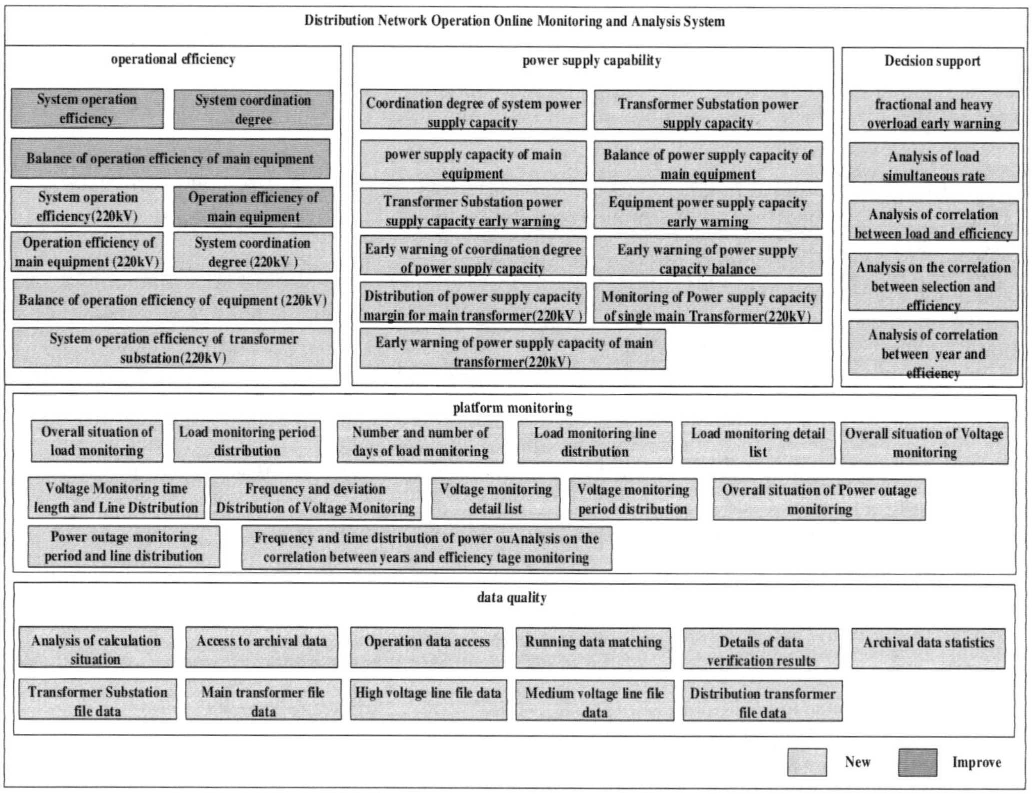

Figure 1. The Business Structure of State Grid Corporation Distribution Network Operation Online Monitoring and Analysis System

2.2. Application Structure

Based on the business structure, it conducted detailed analysis and comprehensively discussed the overall application architecture of the grid operation monitoring business application. The application structure laid the foundation for data architecture design, technology implementation and subsequent development. It is shown in Figure 2.

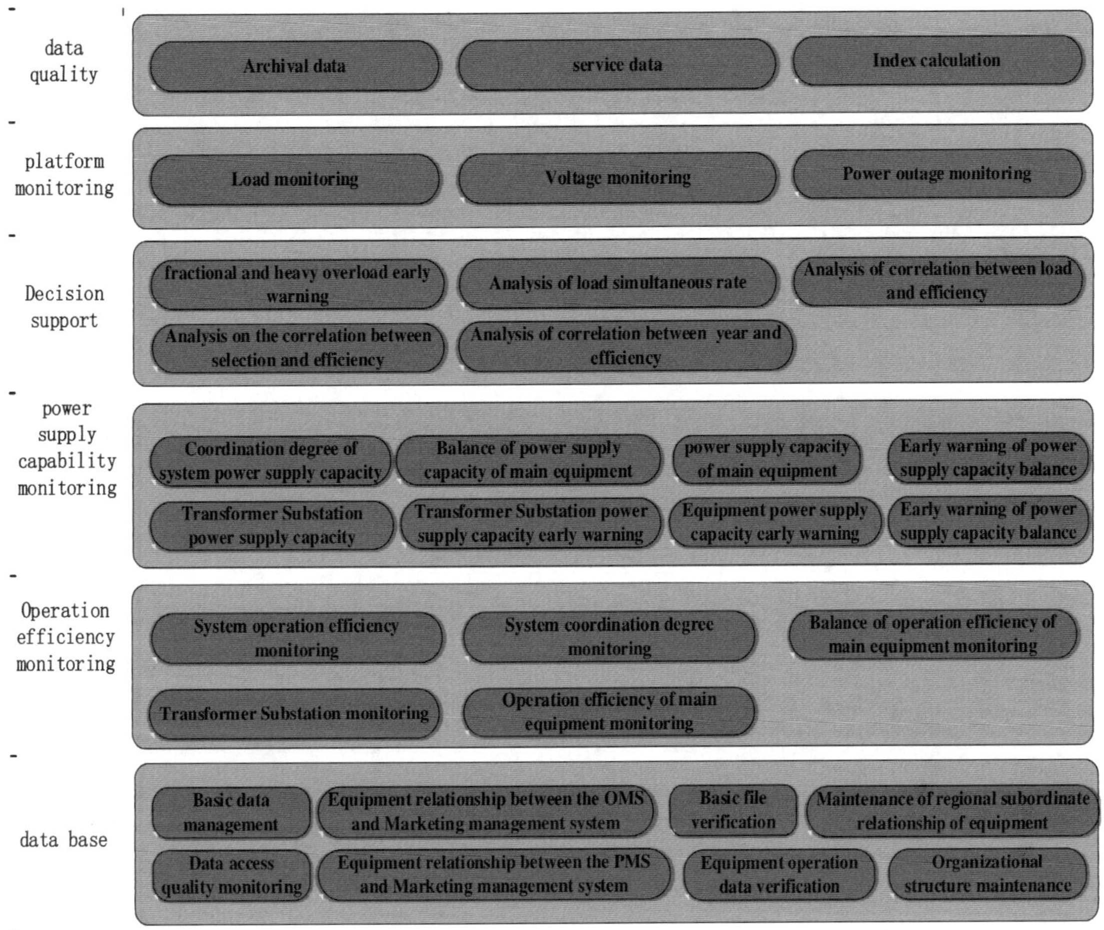

Figure 2. The Application Structure of State Grid Corporation Distribution Network Operation Online Monitoring and Analysis System

2.3. Data Structure

The data source come from Structured Data Platform(SDP) and Massive Data Platform(MDP) , the data about the main transformer, the line, the distribution transformer and other archival data in Structured Data Platform come from the PMS and the Customer Marketing System(CMS), the data about the equipment of running data in MDP come from the OMS and the Power Information Collecting System(PICS). The data structure is shown in Figure 3.

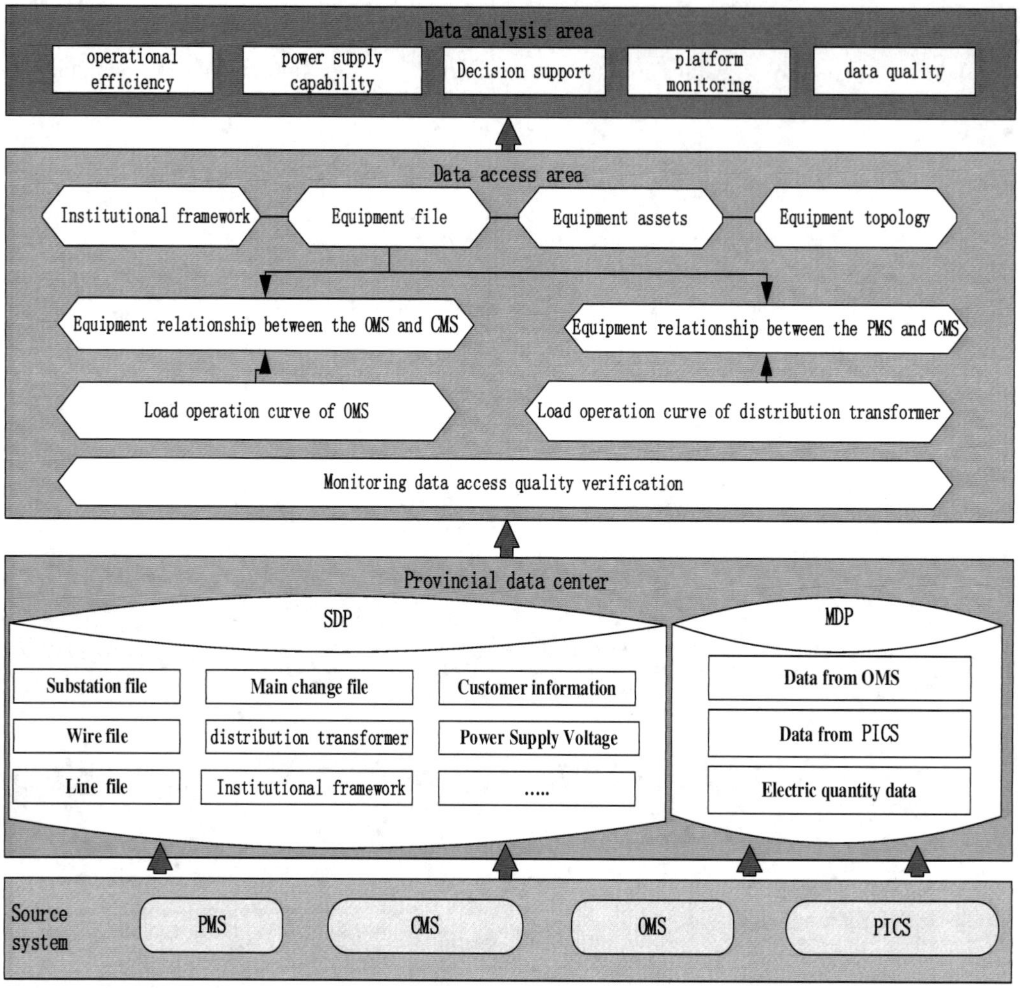

Figure 3. The Data Structure of State Grid Corporation Distribution Network Operation Online Monitoring and Analysis System

2.4. Technical Structure

The online monitoring and analysis system for distribution network operation followed the technical structure of the early stage of the system, based on the SG-UAP platform development, with the Oracle database. The technical structure includes: infrastructure service layer, technical service layer, basic business service layer, application logic layer, and presentation layer as shown in Figure 4.

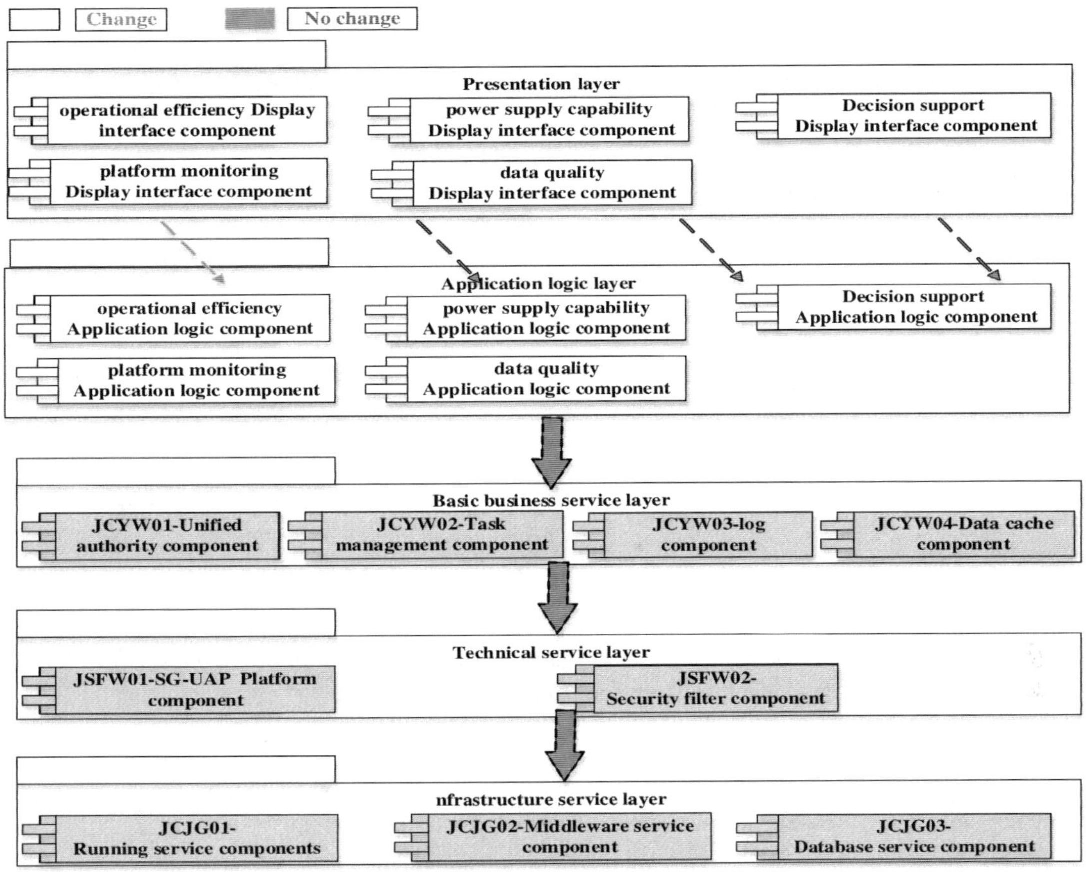

Figure 4. The Technical Structure of State Grid Corporation Distribution Network Operation Online Monitoring and Analysis System

3. Deployment Scheme

The online monitoring and analysis system of distribution network operation adopted the cluster high-availability deployment mode: the intranet application was deployed along the deployment, the server was placed in the production area, the service application included two nodes, and the load balancing is implemented by F5 technology as shown in Figure 5.

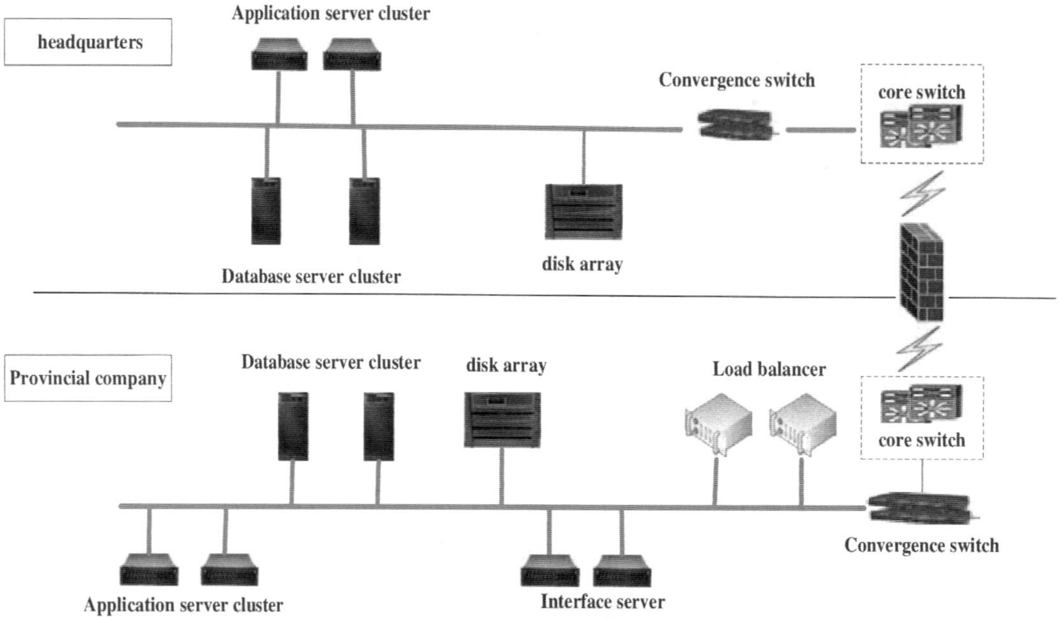

Figure 5. The Deployment Structure of State Grid Corporation Distribution Network Operation Online Monitoring and Analysis System

4. Conclusion

The system has realized the real-time monitoring and comprehensive evaluation of the operating efficiency and power supply capacity of 27 provincial power grid assets [6]. As analysis, the grid efficiency was 0.2538 in 2018, the transmission grid efficiency was 0.3, up 5.64%, and the distribution grid efficiency was 0.2596, down 3.97% year-on-year; 0.2853 in the east, 0.256 in the middle, 0.2413 in the west, and the efficiency in the east-middle-west was gradually decreasing. The load levels of the voltage levels are reasonable, but the regional differences were large. The distribution network high voltage lines, main transformers, medium voltage lines, and distribution transformer power supply margins are 0.4015, 0.2163, 0.3597, and 0.5204, respectively. Among them, the distribution capacity margin can be equivalent to the five-year expansion demand. The comprehensive evaluation results have provided the decision support for the company's power grid planning, construction and production operations, also and promoted the level of grid input and output efficiency as well as the asset lean management level [7].

5. References

[1] Gao Ciwei, Wu Qian, Chen Xihan, Chen Shaojun. Research on Evaluation method of Transmission Network Operation efficiency [J]. East China Electric Power, 2013, 41(03):485-490.

[2] Yang Weihong, Cai Qi, He Yongxiu, Dai Aiying, Wu Liangqi. Evaluation of Coordination between Development and Economic Development of Beijing Power Grid. East China Electric Power, 2009, 10.

[3] Lou Suhua, Wu Yaowu, Li Tingting, Yi Lin. Characteristic Analysis of improving Power Grid Equalization based on demand side Storage. Journal of Electric Power Science and Technology, 2013, 3.

[4] Liu Guozhong. Software flow Optimization Design method and Application of Management Information system in Electric Power Enterprises[D]. North China Electric Power University (Beijing), 2004.

[5] Zhang Lei. Research on Financial intensive Management Mode of Power Grid Enterprises[J]. Guangdong science and technology, 2013, 22(10):204-205.

IOP Conf. Series: Materials Science and Engineering **719** (2020) 012013 doi:10.1088/1757-899X/719/1/012013

[6] Zhou Bowen, Yang Jun. Comprehensive Evaluation Index system and Evaluation Model from the full Perspective of Power Grid Operation efficiency[Papers on the 27th academic year Conference of Power system and Automation Specialty in Colleges and Universities in China], 2011.

[7] Meng Han, Ren Junxia, Tao Ran, Yu Jiehua. Discussion on Lean Management system of equipment (assets) Operation and maintenance in Power Grid Company. Communication world, 2016, 5.

IOP Conf. Series: Materials Science and Engineering **719** (2020) 012014 doi:10.1088/1757-899X/719/1/012014

The Research on SPA Distribution Channel from the Perspective of International Supply Chain Environment

Ling Qiu

Fuzhou University of International Studies and Trade YuHuan Road, Souzhan New District, Changle District, Fuzhou City, Fujian Province, China

Email: 23794648 @qq.com

Abstract. In the context of economic globalization, the environment of distribution industry has changed in essence. The rapid development of e-commerce based on Internet information technology has changed and even subverted the production and sales model of products. Personalized and diversified consumption characteristics have become the new common things. Based on the consumer demand, this paper studied SPA distribution channels and constructed a unique supply chain. On the basis of summarizing previous studies, this research used analytic hierarchy process to confirm the weights and used the empirical study to prove practicality of the index system and model.

1. Introduction

SPA(Specialty retailer of Private label Apparel) model refers to a business model in which clothing enterprises have their own brands, and the head office is responsible for the planning of clothing, product design, purchase of raw materials, production and sales of commodities, which is highly vertical and unified in the supply chain. Specialized stores centered on textiles and clothing are pioneers in the internationalization of retail. Moreover, because the internationalization of specialized stores adopts more standardized ways, it is easy to apply the research methods of internationalization of manufacturing industry.

2. Evauation Indexe Selection of SPA Format in International Supply Chain Environment

Under the background of economic globalization, SPA mode takes meeting consumer demand as the primary goal, which can effectively connect consumers and suppliers, and connect the procurement, production and retail links of enterprises. Through the innovation of supply method and supply chain process, it can achieve rapid response to the market. Under the international supply chain environment, the evaluation index selection of SPA format is the basis of constructing the evaluation system.

In the international supply chain environment, the research of SPA format mainly focuses on consumer demand, procurement, production and retail. In this paper, 35 evaluation indexes with high occurrence frequency are preliminarily selected through the frequency statistics of evaluation indexes of SPA business form research at home and abroad. Professional shops are studied from the perspective of supply chain, and the indexes are modified according to the investigation of enterprise evaluation personnel, in an attempt to establish a set of applicable index system.

This paper conducted two questionnaire surveys. In the first survey questionnaire, namely the expert questionnaire, industry experts, university teachers and leaders of well-known enterprises were invited to score the importance of indicator factors according to the designed expert questionnaire. According to "unimportant, slightly important, relatively important, very important and most important", experts set scores of 1, 3, 5, 7 and 9 respectively for the importance of major indexes. A preliminary survey of the importance of subdivision index under the five major indexes of the clothing SPA model in the international supply chain environment was completed.

In the first survey, 18 questionnaires were issued and 15 were recovered. After statistical sorting of the questionnaire results, 35 indicators were obtained. In this paper, SPSS22 was used to process the data in order to identify the key factors affecting the clothing SPA. It can be determined that these 15 factors are the main factors affecting SPA enterprises in the international supply chain environment, as shown in table 1:

Table 1. Factor Variances and Weights Table

Factor	Characteristic Roots	Variance ratio %	Cumulative Contribution Rate %	The Weight
Business Circle	17.694	17.091	17.091	0.172043638
Brand Management	13.688	13.221	30.312	0.133092196
The Regional Economy	12.409	11.985	42.297	0.120656127
Location	9.583	9.256	51.554	0.09317815
Commodity Combination	9.350	9.031	60.585	0.090912627
Price	7.749	7.485	68.070	0.075345662
Regional Culture	6.772	6.541	74.611	0.065846022
Visual Identification	5.549	5.359	79.971	0.053954456
The Service Level	4.659	4.500	84.471	0.045300741
Image Recognition	4.079	3.940	88.411	0.039661241
Sales Promotion	3.222	3.112	91.524	0.031328394
Experience Marketing	2.969	2.868	94.392	0.028868405
Demographic Factors	2.082	2.011	96.403	0.02024386
Flow of Customer	1.876	1.812	98.215	0.018240865
Concept Identifying	1.165	1.126	99.341	0.011327616

3. The Construction Process of the Evaluation Index System of SPA for SPA Format in the International Supply Chain

The second questionnaire compares the relative importance of primary and secondary indicators. This paper conducted research on the five departments of marketing, production, sales, purchasing and management of two typical private brand chain enterprises, and scored the proportion relationship between SPA indexes in pairs under the international supply chain environment from their perspective.

3.1.Questionnaire Distribution and Recycling

This questionnaire was conducted by the staff of the Marketing Department, production department, sales department, purchase department and management department of private brand chain enterprises. A total of 12 questionnaires were issued and 11 valid questionnaires were collected, reaching 90%.

3.2.The Data Analysis

In order to ensure the effectiveness of the results of the questionnaire, the results of the questionnaire were analyzed.

3.2.1.The reliability analysis

The questionnaire data were analyzed by SPSS22, and Cronbach's alpha was 0.872. The higher Cronbach's alpha coefficient is, the higher the reliability is. When the Cronbach's alpha coefficient is greater than 0.6, the higher the reliability is. It can be seen that this questionnaire has a high reliability and the results are reliable.

3.2.2.Validity analysis

Further statistical test was conducted on the questionnaire. KMO value and Bartlett test showed that KMO=0.736 and KMO value was greater than 0.5, indicating that the questionnaire was effective and suitable for AHP analysis. The significance of Bartlett sphericity test is 0.000, less than 0.001, and this questionnaire is valid.

This questionnaire has reached the design specification, so the data collected and processed according to this questionnaire can objectively reflect the needs of customers. These 15 factors are classified into five categories: retail mix, specialty store management, business environment, specialty store design and specialty store location.

3.2.3.Determination of index weight based on AHP

Based on the analysis results of questionnaire data, this part constructs the evaluation index system of SPA format in the international supply chain environment. Based on AHP, the index weight is calculated to determine the important indexes affecting SPA, providing a basis for the study of SPA formats in the international supply chain environment.

3.2.3.1. Evaluation index system

Based on the analysis of questionnaire data, this part constructs the evaluation index system of SPA format in the international supply chain environment. Based on AHP, the index weight is calculated to determine the important indexes affecting SPA, providing a basis for the study of SPA formats in the international supply chain environment. According to the survey results of SPA indicators and the results of the first expert survey questionnaire, 15 factors were classified and named into 5 categories: retail mix, professional store management, business environment, professional store design and site selection.

3.2.3.2.Evaluation model and construction of judgment matrix

The general target layer A is composed of subsets B1, B2, B3, B4 and B5. B1 is composed of C11, C12 and C13, that is, B1={C11, C12, C13}. Similarly, B1 is composed of B2, B3, B4 and B5. See figure 1.

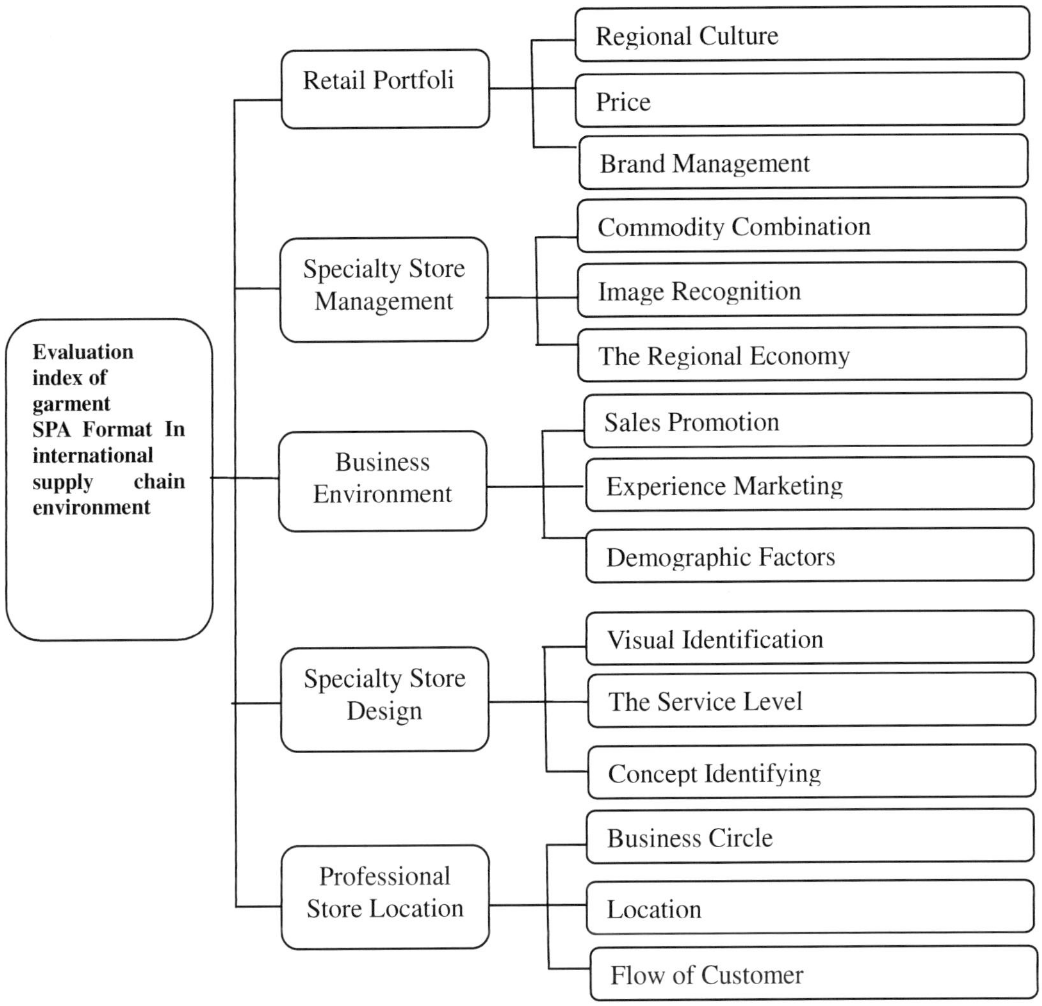

Figure 1. Evaluation Index Level of SPA Format in the International Supply Chain Environment

3.2.3.3.Consistency test and index weight determination

After the scores of 12 experts, a matrix is established. Among them, the consistency ratio of evaluation index A of SPA format in the international supply chain environment: 0.0148; Weight of "evaluation index A of SPA format in the international supply chain environment" : 1.0000; Lambda Max: 5.0341.

According to the weight ordering, the first layer is a very important layer, that is, the weight value is greater than 0.1, and they are C51 and C13 respectively. The second level is the important level, that is, those with weight value between 0.05 and 0.1 are C52, C21 and C11, respectively. The rest are the third level with weight value between 0.05 and 0.02, and the fourth level with weight value less than 0.02, the general level.

Table 2. Ranking Weights of Factors in Scheme Layer to Decision Objectives

Plan	The Weight	The Sorting
Business Circle C51	0.2314	1
Sales Promotion C13	0.1601	2
Experience Marketing C23	0.1232	3
Location C52	0.0712	4
The Service Level C21	0.0611	5
Price C12	0.0534	6
Commodity Combination C11	0.0503	7
Regional Culture C41	0.0446	8
Image Recognition C42	0.0428	9
Brand Management C22	0.0411	10
The Regional Economy C31	0.0373	11
Visual Identification C32	0.0301	12
Demographic Factors C33	0.0223	13
Flow of Customer C53	0.0178	14
Concept Identifying C43	0.0133	15

4. The Empirical Research

X clothing enterprise, with an international advanced clothing production line, mainly operates in the form of professional shops at home and abroad, has great popularity and market space in the international market. In this paper, six specialized stores of the clothing enterprise are summarized and counted according to the operation data that lasts for half a year before and after the countermeasures are taken according to the evaluation index weight of the clothing SPA format.

4.1. Regression Model of Dummy Explanatory Variables

If there is a significant change in sales volume before and after the introduction of dummy variables to explain the change of countermeasures, the explanatory variables have only two mutually exclusive qualitative variables and no regression of quantitative variables. Therefore, the variance analysis model is set, with the type as follows:

$$Y_t = aD_t + b + u_t \tag{1}$$

Where, is the operating quota of X enterprise in month t (yuan), is the dummy variable; And the game changes, the game doesn't change.

The significance of formula (1) is that, assuming that other factors remain unchanged, it studies whether there is a difference in regional express turnover before and after the change of strategy. When the classical hypothesis is satisfied, formula (1) can be written:

Average monthly operation without countermeasures:

$$E\left(Y_t \middle| D_t = 0\right) = b \tag{2}$$

Average monthly operation after the change of countermeasures:

$$E\left(Y_t \middle| D_t = 1\right) = a + b \tag{3}$$

That is, in formula (1), intercept term b gives the average monthly operation amount of the unimplemented game, while another coefficient indicates the part that changes the average monthly operation amount of the game is different from the average monthly operation amount of the unchanged game, which can be obtained from formula (2) and formula (3). The left and right of dummy variable is to change the intercept level of the set model.

To test whether there is a significant difference between the average monthly operation of the unchanged game and the changed game, we can construct a hypothesis that there is no difference between the changed game and the monthly operation. For the regression of formula (3), the acceptance or non-acceptance of hypothesis can be judged according to whether the t-test of the estimated value is significant.

4.2. Regression Analysis of Monthly Operating Income

Regression analysis was made on the sales volume of 9 consecutive months before and after the implementation of countermeasures by 9 professional stores of X enterprise. Since the t test of each coefficient is greater than 2, it indicates that the coefficient of each explanatory variable is significantly different from 0, which indicates that the regression equation of the change in countermeasures to the monthly operating amount is indeed different in statistical significance. Regression results are summarized in table 3:

Table 3. Regression results

parameter	**Speciality Store**					
	A	**B**	**C**	**D**	**E**	**F**
DI	632.5500 t=(2.3147) p=(0.0514)	3013.600 t=(2.3325) p=(0.0214)	1322.750 t=(2.2143) p=(0.0614)	4028.700 t=(3.9935) p=(0.0052)	674.2500 t=(2.1875) p=(0.0416)	4033.250 t=(3.1473) p=(0.0196)
C	5421.300 t=(2.1322) p=(0.0000)	9420.500 t=(18.7188) p=(0.0000)	12131.45 t=(25.1467) P=(0.0000)	15001.70 t=(21.1463) p=(0.0000)	5660.75 t=(30.7427) p=(0.0000)	18132.70 t=(19.3124) p=(0.0000)

Through the above table, it can be concluded that the change of countermeasures does have a positive effect on the monthly operation volume of various regions, and it also has certain reference significance for other professional stores and enterprises.

5. Acknowledgement

This paper was supported by Fujian Social Science Planning Youth Project 2016. The project is No. FJ2016C149. The topic of this project is named with"Research on the innovative development of garment SPA industry in fujian province under the international supply chain environment".

6. References

[1] Chen dandan. Research on the supply chain of changle textile industry in fujian province to join the international SPA clothing specialty store [J].Current logistics and engineering,2015,(5): 214-215.
[2] Zheng linyan. Research on the regional selection of international SPA enterprises to enter the Chinese market [D]. Fuzhou university, 2014.
[3] Zhang hao. The foundation of channel relationship in the process of supply chain construction -- an analysis of Japanese garment industry [J].Journal of Beijing technology and business university (social science edition), 2015, 30(02):56-62.
[4] Nie linpeng. Application of SPA model of clothing enterprises in the new retail environment [J]. Logistics technology and application, 2008, 23(04):147-149.

Route Analysis of Trail Adjustable Wagon Based on Crank Rocker Mechanism

Yinglong Liao[1], Jiangang Yi[1,*], Ziyan Fang[1] and Cuixia Hao[1]

[1] School of Electromechanical and Architectural Engineering, Jianghan University, Wuhan 430056, China
469883553@qq.com

Abstract. The trajectory analysis of a transport wagon is a prerequisite for its rational path planning. In order to analyze the influence of crank and rocker lengths on the trajectory of a transport wagon in a four-bar spatial mechanism, two kinds of trajectories, single "8" and double "8" were simulated. The results show that changing the crank length will obviously change the trajectory shape, and changing the rocker length will mainly change the walking angle. This conclusion provides a basis for the route adjustment and planning of transport vehicles.

1. Introduction

In the increasingly busy workshop transportation production line, the trajectory of the transport wagon is more complex. On the one hand, with the increase of production line function, more requirements are put forward for the traveling path of the transport wagon. On the other hand, in order to reduce the overall production cost, the space utilization rate of the production line workshop is constantly improving, which makes the traveling space of the transport wagon smaller and smaller, so it is necessary to design a more flexible transport wagon. Crank rocker mechanism is a common mechanism in the steering part of transport wagon, and it is the most basic form of four-bar mechanism. It is widely used in feeding mechanism of bullhead planer, radar adjustment mechanism, foot mechanism of sewing machine, compound pendulum jaw crusher and steel conveyor. Aiming at the steering part of the transport wagon, this paper analyses the trajectory change rule and adjustment strategy of the trajectory based on crank rocker mechanism.

2. Steering Structure of the Wagon

As shown in Figure 1, the steering structure of the transport wagon is mainly composed of a hinged four-bar crank-rocker mechanism consisting of a crank and a rocker. Among them, the crank is the active part, driven by the motor or other energy devices, and rotates at the same speed, while the rocker is the follower to swing back and forth at variable speed, and the connecting rod is a plane compound motion. In order to make the spatial rotation of the connecting rod more stable, the joint connecting bearing between crank and rocker adopts universal coupling, thus turning the steering structure into a spatial crank rocker mechanism.

In order to calculate and analyze the trajectory of steering mechanism, assuming that r1 is the length of crank, L is the length of connecting rod, C is the length of rocker, b is the length of crank to origin, and d is the length of rocker to origin, the motion coordinate system of steering mechanism of transport wagon can be established, as shown in Figure 2. Consider the steering wheel of the transport wagon as a starting point. In practical application, in order to facilitate manufacture and installation, the length of crank and connecting rod will not be changed. Therefore, in order to simplify the calculation, the length of crank and connecting rod is set constantly.

Figure 1. Steering Structure of the Wagon

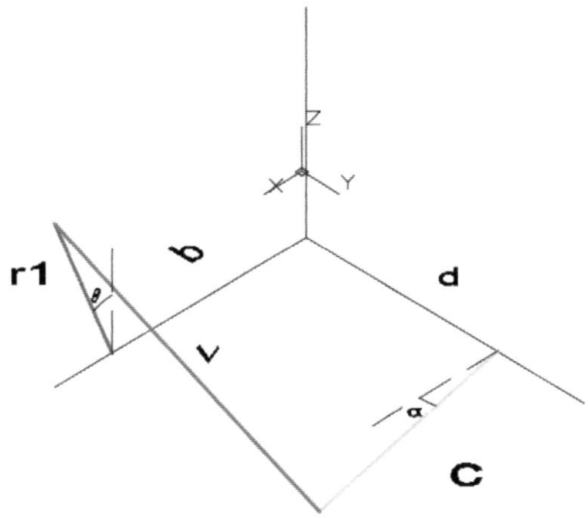

Figure 2. Motion Coordinates of Steering Mechanism

3. Trajectory Change Analysis

In order to compare and analyze the changed paths, the initial trajectory is designed as shown in Figure 3. Starting from the horizontal direction of the X-axis at the right end of the trajectory, the moving wagon moves along the left X-axis at a certain angle to the horizontal direction. During walking, the crank rocker mechanism makes the steering mechanism constantly change its angle. The car moves along the curve to the left end of the X-axis and then continues to move along the curve to the starting point, forming a closed circular curve. The whole trajectory is similar to an double "8" placed horizontally.

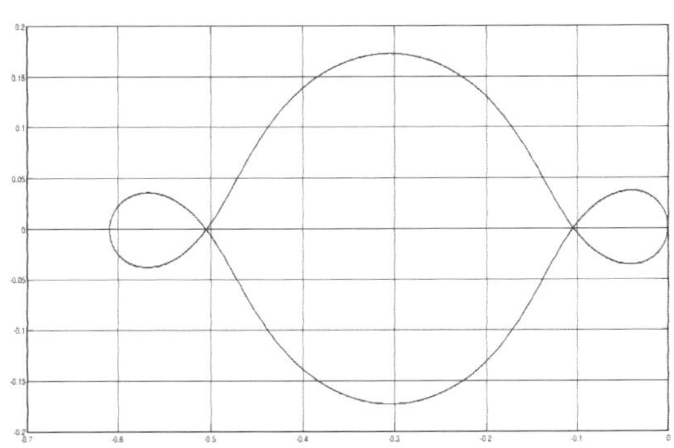

Figure 3. Initial Trajectory

For ease of analysis, it is assumed that the length b of the crank to the origin varies in the range. The limit position values b-3 and b+3 are taken respectively, and the trajectory of the car is shown in Figure 4. Among them, the blue track is the shortest crank-to-origin length; the black track is the same crank-to-origin length; and the red track is the longest crank-to-origin length. From Figure 4, it can be found that changing the b value can change the upward or downward migration angle of the trajectory, but does not change the distance of the trajectory in the horizontal direction. Increasing or decreasing the value of b will affect the downward or upward migration angle of the trajectory, and has no significant effect on the trajectory shape.

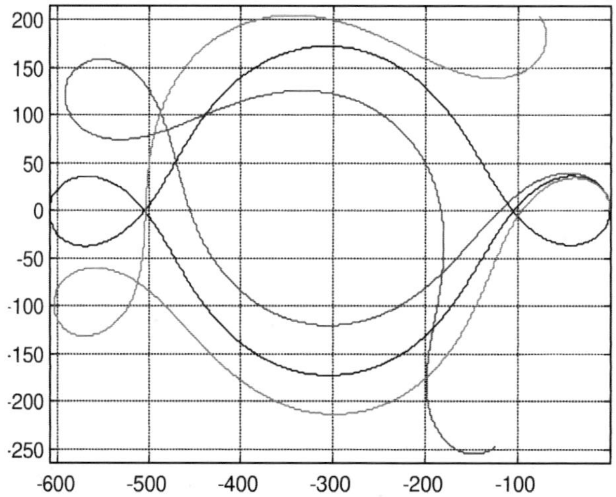

Figure 4. Crank Length Changing in Double "8" Path

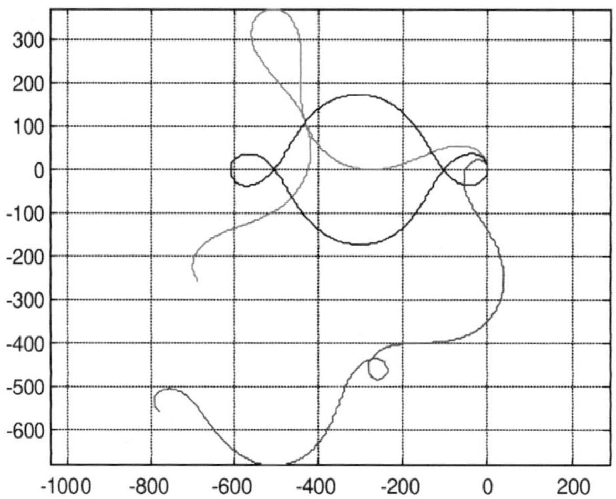

Figure 5. Rocker Length Changing in Double "8" Path

Similarly, assuming that the range of the length C of the rocker is the limit values C-3 and C+3, the trajectory diagram of the wagon is shown in Figure 5. Among them, the blue trajectory is the shortest case of rocker length; the black trajectory is the case of rocker length unchanged; and the red trajectory is the case of rocker length is the longest case. Figure 5 shows that when the length of rocker decreases, the trajectory will rotate counterclockwise, and the route will change from closed to open; when the length of rocker increases, the trajectory will rotate clockwise, and the trajectory will change to "8". Thus, changing the length of the rocker will significantly change the shape of the trajectory. Therefore, when adjusting the angle of steering mechanism, it is not appropriate to adjust the length of rocker.

IOP Conf. Series: Materials Science and Engineering **719** (2020) 012015 doi:10.1088/1757-899X/719/1/012015

4. Path Change Analysis

In order to compare different planning paths, the horizontal double "8" path is changed to the horizontal single "8" path. By changing the length of crank, rocker or connecting rod respectively, the change of trajectory is observed. As shown in Figure 6, the blue trajectory reduces the length of the crank to the origin; the black trajectory keeps the length of the crank to the origin unchanged; and the red trajectory increases the length of the crank to the origin. From Figure 6, it can be found that when the length of the crank to the origin decreases, the trajectory will rotate clockwise, while the shape of the trajectory remains the trajectory "8"; when the length of the crank to the origin increases, the trajectory will rotate clockwise, while the trajectory remains the word "8". Thus, when the angle of the path needs to be changed without changing the shape of the path, the length of the crank to the origin can be moderately increased or reduced.

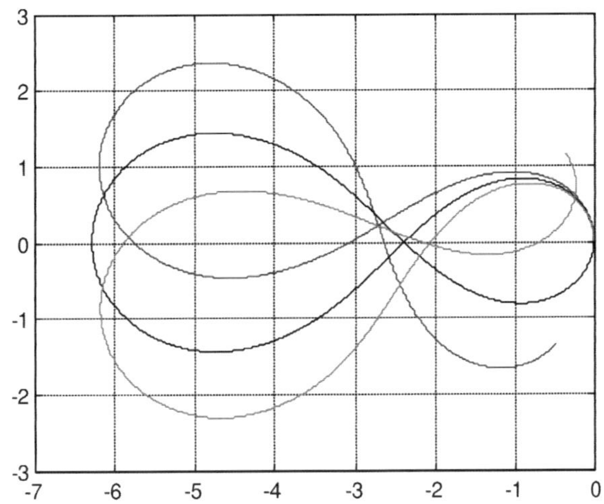

Figure 6. Crank Length Changing in Single "8" Path

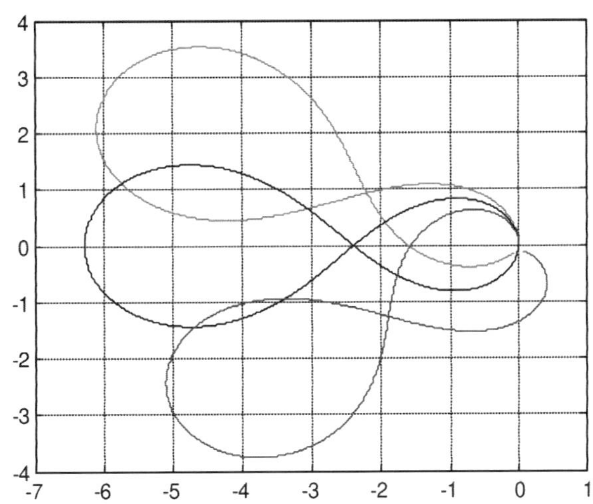

Figure 7. Rocker Length Changing in Single "8" Path

Figure 7 shows the change of steering mechanism path caused by changing the length of rocker in a single "8" path. Among them, the blue trajectory is the shortest case of rocker length; the black trajectory is the case of rocker length unchanged; and the red trajectory is the case of rocker length is the longest case. From Figure 7, it can be found that when the length of rocker decreases, the trajectory will rotate counterclockwise while the shape of the route remains unchanged; when the length of rocker increases, the trajectory will rotate clockwise and the trajectory will remain "8". It can be seen that changing the length of rocker has no obvious effect on the shape of the path, but it will obviously change the deviation angle of the path.

5. Conclusions

By changing the distance from crank to origin and the length of rocker, the trajectories of double "8" and single "8" of steering mechanism of transport car are simulated and analyzed. It is found that changing the distance from crank to origin and the length of rocker will change the trajectory of transport wagon. At the same time, in the change of crank or rocker length in four-bar space, the angle of counter-clockwise rotation of the wagon is larger than that of clockwise rotation. These laws provide a basis for the trajectory planning of transport vehicles.

6. References

[1] Faxin L , Chuanyu W , Liang S . Analysis and test of influence of revolute joint clearance on performance of crank-rocker style transplanting mechanism[J]. Transactions of the Chinese Society of Agricultural Engineering, 2016.
[2] Wu L J , Li A P , Liu X M . Modeling and Simulation of a Slider-Crank Mechanism with Clearance and Lubricant Joint[J]. Advanced Materials Research, 2012, 510:458-466.
[3] Li D U , Jun F , Cheng-Xuan X , et al. Motion Behavior and Simulation Analysis on Coaxial Out-put Impulse Mechanism[J]. Journal of System Simulation, 2009, 21(2):581-585.
[4] Zou H X , Zou X J , Chen Y L , et al. Research on Virtual Design and Simulation System for Manipulators of Multi-Degree of Freedom[J]. Advanced Materials Research, 2010, 97-101:3475-3478.
[5] Tang Y , Liu Z , He B . Optimization Design for Crank-rocker Mechanism Based on Genetic Algorithm[C]// Second International Conference on Intelligent System Design & Engineering Application. IEEE Computer Society, 2012.
[6] Chen L J , Bai X W , Wang J L . Optimization and simulation of crank-rocker mechanism based on Excel[C]// International Conference on Electronic and Mechanical Engineering and Information Technology, EMEIT 2011, Harbin, Heilongjiang, China, 12-14 August, 2011. IEEE, 2011.

IOP Conf. Series: Materials Science and Engineering **719** (2020) 012016 doi:10.1088/1757-899X/719/1/012016

Computer Simulation Design of Joint Manipulator for Handling Pipeline

Peng Zeng[1,*], Ziqiang Fang[2] and Feihu Xing[2]

[1] School of Mathematics and Computer Science, Jianghan University, Wuhan 430056, China

[2] School of Electromechanical and Architectural Engineering, Jianghan University, Wuhan 430056, China

419455942@qq.com

Abstract. The joint manipulator is a transmission device suitable for operation close to the body, which can realize multiple degrees of freedom and move flexibly. In this paper, through the transmission analysis of the big arm, the small arm, the wrist and the other parts of the manipulator, a kind of joint manipulator used in the transportation pipeline is proposed, and a three-degree-of-freedom wrist transmission mechanism is designed. On this basis, the structure of the wrist transmission parts is analyzed and designed, and the mechanical analysis and strength check of the gear connecting shaft of the wrist part are carried out.

1. Introduction

Industrial robots are widely used in production because of their high automation and mechanization. Under the control of the program, industrial robots can complete some humanoid operations. Therefore, for some dangerous actions or repeated movements, industrial robots can be completed by manipulators to reduce the labour intensity of workers and improve production efficiency. At present, industrial robotics technology in China is developing rapidly, but there is still a long way to go compared with some developed countries. Especially for high precision manipulator parts which needs to be solved by import, our country cannot manufacture them independently. In view of this situation, we take the material handling manipulator which is most widely used in industrial production line as an example, and propose a design scheme of three-degree-of-freedom joint manipulator.

2. Site Requirements

According to the requirement of the production site, the industrial robots designed are installed in the middle of two production lines and controlled by motors, which are used to transfer workpieces between the two production lines. The workpieces are approximately a cube with a weight of 5Kg and a side length of 30cm. The two parallel production lines are 1 metre apart, and the maximum elongation of the manipulator is about 1.3 metres. The waist can rotate in space for a week. The bottom is fixed on the ground with bolts, as shown in Figure 1.

IOP Conf. Series: Materials Science and Engineering **719** (2020) 012016 doi:10.1088/1757-899X/719/1/012016

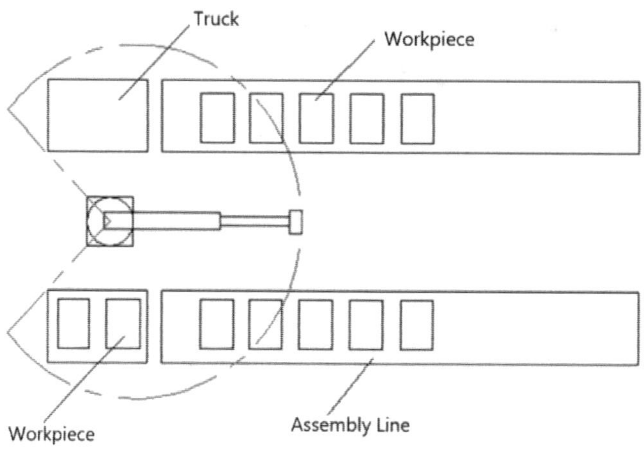

Figure 1. Site Layout

According to the requirement of the production site, the industrial robots designed are installed in the middle of two production lines and controlled by motors, which are used to transfer workpieces between the two production lines. The workpieces are approximately a cube with a weight of 5Kg and a side length of 30cm. The two parallel production lines are 1 metre apart, and the maximum elongation of the manipulator is about 1.3 metres. The waist can rotate in space for a week. The bottom is fixed on the ground with bolts, as shown in Figure 1.

Specific requirements of the industrial robot include:

(1) The workspace is large and the robot occupies a small area.

(2) Flexible handling of goods, faster speed and accuracy.

(3) It can be placed in the box body to facilitate the maintenance and maintenance.

(4) When driving each axis, the torque is small and the energy consumption is small.

3. Project Design

According to the requirements, we decided to adopt joint robots structure. The articulated robot arm consists of two parts: the dynamic joint and the connecting rod. The arm mechanism is a four-link mechanism, whose action can reach any point in the space range. Rotary frame is selected for the machine base. The wrist is located between the arm and the end effector, and its motion determines the action and posture of the end effector. Because of the high requirement for the motion of the end effector, a three-degree-of-freedom articulated manipulator is designed as shown in Figure 2.

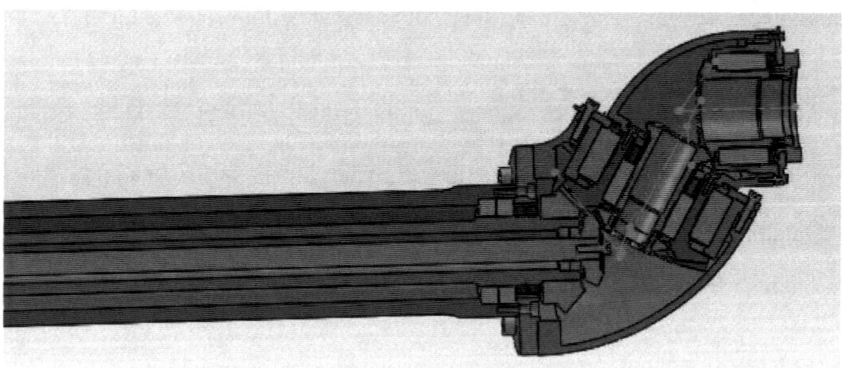

Figure 2. Three-degree-of-freedom Articulated Manipulator

The end of the three-degree-of-freedom robot wrist is connected with the arm of a four-bar mechanism. The control motor is installed on the arm, which can reduce the wrist's weight and facilitate the wrist's flexible rotation. The transmission structure of the small arm joint is shown in Figure 3. The arm can pitch in three-dimensional space, and then determine the position of the wrist. The range of action is from + 130 degree to – 90 degree. The servo motor is mounted on the driving arm, that is, the joint of the big arm. Through the through hole at the bottom of the arm, two connecting rods are used to connect the small arm seat. Through the action of the connecting rod, the pitching action of the small arm seat is realized. The bottom is connected by a flat key, while the upper arm and the bottom pole are connected by a tension sleeve. There are through holes at the bottom of the arm. The arm is connected with the reducer through the through holes. The motor is installed side by side with the reducer. The position of wrist in the plane is determined by the pitching motion of the upper arm and the lower arm. The position of the manipulator in the workspace is realized by the waist rotation and the pitching motion of the arm and the arm. The shaft at the waist has a central shaft hole, which matches the shaft of the motor and transfers the torque of the motor to the shaft by means of keys. The waist transmission structure is worm wheel and worm drive mode. The rotation of the motor drives the turbine to rotate and makes the waist rotate.

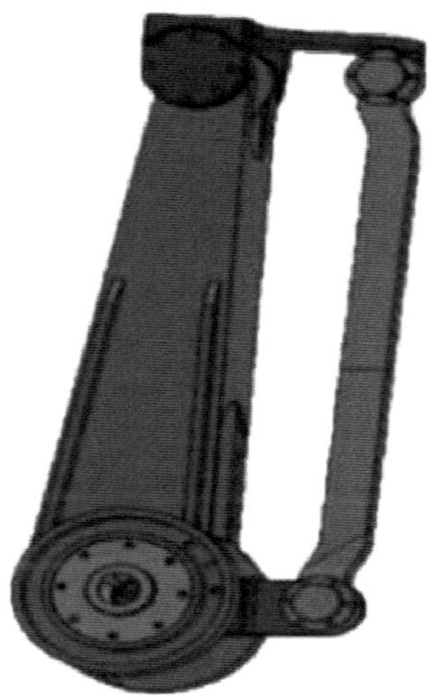

Figure 3. Robot Arm

4. Computational Analysis

According to the requirements of production, it is known that the arm inertia of the workpiece is:

$$J = ms^2/6 = 7.5 \times 10^{-4} kg.m^2 \tag{1}$$

The rotational speed is:

$$w = 330^{\circ}/s$$

The rotational angular acceleration is:

$$\beta = 3300^{\circ}/s^2$$

IOP Conf. Series: Materials Science and Engineering **719** (2020) 012016 doi:10.1088/1757-899X/719/1/012016

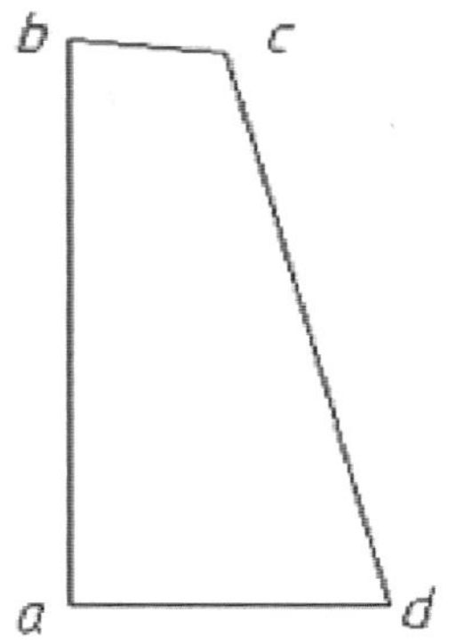

Figure 4. Four-bar Linkage

The crank-rocker mechanism is selected for the four-bar linkage. Among them, in Figure 4, L arm = 1000mm, represented by a-b; L base bar = 400mm, represented by a-d; L rear bar = 1000mm, represented by d-c; L arm = 200mm, represented by b-c. The frame, rocker and crank of four-bar linkage are a-b bar, a-d bar and b-c bar respectively.

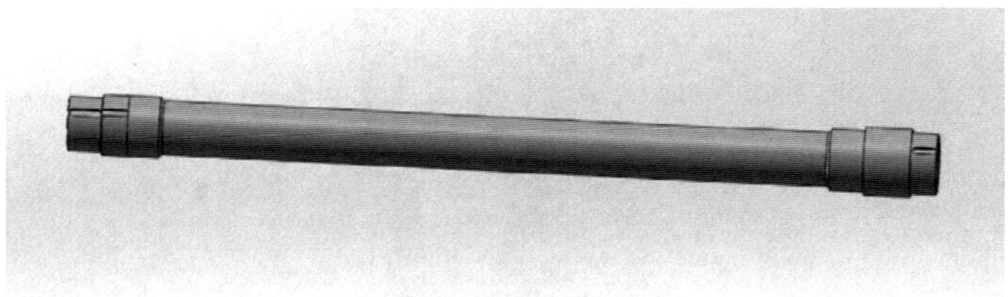

Figure 5. Connecting axle

Because the gear and key are installed at the connecting axle of wrist gear, the force of wrist gear is complex. Therefore, it is necessary to analyze the force of wrist gear axle. As shown in Figure 5, according to the above analysis, it is known that the upper torque of the gear is:

$$T_1 = 9.55 \times 10^6 \times \left(\frac{P_1}{n_1}\right)\eta \tag{2}$$

Then the circumferential force is:

$$F_t = \frac{2T_1}{d_1} = 1260 \tag{3}$$

After checking, the strength of the wrist gear connecting shaft meets the requirements.

5. Conclusions

The structural design of articulated robot wrist is the key and difficult point in the design of the whole structure of robot. In this paper, a mechanical system of robot used in handling assembly line is analyzed, including waist rotation structure, large lower arm rotation mechanism and three-degree-of-freedom wrist structure. On this basis, the structure of the wrist transmission parts is analyzed and designed, and the mechanical analysis and strength check of the gear connecting shaft of the wrist part are carried out. After analysis, the range of motion of the designed robot wrist can meet the requirements of the field. The design scheme also provides a useful reference for similar joint manipulator design.

6. References

[1] Wu B , Guo W , Liu Y , et al. Structural Design of Multi-joint Foldable Robot Manipulator for Remote Inspection in Experimental Advanced Superconducting Tokamak (EAST)[C]// International Conference on Intelligent Autonomous Systems. Springer, Cham, 2017.

[2] Dirk N . Large manipulator having a vibration damping capacity[J]. Journal of the Acoustical Society of America, 2006(121):3263.

[3] Lee K H , Fu D K C , Leong M C W , et al. Nonparametric Online Learning Control for Soft Continuum Robot: An Enabling Technique for Effective Endoscopic Navigation[J]. Soft Robotics, 2017:soro.2016.0065.

[4] Wang A , Deng M . Operator-based robust nonlinear tracking control for a human multi-joint arm-like manipulator with unknown time-varying delays[J]. Applied Mathematics & Information Sciences, 2012, 6(6):459-468.

[5] Tosunoglu S, Monteverde V. Kinematic and Structural Design Assessment of Fault-Tolerant Manipulators[J]. Intelligent Automation & Soft Computing, 1998, 4(3):8.

[6] Zhang L., Yan X., Zhang Q. Design and analysis of 3-DOF cylindrical-coordinate-based manipulator [J]. Robotics and Computer-Integrated Manufacturing, 2018, 52:35-45.

Design of Power Communication Anti-accident Maneuver Simulation System Based on Alarm Deduction

Shengchao Yan[1], Haobo Zhou[2] and Juan Qi[1]

[1] Nari Group Corporation/State Grid Electric Power Research Institute, PRC.
[2] Inner Mongolia Power (Group) Co., Ltd. Information Communication Branch, PRC.

Abstract. In order to improve the effect of the power communication anti-accident maneuver, and to enhance the ability of operation and maintenance personnel to judge and remove network faults quickly and accurately, series of key approaches are adopted, such as root alarm analysis and business analysis. Considering the relationship between the fault and alarm, power communication anti-accident maneuver system based on alarm deduction is realized. Ultimately the purpose of improving the quality and efficiency of the operation and maintenance personnel is achieved.

1. Introduction

With the construction of smart grid, power communication network has played an increasingly prominent role. Carrying out the communication network anti-accident maneuvers simulation activities on a regular basis is an effective means to improve the ability of operation and maintenance work. Therefore, the use of computer technology to study communications network anti-accident simulation system [1-4], can quickly improve the accuracy of communications professionals to determine the fault and accumulated experience in failure to improve business technical knowledge and operation and maintenance management level to ensure that all types of communications networks Business information flow.

Based on the communication resource management and comprehensive monitoring, the simulation system of power communication anti-accident maneuver described in this paper adopts the idea of synchronicity and anti-push between failure and alarm to take the roles of the interviewer, examiner and candidate Functional structure as the basis to simulate the simulation process, using fault alarm deduction, root alarm analysis and business impact analysis and other key approaches to achieve the purpose of anti-accident training simulation training to improve the communication professionals quickly and accurately determine and troubleshoot the ability.

2. System Design

2.1. Demand Analysis

In accordance with the system role to points, communication anti-accident simulation system is often composed of three parts: a copy of the function of the examiner and examinee functions.

The topic and who has three main functions: First, the director Q & A, including new topics and perform maintenance on existing title; the second is director simulation exercises, simulation exercises content that is set to begin setting simulation exercises, simulation exercises time and termination Select the need to participate in the simulation of the staff and so on. The third is to manage users, as well as examiners, examiners distribution system login account.

2.2. Basic Data

The system will make full use of the basic data (including communication sites, transmission equipment, communication optical cables, and bearer services) existing in the communication management system for resource allocation information [5], and obtain the following relevant basis through the interconnection based on the open Web Service interface data:

(1) relay protection, stability control, remote control, dispatching telephone, video conferencing and other important communications services;

(2) circuit (such as transmission channel, 64K circuit) and optical fiber (such as dedicated core) and other resources;

(3) transmission network resources (such as equipment, slots, ports, time slots), cable resources (such as cable segments, fiber core, etc.), wiring resources and space resources.

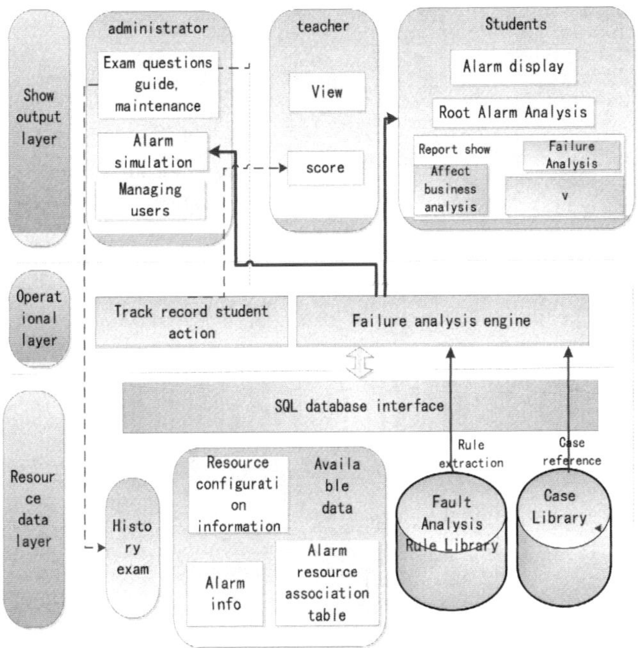

Figure 1. System Architecture

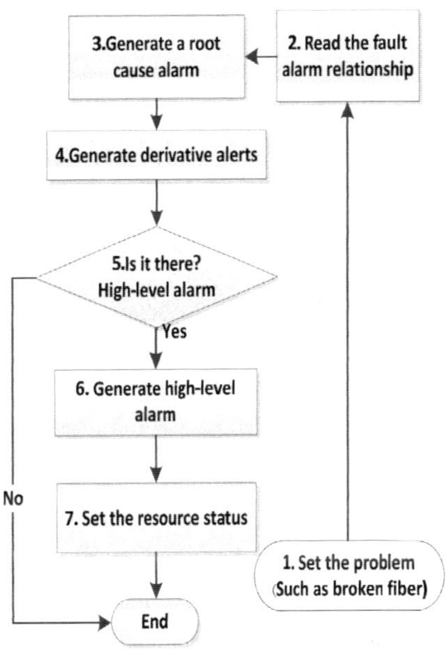

Figure 2. Fault Alarm Flowchart

2.3. System Structure

Based on the above considerations, the design of the system structure shown in Figure 1.

(1) Resource data layer: The data used in the exercise simulation comes from the replication of the real-time database of the communication management system and includes the system's unique rule base, case base and test database.

(2) Operational layer: The fault analysis and processing engine uses the fault analysis rules to combine the resource configuration information and alarm information for alarm presentation, root alarm analysis, service area analysis and correlation analysis, and determines the root failure with reference to the case base.

(3) to show the output layer: According to the questions, examiners, candidates three different roles to show the specific operation.

3. Key Approaches

In order to realize the flexibility, practicability, authenticity and reliability of the anti-accident maneuver simulation system, the key is to realize the alarm deduction management. The system adopts the following key technologies.

IOP Conf. Series: Materials Science and Engineering **719** (2020) 012017 doi:10.1088/1757-899X/719/1/012017

3.1. Fault Alarm Deduction

Based on the correlation between the fault type and the alarm objects, a fault alarm cis-push model is set up to implement the process of generating root alarms, derivative alarms and high-level alarms from faults. The specific method is as follows: establishing a fault alarm relation table according to a specific communication network resource type, and establishing an association model between network resource objects; and according to a fault alarm relationship table and a resource association model, establishing a root type alarm and a derivative alarm and a high-level alarm. The associated model, the final formation of failure alarm deduction model. Based on the above model, the design of fault alarm deduction algorithm, the process shown in Figure 2.

Figure 3. Root Alarm Analysis Flowchart

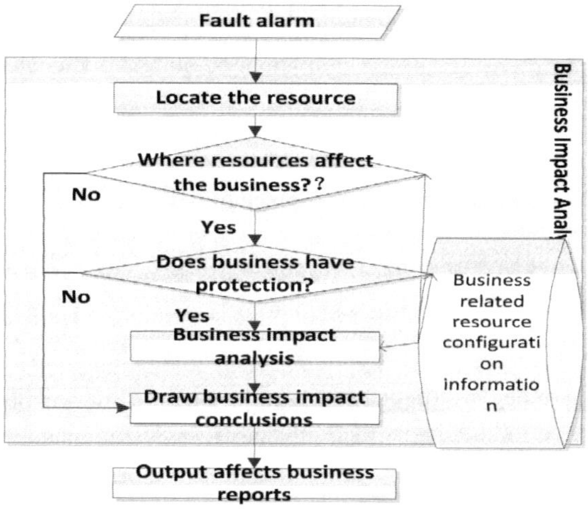

Figure 4. Business Analysis Diagram

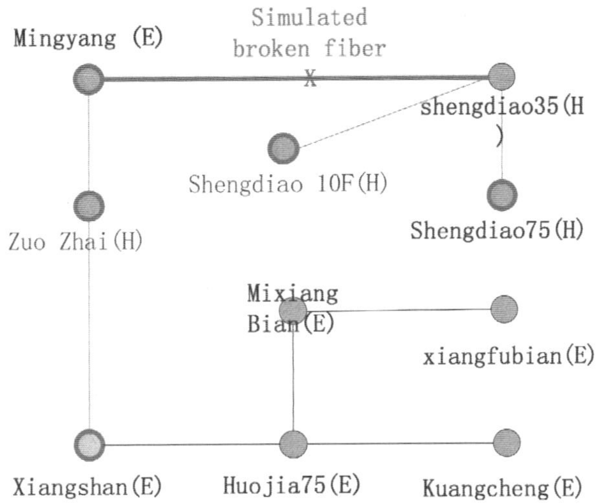

Figure 5. Simulated Fault Case

3.2. Root Alarm Analysis

The root alarm analysis uses the reverse-thinking idea of fault-to-alarm to make a further comprehensive analysis of the alarm information and locate the root cause alarm from the related alarms [6].

The specific implementation method is as follows: all alarms are divided into several correlation analysis and processing queues according to the degree of alarm correlation, and each of the queues performs correlation analysis separately according to conditions such as area, equipment, system and time, filters out homologous and related repetitions, times To, False and other alarms, get the root of each queue alarm group.

Then the rest of the queue alarms are grouped according to their relevancy levels to conduct a correlation analysis.

Finally, the root alarm analysis results are displayed in a hierarchical structure. Process shown in Figure 3.

3.3. Business Impact Analysis

The business analysis model is established by using fault alarm information, resource failure and business related resource configuration information to analyze the affected circuits and business scope when the fault occurs.

The specific implementation method is as follows: taking a simulated fault in a device or a pipeline as an example, generating alarms such as LOS (receiving optical signal loss) and AIS (all 1signal), and determining a root alarm according to a root alarm analysis model and an algorithm. Business analysis as shown in Figure 4.

3.4. Rule-Based Troubleshooting

Failure analysis and processing relies on the alarm correlation rule base and troubleshooting base in the resource data layer. The system provides the maintenance and management of rule bases and case bases, electronically analyzes the operation and maintenance experience of operation and maintenance personnel, and forms rules for alarm analysis and troubleshooting. These rules include: alarm dependency rules, resource dependency rules, alarm and resource association rules, and alarms And resource mapping association table, circuit and business relationship; and the new exercise simulation targeted to form a case for future reference. When a fault occurs, the rules match and a large number of related alarms are generated. Candidates can customize the corresponding rules and reference case base for root alarm analysis, affecting business analysis, correlation analysis and determine root failure [7-8].

IOP Conf. Series: Materials Science and Engineering **719** (2020) 012017 doi:10.1088/1757-899X/719/1/012017

4. Conclusion

The anti-accident maneuver simulation system designed in this paper can effectively improve the ability of operation and maintenance personnel to quickly determine the fault. In the follow-up, the functions of the anti-accident maneuver simulation system will be expanded. Based on the above design method, the simulation system of power communication anti-accident maneuver is developed as a sub-function system of communication management system and successfully applied to a provincial power company shown in Figure 5.And methods can be further studied and extended to analyze the cause of the root cause of the fault and apply it to the communication network fault diagnosis and intelligent analysis of the impact of the alarm business scope, from alarm to failure, intelligent business analysis.

5. References

[1] WANG Hong-jun, LI Zheng-chao, XIE Yong-zhong. Development and application of anti-accident analysis system for electric power communication [J]. Telecommunications for Electric Power System, 2007, 28(180):55-57

[2] WANG Ai-min. Communication Network Anti-accident Maneuver System Based on Multi-Agent Collaboration Technology [D]. Southeast University: 2010.6.12

[3] GONG Ping-ping. Design and implementation of anti-accident maneuver system for communication network [D]. Beijing University of Posts and Telecommunications: 2010.6.12

[4] Anti-accident Exercise Management for East China Power Grid [J]. East China Electric Power, 2007, 35(9):77-79

[5] YAN Sheng-chao, ZHANG Tian-bing, ZHANG Chun-ping. Analysis on The Fault of Power Communication Network Based on Combined Rules [J]. Telecommunications for Electric Power System, 2011, 32(221):34-38

[6] TANG Yi-ze, ZHUANG Yi. Principles and Research on Root Alarm Diagnosing in Power Communication Network [J]. Telecommunications for Electric Power System, 2011, 32(229):9-12

[7] HU Zhong-yu, SHEN Tao, LI Gao-feng, GONG Run-sheng. Interference Finding Expert System Based on Case Reasoning and Rule Reasoning [J]. Computer Engineering, 2009, 35(18):185-190

[8] GUO Feng. Research on Distribution Network Fault Disgnosis Based on Rule-based and Case-based Reasoning [D]. CENTRAL SOUTH UNIVERSITY: 2011. 5. 25

Alarm Filtering Method of Power Communication Network Based on Correlation Set

Shengchao Yan[1] and Lei Li[2]

[1] Nari Group Corporation/State Grid Electric Power Research Institute, PRC.
[2] State Grid Hubei Electric Power Co., Ltd. Information Communication Branch, PRC.

Abstract. In order to find out the root from a large number of alarms of alarm events, we did a good study on the alarm's characters of power communication network equipments, through which we found the alarm relevance sets. On this basis, we present a relevance-set based alarm filtering approach, which make the unnecessary and none-root alarms filtered much more quickly.

1. Introduction

System structure design Alarm correlation analysis is an effective method for root alarm analysis. At present, there are some breakthroughs in the research of alarm correlation analysis technology at home and abroad, such as data mining [1], association rules [2], pattern matching [3], signal flow [4], event tree [5] and other cutting-edge technologies, and some useful results have been achieved. In the literature [6-8], the layered filtering mechanism [7], search tree [8] and other technologies are used to filter a large number of alarms, and some practical results have been achieved in the application.

Through the in-depth study of power communication network equipment alarm, combined with its characteristics, find and sort out the alarm correlation set rules, and realize the alarm filtering technology based on the combination of alarm correlation. So as to realize the alarm filtering of alarm tide events in power communication network, eliminate the secondary and non-fundamental alarms in a large number of alarms, and provide the basis for fault location and processing of power communication network.

2. System Structure Design

The system consists of two parts. The first part: the collection of alarm filtering rules, which can realize the sorting of alarm rules commonly used in communication network, and the application of rules in the specific correlation analysis.

Fig.1 System Architecture

The second part: correlation analysis engine. Call correlation set rules to analyze the correlation of a large number of alarms, and filter the secondary and non-root alarms under the control of time folding window. The structure diagram is shown in Figure 1.

If a large number of alarm information is collected in a certain period of time, it needs to be filtered first and then submitted to the correlation analysis engine for analysis; if a simple and small amount of alarm information is collected, it will be directly sent to the correlation analysis engine for analysis and processing.

3. Rule Set of Alarm Filtering

3.1. Set Model of Alarm Filter

Alarm filter is realized by alarm filter, which is built on a mathematical model. The description of the mathematical model is shown in Formula (1) as follows.

$$[A, B, C, D] \rightarrow A \quad \text{or} \quad [A, p(A) \notin H] \rightarrow \phi \tag{1}$$

Where: a represents the alarm information of a certain group or a certain group that is monitored, P (a) refers to all parameter sets involving alarm a that may be determined as fault event, and H is the set of legal parameters and conditions that constitute the event.

It can be seen from the above formula that if the P (a) value of alarm a does not belong to the legal parameter and condition value set h, but belongs to the empty set, these alarms will be filtered out by the filter, so as to delete the alarms that do not meet the requirements of alarm correlation. The filter function block diagram is shown in Figure 2.

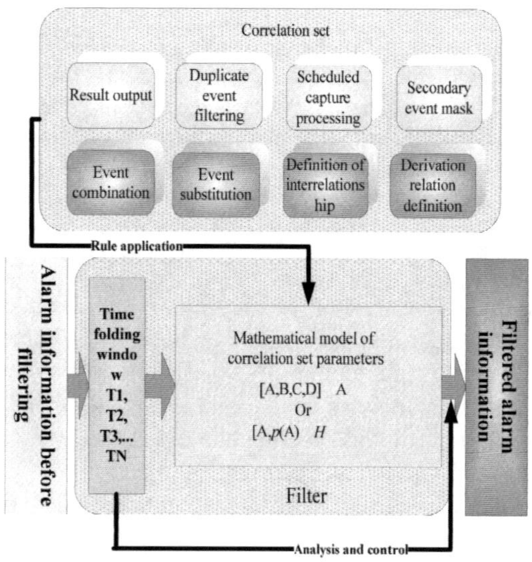

Figure 1. System Architecture

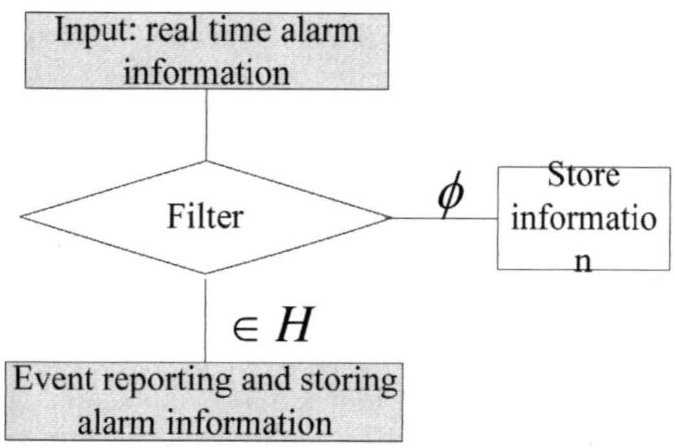

Figure 2. Alarm filtering Diagram

3.2. Alarm Filter Rule Category

The P(a) function is a set of conditions for alarm notification and event reporting. It is constructed by system settings according to various correlations, and it is used to determine which events can be reported or which events can not be reported. The system supports the setting of one or more combined alarm filter conditions. The main filter conditions set in this article are:

(1) low level alarms in pairable alarms (events): events with different alarm levels but the same other parts are called pairing alarms (events). Pairing events can be treated as a combination to clear and filter out the low-level alarms in pairing. For example, there are three alarms in an SDH device, and the alarm levels are: general alarm, main alarm and serious alarm. In this case, general alarms can be filtered, leaving major and serious alarms.

(2) false event filtering: filter mutually exclusive events and ultra short recovery events.

(3) repeat event filtering. Filter recurring events over a period of time.

(4) scheduled event capture: filter alarm events that do not meet user-defined conditions. For example, if the analog threshold of communication DC power supply voltage is set to 54 V, all alarms below 54 V will be filtered.

(5) secondary event shielding: filter the secondary alarm events that do not affect the normal operation of the equipment.

(6) event derivation filter: filter the accompanying alarm. Using the inferential relationship between events, we can filter the accompanying alarms.

The priority of these filters is defined as: false event filtering > repeated event filtering > scheduled capture processing > secondary event screening > Event derivation. Although the filtered alarm information will not be subject to subsequent intelligent analysis, its data still exists in the system database and can be queried and counted when necessary.

After the alarm filter, a large number of low-level, false, repeated, below threshold, secondary and accompanying alarms are filtered. From the actual operation, the filtration ratio can reach 50%. The remaining high-level, true, out of limit and main alarms are sent to the second part of this article: correlation analysis engine to analyze the correlation of alarms.

4. Correlation Analysis Engine

In practical application, if we directly deal with a large number of filtered alarms, then: first, the workload is large and the analysis time is long; second, there is no correlation between some alarm groups and other alarm groups at all, and the combination analysis of them has no effect and affects the analysis accuracy. Finally, the relationship between the source and the cause of the alarm can not be determined in a short time, the analysis is of low intelligence, and the work efficiency is difficult to meet the requirements.

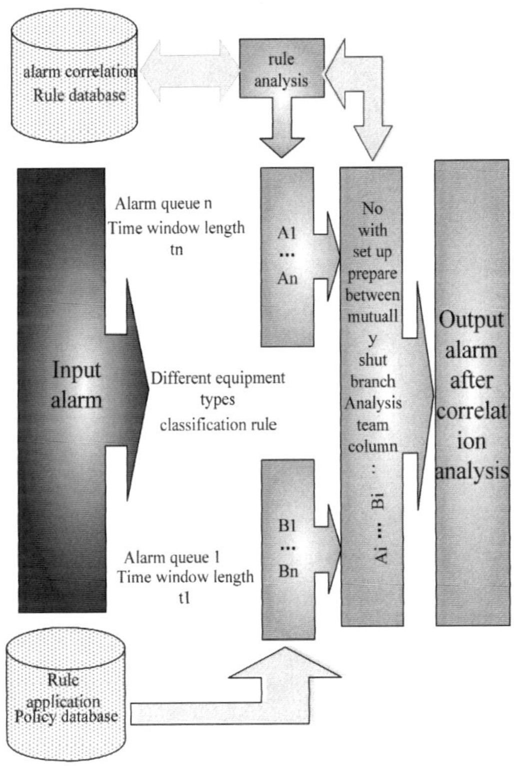

Figure 3. Analysis System Modular Diagram

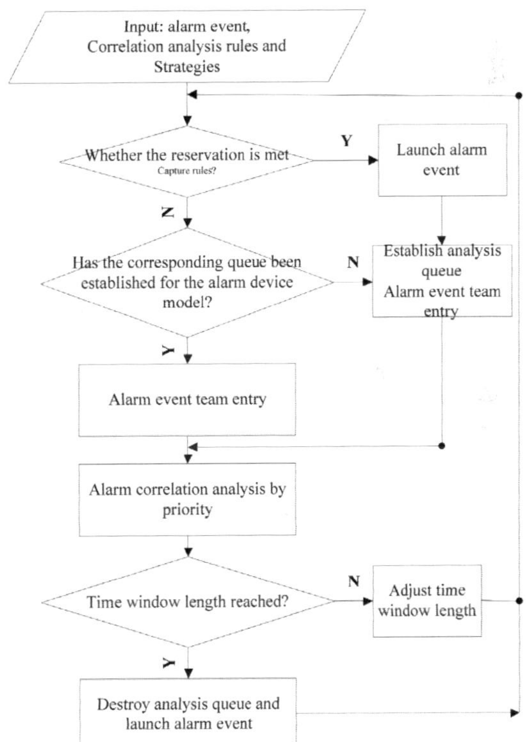

Figure 4. Analysis Flow Diagram

4.1. *Principle of Correlation Analysis*

As for the correlation characteristics, the alarm events belonging to the same area, the same equipment, the same system, the same time and other conditions have a larger correlation, while the alarm events belonging to different areas, equipment, systems, time and other conditions have a smaller correlation, or even no correlation shown in Figure 3.

Therefore, it is considered to divide all alarms into several correlation analysis processing queues according to the alarm correlation degree. Each queue has the maximum correlation according to the conditions such as area, equipment, system and time. The correlation analysis is carried out for each queue respectively to filter out the same origin, related repeated, secondary and false alarms and get the root alarm group of each queue. Then, according to the correlation level, the remaining alarms in each queue are analyzed or output directly. In this way, a large number of alarm analysis without correlation between them is avoided, and the efficiency and processing speed of fault analysis are improved.

4.2. Analysis Process

The flow chart of alarm correlation analysis is shown in Figure 4.

The specific steps are as follows:

(1) step 1: when there is an alarm input, judge whether the queue of the device type of the alarm exists. If not, establish the alarm queue for the device type. Otherwise, the alarm event will be transferred to the team.

(2) step 2: read in the priority and time window length of the correlation processing type from the rule application strategy data table, and read in the rules belonging to the corresponding equipment model of the alarm from the rule data table.

(3) the third step: according to the priority of the correlation processing type, we call the correlation analysis interface and analyze it. Step 4: judge whether the queue has reached the time window length. If so, output all alarms in the queue, destroy the queue, transfer to the start, or transfer to the start, and adjust the window length.

5. Conclusion

Through the realization of the function of this article, it can quickly filter from many complex alarms in the power communication network, and finally determine the root cause of the fault, so as to shorten the decision-making time of fault processing, and improve the intelligent level of the operation and dispatching command of the power communication network.

According to the statistical analysis, the accident probability of electric power communication network is 0.05 times / year. In one year, the average maximum load of a province's power grid is 25 million kilowatts, and the average loss rate of load caused by communication fault affecting the important business of the power grid is 1%. Due to the function realization and practical application of this article in the intelligent analysis and decision-making system for operation and dispatching of electric power communication network, the fault processing time is shortened by 240 hours / year as a whole, and the multi transmission power is 300 million kwh. According to 0.45 yuan / kWh, the new output value is 27 million yuan / year.

6. References

[1] Zheng Qing-guo, Lv Wei-feng. The Alarm Correlation Analysis in Communication Network Management. Computer Engineering and Applications, 2002, 2:11-14

[2] WU Jian, Li Xing-ming. Efficient Distributed Mining Algorithm for Alarm Correlation in Communication Networks. Computer Science, 2009, 36(11):204-212

[3] Lu Xiaobin. Alerts correlation based on intrusion action pattern[J].Microelectronics & Computer, 2005, 24(7):17-20

[4] Li Mingfang. Relativity analysis of optical transmission network alarm based onignal flow[J].Telecommunications Technology, 2005, 22(10):23-26

[5] XU Peng, LU Hai-jun, LI Ming-wei. Alarm Filtering Mechanism for Power Communication Network Based on Event-Tree Model. Power System Technology, 2008, 32(8), 91-94.

[6] SHI Yong-ge, MEI Yu-jie, SHI Feng. Research and application of alarm filtering in network management systems. Computer Engineering and Design, 2008, 29(9):2169-2171.

[7] ZHANG Xian-fei, XU Run-ping, LI Shu-jing. Design of alarms stratification filtering mechanisms in power communication network. International Electronic Elements, 2008, 12:47-48.

[8] WANG Bao-yi, GUO Ya-wei, SHI Zhan-cheng, et al. Research of alarm correlation method based on dependency search tree in electric power communication network. Relay, 2008, 36(6):59-64.

IOP Conf. Series: Materials Science and Engineering **719** (2020) 012019 doi:10.1088/1757-899X/719/1/012019

The Implementation and Application of the MBPP Technology in Aircraft Development Process

Haijun Yin[1,a], Hui Zhang[2,b] and Yingji Shan[3,c]

[1,2,3] Shenyang Aircraft Industry (Group) Co., Ltd., Shenyang, China
[a]yin_hj2010@163.com, [b]skyjz@163.com, [c]qxh20090101@163.com

Abstract. In this paper, the model based technical requirements of MBPP system Implementation plan is put forward by the application requirement analysis of MBD technology in aircraft development process. Then it is described for the application target, application architecture, implementation process and the main business functions of MBPP system. Based on the application example, the application process of the system is also introduces in the development process of certain aircraft, which has some reference value for the development and application of 3D CAPP based on model.

1. Introduction

With the development of CAD technology, MBD(Model Based Definition) technology is now maturely applied on design end of aircraft product at present. With the realization of the design of "three-dimensional model, parametric information, correlation parameters", which supported the rapid development process of aircraft product and improved the efficiency and quality of aircraft product design. With the implementation of digital manufacturing project and the popularization of numerical control equipment on production line on factory manufacturing line, NC machining and NC molding technology has been applied deeply, witch drives other professional processing ability to fast lifting.While CAPP technology witch is the hub and bridge between product design and manufacturing has not taken revolutionary change. It has became a technical bottleneck between product design and manufacturing that 2D CAPP still plays a leading role and has not adequately came into process expression efficiency, which restricted the development of the whole level of manufacturing end. Therefore,it has become an active demand of current aircraft manufacturing enterprise in the process of new generation aircraft development to develop MBPP system based on 3d modeling process design and planning,to play the technology advantage of MBD and establish the process design line based on model.

2. Research Target

MBD is an entity to express complete product definition with the integrated 3D entity model.It adds various manufacturing semantic information such as product size, tolerance and technical requirements,making full use of the expressiveness of 3D model.It is a product information definition method witch is more convenient to understand and more efficient[1-2].

The process design of traditional CAPP system based on two-dimensional CAD environment and two-dimensional graphic information has been unable to meet the requirements of rapid development and production of products[3]. Dou proposed a process oriented plan to express and integrate related informations with process tree based on 3D.The efectiness has been verified[4]. Cezary etc. proposed a process planning procedure based on application of the chain codes for description of the part's selected longitudinal cross sections[5]. Liu proposed a a method to achieve to express and manage the process information simply and efficiently by using working procedure model was proposed[6]. Deja presented a feature-based reasoning approach for generating machining sequences in terms of part setups and the assignment of machine alternatives[7]. The approach has been verified through an illustrative case study using a sample mill-turn part of considerable complexity.

The goal of this paper is to study the construction technology of model-based 3D process system and the BOM reconfiguration,the 3D process planning and visualization of parts and assemblies based on model,forming a MBPP platform based on model driven.Then the key technologies such as EBOM, PBOM and MBOM dynamic evolution, 3D process planning and visualization expression of parts and assemblies in the process of transformation from design model to process model and manufacturing model are solved.

3. Planning of MBPP System

MBPP is a process planning system based on model, using the three-dimensional model as the carrier of process information and the source of information integration and driving the process planning process and the field production process with model[8].

3.1. Overall Framework

On the basis of the existing resource conditions of the enterprise, we can achieve the model driven visualization process through carrying out the research on model-driven 3D process system construction technology and developing 3D MBPP process system independently to support the integration design and manufacturing. Specifically for:

(1)The data between design and manufacturing and based on 3D models is synchronous and unified;

(2)The design, process, manufacturing and inspection for the same data source are concurrent and collaborative;

(3)Organize, contact,transfer and publish the data based on full 3D model;

(4)With 3D BOM as a single data source, BOM reconstruction, 3D process planning and field visualization technology based on model are used to realize 3D digital process planning of full business process;

(5)Multidisciplinary simulation based on full 3D model.

3.2. Requirements of Overall Functional Architecture

(1)Integration of software tools and system encapsulation

Select CATIA software tools and implement seamless integration of software tools and management environment.

(2)Economical and practical

Through the customized development of the system and CATIA software, the function of product structure division and simulation in DELMIA software is realized.

3.3. Service Demand Analysis

(1)Based on model

The result of process design is a related model of comprehensive information, which inherits the design information and expresses the process manufacturing information. The 3D model is the only basis in the manufacturing process.

(2)Cite and extraction of information

The design dimension, tolerance, annotation and annotation information used in process design should be quoted or extracted from the original design model information. Craftsmen do not repeat manual input to ensure the correlation and consistency between the upstream and downstream data, and ensure the accuracy of the changes.

(3)Dominant change

As much as possible to make the implicit information in the design model (such as size) explicit, rather than re-marking the size and tolerance after projection. Use the system function automatic conversion as much as possibly and reduce manual writing and labeling.

3.4. Systemic Architecture and Function

3.4.1. Systemic architecture

Aiming at the application status of CATIA system and CAPP system in the company, this project takes VPM system as the design platform under CAA environment and realizes the model-based working environment supporting the overall process planning and detailed process design through the secondary development, and forms a B/S architecture of three-dimensional MBPP system, which is mainly composed of process data base platform and process design management function component, process business integration management service, client and so on.

The overall architecture of the system is three levels, including data layer, business logic layer and user layer.

Data layer provides data support and access for business logic layer. It is the basis for supporting the operation of the whole system.Using multi-site replication technology, data is copied from the design institute's source database server (VPM server) to the PDM system database on the manufacturing side. The PDM database table on the manufacturing side is consistent with the VPM database table of the design institute, and the process database used for business processing is a multi-instance-based process database designed independently on the basis of reference to the mature design concept of the PDM system and meeting the requirements of the current business processing mode.It is associated with the VPM database on the manufacturing side.

Business logic layer is mainly responsible for the logical implementation of the system.Such as system user verification, login user privilege settings, functional allocation, components of supporting system applications and other compatible system interfaces are implemented in this layer, and they are the supports for user layer to implement business process.

The user layer provides business interface for system users. Users can operate through the functions provided by the interface, so as to complete the related business processing.

3.4.2. System function

On the whole, System can realize EBOM/PBOM/MBOM management, process design management and field visualization management. From the business function, the process of process planning is divided into four work. Firstly, generate factory EBOM synchronously from design data source, secondly, add process routes and process assembly to created PBOM based on the EBOM. Thirdly, according to the assembly position, AO , gradually form the MBOM based on PBOM. Fourthly, compile the corresponding technological procedure.

The first item is using multi-site replication technology to extract the design data of aircraft type (EBOM) from the designing institute's(PDM) system database ,and send it to manufacturing the database server ,and then import the design data of draw to the local database of MBPP system. And ensure the PDM system data of design side and MBPP system data is updated on a regular basis.

3.4.2.1. Model based BOM planning

Based on the model ,System can realize the EBOM to PBOM construction planning and also realize the PBOM to MBOM construction and programming of position . based on the batch process effective management, it realize the associated design change and remind. In the process of lower BOM planning that based on the upper BOM ,it realize the information association and in the process of classification by consumption, it provide a error-proofing remind function.

(1)Building PBOM structure

Based on the EBOM model tree structure, it takes the form of "drag" and attributive adding to form PBOM structure tree. It Contains two main job ,the first is building process combination and split; The second is to add technology division of attribute information .The key technical point is to realize the consumption type division based the model . For parts that have been divided, it reminds on the EBOM according to the color and it can't be second divided. The pipeline packaging pieces of VPM system show on the PBOM clearly after division, and it can automatically generate symmetrical part after analyzing the standard file of the NOTE. The PBOM that has been produced can generate two-dimensional process schedule and related reports according to need.

(2)The top layout of MBOM and AO

Based on PBOM tree structure, it forms the MBOM structure tree in the form of "drag" and attribute adding. it Mainly contains MBOM top stance partition, Parts and standard parts from the PBOM to MBOM stance and AO partition on nodes. Then it form AO matching directory In preparation for the process discipline constitution.The key technical points is to realize the top-down consumption type divide, PBOM information accurate inheritance and associated, and also achieving pokayoke inspection.

(3) The division of the standard parts

In the process of various BOM refactoring, it need to achieve the standard parts of NOTE file accurate resolution ,and ensure the relevance of its upstream and downstream.

In the process of BOM build, Through the design NOTE file parsing, it add the information that division of labor in the form of route to standard parts that based on the dotted line sets increased, and then Divide it into the PBOM and MBOM structure, and Divided to different position, different process, For each standard parts have accurate division of information property.Standard parts that Divided into the AO need to achieve information's quick, automatic label and statistics such as row distance, edge distance, specification, quantity, etc. And it can realize instantiation, which is Used in the assembly process simulation.

3.4.2.2. Model based process planning

The process of assemblies and parts process plan compiling , is the process of refining MBOM content. it forms AO and FO, and finally form MBOM basted on the model.

(1)The compilation of AO/FO

On the node of MBOM components or parts, AO or FO is chosen and created. The system will create PPR file of assembly or parts. Based on the structure tree of PPR file, compilation work is performed about AO/FO content, information for work steps, operating instruction, autoassociated frocks, tools and standard parts.

For matching parts that used in each work step, automatically check the integrity of it item by item. For matching parts that used in each work step, the assembly integrity check is performed automatically item by item. At the same time, the used parts are hidden in the model structure tree, automatic checking is also performed for consistency of parts quantity and parts status among EBOM-PBOM - MBOM. Lastly, the AO/FO editing results are converted into simulation file which is used to perform process simulation to verify the process rationality. Then, it is transmitted to MES for field application.

In the process of FO content establishment, process planning can be generated automatically by recognition part features based on process resource and knowledge base, automatic planning process, and decision-making model can also be formed. It can also be realized for rapid programming, construction of processing unit, selection of cutting tools and related parameters, realizing the automation of whole process of programming.

(2) The design change management

After design BOM have changed, the EBOM of system automatically update, and make changes remind. PBOM and MBOM related personnel according to the prompt for manufacturing BOM change maintenance, and Engineer do the work that changes for AO/FO.Among them, the change description of AO/FO based on model tree structure's parametric expression, and Changes instruction can be output on-demand in the format that two-dimensional procedure change the of the single.

4. The Application Verification of System

The MBPP system that develops based on the above planning ideas has been broadly applied in the course of the development of a new type, and support this type of machine developed by parallel process review, process design, data release, the production site visual production process, providing a good technical support to the type machine's development and satisfy the requirement "good, fast and cheap".

AO/FO and PBOM/MBOM were generated for all the structural parts of the whole machine during the development process. The process data generated in the system maintained an association

IOP Conf. Series: Materials Science and Engineering **719** (2020) 012019 doi:10.1088/1757-899X/719/1/012019

relationship with the upstream data, which could be displayed in 3d form or exported in 2d form for application as required.

Finally, inorder to guide the production process,the finalized MBOM and AO/FO will be released to the ERP system and the field execution system respectively Tools for on-site demonstration of process documents based on true model or light model and two-dimensional view are embedded in MES system. After receiving corresponding work tasks in MES system, operators can view corresponding three-dimensional or two-dimensional process documents and carry out corresponding assembly and processing operations in combination with simulation process.

5. Conclusion

Process planning based on three-dimensional CAD is one of the hot spots in the field of advanced manufacturing technology. According to the specific requirements of model-based development and production in aviation manufacturing enterprises, this paper developed a BPP system for process design and process information integration in three-dimensional environment, realized process design and process collaboration mode based on full three-dimensional information related model, and significantly improved the efficiency of process design. The results of the project have been applied in the development of several new models, which has important reference significance for research and application based on three-dimensional CAPP.

6. References

[1] Fan Yuqing. Model based definition technology and its practices [J]. Aviation Manufacturing Technology,2012(06):42-47.

[2] Yu Zhiqiang,Chen Song,Sun Wei. Application of MBD-based three-Dimensional Module in aircraft manufacturing [J]. Aviation Manufacturing Technology, 2009(25): 82-85.

[3] Jiang Qian,Hou Zhixia,Liang Xuemei. Computer Simulation Technology of Digital Alignment of Large Size Aircraft Large Part[J]. Aviation Manufacturing Technology,2013 (22): 87-89.

[4] Dou Guangji, Mo Rong, Wan Nene. Expression and Integration Technology of Three-Dimensional CAPP Process Information[J]. Advanced Materials Research, 2014(7): 92-95.

[5] Cezary, Grabowik. An attempt to application of chain codes for design similarity evaluation[J]. Advanced Materials Research, 2014(1036):897-902.

[6] Liu, Jinfeng , Ni Zhonghua,Liu Xiaojun. Process information expression and management technology based on 3D process model [J]. Computer Integrated Manufacturing Systems, 2014, 21(7): 1756 -1763.

[7] Deja, Mariusz. Machining process sequencing and machine assignment in generative feature-based CAPP for mill-turn parts[J]. Journal of Manufacturing Systems, 2018, 48(2):49-62.

[8] Zhang Bo. Research on Intellignet Technology of Process Planning Based on 3D Model[D]. Harbin Institute of Technology, 2013.

IOP Conf. Series: Materials Science and Engineering **719** (2020) 012020 doi:10.1088/1757-899X/719/1/012020

SWRL Parallel Reasoning Implementation with Spark SQL

Wan Li, Huaai Kang, Dongbo Ma and Weiwei Wei

Software and Information School, Beijing Information Technology College, Beijing, China.
Email: lee_wan@sina.com

Abstract. With the rapid development of semantic Web and big data technology, ontology data has the characteristics of large-scale, high-speed growth and diversity which big data has. On one hand, the conventional ontology reasoners do not scale well for large amounts of ontologies because they are designed for run on a single machine. On the other hand, the existing scalable reasoners are not perfect enough, for example, to completely support the widely used Semantic Web Rule Language (SWRL) rules. This paper presents an implementation for SWRL scalable parallel reasoning using the Spark SQL programming model, and optimizes and processes some of the problems in the implementation.

1. Introduction

In the field of Semantic Web[1][2], based on RDF [3], RDFS [4] and OWL [5], W3C has introduced SWRL [6] with more logical expression ability combining description logic and rule. With the rapid development of the Semantic Web and big data technologies, ontology data has presented big data characteristics such as large-scale, high-speed growth, and diversity. To this end, researchers have introduced distributed computing technology into the field of Semantic Web research to explore high-efficiency ontology reasoning methods in distributed environments [7][8][9][10].

In our previous work [11], based on the analysis of existing semantic reasoning algorithms, combined with Spark SQL[12] which is a newer and higher level parallel computing platform with structured data processing capabilities, we proposed a SWRL parallel reasoning method based on Spark SQL. This paper is about the specific implementation of the method. We implemented SWRL parallel reasoning based on Spark SQL and optimized the parallel reasoning algorithm to improve performance.

2. Preliminary

2.1. SWRL Semantic Reasoning

A RDF triple, which is a triple of resource <s, p, o>, asserts that the relationship denoted by the predicate is held between the subject and object of the triple. RDF defines a simple graph model to denote relationships between resources using the format of RDF triple.

Based on RDF, both RDFS and OWL define a set of rules respectively with the ability to represent implicit information. SWRL not only includes the RDFS/OWL Horst semantics, but also can be used to express application-specific semantics. Reasoning is the process of deducing implicit information from existing RDF data by using the RDFS or OWL or SWRL reasoning rules. Given an RDF graph G, some new triples denoted as T can be derived from the RDFS or OWL or SWRL rules. Add T to G then we can obtain a bigger RDF graph G'. The process from G to G' is called reasoning [13]. Reasoning is an iterative process that does not end until new results are not derived.

A SWRL rule r has the form (1), where $B = B_0\wedge\cdots\wedge B_m$ is the body of the rule, and $H = H_0\wedge\cdots\wedge H_n$ is the head of the rule. Both B_i and H_j are RDF atoms, which are triples of resources or variables. B_i is a body atom, and H_j is a head atom. To make a rule safe, SWRL constrain that a variable in the head must occur at less one time in the body.

$$B_0\wedge\cdots\wedge B_m \Rightarrow H_0\wedge\cdots\wedge H_n \quad (1)$$

A variable substitution [10][14] $\varphi_t^{?x}$ is an operation that apply to an atom to replace the occurrences of the variable $?x$ with a resource t. A substitution application (SA) $A[\varphi]$ is a triple obtained by applying a substitution φ to an atom A. Composition of substitutions φ_1 and φ_2 is also a substitution and defined as usual in [15]. Given a fact base I and a rule r, the inference result $r(I)$ is the smallest set containing $H_j(r)[\varphi]$ (for each $0\leq j\leq n$) for each substitution φ such that $B_i(r)[\varphi]\in I$ (for each $0\leq i\leq m$).

2.2. Our Previous Work: SWRL Parallel Reasoning Method with Spark SQL

We divide the SWRL rule execution procedure into three stages [11]: (1) finding SA set for each body atom (FindSA), (2) joining SA sets of all the body atoms of the rule to find out the variable substitutions of the rule head (JoinSA) (3) applying the variable substitutions of the rule head to each atom of the rule head (ApplySH). SQL queries are executed by Spark SQL to implement parallelized rule execution.

Take the following rule r1 given in [10] as an example. The reasoning process is shown in figure 1.

r_1: B_0: $(?x, \textit{worksFor}, ?y_1) \wedge B_1$: $(?y_1, \textit{subOrganizationOf}, ?y_2) \Rightarrow H_0$: $(?x, \textit{worksFor}, ?y_2)$

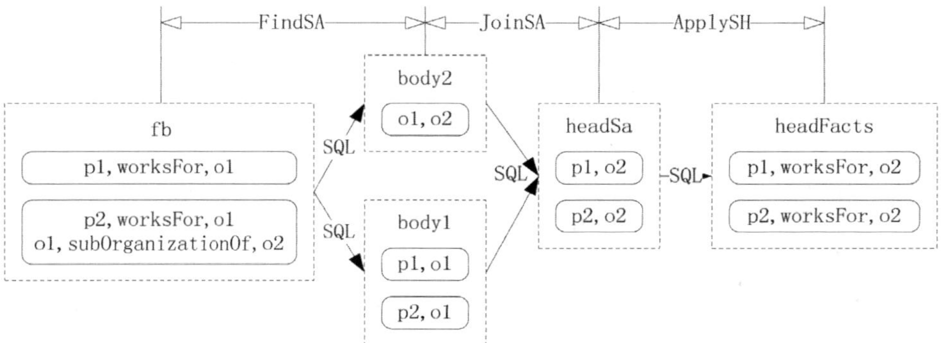

Figure 1. The execution workflow of rule r1

The SQL statements of the three stages are as follows:

FindSA:

```
SELECT t0 AS ?x, t2 AS ?y1 FROM fb WHERE t1='worksFor'
SELECT t0 AS ?y1, t2 AS ?y2 FROM fb WHERE t1='subOrganizationOf'
```

JoinSA:

```
SELECT body0.?x, body1.?y2 FROM body0, body1 WHERE body0.?y1=body1.?y1
```

ApplySH:

```
SELECT DISTINCT ?x AS t0, 'worksFor' AS t1, ?y2 AS t2 FROM headSa
```

3. SWRL Parallel Reasoning Implementation with Spark SQL

3.1. Reasoning Implementation for a Rule

As mentioned above, the reasoning of a rule is to use Spark SQL to execute the SQL statements of the three stages of the rule's reasoning plan (FindSA, JoinSA and ApplySH). The SQL statements for the three stages are already generated in the generateing rule planning step.

Algorithm 1 shows the process of reasoning a rule by sequentially executing these SQL statements of the three stages. Firstly call the Spark DataFrame's *createOrReplaceTempView* function to register the input triples library *tb* as a SQL temporary view. Then call the *findSa* function to find the substitution application (SA) set of each rule body atom. This is done by executing the previously generated SQL statement *findBodyAtomSaSql* to query in the view of *tb*. The SA data sets of each rule body atom are each registered as a SQL temporary view. Then call the *joinSa* function to join the SA sets of all the rule body atoms to find the variable substitution of the rule header. This is done by executing the previously generated SQL statement *joinSaSql* to query in the view of the SA data set of each rule body atom obtained in the FindSA phase. The variable substitution data set of the rule header is registered as a SQL temporary view. Then call the *applySh* function to apply variable substitution to each rule header atom. This is done by executing the previously generated SQL statement *applyHeadAtomSaSql* to query in the view of the variable substitution data set of the rule header obtained in the JoinSA phase. Union the triple sets obtained by each rule header atom to get the rule's reasoning result triple set.

Algorithm 1. Reasoning a rule with Spark SQL

```
Input: rulePlan (reasoning plan of a rule), tb (current triples)
Output: (derived triples by the rule)
def reasoningRule(rulePlan: RulePlan, tb: DataFrame): DataFrame = {
  tb.createOrReplaceTempView(TB_VIEW_NAME)
  findSa(rulePlan.findSaPlan)
  joinSa(rulePlan.joinSaPlan)
  return applySh(rulePlan.applyShPlan)
}
def findSa(findSaPlan: List[FindBodyAtomSaPlan]) = {
  for (findBodyAtomSaPlan <- findSaPlan)
    findBodyAtomSa(findBodyAtomSaPlan)
}
def findBodyAtomSa(findBodyAtomSaPlan: FindBodyAtomSaPlan) = {
  val bodyAtomSa = spark.sql(findBodyAtomSaPlan.findBodyAtomSaSql)
  bodyAtomSa.createOrReplaceTempView(findBodyAtomSaPlan.bodyAtomName)
}
def joinSa(joinSaPlan: JoinSaPlan) = {
  val headSa = spark.sql(joinSaPlan.joinSaSql)
  headSa.createOrReplaceTempView(HEADSA_VIEW_NAME)
}
def applySh(applyShPlan: List[ApplyHeadAtomSaPlan]): DataFrame = {
  var headTriples = createEmptyTripleDataFrame()
  for (applyHeadAtomSaPlan <- applyShPlan) {
    val headAtomTriples = applyHeadAtomSa(applyHeadAtomSaPlan)
    headTriples = headTriples.union(headAtomTriples)
  }
  return headTriples
}
def applyHeadAtomSa(applyHeadAtomSaPlan: ApplyHeadAtomSaPlan): DataFrame = {
  return spark.sql(applyHeadAtomSaPlan.applyHeadAtomSaSql)
}
```

3.2. Reasoning Implementation for Rule Base

The reasoning process of a rule base is divided into two steps: the first step is to parse each rule to generate the reasoning plan in the form of SQL statements with Spark SQL and save it; the second step is to execute the reasoning plan SQL statements with Spark SQL for reasoning. The first step and the second step are separable. The performance of the reasonor is mainly determined by the second step. We did some work to improve the performance of the second step. Here, we explain the implementation of the second step.

To put it simply, the second step of the reasoning process is to execute the rule's reasoning plan one by one, add the derived new triple facts into the fact base, and loop the reasoning until no new facts are derived, then the reasoning ends. However, in practice, the implementation of the reasoning process is more complicated due to the DAG mechanism of the Spark computing framework.

The reasoning of the rule base requires iterative calculations. Therefore, Spark's calculation process DAG will be particularly long. The entire DAG calculation needs to be completed before the result is obtained. During this period, datasets with a large amount of lineage will be generated, which will result out of memory. Therefore, it is necessary to truncate the lineage. It is a possible way to truncate the lineage by Spark's checkpoint mechanism. However, checkpoint stores data in the disk file system, writes many files to the file system every short time, and does not automatically delete the previous checkpoint file, which is detrimental to the file system performance of the entire cluster. Therefore, we use the method of writing and then reading the file in our own program to truncate the lineage. In terms of time efficiency, there is not much difference with Spark's checkpoint mechanism, but it does not generate a lot of files, which improves space efficiency.

If the new triples that are reasoned by a rule are added to the triple library *tb*, and *tb* is applied to the next rule to reason more new triples, then the iterative reasoning process will cause *tb* to generate a large amount of lineage. Since the triple library *tb* may be large, frequently truncating the lineage of *tb* by writing and then reading the file will affect system performance. To resolve this problem, we have improved the algorithm to avoid the large amount of lineage of *tb* by using intermediate results.

Algorithm 2. Reasoning rule base with Spark SQL

```
Input: rulesPlan (reasoning plans of rules), tb0 (original triples)
Output: (derived triples)
def reasoningRules(rulesPlan: List[RulePlan], tb0: DataFrame): DataFrame = {
  var tb = tb0
  var derived = createEmptyTripleDataFrame()
  var tbLoop0 = tb0.union(derived)
  var loopDerived = createEmptyTripleDataFrame()
  var hasDerived = true
  do {
    hasDerived = false
    for (rulePlan <- rulesPlan) {
      val ruleDerived = reasoningRule(rulePlan, tb)
      loopDerived = loopDerived.union(ruleDerived).except(tbLoop0)
      loopDerived = loopDerivedTruncateLineage(loopDerived)
      tb = tbLoop0.union(loopDerived)
    }
    if (!loopDerived.isEmpty()) {
      derived = derived.union(loopDerived).except(tb0)
      derived = derivedTruncateLineage(derived)
      tbLoop0 = tb0.union(derived)
      loopDerived = createEmptyTripleDataFrame()
      hasDerived = true
    }
  } while (hasDerived)
  return derived
}
```

Algorithm 2 shows the iterative reasoning process of the rule base. Where *tb0* is the original triples library before starting the inference, and *derived* is all the new triples that have been reasoned before starting a loop of reasoning on the rule base, *tbLoop0* is the triples library before starting a loop of reasoning on the rule base, *loopDerived* is the new triples reasoned in current loop, *ruleDerived* is the new triples reasoned in current rule. Using *tb0* union *derived* to update *tbLoop0*, using *tbLoop0* union *loopDerived* to update *tb*, using *loopDerived* union *ruleDerived* to update *loopDerived* itself, using *derived* union *loopDerived* to update *derived* itself. To truncate lineage on *loopDerived* and *derived* that avoid the large amount of lineage of *tb*, and the *loopDerived* that is frequently truncated is small, while the larger *derived* (which is a subset of *tb*, smaller than *tb*) is truncated less frequently (once per loop).

3.3. Eliminating Duplicated Facts
It is inevitable that some results of the rules duplicate the facts in the knowledge base during the rule execution process. The duplicated facts not only degrade system performance but also increase management overhead. However, removing duplication also bring additional computation to the reasoner. Existing systems adopt a series of duplication elimination strategies [8][9][10]. We used several strategies to remove all duplicated facts, as follows: (1) removing all duplications before the reasoning process begins; (2) using the DISTINCT clause to remove duplications in the derived data in the ApplySH stage of each rule; (3) after each rule is reasoned, use the *except* method to eliminate the duplications in *loopDerived* and the duplications between *loopDerived* and *tbLoop0*; (4) after each time looping through all the rules, use the *except* method to eliminate the duplications in *derived* and the duplications between *derived* and *tb0*.

3.4. SWRL Reasoning Implementation with RDFS/OWL/SWRL Reasoning Rules
SWRL reasoning is the process of deducing implicit information from existing RDF data by using the RDFS/OWL/SWRL reasoning rules. A rule dependency exists when the evaluation results of a rule trigger another rule execution. Thus improper rule evaluation order will conduct unnecessary job running and bring significant performance degradation [9][10]. We optimize the execution order of RDFS and OWL rules according to the method in [9]. Optimizing the execution order of SWRL rules is what we will do in the future. The SWRL reasoning implementation with the RDFS/OWL/SWRL reasoning rules is shown in Algorithm 3.

Algorithm 3. Reasoning RDFS/OWL/SWRL rule base with Spark SQL

```
Input: rdfsRulesPlan (reasoning plans of RDFS rules), owlRulesPlan (reasoning plans of OWL
      rules), swrlRulesPlan (reasoning plans of SWRL rules), tb0 (original triples)
Output: (derived triples)
def reasoningMixedRules(rdfsRulesPlan: Dataset[RulePlan], owlRulesPlan: Dataset[RulePlan],
      swrlRulesPlan: Dataset[RulePlan], tb0: DataFrame): DataFrame = {
  val rulesPlan = rdfsRulesPlan.union(owlRulesPlan).union(swrlRulesPlan).collect.toList
  return reasoningRules(rulesPlan, tb0)
}
```

4. Conclusion and Future Work
In our previous work, based on the analysis of existing semantic reasoning algorithms, combined with Spark SQL which is a newer and higher level parallel computing platform with structured data processing capabilities, we proposed a SWRL parallel reasoning method based on Spark SQL. This paper is about the specific implementation of the method. We implemented SWRL parallel reasoning based on Spark SQL and optimized the parallel reasoning algorithm to improve performance. In the future work, we intend to further improve the reasoning system and use large data sets to evaluate our system.

5. Acknowledgment
This work was supported in part by the Beijing Municipal Education Commission Science and Technology Plan Project (KM201710857001).

6. References
[1] Tim Berners-Lee. Semantic Web Road Map. September 1998. http://www.w3.org/DesignIssues/Semantic.html.
[2] Tim Berners-Lee, James Hendler, and Ora Lassila. The Semantic Web. Scientific American, 2001, 284(5):35-43.
[3] W3C. Resource Description Framework (RDF): Concepts and Abstract Syntax. http://www.w3.org/TR/rdf-concepts/.
[4] W3C. RDF Schema. http://www.w3.org/TR/rdf-schema/.
[5] W3C. OWL 2 Web Ontology Language: Profiles. https://www.w3.org/TR/owl2-profiles/.
[6] W3C. SWRL: A Semantic Web Rule Language Combining OWL and RuleML. http://www.w3.org/Submission/SWRL/.
[7] Li Ren. Research on Key Technologies of Large-Scaled Semantic Web Ontologies Querying and Reasoning based on Hadoop. Chongqing University PhD thesis. 2013.
[8] Urbani, J., Kotoulas, S., Maassen, J., et al.:WebPIE: a web-scale parallel inference engine using MapReduce. J. Web Semant. 17(44), 59-75 (2012).
[9] Gu, R., Wang, S., Wang, F., et al.: Cichlid: efficient large scale RDFS/OWL reasoning with Spark. In: IPDPS, pp. 700-709 (2015).
[10] Haijiang Wu, Jie Liu, Tao Wang, Dan Ye, Jun Wei, Hua Zhong. Parallel Materialization of Datalog Programs with Spark for Scalable Reasoning. 17th International Conference on Web Information System Engineering (WISE 2016), pp. 363-379.
[11] Wan Li, Huaai Kang, Dongbo Ma and Weiwei Wei. SWRL Parallel Reasoning Method with Spark SQL. 18th IEEE/ACIS International Conference on Computer and Information Science (ICIS 2019), pp. 270-273.
[12] Spark SQL, DataFrames and Datasets Guide. http://spark.apache.org/docs/latest/sql-programming-guide.html
[13] J. Lu, L. Ma, L. Zhang, J.-S. Brunner, C. Wang, Y. Pan, and Y. Yu, “Sor: a practical system for ontology storage, reasoning and search,” in Proceedings of the 33rd international conference on Very large data bases. VLDB Endowment, 2007, pp. 1402–1405.
[14] Abiteboul, S., Hull, R., Vianu, V.: Foundations of Databases. Addison Wesley, Boston (1995).
[15] Ullman, J.D.: Principles of Database and Knowledge-base Systems, vol. I. Computer Science Press, New York (1988).

IOP Conf. Series: Materials Science and Engineering **719** (2020) 012021 doi:10.1088/1757-899X/719/1/012021

Mining and Analyzing "Multi-Meters in One" Comprehensive Energy Data

Zhelong Wang[1], Yan Du[1], Liang Guo [2], Hongxia Guo[1], Zhen Jing[1], Zhe Li [2], Hongxia Zhu[1] and Jie Yang[1]

1. State Grid ShanDong Electric Power Research Institute, Ji Nan, China
2. .State Grid ShanDong Electric Power Company, Ji Nan, China
461849930@qq.com

Abstract. In the present paper, the massive comprehensive energy data was utilized to establish energy consumption analysis model, energy consumption suggestion model, energy consumption loss model, vacant household discrimination, holiday energy consumption analysis model, and so on. The comprehensive energy data was fully excavated, which can guide customers to use energy scientifically, help enterprises to save energy and reduce losses, and provide data support for government decision-making.

1. Introduction

With the wide application of "Internet +", the resident customers urgently need convenient and saving energy consumption, and experience a good sense of using energy to handle business without leaving home. The energy supply enterprises urgently need to solve the problems such as high loss and poor precision service to improve management and service level. and make government decisions, smart city construction and "interconnection". In addition, government decision-making, smart city construction and the development of "Internet + government services" needs timely and accurate energy data support [1].

At present, the number of "multi-meters in one" acquisition and access customers of Shandong Electric Power Company has reached more than 1.1 million. The massive data acquisition is only sent to water, gas and heat enterprises for meter reading and settlement, and the data value cannot be fully exploited. The data of electricity, water, gas and heat are independent. How to establish the relationship between the data of electricity, water, gas and heat so as to realize the analysis of abnormal energy consumption and improve the economic benefits of energy supply enterprises? Based on the electricity, water, gas and heat energy consumption data of customer, how to optimize the customer's energy consumption mode through the electric water gas heat data analysis, give the customer energy recommendations, and reduce the customer's energy cost. The government urgently needs the results of housing vacancy rate analysis to guide the formulation of housing price policy. How to accurately analyze the housing vacancy rate through the correlation analysis of electric water, gas and heat data. The above contents are the focus of this paper [2-3].

2. Referenced Data Resources

This paper is based on the basic archives information, data collection and events of electricity, water, gas and heat. The external data are introduced appropriately according to the needs of analysis dimension and depth. The data sources are mainly divided into the following two parts: internal data and external data.

IOP Conf. Series: Materials Science and Engineering **719** (2020) 012021 doi:10.1088/1757-899X/719/1/012021

2.1. Internal Data

The internal data mainly include the following: 124 kinds of data of "multi-meter in one" customer, such as basic archives information of electricity, water, gas and heat, assets information of meter, topological relationship of meter (mainly including desk area topological relationship, pipeline topological relationship), category of energy use, unit price of energy use, relationship of quantity and price, etc.) ; 26 items of data such as meter's electric energy indication value, opening cover record, voltage phase break etc.; 12 items of data such as water meter status word, accumulated water flow indication value and fault information etc; 13 items of data such as status word, current instantaneous flow velocity and fault information of gas meter, 18 items of data such as instantaneous flow velocity, inlet temperature and return water temperature of heat meter, totalling 450 million items of data.

2.2. External Data

The external data are mainly statistics and technical data issued by price bureau, statistics bureau, meteorological bureau, energy bureau, transportation industry, third party authoritative census structure, research institutes and technical organizations, mainly including 24 items of data such as energy ladder and price, energy category price, natural gas combustion value, energy heating utilization efficiency, heating and refrigeration efficiency for ground source heat (cold) pump and pollutant emissions from various energy sources when they generate 1 kilojoule of heat, etc.

3. Analysis of User Energy Consumption Data

3.1. Analysis of Basic Energy Consumption

The analysis of basic energy consumption includes two parts, that is, the proportion of energy consumption and the composition of energy consumption. The proportion of energy consumption and consumption ranking are calculated according to consumption and unit price. In addition, the electrical heat is converted into thermal energy, and the composition and ranking of energy consumption are calculated from two dimensions of thermal energy consumption and water energy consumption. The energy consumption levels of customers in different regions are displayed through the average and ranking of energy consumption at different levels [4-5].

Taking the energy consumption of a customer in June 2019 as an example, the energy consumption status of the customer is shown in Table 1. The energy consumption proportion of each item for this user can be calculated according to formula (1).

Table 1. Energy consumption status of a customer in June 2019

Category of Energy Consumption	Quantity	Unit Price	Amount of Money (Yuan)
Electricity	248kWh	0.5469Yuan/kWh	135.63
Natural gas	22m^3	3.6Yuan/ m^3	79.20
Water	11 m^3	4.2Yuan/ m^3	46.20
Heat	20kcal	26.7Yuan/ kcal	534.00

$$p_{ci} = \frac{Q_i \times P_i}{\sum_{i=1}^{n} Q_i \times P_i} \times 100\% \qquad (1)$$

p_{ci} :Proportion of energy consumption.

Q_i :Step energy consumption quantity.

P_i :Step unit price.

3.2. Suggestion Model for Energy Consumption

The current energy consumption cost model is established based on the data of step electricity price, step gas price, calorific value of electric energy, calorific value of natural gas, combustion efficiency, heat utilization rate and electro-thermal conversion rate and so on.

$$\begin{cases} \dfrac{p_1}{q_1\eta_1\mu_1} < \dfrac{p_2}{q_2\eta_2\mu_2} \\ \dfrac{p_1}{q_1\eta_1\mu_1} = \dfrac{p_2}{q_2\eta_2\mu_2} \\ \dfrac{p_1}{q_1\eta_1\mu_1} > \dfrac{p_2}{q_2\eta_2\mu_2} \end{cases} \tag{2}$$

Where, p is the energy step unit price, q is the calorific value of energy, η is the energy heat utilization efficiency, μ is the energy combustion efficiency, and this coefficient is 100% for non-combustion energy. When the energy cost of Energy 1 is less than that of Energy 2, Energy 1 is recommended; when the energy cost of Energy 1 is equal to that of Energy 2, Energy 1 and Energy 2 are both acceptable; when the energy cost of Energy 1 is greater than that of Energy 2, Energy 2 is recommended.

Calculate the cost of electricity and gas to produce one kilojoule of heat using electric energy and gas energy respectively. As shown in Table 2, we get the most economical energy consumption decision of customers at present, aiming at the energy source with the lowest cost.

Table 2. Energy consumption recommendation

Gas(m^3) / ecommen dation / (kW·h)	**Step I**	**Step II**	**Step III**
Step I(Peak)	Gas	Electricity	Electricity
Step I(Valley)	Electricity	Electricity	Electricity
Step II(Peak)	Gas	Electricity	Electricity
Step II(Valley)	Electricity	Electricity	Electricity
Step III(Peak)	Gas	Gas	Gas
Step III(Valley)	Gas	Gas	Electricity

For example, a customer's current electricity consumption is in Step I, and the gas consumption is in Step II. The energy consumption is required at 18:00 a.m. on a certain day in June (the valley electricity price during the non-heating period is executed). Table 1 show that it is more economical to choose electricity. For instance, this customer heats 100kg of water, using electricity can save 0.56Yuan and 9.98% of the cost, compared with using gas.

3.3. Annual Comprehensive Energy Consumption Plan

Based on the total amount of electricity and gas energy used by customers in the previous year, the objective function is established aiming at the lowest total energy consumption cost. The total energy consumption of customers is expressed by electric energy. The annual energy utilization plan of electricity and gas energy is obtained (see Table 3) via the equivalent conversion of gas energy into electrical energy (Q Electricity = 5.684 Q Gas). When the customer's actual energy consumption deviates from the given energy plan, the optimal energy consumption scheme can be dynamically generated based on particle swarm optimization according to the current energy consumption.

Table 3. Energy consumption recommendation

Interval Name	Amount of energy consumption (kWh)	Energy consumption plan
Interval I	[0,1227)	Without comprehensive energy consumption plan, the customers can use more gas to reduce energy consumption costs, and the corresponding gas consumption volume is in the range of [0,216 m³].
Interval II	[1227,6027)	The customers use 216 m³ of gas, and the excess part is replaced by electricity. The corresponding total amount of electricity consumption is in the range of [0,4800 kWh].
Interval III	[6027, +∞)	After the customers use 216 m³ of gas and 4800 kWh of electricity, the excess part is replaced by gas. The corresponding total gas consumption volume is in the range of [216 m³, +∞).

For example, a customer in 2017 used 2021 kWh of electricity and 314 m^3 of gas, and the total energy used by this customer was 3805.776 kWh. This energy consumption amount was located in Interval II. According to the comprehensive energy consumption plan, the customer used 216m^3 of gas, the excess part is replaced by electricity, and it is required that 98 m^3 (314 m^3-216 m^3) is equivalently converted into 557.032 kWh of electricity energy, and the total amount of electric consumption is:

$$2021 \text{ kWh} + 557.032 \text{ kWh} = 2578.032 \text{ (kWh)}.$$

Namely, the customer in 2018 will use 216 m^3 of natural gas and 2578.032 kWh of electricity, which cost is lowest. The energy cost of this scheme is 45.72 yuan lower than that of last year. Converting amount of electricity-gas inter-replacing to every month, the monthly plan for using electricity and gas in 2018 can be obtained.

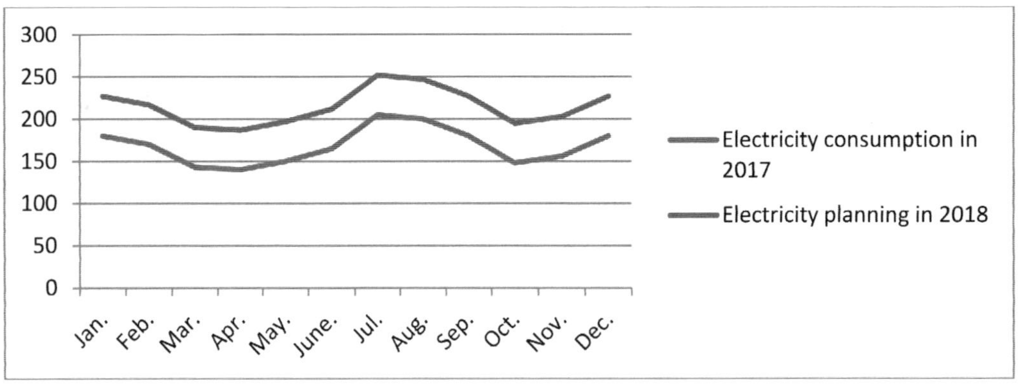

Figure 1. Annual energy curve of a customer

4. Data Analysis of Energy Supply Enterprises

4.1. Analysis of Energy Conservation and Loss Reduction

The loss calculation models are constructed according to electricity, water, heat and gas meters, respectively. The model can be the assessment object of a certain energy supply enterprise or the whole energy supply enterprise. In the present paper, a loss calculation model of "assessment unit - assessment object-input/output metering point". The system first needs to maintain the assessment unit for the assessment of energy supply enterprises, then takes the assessment area as the assessment object, followed by allocating the input/output metering point according to the actual situation on side. After configuring the loss calculation model, calculate the loss by using the loss calculation formula [6-7]. The formula is as follows:

$$\begin{cases} \Delta Q = Q_{input} - Q_{output} \\ Q_{input} = \sum_{i=1}^{n} Q_{iinput} \\ Q_{output} = \sum_{j=1}^{m} Q_{joutput} \\ \delta = \Delta Q / Q_{input} \times 100\% \end{cases} \quad (3)$$

ΔQ :Energy consumption amount

Q_{input} :Total energy input

Q_{output} :Total energy output

Q_{iinput} :Energy consumption amount of input metering point configured in the ith assessment unit

$Q_{joutput}$:Energy consumption amount of output metering point configured in the jth assessment unit

δ :Loss rate

4.2. Analysis of Potential Customers for Electricity Power Replacing

The energy consumption level is divided based on the criteria for classifying the consumption class. By the comparative analysis of the electricity consumption situation among different users, select the users whose total energy cost is in the same class, together with the proportion of electricity cost not less than 20% and more than 60% in the same community, and the electric energy substitution promotion of common household appliances and high-end appliances are carried out respectively.

5. Analysis of the Government Over all Energy

5.1. Analysis of Housing Vacancy Rate

The calculation of vacant households can be applied to value-added services and data sharing. In particular, the value-added services can exploit the non-electric information value of electric power data and develop customer drainage through residential housing use information. In terms of data sharing, it can provide housing use to government management departments, assist government decision-making and improve macro-management ability.

The energy consumption of urban residents is zero or very low for consecutive six months or more. The judgment model is that the various energy consumption monthly is less than or equal to N_i (i = 1,2,3,4; representing the four types of electricity, water, gas and heat energy), or the daily energy consumption is less than or equal to M_i (i=1,2,3,4; representing the four types of electricity, water, gas and heat energy).

N_i is the peak threshold of various energy consumption monthly of vacant households. When the various energy consumption monthly are all less than the corresponding threshold N_i, the user is judged to be vacant household. Otherwise, it is considered that the residential behavior occurred.

M_i is the peak threshold of various energy consumption daily of vacant households. When the various energy consumption daily are all less than the corresponding threshold M_i, the user is judged to be vacant household. Otherwise, it is considered that the residential behavior occurred.

5.2. Analysis of Typical Energy Consumption in Holidays and Festivals

The analysis of typical energy consumption in holidays and festivals is to depict the households who use electricity regularly on fixed holidays, mainly in the form of inflow and outflow. The population mobility of residents in the community is reflected by describing the changes of residents' electricity consumption during the Spring Festival and National Day holidays.

If the average level of electricity consumption before the festival is lower than that during the festival, and the electricity consumption before the festival is at the level of vacant households, it indicates that it is inflow households. If the electricity consumption before the festival is higher than that during the festival, and the electricity consumption during the festival is below the average level of the electricity consumption before the festival, it indicates that it is an outflow family.
The computational model is as follows:

$$I_e = [\frac{\log(1+Q_h)}{\log(1+Q_a)} \quad 1]*100, Q_a \leq Q^* \tag{4}$$

$$O_e = [\frac{\log(1+Q_a)}{\log(1+Q_h)} \quad 1]*100, Q_h \leqslant Q_a \tag{5}$$

I_e :Inflow index

O_e :Outflow index

Q_h :Represent the average daily energy consumption level of residents in a community during a holiday

Q_a : Represent the average daily energy consumption level of residents in the three months before the Ffestival

Q^* : Represent the average daily electricity consumption level of vacant households

The increase in household inflow (out) index indicates that the injection (out) behavior is more obvious. Finally, the comparison is conducted according to the proportion of the inflow and outflow households in each district. If there are more inflow households, the district is an inflow-type residential area; if there are more outflow households, the district is an outflow-type residential area.

5.3. Analysis of Urban Supporting Plan

The consumption level and consumption behavior of customers are evaluated through the establishment of a discrete two-dimensional analysis model of the occupancy rate and the average level of energy consumption for customers in the area, which provides a reference for urban supporting plan. For example, in areas with high occupancy rate and low average energy consumption, the government needs to strengthen the policy for benefiting the people, such as allocating more public transport resources, while in areas with high occupancy rate and average energy consumption, it needs to plan high-end business support and education and medical resources.

5.4. Analysis of Urban Industrial Energy Consumption

The energy consumption and pollutant emission of a single product are calculated based on the industrial productivity and energy consumption, and referring to the emission standards of electricity, gas and heat pollutants. The optimal energy utilization ratio of a single product is further calculated according to the conversion standards of electricity, gas, heat and standard coal, so as to reduce the cost of energy use and decrease the pollutant emission, and provide data support for the government to carry out environmental protection, high energy-consuming enterprise governance or energy replacement plan [8].

6. END

The diverse needs of customers for intelligent energy use is satisfied by mining and analyzing comprehensive energy data in the present paper, which effectively avoids the high-level energy consumption of customers, and reduces the energy consumption cost; automatically identify and accurately locate the abnormalities, faults, energy stealing, crossed meters and other issues of energy meter, so as to avoid customer safety problems caused by leakage of water and air. The real statistical analysis has been achieved, including housing vacancy rate, personnel mobility in holidays and festivals, urban industrial energy consumption, etc., which will provide accurate support for municipal planning, energy regulation, energy conservation and emission reduction, infrastructure construction and "Internet + government services", offering accurate data support for national policy formulation and macroeconomic regulation.

7. References

[1] LI Yang, SONG Tian-li, WANG Zi-jian, "Research on key issues of integrated energy services based on user data deep-mining,", POWER DSM, Vol.20, No.3 May, 2018, PP:1-5
[2] Gong Feixiang, Li Dezhi, Tian Shiming, Zhang Yang, Shi Kun, Zhang Pengfei, Li Junchi , Review and prospect of core technologies of integrated energy system, Renewable Energy Resources, Vol.37 No.8, Aug. 2019, pp: 1229-1235.
[3] Yang Ming, Du Ping-jing, Liu Feng-quan, Hao Xu-peng, Bo Yi-fan, Review of energy consumption and demand forecasting methods, Journal of Shandong University(Engineering Science) :1-8[2019-09-19].http://kns.cnki.net/kcms/detail/37.1391.T.20190909.1738.024.html.
[4] Tan Xue, Guo Qiang, Xu Jian, Song Weihua, Guan Jun, ect. Characteristics of energy use of typical buildings and its behavioral influencing analysis in one university campus in hot summer and cold winter, China, Journal of Nanjing University of Science and Technology, Vol.43 No.1, Feb.2019, pp:101-107
[5] Ma Tian-nan, Wang Chao, Peng Li-lin, Guo Xiao-fan,ect. Research on consumer energy use behavior forecasting and analysis of integrated energy system under multisource heterogeneous data, Smart Power, Vol 46 No.10, 2018,pp: 86-95.
[6] Liu Qingshi, Zou Peng, Zhao He, Wang Yuhong ,Comprehensive energy efficiency evaluation for smart power consumption communities, Electrical Measurement & Instrumentation, Vol. 53 No. 21, Nov. 10, 2016,PP: 59-64
[7] Wang Xingang, Zhu Enguo, Zhu Binruo, Cao Yi , Study on abnormity diagnosis and treatment method of smart meters based on multiple metering system [J], Electrical Measurement & Instrumentation, Vol. 55 No. 2, Jan. 25, 2018, pp: 86-91.
[8] ZHOU Qian, HE Qing, DU Dongmei, Assessment Model of Electrical Energy R eplacement Based on Cost Utility Analysis, Power & Energy , Vol 39 No.04, 2018, pp:538-543.

IOP Conf. Series: Materials Science and Engineering **719** (2020) 012022 doi:10.1088/1757-899X/719/1/012022

The Dynamic Trend and Hotspot in the Clothing Field in China in Recent Five Years——Based on the Visualization of CNKI Mapping Knowledge Domain

Shuchang Zhu[1, a] and Kai Wang[2, b]

1. College of Clothing Engineering,Jilin Engineering Normal University, Changchun, Jilin 130050,China.
2. College of Computer Science and Technology,Jilin University, Changchun, Jilin 130012, China.

[a]Email: 422820115@qq.com, [b]Email: wangkai87@jlu.edu.cn

Abstract. This paper took the recent literature in the field of clothology in China as the object to reveal new researching characteristics and to provide references for relevant industrial development. Based on the mapping knowledge domain, this paper adopts literature research method to analyze the sample literature by co-word analysis and so on. Studies show that, research in the field has drawn tremendous attention and exhibit a steady development. The related researches, meanwhile, are concentrated in three field of textile industry, industrial economy and fine arts. The hot spots could be summarized in three aspects: fashion design, apparel culture and fashion brand. Innovations: basically the relevant researches in this field used the foreign specialized software such as CiteSpace, which relatively more difficult. In this paper we provide a new research approach - a visual analysis embedded in CNKI. It combines the multiple functions of specialized software. Especially its" co-occurrence matrix analysis" and "keywords co-occurrence network" cluster analysis, make ordinary customers could grasp them.

1. Introduction

With the development of society, clothing has become the cultural symbol of modern people's temperament. Modern science have been continuously feeding new elements into clothing. Thus, new products, new materials and new processes are emerging prominently. As a theoretical support of the clothing industry, clothology has been drawing more and more attention, and the research results have been more and more abundant. It is necessary to systematically sort out the studies in this field, which will assist the theoretical study in clothing science and promote the development of the industry.

Clothology first appeared in China as an independent discipline in the 1980s. This is "a systematic knowledge based on the objective of clothing as the main line". The study of clothing and its related phenomena is a interdisciplinary comprehensive discipline combined with engineering, sociology, art, marketing and other disciplines [1]. The establishment of clothology has become a milestone in the field of clothing science in China. With the Opening-up policy, clothing science has achieved a qualitative lea. In the 1990s, China became the world's largest garment exporter. Clothology in China paced up with the international trend rapidly. Clothing science in China is blooming and developing dramatically.

Literature is the crystallization of knowledge and also an important embodiment of a discipline's development. This paper takes the literature in this field as the research object. Based on the searching result in China National Knowledge Infrastructure, we found there are very few articles on the overall study of Chinese clothing science from the perspective of bibliometrics, using knowledge maps. There are mainly two articles: "A study on the literatures for Chinese clothing discipline based on bibliometrics" by Fangping Yin , et al. [2], "Analysis of the research of Chinese clothing discipline from 2005 to 2016", Yan Li , et al. [3]. They both analyze the literature from a very professional perspective to reveal the characteristics of researching domain. The analysis method used by Yin Fangping is a statistical tables. Li Yan used the Cite Space software. Several other related papers are concentrating on a certain aspect of clothology, for example, Huiwei Zhang's "Research on literature metrology of digital garment technology based on scientific knowledge map" [4].

To summarize, this paper selects the literature in China National Knowledge Infrastructure for the past five years, and uses CNKI's embedded automatic statistical analysis function, especially knowledge map, to make an comprehensive analysis of Chinese clothing science research, reflecting the new features in this field. At the same time, it provides a new way for future research in the field - CNKI's embedded hot-spot analysis function.

2. Data Sources and Research Methods

In this paper, the literature is processed by analyzing the time series and co-word analysis by using knowledge maps, in order to study the age distribution, subject distribution of the papers, high frequency word co-occurrence matrix and co-occurrence network structure and to find out research hot-spots in clothology. The tools used are mainly embedded data acquisition and analysis functions of the CNKI platform..

2.1. Data Source

The data source is CNKI "Chinese Academic Journals (online version)" (CAJD), which is a well-recognized full-text library of Chinese journals. The search strategy is:

(Subject = "clothing" or "fashion" or "costume")
Not (Title = "collection" or "essay" or "notice" or "speech")
And (Time = 2014.1.1 – 2018.12.31)
And core journals

The search time is October 26, 2019. In order to ensure that the sample data is comprehensive and accurate, the search field is selected as subject, and the exclusion function of the CNKI search platform is used to directly block the invalid information such as "collection". In order to ensure the data source is academic, the core journal literature is selected. This led to the retrieval of 4,227 valid articles, which was used as the research object.

2.2. Knowledge Map

The broad knowledge map includes: gene map, cognitive map, etc., which is a visual display of knowledge. The narrow knowledge map mainly includes scientific maps, bibliometrics graphs, etc. [5]. The core function of this narrow knowledge map being discussed here, is to display abstract information in image, revealing the network structure and the complex relationship between interaction and intersection among knowledge units or knowledge groups, and intuitively reflecting subjects from various levels', fields' information on the profile of development, knowledge structure, etc., which has been applied to many disciplines.

2.3. Co-word Analysis

The basic principle of the co-word analysis is to use keywords as the analysis object, and use a large number of co-occurrence keywords pairs in the literature to reflect the relationship between the topics. The more the word pairs appear in the same document, the more relevant the two themes are represented. Co-word analysis, is the statistics of co-occurrence of word pairs, which displays the statistics in the knowledge map.

2.4. Research Tools

By searching CNKI we found that CiteSpace and other software are used in this study field. The data needs to be imported specifically for those software, which some need the assistance of third-party

IOP Conf. Series: Materials Science and Engineering **719** (2020) 012022 doi:10.1088/1757-899X/719/1/012022

software such as Ucinet. It is difficult and hard to learn. This paper provides a new approach - the visualization of bibliometrics embedded in CNKI. CNKI platform has powerful functions: multi-grouping, statistics visualization and analysis, and multiple functions of those specialized software, which is easy to operate. The research tool in this paper is the three visualization analyzing parts embedded in CNKI - time, subject and keyword, mainly about the cluster analysis.

3. Research Results and Analysis

3.1. The Age Distribution of theLiterature

Knowing the quantity of annual literature, we can macroscopically grasp the development trend of research in the field. In 2014-2018, the research literature in the field of clothing in China was 885, 852, 802, 942, and 746, as shown in figure 1.

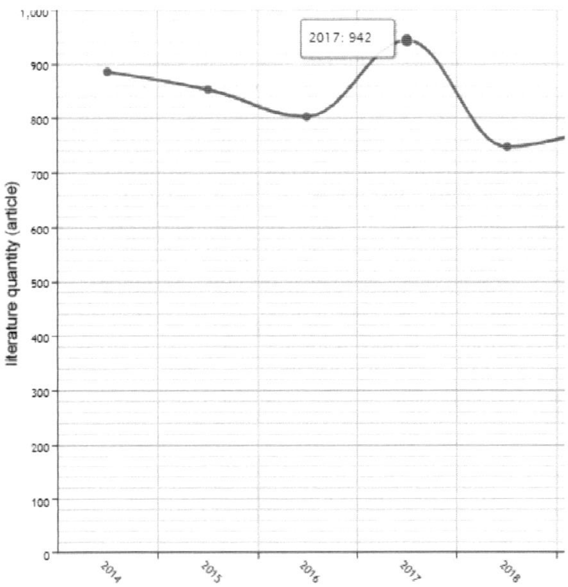

Figure 1. Distribution of literature quantity in the clothing field in recent five years.

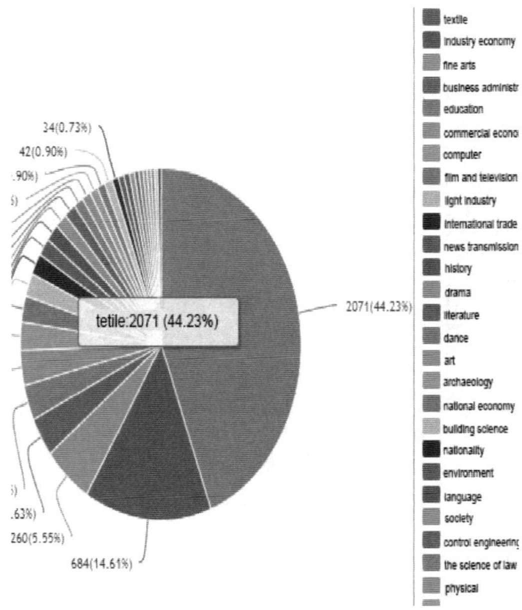

Figure 2. Discipline distribution of literature volume Top30.

It can be seen from Fig.1 that in the past five years, Chinese clothology research can be summarized into three stages: the first stage is 2014-2016, the second stage is 2017, the third stage is 2018, the first and third stages show a downward trend, the research heat reduction. In the second phase of 2017, the amount of literature increased significantly, the research was becoming more active. From the perspective of general trend, it is a wave-like development trend. The total amount of literature in each year is higher than 700, indicating that has been receiving attention. It is in a relatively stable development stage.

3.2. Disciplinary Source Distribution of the Literature

The distribution of disciplines reflects the cross-infiltration relationship between research in this field and other disciplines. The disciplines of Top30 in this field in the past five years is shown in Fig.2.

It can be seen from Fig.2 that the literature in this field is mainly concentrated in the fields of textile, industrial economy and fine arts. The most literature-rich part is textiles which accounts for 44.66%. From the perspective of disciplinary attribution, GB/T 13745-2009 " Discipline Classification and Code List of the People's Republic of China", the subject related to clothing, is the "clothing technology" under the category of textile science and technology, explained the amount of textile literature is the largest. Secondly, the reason for high literature amount in industrial economy may be the huge consumption power in clothing of China. The research in clothology is closely related to the development of industrial economy. The third largest part is fine art. Modern people are more concerned about aesthetic taste so the demands are becoming pluralistic. Therefore, it produces a large number of interdisciplinary literature that intersect with clothing and fine art.

3.3. High Frequency Keyword Co-occurrence

The keyword is the word that can expresses the core content, so the keyword is selected as the research object. The keyword visualization of CNKI searching results shows the highest frequency keywords and their corresponding literature in 4,227 documents according to the frequency. The Top30 high frequency words are shown in Fig.3.

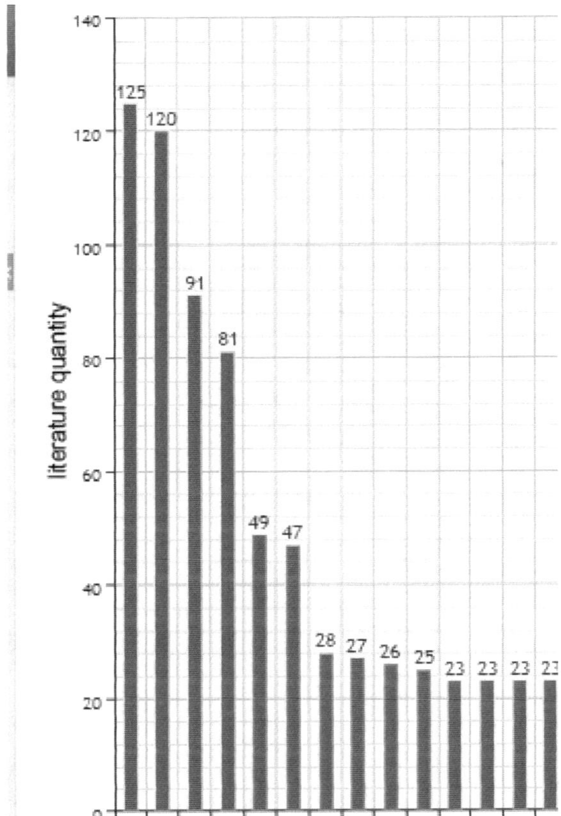

Figure 3. Top30 high frequency keyword

It can be seen from Fig.3 that the keywords with the highest frequency are clothing, fashion design, costumes, etc. These are the main objects in the field of clothing research. However, the statistical comparison of single high-frequency word cannot reflect the essential relationship between the subjects, nor can it effectively reveal the hot-spots of research field. The following will further explore them by using keyword co-occurrence matrix and cluster analysis.

The system randomly selected the top 20 high-frequency keywords (frequency ≥ 20), and performed pairwise statistics to obtain a 20×20 keyword co-occurrence matrix (see figure 4), which shows the most related 20 word pairs, indicating that there is a high degree of cross-infiltration between the themes represented by these high-frequency words.

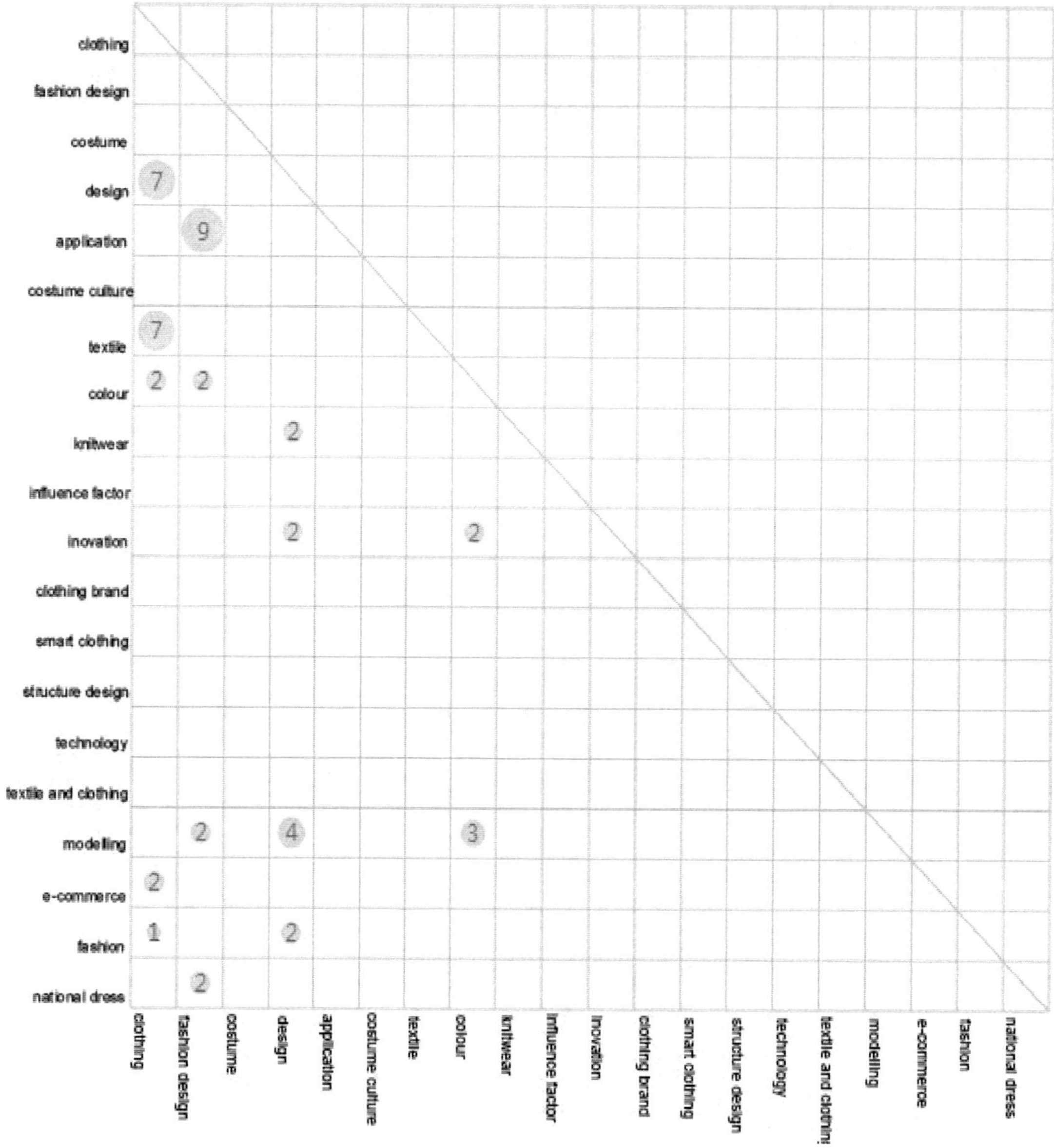

Figure 4. Top20 high frequency keywords co-occurrence matrix.

3.4. High Frequency Word Co-occurrence Clustering

The knowledge map (figure 5) of the high-frequency word co-occurrence matrix (figure 4) is obtained by CNKI embedded visual analysis of searching results.

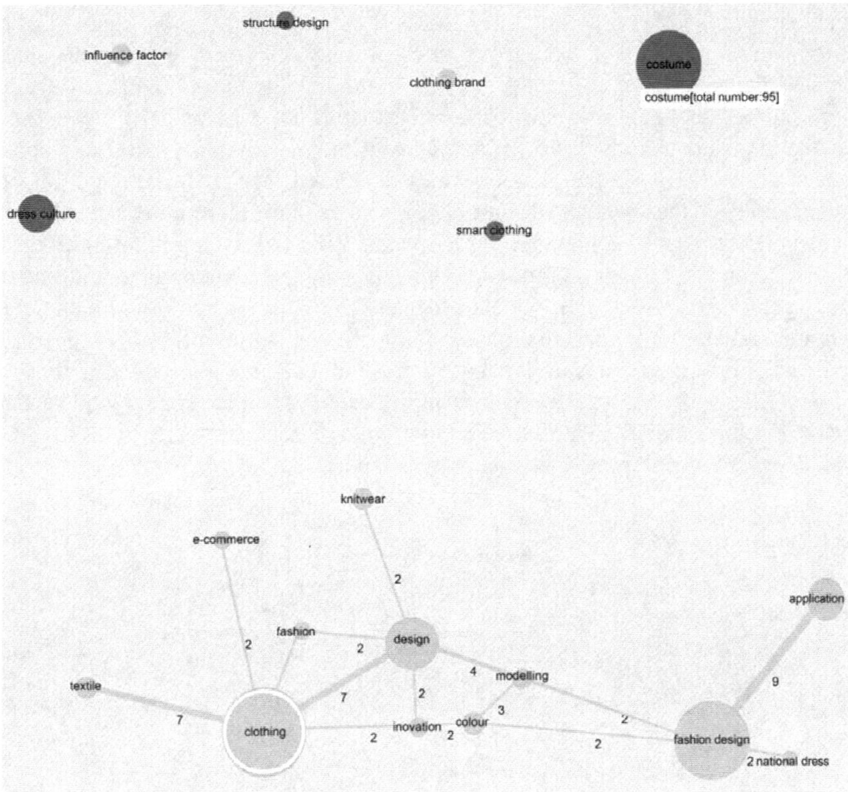

Figure 5. Top20 high frequency keyword co-occurrence map

The keywords in Fig.5 are represented by circular nodes, the size of the nodes increasing with the document amounts. The connection between nodes indicates that there is a co-occurrence relationship between the keywords. The thicker the connection, the closer the two keywords are connected. Different colors represent different word families, the largest word family - "fashion-design" family has 12 keywords, followed by "apparel-culture" family, "fashion-brand" family, "structural design" family. Each of the colors represents a subfamily of multiple collections of documents and multiple keywords. Each family includes a number of keywords and their corresponding documents, as shown in Fig.6.

According to the analysis above, the research hot-spots in the field of clothing in the past five years can be summarized into three domain: fashion design, apparel culture, fashion brand, combining the relevance of the keywords and the subject attributes.

4. Analysis of Research Gotspots in the Field of Clothing

4.1. Fashion Design

This hot spot stems from clustering: "fashion-design" (Fig.5), high-frequency words such as clothing, fashion design, applications, etc. "fashion" and "fashion design" are the two nodes that have the largest amount of literature in entire clustering network. (Fig.5), "fashion design" can be described as a key-spot in hot-spots. China is transforming from clothing manufacturing into clothing creating. The core of creation is innovation. Design is the first step to fulfill innovation. Therefore, it has become the biggest research hot-spot in the field of clothing. As can be seen from Fig.4, in this hot-spot, "application" and "fashion design" have the highest frequency of co-occurrence. The application of various design elements and techniques in fashion design is particularly concerned. In Sunan Xue's "Research of bionic elements and its artistic expression in clothing design"[6] , it pointed out that new bionic technology should be applied to fashion design in time, expounding various bionic elements in shape, color, pattern texture and functional design. In Hongjuan Chen's " Design of pleat knit fabric and its application in knitwear "[7], it describes three process designs of woven knit pleats: four flat strips, positive and negative plain compound, partial braiding. They analyze the texture performance of knitted pleats and its application in knitted garments. It is believed that the pleats formed by this method are stable and natural, which are different from those formed by external forces of woven techniques, which are in line with the characteristics of knitted fabrics. In Fengqiu Guo's "Study on the creative application of the element of Chu phoenix in modern clothing pattern design" [8], they analyze the characteristics of the phoenix bird of Chu Silk fabric, highlighting four methods of applying phoenix elements to contemporary clothing design using modern media and technology: transfer, deconstruction, reconstruction, misplacement. These are the traditional techniques inherited from Chu Culture and further promoting it, enriching the semantics of fashion design. In Yuting Yang's "Creative application of decorative art in modern costume design" [9], they analyze the application characteristics of decorative art in fashion design. The creative method of decorative art has been explained in the paper and the dynamic integration of decorative arts in fashion design is also proved to be an important developing direction in their report.

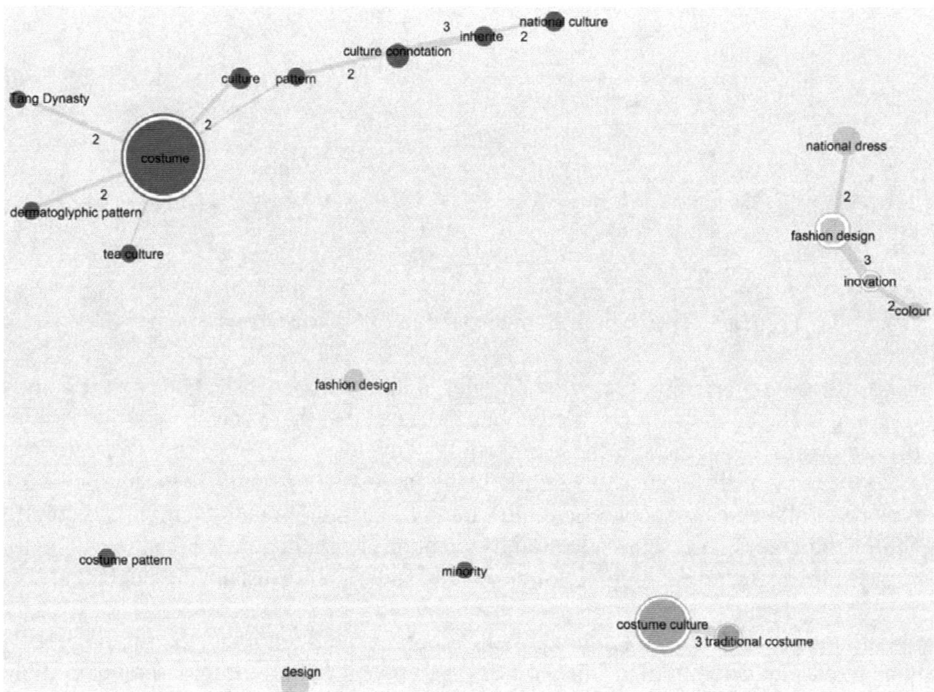

Figure 6. Top20 high frequency keywords co-occurrence map in "dress-culture" family

4.2. Apparel Culture

"Apparel", that is, clothing and its decorations. The hot-spot is derived from the cluster "apparel" and "apparel culture". The main keywords include: apparel, apparel culture, national costumes, etc., reflecting the joint connection between Chinese clothing and culture, national elements. In Ping Yan's "The inheritance and development of the Chinese costume culture" [10] ,it is believed that clothing is an integral part of human culture, reflecting historical and cultural connotations of social politics, economy, folklore, etc., which is a materialized form of aesthetic consciousness as well. In Xiaoyu Wang's, "Discrimination of ho-momorphic patterns on Han's traditional costumes" [11] , it is believed that costume pattern is the most representative image record of regional culture. Homomorphic pattern originates from artistic representation and is essentially the embodiment of cultural connotation. The paper analyzes the elements, artistic representation and cultural connotation of homomorphic pattern, and reveals the subjective and objective reasons for its existence. In Liao Jiangbo, Xiaoming Yang's,

"Investigations on commoner clothing culture based on comparative perspective of hemp and silk" [12] , it is believed that apparel fabrics carry a certain social and cultural connotation. Based on the contrast between cloth and enamel, through the literature and cultural relics, the role of cloth apparel culture in the shape of clothing is examined, revealing the dual cultural connotation of cloth as the wearing apparel and the civilized sprite. In Qiang Xu and other's "Interpretation of Taiwan Paiwan tribe clothing patterns cultural connotation" [13], it analyzes the theme and characteristics of the Paiwan costumes, and summarizes its cultural connotations from six aspects: grade markers, ancestor admiration, reproduction worship, national customs, faith sustenance, myths and legends.

4.3. FashionBrand

The hot-spot stems from clustering "fashion brand" and "influencing factors", related words such as brand, design, e-commerce, etc. In the era of brand economy, brand, such as British BURBERRY, French CHANEL, becomes their ID in the international market. It provides an important guarantee for their stable position in the world economy. In Ying Cheng's "Study on the international strategy of the brand of the textile and garment enterprises"[14] , it puts forward the entry mode for international brand market, emphasizing three types of international marketing strategies of textile and apparel industry brands. In Ling Zhou's "Research on the impact of e-commerce development on brand clothing competitiveness" [15], it analyzes the e-commerce model and its impact on brand apparel competitiveness, and puts forward six principles of branding under the e-commerce model. In Jing Zheng's "Research on the impact of e-commerce development on brand clothing competitiveness" [16] , it emphasizes that the brand style gene should be fully understood as it is a long-term accumulative wealth and should be widely integrated with new styles like series design. It also analyzes three series of clothing design patterns based on brand style genes. In Ling Yu's "Influencing factors and weight distribution of brand garments showcase display image"[17], it divides the influencing factors into two-level index factors and builds the evaluation index system. It turns out that, clothing culture factors, personnel professionalism and attitude are relatively important, and the internal environment construction of the showcase is particularly important. The background design factor of the showcase is the most influential.

5. Conclusion

Based on the principle of bibliometrics, this paper adopts the literature research method, and takes the literature of clothology in CNKI core journals in the past five years as the research object. Cluster analysis of research hot-spots in this field, the conclusions are as follows:

5.1. From the perspective of the distribution of literature volume, the research in the field of clothing science in China has been in a wavy, overall stable and developing state which drew a lot of attention.

5.2. .From the perspective of the distribution of literature subjects, the research in the field of clothing science in China is mainly concentrated in the fields of textile, industrial economy and fine arts, which are determined by the subject property, commodity and sociality of clothing.

5.3. Through the co-occurrence cluster analysis of high-frequency words, the research hot-spots in the field of clothing science in China in the past five years can be summarized into three: clothing design, apparel culture, and clothing brand.

In the era of innovative economy, society calls for innovation. As a major import and export country of clothing, China needs the innovation in clothing. Clothing design is the first step in clothing innovation; apparel culture innovation is the first priority of clothing innovation; high-end clothing branding is the necessary way for Chinese clothing to become international.

6. Acknowledgments

This paper was funded by the 2019 general planning project of the 13th five-year plan of education science of jilin province, subject approval number: GH19329.

7. References

[1] Xiaogang Liu 2016 *Conspectus of Clothing Science: Revised Edition* (Shanghai: Donghua University Press) **p 103**

[2] Fangping Yin, Jingwei Liu and Dawei Zhang 2007 A study on the literatures for Chinese clothing discipline based on bibliometrics *Journal of Xi'an University of Engineering Science and Technology* **(01):66-74**

[3] Yan Li and Jie Li 2017 Analysis of the research of chinese clothing discipline from 2005 to 2016 *Journal of Beijing Institute of Clothing Technology:Natural Science Edition* **37(3):33-45.**

[4] Huiwei Zhang 2017 Research on literature metrology of digital garment technology based on scientific knowledge map (Hangzhou: Zhejiang Sci-Tech University)

[5] Siluo Yang and Ruizhen Han 2013 Studies in the status quo of knowledge mapping application research abroad *In-formation and Documentation Services* **(6):15-20.**

[6] Sunan Xue 2017 Research of bionic elements and its artistic expression in clothing design *Wool Textile Journal* **45(11):51-53**

[7] Hongjuan Chen 2017 Design of pleat knit fabric and its application in knitwear *Shanghai textile science & technology* **45(12):33-36+56**

[8] Fengqiu Guo 2017 Study on the creative application of the element of chu phoenix in modern clothing pattern de-sign *Art & Design* **(12):140-141**

[9] Yuting Yang 2015 Creative application of decorative art in modern costume design *Hundred Schools in Arts* **31(S1):140-142**

[10] Ping Yan and Fu Cheng 2014 The inheritance and development of the chinese costume culture *Jiangxi Social Sciences* **34(9):244-248**.

[11] Xiaoyu Wang and Weidong Gao 2017 Discrimination of ho-momorphic patterns on Han' s traditional costumes *Journal of textile research* **38(01) 116-120**

[12] Jiangbo Liao and Xiaoming Yang 2016 Nvestigations on commoner clothing culture based on comparative per-spective of hemp and silk *Journal of silk* **53(04) 72-77**

[13] Qiang Xu, Dongsheng Chen and Yingjin Gan 2016 Interpre-tation of Taiwan Paiwan tribe clothing patterns cultural connotation *Journal of textile researc*h 37(12) 111-116

[14] Ying Cheng 2015 Study on the international strategy of the brand of the textile and garment enterprises *Shanghai textile science & technology* **43(9)1-2**

[15] Ling Zhou 2017 Research on the impact of ecommerce development on brand clothing competitiveness *Wool textile journal* **45(9) 97-99**

[16] Jing Zheng 2017 Clothing series design method based on brand style gene: taking Dior as an example *Art & design* **(11) 128-129**

[17] Ling Yu , Wenjin Hong and Miao Yu 2018 Influencing factors and weight distribution of brand garments showcase display image *Journal of textile research* **39(06) 162-166**

Implementation and Application of APM Monitoring System under Big Data of Power Grid

Yang Shen[1], Lin Qiao[1], Wei Liu[1] and Chao Yang[1,2]

[1] Information and Communication Branch, State Grid Liaoning Electric Power Co., Ltd., Shenyang Liaoning China.
[2] Dalian Power Supply Company, State Grid Liaoning Electric Power Co., Ltd, Dalian Liaoning China.
Email: liuw@ln.sgcc.com.cn

Abstract. With the rapid development of smart grid application system and the increasing complexity of information system, the difficulty and importance of data center operation and maintenance management are increasingly prominent. In order to further analyze the existing data under the big data of power grid, improve work efficiency and service quality, a set of scientific and efficient application performance management (APM) monitoring system is proposed. Through the real-time monitoring of power system performance, the latest information about the operation of software and hardware system can be ensured. With the performance data of the system under a series of activities and loads, users can define a performance baseline to represent the acceptable performance under typical operating conditions. This baseline provides a reference point for locating problems more easily. Through scoring system, the health degree of information system operation can be displayed intuitively. Once the score drops, the system is in a non-health state, and the location of the performance bottleneck can be roughly determined by the secondary score. The dispatcher can quickly contact the relevant operation and maintenance team to solve the performance problems, greatly reducing the response time of the performance problems.

1. Introduction

The electric power informatization started from the single machine one-way application in the 1990s, and now it has been applied in the whole network business. With the increase of the system traffic and the transformation of the surrounding environment, many application problems of power business system can not be found and solved in the first time by existing means; Meanwhile, due to the limited application monitoring means, application system cannot be monitored in real time. In addition, there is still no product comparable to foreign technology in the application performance monitoring, diagnosis, analysis and optimization technology in China, while IBM, Oracle and other research products in foreign countries do not provide localization services, cannot provide corresponding technical support and guarantee, and have high requirements for technical personnel and expensive product price and service fee. Therefore, it is urgent to build a set of scientific and efficient application performance management system to ensure the stable operation of power information system.

Different from the lagging mode of normal monitoring and post processing, power system performance monitoring can assist dispatching, operation and maintenance personnel to prevent performance bottlenecks and quickly locate performance problems, thus ensuring higher availability and stability of information systems. The research results are mainly used to deal with the performance evaluation, architecture optimization and code analysis diagnosis of business application system. It can bring certain benefits, such as reducing the demand of operation and maintenance personnel, saving labor costs, establishing dashboard performance control, promoting business standardization, etc. In addition, the platform can also provide powerful and dynamic real-time performance monitoring interface and business data analysis interface for application systems of typical industries (such as banking and securities and other financial industries, aluminum electrolysis and other industrial control industries, transportation industry, government data center [1,2]). The big data generated by these industry application systems can be deeply mined and analyzed.

2. Requirement Analysis and Design of the System

Today's power business system is different from the traditional LAN application, which adopts a multi-layer technology architecture. It brings many challenges to the new generation of power business system. So we need to design a product to monitor the performance of the system in real time, to help the staff quickly locate the location of the problem, so as to quickly analyze and solve the problem.

2.1. System Design Requirements

In order to better monitor the performance status of power grid system, the main functions that the system hopes to achieve are as follows: (1) Real-time monitoring. It can achieve the millisecond interval to obtain performance data, real-time monitoring warning data and business code data. (2) JDBC monitoring. For fine-grained monitoring of resources, it can monitor the unreleased code of JDBC resources and exceptions in the process of database interaction. (3) Detailed monitoring of thread pool. For example, the number of threads, real-time thread execution, thread feedback, and the number of rows returned by the current data in the database; in addition, human intervention and thread release operations can be performed on zombie threads, interlocking threads, and threads occupying resources. (4) Monitoring of socket and I/O. (5) Professional monitoring large screen to ensure real-time display of all monitored data. (6) Customization of the monitoring range. .

2.2. Application Performance Management and Acquisition Technology

Gold application performance management (Gold APM) is the next generation of intelligent and easy-to-use application business performance management solution. Through the embedded collection method to achieve code level analysis, SQL scripts, services and other in-depth analysis, to achieve the end-to-end application performance monitoring for the real behavior of users from the user end to the service layer. The working principle of gold APM "application probe" is to analyze the "real production data" in the application. Operate the application from the front-end device, click the button to send the request, and the back-end server receives the request and processes it, such as calling the stack, database or accessing the third party and cache. "Application probe" is collected and processed uniformly, and uploaded to gold APM Data Center for processing display. The deployment mode of gold APM application performance management platform is shown in Fig.1.

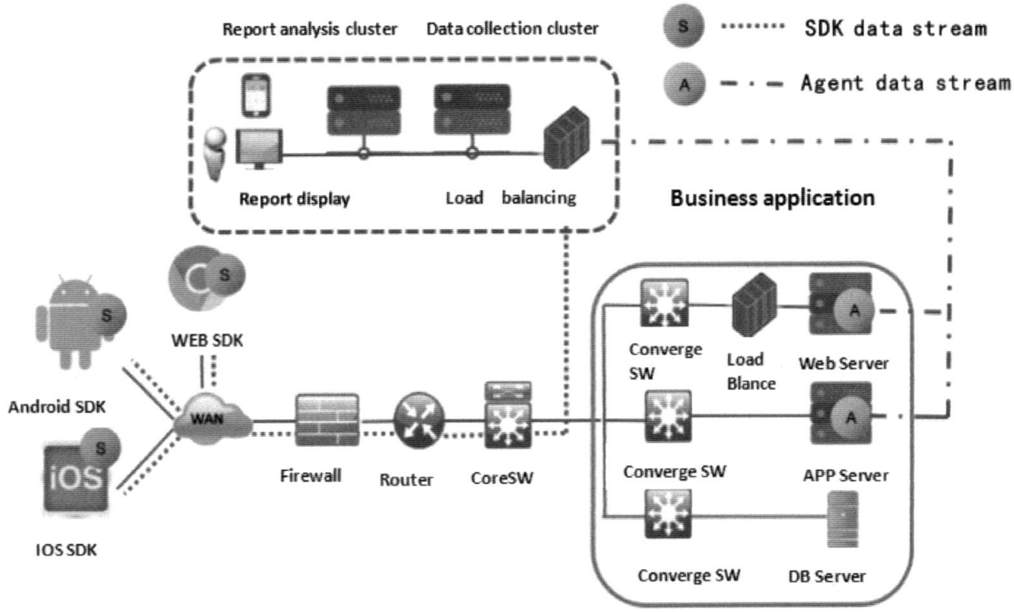

Figure 1. APM deployment mode.

Through APM technology, the data content of the whole life cycle of the application system can be collected, and the results can be achieved as follows: (1) Minimize downtime [3]; (2) Establish integrated dashboard performance control. APM's real-time integrated dashboard provides a powerful and dynamic real-time performance monitoring interface [4]; (3) Ensure the fault corresponding ability. Through continuous system performance monitoring, the possible dangers in the future can be predicted and prevented, and provide automatic alarm function in case of problems [5]; (4) Service satisfaction. The automatic load control (PLC) function can be used to ensure the stability of service, the minimization of downtime and the corresponding ability of failure, so as to realize a more stable and reliable 24-hour×365 day system operation[6]; (5) Reduce the overall cost. TCO can be reduced by minimizing downtime, ensuring fault response capability, optimizing application adjustment performance, maximizing IT resource efficiency and effectively using human resource applications; (6) Ensuring quantitative performance basis data. According to the quantitative data collection and statistical analysis of connected users and load, the system expansion and modification time can be predicted to ensure the quantitative basis data needed in addition and optimization [7,8].

3. Main Research Contents and Technical Solutions

Power application performance diagnosis, analysis and optimization monitoring system needs to provide excellent end-user experience monitoring and early warning, code level performance monitoring and fault handling, application performance bottleneck positioning, application performance problem root cause analysis, external things analysis, etc. Provide end-to-end solutions for end-user perception, application performance on the network and system operation health. Use probe technology to monitor, measure and detect application performance and abnormal conditions, provide detailed real-time view and historical record view, and provide visual reports on business transactions.

3.1. Main Research Contents

(1) Study the high-efficiency storage and reading technology of massive program code; (2) Study the application code collection technology in different operating environments; (3) Develop an efficient code data processing engine to optimize the application program in multi concurrent, distributed and cloud environments; (4) Study the code level program running state collection technology and non intrusive code monitoring technology; (5) Research on stack data tracking monitoring technology of

application system; (6) Research on dynamic link tracking technology; (7) Research on thread pool management tracking monitoring technology.

3.2. Implementation Plan and Technical Solution

The project mainly consists of four modules: data acquisition subsystem, data processing subsystem, data management subsystem and data analysis subsystem.

The data acquisition subsystem is composed of communication acquisition data module, profile data acquisition module [9], I/O data acquisition module, thread stack state acquisition module, JDBC data acquisition module [10, 11]; it is mainly developed by Java and C++ language, and data acquisition is mainly realized by C++ language in the bottom operation of the system. Data processing subsystem [12] consists of data acquisition interface, data integration module, data classification storage module and data preprocessing module. The standard deviation algorithm in statistics is used to process and classify the collected data, and to relate the scattered data. Big data algorithm is developed by C and C++ language, and mining algorithm function library is realized by C++ language in the form of dynamic link library. In the implementation of mining algorithm function library, STL basic container is used to evaluate the development efficiency and operation efficiency, and boost class library is used to provide basic data structure support. The data management subsystem consists of visualization module, warning module, authority management module, utility module and access control module. Be responsible for the management, task scheduling and control in the system, realize the dynamic display of application system operation status and behavior data, and present the analysis results of big data intuitively. Data analysis subsystem is the system architecture of data mining platform based on MapReduce, which is composed of regression analysis module, association analysis module, data classification module, data clustering module, text mining module and web mining module.

Java technology is mainly used to monitor the system operation status, application system behavior and application system business data, which is realized without modifying Java application source code. Some trace codes and profile information output codes used to collect performance data are inserted into Java application class through reflection and mapping technology, so as to realize data monitoring. In the aspect of I/O data and socket stack data monitoring, the method is to modify the program code to realize the output of specific stack information content, and use the Java virtual machine loading mechanism to realize the change of loading mode. In the aspect of profile information monitoring, by inheriting and overloading Java. Lang. object, Java. SQL.*, Java. Io. * and other system base classes, the base classes introduced by the application program will have code offset, so as to realize the corresponding data monitoring.

JDBC related data monitoring is based on the implementation of database link driven, and be divided into three types and makes corresponding configuration, including the connection pool mode, the native JDBC mode and the custom connection pool mode, so as to realize all the existing JDBC mode data monitoring. In the aspect of thread stack and queue status data monitoring, thread tracking identification technology is used. It is mainly used to monitor the thread class, uniquely identify each generated server thread, monitor and trace the thread throughout the process, identify its status, and extend some attributes of the thread, add thread response time factor to achieve the number of threads According to the monitoring of each state and the information of queue depth [13]. In terms of data storage, different storage methods are adopted for different types of data. For the behavioral data and profile data generated by the application system, the product uses the self-developed file data base for storage. The file database is developed based on the ISAM index sequential access method [14]. The index mode is very fast and can be used in the later data analysis and mining. The central integrated management mainly adopts the spring MVC architecture development, follows the MVC mode, and mainly uses H5 technology in visualization; for profile data, uses the transaction end time and the total response time to form a real-time scatter diagram, which can sense the situation of application server processing transactions in time through the shape of the diagram, and deal with the problems of insufficient resources and failures in time. In the aspect of trend display, line chart and column chart are used.

4. Implementation of APM Monitoring System

APM monitoring system is to establish an effective performance monitoring and fault response strategy, which is suitable for Java development application system. Based on the application layer, system performance data monitoring and code level data monitoring are realized, so as to quickly and accurately find the application system fault, and make the application system run more stably and efficiently.

4.1. Intelligent Alarm

One of the core capabilities of monitoring products is intelligent alarm. This section will introduce the intelligent alarm function of the system in detail, including the customized precise alarm and alarm real-time notification function.

4.1.1. Custom precision alarm

The alarm service supports the setting of service index and performance index alarm of a single service, and can also be set according to a class of service. At the same time, the alarm threshold can be set for the indicators of the whole business, and flexible alarm combination can be carried out, such as different indicator combination, different business combination, different alarm mode combination. First, lock the service request through the access path, server IP and message information, and then set the alarm threshold for the service request to achieve the real service alarm based, as shown in Fig.2. Second, the components of alarm strategy are modularized. Alarm monitoring index and threshold are designed as a module, alarm notification as a module, alarm name and enabling status as a module. Each module can be freely combined without mutual influence. This combination of alarm strategies can combine different alarm strategies according to alarm needs to meet the use of different services.

Business name	server	Route
successfully joined cart	192.168.132	/app/cart
get user information	192.168.132	/app/user/get-user
/app/category/list-with-product	192.168.132	/app/category/list-with-product

Figure 2. Business request location figure.

4.1.2. Real-time alarm notification

When the application is abnormal, it triggers the alarm strategy and sends a notice to the user. At the same time, it also records the application information and records the alarm information into the alarm database. The system has developed the client application. By installing the gold APM client, you can know the current status of the application in real time, receive the alarm notice of the application in time, and view the alarm information, as shown in Fig.3.

IOP Conf. Series: Materials Science and Engineering **719** (2020) 012023 doi:10.1088/1757-899X/719/1/012023

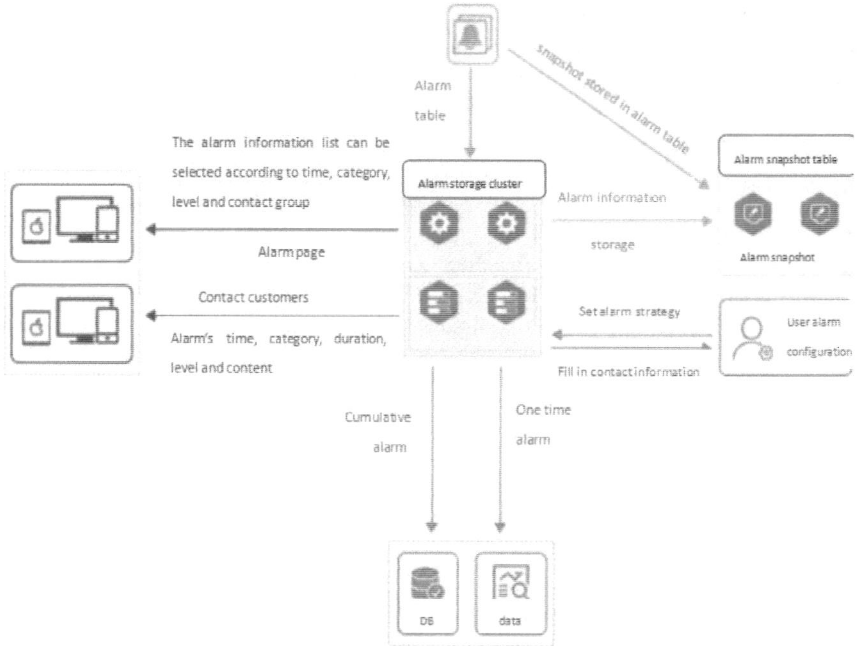

Figure 3. Real time alarm notification chart

After an alarm occurs, users can click the analysis button in the alarm list (both in the alarm overview and alarm details) to enter the alarm analysis page. In the alarm analysis page, the data of 20 minutes before and after the alarm is displayed, and the key indicators affecting the service request are graphically displayed, so that users can analyze the change of service request during the alarm.

4.2. Other Main Functions of the System

Start the server program by running startup.bat under the server installation directory. / bin. Then access the http://server addressr:7900, enter the the system home page. The system lists all commonly used custom monitoring parameters in the centralized monitoring page. For example, it displays the warning information of all monitored servers. The warnings are divided into four types: serious, error, warning and message, which are expressed in different colors. At the same time, the system also has a very professional real-time monitoring screen to ensure the real-time display of all monitored data.

Health score intuitively understands the health degree of the tested application. First, the application performance health index, second, the business health index and how the business performance is. The two indices represent the overall health status of the application. Set the weight of the core indicators of the application, and calculate the index score through the weight relationship, which can directly reflect the current health of the application. The full score is 100 points. Refresh every 5 minutes. The higher the score is, the healthier the application is. The border color of the score will change according to the score value. The platform also has the function of generating report books. By default, monitoring reports are generated on a weekly basis. After selecting the date range and time range, click query to generate a preview report, and print or export the report file.

5. Conclusion

In order to realize the real-time monitoring and early warning of power information system, quickly determine the application performance bottleneck position, and ensure the stable operation of power information system, We proposes a performance monitoring platform based on APM. The system uses probe technology to monitor, measure and detect application performance status and abnormal conditions; it provides the function of viewing detailed real-time and historical records, and generates visual application performance reports. This product is applicable to all mainstream operating systems and various Java versions (from version 1.4). Even a simple jar program without middleware can achieve full function monitoring. In addition, the research results are also applicable to financial, industrial control, transportation, medical and other industries, as well as e-government and other government information business data centers.

6. Acknowledgments

This work is supported by Science and technology project of State Grid Liaoning Electric Power Co., Ltd.(2019YF-58).

7. References

[1] Wu Yuwen 2019 Analysis of innovative forms of bank financial management in the era of big data *Accounting learning* **23** pp 51–52.
[2] Zhang Ling 2019 Research on the change of enterprise human resource management in the era of big data *Intelligence* **23** pp 244–245.
[3] Guo Weihua 2019 Application of ARP system in asset management *China's collective economy* **13** pp 27–28.
[4] Zhou Xuefei, Li Chengyu, Liu Xuefe 2019 How scientific research institutions do well in the connection of government accounting system --Based on the ARP system update of Chinese Academy of Sciences *Economist* **1** pp 93–94.
[5] Qu Guangping 2017 Discussion on free ARP application *Network security and informatization,* **05** pp 124–125.
[6] Guo Qifeng, Zeng Dajun 2018 Research on ARP virus attack and defense in University LAN *Information recording materials* **19** p 69
[7] B. Prabadevi, N. Jeyanthi 2018 TSCBA-A Mitigation System for ARP Cache Poisoning Attacks *Cybernetics and Information Technologies* , **18** p 4.
[8] Zhao Hongli 2018 Application of management accounting of Chinese Academy of Sciences Based on ARP information system *Business accounting* **20** pp 59–60.
[9] Luo Ningning, Liu Xuefei, Li Chengyu 2018 A preliminary study on the solution of bank enterprise interconnection based on ARP system *Economist* **2** pp 64–65.
[10] Wang Xiaoni, Han Jiangang 2018 Research on defense strategy of ARP spoofing attack in campus network based on immune network *Office automation* **23** pp 19–21.
[11] Cheng Yu 2017 Attack analysis and protection of ARP in signal security network *Shanghai Railway Science and technology* **3** pp 140–141.
[12] Zhang Bingbing 2017 Research on discrete event system for sensing LAN attacks *Electronic technology* **30** pp 169–172.
[13] Li Hang 2018 Practice and exploration of salary management information system based on ARP system *National circulation economy* **9** pp 47–48.
[14] Yue Xuanyu, Chen Yuanping, Du Yihua 2017 Construction of mobile application center of distributed system based on "cloud service" *Computer system application* **26** pp 63–67.

IOP Conf. Series: Materials Science and Engineering **719** (2020) 012024 doi:10.1088/1757-899X/719/1/012024

Improved Disease Gene Predication Method

Gerui He[1,b], Zhiming Liu[1,a,*], Lingyun Luo[1] and Yaping Wan[1]

[1]School of Computer, University of South China, Hunan, Hengyang , 421001, China
Email: [a]nhdxlzm@usc.edu.cn, [b]616281384@qq.com

Abstract. The prediction of disease genes has always been a hot topic in the field of bioinformatics. Machine learning methods can effectively dig out the hidden relationship between disease-causing genes and predict disease genes. At present, the prediction algorithm of Gene Ontology (GO) combined with GO annotation has limitations. It is believed that disease genes will only accumulate on the biological process branches of GO, ignoring the cellular components and molecular function branches. Disease gene prediction is performed by considering data from three branches of biological processes, cell components, and molecular functions. Multiple sets of experiments were performed. The data showed that the use of three branches to predict disease genes increased the accuracy from 78% to 91%, indicating that the disease genes not only aggregate on the branches of biological processes but also aggregate on molecular functions and cellular components.

1. Introduction

Some diseases in the traditional concept belong to mental illness, such as schizophrenia, autistic spectrum disorder(ASD), major depression. But the progress of genetic technology has sent a voice of opposition. These diseases are no longer considered to be simple mental illnesses but are also considered to be genetic diseases, and in some cases identify potentially relevant genes [1].

Autism Spectrum Disorder is a complex mental disorder that has a strong genetic influence, multiple causes, and hundreds of different associated genes. A mass of genetic studies related to ASD has identified hundreds of disease-causing genes [2,3]. However, the massive data generated by these large-scale research institutes bring more invalid information while bringing effective information. It has become a challenge and a necessary task to dig out effective information from the vast amount of data and identify the true pathogenic genes. At present, 20%-25% of genetic factors of ASD disease are found, and there is still much room for improvement, which has research significance.

In recent years, machine learning has been applied to various research fields, and the analysis of genomic data sets is no exception [4]. A supervised machine learning approach can identify hidden relationships between disease genes in a data set and then use this information to distinguish disease genes from non-disease genes [5,6]. At present, how to use existing genomic data to reliably predict disease genes and improve the accuracy of prediction accuracy is one of the problems that need to be solved to transform genetic technology into biomedical applications.

2. Related Work

Krishnan used the weighted support vector machine (SVM) to predict the association probability between brain genes and ASD [7]. They use weighted SVM to train on specialized gene networks that integrate gene expression, protein-protein interactions, and regulatory sequences of brain genes. The classifier evaluates on high confidence disease gene of ASD. Besides, there are limitations in the network which is difficult to express weak interactions. Therefore, protein-protein interactions and co-expression network data cannot express weak interactions that limit the prediction range of disease genes.

To overcome the shortcomings of network data not expressing weak interactions, Muhammad used gene ontology to predict gene functional similarity as training data [8]. Genes with similar functions have a similar expression trend. For example, in ASD, a disease-related gene that is disrupted by genetic variation tends to accumulate on specific biological processes [9]. That indicates that disease genes may belong to the same hierarchical path of GO and have a higher functional similarity. Based on this hypothesis, Muhammad used GO predictor gene functional similarity to predict potential ASD candidate genes.

Muhammad believes that disease genes only accumulate in biological processes, ignoring molecular function and cellular components. The effect of the calculation is that some genes have a functional similarity to any gene that is zero, including itself. The reason for this illogical phenomenon is that the term annotating the gene is not on the branch of the biological process. For example, the gene "AAAS", whose annotation terminology is shown in Figure 1, is all located in the branch of the cell component and has no annotations located in the branch of the biological process. The genetic similarity between all genes and "AAAS "is 0, include "AAAS".

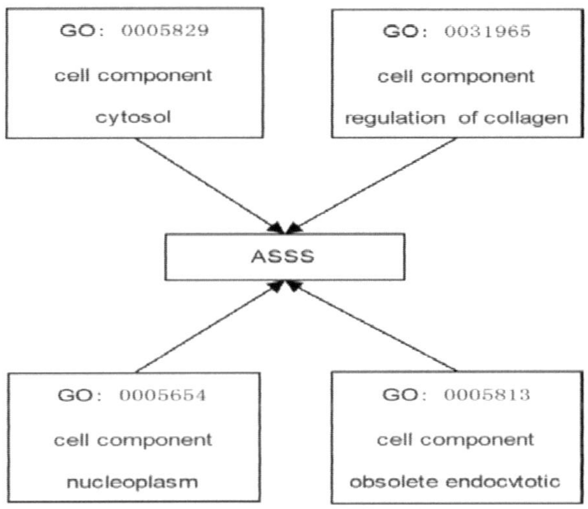

Figure 1. Annotation relationship of Gene Ontology term and gene of AAAS

In response to the above illogical shortcomings, it is believed that disease genes not only accumulate on the biological processes but also aggregate on cellular components and molecular functions. When calculating the functional similarity of genes, considering the three branches of GO. And then the final prediction accuracy is improved.

3. Improved Method

3.1. Overview

The method of Muhammad consider that disease genes will only accumulate on branches of biological processes, Named it biological process aggregation (BPA) methods. The full-branch aggregation (FBA) method in this paper believes that disease genes not only aggregate on biological process branches but also aggregate on molecular functions and cell component branches.

Figure 1 shows the steps of the FBA method. A gene functional similarity matrix is generated for a given data set. Gene ontology annotations for molecular functions and cellular components are no longer ignored when calculating gene functional similarities. Classifier training and testing on the gene functional similarity matrix. Classifiers from machine learning methods used: random forest, naive Bayes, support vector machine. Also, during training and prediction, the undersampling method is used to strictly keep the number of positive class (disease gene) and negative class (non-disease gene) consistent to avoid partial estimation. Each classifier is trained and tested repeatedly for 20 averaging.

3.2. Gene Ontology Semantic Similarity

Gene functional similarity calculated using gene ontology semantic similarity when constructing similarity matrix. This article uses four different semantic similarity method: resnik [10], rel [11], wang [12], Netsim [13]. The resnik and rel algorithms are based on information content. The information content of gene ontology terms is defined as follows:

$$\mathrm{IC(c)} = -\log\big(\mathrm{p(c)}\big) \tag{1}$$

The resnik method is defined as follows

$$\mathrm{Sim_{resnik}(t_1, t_2)} = \max_t \mathrm{IC(t)(t \in T)} \tag{2}$$

Where T is a set of common ancestor terms for t1and t2.

The rel method is defined as follows:

$$Sim_{rel}(t_1, t_2) = \max_t \frac{2 \times IC(t)}{IC(t_1)+ IC(t_2)} (t \in T) \tag{3}$$

Where T is a set of common ancestor terms for t1and t2.

Wang's method is based on semantic contribution. For a given term, calculate the semantic contribution of the ancestor node of the term to the node. The similarity of two terms is expressed by the semantic contribution of the common ancestor to the two terms. The formula is as follows:

$$Sim_{wang}(t_1, t_2) = \sum_t \frac{S_A(t)+ S_B(t)}{SV(t_1)+SV(t_2)} (t \in T) \tag{4}$$

Where T is a set of common ancestor terms for t1, t2.S_A(t) is the semantic contribution of the term t to the term t1.S_B(t) is the semantic contribution of the term t to the term t2.SV(t1) is the sum of the semantic contributions of all ancestor terms of the term t1 to t1. SV(t2) is the sum of the semantic contributions of all ancestral terms to t2 in the term t2.

Netsim is a newer method that combines GO, GO annotation and co-function data. The method performs better on metabolic pathways.

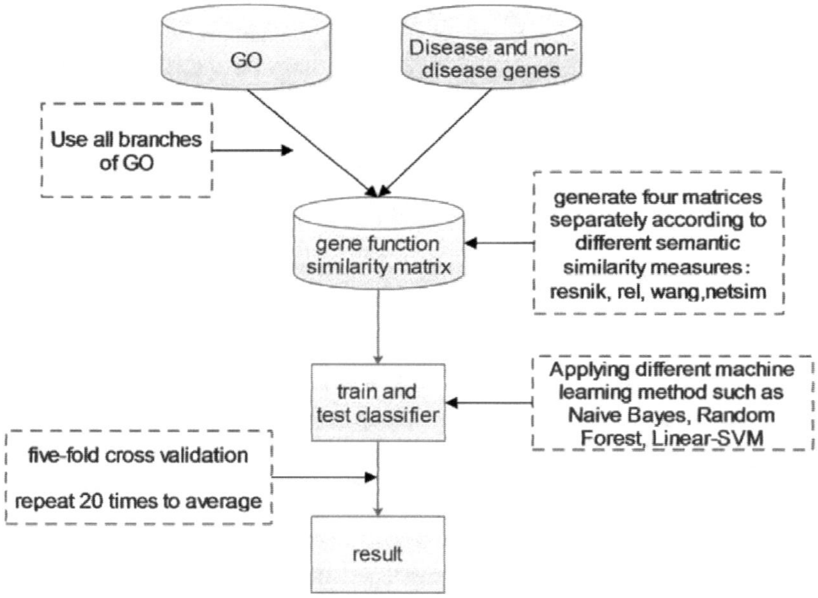

Figure 2. Step of proposed method to predict disease genes

3.3. Gene Function Similarity

The calculation methods of gene functional similarity are divided into two categories: the pairwise comparison method and overall comparison method. Since the overall comparison method does not pass the semantic similarity between gene ontology, the overall comparison method is not considered. In the pairwise comparison method, each gene corresponds to a set of gene ontology terms, and comparing the similarities of the two genes is equivalent to comparing the corresponding term sets. Common methods for integrating the similarity of two-term sets include average, maximum, and sum. This paper chooses the most widely used maximum method.

For a given two genes g1, g2. Through the GO annotation data, the corresponding term set T1, T2 is obtained. For example, in Figure 1, the set of terms corresponding to the gene ASSS is {"GO: 0005829", "Go: 0031965", "GO: 0005654", "GO: 0005813"}

The gene functional similarity of g1, g2 is defined as follows:

$$(g_1, g_2) = max\left((t_i, t_j)\right)\left(t_i \in T_1, t_j \in T_2\right) \quad (5)$$

3.4. Data Set

Disease gene of ASD was obtained from the Simons Foundation Autism Research Initiative (SFARI) gene database(https://gene.sfari.org/)

Genes are classified into 7 categories according to the degree of correlation with the disease in the SFARI data set. The most relevant, reproducible evidence is classified as a class 1 gene. Reproducible under certain constraints is classified as a class 2 gene. Class 3, 4 genes have small research evidence, 5 genes are indirectly related, and 6 genes are completely unrelated. There is another type of gene that does not indicate a score. In this study, only 746 genes of 1, 2, 3, and 4 genes were used, of which 1, 2 were used as High confidence Disease gene HD, a total of 87, 3, 4 As a low confidence disease gene (LD), a total of 649. A total of 1111 non-ASD genes were from Muhammad's paper [13].

In the experiment, by dividing the positive class (ie disease-related genes) into HD and LD, two sub-datasets were generated, one being HD as a positive class and NoASD as a negative class, which is called an HD data set. The other is HD+LD as a positive class and NoASD as a negative class called the HD+LD dataset.

Furthermore, due to the update of SFARI data, 8 of the LD genes intersected with non-ASD. These 8 genes were used as expired invalid data, and these 8 genes were deleted in both non-ASD and LD. The classification of the final gene is shown in Figure 2.

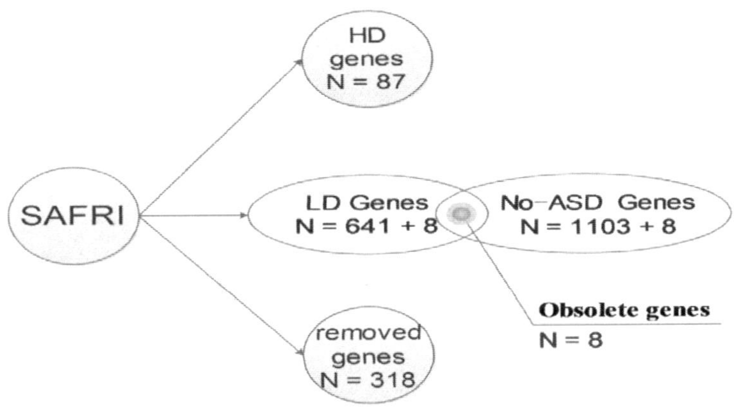

Figure 3. ASD data set gene classification

3.5. *Class Imbalance Problem*

There is a class imbalance in the dataset. The non-ASD gene is more than the ASD gene, especially in the HD data set, reaching a ratio of 87:1103. Therefore, this paper adopts a random undersampling method to solve the problem of class imbalance. At the time of training and testing, random undersampling of non-ASD genes, random acquisition of non-ASD gene construction training sets and test sets consistent with the number of ASD genes. For example, when performing HD gene training and testing, there are 87 genes in the normal HD gene, and the sample size of the negative class is also 87.

However, under-sampling introduces the risk of data specialization, it is easy for the model to learn only a part of the features. Therefore, the experiment was repeated several times and then averaged to avoid the problem of data specialization.

4. Experimental Result

Four different ontology similarity algorithms generate four matrices, which are classified according to HD and HD+LD. The results are shown in Table 1.

Table 1. Experiment result

		HD		HD+LD	
		BPA	FBA	BPA	FBA
Resnik	RF	0.793	0.908	0.759	0.897
	NB	0.701	0.707	0.667	0.718
	SVM	0.793	0.897	0.756	0.885
Rel	RF	0.753	0.891	0.787	0.897
	NB	0.660	0.741	0.655	0.747
	SVM	0.736	0.799	0.736	0.810
Wang	RF	0.787	0.914	0.793	0.897
	NB	0.690	0.724	0.701	0.747
	SVM	0.770	0.862	0.782	0.839
Netsim	RF	0.770	0.793	0.770	0.805
	NB	0.598	0.517	0.506	0.529
	SVM	0.672	0.615	0.684	0.615

In Muhammad's paper, the performance of HD datasets is significantly better than HD+LD, and the experimental data in this paper has the same performance. The BPA method based on resnik similarity has better performance of HD dataset than HD+LD on all classifiers. In rel, wang and netsim methods, the performance on the HD dataset is not much different from the performance of HD+LD.

There is no statistically significant difference in the performance of the classifier using the FBA method on the two data sets. This indicates that the full branch aggregation method has lower quality requirements than the biological process aggregation method.

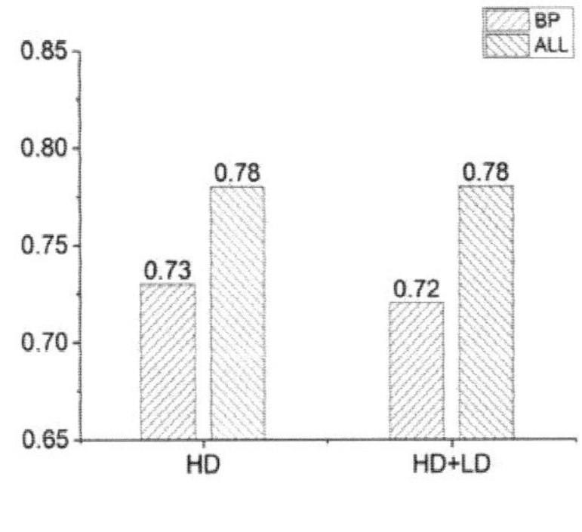

Figure 4. Average performance comparison of data sets

The average classification accuracy is shown in Figure 3 also shows that the HD performance of the algorithm for biological process aggregation is better than HD+LD, while the performance of full-branch aggregation is consistent.

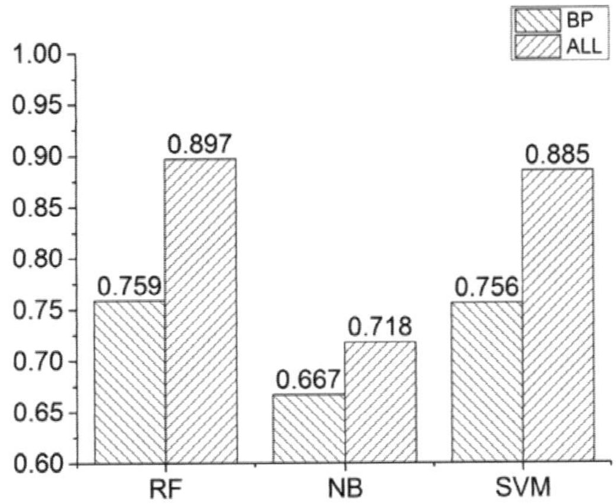

Figure 5. Classification result based on HD+LD in resnik method

The results of the four gene ontology similarity methods, resnik, rel, and wang all showed the same trend: whether in the HD dataset or the HD+LD dataset, the full branch aggregation is better than the biological process aggregation algorithm. The classification accuracy has a 10% improvement. Figure 4 is a comparison of the resnik algorithm on the HD+LD data set.

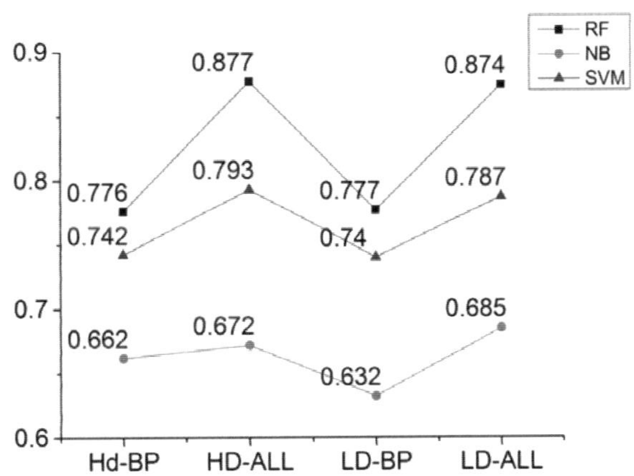

Figure 6. Average accuracy of classification of three classifiers

But the results based on netsim semantic similarity have a different performance. FBA on random forest classifiers has some improvement over BPA. However, there is no improvement in Naive Bayes and SVM classifiers, but it has declined. The main reason for this is that NETSIM uses other data sources. NETSIM not only considers GO and annotation, but also considers gene functional distance, and considers the similarity of gene function when calculating semantic similarity.

In this paper, three classifiers are used. The results are shown in Figure 5. In different cases, the RF performance is higher than SVM, and SVM is higher than NB. The best performing classifier is the RF classifier running on the HD dataset. The matrix is generated using wang's semantic similarity method and FBA with an accuracy of 91.4%.

5. Summary

The improved disease gene prediction method effectively improves the prediction accuracy of identifying ASD disease genes. Besides, disease gene prediction based on full-branch aggregation is superior to biological process branch aggregation. This indicates that disease genes not only accumulate in biological process but also aggregate on molecular functions and cellular components, which is also helpful for predicting other diseases in the future.

6. Acknowledgement

This work is supported by the Hunan Natural Science Foundation Youth Project(2019JJ50520)

7. References

[1] Geschwind D H , Flint J . Genetics and genomics of psychiatric disease[J]. Science, 2015, 349(6255):1489-1494.
[2] Sanders S J . First glimpses of the neurobiology of autism spectrum disorder[J]. Current Opinion in Genetics & Development, 2015, 33:80-92.
[3] Stephan Ripke etc al. Biological insights from 108 schizophrenia-associated genetic loci[J]. Nature, 2014, 511(7510):421-427.
[4] Libbrecht M W , Noble W S . Machine learning applications in genetics and genomics[J]. Nature Reviews Genetics, 2015, 16(6):321-332.
[5] Luo P, et al. Identifying disease genes from PPI networks weighted by gene expression under different conditions[C] //2016 IEEE International Conference on Bioinformatics and Biomedicine (BIBM). IEEE, 2016: 1259-1264.
[6] Radivojac P , et al. An integrated approach to inferring gene-disease associations in humans.[J]. Proteins-structure Function & Bioinformatics, 2010, 72(3):1030-1037.
[7] Krishnan A , Zhang R , Yao V , et al. Genome-wide prediction and functional characterization of the genetic basis of autism spectrum disorder[J]. Nature Neuroscience, 2016. (2016): 1454-1462.
[8] Asif M, et al. Identifying disease genes using machine learning and gene functional similarities, assessed through Gene Ontology[J]. PloS one, 2018, 13(12): e0208626.
[9] Voineagu I, Eapen V. Converging pathways in autism spectrum disorders: interplay between synaptic dysfunction and immune responses[J]. Frontiers in human neuroscience, 2013, 7: 738.
[10] Resnik P. Semantic similarity in a taxonomy: An information-based measure and its application to problems of ambiguity in natural language[J]. Journal of artificial intelligence research, 1999, 11: 95-130.
[11] Schlicker A, et al. A new measure for functional similarity of gene products based on Gene Ontology[J]. BMC bioinformatics, 2006, 7(1): 302.
[12] Wang J Z, et al. A new method to measure the semantic similarity of GO terms[J]. Bioinformatics, 2007, 23(10):1274-1281.
[13] Peng J, Uygun S, Kim T, et al. Measuring semantic similarities by combining gene ontology annotations and gene co-function networks.[J]. Bmc Bioinformatics, 2015, 16(1):1-14.

IOP Conf. Series: Materials Science and Engineering **719** (2020) 012025 doi:10.1088/1757-899X/719/1/012025

Research on Single-Input Multiple-Output Converter for Rotorcraft Unmanned Aerial Vehicle

Yifan Gan[1,2] and Chunxiao Xu[1]

1. Institute of Electrical Engineering, Chinese Academy of Sciences, Beijing, 100190, China
2. University of Chinese Academy of Sciences, Beijing, 100049, China
Email: ganyifan@mail.iee.ac.cn

Abstract. To satisfy the different voltage requirements of different power units in small rotorcraft unmanned aerial vehicle (UAV), we investigated two different single-input and multiple-output converters, three-output flyback converter and single inductor three-output Buck converter. The hardware components and operating principles of the two single-input and multiple-output converters were analyzed, and the control strategies of them were designed. Furthermore, operations of the two single-input multiple-output converters were simulated using Simulink in Matlab. We found that the three-output flyback converter can be used in the design of UAV when its components are settled and the voltage quality requirements are not strict. When the expansibility of component and high-stable voltage are required for the design of UAV, the single inductor three-output Buck converter should be used.

1. Introduction

Due to its low cost, high cost-effectiveness, strong survivability, mobility, ease of use, strong adaptive capacity to environment, and no risk of injury, small rotorcraft unmanned aerial vehicle (UAV) was widely used in transport, detection, investigation, fighting against terrorism, communication, etc. The application area of small rotorcraft UAV will be more and more widely with the increase in UAV performance [1-4], which will result in the increase of electrical units in UAV. However, the voltage of battery carried by the small rotorcraft UAV is constant, and the voltage required for each power unit may be different. Therefore, a single-input multi-output (SIMO) DC-DC converter that can convert a single voltage to multiple voltages is required[4].

In research of single-input multi-output DC-DC converters, the shunt neutralization of single-input single-output (SISO) DC-DC converters was proposed. The combination of multiple SISO converters with different voltages could produce various voltage outputs. However, this led to the increase of system cost and control complexity[5]. SIMO converters can reduce system complexity and make control easier, and were widely used in renewable energy systems and electric vehicles[6, 7]. In research of topology and control of SIMO converter, a shared ZCS (Zero-Current-Switch) was used to lag the branch topology and soft-switching technology to reduce loss. However, the circuit of SIMO converter has three full-bridge converters, which increased component density and cost, reduced efficiency, and was not suitable for small rotorcraft UAV[8]. To reduce the number of switching tubes and the switching loss, a new SIMO Buck converter was proposed. However, it was still has multiple inductors in the circuit topology, which led to complexity of the circuit structure and control strategy[9]. Cavalcanti et al. calculated the switching loss of a single-inductance three-output DC-DC converter model, and analyzed the relation between the power loss of IGBT and circuit and the operating parameters for the direction of specific application [10].

To identify suitable SIMO converter for small rotorcraft UAV, we improved the topology and control strategy of a single-inductor multi-output DC-DC converter, and compared it with a three-output flyback converter.

2. Topology and Control Strategy of Three-output Flyback Converter

The flyback converter that has a simple circuit structure, a wide input voltage range, low circuit loss and stable output is suitable for low power multi-output requirement (Fig. 1). [11]. When IGBT is turned on, the DC power supplies power to the primary winding of the transformer; when IGBT is turned off, the power of the primary winding of the transformer is transferred to the secondary winding, and the load resistance is supplied through the freewheeling diode D and the filter capacitor C.

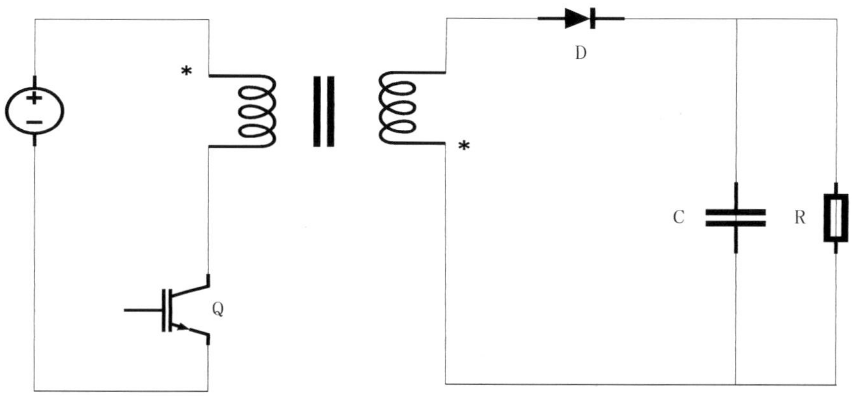

Figure 1. Basic structure of flyback converter

In steady state, the average voltage of the inductor in one cycle is 0 according to the volt-second principle.

$$V_{in}t_{on} + (-\frac{N_1}{N_2}V_o)(T_s - t_{on}) = 0 \tag{1}$$

Here, V_{in} is the input voltage, V_o is the output voltage, t_{on} is the on-time of the switch, T_s is a cycle time, $\frac{N_1}{N_2}$ is the turns ratio of the primary and secondary sides of the transformer.

Therefore, the transformer ratio of flyback converter can be obtained as follows:

$$M = \frac{V_o}{V_{in}} = \frac{N_2}{N_1}\frac{t_{on}}{T_s - t_{on}} \tag{2}$$

Another feature of the flyback converter is that it can add multiple coils to the secondary winding to generate multiple outputs according to the user's requirement. In our research, a three-output flyback converter was used for further analysis (Fig. 2)[12]. The operating principle of this three-output flyback converter is similar to that of a single output flyback converter.

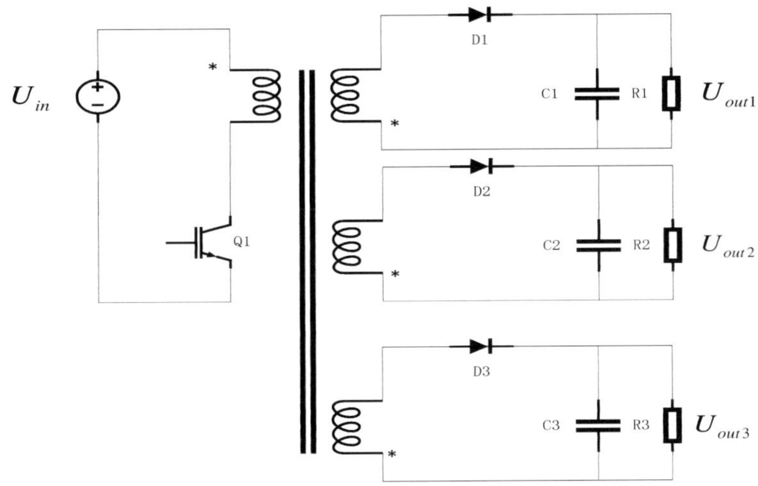

Figure 2. Three-output flyback converter

Similarly, the transformer ratio of the three-output flyback converter can be obtained as follows:

$$\begin{aligned} M_1 &= \frac{U_{out1}}{U_{in}} = \frac{N_1}{N_0}\frac{t_{on}}{T_s - t_{on}} \\ M_2 &= \frac{U_{out2}}{U_{in}} = \frac{N_2}{N_0}\frac{t_{on}}{T_s - t_{on}} \\ M_3 &= \frac{U_{out3}}{U_{in}} = \frac{N_3}{N_0}\frac{t_{on}}{T_s - t_{on}} \end{aligned} \tag{3}$$

Here, M_1, M_2 and M_3 are transformer ratios of different branches; N_0 is the number of turns of the transformer primary winding; N_1, N_2 and N_3 are numbers of turns of the different branches of the transformer secondary winding; U_{in} is the input voltage; U_{out1}, U_{out2} and U_{out3} are output voltages; t_{on} is the switch on time; T_s is a cycle time.

In the three branches, the duty ratios of IGBT are the same. Therefore, only one feedback loop is needed, which greatly simplifies the control loop. According to the above analysis, we designed the control strategy of three-output flyback converter as follows: In the three outputs of the converter circuit, the branch that has the feedback loop was defined as the control branch, and other branches were defined as follower branches. The reference voltage was set in the control branch, and the on/off of IGBT was controlled through the appropriate output of duty ratio from the feedback loop. When the duty ratio of IGBT is constant, the given voltage of the branch output is determined by the turns ratio of the primary winding and the secondary winding of the transformer judged from formula (3). Once the hardware circuit of converter is determined, only the voltage of the control branch can be changed. Voltages of the following branches can only be changed proportionally with the change of control branch voltage.

3. Topology and Control Strategy of Single Inductor Three Output Buck Converter

The basic structure of a single-inductor three-output Buck converter is shown in Fig. 3. [13]. When the IGBT Q_{11} is turned on and the freewheeling diode D_{11} is turned off, the input battery then charges the storage inductor L_{11}, the filter capacitor C_{13} and the load 3. Meanwhile, the filter capacitor C_{11}, the filter capacitor C_{12}, the load 1 and the load 2 are charged when the IGBT Q_{12} and the IGBT Q_{13} are turned on. When the IGBT Q_{11} is turned off and the freewheeling diode D_{11} is turned on, the energy storage inductor L_{11} charges the filter capacitor C_{13} and the load 3. At the same time, the filter capacitor C_{11}, the filter capacitor C_{12}, the load 1 and the load 2 are charged when the IGBT Q_{12} and the

IGBT Q_{13} are turned on. According to the secondary volt balance, the transformer ratio is:

$$M = \frac{U_{out}}{U_{in}} = \frac{t_{on}}{T_s} \tag{4}$$

Here, U_{in} is the input voltage, U_{out} is the output voltage, t_{on} is the switch on time, T_s is a cycle time.

According to the above analysis, the following control strategy was set as follows: the branch of the load 3 was set as the main control branch, and the reference value of output voltage was set as U_{ref3} to ensure that U_{ref3} was the maximum output voltage. U_{ref3} and the actual output voltage U_{out3} were input into the corresponding PID controller, and then the PID controller output to the PWM generating device to control the on/off of the IGBT Q_{11} of the main control branch, and output the required waveform. The load 1 and load 2 branches were follower branches, and the reference values of output voltages of U_{ref1} and U_{ref2} were set by the comparison control method. The actual values of output voltages of U_{out1} and U_{out2} were compared with U_{ref1} and U_{ref2}, respectively When U_{out1} and U_{out2} were less than U_{ref1} and U_{ref2}, the IGBT of the corresponding branch was turned on. When U_{out1} and U_{out2} were greater than U_{ref1} and U_{ref2}, the IGBT of the corresponding branch was turned off. These could ensure the simplicity and feasibility of the control method. Moreover, the control loop was reduced greatly.

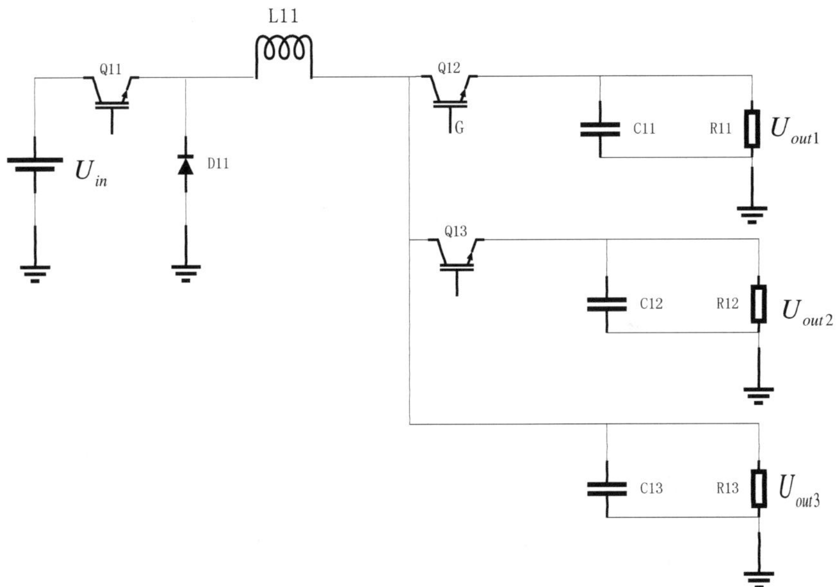

Figure 3. Basic structure of single inductor three output Buck converter

4. Simulation of Single-Input Multi-Output Converters

The three-output flyback converter and the single-inductor three-output Buck converterwere simulated according to the actual situation of the small rotorcraft UAV system. The simulation tool was Simulink in Matlab. The simulation requirements were as follows: the input voltage was 24V; three output voltages were 3.3V, 9V and 12V, respectively; the required voltage fluctuation did not exceed 5%; the voltage fluctuation was required to be less than 5% when any load changed.

4.1. Simulation of the Flyback Converter with Three Outputs

The main branch of the three-output flyback converter was controlled by voltage plus PIDwhose parameters were 30, 0.1 and 0, respectively. The follower branches were open loop controlled. The load was set as the resistive load with 10Ω to verify the stability of the circuit when the load changed, after 10s, the 10Ω resistive load was connected in parallel to the main control branch to reduce the

IOP Conf. Series: Materials Science and Engineering **719** (2020) 012025 doi:10.1088/1757-899X/719/1/012025

load. The simulation circuit was shown in Fig. 4.

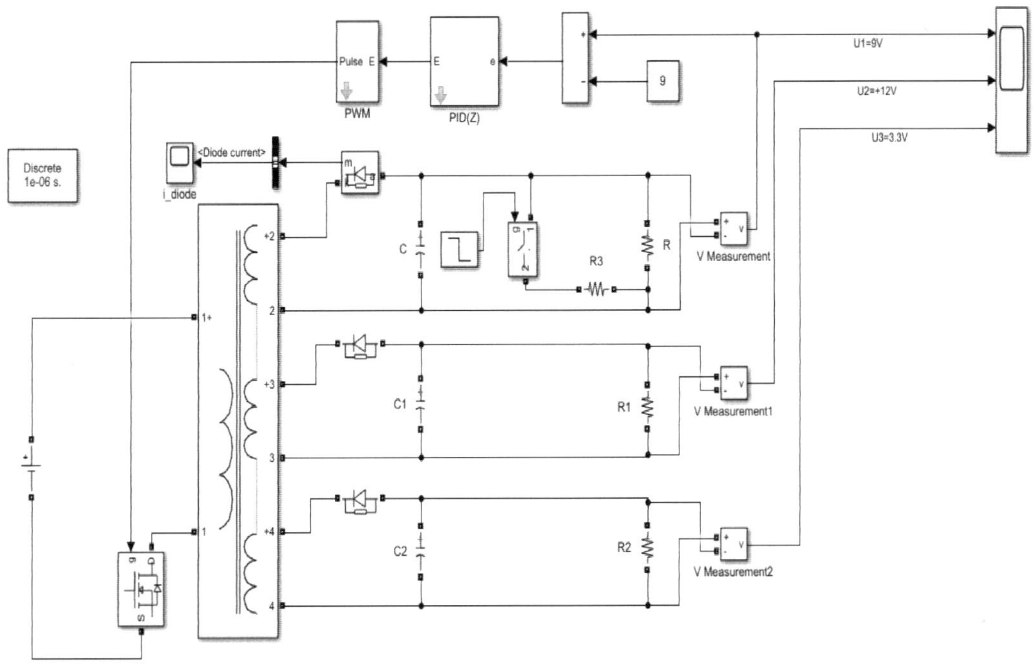

Figure 4. Three-output flyback converter simulation circuit

The simulation results were shown in Fig. 5. The three outputs of the three-way output flyback converter were overshooted, which were larger than 50% of the required output voltages before 0.15s. When the load changed abruptly at 0.5s, the output voltages did not change significantly. After 0.15s, the outputs were stable, and the voltage fluctuation of the three-output flyback converter did not exceed 5%. Before 0.5s, the output voltage of the main control branch was among the range of 8.8V~9.2V. The fluctuation was approximately 2.2%. In the follower loops, the output voltage of the branch with 12V reference voltage was between 11.8 and 12.2V with a fluctuation of approximately 1.7%. The output voltage of the other branch with 3.3V reference voltage was between 3.3V and 3.4V with a fluctuation of approximately 3%. After 0.5 s, the voltage fluctuations became smaller after the load was reduced, which satisfied the requirements.

IOP Conf. Series: Materials Science and Engineering **719** (2020) 012025 doi:10.1088/1757-899X/719/1/012025

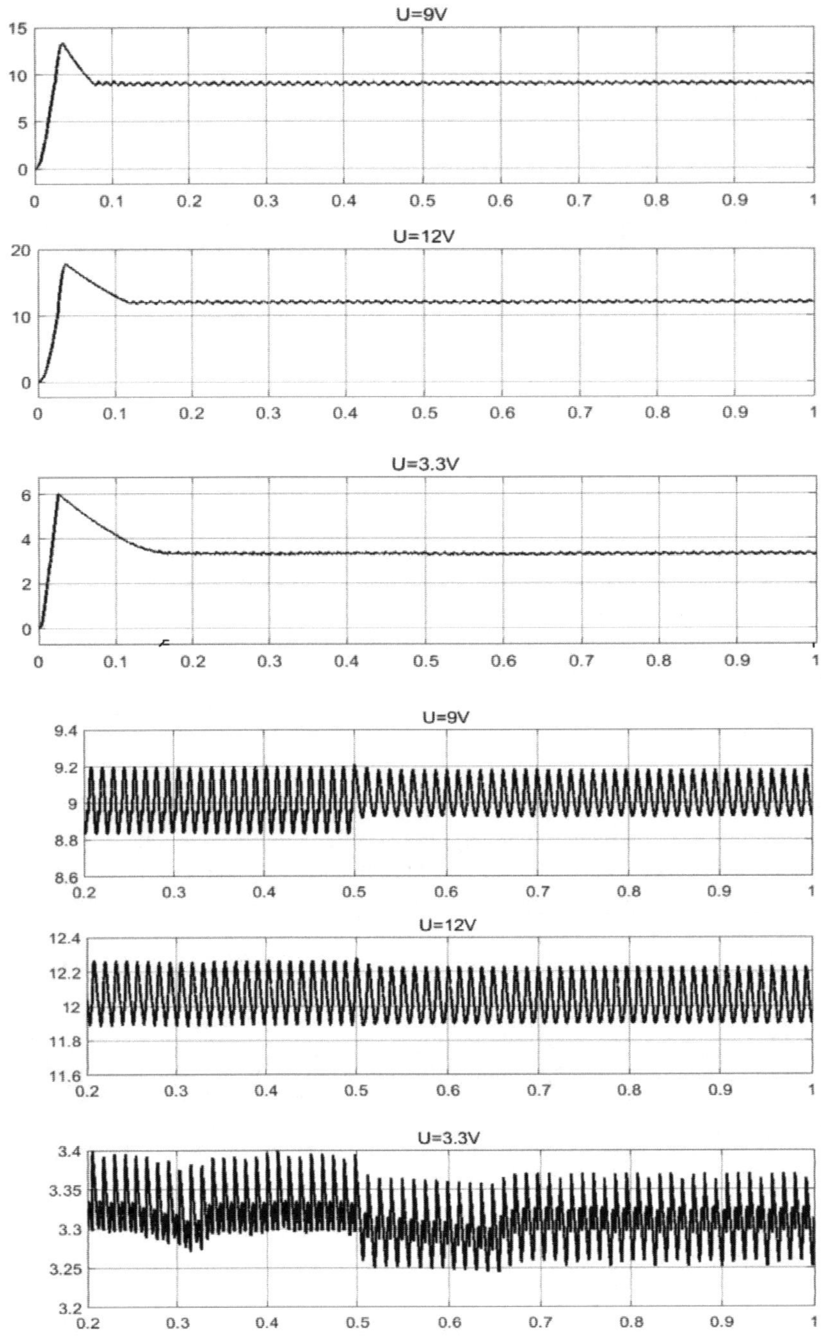

Figure 5. Three-output flyback converter output with partial enlargement

4.2. Simulation of Single Inductor Three Output Buck Converter

In the single-inductance three-output Buck converter, the branch with 12V reference voltage output was set as the main control branch, which was controlled by voltage feedback plus PID. The PID parameters were 150, 0.1 and 1. The branches with 9V and 3.3V reference voltage outputs was set as follower branches, which were controlled by comparison control. The load was set as a resistive load of 10 Ω. A 10 Ω resistive load was in parallel with the main control branch after 0.5 s. The simulation circuit was shown in Fig. 6.

IOP Conf. Series: Materials Science and Engineering **719** (2020) 012025 doi:10.1088/1757-899X/719/1/012025

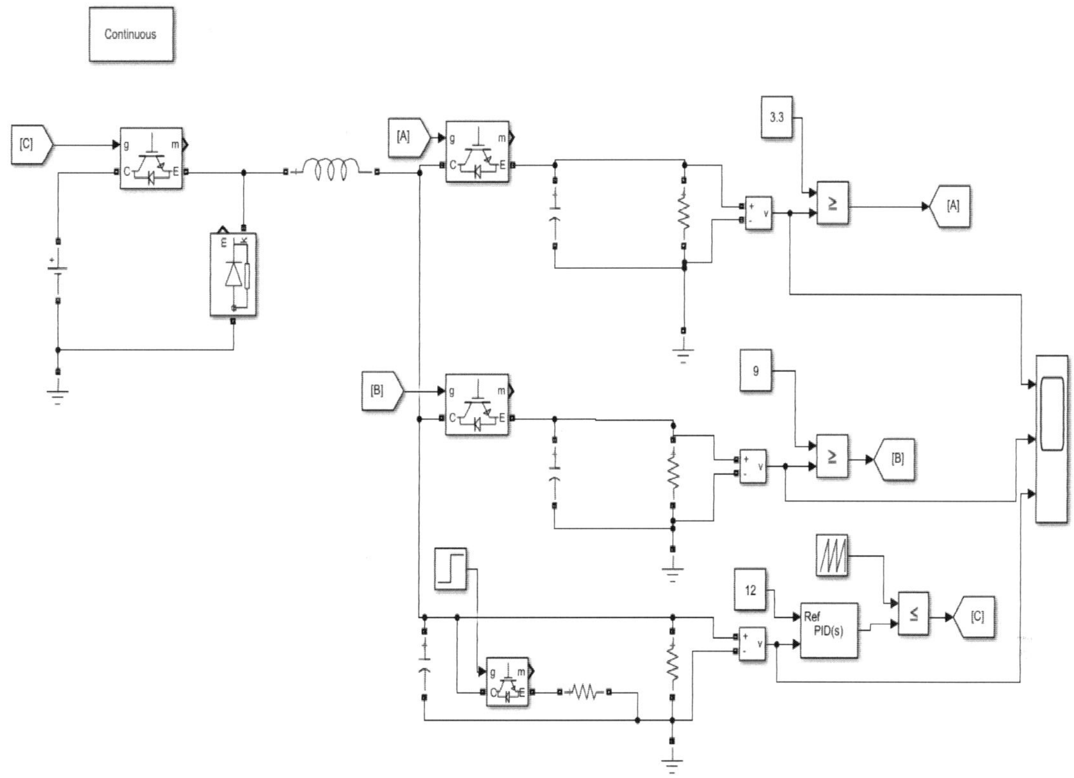

Figure 6. Single inductor three-output buck converter simulation circuit

The simulation results were shown in Fig. 7. The overshoots of the two follower branches were extremely small, while the overshoot of the main control branch was larger than twice of the reference voltage. After 0.05s, the voltages of all branches were stable and the adjustment time was short. When the 0.5s load was reduced, almost no change was found in the output voltage. When the voltage outputs after 0.05s were amplified, we found that fluctuations of the voltage outputs of the three branches did not exceed 5% of the reference voltages. The output voltage of the branch with 3.3V reference voltage was 3.3 V~3.35V, and the fluctuation was about 1.5%. The voltage output of the branch with 9V reference voltage was between 9V~9.1V, The fluctuation was about 1.1%. The actual output value of the main control branch was between 11.7V~12.3V. The voltage fluctuation was about 2.5%. After 0.5s, the voltage fluctuation of the main control loop increased to about 4.2%.

IOP Conf. Series: Materials Science and Engineering **719** (2020) 012025 doi:10.1088/1757-899X/719/1/012025

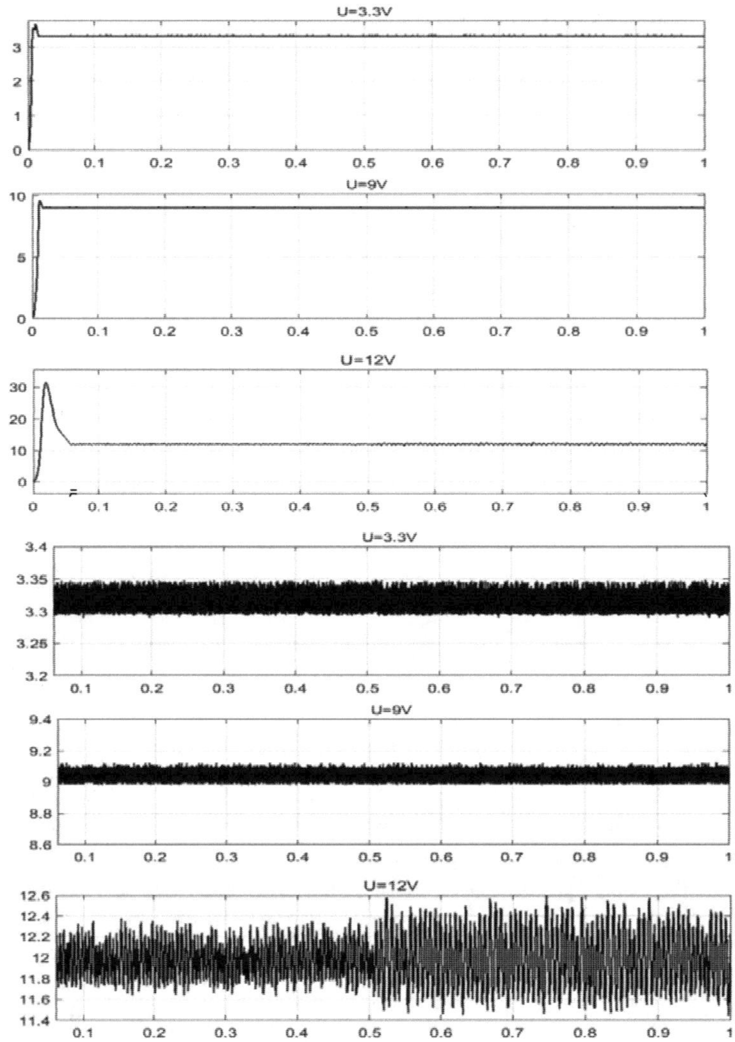

Figure 7. Single inductor three output Buck converter output with partial enlargement

4.3. Comparison of Simulation Results

The follower branch of the single-inductance three-output Buck converter had a very small overshoot which only oscillated at a high frequency above the reference voltage compared with that of the three output flyback converter (Fig.5, Fig.7). However, the overshoot of the three output flyback converter oscillated around the reference voltage with a low frequency. The main control branch of the single-inductance three-output Buck converter had a larger overshoot and a larger voltage oscillation compared with those of the three output flyback converter. The overshoot oscillated around the reference voltage with a high frequency. However, the adjustment time is shorter than 0.05s. The quality of voltage output of the main control branch in the three output flyback converter was higher than that of the single-inductance three-output Buck converter.

The improvement of the voltage output quality of the single-inductor three-output buck converter can be achieved by increasing the follower branch with better voltage output. The main control branch can be connected to a smaller load to improve the quality of voltage output, though this will sacrifice a small amount of power and heat dissipation.

5. Conclusion

The three output flyback converter, which has fewer components, simple circuit structure and wide voltage output, can meet the power supply that the output voltage is higher than the input voltage. However, the voltage output quality of the three output flyback converter is poor compared to the single-inductor three-output Buck converter. In addition, once the circuit is completed, only the reference voltage of the main control branch in the three output flyback converter can be adjusted. The reference voltages of the follower branchs can only be changed proportionally. Furthermore, the power interface cannot be reserved. Therefore, the three output flyback converter is only suitable for the design of small rotorcraft UAV with shaped component and low voltage quality requirement. If extensibility is required for the design of a small rotorcraft UAV, the three output flyback converter is difficult to power the functional components.

The voltage output quality of the single-inductor three-output Buck converter is high. Moreover, the single-inductor three-output Buck converter is easy to expand, and the reference voltage of branch in the single-inductor three-output Buck converter is easy to adjust. Therefore, when the single-inductor three-output Buck converter is used for the design of a small rotorcraft UAV, it can meet the power demand of various functional components in the UAV at the same time.

6. References

[1] G. Ludeno, I. Catapano, A. Renga, A. R. Vetrella, G. Fasano, F. Soldovieri. Assessment of a micro-UAV system for microwave tomography radar imaging[J], Remote Sensing of Environment, 2018, 212:90-102.
[2] E. J. Lee, S. Y. Shin, B. C. Ko, C. Chang. Early sinkhole detection using a drone-based thermal camera and image processing[J], Infrared Physics & Technology, 2016, 78:223-232.
[3] Selvaganapathy S, Ilangumaran A. Design of Quadcopter for Aerial View and Organ Transportation Using Drone Technology [J]. Asian Journal of Applied Science and Technology (AJAST), 2017, 1(3): 311-5.
[4] Hassanalian M, Abdelkefi A. Classifications, applications, and design challenges of drones: a review [J]. Progress in Aerospace Sciences, 2017, 91:99-131.
[5] Deshpande N, Chavhan RD, Deshpande N. Coupled inductor based single-input multiple-outputs boost converter; proceedings of the 2017 Third International Conference on Advances in Electrical, Electronics, Information, Communication and Bio-Informatics (AEEICB), F, 2017 [C]. IEEE.
[6] Ghodke DV, Chatterjee K, Fernandes BG. Modified soft-switched three-phase three-level DC–DC converter for high-power applications having extended duty cycle range [J]. IEEE Transactions on Industrial Electronics, 2011, 59(9): 3362-72.
[7] Ouyang Z, Zhang Z, Thomsen OC, et al. Planar-integrated magnetics (PIM) module in hybrid bidirectional DC–DC converter for fuel cell application [J]. IEEE Transactions on Power Electronics, 2011, 26(11): 3254-64.
[8] Chen Y, Kang Y, Nie S, et al. The multiple-output DC–DC converter with shared ZCS lagging leg [J]. IEEE Transactions on Power Electronics, 2010, 26(8): 2278-94.
[9] Sabbarapu BK, Nezamuddin O, McGinnis A, et al. Single-input multiple-output synchronous DC-DC buck converter; proceedings of the 2016 IEEE Energy Conversion Congress and Exposition (ECCE), F, 2016 [C]. IEEE.
[10] CAVALCANTI M C, DA SILVA E, BOROYEVICH D, et al. A feasible loss model for IGBT in soft-switching inverters; proceedings of the IEEE 34th Annual Conference on Power Electronics Specialist, 2003 PESC'03, F, 2003 [C]. IEEE.
[11] Wang CM. A novel ZCS-PWM flyback converter with a simple ZCS-PWM commutation cell [J]. IEEE Transactions on Industrial Electronics, 2008, 55(2): 749-57.
[12] El Kattel MB, Mayer R, Oliveira SVG, et al. Three-phase flyback/current-fed push-pull dc-dc converter with y-Δ connected transformer; proceedings of the 2015 IEEE 13th Brazilian Power Electronics Conference and 1st Southern Power Electronics Conference (COBEP/SPEC), F, 2015 [C]. IEEE.
[13] Patra P, Patra A, Misra N. A single-inductor multiple-output switcher with simultaneous buck, boost, and inverted outputs [J]. IEEE Transactions on Power Electronics, 2011, 27(4): 1936-51.

A BCI System with Motor Imagery Based on Bidirectional Long-Short Term Memory

Jzau-Sheng Lin and Bing-Hong She

Dept. of Computer Science and Information Engineering, Nat'l Chin-Yi University of Technology, Taichung, Taiwan, R.O.C
Email: jslin@ncut.edu.tw and tombear07@gmail.com

Abstract. A bidirectional long short term memory (BiLSTM) neural network was embedded into a brain-computer interface (BCI) system based on motor-imagery (MI) in this paper. The MI-based electroencephalogram (EEG) signals were used to recognize different imagery actions. The dynamic characteristics of MI signals in EEG are usually low signal-to-noise ratio as non-stationary time series. A lot of strategies have been proposed to clustering MI-EEG signals. However they are not considering the concept of series features of the signal in time domain with forward and backward manners, so the recognition results are not promising. The discrete wavelet transform (DWT) was also used to get the frequency feature from transforming each channel of MI-EEG in this paper. Then the proposed BiLSTM is proposed as a classifying system to identify the MI-EEG data. BiLSTM can extract dependencies of different time points by each recurrent unit with an adaptive manner. Besides the forward manner of time series signals in the LSTM unit, the BiLSTM also puts the output signals into previous layers with backward manner. The BiLSTM system can get more promising results in the classification of MI-EEG than those obtained by other strategies shown as in experimental results.

1. Introduction

The BCI system is a different way of intercourse through EEG signals to support one of the most important aspects. The response of brain wave is translated into activities by using of electroencephalogram (EEG) signals extracted from electrodes contacting on scalp in a BCI system. These EEG signals can be processed to provide a communication channel through system's hardware and software to control several systems such as computer games, electric wheelchair and so on. Being one of a topic in the research field of BCI, motor imagery (MI) emulates a given mental activity, e.g., imaging the motions of the legs [1]. MI refers to the visualization of any moving action which responses the various voltage changes in the connectivity between neurons in the cortex without any actual executing activity that is in either an event related desynchronization (ERD) or event-related synchronization (ERS) of mu rhythms. These effects are resulted from the change of intrinsic membrane properties of local neurons, the change in strength between the interconnections or the change of micro voltage in the chemical synapses between neurons. A lot of strategies were used to control electric wheelchair. An MI-based BCI system to control electric wheelchair was constructed by Lin and Lo [2]. In 2017, Sreeja et al. [3] presented two different techniques and modeling using Gaussian Naive Bayes (GNB) classifier for the purpose of selecting optimal features to construct a machine-learning classification platform for MI-EEG signals. They proved that their strategy can provide improved accuracy than LEA and SVM methods. A classification framework and data reduction method for classifying MI-EEG signals was proposed by Guan et al. [4] in 2019. They used the manifold of covariance matrices in a Riemannian perspective through decision tree framework with filter geodesic minimum distance to recognize MI tasks.Xie et al. [5] proposed a symmetric positive-definite (SPD) covariance matrix for EEG signals to convey important discriminant information for the MI BCI system in 2016. In 2015, some classification strategies for MI-Based BCI system were surveyed by Jois et al. [6]. They pointed out that general features such as band power values in the EEG signals can be extracted by proper classification methods like neural networks, SVM or ensemble classifiers. The different classifiers can obtain different performances and are compared to find the better techniques for using equal number of features. They proved that the neural network strategies can get the most efficient performance. One barrier of the general neural networks for their wide application is the initial conditions must be set carefully. Generally, small initial value could result in the multilayer network un-trainable due to weight diffusion, and large weight values could make wicked local minima [7]. For the purpose of resolving this obstacle and organize higher promising neural networks, new novel model of techniques and methodologies, called deep learning (DL), have been effectively proposed and become dominant in some areas [8]. Recently, one of models in DL such as recurrent neural networks (RNNs) have been certified that they can obtain promising results in many research field [9] especially in time-sequence processing with variable-length input/output. For the problems in EEG signals classification, Petrosian et al. [10] used wavelet transform to extract the compact features and RNN to classify EEG signals into several categories. Because the scalp EEG containing external noises causes not promising results, the RNN is not suitable to classify EEG signals from scalp. Several researchers have proved that the RNN network can results in gradient explosion and gradient vanished when the weights are updated layer by layer. The RNNs with Long short-term memory (LSTM) [11] using the time-series features of signals is an effective deep learning model in several sequential-data applications. Not only the problems in RNN such as gradient explosion and gradient vanished can be solved but also the long time information can be stored by the memory cell in the LSTM-based RNN. The gating mechanism is organized in the LSTM to avoid vanishing gradients. In addition to occupy the LSTM unit, BiLSTM also feedback transfers the output into previous layers and sequence data with variable-length manner can be captured by each recurrent unit adaptively.

In this paper, we proposed a platform that the DWT was used to extract compact data and the BiLSTM was applied to classify the EEG signals. Each channel, the MI-EEG signalt is extracted and converted by a DWT with an effective time-frequency characteristics. Then we calculated the average power spectrum of MI-EEG signals and determined the effective time segment. The experimental results showed that BiLSTM method can make full use of the time-frequency information of MI-EEG as well as time sequence information, and can get promising classification results.

The organization of this paper is listed as follows. The system architecture is described in Section 2, Section 3 discusses the DWT. The LSTM recurrent network is presented in Section 4. The architecture of BiLSTM is shown in Section 5. Section 6 discusses the classification accuracies obtained by the BiLSTM compared with other strategies. Finally, the conclusion is given in Section 7.

2. System Architecture

The Emotiv EPOC chip and g.tec dry electrodes were combined an extracting subsystem to extract EEG signals from electrodes located on C3, C4 and Cz in the proposed BCI system. MI brain signals including right and left hand actions were recognized in the proposed MI-EEG based BCI system respectively. For the purpose of establishing a sampling model, The DWT is used to transform the extracted brainwave signal to get the spectrums in frequency domain. Then the BiLSTM is used to classify the features with frequency manner into different labels. The NVIDIA Jetson TK1 was also embedded in the proposed platform in order to promote the calculating speed. Additionally, the Bluetooth 2.1 was also used to transfer to EEG signals from extracting subsystem to computer with wireless manner. Figure 1 shows the proposed BCI platform.

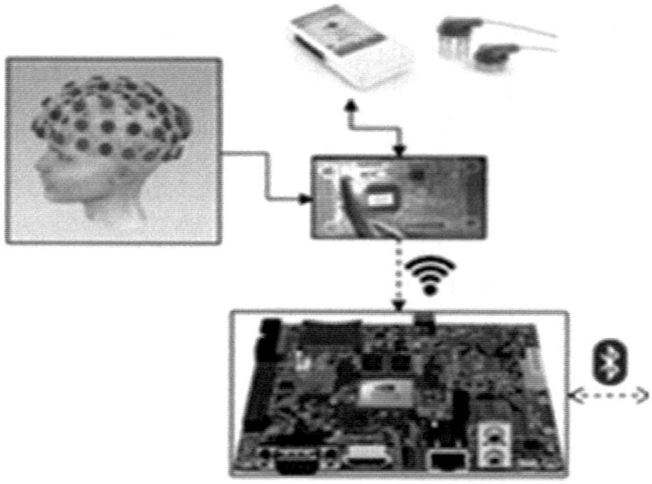

Figure 1. The proposed BCI platform

3. Wavelet Transform

In 1981 Jean Morlet proposed the strategy of wavelet. In applications of noise filtering, digital signal analysis, and signal compression and so on, the wavelet is always used. Several series db with Daubechies wavelet [12] can obtain better results in signal analysis. db4 wavelets were used in order to capture main features from EEG signals in this paper. The single-resolution WT is not easy to get detailed features when a signal has an altitude-manner diversity in a proper field. Instead, the multi-resolution methodology can decompose the lower-degree signals to obtain more message. Therefore, the signals with lower frequency can be continuously decomposed to show more characteristics. However, a lot of decomposition iterations of the signal can results in few number of samples that make less obvious signal features. Therefore, the number of layers for the decomposition of signals is limited. A high-pass filter and a low-pass filter are respectively used to transform the original signal in wavelet decomposition. The consistency of the original signal is retained in the low-pass filter while the variability of the original data is reversed high-pass filter. The wavelet and scale functions can be combined with the DWT transform. It has a lower-frequency resolution and a higher-time resolution in the high frequency. And in the low frequency part, it occupies a high frequency resolution and low temporal resolution.

4. LSTM Recurrent Network

The RNNs are called recurrent module because it performs a work for every element of sequential signals with the output being depended on the previous states in the hidden layers. Being popular networks, the RNNs have been shown great promise in many sequence-manner tasks. More complex types of RNNs have been created to resolve several problems recently. Same as the conventional the neural networks, the training process of RNNs have difficult learning process with long-term dependencies owing to the problems of gradient vanishing and exploding. Without a fundamentally different organization from RNNs, the LSTMs use a different manner to update the states in hidden layer. The memory cell in the LSTMs can be worked as a black box that input the previous state and current input data. The data in memory will be internally decided what to be kept in or erased from memory by this cell. Then the input data, the current memory, and the previous state to be combined to construct a neuron node. It is proved that these neuron node can capture long-term dependencies with a very efficient manner. The architecture of the LSTM is shown as in Figure 2. Eq. (1) shows the state of forget gate f_t which is calculated by a sigmoid function from previous cell state c_{t-1}, previous hidden layer state h_{t-i} and input data x_t .

$$f_t = \sigma(w_{c,f}C_{t-1} + w_{x,f}x_t + w_{h,f}h_{t-i}) + b_f \tag{1}$$

The function of cell state can be derived in Eq. (2), It is calculated by the forget-gate state f_t, previous cell state c_{t-1}, and $i_t * \tilde{C}_t$.

$$C_t = f_t * C_{t-1} + i_t\tilde{C}_t \tag{2}$$

where

$$i_t = \sigma(w_{c,i} * C_{t-1} + w_{x,i} * x_t + w_{h,i} * h_{t-1} + b_i) \tag{3}$$

and

$$\tilde{C}_t = \tanh(w_{x,c} * x_t + w_{h,c} * h_{t-1} + b_c) \tag{4}$$

Then, the states of hidden layer and output gate can be calculated by Eq. (5) and Eq. (6), respectively.

$$h_t = o_t \tanh(C_t) \tag{5}$$

$$o_t = \sigma(w_{c,o} * C_t + w_{x,o} * x_t + w_{h,o} * h_{t-1} + b_o) \tag{6}$$

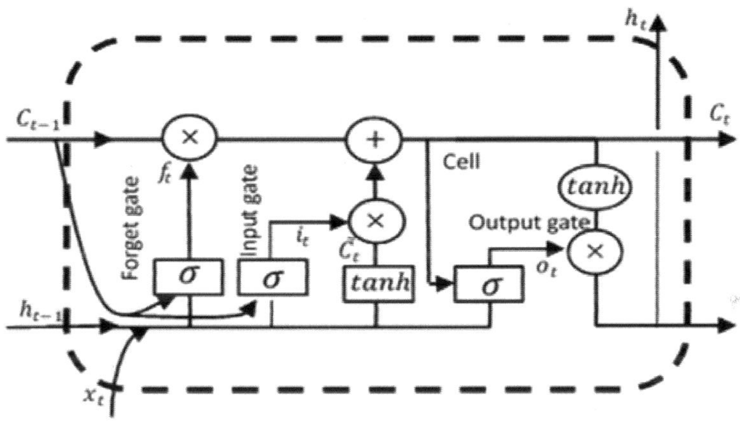

Figure 2. Long-Short term Memory

5. Bidirectional LSTM

A recurrent neural network (RNN) can iteratively compute the hidden vector sequence h_t and output vector sequence y_t for an input sequence x_t with Eqs. (7) and (8).

$$h_t = \sigma(w_{xh}x_t + w_{hh}x_{t-1} + b_h) \tag{7}$$

$$y_t = w_{hy}h_t + b_y \tag{8}$$

where w is the weight matrices, b denotes the bias vectors and σ is the sigmoid function in hidden layer. LSTMs preserve information from inputs that has already passed through it using the hidden state because the only inputs it has seen are from the past. In order to improve model performance on sequence classification problems, LSTMs can be extended as BiLSTMs. Instead of one direction in LSTMs, BiLSTMs are a modification of the conventional LSTMs to process sequences of signals in both directions, one from past to future and the other from future to past. This can provide additional context to the network and result in fuller learning on the problem.

IOP Conf. Series: Materials Science and Engineering **719** (2020) 012026 doi:10.1088/1757-899X/719/1/012026

As shown in Figure 3, combine RNN and LSTM to construct a BiLSTM which uses past and future states to access long-range context in both directions to predict the imaging action of a given EEG signal. As illustrated in Figure 3, the states of forward hidden sequence and the states of backward hidden sequence are calculated by the BiLSTM iteratively. The output sequence y from the backward layer and the forward layer then can be updated. They are derived as following equations.

$$\vec{h}_t = \sigma(w_{x\vec{h}} x_t + w_{\vec{h}\vec{h}} \vec{h}_{t-1} + b_{\vec{h}}) \quad (9)$$

$$\overleftarrow{h}_t = \sigma(w_{x\overleftarrow{h}} x_t + w_{\overleftarrow{h}\overleftarrow{h}} \overleftarrow{h}_{t-1} + b_{\overleftarrow{h}}) \quad (10)$$

$$y_t = w_{\vec{h}y} \vec{h}_t + w_{\overleftarrow{h}y} \overleftarrow{h}_t + b_y \quad (11)$$

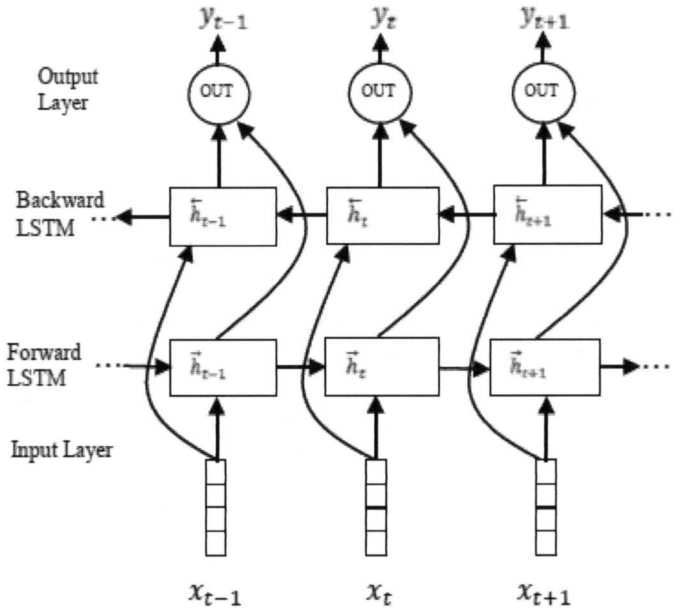

Figure 3. Bidirection Long-Short term Memory

6. Experimental Results

The EEG signals were extracted on locations C3, Cz and C4 in this paper and Emotiv EPOC chip, g.tec dry electrode and Ultracortex helmet are connected to record MI-EEG signals such as imagine left-hand and right-hand movements. It is consumed 9 seconds for each imaginary action to get a data set. The wavelet transform was used to transform EEG signals those were extracted 28 times to obtain their features. Within an interval of two minutes, a data set every 9 seconds was obtained in the experimental data acquisition process. On the first two seconds, the waiting time is set then the testing process is started and a cross sign "+" was displayed for one second after a stimulus signal was sound. Then the right- or left-arrow is shown to hint a subject to imaging right- or left-hand moving. In the acquisition process, the sampling rate is set 128Hz.

Since MI-EEG signals were captured from the electrodes on C3, Cz, and C4 then classified them into several groups. The forward and backward hidden layers in the proposed BiLSTM are all set into seven neurons while three channels are set for the proposed MI-EEG based BCI system. For the purpose of evaluating the classification results and obtain a stable and reliable BCI system, the classification accuracy is estimated by this proposed model executed 500 cross validation. In order to prove a stable and flexible system being obtained, the proposed methods compared to other strategies based on “BCI Competition 2003” [13]. The experimental results are shown in Table 1. From Table I, we can find that the proposed BiLSTM can get better performance than others.

Table 1. The Accuracy Rates of Different Strategies for BCI Competition 2003

Authors	Features	Classifiers	Accuracy rates
Akash Narayana [13]	AR	LDA	84.29%
GAO Xiaorong [13]	ERD	LDA	86.43%
The proposed BiLSTM	DWT	BLSTM	87.14%

7. Conclusions

In this paper, a deep-learning model named BiLSTM was applied to be embedded into a BCI system for MI-EEG signals to identify two imagery movements such as imaging right hand and imaging left hand actions. In the proposed BCI system, the Emotiv EPOC IC with brain waves helmet to capture brainwave signal on C3, Cz, and C4. In this paper, we use the Daubechies wavelet to get feature values on db4 coefficient. The BiLSTM can make each recurrent unit to capture variable-length sequences adaptively. Modified from LSTM, the BiLSTM has both feeding directions, one from past to future and the other from future to past. This can provide additional context to the network and result in fuller learning on the problem. The experimental results show that the BILSTM can get better performance than other strategies.

8. Acknowledgments

In this paper, the research was sponsored by Chin-Yi University of Technology under the Grant NCUT 19-R-CC-010.

9. References

[1] Decety J and Ingvar D H, 1990 *Acta Psychol* **73**, 13
[2] Lin J S and Lo C H 2016 *Int. J. of Computer, Consumer and Control*, **5** 18
[3] Sreeja S R, Joytirmoy R, Nagarjuna K Y, Samanta D, Mitra P and Sarma M 2017 *Proc. Int. Conf. on New Trends in Computing Sciences* (Amman:Jordan) p 61
[4] Guan S, Zhao K and Yang S 2019 *Computational Intelligence and Neuroscience*, **29** https://doi.org/10.1155/2019/5627156
[5] Xie X, Yu Z L, Lu H, Gu Z and Li Y 2016 *IEEE Trans. On Neural Systems and Rehabilitation Engineering* **25** 504
[6] Jois K, Garg R, Singh V, and Darji A 2015 *IEEE Workshop on Computational Intelligence: Theories, Applications and Future Directions* (WCI) DOI: 10.1109/WCI.2015.7495507
[7] Hinton G E and Salakhutdinov R R 2006 *Science* **313** 504
[8] Bengio Y, Courville A, and Vincent P 2013 *IEEE Trans. Pattern Anal. Mach. Intell.* **35** 1798
[9] Graves A, Mohamed A -R, and Hinton G 2013 *Proc. IEEE Int. Conf. on Acoustics, Speech and Signal Processing*(Vancouver, BC, Canada) p 6645
[10] Petrosian D, Prokhorov V, Homan R and Wunsch D C 2000 *Neurocomputing* **30** 201.
[11] Li M, Zhang M, Luo X and Yang J 2016 *Proc. of IEEE International Conference on Mechatronics and Automation* (Harbin, China) p 1971
[12] Daubechies I 1988 *Communications on Pure and Applied Mathematics*, **41** 909.
[13] Blankertz B, Müller K R and Curio G 2004 *IEEE Transactions on Biomedical Engineering* **51**1044

Single Channel Speech Enhancement Algorithm based on BLSTM-DNN Bidirectional Optimized Hybrid Model

Xiaoyue Sun[1], Ruwei Li[1], Tao Li[1] and Dengcai Yang[2]

1. Department of Information and Communication Engineering, University of Beijing Technology, Faculty of Artificial Intelligence, Science building, room 611, Beijing, China;
2. Institute of Science and Technology Development, Beijing University of Technology, Knowledge and Practice building, room 415, Beijing, China.
Email: li_stu_public@163.com

Abstract. The performance of existing speech enhancement algorithms based on deep learning is not ideal in complex noise environment. To improve the problem, a bidirectional optimized hybrid network named BLSTM-DNN is constructed based on bidirectional long-short term memory (BLSTM) network and fully-connected deep neural network (DNN). This structure uses BLSTM to extract high-level information including past and future temporal context of noisy speech. Next, fully-connected DNN fits the high-level information to ideal ratio mask (IRM). Finally, the IRMs estimated by the BLSTM-DNN are used to enhance the noisy speech. Experimental results show that the proposed method can effectively improve the speech quality and intelligibility under unknown noise conditions.

1. Introduction

Recently, with the development of deep learning technology, deep neural networks have been widely used in the field of speech enhancement due to their simple network structure, powerful modeling capabilities and good fitting effects [1]. For example, Wang et al. [2] learn a variety of complementary features under the DNN framework to further improve the performance of speech enhancement. Li et al. use DNN to achieve noise classification and denoising [3]. However, although contextual information can be incorporated into the training of the DNN through frame expansion [2], that cannot be represented by the DNN in the long-term acoustic model. Recurrent neural networks (RNNs) can correctly represent that relationship between the previous frame and the current frame [4]. But because of signal internal structure and the lack of nonlinear activation function, the gradient of RNN will gradually vanish and explode after multi-stage propagation, which makes RNNs difficult to capture the long-term contextual information [5]. Long short-term memory networks may alleviate this problem by introducing a series of gates and the concepts of memory cell to dynamically control the flow of information, which makes the networks have long-term memory [6]. In order to access past and future context information, Graves et al. [7] propose a Bidirectional long-short term memory (BLSTM) networks based on LSTM, which integrate forward and backward contextual information into the model to simulate the timing variation of speech and noise over a longer period of time [8] and improve the speech enhancement performance. However, more network parameters and lower fitting efficiency result in the BLSTM model not easily converging in speech enhancement tasks [9].

In this paper, a novel speech enhancement algorithm named BLSTM-DNN is proposed based on the advantage of BLSTM and DNN to capture the contextual information a while improving the matching efficiency of the networks, which combined the advantages of BLSTM and DNN to jointly optimize them. Bidirectional context information for noisy speech is extracted by BLSTM, which is used by DNN to fit the training target. The effectiveness of the proposed algorithm is verified by experiments from the aspects of enhanced speech quality and intelligibility as well as the waveform and spectrogram.

2. Bidirectional Optimized Speech Enhancement

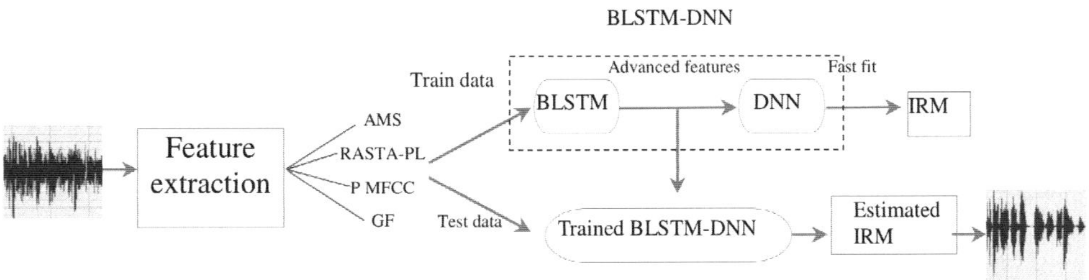

Figure 1. Schematic diagram of the proposed speech enhancement algorithm.

The proposed system is shown in figure 1. In the training phase, the noisy speech is decomposed into time-frequency units, and the Mel-frequency cepstral coefficients (MFCC), amplitude modulation spectrogram (AMS), relative spectral transform and perceptual linear prediction (RASTA-PLP) and gammatone frequency (GF) are extracted from the time-frequency units as the inputs of BLSTM-DNN [10]. High-level abstraction information of noisy features is extracted by BLSTM, which is used to fit the IRM by DNN. In the enhancement phase, the complementary features of the test noisy speech are input into the trained BLSTM-DNN to estimate IRMs. It is used to weight the noisy speech to synthesize the enhanced speech.

2.1.BLSTM-DNN Hybrid Model Construction

In order to further incorporate past and future context into model to simulate the temporal changes of speech as well as noise in a longer time range and improve the fitting ability of the network, the BLSTM-DNN bidirectional optimized hybrid model is constructed. The topology of the model is shown in figure. 2. The input layer $x = (x_1, x_2, \cdots, x_T)$ is a frame-level complementary feature vector, where $T = 246$ represents the feature vector dimension. The first three layers are BLSTM layers, each of which consists of a forward LSTM layer and a backward LSTM layer. The remaining two layers are fully connected DNN layers.

2.2.BLSTM-based Bidirectional Optimization

Figure 2 shows that the BLSTM layer scans the input sequence in two opposite directions and connects to the same output layer. The output layer is updated by the backward hidden layer from t=T iteration to 1 and the forward hidden layer from t=1 iteration to T, which is used to calculate the forward hidden sequence, the backward hidden sequence and the output sequence. That bidirectional optimization process can further simulate the temporal context between speech and noise. This process is implemented by the following formulas:

$$\overrightarrow{h_t} = H\left(W_{x\overrightarrow{h}}x_t + W_{\overrightarrow{h}\overrightarrow{h}}h_{t-1} + b_{\overrightarrow{h}}\right) \quad (1)$$

$$\overleftarrow{h_t} = H\left(W_{x\overleftarrow{h}}x_t + W_{\overleftarrow{h}\overleftarrow{h}}\overleftarrow{h}_{t-1} + b_{\overleftarrow{h}}\right) \quad (2)$$

$$y_t = W_{\overrightarrow{h}y}\overrightarrow{h}_t + W_{\overleftarrow{h}y}\overleftarrow{h}_t + b_y \quad (3)$$

Where $\overrightarrow{h} = (\overrightarrow{h_1}, \cdots, \overrightarrow{h_t}, \cdots, \overrightarrow{h_T})$ denotes the forward hidden layer sequence. $\overleftarrow{h} = (\overrightarrow{h_1}, \cdots, \overrightarrow{h_t}, \cdots, \overrightarrow{h_T})$ denotes the backward hidden layer sequences. $y^l = (y_1, \cdots, y_t, \cdots, y_T)$ indicates the output sequence after bidirectional optimization of the lth (l=1, 2, 3) BLSTM layer. H denotes the activation function of hidden state. W_{xh}, W_{hh} and W_{hy} represent the weight matrix of the input layer-hidden layer, the hidden layer-hidden layer and the hidden layer-output layer, respectively. $b_{\vec{h}}$, $b_{\overleftarrow{h}}$ and b_y represent the bias of the forward and backward hidden sequences and the offset of the output vector, respectively. $t = (1,2,\cdots,T)$ indicates the time index in the iterative process.

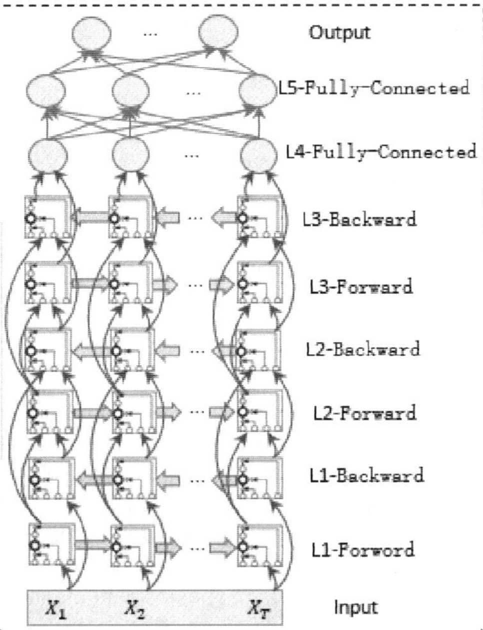

Figure 2. Topology diagram of the BLSTM-DNN

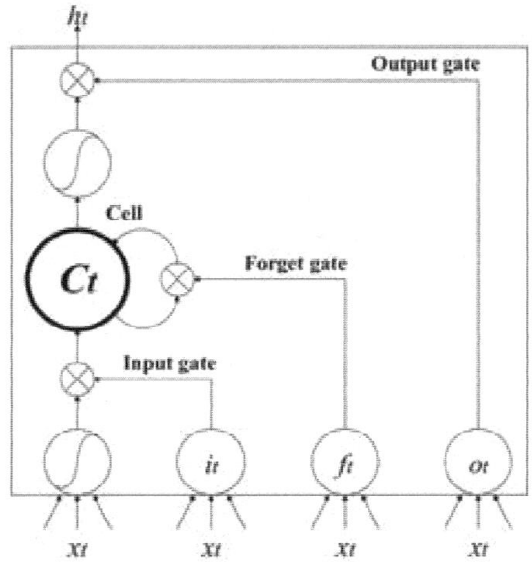

Figure 3. Block diagram of a single memory module in BLSTM

The forward hidden sequence $\vec{h}$ and the backward hidden sequence $\overleftarrow{h}$ for the BLSTM layer are obtained by optimizing the cell state c_t in each memory module (shown in figure 3) as follows:

$$i_t = \sigma(W_{xi}x_t + W_{hi}h_{t-1} + W_{ci}c_{t-1} + b_i) \tag{4}$$

$$f_t = \sigma\left(W_{xf}x_t + W_{hf}h_{t-1} + W_{cf}c_{t-1} + b_f\right) \tag{5}$$

$$c_t = f_t c_{t-1} + i_t tanh(W_{xc}x_t + W_{hc}h_{t-1} + b_c) \tag{6}$$

$$o_t = \sigma(W_{xo}x_t + W_{ho}h_{t-1} + W_{co}c_{t-1} + b_0) \tag{7}$$

$$h_t = o_t \cdot tanh(c_t) \tag{8}$$

Where i_t, f_t, o_t, c_t represent the input gate, the forgetting gate, the output gate and the cell, respectively. The dimension of them is the same as the hidden layer vector h_t. W_{hi} denotes the hidden layer-input gate weight matrix. W_{xo} denotes the input layer-output layer weight matrix. c_t is the cell state at time t. h_t is the hidden layer state at timet, and $\sigma(\cdot)$ is the sigmoid function.

Table 1. Test results of PESQ

Noise	SNR (dB)	DNN	BLSTM	BLSTM-DNN
Babble	10	3.026	3.442	3.483
	5	2.765	3.036	3.188
	0	2.425	2.816	2.841
	-5	1.997	2.467	2.498
White	10	2.832	3.019	3.106
	5	2.659	2.778	2.821
	0	2.316	2.420	2.445
	-5	1.913	2.041	2.058

2.3.DNN-based Fitting Process

The formula for network iteration of the fully-connected DNN layers ($l = 4,5$) is as follows:

$$y^l = \varphi\left(W^l y^{l-1} + b^l\right) \tag{9}$$

Where y^l, W^l and b^lrepresent the output of the activation units, weight matrix and the bias of lth DNN layer, respectively. φ represents the ReLU activation function. The function of fully connected DNN layers is efficiently fit the training target through gradient descent and error backpropagation.

Table 2. Test results of STOI

Noise	SNR (dB)	DNN	BLSTM	BLSTM-DNN
Babble	10	0.944	0.9509	0.9544
	5	0.891	0.9004	0.9155
	0	0.806	0.8112	0.8452
	-5	0.655	0.7336	0.7376
White	10	0.957	0.9580	0.9597
	5	0.921	0.9216	0.9216
	0	0.862	0.8701	0.8715
	-5	0.788	0.7957	0.8016

Table 3. Test results of LSD

Noise	SNR (dB)	DNN	BLSTM	BLSTM-DNN
Babble	10	6.312	4.065	4.062
	5	7.319	4.567	4.543
	0	8.952	6.792	6.514
	-5	11.634	7.540	6.929
White	10	7.821	4.239	4.184
	5	8.932	4.948	4.283
	0	10.271	5.418	5.405
	-5	11.700	6.413	5.552

3. Experimental Results and Analysis

3.1.Experimental Data

288 clean utterances of the experiments come from the NTT corpus. 100 noise types are from TIMIT database in training stage to improve the generalization capacity of unseen environments. This paper selects 200 utterances, which are corrupted with each noise type at 10dB,5dB, 0dB and -5dB to build training set. The remaining 88 utterances are used to construct the test set for each combination of noise types and SNR levels. Babble, white and factory2 from the NOISEX-92 corpus are used to evaluate on unseen noise types. All the signals are sampled at 16kHz rate. The frame length and shift are 320 and 160 samples, respectively. In order to verify the effectiveness of the proposed algorithm, we select on complementary features speech enhancement model based on DNN [2] as the first contrast algorithm, and complementary features speech enhancement model based on BLSTM as the second contrast algorithm. Both DNN and BLSTM model have 4 hidden layers with 1024 nodes for each layer. The Dropout ratio is 0.2, and the number of iterations is set to 50.

3.2.Speech Enhancement Performance Comparison

3.2.1.*Objective performance evaluation.* The test results of Segmental Signal to Noise Ratio (SegSNR), Evaluation of Speech Quality (PESQ), Log-Spectral Distortion (LSD) and Short-time Objective Intelligibility (STOI) are shown in table 1, table 2, table3 and table4, respectively. It can be seen from the tables that the objective test results of the BLSTM-DNN model are better than the BLSTM and

DNN models. Because BLSTM can further incorporate bidirectional temporal context and simulate the temporal variation of speech and noise in a longer time, and makes the model have the high-efficiency fitting ability that the BLSTM model does not.

Table 4. Test results of SegSNR

Noise	SNR (dB)	DNN	BLSTM	BLSTM-DNN
Babble	10	0.864	2.272	2.377
	5	-1.951	0.622	1.967
	0	-6.409	-4.275	1.262
	-5	-14.845	-10.063	-8.509
White	10	3.503	3.942	4.411
	5	1.167	1.354	2.208
	0	-0.870	-0.982	0.074
	-5	-3.237	-2.510	-2.420

3.2.2. Comparison of waveform and spectrogram. In order to compare the speech enhancement effects of DNN, BLSTM and BLSTM-DNN visually, figure 4 shows the waveform and spectrogram of the proposed algorithm and the contrast algorithm with factory2 noise at SNR = -5dB. We can know from the spectrogram that the speech enhanced by BLSTM-DNN is more similar with pure speech, because BLSTM layers could build long-term dependence of noisy speech to better preserve the temporal context information. Besides, the speech enhanced with BLSTM-DNN has less background noise, especially in the non-speech segment. That indicates that the fully connected DNN could efficiently fit speech information processed by BLSTM layers to the training target to achieve better denoising performance. The experimental results show that BLSTM-DNN has better denoising performance than BLSTM and DNN, which is consistent with the objective test results.

4. Conclusion

The proposed algorithm can bi-directionally model the temporal context information of noisy speech, and simulate the timing changes of speech and noise in a longer period through BLSTM. In addition, the high-level information extracted by the BLSTM quickly fits the IRM by using the efficient fitting ability of the DNN. It ensures that the deep neural network has a good fitting ability while modeling the temporal relationship of the speech. Compared with the comparison algorithm, the proposed algorithm improves the quality and intelligibility in unseen noise conditions.

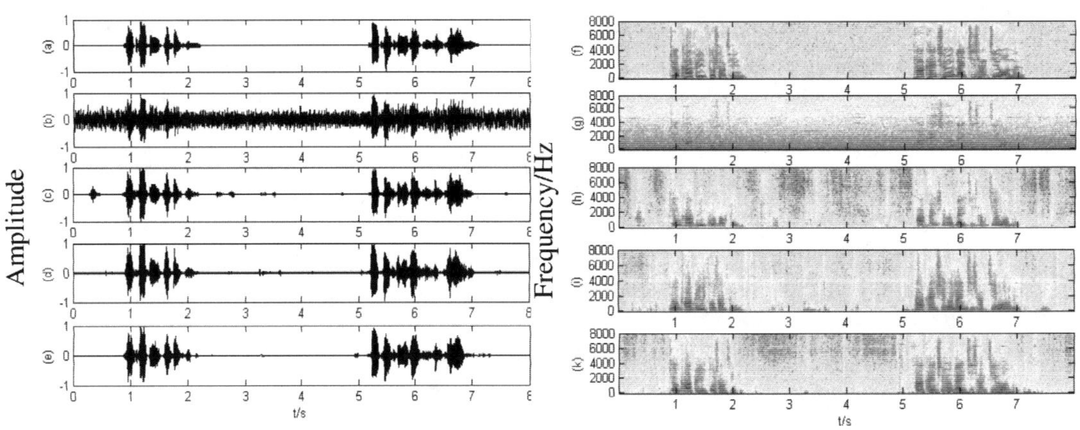

Figure 4. Speech enhancement effect samples with -5dB factory2 noise
(a),(f)Waveform and spectrogram of clean speech (b),(g)waveform and spectrogram of noisy speech(c),(h) waveform and spectrogram of speech with DNN (d),(i) waveform and spectrogram of speech with BLSTM (e),(j) waveform and spectrogram of enhanced speech with BLSTM-DNN

5. Acknowledgments

This work was supported by Beijing Municipal Natural Science Foundation Cooperation Beijing Education Committee (No. 201910005007) and the National Natural Science Foundation of China (No. 61871007).

6. Reference

[1] Li R, Sun X, Liu Y, et al. Multi-resolution auditory cepstral coefficient and adaptive mask for speech enhancement with deep neural network[J]. EURASIP Journal on Advances in Signal Processing, 2019, 2019(1): 22.
[2] Wang Y, Narayanan A, Wang D L. On training targets for supervised speech separation[J]. IEEE/ACM transactions on audio, speech, and language processing, 2014, 22(12): 1849-1858.
[3] Li R, Liu Y, Shi Y, et al. ILMSAF based speech enhancement with DNN and noise classification[J]. Speech Communication, 2016, 85: 53-70.
[4] Mehri S, Kumar K, Gulrajani I, et al. SampleRNN: An unconditional end-to-end neural audio generation model[J]. arXiv preprint arXiv:1612.07837, 2016.
[5] Tang Z, Shi Y, Wang D, et al. Memory visualization for gated recurrent neural networks in speech recognition[C]//2017 IEEE International Conference on Acoustics, Speech and Signal Processing (ICASSP). IEEE, 2017: 2736-2740.
[6] Sun L, Du J, Dai L R, et al. Multiple-target deep learning for LSTM-RNN based speech enhancement[C]//2017 Hands-free Speech Communications and Microphone Arrays (HSCMA). IEEE, 2017: 136-140.
[7] Graves A, Schmidhuber J. Framewise phoneme classification with bidirectional LSTM and other neural network architectures[J]. Neural Networks, 2005, 18(5-6): 602-610.
[8] Cheng G, Huang L, Sun J, et al. Bidirectional LSTM with Extended Input Context[C]//2018 11th International Symposium on Chinese Spoken Language Processing (ISCSLP). IEEE, 2018: 364-368.
[9] Wang Q, Du J, Dai L R, et al. A multiobjective learning and ensembling approach to high-performance speech enhancement with compact neural network architectures[J]. IEEE/ACM Transactions on Audio, Speech and Language Processing (TASLP), 2018, 26(7): 1181-1193.
[10] Chen J, Wang Y, Wang D L. A feature study for classification-based speech separation at low signal-to-noise ratios[J]. IEEE/ACM Transactions on Audio, Speech, and Language Processing, 2014, 22(12): 1993-2002.

IOP Conf. Series: Materials Science and Engineering **719** (2020) 012028 doi:10.1088/1757-899X/719/1/012028

An Adaptive QP Adjustment of Multimedia over Heterogeneous Wireless Networks

Chao He[1,2], Zhidong Xie[1,2] and Chang Tian[1]

[1] College of Communications Engineering, Army Engineering University of PLA, Nanjing 210007, China
[2] National Innovation Institute of Defense Technology, Academy of Military Sciences of PLA, Beijing 100071, China
Email: xzd313@163.com

Abstract. When videos are transmitted to the moving vehicle through heterogeneous networks, the mobility will frequently trigger the vertical handoff of networks. This may cause large fluctuation of channel bandwidth and will affect the Quality of Experience(QoE) of videos. In order to effectively solve the problem, an adaptive video transmission mechanism was proposed to guarantee the video quality. An information feedback link from the vehicle terminal was designed first. After analyzing the queue length of the vehicle receiver buffer and its future variation trend, the terminal fed back the judgment and prediction and the video's encoding Quantization Parameters(QP) were updated at the sender. Simulation results show that the variation of video encoding rates in our algorithm could properly adapt to the bandwidth fluctuation of heterogeneous networks. This mechanism not only makes full use of the wireless bandwidth resource, but also effectively avoids video interruption and ensures the video QoE.

1. Introduction

In the process of driving, live video broadcast can provide drivers with more support for operating judgment and it can also meet the online multimedia viewing and instant video communication requirements of the passengers in the vehicles.

The real-time vehicle video transmission needs to depend on the wireless access networks which are always heterogeneous during driving. Generally speaking, heterogeneous wireless networks often integrate various access modes such as Wireless Local Area Network (WLAN), terrestrial 4G/5G cellular network and satellite communication network[1-2]. When getting access to them, vehicular terminals not only need to have diverse accessing capabilities and compatibility, but also have to consider the impact of vertical handoff between networks, which may bring about inevitable fluctuation of wireless bandwidth. Thus, it has become an urgent problem to dynamically adjust the video encoding rate to adapt to the variation of channel resources and ensure the QoE of video transmission.

The above adaptation problem in video transmission has been extensively studied, for example, the HTTP-based MPEG-DASH (Dynamic Adaptive Streaming over HTTP) standard[3]. Although the original DASH cannot adequately meet the demand of real-time rate adjustment, the basic method that the receiver sends back requests to select different quality video segments at the sender has offered a good idea for adaptive streaming. Many algorithms are based on the feedback mode[4-7] and their difference always lies in the contents of feedback. Generally, they can be divided into three kinds: a method based on channel throughput, a method based on receiver buffer queue, and a hybrid method combining the previous two[8-11]. Among them, the buffer queue-based method directly starts from the results of bandwidth variation, and adopts statistical analysis methods, which has low complexity, good real-time performance and more accurate judgment.

In video encoding algorithms, QP is a kind of parameter which is easy to be adjusted and realized. In order to meet the requirement of video adaptation for the vehicle terminal travelling in heterogeneous networks, this paper proposes a QP adaptive adjustment method based on the receiver buffer prediction. The main contributions of this paper are as follows:

- In order to improve the video QoE, this mechanism adopts the way of changing video bitrates by updating the encoding QP at the sender.
- The receiving buffer queue of the vehicle terminal is analyzed. The video interruptions in the future is regarded as a rare event, and the Buffer Underflow Probability(BUP) is solved by adopting the Large Deviation Principle(LDP).
- A specific algorithm for adjusting the video QP values at the sender is proposed.
- We make the real-time comparison between the average video bitrates and heterogeneous network throughput and find that the proposed algorithm could accurately follow the changes of network bandwidth and provide satisfying video quality.

2. System Model and Problem Formulation

As shown in figure 1, when the vehicle is driven along the red dotted path, it will pass through the coverage areas of the Macro and Pico Cellular Networks. The Vehicle Terminal Equipment(VTE) will get access to one of them continuously. During the moving process, the VTE receives the real-time monitoring video and broadcast it instantly to the passengers in the car. Neither of the two kinds of networks completely covers the trajectory of vehicle, while the combination of the them can seamlessly cover the whole trajectory. Therefore, the vehicle antenna could compatibly receive signals from the two cellular networks and adaptively switch between them.

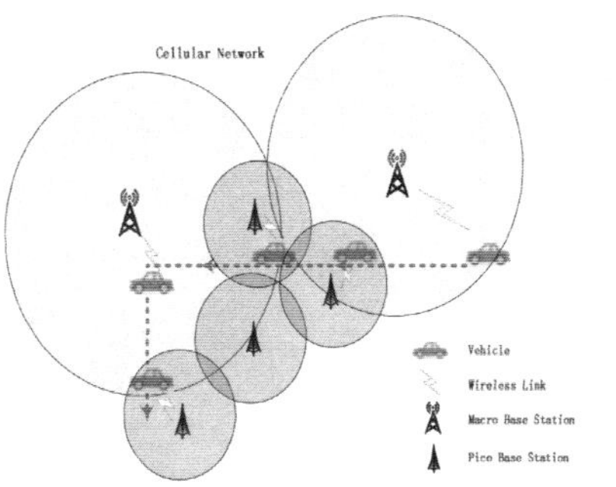

Figure 1. Path of Driving

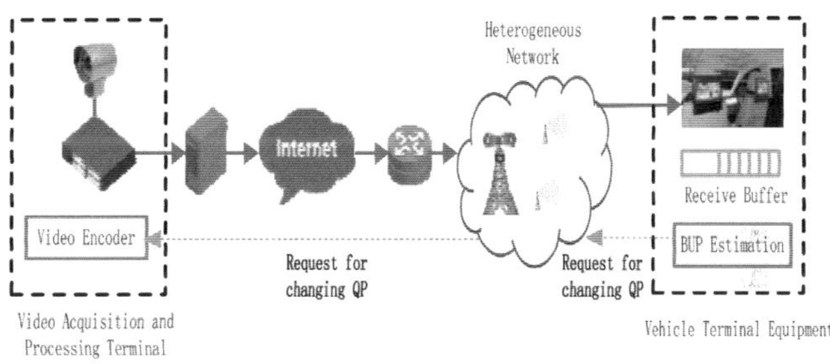

Figure 2. Adaptive Video Transmission System

The system model is shown in figure 2. At the sender of the communication, the Video Acquisition and Processing Terminal(VAPT) mainly takes the tasks of video shooting, and continuously compresses and encodes the video frames. The encoded video streaming is transmitted over the Internet and eventually sent to the VTE through the heterogeneous cellular networks. The VTE first stores the received video data in its receiving buffer, and then decodes and broadcasts it at a certain rate.

In order to better evaluate the transmission effect from the perspective of application layer services, this paper studies the packet arrivals at VTE buffer queue in the unit of frame. Assuming that the video frame rate is f, the duration time of each slot is set as $eachslot = 1/f$. Therefore, the VTE broadcasts a video frame in one slot, and the buffer queue also decreases one frame per slot. Due to the frequent occurrence of vertical handoff, the channel conditions are always changing, which makes the number of frames arriving at the VTE dynamically change in each slot. We continuously observe the size of the receiving buffer and the arrival of new frames at the VTE. According to the observation and prediction, the request of changing the video QP value is fed back to the VAPT through the network. As a result, the video encoding rate is adjusted in real time.

If the encoding rates do not match with the channel condition, the frame arriving rate will be either too fast or too slow, which will result in the buffer overflow or underflow. These need to be avoided because both of them could affect the users' QoE. In the following discussion, we assume that VTE's buffer capacity is large enough and the only disadvantage lies in buffer underflow, which means the real-time throughput of wireless link cannot meet the requirement of video transmission rate.

First, two thresholds are specified. (1) The minimum queue length that the VTE buffer can tolerate is $L_{\min}$. (2) The allowed maximum probability that the actual queue length is less than $L_{\min}$ is $P_{\max}$. The goal of video transmission is to provide users with good QoE. In order to objectively describe QoE, we analyze the Q-STAR model[12] first. From this model we know if the frame rate and resolution of the video at the VTE keeps unchanged, the key factor affecting QoE value depends on the quantization step. As QP is the serial number of the quantization step, so we could change the video QoE by adjusting the QP values, which vary from 0 to 51. The smaller the QP is, the higher the video encoding rate could achieve, the better the video quality is. So the optimization problem can be described as:

$$\begin{aligned} &\min \ \ QP \\ &s.t. \ \ l_i > L_{\min} \\ &\quad P(l_i < L_{\min}) < P_{\max} \\ &\quad P(l_{i+N} < L_{\min}) < P_{\max} \\ &\quad 0 < QP < 51 \end{aligned} \tag{1}$$

The current slot is set as i, the queue length of the current slot is l_i, and the queue length after N slots is l_{i+N}. $P(\cdot)$ represents taking probability values. This expression not only involves the current queue size, but also considers the probability after N slots, which reflects the prediction and estimation of the future bandwidth. When solving this problem, $L_{\min}$ and $P_{\max}$ in equation (1) as the thresholds are known. The queue length of the current slot can be obtained by observation. The problem lies in predicting $P(l_{i+N} < L_{\min})$, which is the BUP after N slots .

3. The Queue Length Prediction

The variations of network bandwidth are the main factor that causes the changes of buffer queue. By observing the length of the receiving buffer and estimating the BUP of $P(l_{i+N} < L_{\min})$, we can infer the condition of the network in the future.

It is assumed that the arrival of video frames is independent and identically distributed. Compared with slot i, the queue length increases Δl_N after N slots, the probability can be transformed as

$$P(l_{i+N} < L_{\min}) = P(l_i + \Delta l_N < L_{\min}) = P(\Delta l_N < L_{\min} - l_i) = P(\Delta d_N > (l_i - L_{\min})) \tag{2}$$

where $\Delta d_N = -\Delta l_N = d_{i+1} + d_{i+2} + ... + d_{i+N}$, and $d_{i+1}, d_{i+2}, ... d_{i+N}$ is the reduction of queue length per slot relative to the previous slot, from slot $i+1$ to $i+N$. It is hoped that video interruption of broadcast is a rare event. Since the LDP provides a probabilistic method for solving these problem[13], we can adopt the theory for solving equation (2). Suppose $\beta = \frac{1}{N}(l_i - L_{\min})$, then

$$\begin{aligned} &P(\frac{1}{N}(d_{i+1} + d_{i+2} + ... + d_{i+N}) > \beta) = \exp[-N \cdot l(\beta) + o(N)] \approx \exp[-N \cdot l(\beta)] \\ &l(\beta) = \sup[\theta\beta - \log M(\theta)] \\ &M(\theta) = E[e^{\theta d_{i+k}}] \quad k = 1, 2, ..., N \end{aligned} \tag{3}$$

The above expressions are based on the precondition of $E[d_{i+k}] < \frac{1}{N}(l_i - L_{\min})$. Suppose, at slot $i+k$,

j_k frames arrive at the VTE buffer queue and one frame is broadcast, so $d_{i+k} = 1 - j_k$. We consider that the arrival process of video frames obeys Poisson distribution and suppose the frame arriving rate is λ, then

$$M(\theta) = E[e^{\theta d_{i+k}}] = E[e^{\theta(1-j_k)}] = E[e^{\theta}] \cdot E[e^{-\theta j_k}] = \exp(\lambda e^{-\theta} - \lambda + \theta)$$
$$L(\beta) = \sup[\theta\beta - \log M(\theta)] = (\beta - 1)\ln\frac{\lambda}{1-\beta} + \lambda + \beta - 1 \quad (4)$$

In equation (4), λ can be gained from the historical observation values of the queue and calculated with the method of a sliding window.

4. The QP Adjustment Algorithm

Based on the above judgment and prediction, we can design a real-time QP adjustment algorithm for video parameters. The overall flow of the algorithm is shown in figure 3. The QP of the current slot i is QP_i and the one for the next slot $i+1$ is QP_{i+1}, and A, B, C, D are natural numbers. In order to make the VTE respond to the change of channel bandwidth in time, this paper adopts a scheme similar to Additive Increment and Multiplication Decrease(AIMD) to change the QP values.

Firstly, all QP values in the encoding process are gathered and sorted in descending order, with the serial numbers ranging from 1 to $level_{MAX}$. The serial number of QP value corresponding to current slot i is recorded as $level_i$. When the video interruption occurs as a certain event, the serial number will be decreased multiplicatively. If there is a possibility of video interruption after N slots, we only subtract 1 from the corresponding serial number. Also, we adopt a conservative way to increase the QP value.

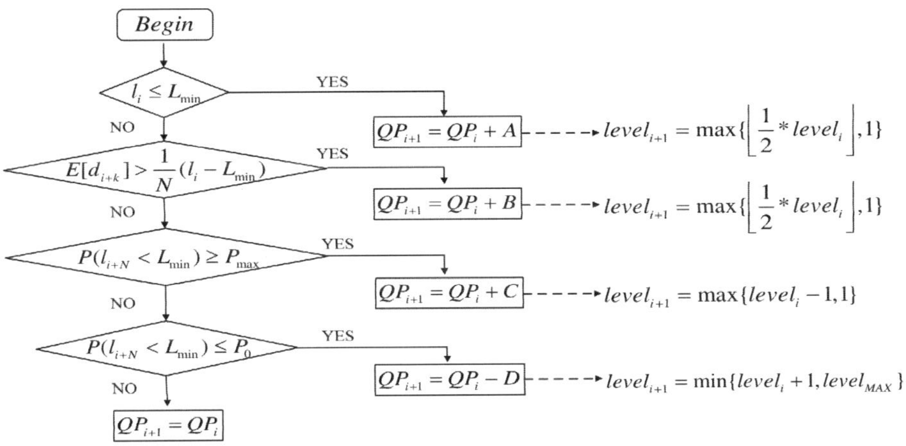

Figure 3. The Flow of the Algorithm

5. Simulations and Analyses

The simulations in this paper are based on the platform of MATLAB R2017a. We uses one of the video sequences in [14] named "Big Buck Bunny". The basic features of the video are listed in table 1. The considered future slot is set as $N = 100$. The number of pre-stored video frames in VTE is 20. The lower queue threshold of the buffer is L_{min}=6. The threshold of BUP is $P_{max}=3\times10^{-3}$, and the threshold $P_0=10^{-7}$. Table 2 lists the average encoding rates of the video corresponding to different QP values. In the simulation, VTE maintains the buffer queue, calculates the length and predicted value in each slot according to the proposed algorithm, and gives the adjustment strategy of QP. The sender dynamically adjusts the QP value according to the feedback adjustment strategy of the receiver, and finally changes the transmission rate and arrival rate of the video frames. Then, it goes on to the next slot and repeats the process until the end of the simulation. The encoding rate of video recorded in each slot is analyzed after simulation.

Table 1. Essential Features of the Video Sequence

Sequence: Big Buck Bunny
Resolution: 1920x1080
FPS: 24
Encoder: H.265/HEVC 10.1
GoP pattern: G24B7
Number of Frames: 14280

Table 2. Encoding Parameters of the Video Sequence

level	QP	Average Encoding Rate(bps)
1	45	154635.4
2	40	288554.6
3	35	524949.9
4	30	970807
5	25	1841115
6	20	3620199
7	15	6651106.7
8	10	12628692.3

In order to analyze the adaptability of the proposed algorithm to the change of channel bandwidth, the average video encoding rate achieved by VAPT transmitter is investigated first. When the vehicle is moving, it is assumed that the bandwidth of heterogeneous wireless networks obtained by VTE varies from 100 kbps to 18 Mbps, and there is a random jitter of 100 kbps to describe the channel fading. The channel bandwidth change over time is shown in the red dotted line in figure 4. It can be seen that there are many times when the channel bandwidth suddenly decreases or increases. This means that in the process of vehicle driving, the vertical handoff is triggered in different networks. The blue curve in figure 4 is the actual video encoding rate corresponding to each slot by our proposed algorithm. Overall, when the bandwidth provided by heterogeneous networks increases, the algorithm can reduce the QP value of the video at the sender and increase the encoding rate. When the heterogeneous network throughput decreases, requests to change QP will also reduce the bit rate of the sender. Compared with the change of throughput, there is no significant delay in the change of average encoding rate.

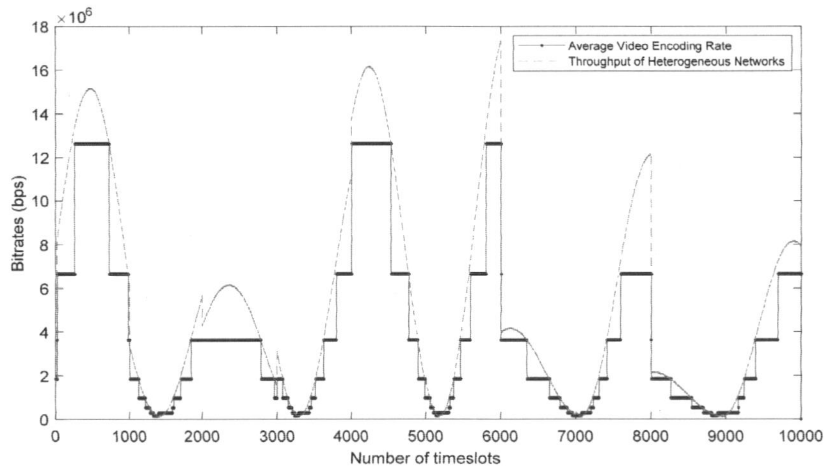

Figure 4. The Variation of Average Encoding Rate and Throughput of Network

The change of network throughput, or called bandwidth, not only causes the change of video encoding rate, but also further affects the QoE video transmission. We analyze the relationship between average network throughput and average Peak Signal to Noise Ratio (PSNR) of the video, as shown in figure 5. The blue circle curve represents the ideal situation where the channel bandwidth is fully utilized. That means the average throughput value of the horizontal axis is exactly equal to the video encoding rate. We get their corresponding average PSNR. The red asterisk curve shows the average PSNR that the proposed algorithm can achieve under different network throughput. It can be seen that the algorithm proposed in this paper is close to the ideal situation. Under the condition of heterogeneous networks with frequent channel bandwidth switching, it can provide good video quality for mobile vehicle users.

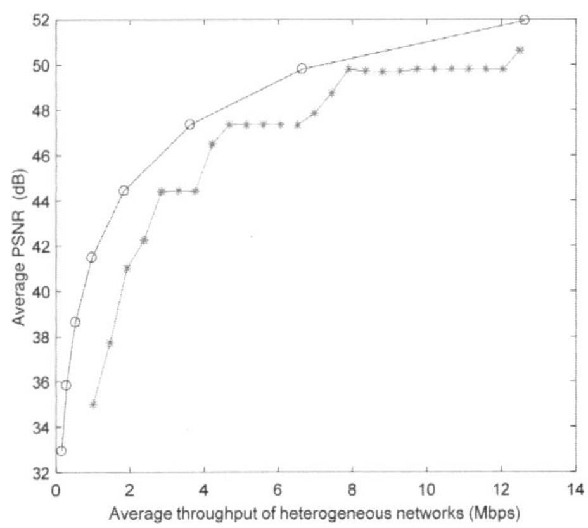

Figure 5. Video Quality vs. Average Throughput

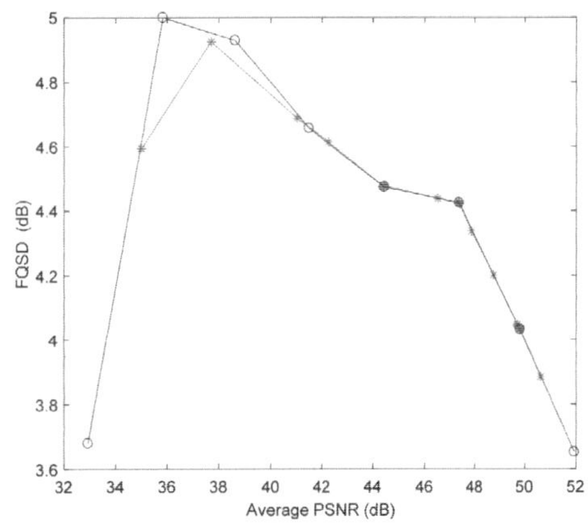

Figure 6. FQSD vs Average PSNR

In addition to video interruption, the flicker effect also needs to be concerned. Frame Quality Standard Deviation (FQSD) could reflect the offset of video quality relative to the mean value. When it is large, video flicker is apt to happen. In order to effectively evaluate the flicker effect caused by the our adjustment algorithm, the corresponding relationship between FQSD and average PSNR is analyzed in this paper. In figure 6, the blue curve is still the FQSD corresponding to the ideal case. The red curve is the simulation result of our algorithm. Because of the vertical handover of heterogeneous networks, the changes of networks themselves are full of unknown and mutation. Therefore, when using FQSD to evaluate video flicker effect, this ideal situation is not an optimal reference. On the contrary, the frequent switching of the network can easily lead to the occurrence of video flicker. The simulation results show that the proposed algorithm is basically the same as the ideal one, and in some cases it is slightly lower than the ideal FQSD value. So it does not trigger more serious video flicker phenomena.

6. Conclusion

When mobile vehicle users receive video in real time through heterogeneous networks, frequent fluctuations of channel bandwidth will affect the quality of video viewing. In order to effectively solve this problem, an adaptive QP adjustment mechanism is proposed in this paper. The mechanism estimates and predicts the current and future changes of channel bandwidth resources in the heterogeneous networks from the perspectives of the queue length of the buffer at the VTE. The decision results of the terminal buffer queue are fed back to the video sender in time. By the feedback information, the sender updates the QP of the video based on the basic idea of AIMD. Thus, the bit rate of video is changed, so that the transmission can follow up the bandwidth changes of heterogeneous networks adaptively. The simulation results show that the mechanism has sufficient timeliness and accuracy in tracking channel bandwidth variation. It can not only make full use of bandwidth, but also effectively avoid video interruption, which provides ideal video QoE for vehicle terminal users.

7. Acknowledgments

This work was funded by the Project of Natural Science Foundations of China (No. 91738201 and 61401507) and China Postdoctoral Science Foundation (No.2017M613403).

8. References

[1] J. Wu, C. Yuen, M. Wang and J. Chen, "Content-Aware Concurrent Multipath Transfer for High-Definition Video Streaming over Heterogeneous Wireless Networks," in *IEEE Transactions on Parallel and Distributed Systems*, vol. 27, no. 3, pp. 710-723, 1 March 2016.

[2] M. Liu, J. Yang and G. Gui, "DSF-NOMA: UAV-Assisted Emergency Communication Technology in a Heterogeneous Internet of Things," in *IEEE Internet of Things Journal*, vol. 6, no. 3, pp. 5508-5519, June 2019.

[3] ISO/IEC. Information technology dynamic adaptive streaming over HTTP(DASH) http://standards.iso.org/ittf/PubliclyAvailableStandards/c057623_ISO_IEC_23009-1_2012.zip

[4] C. Zhou, C. Lin and Z. Guo, "mDASH: A Markov Decision-Based Rate Adaptation Approach for Dynamic HTTP Streaming," in *IEEE Transactions on Multimedia*, vol. 18, no. 4, pp. 738-751, April 2016.

[5] Xiao, J. Liu, Y. Li, Q. Song and N. Ge, "Two-phase rate adaptation strategy for improving real-time video QoE in mobile networks," in *China Communications*, vol. 15, no. 10, pp. 12-24, Oct. 2018.

[6] S. Han, Y. Go, H. Noh and H. Song, "Cooperative Server-Client HTTP Adaptive Streaming System for Live Video Streaming," *2019 International Conference on Information Networking (ICOIN)*, Kuala Lumpur, Malaysia, 2019, pp. 176-180.

[7] J. Yang, R. Qiao, R. Ma and F. Li, "Cross-layer design of resource allocation and segment adaptation based on CPIQ model for DASH clients over LTE networks," in *IET Communications*, vol. 13, no. 10, pp. 1405-1414, 25 6 2019.

[8] BOUTEN N, SCHMIDT R D O, FAMAEY J, et al. “QoE-driven in-network optimization for adaptive video streaming based on packet sampling measurements”. *Computer Networks*, 2015, 81, 96-115.

[9] Huang, Te-Yuan; Johari, Ramesh; McKeown, Nick *et al.*, “A buffer-based approach to rate adaptation”, *Proceedings of the 2014 ACM conference on SIGCOMM*, 2014, pp.187-198.

[10] Shuangwu Chen, JianYang,Yongyi Ran, and Enzhong Yang, "Adaptive Layer Switching Algorithm Based on Buffer Underflow Probability for Scalable Video Streaming Over Wireless Networks," *IEEE Trans. Circuits Syst. Video Technol.*, vol. 26, no. 6, pp. 1146–1160, June 2016.

[11] D. J. Vergados, A. Michalas, A. Sgora, D. D. Vergados and P. Chatzimisios, "FDASH: A Fuzzy-Based MPEG/DASH Adaptation Algorithm," in *IEEE Systems Journal*, vol. 10, no. 2, pp. 859-868, June 2016.

[12] Ou, Y.; Xue, Y.; Wang, Y. “Q-STAR: A Perceptual Video Quality Model Considering Impact of Spatial, Temporal, and Amplitude Resolutions”. *IEEE Transactions on Image Processing*, 2014, 23, 2473–2486. 441.

[13] M. Mandjes, “Large Deviations for Gaussian Queues: Modelling Communications Networks”. New York: Wiley, 2007.

[14] Patrick Seeling and Martin Reisslein. “Video Transport Evaluation With H.264 Video Traces”. *IEEE Communications Surveys and Tutorials* 14(4): 1142-1165, Fourth Quarter 2012.

IOP Conf. Series: Materials Science and Engineering **719** (2020) 012029 doi:10.1088/1757-899X/719/1/012029

Timing Modeling Technology of DICE SRAM Based on SiliconSmart

Wu Yang, Yujing Li and Yi Tao

Sichuan Institute of Solid State Circuits, Chongqing, P.R. China.
Email: yangwu3101@163.com

Abstract. The rapid development of aerospace industry and semiconductor technology has put forward higher requirements for the radiation hardening of microelectronic devices. At the same time, with more and more applications of SRAM in SOC circuits, the anti-single event reversal effect ability of SRAM is more and more important. In this paper, the DICE structure is used to design SRAM to resist single event reversal effect. The simulation and test results show that the SRAM can resist single event reversal. In order to make SRAM with DICE structure be used in the digital design flow of SOC, this paper uses SiliconSmart to create timing model and successfully applies it to the design flow in the SOC project.

1. Introduction

With the development of integrated circuit technology to deep submicron, single event effect has put forward higher requirements for the radiation hardening of chip. As the process size decreases, the critical charge required for the inversion of the node level decreases, so the sensitivity of deep submicron integrated circuits to the single event effect will increase.

Single event effects mainly include single event transient effect (SET), single event latch-up effect (SEL) and single event reversal effect (SEU). The single event reversal effect (SEU) is the main source of single event effect in integrated circuits. Single event reversal (SEU) refers to the phenomenon that a single high-energy particle radiation causes the logical state of the circuit to be reversed, referred to as SEU. SEU effect will not cause irreversible damage to circuit function, and can restore its original function through system reset and re-charging, so SEU effect is also called soft error. Since there is no feedback signal in the combined circuit, when the duration of SEU effect is less than 1ns, most of the combined circuits can restore their original logical state after the high energy particle radiation is over. However, due to the feedback signal contained in the sequential circuit(such as SRAM), when the internal level of a node flips, the wrong data after the node flips may be locked, thus affecting the function of the circuit.

In this paper, we use DICE structure to design SRAM memory which can resist SEU effect. At the same time, SRAM with anti-SEU capability needs timing model to be integrated into the automatic place and routing flow of VLSI design. Therefore, this paper uses SiliconSmart to model the timing library of SRAM with DICE structure, so it can be used in the digital design flow.

2. Design of SRAM with DICE Structure

2.1. Theory Analysis of DICE SRAM

At present, the reinforcement technology against SEU effect is relatively mature, including Error Correction Codes, Triple Redundancy (TMR) and Double Interlocking Memory Cell (DICE) and other

reinforcement technologies. Considering the influence of area, speed and power consumption, SRAM with DICE structure is a widely used and dominant reinforcement technology in anti-SEU design.
The DICE SRAM storage cell in this paper uses 12-tube DICE structure to reinforce the anti-SEU ability. The circuit structure is shown in Figure 1.

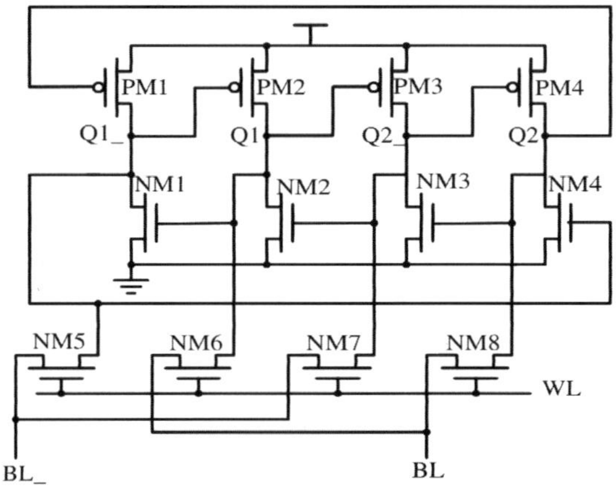

Figure 1. 12-tube DICE cell.

Compared with 6-tube element, 12-tube DICE structure has two-node feedback mechanism. During SEU effect, the destroyed node information is restored by the undestroyed node information, so as to achieve the anti-SEU effect. At the same time, the value of the corresponding node is determined only by two adjacent nodes, and there is no direct relationship between the non-adjacent nodes.

In Figure 1, when the DICE cell is logical "1", that is Q1 and Q2 are "1", and Q1_and Q2_are "0". Four storage nodes have two pairs of storage information, read and write operations is done through the transmission gate NM5-NM8. NM1, PM2, NM3 and PM4 are turned on and combine to form two latches: NM1 and PM2, NM3 and PM4. PM1, NM2, PM3 and NM4 are cut off and these four transistors constitute two bidirectional feedback circuits of NM2 and PM3, NM2 and PM3,so that they are separated from the latches. Eight single transistor inverters form two feedback loops, PMOS and NMOS. This feedback mechanism enables the correct node information to correct the wrong node information.

The 12-tube DICE structure has the same read-write reliability as the 6-Tube structure. In order to ensure the read-write reliability of the circuit, the width-length ratio of the DICE cell designed in this paper is shown in Table 1.

Table 1. Width-Length Ratio of DICE cells.

MOS tube	Width-Length Ratio
NM1~NM4	250n/190n
PM1~PM4	220n/190n
NM5~NM8	220n/290n

2.2. Radiation Simulation of DICE SRAM

Initial storage nodes store information as Q1 = Q2 = "1", Q1 = Q2 = "0", so NM1 and NM3 are turned on, NM2 and NM4 are cut off, so sensitive nodes are Q1 and Q2.

In Q1 node, we added double exponential current source, the simulation results are shown in Figure 2. Only Q1 flips during the current period. After the current is over, the positive feedback of the other three nodes restores Q1 to its original value. Q2 node is also a sensitive point. The effect of adding double exponential current source in Q2 node is the same as that of adding current source in Q1 node. It can be seen that the structure of DICE has the function of hardening single event reversal.

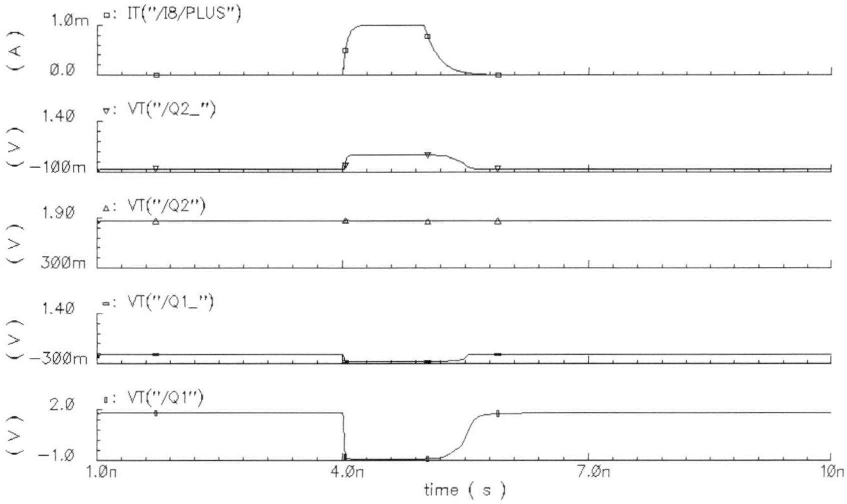

Figure 2. Simulation results of Q1 plus current source.

3. Create Timing Model with SiliconSmart

SiliconSmart is an EDA tool provided by Synopsys for creating timing model. SiliconSmart is the leading characterization and modeling solution tool focused on creating high quality cell libraries for today's advanced nanometer processes. It has been designed from the ground up to handle a full range of cells, from high-performance standard cells to advanced I/O cells implementing the latest communications protocols, and embedded memories, including RAMs, ROMs, and register files. It is the mainstream tool that can simulate SRAM with high speed and high precision and build the timing model. Using advanced algorithms, SiliconSmart can automatically identify the functions of standard cells and complex circuits and model them to generate effective simulation excitation sets for the timing model.

Using SiliconSmart to creating timing model of SRAM, the input data and the flow are shown in figure 3.

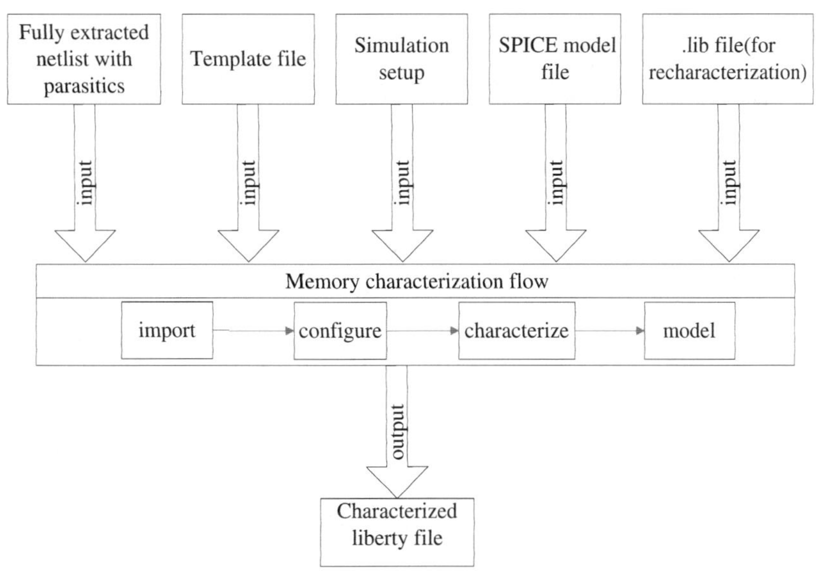

Figure 3. The input data and flow of SilconSmart.

3.1. Input Data

3.1.1. Spice model file. The spice model parameters of basic circuit components (such as NMOS and PMOS) provided by foundry manufacturer. It includes device models of various process and different temperature.

3.1.2. Spice netlist of DICE SRAM. The spice netlist with parasitic parameters is extracted from the layout of DICE SRAM. According to the accuracy of simulation, the parasitic parameter categories that need to be extracted are selected. The parasitic parameter accuracy of R+C is relatively low, but the simulation speed to create timing model is fast. The parasitic parameter netlist of R+C+CC has high accuracy, but the simulation speed is slow.

3.1.3. Template file. The template file contains basic information of the memory. It includes the type of memory, whether there is BIST mode, the function definition of each pin and so on. The structure and pin definition of a dual-port SRAM are shown in Figure 4, and the corresponding template file is shown in Figure 5.

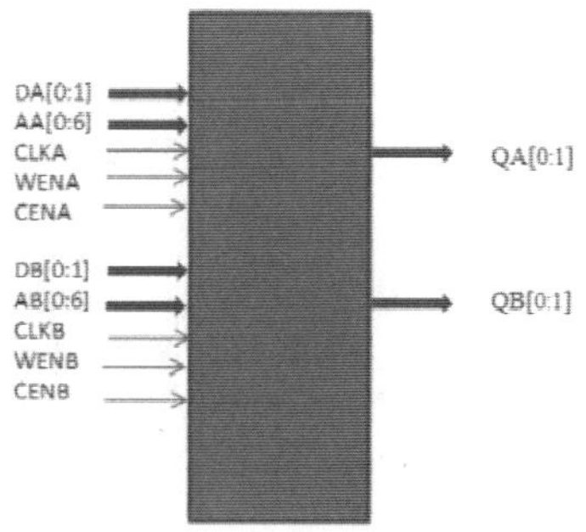

Pin	Description
AA, AB	Addresses: (A[0] = LSB)
DA, DB	Data Inputs: (D[0] = LSB)
CEA, CEB	Chip Enable
WEA, WEB	Write Enable/Read Enable
CLKA,CLKB	Clock
QA, QB	Data Ouputs: (Q[0] = LSB)

Figure 4. Structure and pin definition of dual-port SRAM.

```
set_memory_type multi_port_ram
set_memory_name dpram018
set_bypass_mode OFF
set_pipeline_mode OFF
set_writethrough_mode ON
set_bist_mode OFF
set_extramargin_adjustment_mode OFF

create_readwrite_port A
set_clock CLKA -active r -port A
set_address_bus AA -width 9 -port A
set_data_bus DA -width 16 -port A
set_read_enable WENA -active H -port A
set_write_enable WENA -active L -port A
set_output_enable  OENA  -active L -port   A
set_chip_enable CENA -active L -port A
set_data_output QA -width 16 -port A

create_readwrite_port B
set_clock CLKB -active r -port B
set_address_bus AB -width 9 -port B
set_data_bus DB -width 16 -port B
set_read_enable WENB -active H -port B
set_write_enable WENB -active L -port B
set_output_enable  OENB  -active L -port  B
set_chip_enable CENB -active L -port B
set_data_output QB -width 16 -port B
```

Figure 5. Template file example of SRAM.

3.1.4. Configure file. The file for configuration of simulation parameters include the selection and configuration of simulators and the definition of the header of timing model. In addition to selecting

IOP Conf. Series: Materials Science and Engineering **719** (2020) 012029 doi:10.1088/1757-899X/719/1/012029

and configuring the simulator, the slew and load lookup tables to creating timing model must be specified in the configure.tcl file for simulator.

The example of slew values is shown below (in s).

set numsteps_slew 5
set constraint_numsteps_slew 5
set smallest_slew 20e-12
set default_slew 200e-12
set largest_slew 3e-9

The example of the load value is shown below (in f).

set autorange_load 0
set numsteps_load 5
set smallest_load 20e-15
set default_load 400e-15
set largest_load 1e-12

3.2. The Flow of Creating Timing Model

The flow using SiliconSmart to create a timing model for DICE SRAM is shown in the figure 6.

```
set charpoint test_sram_1022
create $charpoint
set_log_file $charpoint/sis.log
set_location $charpoint
set cells {dpram018_512x16}
exec cp configure.tcl $charpoint/config/configure.tcl
exec ln -sf [pwd]/netlists/${cells}.pex.netlist.pex ${charpoint}/netlists/.
exec ln -sf [pwd]/netlists/${cells}.pex.netlist.${cells}.pxi ${charpoint}/netlists/.

set_config_opt auto_fix 0
set_config_opt char_engine_max_lifespan 12000
set_config_opt cdpl_task_max_lifespan 12000

set_config_opt -cell dpram018_512x16 -from OENB -reference CLKB state_partitions none

import -template ./dpram512x16_template.tcl -netlist_dir ./netlists -ext .pex.netlist $cells -overwrite

set_config_opt -cell dpram018_512x16 -from OENA -reference CLKA state_partitions none
set_config_opt -cell dpram018_512x16 -from OENB -reference CLKB state_partitions none

find_internal_nodes_for_constraint
configure -timing -power $cells
characterize $cells
model -output dpram018_512x16 $cells
```

Figure 6. The flow of creating timing for DICE SRAM.

3.2.1. Simulation option setting. Configure some tool parameters and timing modeling conditions.

3.2.2. Import. Import spice netlist to generate cell.inst file. SiliconSmart can identify the functions of the SRAM from the spice netlist and the template file.

3.2.3. Configure. Configure to build SPICE simulation files for each arc. Add input slew and output load to the SRAM spice netlist to generate simulation files.

3.2.4. Characterize. SiliconSmart calls HSPICE to simulate for each arc of the timing model, and determines whether the arc is successful simulated.

3.2.5. Model. Generates NLDM or CCS timing/power lib libraries as needed which can used in the design flow of SOC.

IOP Conf. Series: Materials Science and Engineering **719** (2020) 012029 doi:10.1088/1757-899X/719/1/012029

4. Conclusion

In this paper, the timing model of DICE SRAM is created by SiliconSmart, which is successfully used in the SOC project design flow. Functional tests show that the timing model can ensure the timing correctness of SOC.

Under the $LET_{TH} \geq 75 MeV \cdot cm^2/mg$ ion source condition, the SEU effect experiment is performed on the SOC with DICE SRAM. The result of the experiment is shown in table 2.

Table 2. A table with headings spanning two columns and containing notesa.

Sample Num.	Fluence rate(ions/cm^2.s)	Fluence(ions /cm^2)	Phenomenon description
1	1.05×10^4	10^7	Output flips 13 times
2	1.05×10^4	10^7	Output flips 15 times
3	1.05×10^4	10^7	Output flips 8 times

From the experimental result of table 2, all the three circuits do not flip more than 15 times, it indicates that these circuits is not sensitive for SEU effect at this LET_{TH} value. The SEU irradiation experiment shows that SRAM with DICE structure has the function of anti-SEU.

5. References

[1] Lingyu Zhang, Yuming Jia, Lei Li. Radiation Resistance SRAM Design Based on DICE Structure. Microelectronics, 2011.4

[2] Jing Shen, Haiwei Xue. Radiation hardening design of SRAM based on DICE structure. ELECTRONICS & PACKAGING, 2016.3

[3] SiliconSmart ACE User Guide [M]. SYNOPSYS, 2014.09

IOP Conf. Series: Materials Science and Engineering **719** (2020) 012030 doi:10.1088/1757-899X/719/1/012030

Analysis of Sound Field Transfer Loss in Inlet Pipeline of Industrial Gas Turbine

Jiangang Yi[1,*] and Siwei Ju[1]

[1] School of Electromechanical and Architectural Engineering, Jianghan University, Wuhan 430056, China
469883553@qq.com

Abstract. Taking an industrial gas turbine intake pipeline as the research object, the acoustic field of the intake pipeline is simulated and analyzed by using acoustic finite element method, and the internal acoustic field and transmission loss of the intake pipeline are calculated. Through simulation analysis, it is found that the transmission loss curve of gas turbine intake pipeline fluctuates greatly, and the frequency bandwidth of the peak muffler is narrow. The conclusions obtained provide a theoretical basis for optimizing the structure of industrial gas turbine intake pipeline.

1. Introduction

Industrial gas turbine (IGT) is a kind of internal combustion power machine, which uses continuous flowing gas as working fluid to drive the impeller to rotate at high speed and convert the energy of fuel into useful power. As one of the most important components of chemical gas turbine, intake pipeline, which makes air flow through the intake system of gas turbine, will form contraction, expansion and other flow, resulting in the change of air density and pressure in the pipeline. It can produce pressure wave. In the process of pressure wave propagation, air noise is formed. It not only seriously reduces the production efficiency, but also brings noise pollution. In order to solve this problem, this paper takes the sound field of the industrial gas turbine inlet pipeline as the research object, and analyses its transmission loss status, so as to provide conceptual support for energy saving and noise reduction.

2. Working Process

Figure 1 shows a schematic diagram of an industrial gas turbine. Air flows into the combustion chamber from the inlet of the intake pipeline through the contraction section and the compressor impeller of the gas turbine. The gas is discharged by the gas turbine after full combustion. The intake pipe is composed of two straight pipes and two tapered shrinkage pipes. The inlet diameter is 6.8m, the outlet diameter is diameter is 2.8m, the total length of the pipe is 4.42m, and the wall thickness of the pipe is 5mm.

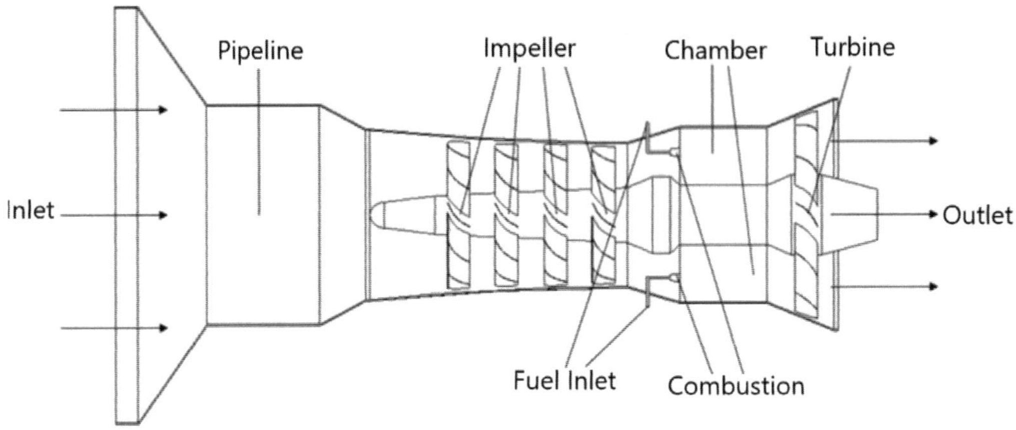

Figure 1. Industrial Gas Turbine

As shown in Figure 2, the length, width and height of the fluid microelement at the entrance of the block are dx, dy, dz. Respectively, the mass of micro-elements flowing into X direction is $\rho u_x dydz$, and the mass of micro-elements flowing out is $\rho(u_x + \frac{\partial u_x}{\partial x}dx)dxdy$. The the mass of micro-elements in X direction caused by medium flow changes to $-\rho\frac{\partial u_x}{\partial x}dxdydz$, and the mass of micro-elements in Y direction and Z direction is $-\rho\frac{\partial u_y}{\partial y}dxdydz$ and $-\rho\frac{\partial u_z}{\partial z}dxdydz$. The microelement obeys the law of conservation of mass, so the variable of mass is equal to the rate of change.

$$-\rho\left(\frac{\partial u_x}{\partial x} + \frac{\partial u_y}{\partial y} + \frac{\partial u_z}{\partial z}\right)dxdydz = \frac{\partial}{\partial t}(\rho dxdydz) \quad (1)$$

Assume $\nabla = \frac{\partial}{\partial x}i + \frac{\partial}{\partial y}j + \frac{\partial}{\partial z}k$, then the continuous equation of the sound wave obtained is as following.

$$\frac{\partial \rho}{\partial t} + \nabla \rho u = 0 \quad (2)$$

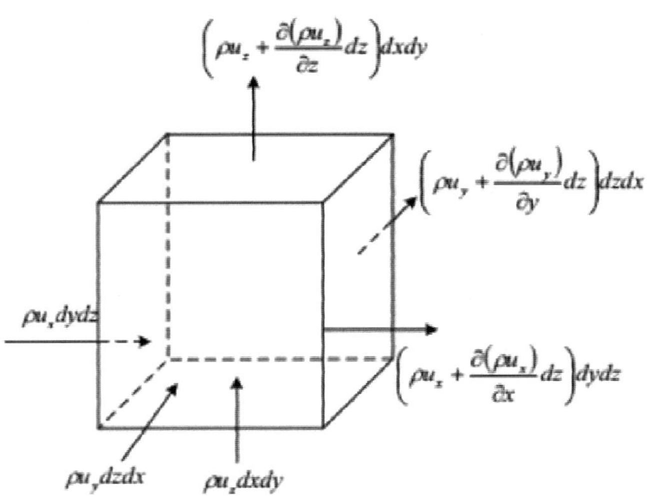

Figure 2. Fluid Microelement

3. Simulation and Calculation

There are three main evaluation indicators of acoustic performance: transfer loss, end noise reduction parameters and insertion loss. Among them, the calculation of insertion loss and other indicators needs to define the sound source and acoustic impedance characteristics. Generally, the parameters of sound source and acoustic impedance characteristics are obtained from physical experiments. However, the transmission loss is the acoustic characteristic of the research object itself, which does not need the definition of sound source and acoustic impedance characteristics. Due to the limitation of conditions, this paper uses the transmission loss to analyze and evaluate the acoustic performance of gas pipelines.

Figure 3 is schematic diagram of the theoretical model of transmission loss of sound field in the intake pipeline. The relationship between the incident sound pressure at x=0 and the inlet sound pressure is as following.

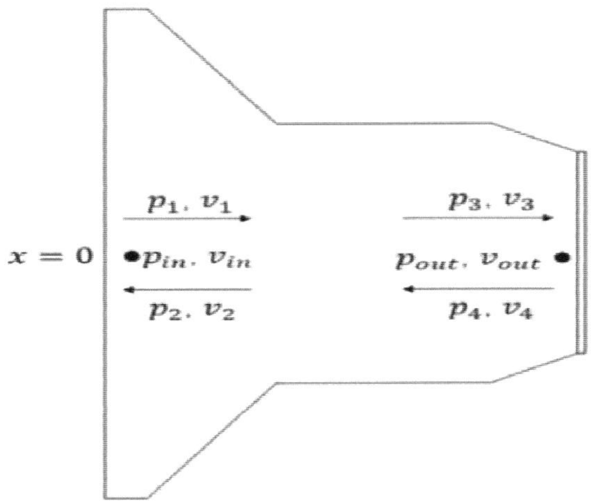

Figure 3. Landing Time Versus ξ

$$p_1 = \frac{1}{2}(p_{in} + \rho c) \tag{3}$$

When the area of the entrance section is A_{in} and the area of the exit section is A_{out}, the sound power of the entrance plane sound wave and the exit plane sound wave is as following:

$$W_{in} = \frac{p_1^2 A_{in}}{\rho c},\ W_{in} = \frac{p_3^2 A_{out}}{\rho c} \tag{4}$$

The formula for calculating the transmission loss defined is as following:

$$TL(db) = 10log\left(\frac{p_1^2}{p_3^2}\frac{A_{in}}{A_{out}}\right) \tag{5}$$

4. Simulation and Analysis

In the acoustic simulation calculation, the unstructured linear mesh generation method is used, as shown in Figure 4. The velocity of particle vibration at the entrance of airflow is known. Because the wall of intake pipeline is rigid and the incident sound wave is plane wave, it is impossible to transmit the internal sound wave to the outside world, so the sound wave will completely reflect at the wall of the pipeline, only from the outlet of the pipeline to the outside space. Considering the influence of the flow of sound transmitting medium on the acoustic performance, it is necessary to calculate the average flow velocity in the pipeline before calculating the acoustic transmission loss at the entrance and exit and the radiated noise at the exit. Flow boundary condition is used to calculate the average flow velocity, in which the inlet flow velocity of the intake pipe is given and the average distribution is given. Because the normal direction of the acoustical grid points to the outside of the grid, the inlet velocity is set to -3m/s and the flow velocity type is set to Flow Velocity. The results of the flow field inside the intake

IOP Conf. Series: Materials Science and Engineering **719** (2020) 012030 doi:10.1088/1757-899X/719/1/012030

pipeline are obtained by calculation. The internal velocity vector distribution is shown in Fig. 5. The maximum particle velocity of the flow field is 27m/s. The center of the inlet and outlet face of intake pipeline is defined as the acoustic input point and the acoustic output point (Output Point). The acoustic pressure response data of the inlet and outlet points are obtained. Based on the transfer loss theory, the transfer loss at the inlet and outlet ends of intake pipeline is calculated by using the formulas mentioned above, as shown in Figure 6. From the figure, it can be seen that the transmission loss curves of 260-700Hz and 820-1000Hz fluctuate greatly, and there are many peaks with narrow bandwidth, among which the maximum value of transmission loss occurs at 980Hz, which is 31.434dB.

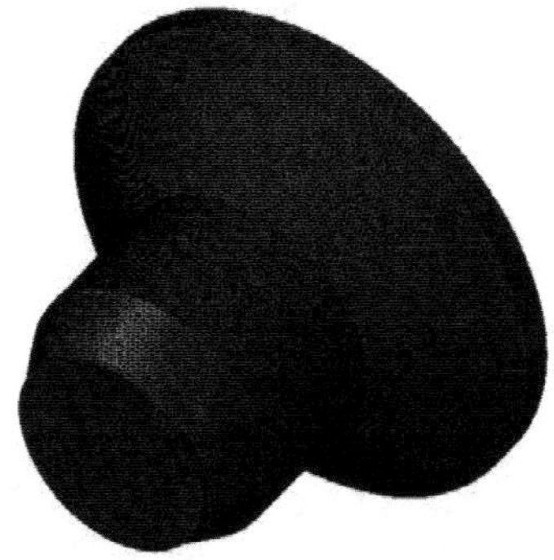

Figure 4. Sound Field Model of Intake Pipeline

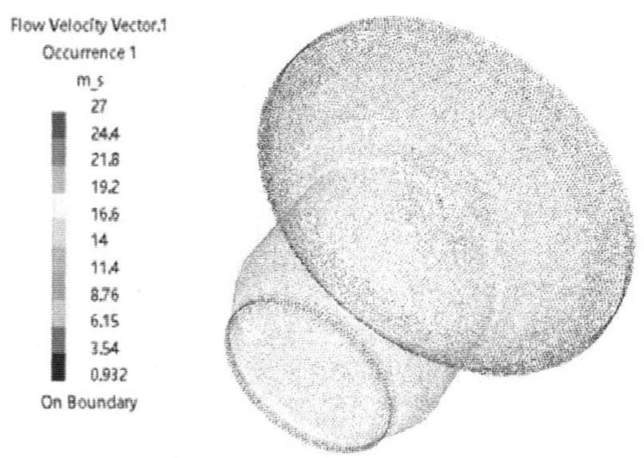

Figure 5. Velocity Vector Distribution of Sound Field in Intake Pipeline

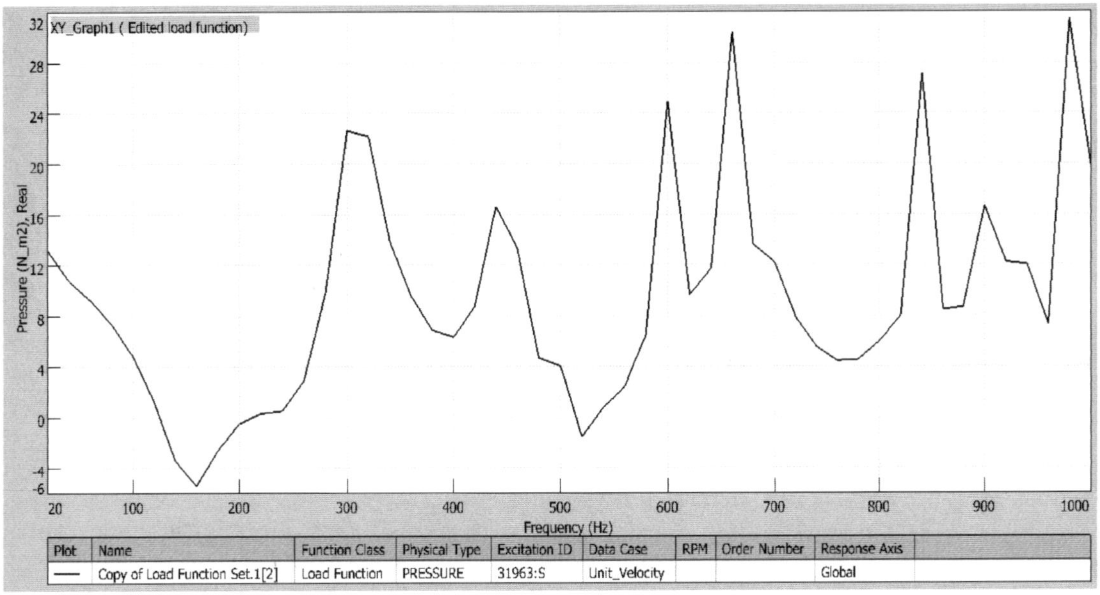

Figure 6. Acoustic Field Transmission Loss of Intake Pipeline

5. Conclusions

Gas turbine is the core equipment in industry, and the design of its intake pipeline is the key link of the overall design. By simulating the sound field of gas turbine intake pipeline, this paper analyses the sound field characteristics of intake pipeline and calculates the transmission loss of sound field. The calculation model is simulated and analyzed by using the finite element tool of sound field, and the transmission loss curve of sound field is established. The conclusion provides a reference and comparison basis for the optimization of pipeline structure. It is of great significance to understand the distribution of noise in intake pipeline and take pertinent noise reduction measures.

6. Acknowledgements

This work is supported by the Industry-university-research project of Wuhan Education Bureau (Granted No: cxy02), China. The authors also gratefully acknowledge the helpful comments and suggestions of the reviewers, which have improved the presentation.

7. References

[1] Guedra M, Penelet G, Lotton P, et al. Theoretical prediction of the onset of thermoacoustic instability from the experimental transfer matrix of a thermoacoustic core[J]. Journal of the Acoustical Society of America, 2011, 130(1):145.

[2] Seyf H R, Feizbakhshi M. Computational analysis of nanofluid effects on convective heat transfer enhancement of micro-pin-fin heat sinks[J]. International Journal of Thermal Sciences, 2012, 58(4):168-179.

[3] Li J, Chen W, Fu Z, et al. Explicit empirical formula evaluating original intensity factors of singular boundary method for potential and Helmholtz problems[J]. Engineering Analysis with Boundary Elements, 2016, 73:161-169.

[4] Habib M A, Anwar A S, Patwari M A U. An innovative concept to improve the muffler performance using automated mechanical IRIS[J]. Journal of Engineering & Applied Sciences, 2016, 11(2):1035.

[5] Shao Y. A result on the acoustic characteristics of the Mixture of Counter-phase Counteract and Split-gas Rushing muffler[J]. 2016, 40:02019.

[6] Zou R, Yang X, Cai Q, et al. Analysis of flow field and pressure loss of Resistance Muffler Based on FLUENT[J]. Design & Manufacture of Diesel Engine, 2016.

IOP Conf. Series: Materials Science and Engineering **719** (2020) 012031 doi:10.1088/1757-899X/719/1/012031

Implementation of SAR Echo Simulation Algorithm on Heterogeneous Embedded Computing Platform

Xiaoyu Hou[1], Daying Quan[1], Wei Fan[1], Xiaoping Jin[1] and Hengliang Liu[2]

[1]Key Laboratory of Electromagnetic Wave Information Technology and Metrology of Zhejiang Province, College of Information Engineering, China Jiliang University, Hangzhou, China

[2]Hangzhou Jianpu Information Co. Ltd., Hangzhou, China

Email: 862376302@qq.com; qdy@cjlu.edu.cn; l312361206@163.com

Abstract. The Synthetic Aperture Radar (SAR) echo simulation is to obtain the SAR original echo signal by performing reverse operation on the pre-set SAR image. The traditional platforms used to implement the simulation algorithms like central processing unit (CPU) + graphic processing unit (GPU) and digital signal processor (DSP) + field programmable gate array (FPGA) have disadvantages such as high power consumption or complicated programming. In order to make up for these shortcomings, the implementation structure of the SAR echo simulation algorithm was improved to be applicable on the heterogeneous embedded platform with the basic SAR simulation algorithms digitized and decomposed. Base on the improved structure, a mobile GPU based heterogeneous computing platform with one multiprocessor system-on-chip (MPSoC) and multiple GPUs was designed to implement SAR echo simulation algorithm. The platform can simultaneously utilize the real-time nature of register transfer level (RTL) design and the ease of programming on GPU. It can achieve relatively faster computing power at lower power consumption and has the characteristics of miniaturization and mobility.

1. Introduction

SAR is a kind of imaging radar used for remote sensing to create two-dimensional or three-dimensional images of scene objects. Its high resolution, strong transmissivity and long detection range make it highly military and civil value. However, the acquisition of the real SAR echo data requires the support of airborne or spaceborne radar, which requires high cost so that it is necessary to develop a SAR echo simulator.

When the traditional platform performs SAR echo simulation algorithm based on general-purpose computer, it takes too much time to simulate the scene with large amount of data, so it is difficult to realize real-time echo simulation. Most computing platforms that can be used for real-time echo simulation are based on FPGAs and DSPs. As a programmable device, FPGA has a wealth of logical computing resources and has the advantage of flexible configuration. A large number of independent multiply operations in echo simulation can be implemented in a highly parallel manner by means of hardware resources such as dedicated multipliers in the FPGA. Wang et al [1] designed a platform of four FPGAs to realize the generation of SAR echo signals and optimized the implementation of the slant range calculation, which is the most computationally intensive part of the entire algorithm. Although implementing the algorithm with FPGA can achieve its real-time parallel computing ability and flexibility, it has the disadvantages of large programming workload and high logic design difficulty. Chen [2] used DSP and FPGA to realize the fast simulation of radar echo. The DSP chip is responsible for the process of range direction information integration. The FPGA chip mainly completes the process of slant range calculation, FFT, complex multiplication and IFFT. The cooperation between different computing units is realized by the heterogeneous form, but the multi-core parallel computing capability of the DSP chip is insufficient, so there is still room for improvement in the process of range direction information integration.

The rapid development of GPU provides a promising and efficient computing platform for SAR echo simulation. Li [3] realized SAR echo simulation algorithm on the GPU, but only stayed on a single GPU for optimization. Jing et al [4] implemented SAR echo simulation algorithm on dual GPUs. The SAR echo simulation using GPU platforms described above is still based on a desktop or server platform and has the problems of large volume and high power consumption, which is not suitable for scenarios with high mobility requirements. Therefore, it is necessary to design a SAR echo simulation platform that combines high performance, low power consumption and small size for these scenarios. On the other hand, with the demand of machine learning and deep learning, embedded GPU processors have been greatly developed, such as NVIDIA's Tegra series SoC.

In order to further improve the integration and mobility of the SAR echo simulators, this paper proposed an embedded heterogeneous computing platform based on one MPSoC and multiple embedded GPU processors. By arranging different computing tasks on different processors, the real-time SAR echo simulation can be realized on the embedded platform with high integration, small size and low power consumption.

2. SAR Echo Simulation Algorithm and it's Embedded Design

2.1. SAR Echo Simulation Algorithm

The most commonly used detection pulse in SAR is the linear frequency modulation (LFM) pulse, which can be written as

$$s_r(\tau) = \omega_r(\tau)\exp(j2\pi f_0\tau + \pi K_r\tau^2) \tag{1}$$

where τ is the time variable in seconds, K_r is the LFM rate in hertz per second, f_0 is the carrier frequency in hertz and $\omega_r(\tau)$ is the window in time domain or the pulse envelope. Suppose a single point target is located in the beam footprint, the baseband echo of the single point target can be written as [5]

$$s_0(\tau,\eta) = A_0\ \omega_r(\tau - 2R(\eta)/c)\ \omega_a(\eta-\eta_c)\exp\{-j4\pi f_0 R(\eta)/c\}\exp\{j\pi K_r(\tau-2R(\eta)/c)^2\} \tag{2}$$

where $A_0 = A_0^{'}\exp(j\psi)$ is the complex backscatter coefficient, τ and η are the fast time variable and slow time variable, respectively, in seconds, $\omega_a(\eta)$ is the antenna pattern, c is the speed of light in meter per second, and $R(\eta)$ is the slant range of the point target in meter.

In a more general scenario, there should be continuous targets or scatter points which make up the whole surface target reference scene in the beam footprint. Thus the backscatter coefficient becomes to be a three dimensional function with two dimension according to the scatter surface and the other in the slow time η, and we call it as range direction scatter function (RDSF). With this function denoted as $g(\tau,\eta) = g'(\tau,\eta)\exp\{j\psi(\tau,\eta)\}$, the baseband echo of the surface targets can be written as

$$s_{bb} = g(\tau,\eta)\otimes h(\tau,\eta) \tag{3}$$

where $h(\tau,\eta)$ is the impulse response of the unified single target and it is formulated as [5]

$$h(\tau,\eta) = \omega_r(\tau - 2R(\eta)/c)\ \omega_a(\eta-\eta_c)\exp\{-j4\pi f_0 R(\eta)/c\}\exp\{j\pi K_r(\tau-2R(\eta)/c)^2\} \tag{4}$$

It is difficult to directly implement equation (3) and equation (4) since they are the continuous form. Another problem in implementing equation (3) and equation (4) is that the parameters in them are coupled. For example, both $g(\tau,\eta)$ and $R(\eta)$ is not only space variant but also slow time variant.

In order to simplify the problem, for η can be seen as a discrete variable, which indicates the sampling time in the azimuth direction, equation (3) and equation (4) can be rewritten as [5]

$$\begin{aligned} h(\tau,\eta_i) &= \omega_r(\tau-2R(\eta_i)/c)\,\omega_a(\eta_i-\eta_c) \\ &\quad\times\exp\{-j4\pi f_0 R(\eta_i)/c\}\exp\{j\pi K_r(\tau-2R(\eta_i)/c)^2\}, i=0,1,...,I-1 \end{aligned} \tag{5}$$

$$\begin{aligned} s_{bb} &= g(\tau,\eta_i)\otimes h(\tau,\eta_i), \\ &= (g(\tau,\eta_i)\,\omega_a(\eta_i-\eta_c)\exp\{-j4\pi f_0 R(\eta_i)/c\}) \\ &\quad \otimes \omega_r(\tau-2R(\eta_i)/c)\exp\{j\pi K_r(\tau-2R(\eta_i)/c)^2\} \\ &= g_t(\tau,\eta_i)\otimes\omega_r(\tau-2R(\eta_i)/c)\exp\{j\pi K_r(\tau-2R(\eta_i)/c)^2, i=0,1,...,I-1 \end{aligned} \tag{6}$$

Where

$$g_t(\tau,\eta_i) = g(\tau,\eta_i)\,\omega_a(\eta_i-\eta_c)\exp\{-j4\pi f_0 R(\eta_i)/c\} \tag{7}$$

With a dedicated i, equation (6) can be regarded to execute a two dimensional convolution for there is a two dimensional scatter surface. Moreover equation (5) represents the i-th sampling along the slow time and meanwhile it is the i-th range direction echo.

There are three important implications in equation (3)-(6):

(1) With different slow time (azimuth) sampling point, equation (5) and equation (6) are independent. That means although slow time sampling η_i is sequencing in the real world, it doesn't matter that the k-th sample is calculated earlier than the l-th one, even if $k > l$. It is also doesn't matter that they are calculated at the same time, thus range direction echo can be calculated in parallel.

(2) Although $g_t(\tau,\eta_i), i=0,1,...,I-1$ in equation (7) is two dimensional in the scatter surface for a dedicated η_i, dimension reduction could be employed via a digitization of equation (6) and equation (7) in order to ease the implementation.

(3) In equation (6), the convolution calculation requires massive calculation resources.

As mentioned above in implication (2), the discrete sampling in range (resulting in range cells) and discrete scatter surface are introduced so that RDSF in equation (7) can be approximately expressed as

$$g_t(\tau_{n\_rc},\eta_i) = \sum_{n\_rc} g(\tau_{n\_rc},\eta_i)\,\omega_a(\eta_i-\eta_c)\exp\{-j4\pi f_0 R_{rc}(\eta_i)/c\} \tag{8}$$

where n_rc denotes the index of the scatter points located in the same range cell. According to the calculation of the RDSF in equation (8), all of the scatter points in the whole two dimensional scatter surface should be traversed and there are 4 steps of calculation for each dedicated azimuth sample:

(1) for every scatter point, the slant range is calculated and then it is approximate to a proper range cell,

(2) for every scatter point, the phase modulation term $\exp\{-j4\pi f_0 R_{rc}(\eta_i)/c$ is calculated,

(3) for every scatter point, the backscatter coefficient $g(\tau_{n\_rc},\eta_i)$ and the antenna pattern modulation $\omega_a(\eta_i-\eta_c)$ are multiplied, and finally,

(4) for every range cell, all scatter points located in are integrated.

For the 4 steps above to calculate the RDSF in equation (8), the range cells and scatter points are located in the range-azimuth two dimensional surface, and it doesn't matter which range cell or which scatter point is calculated first. That means they can be calculated simultaneously, and the only requirement is that equation (8) is calculated before equation (6).

As summarized, to calculate equation (3)-(8) an embedded architecture can be proposed.

2.2. Embedded Design of SAR Echo Simulation Algorithm

The embedded architecture of SAR echo simulation algorithm is presented in Figure 1. For the parallel nature of calculation along the azimuth samples, the whole task is firstly partitioned into sub routine for several azimuth sample subsets, which could be parallel assigned to separate processing units, thus all the range direction echoes of these subsets are computed in parallel. Then inside the procedure in every separate processing unit, there is a repetition to traverse all azimuth points in the corresponding subsets. Meanwhile there are two kinds of different requirements on calculation in these procedures: one is RDSF calculation which requires parallel calculation along all scatter points; the other is the convolution of scatter function and the detection pulse which requires massive and real-time computing.

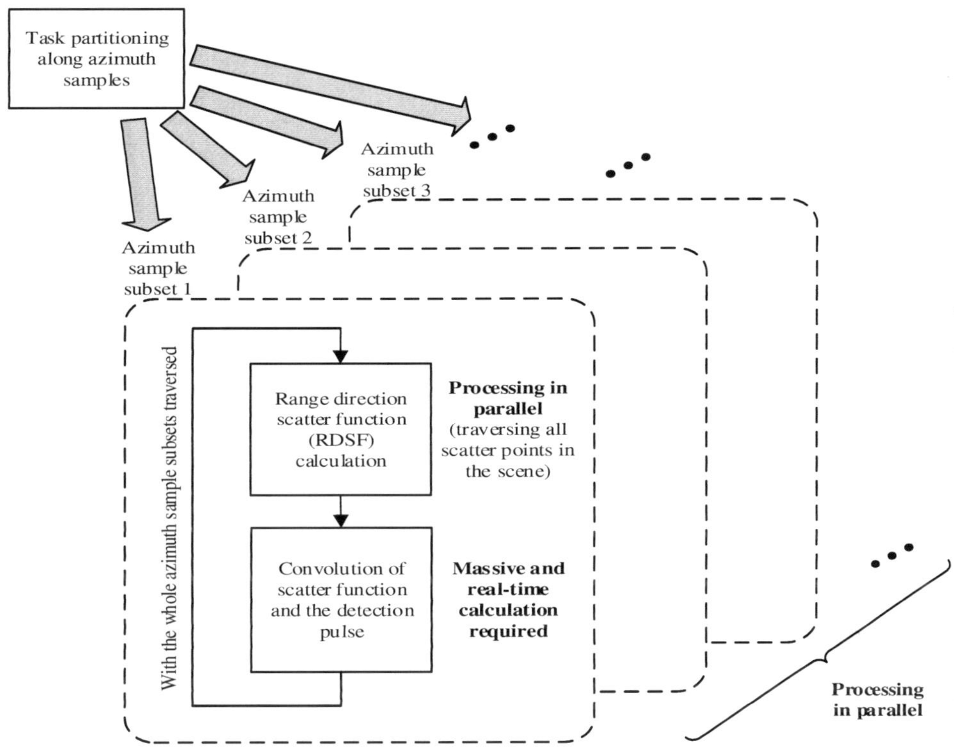

Figure 1. Embedded architecture of SAR echo simulation algorithm

With the different requirements of the calculation tasks depicted in Figure 1, different kinds of processing units could be properly adapted. For task partitioning, small amount of computing but relatively complex controlling is required, so an ARM processor may be suitable to be employed. For RDSF calculation, because the two-dimensional massive scatter points need to be processed in parallel, GPUs are suitable to be used. And finally, due to massive and real-time calculation of convolution, FPGAs are most suitable for it.

3. Prototype Design of Heterogeneous SAR Echo Simulator

3.1. Hardware Design

Base on the architecture depicted in Figure 1 and the analysis presented above, a heterogeneous computing platform featuring in one MPSoC and four GPUs is proposed in Figure 2. The heterogeneous platform is mainly composed of an MPSoC as the central control node, GPU computing nodes, switching nodes and a storage system. GPU computing nodes and switching nodes are connected through PCIe interface. Switching nodes and the central control node are connected through SRIO interface.

The central control node uses the XILINX's Zynq-7000 series MPSoC. GPU computing nodes use the NVIDIA's JETSON Tegra X2 modules. Switching nodes use the XC7K70T from XILINX's Kintex-7 series FPGA, which is used to implement the communication interconnection between the central control node and all GPU computing nodes. The fabricated platform prototype is shown in Figure **3**.

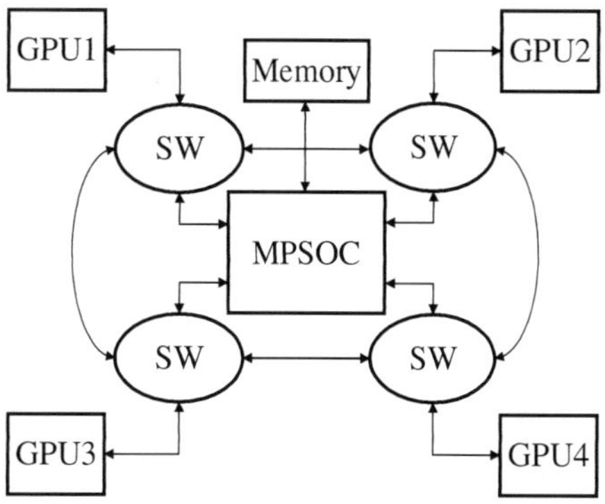

Figure 2. MPSoC+4 GPUs platform structure

Figure 3. The fabricated simulator prototype

3.2. Software Design

3.2.1. Software architecture. Porting the architecture in Figure 1 into hardware prototype shown in Figure 2 and Figure *3*, the proposed software architecture for implementation is presented in Figure 4. PS of Zynq SoC is used to partition calculation tasks for different azimuth sample subsets and assign them to GPUs. GPUs are employed to calculate the RDSF. Finally PL of Zynq SoC is mainly responsible for convolution operations.

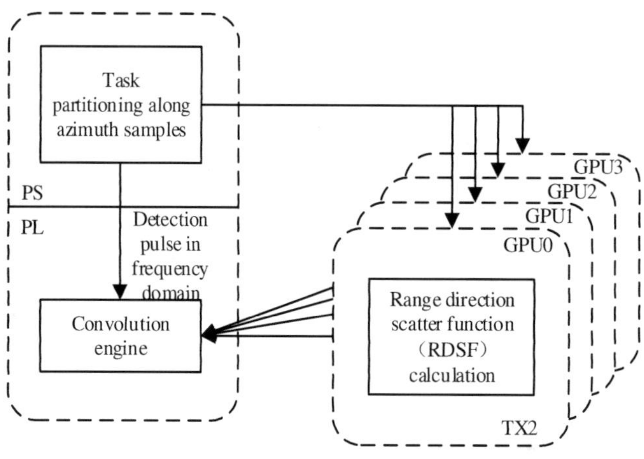

Figure 4. Software architecture of SAR echo simulation algorithm

3.2.2. Zynq SoC software. When PS of Zynq SoC initializes the tasks and scene data, it sends the data to GPU0, GPU1, GPU2 and GPU3, so that they can calculate the point target echoes at the sampling time of 4N, 4N+1, 4N+2 and 4N+3 respectively. Thus the azimuth samples are equally partitioned into four subsets and an average payload along the four GPUs are achieved.

When GPUs send the RDSF according to an azimuth sampling time to PL of Zynq SoC, the PL adopts a convolution engine to perform the convolution in frequency domain, with the detection pulse in frequency domain preset by PS. The convolution in frequency domain performs the procedure of FFT on RDSF, multiplication of detection pulse in frequency domain and finally IFFT operation, which decrease the computation complexity. PL mainly implements the convolution engine and multiplexes along all azimuth samples, so the workload of RTL programming is highly reduced. When PL of Zynq SoC completes convolution, the final result is summarized on PS of Zynq SoC.

3.2.3. GPU software. RDSF in equation (8) is calculated by a four-step procedure as discussed in chapter 2.1, which can be divided into two stages with high parallelism respectively. Two kernel functions are opened for these two stages in each GPU. In the first kernel function, each point target corresponds to one thread. In each thread, the calculation of slant range, range cell, phase modulation term, backscatter coefficient and antenna pattern modulation are implemented. In the second kernel function, different threads in the same block correspond to the product term in Equation (8) of different point targets in the same range cell. In each block, the accumulation of RDSF is implemented by using the reduction algorithm. For a dedicated azimuth sample, the procedure to calculate RDSF is presented in Figure 5.

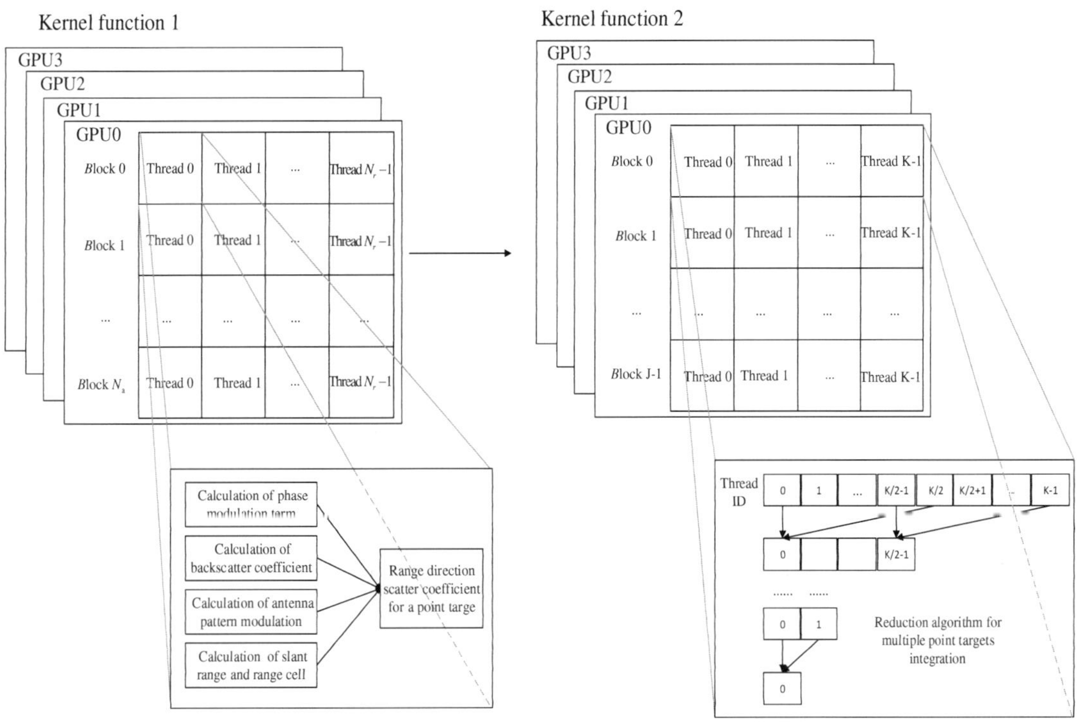

Figure 5. GPU kernel function operation diagram

In the first kernel function, it is assumed that the size of surface target reference scene is $N_a \times N_r$ pixels, where N_a is the sample number of azimuth direction and N_r is the sample number of range direction. At each sampling instant, $N_a \times N_r$ threads are opened, where N_a is the number of blocks and N_r is the number of threads in each block. In each thread, slant range, phase and range cell of the corresponding point target are calculated.

In the second kernel function, it is assumed that there are J range cells in the result of the former and there are $K_1, K_2, ..., K_J$ points in different range cells. Suppose that $K >= \max\{K_1, K_2, ..., K_J\}$ and K is integer powers of 2. $J \times K$ threads are opened. When the number of points in a range cell is less than K, the value in the corresponding thread is assigned to 0. Then, the accumulation of product term in Equation (8) in the same range cell is realized by reduction algorithm. In the reduction algorithm, at the first step, the algorithm simultaneously implements the addition of the 0th and the

$(K / 2)$ th, the 1st and the $(K / 2+1)$ th, ..., thus obtaining $K / 2$ addition results. After $\log_2 K$ steps, we get the RDSF of the corresponding rang cell.

4. Test Results and Analysis

In the echo simulation experiment, the simulation platform parameters are shown in Table 1. The echo simulation of surface target reference scene used in this paper is shown in Figure 6 and its size is 1024 × 1024 pixels. The echo simulation is performed by the heterogeneous computing platform and then the generated echo data is imaged by the Rang-Doppler algorithm, which is shown in Figure **7**.

Table 1. Simulation platform parameters.

Parameter	Value
CUDA version	9.0
CUDA core number	256*4
Memory size	8GB
Memory bandwidth	58.3GB/S

Figure 6. Original image of the simulated surface target

Figure 7. The simulated SAR image

In order to evaluate the performance of the architecture designed in this paper, the paper compares the time consumption and typical power consumption with literature [1], [3] and [6]. The test results are shown in Table 2. Compared with the traditional CPU+GPU architecture, the simulator proposed in this paper not only takes less time but also consumes less power. The time consumption of literature [3] is 114 times that of this paper, while the power consumption is 1.5 times that of this paper. The time consumption of literature [6] is 44 times that of this paper, while the power consumption is 6.8 times that of this paper. Compared with FPGA architecture, there is still a certain gap in computing performance, but the GPU modules used in this paper are low-power products and totally the proposed simulator prototype typically consumes less power.

Table 2. Comparison to results of different platforms.

Platform architecture	Main device	Main device	Time consuming (ms)	Typical power consumption(W)
Zynq SoC+GPU (This paper)	XC7Z100 + 4*Tegra X2	1024*1024	3.87	45
CPU+GPU[3]	GeForce GT620M	1024*1338	441.6	70
CPU+GPU[6]	Tesla K20C	1024*1024	172	310
FPGA[1]	4* StratixII	1000*1000	2.0	80

The detailed time-consuming decomposition of the simulator prototype proposed in this paper is shown in Table 3. It can be seen that it takes most of the time for data transmission, which means wider bandwidth of data interaction among multiple devices is required.

Table 3. Time-consuming composition.

Component	Time consuming(ms)
Calculation	0.73
Data transmission	3.14
Total	3.87

5. Conclusion

This paper digitized and decomposed the basic SAR simulation algorithm, proposed an embedded implementation structure, and designed a heterogeneous computing platform based on one MPSoC and multiple GPUs to implement the proposed algorithm. Compared with traditional platforms, the proposed platform features in higher integration, smaller size, better mobility, and higher energy efficiency.

6. Acknowledgements

This research was supported in part by Natural Science Foundation of Zhejiang Province, China, under Grants No. LY17F010012.

7. References

[1] Hongxian Wang, Yinghui Quan, Mengdao Xing and et al, 2010 Fast Implementation Method of SAR Echo Simulation Based on FPGA *Systems Engineering and Electronics* **32** 2284-89

[2] Lulu Chen, 2014 SAR Echo Signal Simulation Fast Parallel Algorithm and Engineering Implementation Method *Xidian University*

[3] Bo Li 2014 Research and Implementation of GPU-based SAR Echo Signal Simulation *Xidian University*

[4] Guobin Jing, Yunji Zhang and Zhenyu Li, 2016 Efficient Implementation Method of GPU-based SAR Echo Simulation *Systems Engineering and Electronics* **38** 2493-98

[5] Cumming, Ian G. and Frank H. Wong 2005 *Digital processing of synthetic aperture radar data: Algorithms and implementation* (Boston: Artech House) pp 141-147

[6] Yunji Zhang, 2015 GPU-based SAR Echo Simulation and Imaging Method *Xidian University*

IOP Conf. Series: Materials Science and Engineering **719** (2020) 012032 doi:10.1088/1757-899X/719/1/012032

Research on Automatic Discharge System of Stamping Line Tail

Jiangang Yi[1,*], Wen Bo[1] and Zibo Wang[1]

[1] School of Electromechanical and Architectural Engineering, Jianghan University, Wuhan 430056, China
469883553@qq.com

Abstract. The automatic discharge system of stamping line tail is an important part of automatic logistics production line, which can solve the problem of disorderly placement and management of end workpieces. In this paper, an intelligent automatic discharge system for stamping line tail is proposed, focusing on the design of horizontal discharging mechanism, vertical lifting mechanism and conveying mechanism. Through reasonable selection and calculation analysis, the mechanism composition of each part is given. On this basis, the key components are modelled and simulated by finite element method.

1. Introduction

The automatic production line is mainly divided into automatic feeding at the beginning of the line, automatic transfer of semi-finished products between the lines and automatic discharging at the end of the line. The improvement of automation and flexibility of stamping line can not only optimize the rhythm of production line, but also avoid the faults of production line workers in high-tempo production operation and improve the productivity and safety of production line. The traditional tail discharging process of stamping production line mainly depends on manual control, which is difficult to meet the needs of high-speed, flexible and intelligent production. At present, foreign factories have begun to use intelligent logistics warehouse transport robots to achieve material transportation in stamping production line. China's research in this area is relatively backward. Therefore, this paper proposes a flexible automatic discharging system for stamping line tail, discusses its composition scheme and mechanism, and uses it to improve production efficiency and production safety.

2. System Composition

In stamping production line, compared with manual handling, the safety and efficiency of automatic palletizing by manipulator are very high. Therefore, the designed discharging system is used in conjunction with the parallel manipulator on the production line. The discharging system includes conveying mechanism, transverse discharging mechanism and vertical lifting mechanism.

Because of the lower weight of stamping parts, the load on the conveyor is smaller. Therefore, flat belt conveyor can be used for conveying mechanism. Because flat belt conveyor has the characteristics of simple structure and low design cost, it is more suitable for small load workpiece transportation, which has higher flexibility and simpler control mode. In addition, the various parts of flat belt conveyor have been standardized, suitable for mass production, so that the design and manufacture cycle can be greatly reduced. The conveyor is used to transport the workpiece produced from the stamping line to the grasping position of the parallel manipulator, providing a platform for grasping recognition of the parallel manipulator, as shown in Figure 1.

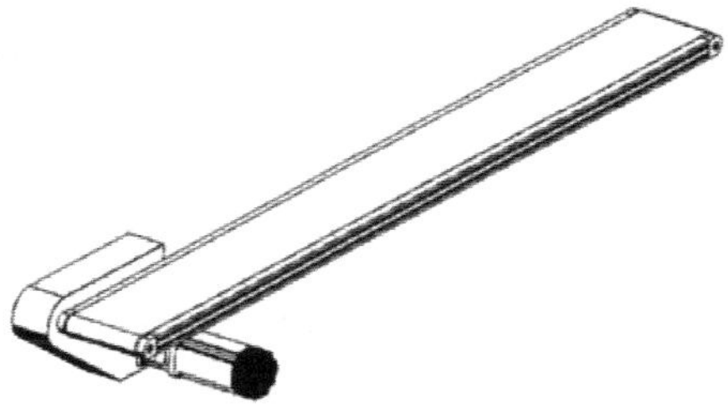

Figure 1. Conveyor Mechanism

The transverse discharging mechanism is used to realize the rapid delivery of the workpiece. The trajectory is high-speed reciprocating motion. Considering the load weight of the stacked workpiece and the stability of high-speed transmission, the reciprocating high-speed linear module composed of synchronous belt drive and linear guide is selected. Using medium-load linear guide can make the life of linear high-speed module in a longer state. Selecting smaller high-torque synchronous pulley and synchronous belt can reduce the manufacturing cost and the size of the overall mechanism. By installing synchronous belt base fittings with slider of linear guideway, high efficiency and high speed linear motion can be guaranteed. The slider holder needs to pass through the loose and tight edges of the synchronous belt, so it needs to be designed.

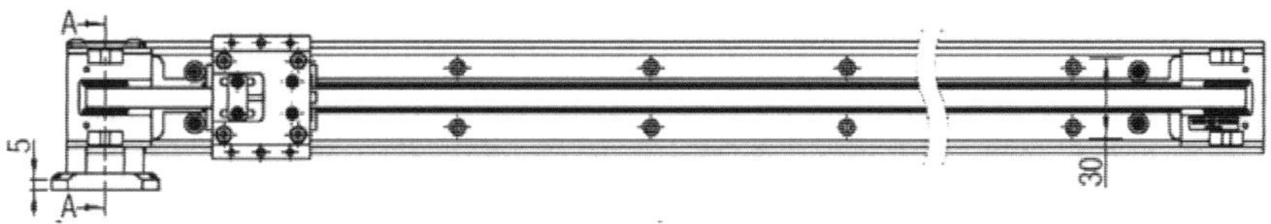

Figure 2. Transverse Discharging Mechanism

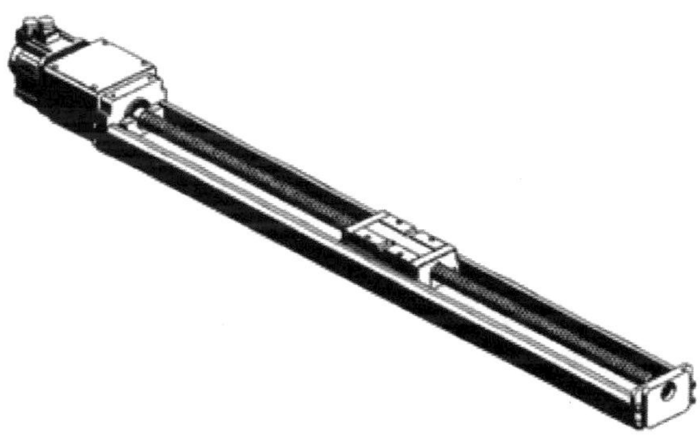

Figure 3. Vertical Lifting Mechanism

The function of vertical lifting mechanism is lifting tray. After the manipulator grabs the workpiece on the flat belt conveyor, the feature points of the workpiece are identified by the camera, and then the workpiece is put into the pallet. For each workpiece, the motor drives the ball screw to drop one unit vertically, waiting for another workpiece until the tray is full. The tray with full workpiece is sent out by the transverse discharging mechanism. The linear module of ball screw is driven by stepping motor, which can achieve better servo control. Because of the need to achieve multi-point residence and infinite speed regulation, the ball screw nut actuator driven by stepping motor is selected.

3. Computational Analysis

Flat belt conveyor has a long conveyor length. In order to maintain the normal operation of the conveyor so that the loose edge of the belt will not slip, it is necessary to set a higher belt tension. Therefore, the conveyor needs to produce higher torque. Ordinary servo motor can not be directly connected with the conveyor driving wheel. Considering the economy of the conveyor, it is necessary to install a reducer before driving the driving wheel. The function of the reducer is to improve the output torque of the transmission shaft and adjust the speed of the motor to a suitable range. According to the output torque of the reducer and the transmission efficiency of the reducer, the minimum output torque required by the motor can be calculated in reverse.

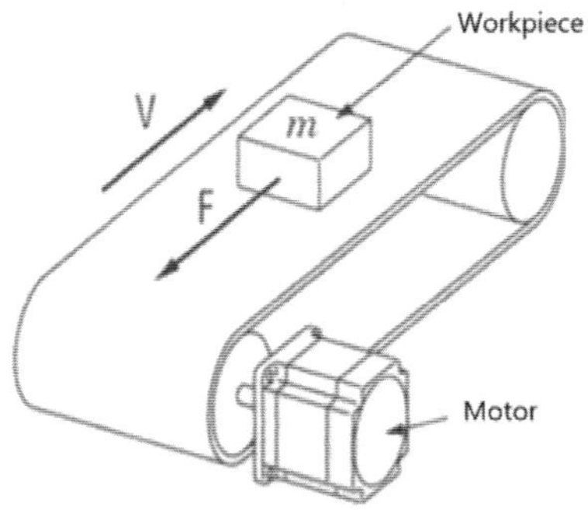

Figure 4. Driving Mode

Considering the size of the motor and the conventional installation method, the transmission mechanism is set to be installed in parallel with the bottom external, and the synchronous belt is used to connect the motor power to drive the driving wheel. The motor drives the roller of the conveyor to rotate. The friction provided by the inner side of the conveyor belt makes the belt move in a straight line, and then drives the driven roll on the other side to roll. Because the movement precision required by the conveyor is not high, only light parts can be transported. In the course of running, the slider will be affected not only by the resistance but also by the moment. The direction of the action of the moment varies with the installation mode. Here, the straight-line guide is installed horizontally, and the load can be calculated by Equation (1).

$$P = F + \frac{C_0}{M_C}(F \times L_R) + \frac{C_0}{M_B}(F \times L_Y) \tag{1}$$

The linear guideway is subjected to the gravity of the loaded object, it also needs linear reciprocating motion, so the position of the centre of gravity and the necessary thrust of the stamping part and the starting, stopping, accelerating and decelerating of the motor drive will cause the load acting on the linear guideway to change, as shown in Figure 5.

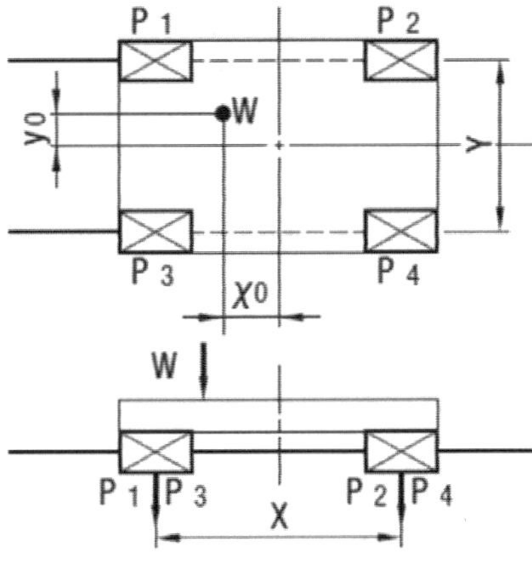

Figure 5. Force Distribution on Guideway

4. Design of Slider Connector Bracket

The slider connector bracket is used to fix the bracket. During operation, the bracket lifts the bracket, and after the machine hand locates and identifies the workpiece, grabs the workpiece from the conveyor and transfers it to the bracket. Since the lifting workpiece is a simple load, the bracket can be simplified to a cantilever beam subjected to uniform load, as shown in Figure 6. In order to reduce its deflection, reinforcement is added to the main load-bearing part. In order to ensure that the connector bracket can withstand the maximum load weight and not deform, it is necessary to carry out mechanical analysis.

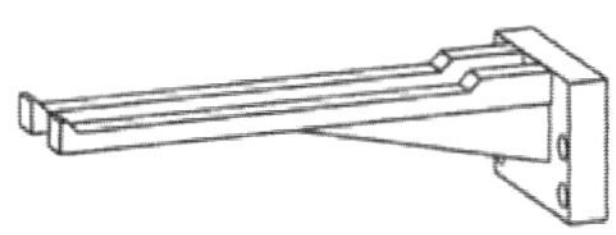

Figure 6. Slider Connector Bracket

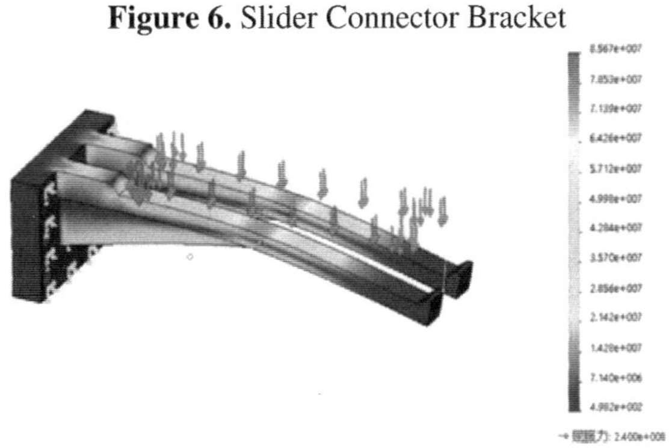

Figure 7. Finite Element Analysis

According to the operating conditions, it is necessary to calculate the deformation of the bracket at full load. If the deformation is too large, the bottom plate of the load will be inclined, resulting in irregular workpiece. Using finite element simulation software, the bracket model as shown in Figure 7 is established. Deformation analysis in static analysis is selected. Under full load, the bracket is subjected to the uniformly distributed pressure of the weight and its own gravity. The analysis shows that the maximum stress is located at the end of the reinforcing bar, and the deformation meets the standard. Therefore, the design of the bracket is reasonable and meets the application requirements.

5. Conclusions

The automatic discharging system of stamping line tail is an intelligent equipment used to improve automatic palletizing of production line. With the development of manufacturing industry, the demand for automated production is getting higher and higher. The designed system includes vertical lifting mechanism composed of ball screw, horizontal discharging mechanism composed of synchronous belt and conveyor composed of flat belt conveyor. Through the structural design, parameter selection and force analysis of the main components, the transmission scheme and overall structure design of the three mechanisms are given. Finally, the strength of the slider connector bracket is checked by finite element software, and the feasibility analysis of the design scheme is verified.

6. Acknowledgements

This work is supported by the Industry-university-research project of Wuhan Education Bureau (Granted No: cxy02), China. The authors also gratefully acknowledge the helpful comments and suggestions of the reviewers, which have improved the presentation.

7. References

[1] Zhen Z, Shan Z D, Chao J, et al. Study on automatic feeding in hot stamping production line[C]// International Conference on Advanced Technology of Design & Manufacture. 2012.

[2] Cao Q X, Xiao, Xiao Ting. Development of Stamping Process CAD System on Automatic Production Line for Robotic Manipulation[J]. Applied Mechanics & Materials, 2014, 556-562:2256-2261.

[3] Xue F W, Zhou, Ji Ping. Simulation Analysis and Structure Optimum Design of the High Speed Conveying Manipulator Based on ADAMS[J]. Advanced Materials Research, 2014, 945-949:121-126.

[4] Li S, Chen Y, Zhang W, et al. Design of automatic stamping line with centralized control mode[J]. Journal of Nanjing University of Science & Technology, 2015.

[5] An C M, Li, Yin Xing. Research on Crash Simulation of Car Stamping Results Based-System Management[J]. Applied Mechanics & Materials, 2013, 321-324:1570-1573.

Adaptive Audio Watermarking Algorithm Based on Wavelet Transform

Hanyu Lu[1], Mingjing Zhong[1], Haixiang Lan*[2], Ruchong Hou[1] and Zhengjiang Hu[1]

[1] College of Big Data and Iinformation engineering, GuiZhou University, Guiyang, GuiZhou.
[2] Guizhou Lichuang Science and Technology Development Co., Ltd. Guiyang, GuiZhou.
luhanyu@163.com

Abstract. Aiming at the problem of small capacity and poor robustness of audio watermarking information, an adaptive digital audio blind watermarking algorithm based on wavelet transform is proposed. In this algorithm, each frame of the audio signal is converted into a discrete transform by using Haar wavelet, and the fine and approximate components are obtained, and the component masking threshold is calculated. Finally, the watermark embedding position is selected by the adaptive algorithm, quantized and modulated in the audio. The intermediate and low frequency coefficients of the signal. The simulation results show that the proposed algorithm has some resistance to common signal processing attacks and is robust to resampling and low-pass filtering attacks.

1. Introduction

In recent years, the rapid development of digital multimedia technology and Internet technology has made image, video and audio transmission convenient and widely used [1-10]. Digital watermarking technology is a specific process meaning and easy to extract information embedded in the original audio signal. In this paper, an adaptive digital watermarking algorithm based on wavelet transform is proposed, which can realize blind detection. The proposed algorithm can adaptively select the embedded noise location with small noise sensitivity. The audio watermark is embedded in specific information and easy to extract. The original audio quality is not affected during the embedding process.

2. Audio Watermarking Algorithm

There are usually two algorithms for implementing audio watermarking: time domain watermarking algorithm and transform domain watermarking algorithm. Time domain audio watermarks are easier to implement than transform domain watermarks but have lower computational complexity and are less able to withstand attack power.

2.1. Masking Threshold

There is a masking effect on the human ear. When there are some energy differences between two or more excitation sounds at the same time, If the watermark strength is limited below the masking threshold, then the unperceivable watermark is reflected. The frequency resolution of the human ear is nonlinear [4].

2.2. Audio Watermarking Method Flow Based on Wavelet Transform

Thus supporting different frequency ranges. Wavelet transform is a very suitable tool for constantly changing the frequency of an audio signal. The audio watermark embedding uses a wavelet transform-based method. The flow is: the original audio signal is converted into audio pre-processing, then the DWT is performed, then the watermark embedding is performed, and the digital watermark is pre-processed in this link, then IDWT, and finally the watermark audio is embedded. .

The wavelet transform decomposes the signal into a set of orthogonal bases by translation and pressure expansion, if the Fourier transform of and is equal to equation (1):

$$\psi(\omega) = \int_{-\infty}^{+\infty} \psi(t) e^{-j\omega t} dt \tag{1}$$

So let wavelet or mother wavelet of $\varphi(t)$, transform or stretch the mother wavelet to get formula (2).

$$\psi_{a,b}(t) = \frac{1}{\sqrt{a}} \psi_{a,b}(\frac{t-b}{a}), (a,b \in R, a \neq 0) \tag{2}$$

For any function $f(t) \in L^2(R)$,the continuous wavelet transform is defined as:

$$W_f(a,b) \leq f(t), \psi_{a,b}(t) \geq \int_{-\infty}^{+\infty} f(t) \overline{\psi_{a,b}(t)} dt \tag{3}$$

The inverse transformation is defined as:

$$f(t) = \frac{1}{c_\psi} \int_0^{+\infty} \int_{-\infty}^{+\infty} \frac{1}{|a|^2} W_f(a,b) \psi_{a,b}(t) dbda \tag{4}$$

3. Audio Watermark Embedding and Extraction Process

In general, digital audio watermarks should have the following basic characteristics: invisibility, robustness, security, certainty, and recovery. The watermarking system is usually implemented in three steps: the embedding and extraction process of the watermark, and the detection of the watermark.

3.1. Adaptive Embedding Watermarking Step

Since the audio signal is a non-stationary signal, not all audio segments are suitable for embedding watermark information. Therefore, this paper conceives an adaptive selection scheme that can improve the performance of watermarks. After the discrete wavelet transform is performed, the approximate component is extracted[6]. Since most of our audio signal energy is medium and low frequency coefficients[7], combined with the robustness of the algorithm, we concentrate the watermark information on the intermediate frequency and low frequency coefficients.

1)The audio signal is segmented into frames, segments, and the original audio signal is segmented into frames containing the same sample points. Each frame contains the same sample point segment.

2)Performing a discrete wavelet transform on each frame of the audio signal $A_1^i A_2^i$ to obtain a fine component $A_{1d}^i A_{2d}^i$ and an approximate component $A_{1a}^i A_{2a}^i$:

$$(A_{1d}^i, A_{1a}^i) = DWT(A_1^i) \tag{5}$$

$$(A_{2d}^i, A_{2a}^i) = DWT(A_2^i) \tag{6}$$

For the approximation component, we don't need to process it, and the embedding watermark is done in the fine component.

3)Calculate the energy of the critical band of the approximate components of the first two segments of each frame E_1^i, E_2^i:

$$E_1^i = \sum_{i=1}^{s} \left| A_{1d}^i(t) \right| \tag{7}$$

$$E_2^i = \sum_{i=1}^{s} \left| A_{1d}^i(t) \right| \tag{8}$$

Where **s** is the number of the fine component of each segment.

4)Since the masking threshold calculation is related to the noise characteristic of the audio, the masking threshold is modified according to the noise characteristic factor. If pure tone masks noise, the masking threshold is reduced by 14.5 dB. Conversely, if the noise masks pure tones, the threshold is reduced by 5.5 dB. The degree to which the audio signal is similar to noise can be represented by the noise characteristic coefficient a(z).

$$a(s) = \min[10\lg(G/A)/E_{\max}, 1] \tag{9}$$

Where A and G are the arithmetic mean and geometric mean of the power spectrum, respectively. If the audio signal is pure tone, then a = 1, if the audio signal is white noise, then a = 0. The masking threshold after considering the noise characteristic coefficient is corrected as follows:

$$O(z) = a(14.5 + z) + 5.5(1 - a) \tag{10}$$

We use the larger value of each fine component as the final masking threshold[8].

5)Reconstructing the audio signal: applying the discrete wavelet inverter to $A_{1d}^{i'}, A_{1a}^{i}, A_{2d}^{i'}, A_{2a}^{i},$, reconstructing the audio signal $A_1^{i'}, A_2^{i'}$:

$$A_1^{i'} = IDWT(A_{1d}^{i'}, A_{1a}^{i}) \tag{11}$$

$$A_2^{i'} = IDWT(A_{2d}^{i'}, A_{2a}^{i}) \tag{12}$$

Finally, the watermarked audio signal A' needs to be reconstructed.

3.2. Watermark Extraction Step

1) Divide A' into frames and segments by embedding.

2) For each frame of the first two segments of the audio signal $A_1^{i'} A_2^{i'}$, Haar discrete wavelet transform is performed separately to obtain a fine component $A_{1d}^{i'} A_{2d}^{i'}$ and an approximate component $A_{1d}^{i'} A_{2d}^{i'}$:

$$(A_{1d}^{i'}, A_{1a}^{i'}) = DWT(A_1^{i'}) \tag{13}$$

$$(A_{2d}^{i'}, A_{2a}^{i'}) = DWT(A_2^{i'}) \tag{14}$$

3) After the calculation, the first second fine component $E_1^{i'} E_2^{i'}$ of each frame is compared, and then the watermark sequence is extracted, as follows: if $E_1^{i'} < E_2^{i'}$, the watermark is extracted as "1"; if, the watermark is extracted as "0".

$$W_s = \{W_s(i, j) = W'(k), 1 \le i \le M_1, 1 \le j \le M_2, k = i \times M_2 + j\} \tag{15}$$

Finally, $W'(i)$ is converted to detect the position of the watermark in the binary watermark image, and the watermark image is restored by increasing the dimension.

3.3. Robust Detection

Robustness detection uses the normalized correlation coefficient (NC) to calculate the similarity between the extracted watermark image and the original watermark image[10]. Let bt be the original watermark sequence, bt' is the extracted watermark sequence, and N is the watermark length. The normalized correlation coefficient is calculated as:

$$NC = \frac{\sum_{i=1}^{N} bt(i)bt'(i)}{\sqrt{\sum_{i=1}^{N} bt(i)^2}\sqrt{\sum_{i=1}^{N} bt'(i)^2}} \tag{16}$$

4. Experimental Simulation and Data Analysis

In the experiment, a 32×32 pixel binary image was used as the watermark, and four different types of carrier audio speech signals (classical style, jazz style, classic movie theme song and popular style) were selected, and their sampling frequency was 44.1. KHz, the playback precision is 16 bits, and the playback files are not in time. Taking the classical speech signal as an example, the audio signal is decomposed by Haar wavelet base and 3-level wavelet decomposition. Each decomposed component is calculated by the algorithm to a larger masking threshold, and the determined fine component is located. PSNR=84dB in the audio signal embedded in this position. The original signal is shown in Figure 1, the original image is shown in Figure 2, and the watermark signal is shown in Figure 3.

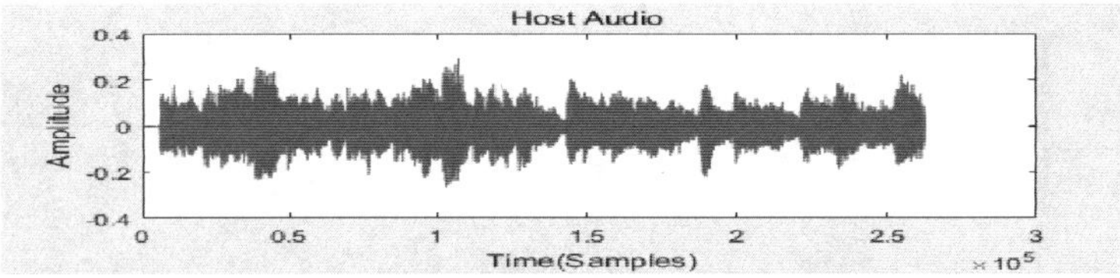

Figure 1. Original audio signal waveform

Figure 2. Original image

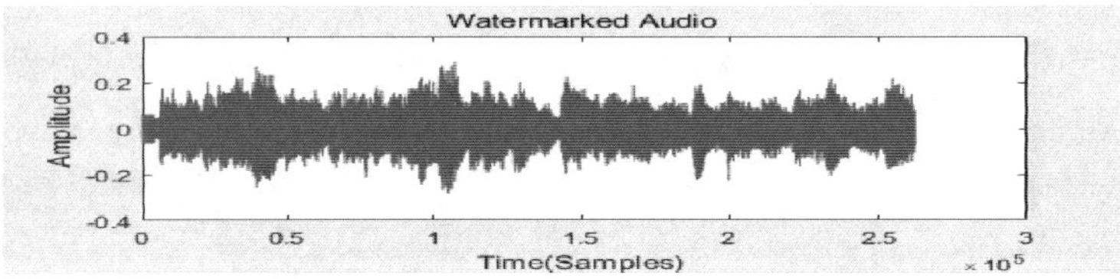

Figure 3. Watermark signal waveform

In order to detect the robustness of the algorithm, it is necessary to use MATLAB and other software to resample the audio signal, low-pass filter, add noise, shear compression, etc. The specific practices are as follows:

4.1. Resampling

The watermark signal is resampled with 48 kHz to obtain a waveform as shown in Figue4, and then the watermark shown in Figue5 is extracted by inverse wavelet transform.

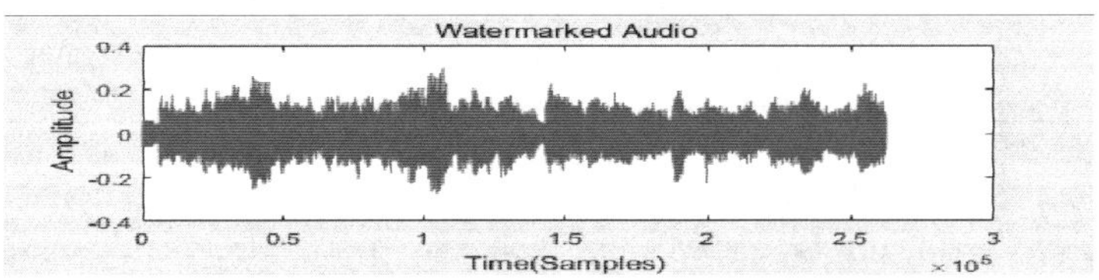

Figure 4. Resampled watermark signal

IOP Conf. Series: Materials Science and Engineering **719** (2020) 012033 doi:10.1088/1757-899X/719/1/012033

Figure 5. Extracting a watermark from a resampled signal

4.2. Low Pass Filtering

The watermark signal is filtered by a Butterworth low pass filter with N=20, wp=0.1, ws=0.9, rp=0.6, rs=0.9. The filtered waveform is as shown in Figure 6, and the watermark extracted from the signal by the low pass filter is shown in Figure 7.

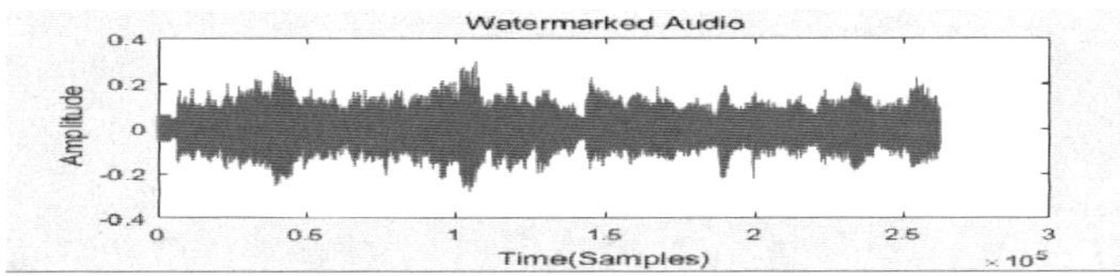

Figure 6. Low pass filtered watermark signal

Figure 7. Signal watermark through low pass filter

4.3. Gaussian Noise

The Gaussian white noise is added to the watermark signal of the Y=AWGN(X, SNR) function, and the watermark signal with Gaussian noise is shown in Figue8. The watermark extraction of the Gaussian noise signal is shown in Figue 9.

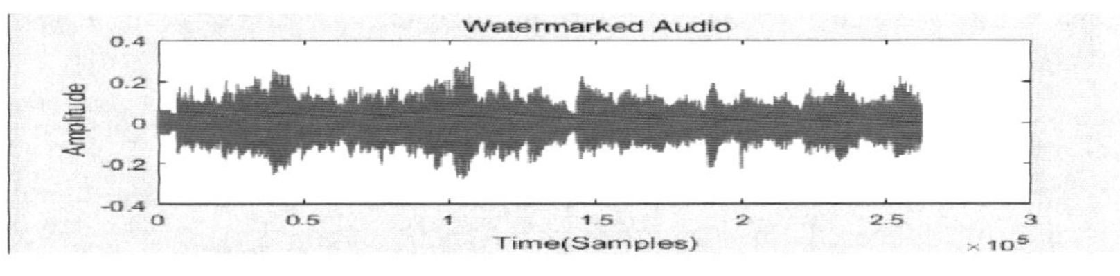

Figure 8. Watermark signal with increased Gaussian noise

Figure 9. Signal watermark with Gaussian noise

4.4. Compression

The audio editing software is selected to compress the entire watermark audio signal, and the compression is 1 dB, and the compression ratio is 89.13%, and the compressed signal waveform as shown in Figue10 and the watermark shown in Figue11 are obtained.

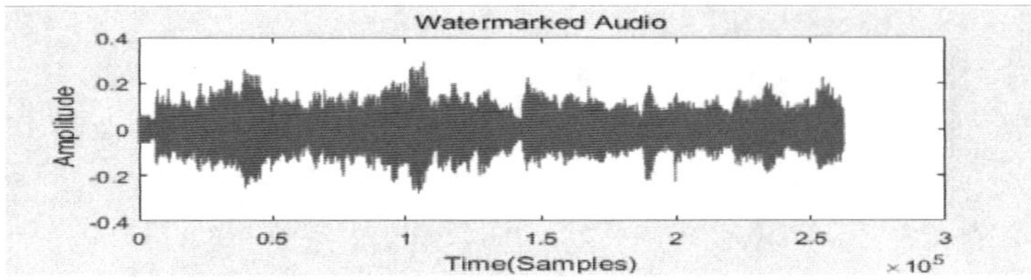

Figure 10. Compressed watermark audio signal

Figure 11. Compressed watermark

4.5. Normalized Correlation Coefficient

In the MATLAB simulation process, in order to detect the robustness of the algorithm, white noise attacks and resampling attacks are performed on the audio files embedded in the watermark respectively. Table 1 shows common voice attack tests, such as low-pass filtering attacks. The extracted watermark, and the corresponding normalized correlation coefficient NC value. The correlation of the watermark is evaluated using the normalization coefficient. For the embedded watermarked audio signal: Table 1 is the unprocessed original image and the extracted watermark NC value. After the original image is resampled and the watermark is extracted, white noise is added to the original image and the watermark is extracted, and the original image is low. Filtering, listening, and compressing, extracting the NC value after watermarking.

Table 1. NC value

Unprocessed	resampling	noise	Low pass filtering	Compression
0.9749	0.9527	0.9458	0.9516	0.8963

In the Matlab simulation process, the normalization function is used: NC= nc(ImageA, ImageB). In this function, ImageA is the original watermark image; ImageB is the extracted watermark image; use this function to get the NC value, which is used for Test the correlation of the watermark extracted from the original image. Referring to the function: x = psnr (a, b), this function is used to calculate the peak signal to noise ratio. In this function, "a" is the original audio signal; "b" is the embedded audio signal, and the similarity between the two signals can be easily obtained by calling the function. It can be seen from the results of Figure 7 and Table 1, the algorithm has strong anti-attack capability.

5. Conclusion

In this paper, image watermarks embedded in audio signals are implemented by adaptive algorithms using human auditory characteristics and discrete wavelet analysis. The algorithm has certain reliability and specific characteristics. Watermark extraction belongs to blind watermark extraction and does not require real-time carrier audio signals. The simulation results show that the algorithm has certain transparency and is robust to resampling and low-pass filtering attacks. Therefore, the audio watermarking technology based on wavelet transform is applied to the field of audio information security, and it is convenient to embed and extract specific meaning information. Since this article only deals with a small frame of audio, the impact of the watermark on this will be considered in future research and more comparisons will be made.

6. Acknowledgments

Thank you for Guizhou Science and Technology Plan Project (Guizhou Kehe [2017] 5305; 5788; 2816, [2016] 5604, [2015] 3054, [2014] 7633);

7. References

[1] Chen Yanli, Wang Haitao, Wang Juan. Digital watermark analysis of color image based on discrete wavelet transform[J].*Fujian Computer*,2018,1:14-15.

[2] Hua G, Huang J, Shi Y Q,et al. Twenty years of digital audio watermarking a comprehensice review[J].*Signal Processing*,2016,128:222-242.

[3] SUN Rui, JIANG Jian-guo, YANG Xing-ming.Audio watermarking algorithm based on wavelet packet and psychoacoustic model[J].*Journal of System Simulation*,2007,19(10):2248-2251.

[4] Bai Yubao, Bai Sen, Liu Chenghao. Digital audio watermarking algorithm based on wavelet and cepstrum domain[J]. *Journal of Computer Applications and Software*, 2012(3):163-167.

[5] W ang Wei, Wang Ji. Research on digital video watermarking technology[J].*Jiangsu Science and Technology Information*, 2018, 8:41-43.

[6] Zhao Cui. Research on digital watermarking technology of audio signal based on discrete wavelet transform[J]. *Radio and TV Technology*,2019.

[7] Yang Yuting. Research on digital watermarking algorithm based on wavelet transform[D]. *Chengdu University of Technology*, 2017.

[8] Jonston J D.Transform coding of audio signals using perceptual noise criteria[J].*IEEE Journal on Selected Areas Communications*,1988,6(2):314-323.

[9] Li Jingli.Research on the implementation of audio digital watermarking [D]. *Zhengzhou University*, 2010.

[10] Liang Jiadong, Yang Shuguo. A digital Image watermarking technique based on wavelet analysis[J]. *Modern computer*,2016:71-74.

IOP Conf. Series: Materials Science and Engineering **719** (2020) 012034 doi:10.1088/1757-899X/719/1/012034

Extraction and Recognition of Fingerprint Characteristic of Mobile Terminal's Transient Signal

Fucai Luo, Fei Wu, Hongfa Li, Ting Li, Jindong He and Qian Chen

State Grid Fujian Electric Power Company, 350000, China.
Email: Fucai Luo. 519594474@qq.com

Abstract. The fingerprint characteristic of mobile terminal's signal is unique and it can be used to identify the source of the signal. In all the characteristics, the characteristic of transient signals has been more favoured because of the greater diversity. However, the duration of transient signal is extremely short and difficult to accurately detect. Therefore, in order to successfully obtain the fingerprint characteristic, most of the research results are based on the laboratory environment. In this paper, through the methods of differential constellation trajectory and neural network, we realize the extraction and recognition of fingerprint characteristic of mobile phone's transient signal in the real environment. Meanwhile, by controlling the distance between the terminal and the base station, we also studied the recognition of fingerprint characteristic under different noise conditions.

1. Introduction

With the development of smart phones, wireless communication plays an increasingly important role in people's lives. Especially since this year, 5G will gradually become the new mainstream communication standard. However, no matter how the communication standard is updated, GSM has a status that cannot be ignored in the history of communication. In this paper, we will mainly consider the fingerprint characteristic of GSM signal, especially the GSM 900M frequency band (Figure 1 is the corresponding band diagram).

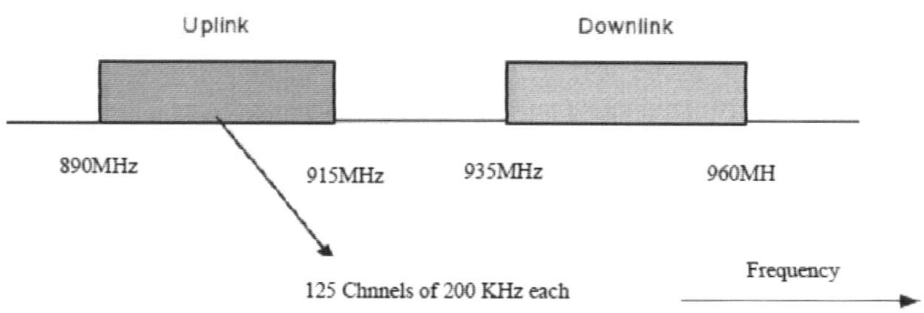

Figure 1. GSM 900M band diagram

Since the first day of wireless communication, electronic attack and defence at the information level has followed. With the development of technologies such as mobile payment, wireless network terminals are increasingly vulnerable to various information intrusion methods: WIFI phishing attacks, data sniffing attacks, etc. to overcome these problems, people usually need to perform various encryption and authentication on the transmitted information. In this process, people usually need to complete operations such as key generation, key agreement, and key distribution. This will cause a lot of waste on communication resources, and will bring a series of security problems. The uniqueness of the fingerprint characteristic of mobile terminal's signal provides the possibility to solve the above problems.

RF fingerprint characteristic is the characteristic vector or vector set extracted from the received wireless signal that uniquely identifies the transmitter individual and embodying the hardware characteristics of the wireless device transmitter. The characteristic has a certain stability; the characteristics are the same when the detection conditions are the same. Which means if the conditions of the two communication terminals do not change, the fingerprint characteristic of them will not change. If the adversary joins this communication, the characteristic will change because the conditions of communication have changed.

In 2003, Hall et.al. firstly propose RF fingerprinting in the identification of wireless device terminals [1]. RF fingerprinting uses the characteristic information extracted by the transient part of the mobile terminal transmitter signal to uniquely identify the terminal, and builds an anti-intrusion detection system that recognizes the RF fingerprint of IEEE 802.11b devices and successfully countered attacks such as MAC address cloning [2]. The characteristic is unique when the terminal is powered on, and divide terminal fingerprint characteristic recognition process into three stages: the transient starting point detection stage, the signal characteristic extraction stage and the signal characteristic recognition stage [3]. This concept is also applied to the field of ZigBee, Bluetooth and GSM terminals [4-9].

According to the timing of signal acquisition, the RF fingerprinting can be divided into transient RF fingerprint features and steady-state RF fingerprint features. Transient RF fingerprint refers to the characteristics of electromagnetic waves when the transmitter is turned on and off. This signal is determined by the hardware itself. Steady-state characteristics are characteristics of the signal itself, such as frequency offset, amplitude error, phase error, IQ offset, IQ imbalance, etc. Transient characteristic duration is very short, need to determine the starting point, difficult to collect. Steady-state characteristic has lower requirements for acquisition equipment, easy to collect. However, the steady-state characteristic may change, have lower difference, and it is not possible to extract identifiable characteristic in a signal. Therefore, transient characteristic better reflect terminal differences. And in transient signal-based RF fingerprinting, detecting the starting point of a transient signal is a critical step.

In order to more clearly depict the characteristics of transient signals, a method based on differential constellation trajectory map is used for RF fingerprint characteristic extraction [10]. In [10], the USRP equipment is used to draw the differential constellation using the signals generated by QPSK, BPSK, MSK, etc. The corresponding graphics are extracted, and the relevant characteristic are extracted to verify the reliability and practicability of the proposed method. It is also confirmed that in the device recognition process, the wireless terminals can be identified without extracting the a priori information of the transmitter.

The rest of the article is structured as follows: Section 2 gives the introduction of fingerprint characteristic extraction method, the differential constellation trajectory. Section 3 gives the introduction of fingerprint characteristic recognition method, the neural network and machine learning. Section 4 gives the specific experimental methods and experimental results.

2. Differential Constellation Trajectory

In digital communication, the constellation trajectory is directly processed by processing the sampled signal in the complex plane. It can provide a convenient way to study the relationship between the I/Q signals and visually represent the relationship between the signals.

When using oversampling to collect the radio frequency fingerprint characteristic, the constellation trajectory can be obtained by drawing the constellation map through the determined sampling points. This is because the nonlinear response of the amplifier and the interference factors from the filter are expressed in the constellation trajectory map, and the received signal characteristics are more closely distinguished in the transmitter signal transmission.

Under normal circumstances, there will exists errors such as frequency offset between the transmitter and the receiver, which will lead to instability of the constellation trajectory map. For a giving transmit signal

$$\boldsymbol{S}(t) = \boldsymbol{X}(t)e^{-2\pi jtf_{ct1}} \tag{1}$$

Then, in the ideal experimental case, the receiving signal is

$$\boldsymbol{Y}(t) = \boldsymbol{S}(t)e^{2\pi jt+\varphi f_{ct2}} \tag{2}$$

If there is a deviation between the receiving end and the transmitting signal, for example, f_{ct1} and f_{ct2} have a deviation Δf, then

$$\boldsymbol{Y}(t) = \boldsymbol{X}(t)e^{2\pi jt+\varphi\Delta f} \tag{3}$$

Obviously, the constellation trajectory map is a time-dependent function. It will rotate over time t, and makes us difficult to compare different images through the constellation. To overcome this problem, we need to differentially process the data under the premise of retaining the frequency offset. Then

$$\boldsymbol{D}(t) = \boldsymbol{Y}(t) * \boldsymbol{Y}^*(t+n) = \boldsymbol{X}(t) \cdot \boldsymbol{X}(t+n)e^{-2\pi j\theta n} \tag{4}$$

Although the image after the difference still rotates, the rotation is a stable value and is related to the frequency offset, so that we can better observe the characteristics of the frequency offset.

3. Neural Network and Machine Learning

Since the 1980s, neural networks have developed very rapidly. In our experiment, we mainly used 3 different neural network schemes. We will introduce them separately.

3.1. BP Neural Network

BP neural network is currently the most widely used neural network architecture for its ability to learn complex multidimensional mappings. It is a nonlinear dynamic systeme. It is made up of many simple processing units, and the larger the number, the better the recognition effect.

The main feature of BP neural network is that the signal is forward propagating and the error is back propagating. The model of BP neural network is shown in the figure 2 below:

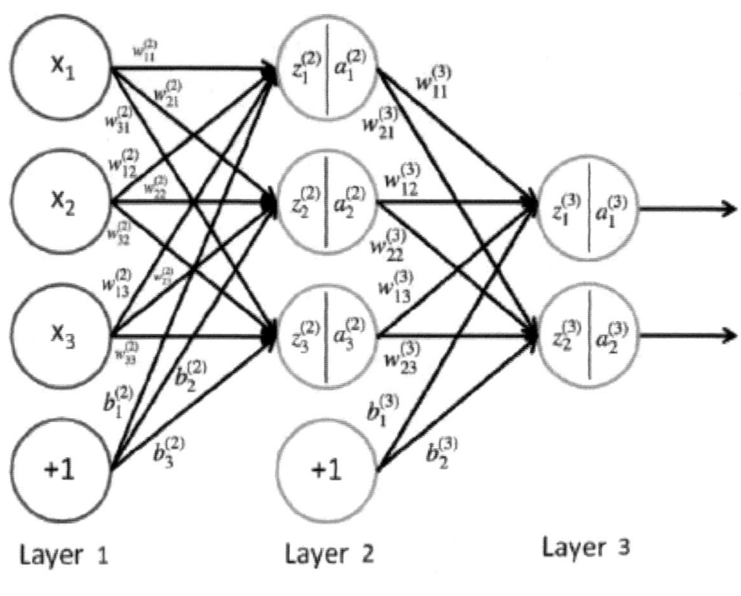

Figure 2. 3-layer BP neural network

The process of BP neural learning network is mainly divided into two stages, the first stage to the forward propagation of the output layer; the second stage is the error return to the input layer back propagation. The input to the neuron includes the sum of its deviation and the weighted input. The output of a neuron depends on the input of the neuron and its transfer function. The network implements the calculation by mapping the input values to the output values. The specific mapping problem to be performed determines the number of inputs and the number of outputs from the network. Therefore, we need to choose the best BP neural network structure according to the type of problem.

3.2. ALEXNET Neural Network

Compared with BP neural networks, ALEXNET neural networks are more mature and more complex. AlEXNET has five layers of convolution, three layers of fully connected network (see Figure 3). At the same time, ALEXNET can use two GPUs for calculation at the same time, which greatly improves the computational efficiency.

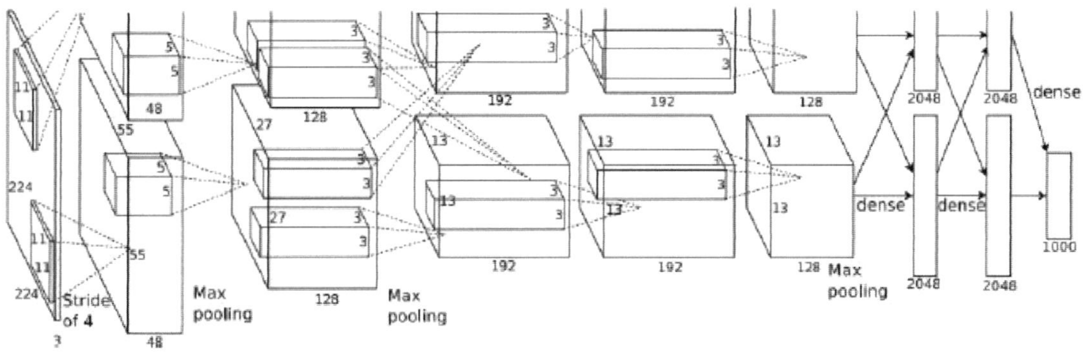

Figure 3. ALEXNET neural network training structure

3.3. Support Vector Machines

Support Vector Machine (SVM) is a new machine learning algorithm based on statistical learning theory with the goal of minimizing structural risk. It is mainly composed of 3 parts: Preprocessor, SVM classifier, and decision system. The preprocessor is to process the data into the form of SVM. The SVM classifier is the core component of the SVM system and needs to be used to train data training to achieve performance. Finally, the judgment equations in the decision system are pained to determine the category in which the object is located.

SVM was originally designed for binary classification problems. It need to construct a suitable multi-class classifier when dealing with multiple types of problems. In this paper, we combine our SVM classifier with the histogram of oriented gradients (HOG) feature to complete our experiments.

In an image, the appearance and the shape of the object can be described by gradient of the shape and the colour. Therefore, we can divide the image into individual cell elements, and analyze every individual cell element. For easy calculations, we can also normalize the date in each cell. Normalization can also achieve better results for changes in the color of image. Then, we can process the graphical gradients and features between cell and cell. According to all these results, we can draw the direction histogram to form the characteristic of each image for comparative recognition. Figure 4 gives the implementation process of HOG.

IOP Conf. Series: Materials Science and Engineering **719** (2020) 012034 doi:10.1088/1757-899X/719/1/012034

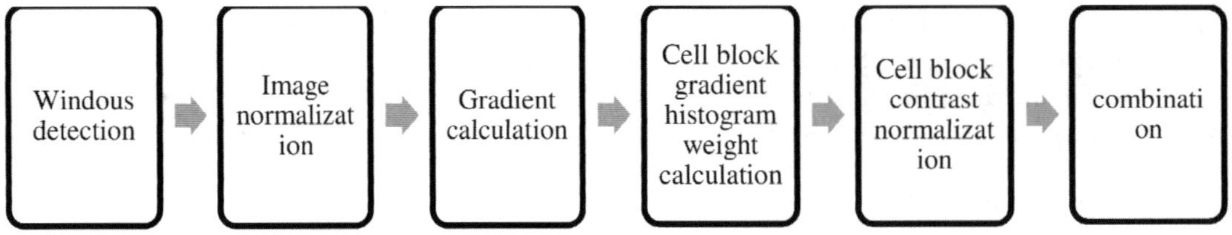

Figure 4. The implementation process of HOG

4. Our Experiments

The experimental equipment we experimented with includes hardware devices and software devices. On the software device, we chose GNU Radio. We use it to write data processing modules, and integrated the modules to form a complete wireless communication system model. On the hardware devices, we use Hack Rf and USRP devices as our transmitter and receiver, respectively. The Hack Rf (see Figure 5) is a software radio external device that mainly use to transmit and receive signals, and to mix and sample the signals. The USRP (Figure 6) is also a general-purpose software radio peripheral. It handles high-speed general-purpose operations such as up-and-down conversion, interpolation, and extraction.

Figure 5. The Hack Rf

Figure 6. The USRP

After we obtain the signals by the Hack RF. We first perform GMSK modulation on the received GSM signal. Then, we analyse the modulation results by means of spectrum analysis, frequency finding and filtering. So as to facilitate our SDCCH detection. With SDCCH, we can extract the transient characteristics of the signal, and then draw a differential constellation trace. Finally, by the neural networks and machine learning algorithms, we identify and classify the signals. Figure 7 gives the process of our experiment.

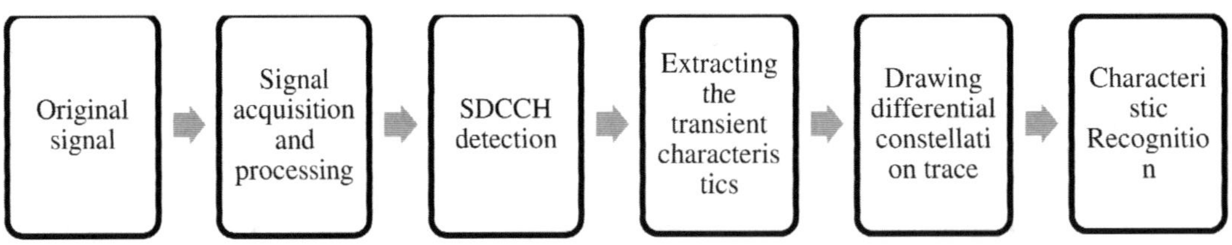

Figure 7. The process of fingerprint characteristic of signal

4.1. Characteristic Extraction

Before processing the acquired data, we use the GNUradio device to map the differential constellation traces of the GMSK modulated signals under ideal conditions. After the GMSK modulation of the transmitted signal, it is received by the USRP and then processed directly by GNUradio, and the interference caused by noise and multipath can be almost ignored. Set a delay amount during the receiving process, and delay and differentially calculate the signal each time. Then a differential constellation trajectory map in an ideal state can be obtained.

For the single GSM signal that has been acquired, we use Matlab to draw the differential constellation for processing. Then we select the appropriate differential value to directly process the data differentially. By this method, a differential constellation trajectory of single GSM signal with obvious features can be obtained.

In order to better reflect the differential constellation, we performed a visualization process to mesh the differential constellation trajectory map, by giving red, green and blue colors according to the density. The higher the density, the closer the color is to red. Note that, when drawing differential constellation trajectories, there will be different results with different interval difference. Figure 8 gives some example of differential constellation trajectory maps with different interval difference.

When choosing the appropriate differential interval, the differential constellation trajectory map will have a better contour, and it will also have a better shape on the curve near the center inside the contour. When the difference interval is large, the image contour will form a wide range of distortion, and it will not exhibit a good range of polymerization. The center curve will become blurred and cannot be a good recognition standard. When the difference interval is too small, it can be found that the constellation trajectory maps no longer exhibits a circular range that can be clearly observed, but the range appears to decrease, and the aggregation is too obvious. Furthermore, When the difference interval is close to 1, the image becomes a single line and no longer provides any feature information. Therefore, appropriate differential spacing is a great help for getting good fingerprint features.

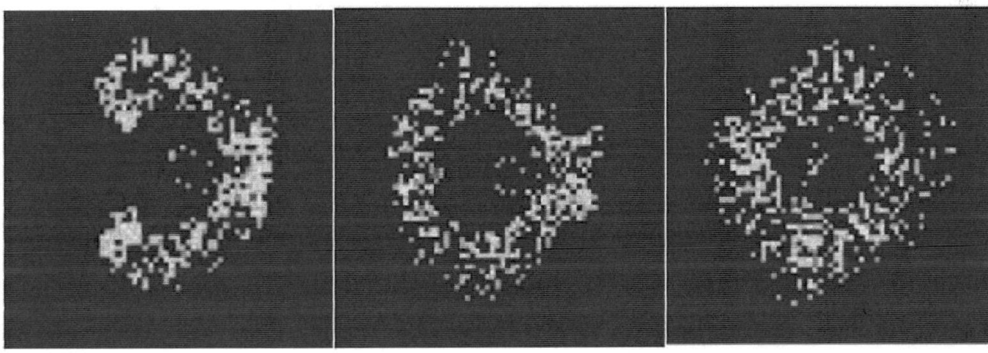

Figure 8. Differential constellation trajectory maps with different interval difference (From left to right, followed by smaller intervals, appropriate intervals, larger intervals)

From the differential constellation, we can observe both steady-state and transient characteristics. Due to the slot structure of GSM, the transient characteristics of the transmitter at the guard interval are fully reflected in the curve near the center of the contour of the differential constellation trajectory map. It is generated due to the differential data of the fading data of the guard interval, and the differential data of the fading data of the partial guard interval and the header bits of the frame, which represent the characteristics of the transmitter at the transmission interval of the base station. Compared with the traditional wavelet transform and instantaneous frequency, envelope and other methods, the differential constellation trajectory map directly shows these data in a curve, which is more intuitive.

After the differential constellation trajectory map is drawn, it is easy to find that in the differential constellation trajectory map of different devices, there will be a certain difference in the contour curve. And this is the key to the fingerprint recognition of mobile devices. Through the difference between the contour and the curve within the contour, we can effectively classify different terminal devices.

In addition, we found that there are still differences in the differential constellation traces of different terminal signals over longer distances. This means that the differential constellation trajectory can still be used as an important feature to identify the terminal signal under certain interference conditions.

4.2. Characteristic Recognition

In this chapter, we will realize the characteristic recognition of terminal signals from different angles through three different methods. They are BP neural network, ALEXNET neural network and Support Vector Machine, respectively.

4.2.1. BP Neural Network. This experiment uses the neural network tool system nntool that comes with Matlab to realize the construction of BP network. We only need to input the length of the vector and the number of output classifications, and then we can get a good neural network after training.

First, we perform binary recognition on the obtained 41*41 differential constellation trajectory map and convert the image into a matrix. Then, we input the obtained matrix into the BP network training box as the basic, and train and tested autonomously another part of the samples.

Here, we set up three layers of hidden layers, which are 25, 12, and 1 neuron, as shown in the figure 9. It can be seen that the BP neural network responds very well to our data, and only with a short training layer number, we have obtained considerable classification results, as shown in the figure 10.

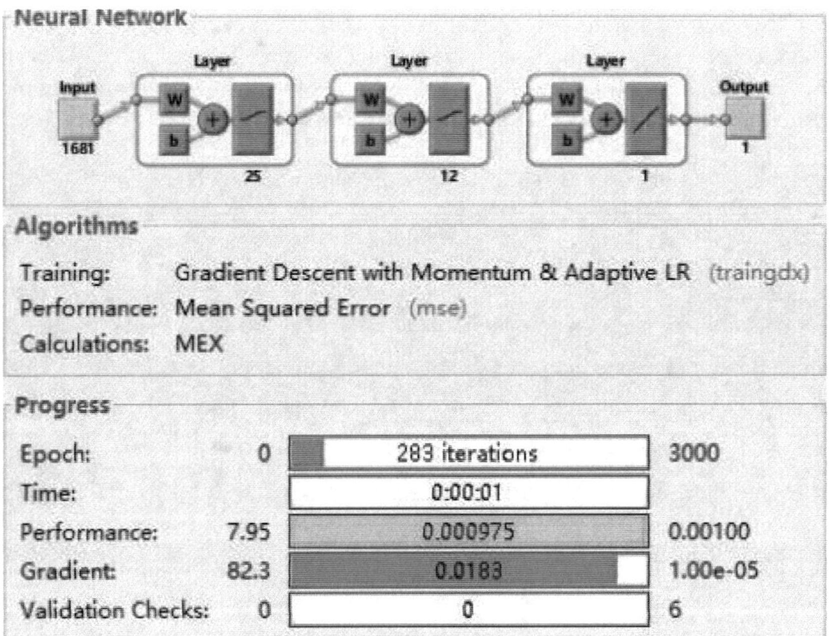

Figure 9. BP Neural Network

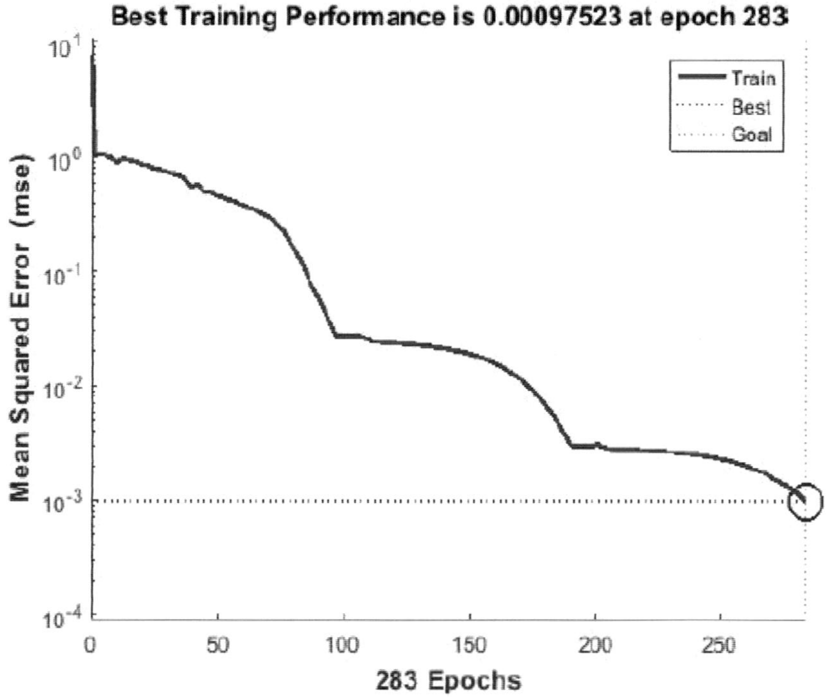

Figure 10. Training performance of BP

We also process the RF fingerprint of the terminal equipment tested in a noisy environment, and identify the single picture when the trained model has been deployed to determine whether it can successfully match the data in the sample library. Although the vast majority of samples have been identified. Unfortunately, one sample was misidentified. The reason is because BP needs to convert the original differential constellation trace map into binary data when inputting data. This can cause some data distortion and even loss of critical data.

4.2.2. ALEXNET neural network. ALEXNET can directly process the differential constellation trace map (41*41*3) without binarization. This means that the image can be fully recognized for recognition without losing some of the key information that can be avoided.
First, we input the differential constellation trace map drawn by the close-range terminal GSM data into the ALEXNET neural network system. Training parameters are shown in Table 1.

Table 1. Training parameters of AVERNET

Training Parameter	Value
Training Sample Set	200
Minimum Batch Size for Each Training Iteration	30
Maximum Number of Training Cycles	30
Initial Learning Rate	0.0001
Network Verification Frequency (Iterations)	30
Data for Verification during Training	Auto

The training results (as shown in Figure 11) show that the training speed is very fast, and all test samples can be correctly classified, and excellent experimental results are obtained (Root mean square error (RMSE) and Loss Value Function (LOSS) are both reduced to a very low range, as shown in Figure 12).

Epoch	Iteration	Time Elapsed (hh:mm:ss)	Mini-batch Accuracy	Mini-batch Loss	Base Learning Rate
1	1	00:00:00	13.28%	3.0845	1.0000e-04
1	50	00:00:02	64.84%	1.0945	1.0000e-04
2	100	00:00:03	74.22%	0.7260	1.0000e-04
3	150	00:00:05	83.59%	0.4741	1.0000e-04
4	200	00:00:06	91.41%	0.3089	1.0000e-04
5	250	00:00:08	92.97%	0.2333	1.0000e-04
6	300	00:00:09	97.66%	0.1539	1.0000e-04
7	350	00:00:11	97.66%	0.1317	1.0000e-04
7	400	00:00:12	96.09%	0.0944	1.0000e-04
8	450	00:00:14	98.44%	0.0662	1.0000e-04
9	500	00:00:15	99.22%	0.0458	1.0000e-04
10	550	00:00:17	100.00%	0.0545	1.0000e-04
11	600	00:00:18	99.22%	0.0660	1.0000e-04
12	650	00:00:19	100.00%	0.0338	1.0000e-04

Figure 11. The training results of AVERNET

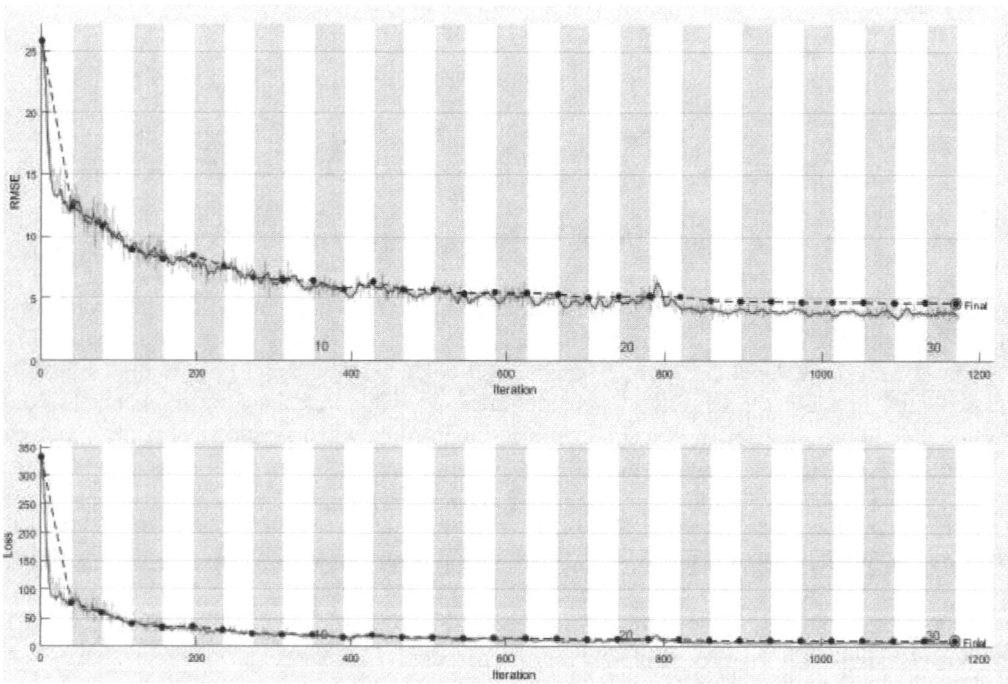

Figure 12. The RMSE and LOSS of training

After that, we imported the trained neural network and input the differential constellation trace map drawn under the loud noise into the SOFTMAX with the connection layer, which obtained a good recognition effect. The results show that the recognition of the constellation map in the ALEXNET training network has achieved a recognition rate of 98.5%.

4.2.3. Support Vector Machine.
As mentioned earlier, the SVM system is more about the binary classification problem. In order to use the SVM system, we took the fitcecoc function carried by MATLAB to complete our experiment. We input the extracted HOG features and the required classification labels through fitcecoc. Then, we can classify them with SVM, create sample sets, and predictively identify them by predict function.

As with the ALEXNET experiment, we used SVM for machine learning processing with a sample set of 200 different constellation trajectory maps and tested the test set. The results show that the recognition of the constellation map in the SVM has also achieved a recognition rate of minimum 85%.

5. Conclusion
In this paper, by analysing the terminal GSM transient signal, the differential constellation trajectory representing the signal fingerprint characteristics are obtained. Then, according to these fingerprint characteristics, the characteristic recognition of the wireless terminal device is realized by means of neural network and machine learning. Our experiments show that BP neural network, AVERNET neural network and SVM have excellent recognition ability for differential constellation trajectory feature recognition under reasonable training conditions. Moreover, the recognition of AVERNET neural network and SVM is more accurate than BP neural network.

6. References
[1] J. Hall, M. Barbeau, and E. Kranakis, "Detection of transient in radio frequency fingerprinting using signal phase," Wireless and Optical Communications, pp. 13–18, 2003

[2] Guyue Li, Jiabao Yu, Yuexiu Xing, and Aiqun Hu. Location invariant physical layer identification approach for WiFi devices. *IEEE Access*, pages 1–1, 2019.

[3] Linning Peng, Aiqun Hu, Junqing Zhang, Yu Jiang, Jiabao Yu, and Yan Yan. Design of a hybrid RF fngerprint extraction and device classifcation scheme. IEEE Internet Things J., 6(1):349–360, Feb. 2019.

[4] Yuexiu Xing, Aiqun Hu, Junqing Zhang, Linning Peng, and Guyue Li. On radio frequency fngerprint identifcation for DSSS systems in low SNR scenarios. *IEEE Commun. Lett.*, 22(11):2326–2329, Nov. 2018.

[5] J. Hall, M. Barbeau, and E. Kranakis, "Detecting rogue devices in Bluetooth networks using radio frequency fingerprinting," in *Proc. IASTED Int. Conf. Commun. Comput. Netw. (CCN)*, Lima, Peru, Oct. 2006, pp. 108–113.

[6] A. M. Ali, E. Uzundurukan, and A. Kara, "Assessment of features and classifiers for bluetooth RF fingerprinting," IEEE Access, vol. 7, pp. 50 524–50 535, 2019.

[7] Jiabao Yu, Aiqun Hu, Guyue Li, and Lin Ning Peng. A robust RF fngerprinting approach using multi-sampling convolutional neural network. *IEEE Internet Things J.*, pages 1–1, 2019.

[8] Xinyu Zhou, Aiqun Hu, Guyue Li, Linning Peng, Yuexiu Xing, and Jiabao Yu. Design of a robust rf fngerprint generation and classifcation scheme for practical device identifcation. In *Proc. IEEE Conf. Commun. Netw. Secur. (CNS)*, pages 1–9, Washington, DC, USA, Jun. 2019.

[9] G. Baldini, C. Gentile, R. Giuliani, and G. Steri, "Comparison of techniques for radiometric identification based on deep convolutional neural networks," *Electron. Lett.*, vol. 55, no. 2, pp. 90–92, Jan. 2019.

[10] Linning Peng, Aiqun Hu, Yu Jiang, Yan Yan, and Changming Zhu. A differential constellation trace fgure based device identifcation method for ZigBee nodes. In *Proc. Int. Conf. Wireless Commun. Signal Process. (WCSP)*, pages 1–6, Yangzhou, China, Oct. 2016.

DOA Estimation for Unfolded Coprime Arrays: Successive-MUSIC Algorithm

Chang Wu[1, 2] and Changbo Ye[1]

1. Department of Electronic Information Engineering, Nanjing University of Aeronautics and Astronautics, Nanjing, China
2. China Electronic Technology Group 55th Institute
Email: [1]wuchang818@126.com, [2]2714985990@qq.com

Abstract. In this paper, we investigate the problem of direction of arrival (DOA) estimation of unfolded coprime arrays with successive-MUSIC algorithm which cascades multiple signal classification (MUSIC) algorithm and rotational invariance techniques (ESPRIT).We propose An algorithm with low computational complexity and high estimation accuracy, which first use ESPRIT algorithm for initial estimation, then according to the results of initial estimation narrow the search scope, finally, more accurate estimation results are obtained by MUSIC algorithm. Compared with the traditional MUSIC algorithm, this algorithm only needs to conduct a small range of angle search, which greatly reduces the algorithm complexity. Compared with the traditional ESPRIT algorithm, this algorithm also effectively improves the angle estimation performance of the algorithm. Moreover, comparing with traditional coprime linear arrays, the enhanced degrees of freedom (DOFs) can be achieved due to larger array aperture, Simulations are presented to validate the superiority of the proposed algorithm.

1. Introduction

In recent years, a new array structure coprime array [1-3] has aroused wide interest, its advantage is to provide a larger array aperture [4], improved estimation performance [5] and enhanced DOFs [6]. A large number of effective algorithms are applied to DOA estimation in coprime array. Such as ESPRIT, MUSIC [7], PARAFAC algorithm as well as others. The effective estimation algorithms of coprime array are mainly defuzzification methods and spatial smoothing algorithms [8]. The DOA estimation of two submatrices is carried out by defuzzification method respectively, and the only DOA estimation value is obtained by comparing the estimation results with the mutual quality characteristics of subarray elements. Although the method is easy to implement, the space degree of freedom is greatly reduced by the method of estimating two submatrices separately.

In this paper, we investigate the problem of DOA estimation of unfolded coprime arrays. The unfolded coprime array is obtained by reversing one of the coprime linear arrays (CLA) to the right of the origin. We proposed a low complexity and high DOA estimation precision of algorithm, by combining the ESPRIT - MUSIC cascading algorithm and unfolded coprime array. on the one hand the proposed algorithm can play coprime array on the advantages of effectively improve spatial degrees of freedom, on the other hand, the proposed algorithm can fully combine the advantages of MUSIC algorithm and ESPRIT algorithm, improve the estimation accuracy of the algorithm while effectively reducing the complexity of the algorithm. numerical simulation verified the effectiveness of the proposed algorithm.

2. Data Model

Consider unfolded coprime arrays configuration which consists of two uniform linear subarrays, The spacing of the subarray to the left of the origin is $d_1 = N\lambda/2$, the number of subarray elements is M, the spacing of the subarray to the left of the origin is $d_2 = M\lambda/2$, the number of subarray elements is N, where M and N are coprime integers and λ denotes the signal wavelength. Its array topology is shown in figure 1.

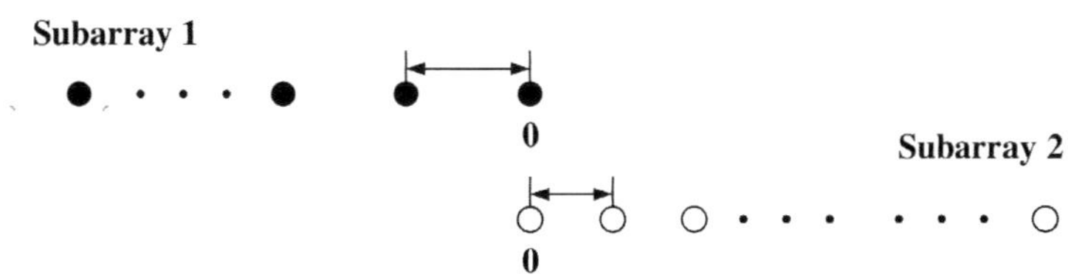

Figure 1. Unfold coprime arrays structure

The unfolded coprime arrays composed of subarray 1 and subarray 2 overlaps at the origin, and the positions of other subarrays elements do not overlap except at the origin. The position of each array element on the expanded matrix can be expressed as

$$Ls = \{(-m_1 d_1, 0), |m_1 = 0,1,2 \cdots M-1\} \cup \{(0, m_2 d_2), |m_2 = 0,1,2 \cdots N-1\} \quad (1)$$

The origin of the arrays can be Shared by two subarrays, so the number of unfolded coprime arrays is $M + N - 1$, but can provide $O\ (MN)$ degrees of freedom.

Considering unfolded coprime arrays, it is divided into subarray 1 and subarray 2. The array manifolds of subarray 1 and subarray 2 are defined as follows

$$\boldsymbol{a}_1(\theta_k) = [e^{j2\pi(M-1)d_1 \sin\theta_k/\lambda}, \cdots, e^{j2\pi d_1 \sin\theta_k/\lambda}, 1]^T \quad (2)$$

$$\boldsymbol{a}_2(\theta_k) = [1, e^{-j2\pi d_2 \sin\theta_k/\lambda}, \cdots, e^{-j2\pi(M-1)d_2 \sin\theta_k/\lambda}]^T \quad (3)$$

The direction matrix of subarray i is defined as

$$\boldsymbol{A}_i = [\boldsymbol{a}_i(\theta_1), \boldsymbol{a}_i(\theta_2), \cdots, \boldsymbol{a}_i(\theta_k)] \quad (4)$$

The source matrix is defined as

$$\boldsymbol{S} = [\boldsymbol{s}_1, \boldsymbol{s}_2, \cdots, \boldsymbol{s}_K]^T \quad (5)$$

Among, $\boldsymbol{s}_k = [s_k(1), s_k(2), \cdots, s_k(J)]$, $(k = 1,2,\cdots,K)$, J represents snapshots.
Then, the received signal of the subarray i is defined as

$$\boldsymbol{X}_i = \boldsymbol{A}_i S + N_i \quad (6)$$

Among, $\boldsymbol{N}_i$ represents noise matrix.

Considering the whole arrays, it is no longer divided into subarray 1 and subarray 2. The array manifold of the unfolded coprime arrays is defined as

$$\boldsymbol{a}(\theta_k) = [e^{\frac{j2\pi(M-1)d_1 \sin\theta_k}{\lambda}}, \cdots, e^{\frac{j2\pi d_1 \sin\theta_k}{\lambda}}, 1]^T 1, e^{-\frac{j2\pi d_2 \sin\theta_k}{\lambda}}, \cdots, e^{-\frac{j2\pi(N-1)d_2 \sin\theta_k}{\lambda}}]^T \quad (7)$$

The direction matrix is defined as

$$\boldsymbol{A} = [\boldsymbol{a}(\theta_1), \boldsymbol{a}(\theta_2), \cdots, \boldsymbol{a}(\theta_k)] \quad (8)$$

Then, the received signal of the arrays is defined as

$$\boldsymbol{X} = \boldsymbol{AS} + \boldsymbol{N} \quad (9)$$

Among, $\boldsymbol{N} \in C^{(M+N-1)\times J}$ represents noise matrix.

Get J snapshot to obtain unfolded coprime arrays covariance matrix $\widehat{\boldsymbol{R}}$ estimated

$$\hat{\boldsymbol{R}} = \frac{1}{J}\sum_{j=1}^{J} \boldsymbol{X}(j)\boldsymbol{X}^H(j) \tag{10}$$

The eigenvalue decomposition of the signal covariance matrix can be expressed as

$$\hat{\boldsymbol{R}} = \boldsymbol{E}_S \boldsymbol{D}_s \boldsymbol{E}_S^H + \boldsymbol{E}_N \boldsymbol{D}_N \boldsymbol{E}_N^H \tag{11}$$

3. The Proposed Algorithm

In this section, we first use ESPRIT algorithm to make the initial estimation, and then use the result of the initial estimation to determine the angle search range. Finally, we use MUSIC algorithm to complete the signal angle parameter estimation.

3.1. Initial Estimation

Firstly, the direction matrix $\boldsymbol{A}_i$ and receiving signal matrix $\boldsymbol{X}_i$ of subarray i are divided into equations (12) and (13) respectively

$$\boldsymbol{A}_i = \begin{bmatrix} a_{i1} \\ \boldsymbol{A}_{i2} \end{bmatrix} = \begin{bmatrix} \boldsymbol{A}_{i1} \\ a_{i2} \end{bmatrix} \tag{12}$$

$$\boldsymbol{X}_i = \begin{bmatrix} x_{i1} \\ \boldsymbol{X}_{i2} \end{bmatrix} = \begin{bmatrix} \boldsymbol{X}_{i1} \\ x_{i2} \end{bmatrix} \tag{13}$$

Among, $\boldsymbol{A}_{i1}$ and $\boldsymbol{A}_{i2}$ are matrices composed of $M_i - 1$ rows before $\boldsymbol{A}_i$ and $M_i - 1$ rows after $\boldsymbol{A}_i$ respectively. $\boldsymbol{X}_{i1}$ and $\boldsymbol{X}_{i2}$ are matrices composed of $M_i - 1$ rows before $\boldsymbol{A}_i$ and $M_i - 1$ rows after $\boldsymbol{A}_i$ respectively. Then the construction matrix $\boldsymbol{Z}_i$ can be obtained by combining the output of $\boldsymbol{X}_{i1}$ and $\boldsymbol{X}_{i2}$.

$$\boldsymbol{Z}_i = \begin{bmatrix} \boldsymbol{X}_{i1} \\ \boldsymbol{X}_{i2} \end{bmatrix} = \begin{bmatrix} \boldsymbol{A}_{i1} \\ \boldsymbol{A}_{i2} \end{bmatrix} \boldsymbol{S} + \boldsymbol{N}_i \tag{14}$$

Among, $\boldsymbol{S}$ is source matrix, $\boldsymbol{N}_i$ represents noise matrix.$\boldsymbol{A}_{i2} = \boldsymbol{A}_{i1}\boldsymbol{\Phi}_i$,$\boldsymbol{\Phi}_i = diag\{e^{-j\mu_{i1}}, \cdots, e^{-j\mu_{ik}}\}$. $\boldsymbol{\Phi}_i$ represents the rotation operator matrix.

To estimate direction of arrival (DOA) $\hat{\theta}_k^{\,ini}$ which is based on matrix $\boldsymbol{Z}_i$. It is necessary to estimate $\boldsymbol{\Phi}_i$.The signal subspace composed of $\boldsymbol{X}_{i1}$ and $\boldsymbol{X}_{i2}$ rotates phase μ_{ik}.
The covariance matrix $\boldsymbol{R}_{iz} = \boldsymbol{Z}_i \boldsymbol{Z}_i^{\,H}$ is constructed for $\boldsymbol{Z}_i$, and the signal subspace $\boldsymbol{E}_{iz}$ is obtained by eigenvalue decomposition. Then there is a matrix $\boldsymbol{T}_i$ that makes the following equation valid

$$\boldsymbol{E}_{iz} = \begin{bmatrix} \boldsymbol{E}_{i1} \\ \boldsymbol{E}_{i2} \end{bmatrix} = \begin{bmatrix} \boldsymbol{A}_{i1} \\ \boldsymbol{A}_{i1}\boldsymbol{\Phi}_i \end{bmatrix} \boldsymbol{T}_i \tag{15}$$

Among, $\boldsymbol{E}_{i1}$ is composed of lines one to $M_i - 1$ of $\boldsymbol{E}_{iz}$,and $\boldsymbol{E}_{i1} = \boldsymbol{A}_{i1}\boldsymbol{T}_i$, $\boldsymbol{E}_{i2}$ is composed of lines M_i of $\boldsymbol{E}_{iz}$ to the last line, and $\boldsymbol{E}_{i2} = \boldsymbol{A}_{i1}\boldsymbol{\Phi}_i\boldsymbol{T}_i$ without considering the influence of noise

$$\boldsymbol{E}_{i2} = \boldsymbol{E}_{i1}\boldsymbol{T}_i^{-1}\boldsymbol{\Phi}_i\boldsymbol{T}_i \tag{16}$$

In this case

$$\boldsymbol{T}_i^{-1}\boldsymbol{\Phi}_i\boldsymbol{T}_i = \boldsymbol{E}_{i1}^{+}\boldsymbol{E}_{i2} \tag{17}$$

Comparisons between $\hat{\theta}_{k1}^{\,ini}$ and $\hat{\theta}_{k2}^{\,ini}$ obtained by subarray 1 and subarray 2 of the unfolded coprime arrays. then get initial estimation $\hat{\theta}_k^{\,ini}$.

3.2. Accurate Estimation

After get $\hat{\theta}_k^{\,ini}$, through the use of MUSIC algorithm space spectrum function in the interval$[\hat{\theta}_k^{\,ini} - \Delta, \hat{\theta}_k^{\,ini} + \Delta]$ (Δ is a small value) within the search theta or get more precise $\hat{\theta}_k$.

$$\mathrm{P_{music}}(\theta_k) = \frac{1}{\boldsymbol{a}^H(\theta_k)\boldsymbol{E}_N\boldsymbol{E}_N^H\boldsymbol{a}(\theta_k)} \tag{18}$$

Among, $\boldsymbol{a}(\theta_k)$ is obtained by $\boldsymbol{E}_N$ according to (11). By looking for the crest to estimate the direction of arrival (DOA), The maximum K peaks relative to $\hat{\theta}_k$ are the Angle parameters of K sources to be estimated.
Remark 1. The number of source searches are required in the case of multiple sources.

3.3. Detailed Steps of the Proposed Algorithm
Step 1. Compute $\hat{\boldsymbol{R}}$ and perform eigen decomposition of $\hat{\boldsymbol{R}}$ to obtain $\hat{\boldsymbol{E}}_n$.
Step 2. Compute $\hat{\theta}_{k1}{}^{ini}$ and $\hat{\theta}_{k2}{}^{ini}$ according ESPRIT algorithm.
Step 3. Obtain source angle initial estimates of $\hat{\theta}_k{}^{ini}$ according to the subarray 1 and subarray 2 coprime relations.
Step 4. Through one-dimensional MUSIC algorithm in $[\hat{\theta}_k{}^{ini} - \varDelta, \hat{\theta}_k{}^{ini} + \varDelta]$ for spectral peaks searching.
Step 5. Obtain $\hat{\theta}_k$ by the maximum K peaks.

4. Performance Analysis

4.1. Complexity Analysis
In this part, we evaluate the complexity calculation of the proposed algorithm and compare it with other related methods. For ESPRIT algorithm, the major complexity calculation is caused by the covariance matrix calculation, eigenvalue decomposition. Suppose that the number of elements of subarray 1 is M , the number of elements of subarray 2 is N, the covariance matrix calculation needs $O\{J(M^2 + N^2)\}$ and eigenvalue decomposition of the covariance matrix requires $O\{M^3 + N^3\}$. The complexity required to obtain $\boldsymbol{\Psi}_i$ is $O\{2K^3 + 3K^2(M + N - 2)\}$, and eigenvalue decomposition of $\boldsymbol{\Psi}_i$ requires $O\{2K^3\}$. Consequently, the total complexity of ESPRIT is given by $\{4K^3 + M^3 + N^3 + (M^2 + N^2)J + 3K^2(M + N - 2)\}$. For the proposed algorithm, the major complexity calculation is caused by using ESPRIT for initial estimation and MUSIC for spectral peak searching. The required complexity of local range MUSIC algorithm search is $O\{n_1[(P - K)(2P + 1)]\}$, so the total complexity of proposed algorithm is given by $O\{n_1[(P - K)(2P + 1)] + 4K^3 + M^3 + N^3 + (M^2 + N^2)J + 3K^2(M + N - 2)\}$.For MUSIC algorithm, the calculation of covariance, eigenvalue decomposition and spectral peak searching requires $\mathrm{O}\{n[(P - K)(2P + 1)] + P^3 + P^2J\}$.Among, total number of elements P equals $M + N - 1$, n_1, n is the number of peak searches, K is the number of sources, J represents the number of snapshots. where $n_1 = 60/\Delta$ and $n = 2K/\Delta$ represent the search times corresponding to finding peaks and the search interval of spectral peak is $\varDelta = 0.01$.

We summarize the computational complexity of above algorithms in Table 1. Moreover, Figure 2 depicts the complexities of the different algorithms versus different N, where $M = 4$, $K = 3$, $J = 500$. The complexity comparison versus snapshots is given in Fig. 3, where $M = 4$, $N = 5$ and $K = 3$. It is observed from Figs. 2 and 3 that the complexity of the proposed algorithm is Significantly lower than MUSIC algorithm but is slightly higher than ESPRIT algorithm. It can be concluded that the complexity of the algorithm mainly comes from peaks search. Because the proposed algorithm greatly reduces the number of peaks searches, the complexity is greatly reduced.

Table 1. Complexity of different methods.

Algorithm	Computational complexity
Proposed	$O\{n_1[(P - K)(2P + 1)] + 4K^3 + M^3 + N^3 + (M^2 + N^2)J + 3K^2(M + N - 2)\}$
Esprit	$O\{4K^3 + M^3 + N^3 + (M^2 + N^2)J + 3K^2(M + N - 2)\}$
MUSIC	$\mathrm{O}\{n[(P - K)(2P + 1)] + P^3 + P^2J\}$

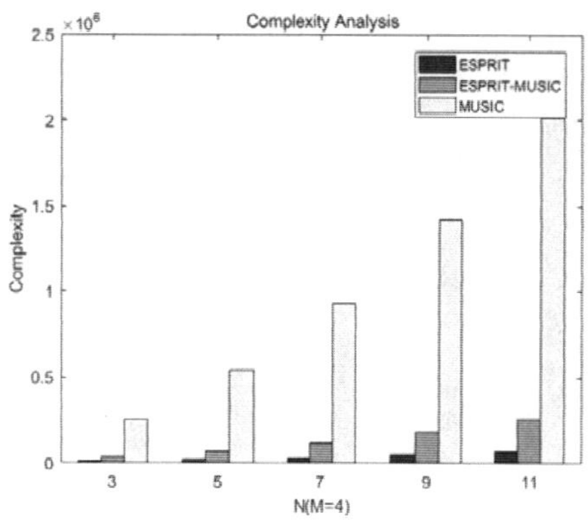

Figure 2. Complexities versus different N

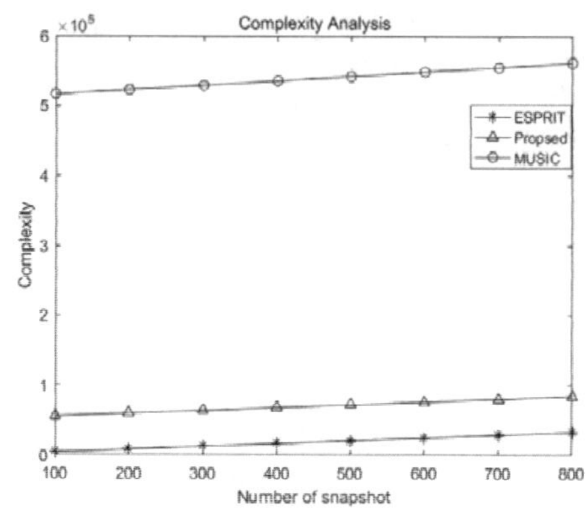

Figure 3. Complexities versus different snapshots

4.2. Advantages

The main advantages of the proposed algorithm are summarized as follows:

1) The proposed algorithm can achieve high resolution one-dimensional DOA estimation.
2) The proposed algorithm compared with the MUSIC algorithm, the complexity is greatly reduced. The performance of DOA estimation is better than ESPRIT algorithm, and very close to the high complexity of the MUSIC algorithm.
3) With the same array element number, the unfolded coprime arrays have a larger array aperture than that of CLA and ULA.

5. Simulation Results

The DOA estimation performance of the MUSIC algorithm, ESPRIT algorithm and proposed algorithm in unfolded coprime arrays was evaluated by 1000 Monte Carlo simulations. The Root Mean Square Error (RMSE) is defined as

$$RMSE = \frac{1}{K}\sum_{k=1}^{K}\sqrt{\frac{1}{1000}\sum_{l=1}^{1000}\left[\left(\hat{\theta}_{k,l} - \theta_k\right)^2\right]} \tag{19}$$

Among, $\hat{\theta}_{k,l}$ is the estimated value of θ_k in Monte Carlo simulation. Assuming that there are three sources in space, the angular information: $\theta_1 = 10°, \theta_2 = 30°$, $\theta_3 = 50°$, M and N represent the number of elements of subarray 1 and subarray 2, $d_1 = N\lambda/2$,$d_2 = M\lambda/2$. d_1 and d_2 represent the spacing of elements of subarray 1 and subarray 2, J represents the number of snapshots, respectively.

5.1. RMSE Results of the Proposed Algorithm

Figure 4 gives the parameter estimation performance of the proposed algorithm with unfolded coprime arrays versus snapshots, where $N = 5, M = 4$. As depicted in Fig. 4, The number of snapshots increases, the sampling data increases. The DOA estimation performance of the algorithm becomes better as the snapshots increases. Figure 5 gives the parameter estimation performance of the proposed algorithm with unfolded coprime arrays versus N, where $J = 200, M = 4$. As depicted in Fig. 5, Array element number increases, namely diversity gain increases. The DOA estimation performance of the algorithm becomes better as the array number increases.

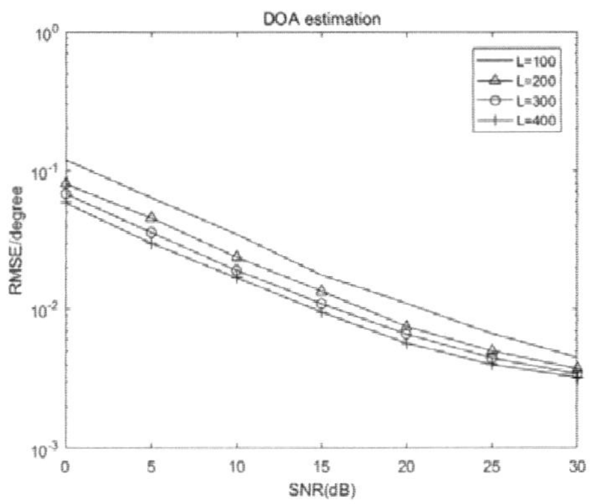

Figure 4. RMSE performance versus snapshots

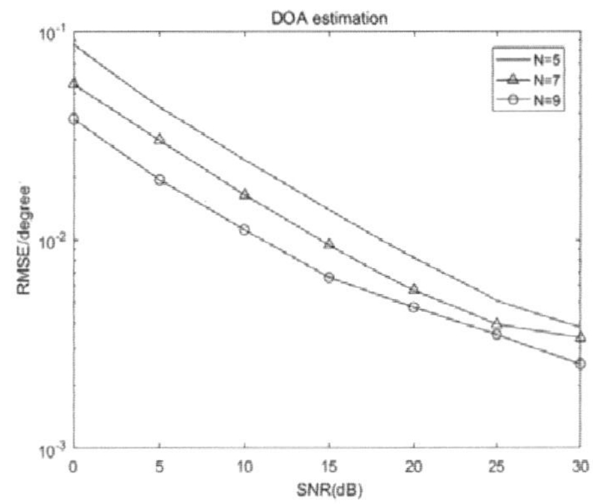

Figure 5. RMSE performance versus *N*

5.2. RMSE Results of Different Methods

Figure 6 gives the parameter estimation performance of the proposed algorithm, MUSIC algorithm and ESPRIT algorithm with unfolded coprime arrays under the same conditions, where $J = 200, M = 4, N = 5$. As depicted in Fig. 6, Array element number increases, namely diversity gain increases. The DOA estimation performance of the algorithm becomes better as the array number increases. the parameter estimation performance the proposed algorithm is very close to MUSIC algorithm but is Significantly higher than ESPRIT algorithm.

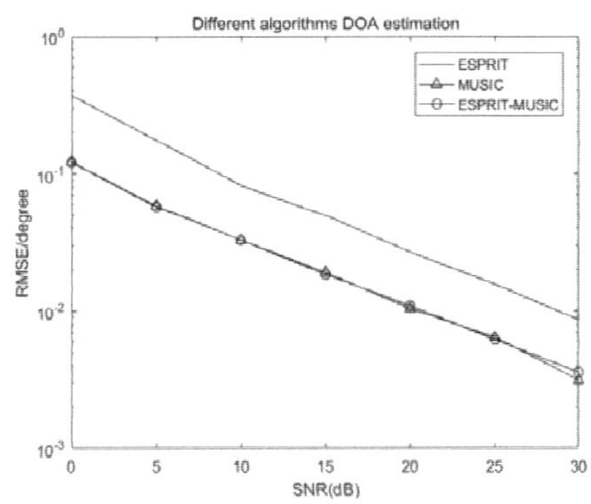

Figure 6. RMSE performance of different algorithms versus SNR

6. Conclusion

In this paper, we propose a low computational complexity and high estimation accuracy algorithm which cascades MUSIC and ESPRIT with unfolded coprime arrays. From the perspective of DOA estimation performance, this algorithm uses spectral peak search in a small range, so the DOA estimation performance of this algorithm is very close to MUSIC algorithm, and better than ESPRIT algorithm using eigenvalue decomposition. In terms of the complexity of the algorithm, since the algorithm uses local search instead of traversing the whole angle range, the complexity of the algorithm is far lower than that of the MUSIC algorithm, but only slightly higher than that of ESPRIT algorithm. Therefore, the algorithm is of great significance for the research on DOA estimation of unfolded coprime arrays. Numerical simulations corroborate the superiority of the algorithm in terms of computational complexity and estimation performance.

7. Acknowledgements

This work is supported by China NSF Grants (No. 61371169, No. 61601167).

IOP Conf. Series: Materials Science and Engineering **719** (2020) 012035 doi:10.1088/1757-899X/719/1/012035

8. References

[1] Vaidyanathan P P, Pal P. Theory of sparse coprime sensing in multiple dimensions. IEEE Transactions on Signal Processing, 2011, 59(8): 3592-3608.
[2] Zhou C, Gu Y, Fan X, et al. Direction-of-arrival estimation for coprime array via virtual array interpolation. IEEE Transactions on Signal Processing, 2018, 66(22): 5956-5971.
[3] Shi Z, Zhou C, Gu Y, et al. Source estimation using coprime array: A sparse reconstruction perspective. IEEE Sensors Journal, 2016, 17(3): 755-765.
[4] Pal P, Vaidyanathan P P. Coprime sampling and the MUSIC algorithm. IEEE Digital signal processing and signal processing education meeting, 2011: 289-294.
[5] Tan Z, Eldar Y C, Nehorai A. Direction of arrival estimation using co-prime arrays: A super resolution viewpoint. IEEE Transactions on Signal Processing, 2014, 62(21): 5565-5576.
[6] Vaidyanathan P P, Pal P. Sparse sensing with co-prime samplers and arrays. IEEE Transactions on Signal Processing, 2010, 59(2): 573-586.
[7] Yuan Q, Chen Q, Sawaya K. MUSIC based DOA finding and polarization estimation using USV with polarization sensitive array antenna. IEEE Radio and Wireless Symposium, 2006: 339-342.
[8] Pal P, Vaidyanathan P P. Coprime sampling and the MUSIC algorithm[A]. In: Digital Signal Processing Workshop and IEEE Signal Processing Education Workshop, IEEE, 2011, 289-294.

Design of Fast Detection and Tracking System Based on FPGA

Feng Tian[1,*], Lei Wang[1] and Yan Xiong[1]

[1]Computer and Information Engineering College, Guizhou University of Commerce, Guiyang, 550014, China
[*]fengtianxql@163.com

Abstract. Aiming at the problem that the classical Sobel edge detection algorithm is sensitive to noise and has a large amount of computation, an adaptive fast edge detection algorithm is proposed. Firstly, some background information which we are not interested in is removed through adaptive window to reduce the computational complexity. Then, Sobel operator is used to detect the real-time edge of the image in the window, and then morphological optimization is carried out. The fast detection algorithm in this paper is implemented on the FPGA. The results show that this method can effectively detect image edges and make the image edges clearer. At the same time, it can greatly reduce the noise and computational load in the image and reduce the power consumption.

1. Introduction

Image is an important feature source for human eyes and computer to recognize image and obtain information[1]. Meanwhile, the edge of image is the most basic feature of image, which can help computer to recognize and understand image automatically. Image edge detection has been widely applied in the fields of aviation, military, medicine, artificial intelligence and other fields that need image processing. Because of the continuity of real-time target movement in image tracking, and we are usually not interested in the related aspects, we can reduce the computational complexity by reserving a search window and discarding other pixels after finding the target in the initial state, which not only reduces the consumption of hardware resources, but also reduces the power consumption.

Mathematical morphology is based on geometric theory and defines a set of transformations based on the set of structural elements. In image processing filtering, morphological filters filter the input image according to the shape of the object in the image. The relatively simple operation and moderate storage requirements of morphological filters make them the most frequently implemented image processing filters.

Traditional real-time image detection and tracking only use Sobel operator for edge processing[2]. However, whether it is to find a larger object or a smaller image object in a larger image, it needs to be calculated at all possible locations, and there is strong background noise. In this paper, Sobel edge detection [3][4] is used for real-time image acquisition by using FPGA to find the edge of the initial state, and then the size of the search window is automatically adjusted by the program to optimize the morphological filtering, and the edge image with clear edge is obtained.

2. System Overall Framework Design

The theory of linear filters is complete and relatively easy to implement, but these filters have difficulties in distinguishing the reasonable transformation of pixel values (such as edges) and abnormal changes caused by noise. But the nonlinear combination of linear filters can also provide more useful output than a simple filter. Sobel edge detection filter is a common and basic non-linear filter in the field of image processing. By utilizing the separability of the filter, a simpler implementation structure can be designed.

In order to get more fine edges, reduce noise interference, suppress background information, reduce computation and reduce power consumption, this paper uses the method of automatically adjusting the window size to determine the frame area which needs to calculate edges, and then carries out morphological optimization. The fast detection and tracking system based on FPGA is shown in Figure 1.

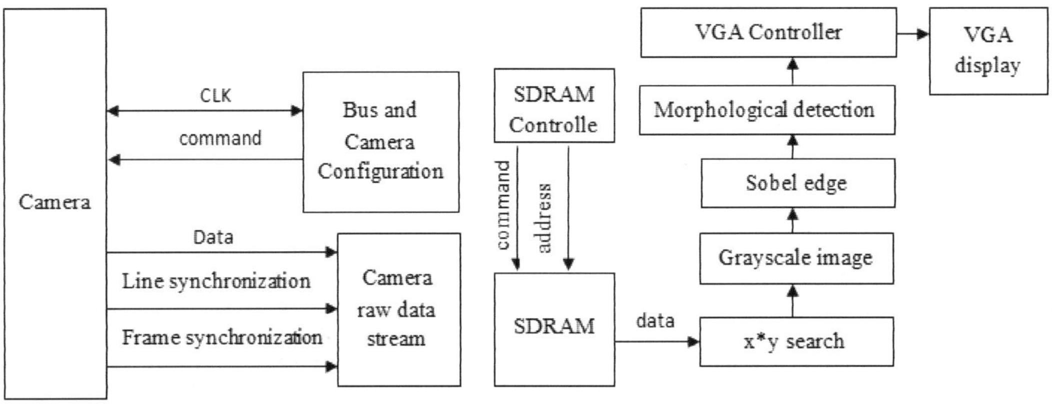

Figure 1. Design block diagram of fast detection and tracking system based on FPGA

3. Mathematical Morphology and Its Basic Operations

Based on set theory, mathematical morphology defines a set of transformations based on the set of structural elements. In image processing filter, the set is the pixel value of the image, and the structural element is considered as an arbitrary window. In binary images, pixels can be divided into two categories: target pixels and background images. Usually the target pixel is represented by 1, while the background pixel is represented by 0.

The basic operation of morphological filtering is corrosion and expansion. For corrosion, a target pixel is retained when the structural unit can be fully contained in the target. The structural unit is regarded as a window, and the output is the target pixel when all inputs are 1. Therefore, corrosion is the logic and function of the pixels in the window.

$$\begin{aligned} Q_{erosion}[x,y] &= I \Theta S \\ &= \wedge_{i,j \in s} I[x+i, y+j] \end{aligned}$$

It is called corrosion because the size of the target decreases as a result of the treatment.

For expansion, each input pixel is replaced by the shape of the structural unit in the output image. This is equivalent to the output of the target pixel when the flipped structure unit covers the target pixel in the input image. Expansion is the logic or the logic of the logical pixels in the window. Compared with corrosion, expansion enlarges the target and reduces the background. Corrosion and expansion are complementary, and the expansion of the target is equivalent to the corrosion of the background.

$$\begin{aligned} Q_{dilation}[x,y] &= I \oplus S \\ &= \vee_{i,j \in s} I[x-i, y-j] \end{aligned}$$

4. Implementation of Adaptive Window and Orphological Optimization

The improved algorithm first detects the Sobel edge based on the collected image which removes part

IOP Conf. Series: Materials Science and Engineering **719** (2020) 012036 doi:10.1088/1757-899X/719/1/012036

of the background information, calculates the next window size, and optimizes the morphological corrosion and expansion of the corresponding edges of the large image. As shown in Figure 2.

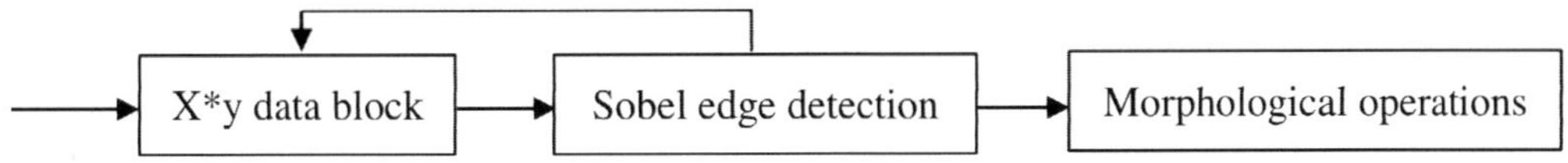

Figure 2. The block diagram of the algorithm in this paper

According to Sobel edge detection data, find out X_{max}, X_{min}, Y_{max}, Y_{min}. In this paper, CMOS outputs VGA(640×480 Resolution)At the time of Initialization, X_{max} =480, X_{min} =1, Y_{max} =640, Y_{min} =1. Since the edge detected by the edge detection operator has an error, and the target may be motion, in order to improve the accuracy, the window size may be preset, that is, the target tracking with the size of the target current position center x×y in the original image of size a×b is selected(a>x,b>y). And the detection, the size of the W sampled data block [5] is obtained, and the image data of the Sobel edge detection is required as the next frame image. The RTL-level schematic of the x×y adaptive window implementation is shown in Figure 3.

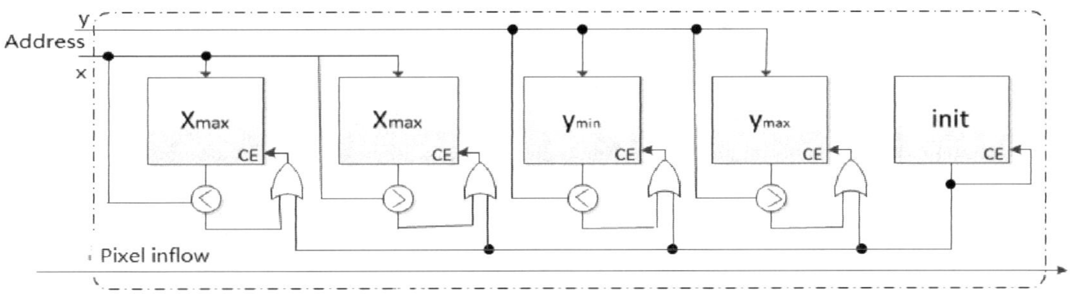

Figure 3. RTL-level schematic diagram of x×y adaptive window implementation

According to Xmax, Xmin, Ymax, Ymin, calculate the center position (centre) and size of the target, where

$$center = (\frac{x_{max} + x_{min}}{2}, \frac{y_{max} + y_{min}}{2})$$

$$size = (x, y) = (x_{max} - x_{min} + z, y_{max} - y_{min} + z)$$

In order to improve the accuracy and consider the motility of the target, this paper chooses Z=8.

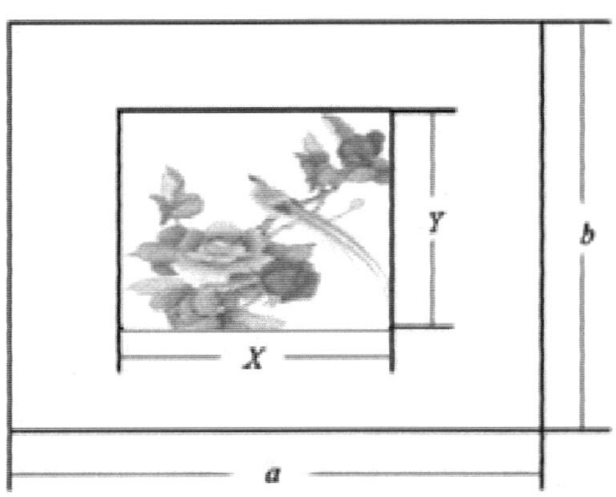

Figure 4. Schematic diagram of adaptive window downsampling

Downsampling the data block removes the background of the target image that we are not paying attention to, and the amount of data is greatly reduced, which is equivalent to the frame drawing and reduction processing of the metadata. The downsampled data block is grayed out and sent to the Sobel edge detection module. In the operation, the Sobel operator is used to convolute the image, and then the gradient formula is used to calculate the image gradient [6]. The multiplication and addition in the convolution operation is performed in parallel with the (programmable multipath parallel adder) in the macroblock library in the software [7].

After the Sobel operator detects the edge of the image, the data is sent to the morphological optimization module. P1~P9 are the edge data to be processed in the 3×3 operation template of the expansion or corrosion module. An implementation module for a morphological filter (corrosion or expansion operation) containing programmable structural elements is shown in Figure 4.

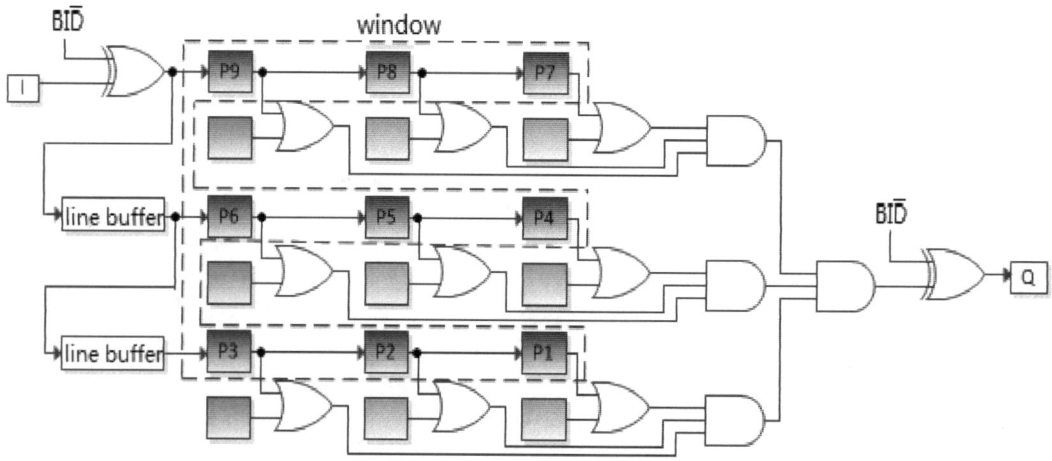

Figure 5. Schematic diagram of the corrosion and expansion morphology filter module

The complementarity of corrosion and expansion enables them to be realized in the same circuit. A control signal is used to compensate the input and output, and the control signal is used to select the operation of corrosion or expansion. Use an additional register associated with each window element to control the window element. When the window element is not a data block, the low level is used to shield the window data. For programmable windows, it is easier to share the control bits of the structural units than to operate the two window filters separately because of the morphological filtering of corrosion and expansion after edge detection.

5. Results and Analysis of Adaptive Window and Morphology Optimization

Figure 6 is the original image of the binary image, Figure 7 is the image detected by the Sobel algorithm, Figure 8 is the image optimized by the Sobel algorithm and then morphologically optimized, Figure 9 is the image detected by the algorithm, ie the improved algorithm the result of. From the comparison of Fig. 7, 8and 9, the algorithm is practical, the noise is suppressed after optimization, the edge is clearer, and the detection accuracy is obviously improved.

Figure 6. Original image before edge detection

Figure 7. Sobe edge detection

Figure 8. Sobel+ morphology optimization

Figure 9. Edge detection method

In this paper, the adaptive window is used to reduce the amount of data to be processed. Sobel edge detection is used to optimize the corrosion and expansion morphology. Finally, it is implemented on the FPGA platform with EP2C35F484C8N as the core. Compared with the traditional edge detection system, this method further reduces the calculation amount of background information, improves the processing speed, reduces the system power consumption, improves the accuracy of edge detection, and suppresses the background or is caused by the camera process. Part of the noise. It can be applied to areas such as monitoring, tracking, and aviation with high real-time performance, large amount of tasks, and large amount of calculation.

6. Acknowledgements

This work is supported by First-class project at school level of Guizhou University of Commerce: 2018YYLKC15 & Young Talent Growth Project of Education Department of Guizhou Province: Qianjiaohe KY Zi [2018]271.

7. References

[1] Yuwen Xia. Verilog Digital System Design Tutorial[M].3. Beijing: Beijing Aerospace University Press, 2013.

[2] Haoli Zheng. VerilogHDL Digital Logic Design Tutorial-Verilog[M]. Beijing: Publishing House of Electronics Industry, 2010.

[3] Chen Xue. Algorithm and application research of video target tracking technology in complex environment[D]. Beijing: Graduate School of Chinese Academy of Sciences, 2010.

[4] Chuang Zhang. Color Image Denoising Based on Edge Detection and Bilateral Filtering[J]. Acta Electronica Sinica, 2010, 8(8).

[5] Bergen He. Design of a fast video tracking system based on TMS320C6455+FPGA+ SDRAM [J]. Chinese Journal of Liquid Crystals and Displays, 2014, 6(29): 1111-1116.
[6] Lusheng Gu. Research on real-time edge detection system based on SOPC[J]. Application of electronic Technique. 2019, 35 (8): 47-75.
[7] Ming Li. Implementation of Sobel Edge Detection Based on FPGA[J]. Modern Electronics Technique, 2019, 33 (16): 44-50.

IOP Conf. Series: Materials Science and Engineering **719** (2020) 012037 doi:10.1088/1757-899X/719/1/012037

A Soft Decoding Algorithm with Hard Decision Boundary Factor for High-Order M-APSK Signal

Yanshuang Cai[1,*], Tingting Lv[1], Kai Wang[2*] and Jianjun Wu[1]

[1]School of Electronics Engineering and Computer Science, Peking University, Beijing, China, 100871.
[2]Institute of War Studies, Academy of Military Sciences
[*]Email: 1601214388@pku.edu.cn

Abstract. High-order amplitude phase shift keying is adopted in the latest digital satellite broadcasting system, which presents new challenges to the design of a receiver demodulation scheme. This study focuses on the soft decision decoding algorithm. Traditionally, maximum likelihood soft decision based on log likelihood ratio estimation is often used as a decoding scheme. Accordingly, a soft decision algorithm based on hard decision boundary is designed and the corresponding soft decision estimation formula is derived. Numerical analytical results show that the proposed soft decision algorithm can improve the error performance of a system and substantially reduce the complexity of implementation.

1. Introduction

The digital satellite broadcasting is an important part of satellite communication, the industry and its demands have considerably changed over the years. Ultra-high definition television, high-speed IP data dissemination, increase in the number of users, and improvement of service quality requirements have driven continuous innovation in satellite broadcasting technology. These factors have resulted in breakthroughs in high access probability, high spectrum efficiency, and enhanced communication services.

In order to improve the spectrum efficiency of satellite communication systems and performance of the system at extremely low signal-to-noise ratio and expand the conversion efficiency of bandwidth and capacity of satellite communication systems, the DVB standard-setting organization ETSI released the DVB-S2 Extensions standard on March 4, 2014, its main body follows the DVB-S2 standard and is also compatible with the DVB-S[1]and DVB-S2[2] standards. In particular, DVB-S2X[3] provides additional options in coefficients, modulation schemes, and forward error correction, including high-order amplitude phase shift keying(APSK), small FIR filter roll-off factors, and considerably advanced filtering techniques. These modifications immensely improve the efficiency of satellite transmission channels. Moreover, the spectrum efficiency of DVB-S2X can be effectively increased by 20%–30%.

The receiver processing algorithm should be improved according to the new technical features of DVB-S2X, particularly the adoption of high-order M-APSK(up to 256APSK).This research introduces a soft decoding algorithm with hard decision boundary factor for the high-order M-APSK signal.

2. System Model

In a DVB-S2X receiver, the demodulation module is consisted by matched filtering, timing offset correction, down sampling, frame synchronization, frequency offset estimation and correction, coarse phase offset estimation and correction, constellation decoding, forward error correction decoding, descrambling, and cyclic redundancy (CRC) check[3]. The receiver processing block diagram is shown in Figure 1.

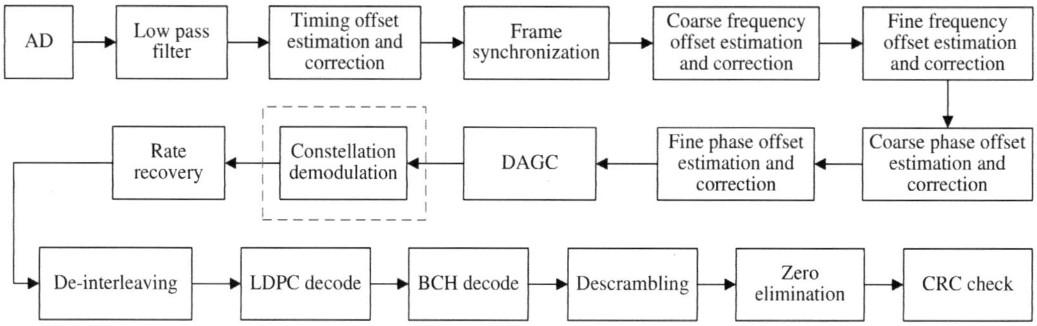

Figure 1. Block diagram of the DVB-S2X Receiver Processing

For low-order modulation, such as QPSK and 8PSK, the small phase and radius deviation in the demodulation process and processing the fine phase offset estimation and correction and DAGC are no longer necessary. We can use the hard decision method to accomplish the decoding given the sparse constellation points and large minimum Euclidean distance between them. However, the following properties are manifested in high-order modulation, particularly for 256 APSK: (1) eight rings are present in the constellation map, (2) the radius of each ring slightly varies, (3) each ring has 32 constellation points, (4) the phase difference between the adjacent constellation points is merely $\pi/16$. Therefore, the residual phase offset should be corrected after phase synchronization and the amplitude automatic control should be considered. The soft decision method should be used in constellation decoding to improve the demodulation performance.

In the high-order demodulation, the residual phase offset must be corrected (after the preceding processing of the coarse frequency offset, fine frequency offset is estimated and corrected and the phase offset estimated and corrected) using the Q-power loop algorithm. Bit decoding (soft demodulation algorithm is designed to improve the system performance for APSK modulation), de-interleaving, LDPC decoding, BCH decoding, descrambling, and CRC decoding are needed upon the completion of the processing in the front-end module of the receiver. Thereafter, the bit and frame error rates are calculated through a comparison with the transmitter.

3. Low Complexity M-APSK Soft Decoding Algorithm

In a communication system, the transmitter converts the encoded data into I and Q real signals through constellation mapping. In the receiver, the I and Q real signals should be decoded into bit streams in the hard demodulation. By contrast, the I and Q real signals are decoding into real sequences corresponding to each bit, and then decoded into bit streams through channel decoding in the soft demodulation. Compared with hard demodulation, soft demodulation retains more channel information, and it can improve the posterior probability of data decoding and reduce the bit error rate. The advantage of soft decision is performance improvement, but high complexity follows.

3.1. Maximum Likelihood Soft Decision Based on Logarithmic Likelihood Ratio Estimation

The commonly used method of soft decision is the estimation method based on logarithmic likelihood ratio. The maximum likelihood soft decision is used to solve the logarithmic likelihood ratio and retain the soft information of each bit. The transmission probability of 0/1 at the sender is the same. That is, P (bi = 0) = P (bi = 1) = 0.5. Thereafter, LLR can be written as follows:

$$L(b_i|y) = \log\left(\frac{\sum_{s:b_i(i)=1} \exp\left(-\frac{1}{2\sigma^2}\|y-h_s\|\right)}{\sum_{s:b_i(i)=0} \exp\left(-\frac{1}{2\sigma^2}\|y-h_s\|\right)}\right) \quad (1)$$

where S represents all symbols in the M-APSK modulation. Although the bit error rate (BER) of this method is low, and the performance is excellent, the computational complexity will exponentially increase with an increase in M. Therefore, this method will generally be replaced by the maximum logarithmic likelihood ratio method[4]:

$$L(b_i|y) \approx \ln\left(\frac{\max_{s::b_i(i)=0}\exp\left(-\frac{1}{2\sigma^2}\|y-h_s\|\right)}{\max_{s::b_i(i)=1}\exp\left(-\frac{1}{2\sigma^2}\|y-h_s\|\right)}\right) \tag{2}$$

which can be simplified as follow[5]:

$$L(b_i|y) \approx \frac{1}{2\sigma^2}\left[\min_{s::b_i(i)=0}\|y-h_s\| - \min_{s::b_i(i)=1}\|y-h_s\|\right] \tag{3}$$

This estimation method is simple and has good performance for low-order modulation. However, a disadvantage of the high-order modulation M-APSK is the substantial increase in complexity and additional resources will be costed[6]. The following subsection discusses an improved algorithm (i.e., soft decision algorithm based on hard decision boundary) for the MLR estimation.

3.2. Soft Decision Algorithm Based on Hard Decision Boundary

A simple BPSK modulation mode is considered. Without loss of generality, we assume that the received signal is composed of I-channel and Q-channel signals. Thereafter, the received signal after channel equalization is provided as follows:

$$r = (h * \cdot y) / \|h\|^2 \tag{4}$$

The soft decision result of b1 based on the LLR maximum likelihood estimation method can be expressed as follows:

$$L(b1 \mid y) = \rho R(r) \text{ , } \rho=2\|h\|^2 / y \tag{5}$$

This logarithmic likelihood ratio can be considered ρ multiplying the boundary of a hard decision: R(r) = 0[7] [8]. We can obtain the soft decision schemes on the basis of the hard decision boundary of 16, 32, 64, 128 and 256 APSK by extending this idea. Thereafter, we analyse the following specific scheme using the 64APSK as an example.

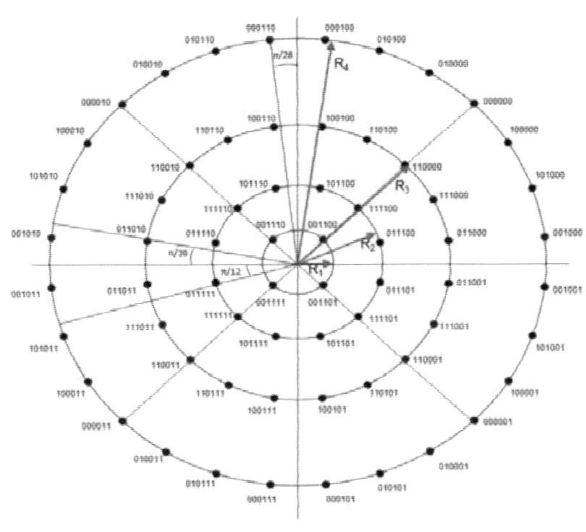

Figure 2. 64APSK signal constellation (4+12+20+28)

The 64 APSK signal constellation is shown in Figure 1. We assume that b1, b2, b3, b4, b5, b6, b7 and b8 represent the constellation points. Given that b7 and b8 determine the quadrant of the constellation points, the hard decision boundary of b7 is R(r) = 0 and that of b8 is I(r) = 0. The constellation points of the second/third/fourth quadrant can be analysed by that in the first quadrant as they are gray codes. Thus, the angle should be normalized to the first quadrant and the conversion of the angle scan be completed using $\theta_r = \arctan(|I(r)/R(r)|)$.

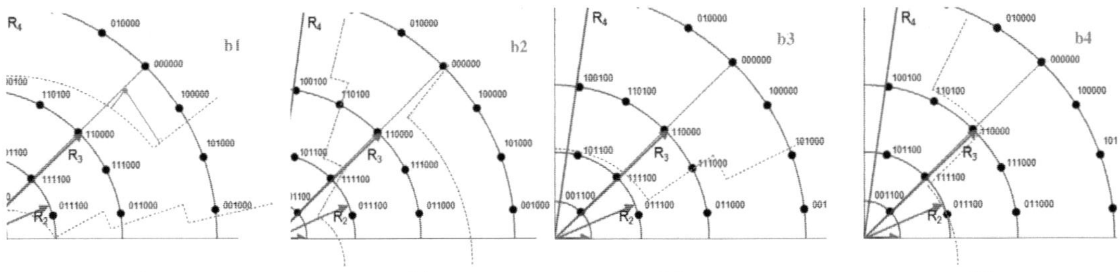

(a) (b) (c) (d)

Figure 3. Boundaries of *b1, b2, b3* and *b4*

Figure 3 is the hard decision boundaries of b1–b2 and b3–b4, respectively. Figure 2 (a) shows that the dotted line is the boundary line of the hard decision of the first bit. We can assess the soft information of the first bit, which is represented by the symbol bit of soft information multiplied by the minimum distance between the signal and the boundary line, according to the different positions of the received signal in the first quadrant. When the outputs of the hard decision are 0 and 1, the symbols are −1 and 1, respectively. For example, the sketch point in the above figure is located in the range of $R \geq (R3+R4)/2 \cap \theta_r \geq 3\pi/14$ and its boundary is an arc $R \geq (R3+R4)/2$ and a ray $\theta \geq 3\pi/14$. Thus, the minimum distance between the point and the two boundaries is as follows:

$$\min\{\mathrm{R}-\frac{(R3+R4)}{2}, R\sin(\theta_r-\frac{3\pi}{14})\} \tag{6}$$

Since that the hard decision output in this range is 0, the final soft decision output is as follows:

$$-\min\{\mathrm{R}-\frac{(R3+R4)}{2}, R\sin(\theta_r-\frac{3\pi}{14})\} \tag{7}$$

The soft information formulas for all bits of 16, 32, 64, 128, and 256 APSK can be similarly derived on the basis of the preceding method. The formula of 256 APS is provided as follows:

$$\bullet LLR\_b1 = R-\frac{(R4+R5)}{2}$$

$$\bullet LLR\_b2 = \begin{cases} \mathrm{R}-\frac{(R2+R3)}{2}, R<\frac{(R2+R3)}{2} \\ \min\{\mathrm{R}-\frac{R2+R3}{2}, \frac{(R6+R7)}{2}-R\}, \frac{R2+R3}{2} \leq R < \frac{(R6+R7)}{2} \\ \frac{(R6+R7)}{2}-R, \mathrm{R} > \frac{(R6+R7)}{2} \end{cases}$$

$$\bullet LLR\_b3 = \begin{cases} -\min\{\mathrm{R}-\frac{(R3+R4)}{2}, \frac{(R5+R6)}{2}-R\}, \frac{(R3+R4)}{2} \leq R < \frac{(R5+R6)}{2} \\ \min\{\mathrm{R}-\frac{(R1+R2)}{2}, \frac{(R3+R4)}{2}-R\}, \frac{(R1+R2)}{2} \leq R < \frac{(R3+R4)}{2} \\ \min\{\mathrm{R}-\frac{(R5+R6)}{2}, \frac{(R7+R8)}{2}-R\}, \frac{(R5+R6)}{2} \leq R < \frac{(R7+R8)}{2} \\ \mathrm{R}-\frac{(R1+R2)}{2}, R<\frac{(R1+R2)}{2} \\ \frac{(R7+R8)}{2}-R, R \geq \frac{(R7+R8)}{2} \end{cases}$$

$$\bullet LLR\_b4 = -imag(r)$$

$$\bullet LLR\_b5 = -real(r);$$

$$\bullet LLR\_b7 = \begin{cases} \mathrm{Rsin}(\theta_r-\frac{\pi}{8}), \theta_r < \frac{\pi}{8} \\ \min\{R\sin(\theta_r-\frac{\pi}{8}), \mathrm{R}\sin(\frac{3\pi}{8}-\theta_r)\}, \frac{\pi}{8} \leq \theta_r < \frac{3\pi}{8} \\ \mathrm{Rsin}(\frac{3\pi}{8}-\theta_r), \theta_r \geq \frac{3\pi}{8} \end{cases}$$

$$\bullet LLR\_b8 = \begin{cases} \mathrm{Rsin}(\theta_r-\frac{\pi}{16}) & \theta_r < \frac{\pi}{16} \\ \min\{R\sin(\theta_r-\frac{\pi}{16}), R\sin(\frac{3\pi}{16}-\theta_r)\}, \frac{\pi}{16} \leq \theta_r < \frac{3\pi}{16} \\ -\min\{R\sin(\theta_r-\frac{3\pi}{16}), R\sin(\frac{5\pi}{16}-\theta_r)\}, \frac{3\pi}{16} \leq \theta_r < \frac{5\pi}{16} \\ \min\{R\sin(\theta_r-\frac{5\pi}{16}), R\sin(\frac{7\pi}{16}-\theta_r)\}, \frac{5\pi}{16} \leq \theta_r < \frac{7\pi}{16} \\ R\sin(\frac{7\pi}{16}-\theta_r), \theta_r \geq \frac{7\pi}{16} \end{cases}$$

$$\bullet LLR\_b6 = \mathrm{Rsin}(\theta_r-\frac{\pi}{4})$$

4. Numerical Analysis

The system performance of the hard and soft decision schemes based on the hard decision boundary of 16, 64, 128 and 256APSK are tested through MATLAB simulations. In the simulations, the minimum number of error bits is 200, the minimum number of simulation frames is 50, the maximum number of simulation frames is 1000, and the LDPC bit rate is 2/3.

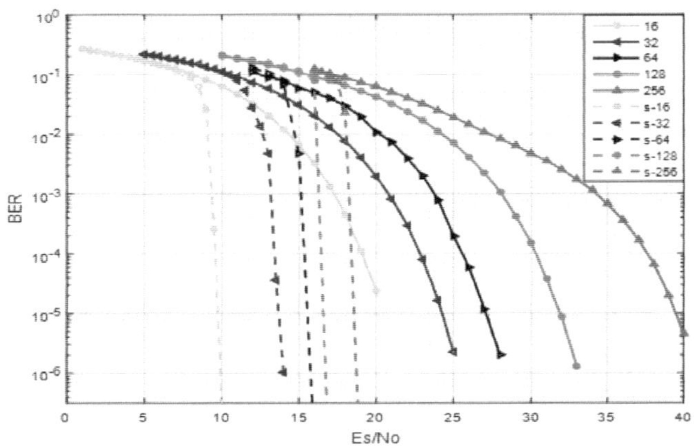

Figure 4. Performance comparison between the M-APSK soft and hard decisions

The advantages of soft demodulation are evident with the increase of SNR (see Figure 4). The BER curves of the soft demodulation algorithms based on the hard decision boundary show rapid fading when SNRs are 8.5, 13.5, 15, 17, and 19dB in 16, 32, 64, 128, and 256APSK, respectively. When BER=10^{-5}, the decreases in SNR of the soft demodulation algorithm based on the hard decision boundary are 10, 10.5, 11.5, 15.5, and 20 dB in 16, 32, 64, 128, and 256APSK, respectively, compared with those based on the hard decision boundary. This finding shows substantial improvement in the system performance.

The performance comparison of the soft demodulation based on the hard decision boundary with the soft decision schemes based on the logarithmic likelihood ratio is shown as follows:

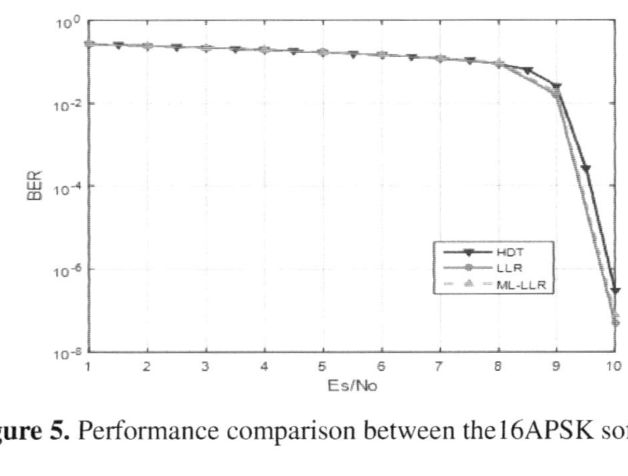

Figure 5. Performance comparison between the16APSK soft decision schemes

The HDT performance is nearly the same as those of the logarithmic likelihood ratio estimation method and maximum likelihood soft decision method based on the logarithmic likelihood ratio (see Figure 5). When BER reaches 10^(−6), the SNR difference is below 0.5dB.

The simulation results manifest that the soft decision algorithm based on the hard decision boundary can immensely improve the system performance. Compared with the maximum likelihood method based on the logarithmic likelihood ratio estimation, the HDT soft demodulation can substantially reduce the complexity of the system at the expense of small system performance. The required adder and multiplier resources are provided as follows:

Table 1. Resource comparison of the three soft decision methods

	Adder		Multiplier	
	ML-LLR	HDT	ML-LLR	HDT
16APSK	192	5+m	128	4
32APSK	480	6+m	320	5
64APSK	1152	12+m	768	12
128APSK	2688	10+m	1792	9
256APSK	6144	5+m	4096	5
16APSK	192	5+m	128	4

Table 1 shows that m is a fixed number of the adder resources occupied by the IQ2AP module. By comparison, the soft decision method based on the hard decision boundary can change the computational complexity from O (MlogM) to O (logM). Accordingly, the consumption of hardware resources is substantially reduced.

5. Summary

This research studied the decoding algorithm for high-order modulation signal. A soft decision algorithm based on hard decision boundary was designed in accordance with the traditional soft decision decoding algorithm, namely, maximum likelihood soft decision algorithm based on logarithmic likelihood ratio estimation. Accordingly, the corresponding soft decision estimation formula was deduced. The numerical results show that the proposed soft decision algorithm can improve the BER performance of the system and substantially reduce the complexity compared with the maximum likelihood soft decision method.

6. Acknowledgment

This paper is supported by the National Natural Science Foundation of China (No.61371073).
Corresponding author: Kai WANG; Email: 1801213629@pku.edu.cn.

7. Reference

[1] ETSI EN 300.421. V1.1.2(1997-08) Digital Video Broadcasting(DVB);Framing structure, channel coding and modulation for 11/12 GHz satellite service.
[2] ETSI EN 302 307-1 V1.4.1, Digital video broadcasting(DVB);second generation framing structure, channel coding and modulation systems for Broadcasting, Interactive services, news gathering and other broadband satellite applications (DVB-S2).2014.
[3] White Paper: DVB-S2X demystified, Business Insight on DVB-S2X March 2014.
[4] C. Pan, J. Huang, K. Peng and L. Fan, "Simplified Soft Demapping Algorithm for Gray-APSK," 2012 8th International Conference on Wireless Communications, Networking and Mobile Computing, Shanghai, 2012, pp. 1-3.
[5] M. Zhang and S. Kim, "Efficient soft demapping for M-ary APSK," ICTC 2011, Seoul, 2011, pp. 641-644.
[6] Gaudenzi R D, Fabregas G A, Martinez A. Turbo-coded APSK modulations for satellite broadband Communications-Part II: End-to-end performance[J]. 2004.
[7] Z. Liu, Q. Xie, K. Peng and Z. Yang, "APSK Constellation with Gray Mapping," in IEEE Communications Letters, vol. 15, no. 12, pp. 1271-1273, December 2011.
[8] K. Cho, K. Hyun, D. Yoon, S. Park and S. Cho, "An Approximated Soft Decoding Algorithm of 16-APSK Signal for DVB-S2," 2007 Digest of Technical Papers International Conference on Consumer Electronics, Las Vegas, NV, 2007, pp. 1-10

Dynamic Frequency Tracking Method for Anti-offset Wireless Power Transfer System Based on Phase Comparison

Xiaokang Wu[1], Cheng Jiang[1], Bin Wei[1], Fan Zhang[2*] and Changsong Cai[2],

1 China Electric Power Research Institute, Beijing, China
2 School of Electrical Engineering and Automation, Wuhan University, Wuhan, Hubei Province, China
Email: frankozhang@foxmail.com

Abstract. In this paper, a dynamic frequency tracking method for the anti-offset wireless power transfer system based on phase comparison is proposed. The anti-offset characteristics of the system are realized by the structural design of the transmitting coil. The proposed anti-offset coil is verified in both simulation and experiment. However, in order to solve the frequency shift caused by the change of system parameters, a dynamic frequency tracking method is proposed in this paper, which continuously captures and calculates the phase difference between the current and voltage at the transmitting unit through the STM32 chip, adjusts the output frequency in time, and finally keeps the working frequency near the slightly inductive area. This paper establishes an experimental prototype and achieves an efficiency improvement of 6% to 8% through the proposed control method.

1. Introduction

In recent years, with the widespread use of various types of portable electronic products, the drawbacks of the wired charging method have gradually emerged, and people began to pursue a wireless way to obtain energy. Due to its safety and convenience, the magnetic coupling technology of magnetic coupling resonance has been gradually applied to electric wireless charging vehicles, transmission line equipment and biological implant equipment [1-3], even if there are some obstacles in the transmission channel, the power transfer will not be affected, ensuring the stability and reliability of the system [4].

However, the research on the wireless power transfer technology of the magnetic coupling resonance mode is mostly established in the case where the transmitting coil and the receiving coil are symmetrical, but in most cases, especially in the wireless power transfer of portable electronic products, limited by the size of the receiving device, the absolute symmetry of the transmitting coil and the receiving coil cannot be achieved [5].

For asymmetric coupling, due to the asymmetry in the size of the transmitting coil and the receiving coil, once the offset between the transmitting coil and the receiving coil occurs, the transmission effect of the wireless power transmission will be greatly affected. Therefore, it is necessary to increase the offset resistance of the transmitting coil. In this regard, there are mainly two ways to achieve, one is to design the structure of the transmitting coil to produce a uniform magnetic field [6,7]; the other is to control through changing the topology of the system [8].

For the wireless power transfer of the magnetic coupling resonance mode, there is an inherent resonant frequency, and the system has the largest transmission power at its inherent resonant frequency. However, during power transmission, the resonant frequency may change for various reasons, such as changes in parasitic parameters, offset of the receiving coil, rise in coil temperature, and so on. The above phenomenon is called detuning. Once detuning occurs, the efficiency of the system drops rapidly, and the input frequency must be changed by the frequency tracking module to accommodate changes in the resonant frequency.

2. Analysis of System Model

2.1. Analysis of Coil Anti-Offset Characteristics

For the conventional coil, a horizontal spiral type winding method is often used, so that the coil inductance can be calculated according to an empirical formula:

$$L = 2.15\times10^{-6}\frac{R_1+R_2}{2}N^{\frac{5}{3}}\ln\frac{8a}{R_2-R_1} \tag{1}$$

where R_1 is the inner radius of the planar spiral coil and R_2 is the outer radius of the planar spiral coil.

However, at the same size, if the winding is performed in a conventional way, more turns will be required, thereby increasing the inductance of the coil, resulting in a small matching resonant capacitor, which seriously affects the system. However, if the number of winding turns of the coil is reduced in order to reduce the inductance of the coil, the coil will form a "concave" magnetic field distribution in the central region, which cannot achieve the uniform of the magnetic field [9].

In order to make the magnetic field distribution on the coil surface uniform, the current density distribution of the coil should satisfy:

$$\sigma(r,R) = i_0 r\left(R^2-r^2\right)^{-\frac{1}{2}}H(R-r) \tag{2}$$

where R is the radius of the transmitting coil, r is the distance between the center point of the receiving coil and the center point of the transmitting coil, h is the vertical distance, the magnetic field on the surface of the coil is:

$$B_z = \frac{\mu_0 i_0 \pi}{4}(r<R) \tag{3}$$

After determining the target magnetic field distribution, the current density distribution of the coil can be reversed by the magnetic field distribution. The area under the current density curve is divided into N equal parts, and each part is replaced by one turn. The turn position f_n is determined by the coefficient a ($0 \le a \le 1$). When $a = 1$, the turn is placed at the rightmost place, and when $a = 0$, the turn is placed at the leftmost place [10].

The expression for the position where the nth turn is placed is:

$$\frac{r_n}{R} = a\sqrt{1-\left(\frac{N-n}{N}\right)^2} + (1-a)\sqrt{1-\left(\frac{N-n+1}{N}\right)} \tag{4}$$

Through the above calculation, the structural distribution of the system transmitting coil can be obtained, and the anti-offset characteristics of the coil can be realized.

2.2. Analysis of Frequency Tracking

For the wireless power transfer system, the basic series-series compensation topology is shown in figure 1.

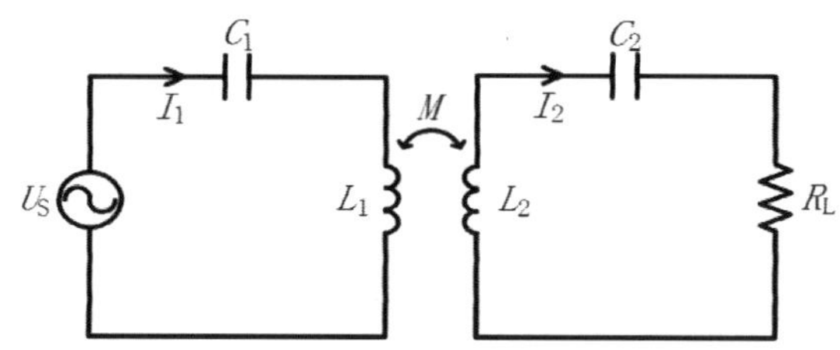

Figure 1. Basic topology diagram of the wireless power transfer system

For figure 1, the mesh current equation can be listed as follows:

$$\begin{cases} \dot{U}_S = \left(j\omega L_1 + \dfrac{1}{j\omega C_1} \right) \dot{I}_1 + j\omega M \dot{I}_2 \\ 0 = j\omega M \dot{I}_1 + \left(j\omega L_2 + \dfrac{1}{j\omega C_2} + R_L \right) \dot{I}_2 \end{cases} \tag{5}$$

Assume that the system operates at the resonant frequency. At this time, the primary and secondary currents are:

$$\begin{cases} \dot{I}_1 = \dfrac{R_L}{\omega^2 M^2} \dot{U}_S \\ \dot{I}_2 = \dfrac{\dot{U}_S}{j\omega M} \end{cases} \tag{6}$$

Taking the power supply voltage as the reference phasor, we can see from the above equation that when the wireless power transmission system operates at the resonant frequency, the primary current phase will be the same as the power supply voltage phase; and once the operating frequency of the system shifts, when $\omega > \omega_0$, the system will be in the inductive zone, at which time the voltage will lead the current; and when $\omega < \omega_0$, the system will be in the capacitive region, at which time the voltage will behind the current.

From the above, we can know that when the operating frequency of the wireless power transfer system deviates from the resonant frequency of the system, it will cause a phase difference between the current and the voltage of the primary side of the system. Based on this, the primary side voltage and current phase of the wireless power transmission system are measured, thereby implementing frequency tracking of the wireless power transmission system.

3. Design and Simulation of the Coil Structure

As described above, the design of the transmitting coil structure is mainly to reasonably distribute the pitch of the coils so that the current density distribution satisfies the obtained current density distribution, thereby realizing a uniform magnetic field of the coil and improving the anti-offset characteristics of the coil.

The transmitting coil designed in this paper is a square coil with a side length of 15 cm. The coil is uniformly wound with a Litz wire having a wire diameter of 1.6 mm, and mainly comprises three parts, wherein part I is the outermost 4 turns, which generates a "concave" type magnetic field; part II is the middle 9 inch, by which fill the magnetic field depression; part III is the innermost 3 turns, which further fills the magnetic field depression.

Taking into account the volume of the portable electronic product, the receiving coil has a circular design with a radius of 1 cm, a wire diameter of 1 mm, and a tight winding of 10 turns.

The parameters of the transmitting coil and receiving coil used in this paper are shown in table 1.

Table 1. Parameters of the coils in the experiment

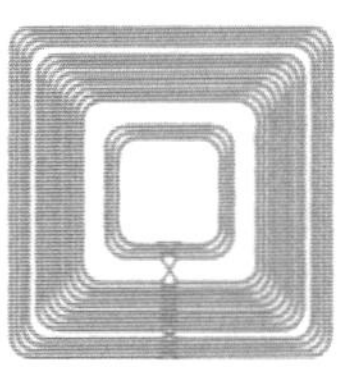	Type of coil	Transmitting coil
	Number of turns	16
	Outer radius / turns of part I	7.5cm/4
	Outer radius / turns of part II	6.5cm/9
	Outer radius / turns of part III	3cm/3
	Diameter of the wire	0.16cm
	Turn spacing	0.1cm
	Number of layers	2
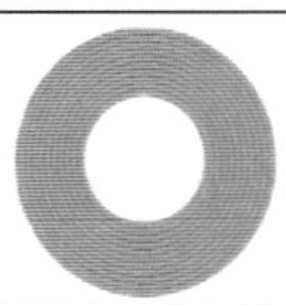	Type of coil	Receiving coil
	Number of turns	10
	Outer radius	2cm
	Diameter of the wire	0.1cm
	Number of layers	2

IOP Conf. Series: Materials Science and Engineering **719** (2020) 012038 doi:10.1088/1757-899X/719/1/012038

The simulation result of the coupling coefficient k with the horizontal displacement of the receiving coil is shown in figure 2. For comparison, three sets of control simulation groups were made at the same time, that is, only the part I of the transmitting coil, only the part I and part II of the transmitting coil, transmitting coil is wound 24 turns directly.

We can see that the coupling coefficient of only part I of the transmitting coil and only the part I and part II of the transmitting coil is low from figure 2, and depression occurs with the horizontal offset so that the anti-offset cannot be realized. When the transmitting coil is wound 24 turns, the coupling coefficient is too large when the offset is small, so the anti-offset effect is not ideal. However, the coil designed in this paper has better anti-offset characteristics and a high coupling coefficient.

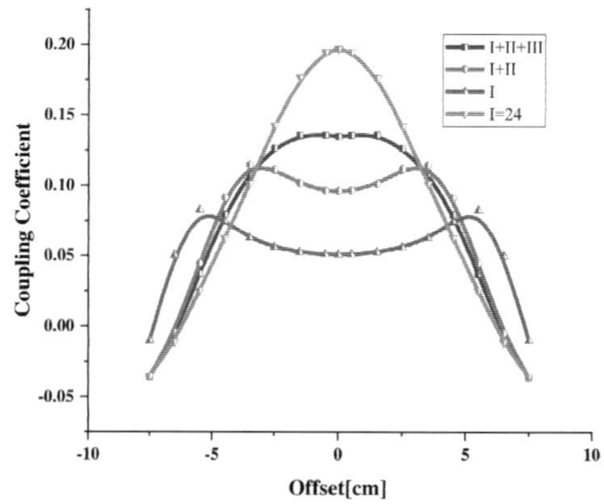

Figure 2. Diagram of coupling coefficient and horizontal offset relationship

4. Model Experiment Verification

According to the coil model designed in the previous section, the Litz wire is used to wind the transmitting coil and the receiving coil. The coils wound with the Litz wire can effectively reduce the effects of the proximity effect and the skin effect.

In order to verify the anti-offset of the proposed coil and the improvement of the frequency tracking to the whole system, the experimental prototype is set up as shown in figure 3.

The transmission distance is set to 5 mm, and the relationship between the coupling coefficient of the actual coil and the horizontal offset is measured as shown in figure 4. It can be seen that the coupling coefficient in the actual measurement is basically consistent with the variation of the coupling coefficient in the simulation.

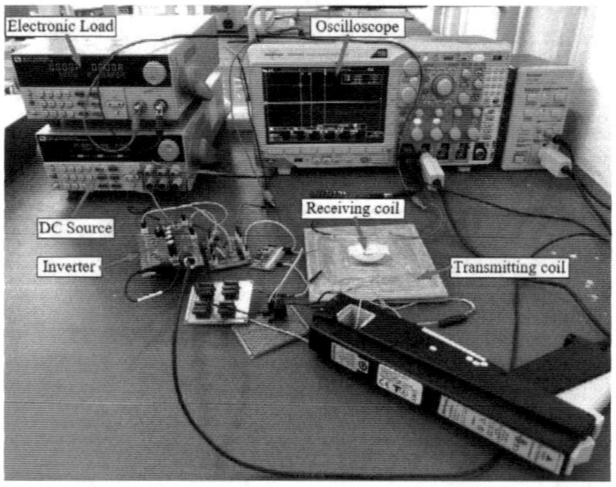

Figure 3. Diagram of the experimental prototype

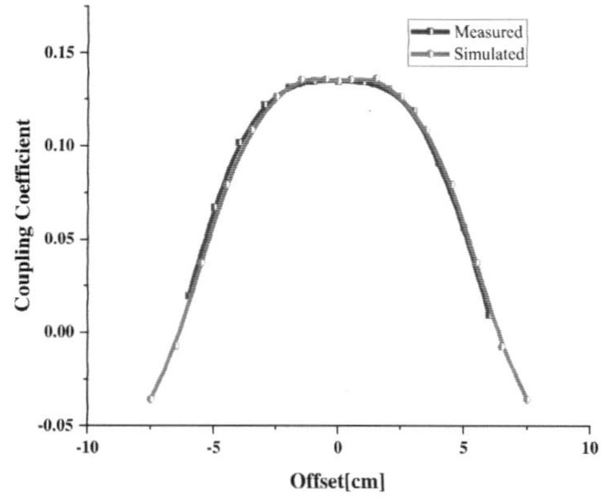

Figure 4. Diagram of coupling coefficient and horizontal offset relationship

Set the DC power supply voltage to 20 V, and the initial operating frequency of the system is 310 kHz. Take the case of no horizontal offset, the voltage and current waveforms of the system's transmitting unit are shown in figure 5(a). After the phase difference between voltage and current is captured and calculated by STM32, the output frequency is adjusted, and the adjusted voltage and current waveforms of the transmitting unit are shown in figure 5(b). At this time, the frequency is about 304 kHz, and the current becomes significantly larger, which is caused by the fact that the equivalent impedance of the LC loop is becoming smaller during the process of shifting to the resonance point. In order to prevent an excessive current from occurring during resonance, the final frequency stays in a slightly inductive region near the resonant point.

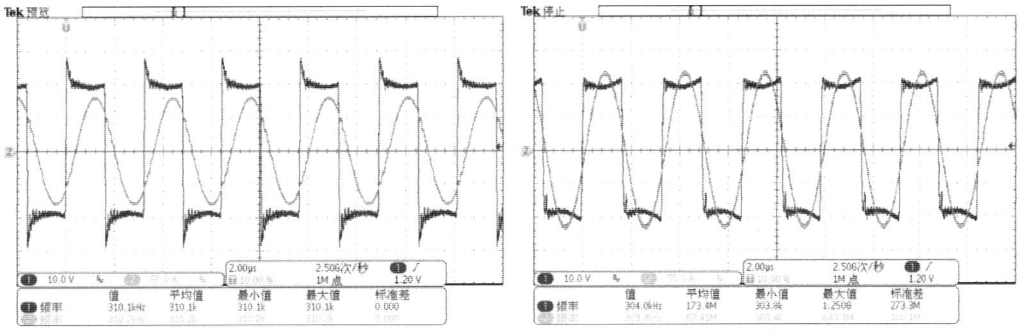

(a) (b)

Figure 5. Waveforms of the voltage and current

The entire frequency tracking process is shown in figure 6. Firstly, the initial operating frequency f0 of the inverter is given, and then the voltage and current phase of the transmitting unit are compared, thereby judging that it is in the inductive region or the capacitive region. If it is in the capacitive region, the frequency will be increased by Δf, and if it is in the inductive region, the frequency will be reduced by Δf until the phase difference meets the accuracy. Finally, the frequency is fine-tuned to operate the system in a slightly inductive region.

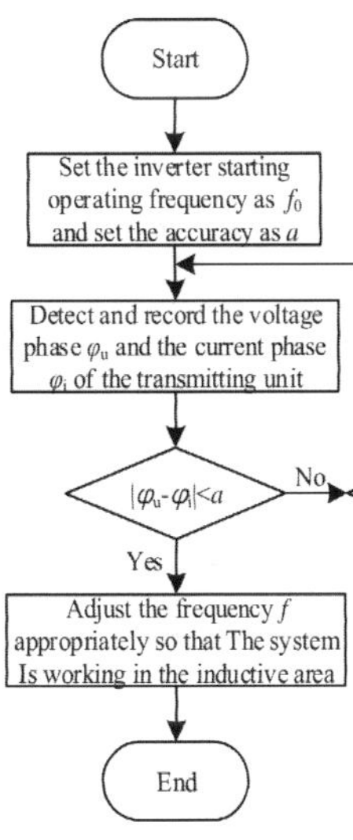

Figure 6. Workflow of the frequency tracking system

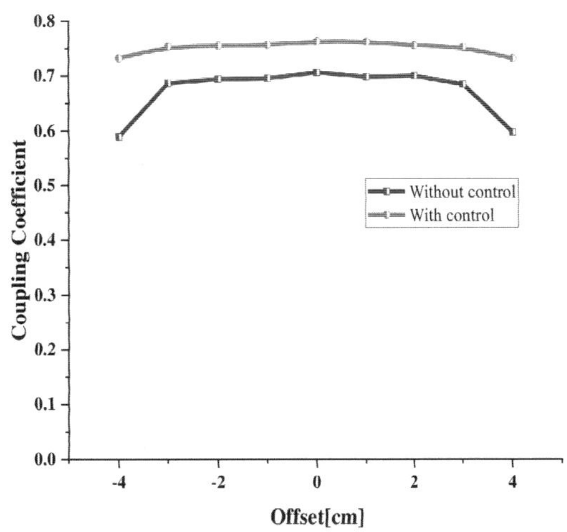

Figure 7. The transmission performance of the promoted wireless power transfer system

The transmission performance for a wireless power transfer system with or without frequency tracking control is shown in figure 7. The DC power supply is 20V. From the figure, we can see that under the effect of frequency tracking control, the transmission efficiency of the wireless charging system has been significantly improved. The transmission efficiency increases by about 6% to 8% on average, while the transmission efficiency does not change a lot in the effective charging area.

5. Conclusion

In this paper, a transmission coil design with anti-offset performance is proposed in order to solve the problem of the transmission efficiency reduce caused by the offset between the transmitting coil and receiving coil in the asymmetric wireless charging system, and the proposed transmitting coil is verified in simulation and experiment. Secondly, a frequency tracking control method based on voltage and current phase comparison is proposed for the problem of system parameter changes. The system can effectively improve the anti-offset characteristics of the wireless charging system, and at the same time ensure high transmission efficiency while ensuring anti-offset characteristics, and improve the flexibility and adaptability of the wireless charging system.

6. Acknowledgments

This work was supported by the Science and Technology Project of State Grid Corporation of China under Project of "Research on Wireless Charging Technology of Substation Patrol Robot Based on Magnetic Coupling Resonance".

7. References

[1] Cai C, Wang J, Liu R, Fang Z, Zhang P, Long M, Hu M, Lin Z, Resonant Wireless Charging System Design for 110-kV High-Voltage Transmission Line Monitoring Equipment[J]. IEEE Transactions on Industrial Electronics, 2019, 66(5): 4118–4129.

[2] Wang J, Cai C, Li L, Zhang P, Liu Q, Zhang F, Fang Z, Extended efficiency control method for WPT systems in smart grid under loose coupling extremes[J]. IET Power Electronics, 2019, 12(10): 2523-2533.

[3] Cai C, Wang J, Fang Z, et al. Design and Optimization of Load-Independent Magnetic Resonant Wireless Charging System for Electric Vehicles[J]. IEEE Access, 2018, 6: 17264-17274.

[4] Jiang, J., Zhu, C., Song, K., Wei, G., Zhang, Q.. (2016). Novel power receiver for dynamic wireless power transfer system. *Conference of the IEEE Industrial Electronics Society*. IEEE.

[5] Lee G, Waters B H, Shi C, et al. Design considerations for asymmetric magnetically coupled resonators used in wireless power transfer applications[C]// Power Amplifiers for Wireless & Radio Applications. IEEE, 2013.

[6] Kim J, Kim D H, Kim K H, et al. Free-positioning wireless charging system for hearing aids using a bowl-shaped transmitting coil[C]// Wireless Power Transfer Conference. IEEE, 2014

[7] Jolani F, Yu Y Q, Chen Z. A planar positioning-free magnetically-coupled resonant wireless power transfer[C]// Wireless Power Transfer Conference. 2015.

[8] Zhang W, Zhang T, Guo Q, et al. High-efficiency wireless power transfer system for 3D, unstationary free-positioning and multi-object charging[J]. IET Electric Power Applications, 2018, 12(5):658-665.

[9] Li S, Niu P, Yu L, Mao R, et al. Design method of primary transmitting coil for realizing large uniform magnetic field distribution[C]// 3rd International Conference on Mechanical, Control and Computer Engineering. 2018.

[10] Waffenschmidt E . Homogeneous Magnetic Coupling for Free Positioning in an Inductive Wireless Power System[J]. Emerging & Selected Topics in Power Electronics IEEE Journal of, 2015, 3(1):226-233.

Research on Mutual Inductance Prediction of Resonant Magnetic Coupling Wireless Power Transmission System Based on Recursive Least Square Method

Songcen Wang[1], Bin Wei[1], Hao He[1], Jianxue Mu[2] and Changsong Cai[2]

1China Electric Power Research Institute, Beijing, China
2School of Electrical Engineering and Automation, Wuhan University, Wuhan, Hubei Province, China.
Email: 1071831864@qq.com

Abstract. In this paper, based on recursive least square algorithm and resonant magnetic coupling wireless power transmission technology, the magnetic coupling mutual inductance value, which is a significant factor affecting the transmission power and efficiency of wireless transmission system, is predicted. According to the characteristics of the system, a mathematical model of LCC-P-type topology structure is established. Through mathematical deduction and analysis, combined with MATLAB/Simulink simulation results are obtained, and its feasibility is verified by experimental results.

1. Introduction

Wireless Power Transmission (WPT) is a technology that can obtain electric power without direct contact.

The resonant magnetic coupling (RMC) WPT studied in this paper is a hot way of WPT at present. Because of its high efficiency, long distance, low loss, low radiation, selective transmission and other advantages, this method has been widely applied in the field of WPT. In the RMC WPT system, monitoring of transmission status can greatly ensure the safe operation and coordinated control of the system. However, the detection process is cumbersome and the cost is high. Therefore, the purpose of this paper is to find out the key electrical parameters of the primary side which can reflect the overall parameters of the system by studying the equivalent circuit model of the RMC WPT system, and finally realize the monitoring of the electrical parameters of the RMC WPT device measurement.

2. WPT System Based on LCC-P Topology

2.1. Selection of Topological Structure

Generally, RMC WPT systems have five typical topologies: SS-type topology, SP-type topology, PS-type topology, PP-type topology and LCC-P-type topology. In practical experiments, it is found that when the frequency of the WPT system is close to the resonant frequency, the series (S) structure will lead to a large primary current, which is often unable to work at the resonant frequency. The LCC-P-type topology, because of the existence of series inductance L_3, can produce stable power transfer even if the coil deviates or the load changes. It generates a constant current in the transmission coil [1], so that the current in the transmission coil does not rise very much. Therefore, it can make the WPT system work at a frequency that is very close to the resonance point, and in the theoretical calculation, it can also produce stable power transfer. It can be assumed that the system works in a fully resonant state, which greatly simplifies the theoretical calculation. Therefore, this paper takes the WPT system with LCC-P topology as an example for structural analysis.

The equivalent circuit model of WPT system with LCC-P-type topology is shown in figure 1. u_p is a high frequency AC voltage source. L_1, C_1, L_3, C_3, L_2 and C_2 constitute primary and secondary side resonance networks respectively. R_1, R_2 and R_3 are the internal resistances of L_1, L_2 and L_3 respectively (because the conductance of capacitors is very small, the influence of capacitance conductance on the model is neglected here). M is the mutual inductance between primary and secondary coils. R_L is the load resistor that is connected to simulate the charging of different types of wireless charging devices in the resonant circuit. I_1 and I_2 are the primary and secondary side high frequency resonance current, respectively.

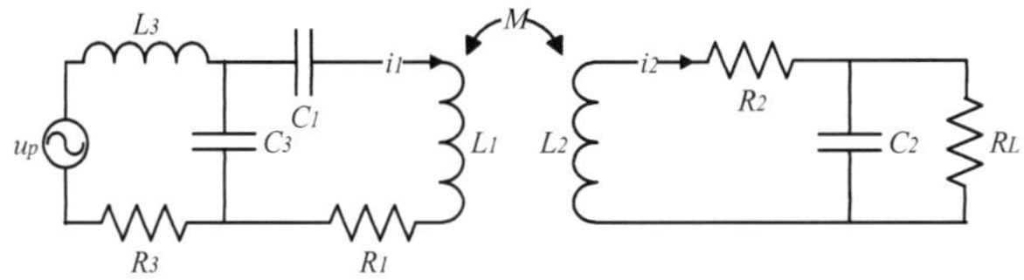

Figure 1. The equivalent circuit model of WPT system with LCC-P-type topology

2.2. Mapping Relation of Electrical Parameters between Primary Side and Secondary Side

Before theoretical analysis, some assumptions need to be made:

1) Circuit parameters L_1, L_2, L_3, C_1, C_2, C_3, R_1, R_2 and R_3 are known quantities measured.

2) Under normal operating conditions, these circuit parameters do not change much, so they are generally considered constant [2-3].

3) The internal resistance R_1, R_2 and R_3 of the coil inductances L_1, L_2 and L_3 of the primary and secondary sides are very small relative to the inductance value, and their influence on the resonance frequency can be neglected. When the system works in the full resonance state, it can be considered that the frequency of the high frequency AC voltage (the frequency used in the WPT system in this study is about 300 kHz) is satisfied

$$f=\frac{1}{2\pi\sqrt{L_1C_1}}=\frac{1}{2\pi\sqrt{L_2C_2}}=\frac{1}{2\pi\sqrt{L_3C_3}} \tag{1}$$

After the above hypothesis analysis, in the equivalent circuit model of the new LCC-P topology, the L_1 and C_1, L_3 and C_3 of the primary side are fully resonant, while the L_2 and C_2 of the secondary side are fully resonant. The theoretical analysis based on this is as follows.

In this WPT system, if the effective value phasor of high frequency AC voltage u_p is set to $\dot{U}_p = U_p\angle 0°$, that is, the effective value of u_p is U_p, and the initial phase angle is 0°, the loop equation can be formulated according to the equivalent circuit model

$$\begin{cases}(R_1+R_3)\dot{I}_1 - j\omega M\cdot\dot{I}_2 = \dot{U}_p \\ -j\omega M\cdot\dot{I}_1 + (R_2+R_L)\dot{I}_2 = 0\end{cases} \tag{2}$$

Among them, $\dot{I}_1$ and $\dot{I}_2$ are the effective phasors of the primary and secondary AC currents i_1 and i_2 respectively.

Solute the (2) equation yields a current parameter expression of the follows for the primary side and the secondary side:

$$\begin{cases} \dot{I}_1 = \dfrac{R_2 + R_L}{\omega^2 M^2 + (R_1 + R_3)(R_2 + R_L)} \dot{U}_p \\ \dot{I}_2 = \dfrac{j\omega M}{\omega^2 M^2 + (R_1 + R_3)(R_2 + R_L)} \dot{U}_p \end{cases} \tag{3}$$

From the deduction of formula (2) ~ (3), it can be seen that the current parameters of the primary and secondary sides are not only related to the parameters of WPT system itself, but also to the high frequency AC voltage. However, the key electrical parameters of the primary side, which can reflect the overall parameters of the system, should not change with the change of high frequency AC voltage, so these parameters are not the key electrical parameters we are looking for. In the next section, we will look for the primary key electrical parameters that can reflect the overall parameters of the system.

2.3. Key Electrical Parameters of Primary Side

In the actual experiment, the high frequency AC voltage is realized by inverters. The circuit structure is shown in figure 2. U_d is the input DC voltage of the primary side, C_d is the large capacitor parallel to U_d, S_1~S_4 is the full bridge inverter, D_1~D_4 is the feedback diode (or continuous current diode), and u_p is the equivalent input voltage of the resonant circuit, that is, the resonant circuit.

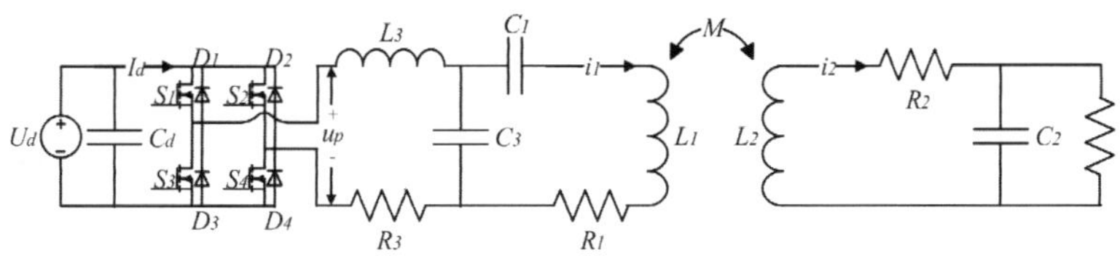

Figure 2. LCC-P-type Topological Equivalent Circuit Model with Inverter

When the phase difference between positive and negative voltage of u_p is 180°, the Fourier series expansion shows that the fundamental effective value of high frequency AC voltage on the primary side has the following quantitative relationship with the primary input DC voltage U_d:

$$U_d = \frac{\sqrt{2}\pi}{4} U_p \tag{4}$$

According to the law of conservation of power, the input power of the inverters should be equal to the output power, ignoring the power loss caused by the inverters. From this we can get:

$$I_d = \frac{2\sqrt{2}}{\pi} I_1 \tag{5}$$

Because the WPT system is in full resonance state at this time, and the dimension of $\dfrac{\omega^2 M^2 + (R_1 + R_3)(R_2 + R_L)}{R_2 + R_L}$ is Ohm (Ω), i.e. $\dot{U}_p$ and $\dot{I}_1$ are in phase, the relationship between DC voltage U_d and current I_d can be obtained by combining formula (3), (4) and (5):

$$U_d = \frac{\pi^2}{8} \cdot \frac{\omega^2 M^2 + (R_1 + R_3)(R_2 + R_L)}{R_2 + R_L} I_d \tag{6}$$

Let $R_d = \dfrac{U_d}{I_d}$, which is called the equivalent input resistance on the DC side, therefore, M can be represented by R_d and R_L as follows:

$$M = \frac{\sqrt{(\frac{8}{\pi^2}R_d - R_1 - R_3)(R_2 + R_L)}}{\omega} \tag{7}$$

In the above formula, R_d is not affected by the high frequency AC voltage. It is the key electrical parameter of the primary side required. The load R_L can be directly known by wireless frequency identification technology [4]. So only M cannot be measured directly from the primary side.

In the follow-up simulation and experiment, because of the inevitable error, M obtained each time is not the real value, which need to be measured repeatedly through a large number of experiments. By changing the phase difference between the two control signals, a series of comparative experiments can be obtained, so that a series of M can be obtained in different ways. A large amount of data measured by the above experiments can be processed by recursive least square (RLS) algorithm.

When the phase difference of the two control signals is adjusted to change, for example, when it is set to be θ, formula (7) will be expressed as follows:

$$M = \frac{\sqrt{[\frac{4(1-\cos\theta)}{\pi^2}R_d^* - R_1 - R_3][R_2 + R_L]}}{\omega} \tag{8}$$

The R_d* is the equivalent input resistance of the DC side measured in the case of the corresponding phase difference θ.

3. Simulation Design Based on RLS Algorithms

3.1. WPT System Simulation Design

A WPT system with LCC-P topology is built on Simulink interface. The simulation wiring is shown in figure 3, and the parameters of the components are shown in table 1.

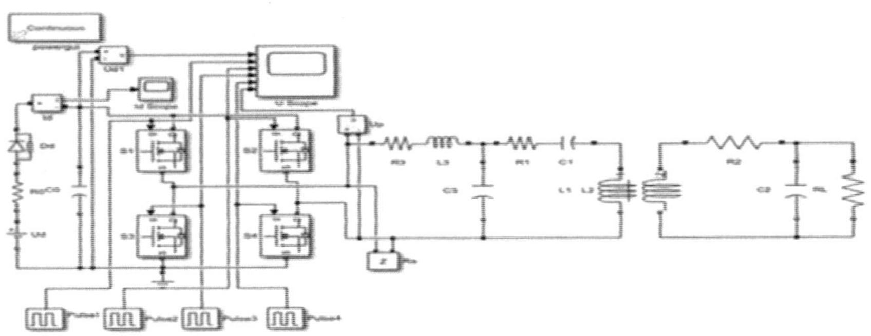

Figure 3. WPT System Simulation Wiring Diagram of LCC-P Topology

Table 1. Component Parameters of Simulation System

Component Number	U_d	R_0	C_0	L_1	L_2	L_3	C_1
Parameter Values	12V or 24V	0.00001Ω	1mF	112μH	14.4525μH	6μH	2.513nF
Component Number	C_2	C_3	R_1	R_2	R_3	R_L	M
Parameter Values	19.474nF	46.9nF	1.55Ω	0.2Ω	0.083Ω	50Ω or 25Ω	8.5μH

IOP Conf. Series: Materials Science and Engineering **719** (2020) 012039 doi:10.1088/1757-899X/719/1/012039

From the above table, the coefficient of M of primary side and secondary side is set at 8.5 μH. The simulation results show that the value fluctuates from 8 μH to 9 μH with an average of 8.57 μH.

3.2. RLS Algorithm

The RLS parameter identification is that when the recognition system runs, with the introduction of new recursive observation data, the parameters are estimated one after another until the estimated value reaches satisfactory accuracy. The basic idea of RLS algorithm can be summarized as follows: the new parameter value to be estimated $\hat{\theta}(k)$ = the old parameter value to be estimated $\hat{\theta}(k-1)$ + the correction term [5].

The comparison of the recursive value of M and the stability value of M is shown in figure 4.

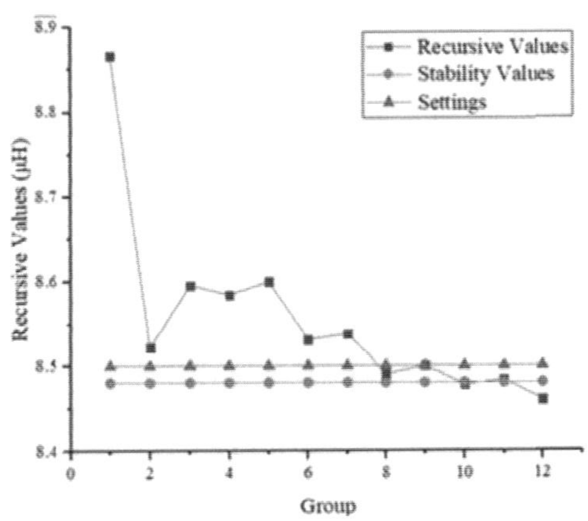

Figure 4. Comparisons among Recursive Values, Stable Values and Settings of Mutual Inductance

From figure 4, it can be seen that after 8 recursions, the M is almost completely stable around 8.48 μH, which shows that the convergence speed of the RLS algorithm is very fast.

According to the results of the above simulation combined with the algorithm, the M is about 8.48 μH, while the actual M set in the simulation experiment is 8.50 μH, the error is about 0.2353%, which is much smaller than the average value error. Through simulation and RLS algorithm, this method can detect the mutual inductance between coils only by measuring the DC side voltage and current of the primary side, which is feasible and has high accuracy.

4. Physical Construction and Actual Measurement

In order to verify the feasibility of the design of the parameter sampling and control circuit in Chapter 3, a practical WPT system circuit is made, as shown in figure 5.

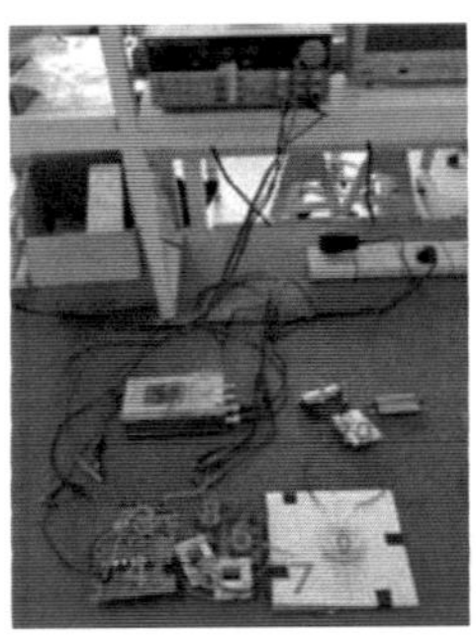

Figure 5. WPT System Circuit

Among them, label 1 is power supply, 2 is control signal generator, 3 is drive circuit board, 4 is inverted circuit board, 5 is inductor L_3 and its internal resistance R_3, 6 is primary capacitor C_3, 7 is primary coil L_1, 8 is primary capacitor C_1, 9 is secondary coil L_2, 10 is secondary capacitor C_2, 11 is load resistance R_L.
The parameters of the components used in the experiment are as shown in table 2.

Table 2. Parameters of Components Used in Experiments

Component Number	U_d	L_1	L_2	L_3	C_1	C_2
Parameter Values	12V or 24V	126.46μH	14.41μH	6.112μH	2.1618nF	19.537nF
Component Number	C_3	R_1	R_2	R_3	R_L	/
Parameter Values	47.37nF	0.046837Ω	0.02Ω	0.00842Ω	50Ω or 25Ω	/

Based on the above parameters, the mutual inductance between the two coils is measured, as shown in figure 6.

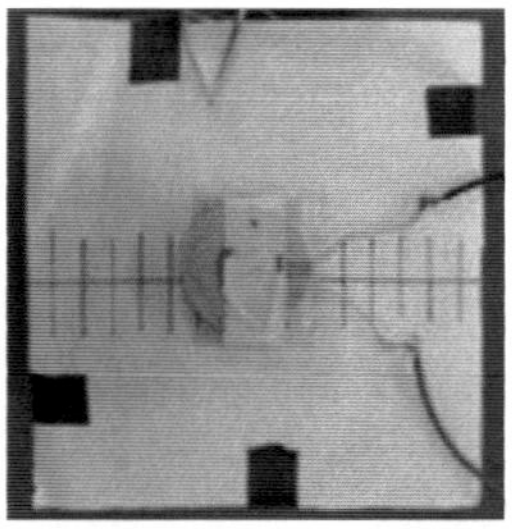

Figure 6. Mutual inductance measurement between two coils

Combining the results with RLS algorithm and importing the data into the program, we can get that the M of the two coils is stable at 7.094 μH.

For the actual M between the primary coil L_1 and the secondary coil L_2, it is impossible to get the exact value, but the "Series Decoupling" method can be used to calculate the very similar value.

The measured value of M between primary and secondary coils is 6.78μH by the above method, and the error between the measured value and the experimental value of 7.094μH is about 4.626%. The error is less than 5%, it is not very large. So it is feasible to monitor the mutual inductance between coils only by measuring the DC side voltage and current of the primary side, and the accuracy is relatively high, which is basically consistent with the simulation results.

The comparison diagrams between the recursive estimates of mutual inductance, the stability values and the measured values are shown in figure 7.

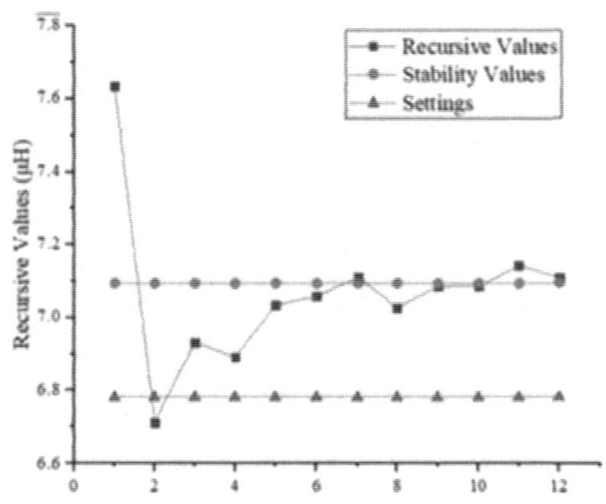

Figure 7. Comparisons among Recursive Values, Stable Values and Settings of Mutual Inductance

5. Conclusion

In this paper, through research and design, the key electrical parameters of the primary side which can reflect the overall parameters of the system are found. Thus, the key electrical parameters of the primary side can be adjusted timely and accurately according to the changes of the electrical parameters corresponding to the secondary side. The monitoring of the electrical parameters of the RMC WPT device is realized. The experimental results show that the error is within the allowable range, the accuracy of the RLS algorithm is higher, and the convergence speed is faster. Therefore, the safe operation and coordinated control of RMC WPT system are greatly guaranteed.

6. Acknowledgments

This work was supported by the Science and Technology Project of State Grid Corporation of China under Project of "Research on Wireless Charging Technology of Substation Patrol Robot Based on Magnetic Coupling Resonance".

7. References

[1] A. Zaheer, H. Hao, G. A. Covic and D. Kacprzak. Investigation of Multiple Decoupled Coil Primary Pad Topologies in Lumped IPT Systems for Interoperable Electric Vehicle Charging[J]. IEEE Trans. Power Electron., vol. 30, no. 4, pp. 1937-1955, Apr. 2015.

[2] Z. Wang, Y. Li, Y. Sun, C. Tang, and X. Lv. Load detection model of voltage-fed inductive power transfer system. IEEE Trans Power Electron., vol. 28, no. 11, pp. 5233–5243, Feb. 2013.

[3] J. Yin, D. Lin, C. Lee, and S. Y. R. Hui. A systematic approach for load monitoring and power control in wireless power transfer systems without any direct output measurement. IEEE Trans. Power Electron., vol. 30, no. 3, pp. 1657–1667, Apr. 2015.

[4] Anonymous. Wireless Frequency Identification (RFID)[J]. Software, 2018, v.39; No.457 (05): 236-237.

[5] Qiang Minghui, Zhang Jing'e. Recognition and Simulation of RLS method based on MATLAB[J]. Automation and instrumentation, 2008 (6).

[6] C. Cai, J. Wang, R. Liu, Z. Fang, P. Zhang, M. Long, M. Hu, and Z. Lin, "Resonant Wireless Charging System Design for 110-kV High-Voltage Transmission Line Monitoring Equipment," IEEE Transactions on Industrial Electronics, vol. 66, no. 5, pp. 4118–4129, 2019.

[7] J. Wang, C. Cai, L. Li, P. Zhang, Q. Liu, F. Zhang, Z. Fang, "Extended efficiency control method for WPT systems in smart grid under loose coupling extremes", IET Power Electronics, vol. 12, no. 10, pp. 2523-2533, 2019.

[8] C. Cai, J. Wang, Z. Fang, P. Zhang, M. Hu, J. Zhang, L. Li, and Z. Lin, "Design and Optimization of Load-Independent Magnetic Resonant Wireless Charging System for Electric Vehicles," IEEE Access, vol. 6, pp. 17264–17274, 2018.

Application Design of the Iris Recognition Technology in the Access Control Management System

Lijun Yu [1], Kefeng Li [2] and Jian Zheng[3]

[1]College of Engineering, Shanghai Second Polytechnic University, Shanghai, China.
Email: ljyu@sspu.edu.cn
[2]College of International Vocational Education, Shanghai Second Polytechnic University, Shanghai, China.
Email: kfli@sspu.edu.cn
[3]College of Engineering, Shanghai Second Polytechnic University, Shanghai, China.
Email: zhengjian@sspu.edu.cn

Abstract. This design is a method of implementing the iris-recognition-based access control management system, including the identity collection device, database server, infrared range finder, iris recognition device, and access controller. The identity collection device collects the target personnel's identity signals on site and outputs iris information. When a target person attempts to enter a door, the infrared range finder needs to work with the identity collection device to start the iris recognition device. The identity collection device will promptly and accurately collect the iris information, and then send the extracted iris information to the database server. The database server compares the received iris information with that previously stored in the database and quickly determines whether the iris information is legal based on the comparison results. If it is legal, the access controller sends a signal to the electric control lock to open the door and the person can enter the door. Otherwise, the person cannot enter the door. This design integrates the infrared sensing technology, human-computer interaction technology, data transmission technology, and communication technology. It utilizes the iris recognition algorithm in the iris recognition technology to improve the precision of recognition. Therefore, this design can efficiently identify and authenticate personal identity. In addition, this design features high reliability, low cost, small volume, complete functionality, and strong scalability on information collection.

1. Introduction

In the field of biometrics recognition, in addition to the application of fingerprint recognition and face recognition, people have applied the iris recognition technology to the access control management system that demands higher security. Iris is a visual characteristic of a human eye, which generally does not change after birth. It is unique for each person and features strong stability. The application of iris recognition in the field of access control will undoubtedly and significantly enhance the security performance of access control.

The iris-based access control management system fully realizes the automatic management of personnel who attempt to pass the access control system by using the data stored on the iris recognition device for comparison and using the computer as the backend processing tool. [1]

2. System Working Principle

The iris-recognition-based access control system is used to identify people by checking their iris. First of all, iris information is registered for each internal staff member, and their authorities are divided according to their positions and departments. When a target person attempts to enter a door, the person needs to place his or her left and right eyes (the right eye is preferred) close to the identity collection device. The identity collection device will promptly and accurately collect the iris information, and then send the extracted iris information to the database server. The database server compares the received iris information with that previously stored in the database and quickly determines whether the iris information is legal based on the comparison results. Then, the system determines whether the person has authority to access the door based on the preset authority of the person. If it is legal, the access controller sends a signal to the electric control lock to open the door and the person can enter the door. Otherwise, the person cannot enter the door. [2]

3. System Architecture

The iris-recognition-based access control system consists of the identity collection device, database server, infrared range finder, iris recognition device and access controller. The identity collection device is connected to the database server by using a cable to recognize the identity of target personnel. The database server is connected to the iris recognition device by using a cable to store standard iris images of personnel. The infrared range finder is connected to the iris recognition device by using a cable to measure the distance between the target person's eyes and the iris recognition device. The iris recognition device is connected to the access controller by using a cable to automatically control whether the target person is permitted to pass the door after the iris image of the target person is compared. Figure 1 shows the architecture of the iris-recognition-based access control system.

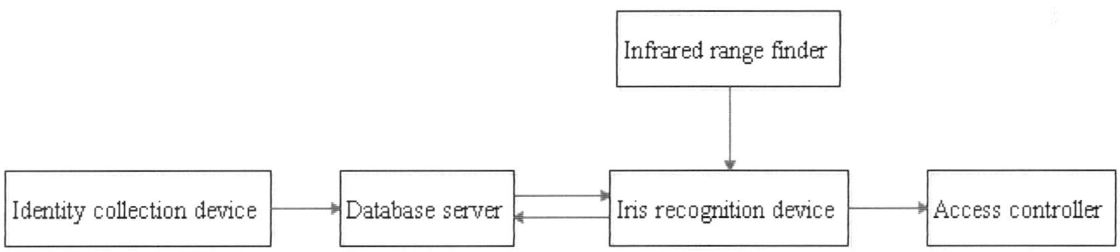

Figure 1. System architecture

The identity collection device is a non-contact smart card recognition device with the IC card function. The model chosen here is ACR122U-A9 NFC card reader. ACR122 is a smart card reader developed on the radio frequency of 13.56 MHz, which complies with the ISO/IEC18092 (NFC) standards.

To improve safety, it is optional to configure a SAM card slot in ACR122. Ensure that the SAM card slot complies with the ISO7816-3 standards. In addition, the ACS can also provide the ACR122 module for easy integration into large devices, such as POS terminals, access control devices and advertising machines. In the infrared range finder, there is a frosted glass ring plate on the first round table of the inner wall of the shell. Two infrared emitters and two light-emitting diodes are installed symmetrically on the second round table of the inner wall of the outer shell. The lower end of the shell is connected with a CCD camera. The infrared range finder is equipped with red and green LED lights. When the distance between the target person's eye and the iris recognition device is within the range of 25 to 35 cm, the green LED light is on, and the iris recognition device is working. When the distance is not within the range of 25 to 35 cm, the red LED light is on, and the iris recognition device is not working. The upper end of the device shell can be buckled onto the eyes of the target person. Under the illumination of the infrared emitter and light-emitting diode, the iris images are collected by the CCD camera and stored in the database server.

The management platform of the iris-recognition-based access control management system is installed on the database server. The management platform provides various functions such as user management, data storage, and access control management.

The iris recognition device consists of the infrared camera, positioning processing module, feature coding module, and feature comparison module. It is the core component of the iris-recognition-based access control management system. [3]

The access controller adopts the main control and sub-control structure, and the chip can use the ARM 32-bit CoreTile boards.

4. Design of Positioning Processing Module in the Iris Recognition Device

The workflow of the iris recognition device is shown in Figure 2.

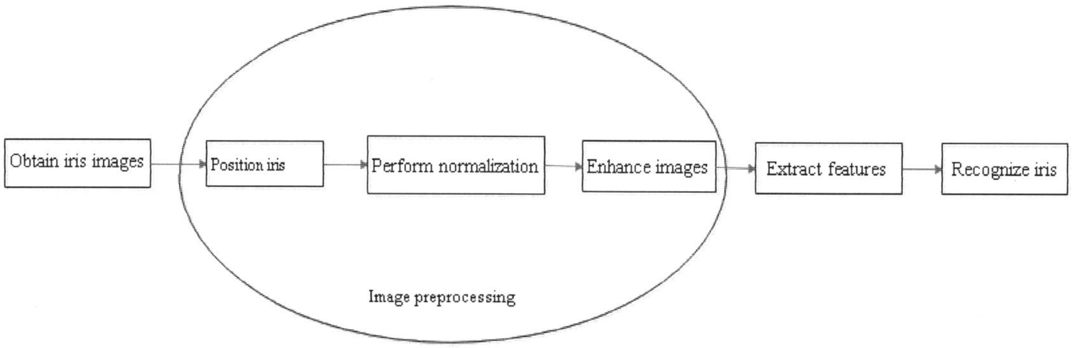

Figure 2. Workflow of the iris recognition device

The iris recognition device obtains iris images from the iris collection device, and then position, normalize and enhance the iris images through the process of image preprocessing. The processed iris images are then sent to the iris feature extraction module to extract the relevant characteristics. Then, the iris recognition device matches the iris characteristics and the iris data previously stored in the database server and obtains the final result.

The iris recognition device mainly includes infrared camera, positioning processing module, feature coding module, and feature comparison module, among which the design of the positioning processing module is particularly important. This is because the accuracy of positioning directly affects the subsequent feature extraction and recognition rate. The inaccuracy of positioning may further lead to the failure of the recognition algorithm.

The inner iris is adjacent to the pupil, and the outer iris is adjacent to the sclera. There is a clear boundary between the iris and the pupil, and the gray-level gradient between them changes significantly. Therefore, the inner edge of the iris can be positioned accurately. The outer edge of the iris is connected with the sclera, but the gray-level gradient between them is not obvious, so it cannot be directly processed. However, a larger gray-level gradient can be formed through the contrast enhancement for the outer edge of the iris.

The positioning processing module is designed to quickly and accurately position iris images. Based on the analysis of the characteristics of iris images and with reference to a few articles about fast iris positioning, this paper proposes a fast positioning algorithm, that is, enhancing iris edge contrast before positioning while suppressing the effect of the iris texture. During the process of iris positioning, use a simple thresholding and the improved Hough transformation. [4]

4.1. Positioning of the Iris Internal Contour

The inner edge of the iris is the circular border between the iris and the pupil, which is much darker than the rest of the image and occupies a different gray-level range. Therefore, as long as an appropriate grayscale threshold is selected, the binary image of the pupil can be easily and quickly obtained. Then you can find out the edge points of the pupil in the binary image and estimate the circular parameters of the pupil edge by using the least square method. In this way, you can accurately position the inner edge of the iris.

The parameters of the iris inner contour can be obtained by the parameter fitting of separated pupils. It is assumed that the iris is a circle. The parameter form of the circle is shown in equation 4-1.

$$(x-x_c)^2+(y-y_c)^2=r^2 \tag{4-1}$$

Where, (x_c, y_c) is the center of the circle, and r is the radius. Suppose there are N points at the edge of the pupil. They are (x_0, y_0) (x_1, y_1) (x_2, y_2) ... (x_n, y_n). Use equation 4-2 to calculate the center and radius of the pupil:

$$AX=B \tag{4-2}$$

Where,

$$X=\begin{bmatrix} 2x_c \\ 2y_c \\ r^2-x_c^2-y_c^2 \end{bmatrix},\quad A=\begin{bmatrix} x_0 & y_0 & 1 \\ x_1 & y_1 & 1 \\ \cdot & \cdot & \cdot \\ \cdot & \cdot & \cdot \\ \cdot & \cdot & \cdot \\ x_n & y_n & 1 \end{bmatrix},\quad B=\begin{bmatrix} x_0^2+y_0^2 \\ x_1^2+y_1^2 \\ \cdot \\ \cdot \\ \cdot \\ x_n^2+y_n^2 \end{bmatrix},\qquad n>5$$

Equation 4-3 is to use the least square method to calculate the circular parameters at the pupil edge:

$$X_{LS}=(A^TA)^{-1}A^TB \tag{4-3}$$

4.2. Positioning of the Outer Edge of the Iris

On the basis of accurate positioning of the inner iris edge, the upper and lower eyelid parts in the image are removed according to the inner edge radius and the center of pupil. According to the scope of the outer radius, the edge points that are too far from the center of the pupil in the horizontal direction are removed, and those that are less than the radius of the inner iris edge at the center of the pupil are also removed.

In order to improve the signal-to-noise ratio of the image at the edge, it is necessary to preprocess the collected image before detecting the outer edge. The preprocessing process involves suppressing noise, reducing the contrast in the iris region, and enhancing the edge contour of the iris region for accurate positioning. The specific method is to shrink the image size to 40%. By extracting vertical edge gradient information from the shrunk image, the system removes some useless edge points. Hough circle detection is conducted within the radius of the outer circle to accurately position the outer edge.

The method of neighborhood averaging and grayscale stretching is used in preprocessing. Neighborhood averaging smoothes the image, eliminates and suppresses the isolated points and line noise in the image, reduces the contrast in the iris region, and enhances the outer edge of the iris. [5]

The solution matrix of circle parameters in Hough circle detection is shown in equation 4-4.

$$\text{When } det(A)\neq 0,\ X=A^{-1}B \tag{4-4}$$

Where,

$$A=\begin{bmatrix} x_0 & y_0 & 1 \\ x_1 & y_1 & 1 \\ x_2 & y_2 & 1 \\ x_3 & y_3 & 1 \\ x_4 & y_4 & 1 \end{bmatrix},\quad B=\begin{bmatrix} x_0^2+y_0^2 \\ x_1^2+y_1^2 \\ x_2^2+y_2^2 \\ x_3^2+y_3^2 \\ y_4^2+y_4^2 \end{bmatrix}$$

(x_0, y_0) (x_1, y_1) (x_2, y_2) (x_3, y_3) (x_4, y_4) are the five points randomly selected at the edge of the pupil. [6]

5. Implementation Process and Steps of the Iris-Recognition-based Access Control Management System

The contactless smart card recognition device collects the iris samples of the target personnel and stores them in the database server. The database server management platform manages user information and conducts system maintenance and update.

Figure 3 shows the flow chart of implementing the iris-recognition-based access control management system.

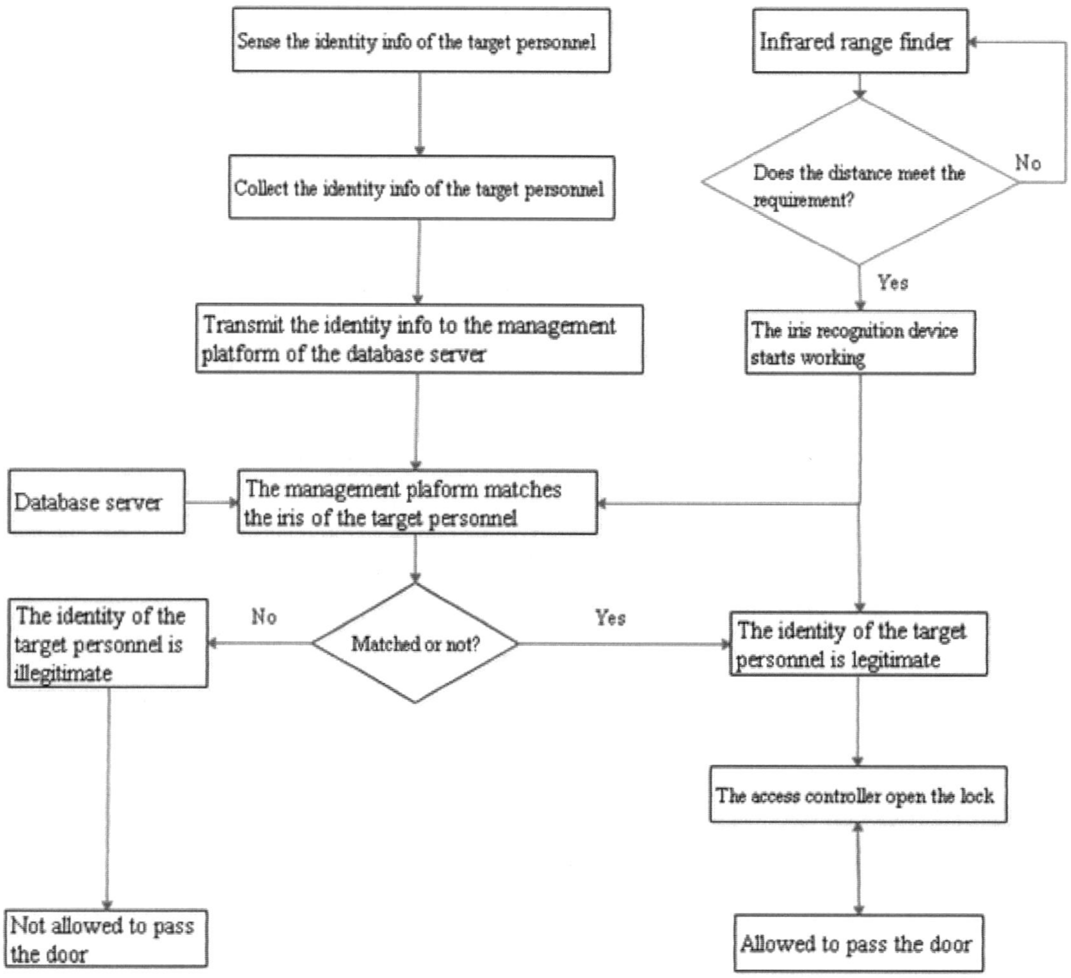

Figure 3. Flow chart of implementing the iris-recognition-based access control management system

The system is implemented in the following steps:

Step 1: The contactless smart card device detects in real time whether there is a target personnel at the access control. When sensing a visitor, the system is triggered to start.

Step 2: The contactless smart card device collects the identity information of the target personnel on site and transmits it to the access control management platform over the network;

Step 3: When the distance between the target personnel's eye and the infrared range finder is within the range of 25 to 35 cm, the green LED light is on and the iris recognition device starts to work.

Step 4: The management platform matches the iris information of the target personnel and compares it with the original iris samples of the target personnel stored in the database server;

Step 5: The management platform determines whether the iris identity of the target person collected on site matches the iris samples stored in the database server. If so, the target person's identity is considered legitimate and you can jump to step 6. If not, the target person's identity is considered illegitimate and the access control is not opened.

Step 6: The management platform judges that the iris identity of the target person is legitimate, and the system then opens the access control through the access controller.

6. Summary

Iris recognition, as an key trend of biometrics recognition, is an important branch of artificial intelligence. Compared with the access control systems based on the existing technologies, the iris-recognition-based access control management system has a lot of advantages. Its design integrates the infrared sensing technology, human-computer interaction technology, data transmission technology, and communication technology. It utilizes the iris recognition algorithm in the iris recognition technology to improve the precision of recognition. Therefore, this design can efficiently identify and authenticate personal identity. In addition, this design features high reliability, low cost, small volume, complete functionality, and strong scalability on information collection.

7. Acknowledgments

This paper was supported in part by the National Natural Science Foundation of China under Grant No.61272036, and the key Disciplines of Computer Science and Technology of Shanghai Polytechnic University under Grant No.XXKZD1604.

8. References

[1] Beijing Kaiping Aisen Information Technology Co., Ltd. Application of Iris Recognition Technology in Access Control System [J]. China Automatic Recognition Technologies. 2007(5): 72-73.
[2] Kang Cheng, Hao Meng. Design of Iris Recognition Access Control System [J]. Science and Technology Information. 2011(9): 49-50.
[3] Ling Yang. A Kind of Access Control System Based on Biological Characteristics [P]. China: 201810052494.6.2018.7.27.
[4] Ningna Dang. Research on Iris Recognition Technology [D]. Northwestern Polytechnical University. 2006: 18-20.
[5] Qichuan Tian. Iris Recognition Principle and Algorithm [M]. Beijing: National Defense Industry Press. 2010: 39-60.
[6] Lili Pan. Research and Implementation of the Iris Recognition Algorithm [D]. University of Electronic Science and Technology of China. 2007: 34-42.

IOP Conf. Series: Materials Science and Engineering **719** (2020) 012041 doi:10.1088/1757-899X/719/1/012041

Design of Power Life-Span Testing System for Mobile Phone Loudspeaker

Zili Gao[1] and Yongmei Huang[1,2]

[1]College of Metrology and Measurement Engineering, China Jiliang University , Hangzhou, Zhejiang , China

[2] E-mail: ymhuang@cjlu.edu.cn

Abstract. In order to obtain the power life-span of mobile phone loudspeaker, a mobile phone loudspeaker test system was designed. The C8051 was taken as the main control core.The sinusoidal signal which was provided by AD9833 signal generating module was amplified by power amplifier module for power life-span test of mobile phone loudspeaker. At the same time, the feedback voltage signals of the system were collected by AD7366. The test software was designed to realize the real-time monitoring and analysis of the power life-span test process of the mobile phone loudspeaker. After many tests, the results manifest that the system runs stably and reliably, and the measurements of each channel are accurate.

1. Introduction

The National standard "GB/T 12060.5-2011. Methods of measurement for Characteristics of loudspeakers", the power life-span test requirement for mobile phone loudspeaker is: A sine wave signal of a specified degree is apply to the loudspeaker, and Dynamically monitor the electrical performance changes of each speaker in real time, during the whole test [1].

According to test requirements, a sinusoidal excitation signal based on DDS (Direct Digital Frequency Synthesis Technology) was designed to provide power life-span test to loudspeakers. The voltage and current through the speaker which continue working for the rated time does not cause a large deviation, then the speaker is determined to pass the power life-span test; when the deviation is too large or almost 0, it is determined that the speaker does not pass the power life-span test.

2. The Overall Design of Test System

The test system is designed to be modular entirety. Mainly divided into MCU control unit, signal generator, power amplifier and voltage and current detection module. The high-precision programmable DDS chip AD9833 is used to provide a sinusoidal excitation signal in the signal generation module for power life-span test. The range of output frequency is20Hz-20kHz, and the accuracy of frequency is ±0.05%. The test system output voltage can be adjusted from 0.5V to 12V. The power amplifier module has a maximum output power of 15W per channel, a maximum total output power of 300W, and a load impedance of 4Ω-32Ω. The voltage error across the speaker is less than±0.5 dB which detected by the voltage and current detection module with reference to the calibration specification[2].

The designed test system is divided into AB groups, 20 channels in total, which can be connected with20 loudspeakers for test to improve test efficiency. The overall structure block diagram of the system is shown in Figure 1. First, the MCU control module C8051 controls the signal source AD9833 chip to generate a sine signal, which is amplified and output to the AB two-stage power amplifier module, which is provided to the loudspeaker power test. During the test, the voltage and current signals at both ends of the speaker are fed back to the MCU control unit through the voltage and current detection module for real-time monitoring to meet the power life-span test requirements of the loudspeaker.

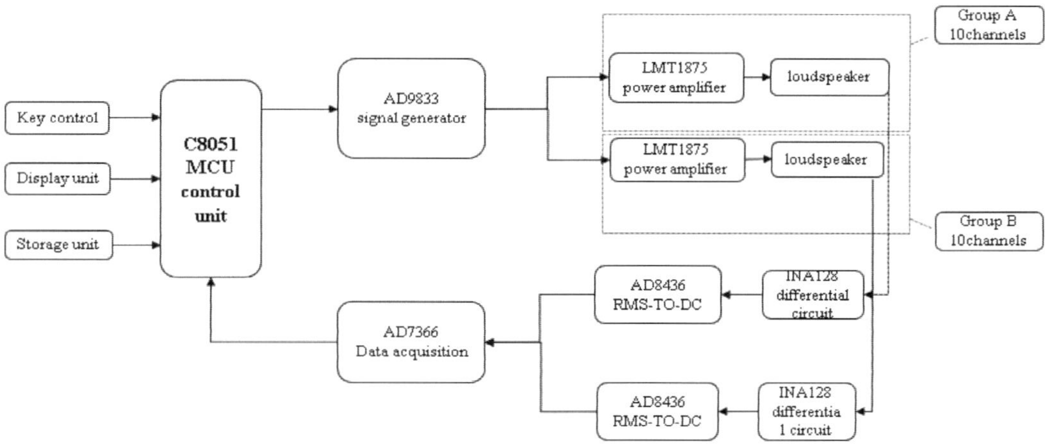

Figure 1. The overall structure diagram of the system.

3. The Hardware Design of Test System

3.1. The MCU Control Unit Module

The MCU control unit module is the control core of the entire power test system. As shown in Figure1, the chip manufactured by Silicon LAB is used, which is a fully integrated mixed-signal, system-on-a-chip MCU with a high-speed 8051-compatible CIP-51 core (100 MIPS). The 128K on-chip Flash memory also has the ability to reprogram in- system, and its function fully meets the test requirements [3]. It mainly communicate with the unit man-machine test interface of Figure 1, which accept the command of the button control unit and performs corresponding drive control on the test system signal generation module according to the instruction, and finally returns the parameter collection result of the data acquisition module to the test interface.

3.2. Signal Generation Module

The signal in the test system is provided by the chip's AD9833 high-precision programmable waveform generator. The AD9833 is a low power, DDS device that requires no external components. The output frequency and phase are software programmable and easy to adjust. The frequency register is 28 bits wide[4]. The output frequency of the AD9833 is:

$$fout = M \cdot f_{MCLK} /228 \quad (1)$$

The f_{MCLK} is the clock frequency; M is the frequency control word, given by external programming.

Since the output signal frequency is in the range of 20Hz~20kHz, the reference clock uses a 1MHz active clock. According to Nyquist's sampling law, the clock source meets the actual demand, The frequency resolution is calculated as follows:

$$\Delta f = 1MHz/228 = 0.004Hz \quad (2)$$

The sinusoidal signal generation module is shown in Figure 2. The MCU control unit C8051 can write data to the AD9833 through a 3-wire SPI interface to control the generation of a sine wave signal. The FSYNC pin in the AD9833 is an enable pin, level-triggered. When serial data is transmitted, FSYNC is asserted low and is biased, amplified, and filtered to produce a sine wave signal SinA[5].

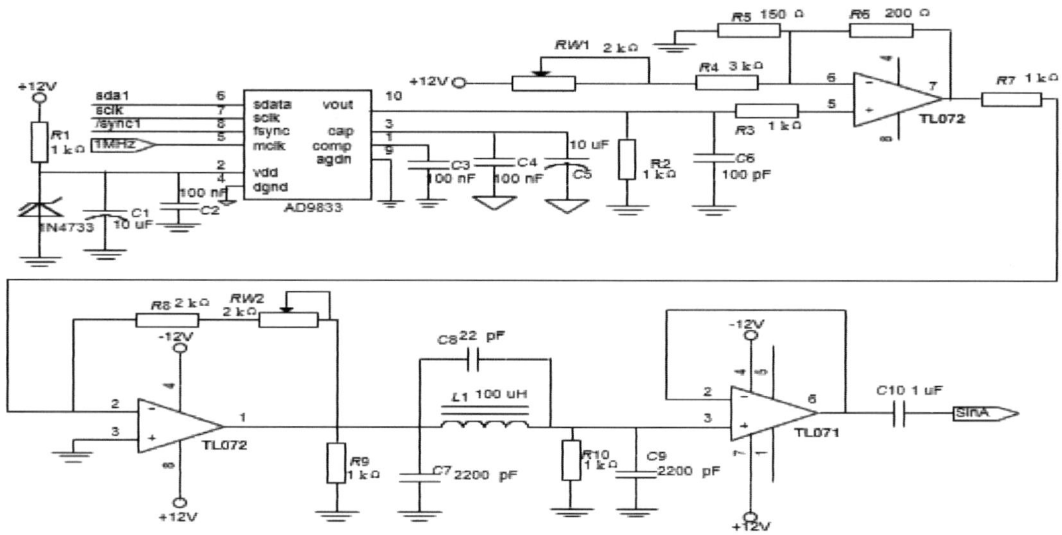

Figure 2. Group A signal generation circuit.

3.3. Power Amplifier Module

The signal generated by the signal distribution module is transmitted to the power amplifier module to drive the loudspeaker to work. One of the power amplifier modules is shown in Figure 4, Which is composed of the operational amplifier NE5532 and the LM1875T power amplifier circuit produced by American Semiconductor Device company. The LM1875T power supply voltage is ±25V.The undistorted power is 20W (THD=0.08%), and the distortion is only 0.015% at 1 kHz, 20 W. At the same time, in order to protect the amplifier from the low resistance in the high frequency state, the resistor is connected in parallel with the inductor at the output of the LM11875T. The amplification gain of LM1875 is determined by the ratio of resistors R20 and R19[6].

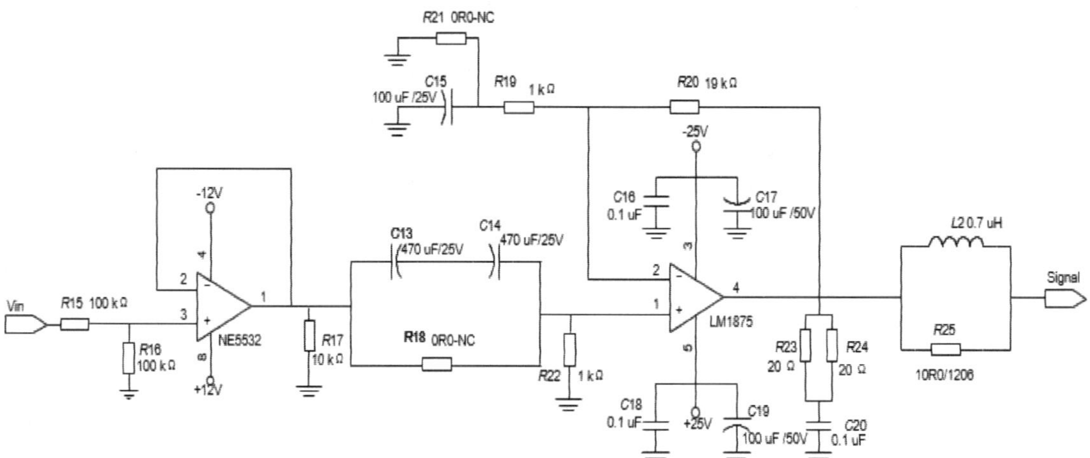

Figure 3. One of the power amplifier circuit.

3.4. The Signal Detection Module

The voltage detection is mainly to collect the voltage across the speaker. Since the voltage across the speaker will be relatively large, in order to protect the chip in the next level circuit,

The voltage detection is first divided by a high-resistance R27 resistor to obtain a smaller voltage Vu.

The current detection mainly uses a differential circuit to detect the loop current of the speaker unit. The loop current is indirectly obtained by the voltage across the high-precision small resistor R26 (0.1Ω/5W) in the acquisition loop. The differential circuit uses a low-power, high-precision instrumentation amplifier, INA128. One of the differential amplifier circuit is shown in Figure 4, which has an RG resistance of 2.63KΩ, so the amplification gain is

$$G=1+50K\Omega/RG=1+50K\Omega/2.63K\Omega=20 \quad (3)$$

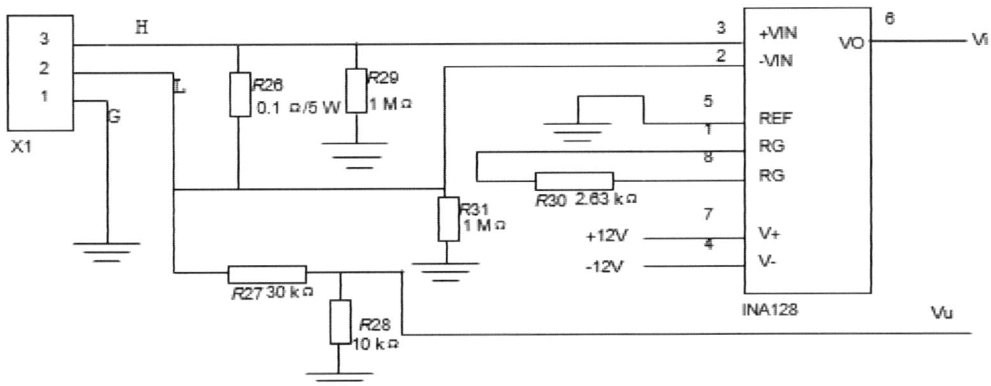

Figure 4. One of the differential amplifier circuit.

3.5. *Signal True RMS Conversion Module*

After the multi-channel sorting and voltage boosting of the respective channel voltages obtained by the signal detection module, the true effective value of the voltage is detected by the AD8436 true RMS chip, and the effective value of the AC voltage is obtained through internal calculation. The conversion circuit is shown in Figure 5. The AD8436 accurately calculates the RMS value of the AC waveform. The RMS conversion accuracy depends on the input signal frequency and the external capacitor CAVG value. At the same input frequency, the larger the CAVG, the smaller the conversion error, and the higher the frequency of the input signal of the same CAVG value, the higher the accuracy. The minimum value of CAVG can be determined based on the frequency of the input signal [7].

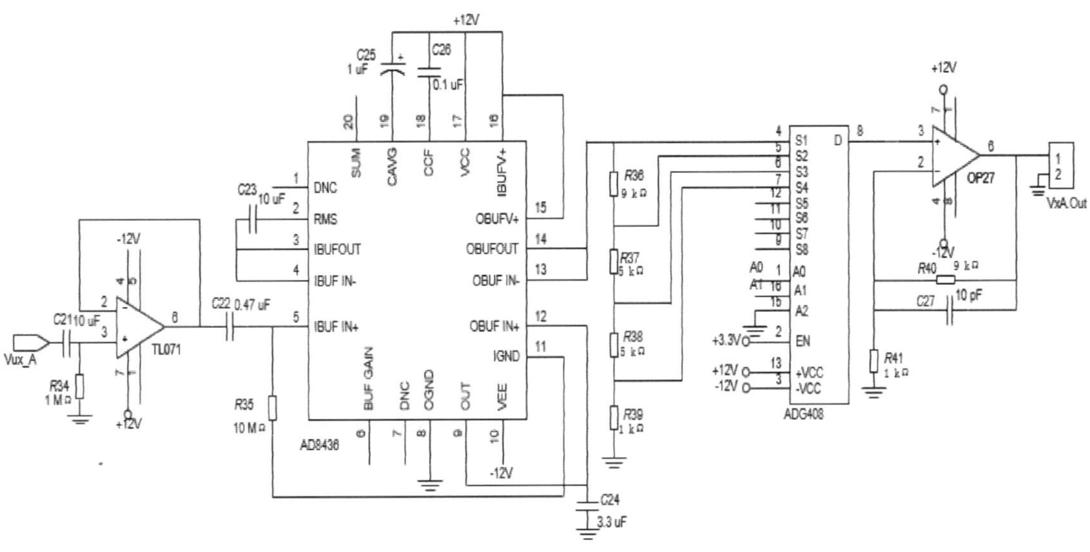

Figure 5. Group A RMS-to-DC circuit.

3.6. Signal Acquisition Control Module

The voltage and current signals (VxA, IxA, VxB, IxB) in Figure 6 are transmitted to the analog-to-digital converter AD7366 via LM358 operational amplification. The AD7366 is a dual-core, 12-bit, high speed, low power, successive approximation analog-to-digital converter (ADC) with a maximum sample rate of 1 MSPS. Two ADCs are built in, and two sets of dual channel multiplexers and a low noise, wide bandwidth sample-and-hold amplifier are available for simultaneous sampling and conversion of two channels [8]. After the data conversion and processing of the AD7366, the data is transmitted to the MCU control system and the real-time monitoring data is displayed through the LCD interface.

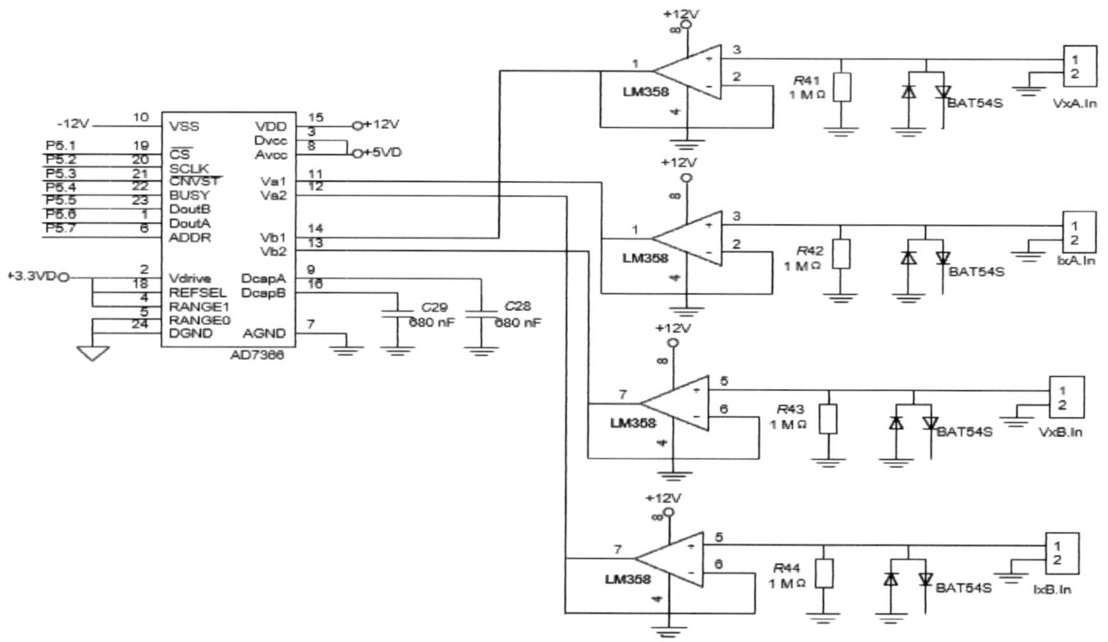

Figure 6. Signal acquisition circuit.

4. The Software Design of Test System

The software of the test system is mainly designed in C++ language, realizing the real-time monitoring of multiple parameters (voltage or current, impedance) and test time of multiple sets of loudspeaker units. According to the actual needs of the test system, the software of the whole system includes an initialization module, a main program module (including a clock module, a display module, a timer interrupt module), a timer interrupt module (including a measurement module, a timing module, a communication module), and a key interrupt module (Contains start-stop module, setup module), etc [9].

After the test system is powered on self-test, the initialization of each component is completed, and then sets the state environment (frequency range, voltage) required by the power life-span test through the button interrupt module. etc.), as well as the start and manual stop of the test. Finally, parameters such as monitoring voltage, current, resistance and test time are displayed on the display panel.

5. The Test Results

The test results shows the real-time voltage monitoring data of the sine sweep test of 20 speakers of the same model and the same batch, and also includes the true RMS value of the voltage across the loudspeaker measured by the Fluke287C digital multimeter, and compares two sets of data. The test type is a sinusoidal signal sweep test. The rated output voltage is 3.86V, the test time is set to 50h, and the rated impedance is 8Ω.

From the data that after the time required for the loudspeaker to continue working under rated power conditions, the error between the real-time voltage and the true RMS value monitored during the test is small (voltage error is less than ±0.5 dB), and the channel consistency is very high. According to the <JJF1203-2008 Calibration Specification for Electro-acoustic Products (Loudspeakers) Power Life-span Measurement Equipments>, the set of loudspeakers passed the power life-span test. The test results show that the speaker power life-span test system meets the design requirements and meets the requirements of the enterprise for power life-span test of the speaker.

6. Conclusion

The mobile phone speaker test system was designed in this paper adopts the modular design concept, which solves the problems of complicated structure of previous experiments, real-time data cannot be recorded and monitored in real time, and low efficiency. The design structure is simple, the system is easy to operate, and the test data can also be saved in the software for subsequent tracking of the product. To meet the requirements of the company's mobile phone speaker power life-span test, you can better promote your own products.

7. References

[1] Han Jie, Zhang Po, You Guolei. An Introduction of National Standard" Methods of measurement for Characteristics of loudspeakers". *Audio Engineering* 2012, 36(11):86-90(in Chinese).

[2] Jian G, Dong P. Design of Dual Phase Signals Generator Based on AD9833 *Advances in Electronic Engineering, Communication and Management* Vol.1. 2012.

[3] ZHU Si-rong, ZHOU Wan-li, BI Chun-yuan. Serial port expansion and communication design of MCU C8051F120. *Shandong Science* 2013, 26(1):69-73.

[4] Zhu W H, Shen Y, Huang G B. Digital Frequency Synthesis Using DDS-Based Approach. *Advanced Materials Research* 2012, 630:226-230.

[5] Zhang Ping. Design of DDS Based Low-Power Digital Signal Generator. *Informatization Research* 2016, 42(5):66-69(in Chinese).

[6] WANG Gang, CHEN Zhuo-yan. Design and Production of Audio Power Amplifier Based on NE5532 and LM1875T. *Science & Technology of West China* 2015.

[7] Staff E. Analog Devices AD8436 RMS-to-DC converter: RMS-to-DC converter measures low signal levels with high accuracy. *Datasheets Com.*

[8] Xing J, Ping H, Wang Z, et al. Design of Data Acquisition System Based on AD7367 and TMS320F2812 *International Conference on Computational & Information Sciences* 2011.

[9] Du Hailong, Wang Qi, LiJuan. Design of Comprehensive Experimental Platform Based on C8051F SCM.*Experimental Technology and Management* 2018, 35(6) (in Chinese).

IOP Conf. Series: Materials Science and Engineering **719** (2020) 012042 doi:10.1088/1757-899X/719/1/012042

Urban Subway Vehicle Dynamic Modelling and Simulation

Xing Cui[1], Runnan Zou[2], Min Gu[2] and Yuan Zou[2]

1. China North Vehicle Research Institute, 4 Huaishuling, Changxindian, Fengtai District, Beijing, 100072
2. National Engineering Lab for Electric Vehicles, School of Mechanical Engineering, Beijing Institute of Technology, Beijing, 100081

Email: shugu_bit@126.com, zouyuanbit@vip.163.com

Abstract. With the continuous development of urban rail transit, urban rail transit project needs to be analysed under different factors. Accurately and quickly calculate the rain real-time operating results could support the research, planning, construction and operation of urban rail transit. This paper used Matlab/Simulink programming language to develop the transit dynamics model and UI design.

1. Introduction

International study on railway operation simulation system started in the 1970s and a quite few companies have launched relatively mature railway operation software till now. [1] For instance, RALSIM software of the United States has set TPC (Train Performance Calculator) as its core function; RAILSYS software of Germany was developed on the basis of micro simulation for rail road net; Japan has developed UTRAS system with the purpose of generality. China has also developed GTMSS system that accords with "train operation calculation regulations" recently [2][3]. These applications lay emphasis on different aspects, but they are all composed of the following three components: the vehicle and route data, the operation process calculation, and the result analysis. Therefore, this paper focuses on these three factors.

2. Urban Railway Train Modelling

2.1. Urban Railway Train Dynamics Modeling

The train moves due to the forces acting on it through the wheel-rail interaction and the forces from the external environment, such as the wind resistance and gravity. In the paper the train is takes as the single particle with all the concentrating mass for during the modelling and all forces acting on the particle change as the train moves regardless of internal force of the train [4]. The external forces act on the particle mainly consist of the train, tractive effort, braking force, basic resistance, and additional resistance.

Traction pushes forward the train via the reactive force of the rail when adhesive force is generated by connecting the driving wheel with the rail through mechanical energy conversion device. In practical traction calculation, it can be worked out through traction characteristic curve. When a train speed is given, the particularly tractive effort can be calculated under linear interpolation method from the curve graph.

$$w_x = w_1 + \frac{(v_x - v_1)(w_2 - w_1)}{v_2 - v_1} \quad (1)$$

where w_x is the traction force of single motor at the required point, kN. Calculation method for traction of the entire motor train unit is:

$$W_x = n_m \times w_x = n_m \times \left[w_1 + \frac{(v_x - v_1)(w_2 - w_1)}{v_2 - v_1} \right] \quad (2)$$

Where n_m is the number of motors in motor train unit and W_x is the traction force of motor train for the required point.

Migration force of unit weight is:

$$w_x^{'} = \frac{W_x \times 1000}{(n_m m_m + n_t m_t) g} \quad (3)$$

Where $w_x^{'}$ is the traction per unit weight at the required point, m_m is the weight of motor train, m_t is the weight of trailer, n_t is the number of trailers in the motor train; g is the acceleration of gravity, m/s^2

2.2. Resistance Model

Resistance of motor train unit during operation includes basic resistance and additional resistance: basic resistance is caused by running resistance between the bullet train and its components, air resistance during operation, friction between the wheel and the rail; additional resistance is caused by route, bridge and tunnel, including additional resistance for gradient factor, additional resistance for curve factor, and additional resistance for tunnel factor. Basic resistance is based on the resistance formula raised by Davis:

$$F_{r0} = A + B \cdot v + C \cdot D \cdot v^2 \quad (4)$$

A, B, C are parameters of this function. v is the velocity of motor train.

Additional resistance is caused due to comprehensive influence of ramp, camber and tunnel. It can be divided into additional resistance for gradient, additional resistance for curve, and additional resistance for tunnel.

In the practical route, included angle θ of the ramp is small, often less than 60‰. Therefore, $\tan\theta$ can be treated as $\sin\theta$, thus the calculation method for gradient resistance is:

$$w_i = \frac{W_i}{(n_m m_m + n_t m_t) g} \times 1000 = 1000 \sin\theta = 1000 \tan\theta = i \quad (5)$$

According to Regulations on Railway Train Traction Calculation [5], the calculation formula of additional resistance for curve of upper curve central angle and curve length is:

$$w_r = \frac{600}{R} = \frac{600}{(\frac{360}{2\pi})(\frac{L_r}{\alpha})} = \frac{10.5\alpha}{L_r} (N / kN) \quad (6)$$

w_r is the additional resistance for curve units, R is the curve radius, m.

Empirical formula is adopted to calculate additional resistance for tunnel of the train [6]:

$$w_s = 0.00013 L_s \quad (7)$$

Where L_s is the tunnel length, m.

2.3. Braking Model

At present, braking method of urban railway often choose electric-pneumatic breaking, and its data can be gained via interpolation method by referring to braking force curve. Suppose that (v_1, w_1) and (v_2, w_2) are two given points on the braking characteristic curve, (v_x, w_x) is point to be calculated between these two points, the braking effort can also be calculated when the velocity v_x is given:

$$w_x = w_1 + \frac{(v_x - v_1)(w_2 - w_1)}{v_2 - v_1}(kN) \tag{8}$$

2.4. Train Dynamics Model

When the train is regarded as rigid system, operation state of the train can be obtained via theorem of kinetic energy. Kinetic energy of the train includes two parts: one is kinetic energy during linear motion, and the other one is the rotational kinetic energy of the rotating part:

$$E_k = \frac{mv^2}{2} + \sum \frac{I\omega^2}{2} = \frac{mv^2}{2} + \sum \frac{Iv^2}{2R^2} = \frac{mv^2}{2}(1 + \sum \frac{I}{2mR^2}) = \frac{mv^2}{2}(1+\gamma) \tag{9}$$

m is the weight of the whole train, v is the speed of train operation, I is the moment of inertia of the rotating part, ω is the angular velocity of the rotating part, γ is the coefficient of rotational mass.

It can be gained through differentiation for (9):

$$dE_k = m(1+\gamma)dv \tag{10}$$

The calculation formula for acceleration can be gained by transforming (10):

$$a = \frac{dv}{dt} = \frac{C}{m(1+\gamma)} \frac{W_x - W_r - W_j}{m(1+\gamma)} \tag{11}$$

Formula for train velocity can be obtained by integration for (11):

$$v = \int \frac{dv}{a} = \int \frac{m(1+\gamma)}{C} dv \tag{12}$$

Formula for operating range of the train can be gained by integration for (12):

$$S = \int dS = \int \frac{v}{C/m(1+\gamma)} dv = \int \frac{v}{a} dv \tag{13}$$

2.5. Calculation for Traction Energy Consumption of the Train

The train will consume power network energy during actual operation and the train is pushed forward by converting electric energy into mechanical energy. Energy consumption during train operation is related to engine attribute, route attribute, and other relevant factors [7][8]. Its energy consumption formula can be simplified into (14) and (15):

$$P = F \cdot V \cdot \eta \tag{14}$$

$$I = P / U \times \cos\varphi \tag{15}$$

In the formula: η is the energy conversion efficiency of the train; $\cos\varphi$ is power factor and often set as 0.95.

Energy consumption for traction during operation of the train can be gained via the energy formula by multiplying traction voltage by traction current by time. Therefore, the calculation formula for total energy consumption is:

$$W = \int Pdt = \int U \cdot Idt \tag{16}$$

2.6. Route Model

Track parameter data refer to route data, covering serial number of sections, accumulated journey, track speed, distance between the two stations, route plane data, vertical data, and tunnel data. Plane data include curve length and curve central angle. Vertical section includes gradient value and gradient distance. Tunnel data include serial number of tunnel and tunnel length. For convenient input, the same variable was adopted for both curve length and ramp length. The specific value was determined according to curve central angle and gradient size. For instance, if there are two different curvatures at a certain gradient, we can suppose that the gradient size is fixed while curve length and central angle vary, and total length of the curve is equal to the sum of the two curve lengths. Refer to the following table for specific variable names.

Table 1. Track parameters table.

Parameter name	Variable
Accumulated journey	L_m
Track speed	V_{limit}
Distance between the two stations	L
Curve length of plane route	L_r
Curve central angle of plane route	α
Vertical section gradient (positive at upslope)	θ
Vertical section slope length	L_s
Tunnel length	L_s

3. Urban Rail Transit Calculation Program

3.1. Traction Calculation Program

The calculation application should include data input module, simulation calculation module, and result output module. The train data and route data will input in graphical display interface, simulation calculation module will feed back data to result output module during simulation, and states of train will present through the graphical interface.

Simulation calculation module is the heart of the entire metro simulation operation software. Its calculation logic is to form a virtual track via the preloaded track data and then determine the working condition of train braking unit and traction unit by detecting the condition of acceleration switch and brake switch; to obtain acceleration and operation state of the train through resultant force calculation module; to gain real-time velocity of the train through velocity calculation module; then to gain the current running distance of the train via distance calculation module. By directing at difference between simulation period and practical period of Matlab, Level 2 S-Function in Matlab language was utilized to write a timer program "sfun_time" and realize the function of real-time simulation via software.

Simulation calculation module in Simulink is presented in the following figure:

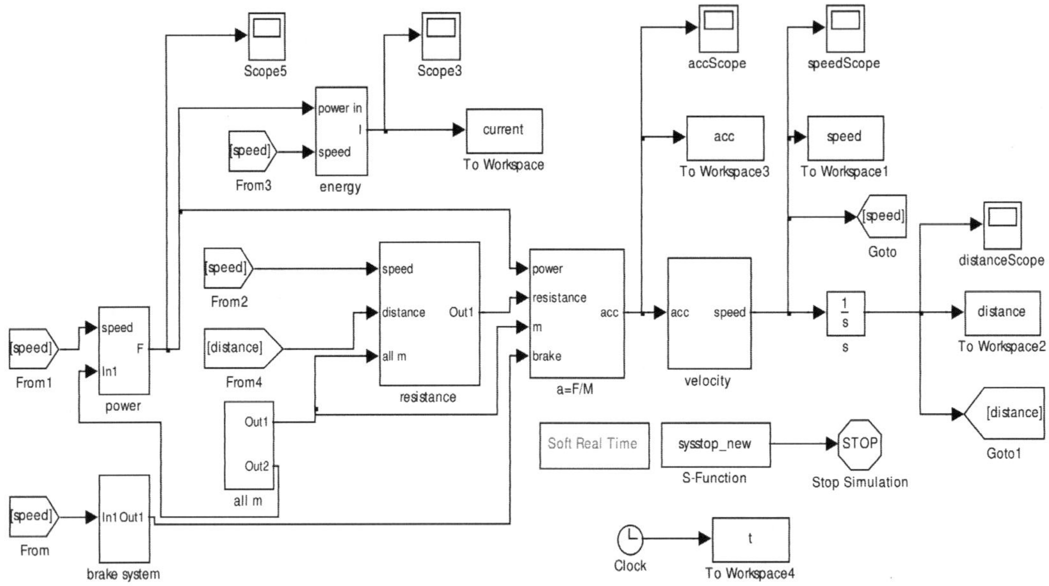

Figure 1. Traction calculation model based on Matlab/ Simulink.

3.2. UI Design

Simple style was adopted for interface of metro train calculation simulation software which is composed of one main window and two secondary windows. The two secondary windows are track parameter loading window and train parameter loading window.

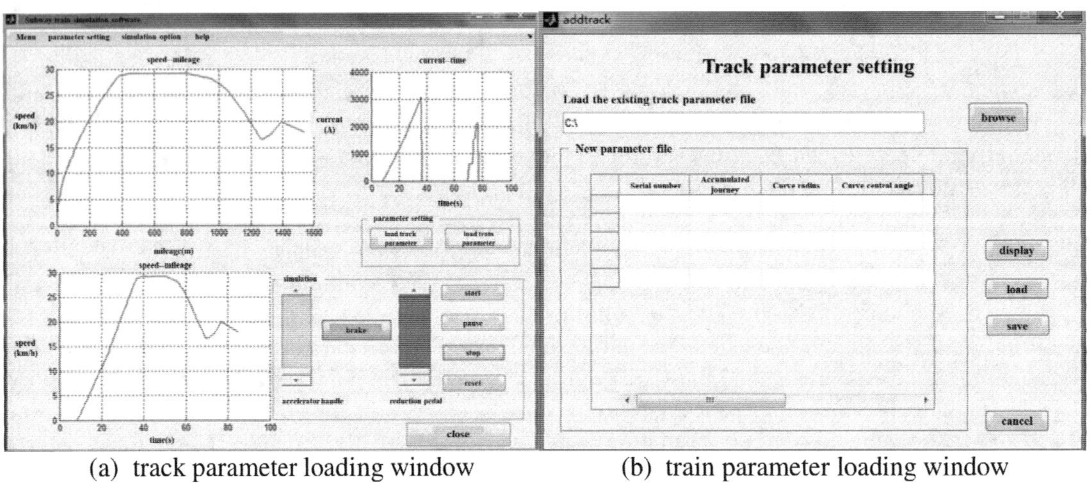

(a) track parameter loading window (b) train parameter loading window

Figure 2. Interface of the application.

During operation, the train can be controlled through the button of accelerator and decelerator, real-time display of the train operation state can be found in the main interface. The track parameter interface can be opened through parameter setting - load track parameter or by pressing the key button of track parameter. The interface of loading track parameter is for loading route data. The loaded file can be browsed or modified in the table interface. Meanwhile, the interface of loading train parameter is for loading train data. The loaded file can be browsed or modified in the table interface, too.

4. Simulation

Basing upon the planned metro route data, we simulated with type B metro and set the highest velocity is 80km/h [9][10]. The parameters are as follows:

Table 2. Vehicle parameters table.

Parameter name	Variable
Bullet train mass (kg)	49800
Bullet train quantity	4
Trailer mass (kg)	44500
Trailer quantity	2
Designed max speed (km/h)	80
Working voltage (V)	1000
Energy conversion efficiency	0.98
Rotary mass coefficient	0.2
Basic resistance parameter 1	2.43
Basic resistance parameter 2	0.02997
Basic resistance parameter 3	0.00078

Table 3. Traction data table.

Velocity(km/h)	Traction force（kN）
0	93.25
42	93.25
50	83.75
60	70
70	59.5
80	45

Table 4. Braking force data table

Velocity（km/h）	Braking force（kN）
0	-150
5	-150
6	-335
64.5	-335
80	-275

After simulation, the total journey is 4,590.97m. The train starts from the origin station and reaches the first stop after 91 seconds. The station dwell time is 51 seconds. Then after 87.5 seconds, it reaches the second stop, with the station dwell time of 49 seconds. It takes 82.5 seconds from the second stop to the third stop. The total operation time is 361 seconds in which the running time is 261 seconds. The average velocity is 44.78km/h, in which the average running velocity is 61.9km/h. The peak current is 4,863 amperes and the total energy consumption is 38.24kwh.

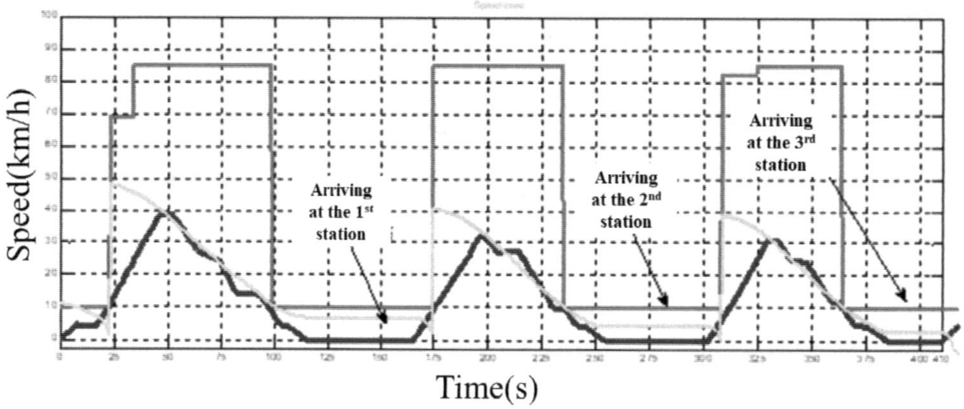

Figure 3. Fastest strategy simulation results.

If energy consumption is considered, the economic targets and velocity targets cannot be paid equal attention to. Therefore, in practical operation of the train, the driver will consciously try to turn the running train into an idle running state within the time allowed by the running schedule, but this will inevitably reduce running velocity of the train. If blocking is further considered, the driver will have no space to consider the economic targets. Therefore, this paper suggests that the economic targets should be brought into the train schedule.

The second group of data shows that the train starts from the origin station and reaches the first stop after 111.3 seconds. The station dwell time is 28.5 seconds. Then after 90.6 seconds, it reaches the second stop, with the station dwell time of 39.5 seconds. It takes 92.5 seconds from the second stop to the third stop. The total operation time is 362.4 seconds in which the running time is 294.4 seconds. The average velocity is 45.60km/h, in which the average running velocity is 56.13km/h. The peak current is 4,206 amperes and the total energy consumption is 22.4kwh.

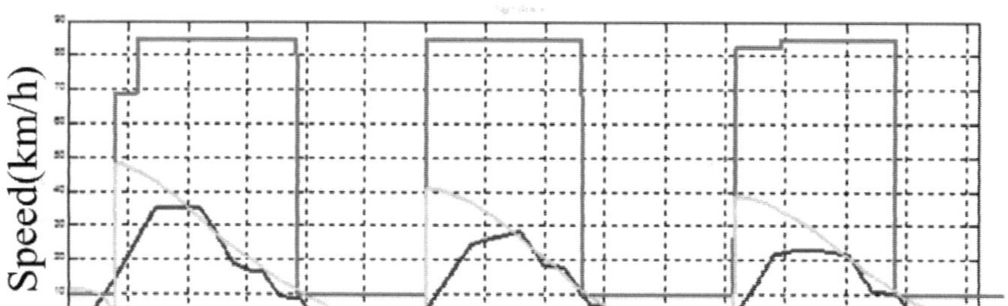

Figure 4. Simulation result based on considering energy costs.

After comparison between these two groups of data, the average running velocity of the second group has decreased by 10.28% when compared with the first group, its energy consumption has reduced by 70.71%, and the peak voltage has decreased by 13.5%.

5. Conclusion

In single-particle model algorithm, metro vehicle simulation software calculates the entire train formation as a whole. The result has proven that the program successfully achieves precise control of a train; it also possesses high simulation efficiency and can be effectively applied to train operation real-time simulation calculation of urban railway system.

6. References

[1] Nicholson G L, Kirkwood D, Roberts C, et al. Benchmarking and evaluation of railway operations performance[J]. Journal of Rail Transport Planning & Management, 2015: S2210970615300159.

[2] Baohua Mao, Wenzheng Jia, Shaokuan Chen and Jianfeng Liu. A computer-aided multi-train simulator for rail traffic[J]. Proceedings of the 2007 IEEE International Conference on Vehicular Electronics and Safety. 108-112.

[3] Zhaoxia Yang, Baohua Mao, Tianjian He. Sitution and Development of Railway Transportation System Simulators[J]. Journal of northern Jiao Tong university. 2002(5): 1-8.

[4] Jing F, Xu G, Xu Y, et al. Research on Optimal Control Strategy of Coupler Force in Urban Rail Transit[J]. Electric Drive for Locomotives, 2017.

[5] TB/T 1407-1998. Train traction calculation procedures[S]. Beijing: China standards Press, 1998.

[6] Xiang Gui. Research and development of urban rail transit traction simulation system[J]. Gansu Science and Technology. 2011(7): 124-128.

[7] E. Pilo, L. Rouco, A. Fernandez, A. Hernandez-Velilla, A simulation tool for the design of the electrical supply system of high-speed railway lines. IEEE PES Summer meeting 2000, Seattle,2000.

[8] Xiaofeng Y, Hao X, Zheng T Q. Stray Current and Rail Potential Dynamic Simulation System Based on Bidirectional Variable Resistance Module[J]. Diangong Jishu Xuebao/Transactions of China Electrotechnical Society, 2019, 34(13):69-81.

[9] Aksjonov A, Vodovozov V, Augsburg K, et al. Design of Regenerative Anti-Lock Braking System Controller for 4 In-Wheel-Motor Drive Electric Vehicle With Road Surface Estimation[J]. International Journal of Automotive Technology, 2018, 19(4):727-742.

[10] Marrazzo M, Manera N, Otta G, et al. Design of a set of rules for the expert control of the disconnection — Reduction of generation in an electrical power system[C]// 2017 XVII Workshop on Information Processing and Control (RPIC). IEEE, 2017.

Test Results for 5G Indoor Pico Base Station

Haitao Liu and Yang Liu

China Telecom Research Institute, Future Science Technology City, Beijing.
Email: liuht@chinatelecom.cn

Abstract. The fifth generation (5G) New Radio (NR) has attracted a large amount of interest with its improvement of user experience and capacity. But the anticipated use of the new technology brings serious challenges to existing solutions for its high spectrum and large bandwidth. Among them, indoor coverage is an important issue before commercial deployment. However, there are few evaluation targeting the pico base station. A new distributed antenna system (DAS) solution towards 5G indoor coverage is introduced. In this paper, the new indoor picocell with different RF configuration is introduced to build more flexible, efficient and evolving network. Besides, A series of trials are carried out to verify the RF performance of two types of radio frequency units (RRU).

1. Introduction

In the approaching 5G era, 70% of new business will occur in indoor scenes, and 80% of high-value customers work in indoor environments. Therefore, indoor coverage has become the key to operators' development of 5G services. However, the 5G deployment frequency band is generally high, which leads to significant increase on path loss and penetration loss. Taking 3.5GHz as an example, it is difficult to cover indoor areas with the outdoor base station. (3GPP TR 38.901 formula) Besides, the existing distributed antenna systems (DAS) can't be updated and reused, since the passive components does not support 3.5GHz frequency band. Therefore, the specific focus of 5G indoor coverage research is pico base station (BS) based on digital DAS, which is one of the most important deployment modes in the future indoor scenarios.

There are various picocell configurations for different requirements. For indoor hotspot scenes, high-capacity and high-density are required to carry out the new 5G services. The picocell with four antennas (4T4R) is more in line with the demand. On the contrary, ensuring coverage at a lower cost is the primary focus of the low-traffic scenario. To reduce the deployment costs, low specification small cells with double antennas (2T2R) will be one of the choices. In this paper, the two types of 3.5GHz picocells is studied by means of a few RF performance trial with a testbed.

In Section 2, we introduce the new DAS solution for 5G and describe its architecture. Section 3 gives an overview of the testbed used and field-trial environments. Section 4 presents the results. Finally, the conclusions are drawn in Section 5.

2. 5G Indoor Coverage

2.1. Digital DAS System

The new digital DAS is seen as the main solution in specific indoor scenarios. As shown in figure 1, The pico BS is mainly composed of a baseband unit (BBU), an extension unit (EU), and several radio frequency units (RRU). The BBU is mainly responsible for baseband signal modulation and demodulation, wireless resource management, mobility management, physical layer processing, equipment state monitoring and other functions. The EU is used for the extension of the wireless baseband signal to enhance the indoor coverage and system capacity. The RRU realizes the radio frequency processing and the radio signal transmission and reception. However, only BBU and RRU are needed to complete the tests.

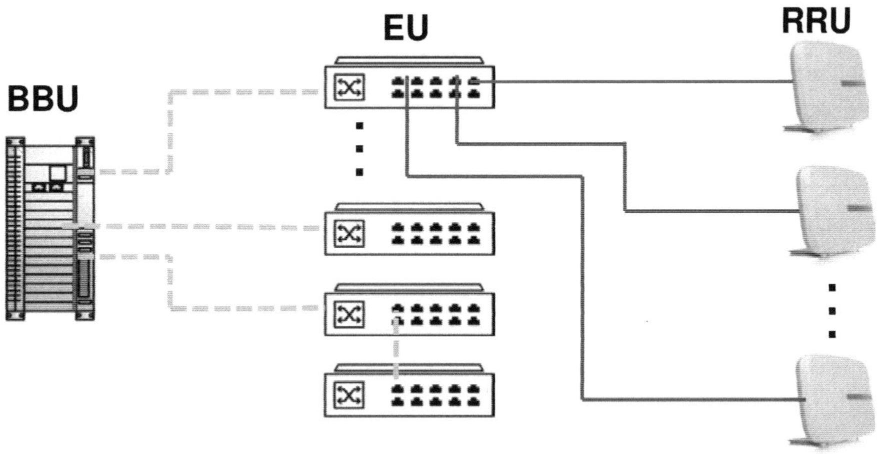

Figure 1. Pico base station system architecture.

The new system architecture has many advantages over traditional passive DAS. At the BBU side, capacity requirements of different scenarios can be dynamically met through cell merging and splitting. On the other hand, benefiting from the active digital device, new functions can be realized such as precise positioning and fault monitoring, which will certainly be of much help RRU maintenance and management. Currently, the active digital DAS has become the mainstream indoor 5G scheme for multiple vendors.

2.2. 4T4R and 2T2R

Considering the differences in coverage and capacity for different scenarios, there are two types of RRU with four or two antennas. From the perspective of coverage, 4T4R of 3.5 GHz can reach the same coverage as LTE of 1.8 GHz. As for system capacity, 4T4R can maximize the system capacity. In addition, 5G terminals support 2T4R, the picocell of 4T4R also matches the mobile phones' capabilities.

However, deployment costs are a huge challenge, and it is easy to waste resources in sparsely populated areas. To meet different scenarios, the 2T2R RRU is more suitable for the lower capacity requirements. The coverage will be lower than 4T4R if the output power is still 24 dBm for each antenna. To compensate for the gain of antenna number loss, the output power is increased to 27 dBm. Table 1 shows the link budget of two configurations. When the transmission power is increased by 3 dB, the downlink coverage are improved as much as uplink coverage. Moreover, the RRU cost is reduced by saving components and lowing the hardware configurations.

Table 1. 3.5 GHz link budget.

Item	Parameter		
Channel	PUSCH	PDSCH	PDSCH
Carrier freqence (GHz)	3.5	3.5	3.5
Sub carrier spacing (kHz)	30	30	30
RB number	100	272	272
Transmit antenna	2	2	2
Output power (dBm)	23	24	27
Receive antenna	2	4	4
Cover range (m)	32.42	26.95	33.48

IOP Conf. Series: Materials Science and Engineering **719** (2020) 012043 doi:10.1088/1757-899X/719/1/012043

3. Test Environment

3.1. Testbed Environment
The RF conduction test of the system was carried out in the electromagnetic shielding room. Figure 2 and figure 3 show the working testbed. NR signals generated by BBU are emitted through RRU. RRU can be powered by PoE network cable from BBU or direct current transformer. For transmitter test, the RRU output is connected to the signal analyzer through a feeder. A protective device is used between the RRU output and the analyser. A N9030B signal analyzer is used for high-performance research and development. The instrumentation can analyze various signals with a wide frequency range of 2 Hz to 50 GHz, and the analysis bandwidth is up to 510 MHz, which completely meet our equipment requirements. For receiver test, A N5121B RF analog signal generator is used to produce standard NR signal.

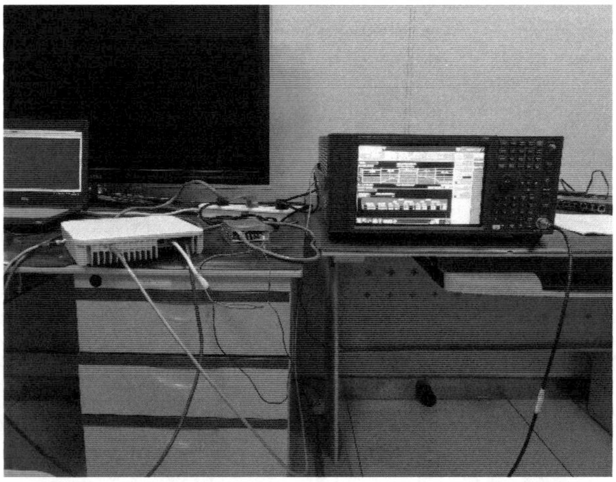

Figure 2. Testbed environments with 2T2R RRU.

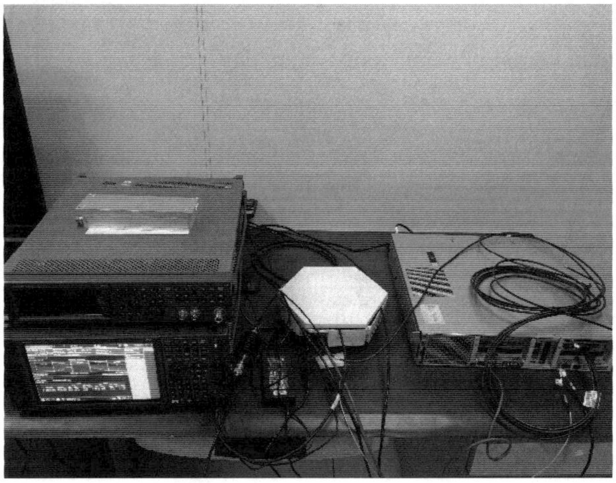

Figure 3. Testbed environments with 4T4R RRU.

3.2. System Configuration
The 5G testbed radio interface has the same main structure according to 3GPP NR standard. As shown in table 1, the duplex mode is time division duplex (TDD). The testbed supports up to 256 QAM modulation in the downlink (DL) and up to 64 QAM in the uplink (UL). The bandwidth is 100 MHz, centered at 3.45 GHz. The total output power is 30 dBm, that means 24 dBm/port for 4T4R RRU and 27 dBm/port for 2T2R RRU.

IOP Conf. Series: Materials Science and Engineering **719** (2020) 012043 doi:10.1088/1757-899X/719/1/012043

Table 2. Main system parameters.

Parameter	Value
Carrier frequency	3.45 GHz
Bandwidth	100 MHz (3.40 – 3.50 GHz)
Subcarrier spacing	30 kHz
Duplex	TDD
Frame structure	2.5 ms double cycle
Duplex scheme	DDDSU DDSUU
Maximum modulation	256 QAM in DL, 64 QAM in UL
Antenna	2, 4
Total output power	30 dBm

4. Results

According to related standards and laboratory conditions, a total of 9 transmitter characteristics has been tested including output power, transmitter off power, transmitter transient period, error vector magnitude (EVM), occupied bandwidth (OBW), adjacent channel leakage power ratio (ACLR), operating band unwanted emissions (OBUE), and transmitter spurious emissions. The receiver indicator was only tested for reference sensitivity level. The detailed RF test results are listed in the table 2. Partial test screenshots are given in figure 4 and figure 5.

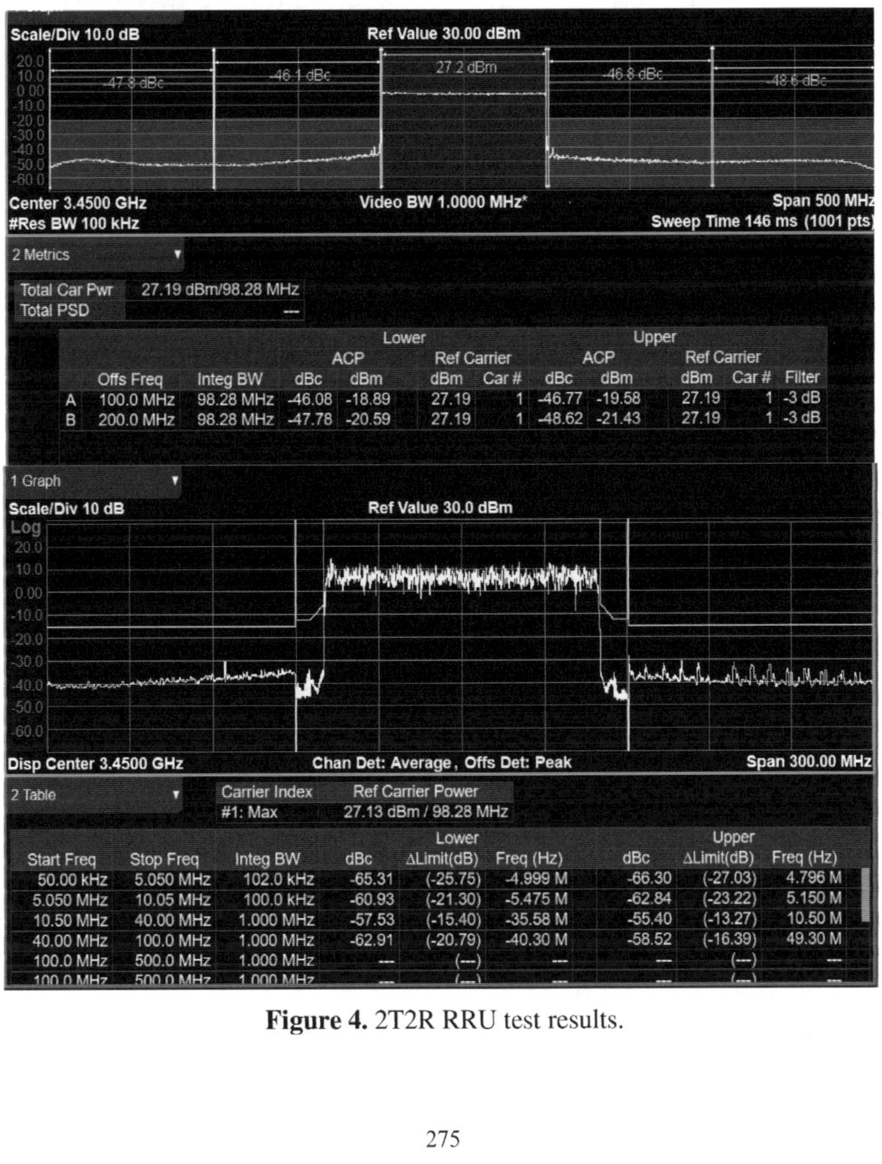

Figure 4. 2T2R RRU test results.

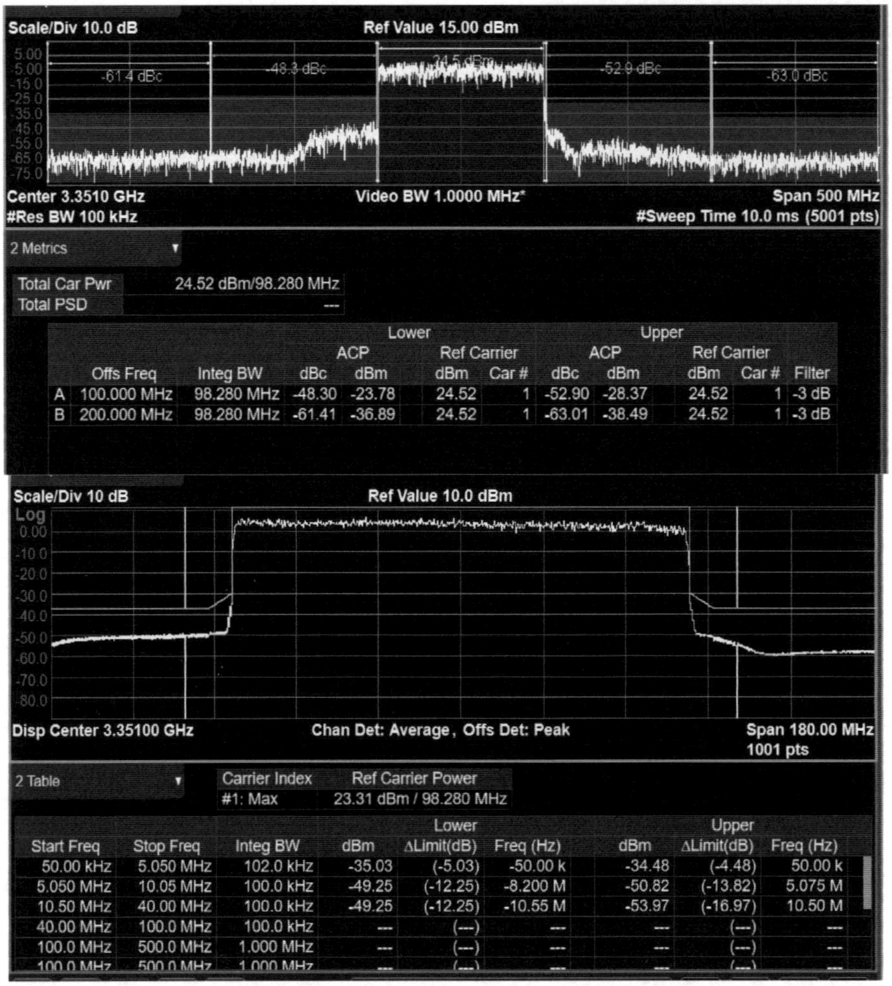

Figure 5. 4T4R RRU test results.

Table 3. RF test results.

Item		2T2R RRU	4R4R RRU	3GPP
Output power		27.19 dBm	24.52 dBm	24 dBm
Transmitter off power		-99.84 dBm/MHz	-98.88 dBm/MHz	-89 dBm/MHz
Transmitter transient period		< 0.5 us	< 2.7 us	10 us
EVM	QPSK	1.94%	1.4%	17.5%
	16QAM	1.74%	1.56%	12.5%
	64QAM	1.6%	1.67%	8%
	256QAM	1.73%	1.8%	3.5%
OBW		97.3 MHz	97.2 MHz	100 MHz
ACLR		-46.1 dBc	-48.3 dBc	-45 dBc
Reference sensitivity level		-94.5 dBm	-93 dBm	-87.6 dBm

As can be seen from table 3, the two kinds of RRU pass all the RF tests, and most of results are better than 3GPP standard requirements. Although 27 dBm exceeds the norm by 3 dB for 2T2R RRU, the total output power is still 30 dBm as same as that of 4T4R RRU. The OBUE cases in figure 4 and 5 show good NR waveforms generated by the tested device. The basic limit of transmitter spurious emissions -36 dBm under 1 GHz and -30 dBm above 1 GHz, and the test results are in line with the requirements. For 2T2R RRU, it must also be mentioned that the digital pre-distortion (DPD) algorithm is used to enhance the efficiency of power amplifier, which is one of the crucial factors of qualified RF performance.

5. Conclusion

In 5G era, the picocell based on the active digital DAS is expected to provide high quality service in specific indoor scenarios while reducing deployment costs. The 2T2R RRU with output power of 27 dBm presents a wide range of possibilities for the further development of indoor solution. A NR testbed is established and a series of RF tests are carried out in this paper. At this stage, the testbed has only examined the RF conducted characteristics, and all tested items of the two types of RRU meet the requirements of 3GPP. In the future, we will continue to verify the performance of pico base stations for indoor coverage, especially of the inexpensive solutions.

6. References

[1] B. Halvarsson, A. Simonsson, A. Elgcrona, R. Chana, P. Machado and H. Asplund, "5G NR Testbed 3.5 GHz Coverage Results," *2018 IEEE 87th Vehicular Technology Conference (VTC Spring)*, Porto, 2018, pp. 1-5.

[2] ``NR; Radio Link Control (RLC) protocol specification," 2018 3GPP TS 38.322 version 15.4.0

[3] "NR Base Station (BS) radio transmission and reception," 2018 3GPP TS 38.104: version 15.4.0

[4] Adrian S, Ruben M, Andreas B, "3.5 GHz Coverage Assessment with a 5G Testbed," 2019, *Vehicular Technology Conference, (VTC2019-Spring)*, pp. 1-6, 2019

[5] 3GPP TS 38.141-1: "NR; Base Station (BS) conformance testing; Part 1: Conducted conformance testing," 2018 3GPP TS 38.141-1 version

IOP Conf. Series: Materials Science and Engineering **719** (2020) 012044 doi:10.1088/1757-899X/719/1/012044

A QoE-Oriented Network Selection Mechanism in Heterogeneous Wireless Networks

Chao He[1,2], Zhidong Xie[1,2] and Chang Tian[1]

[1] College of Communications Engineering, Army Engineering University of PLA, Nanjing 210007, China

[2] National Innovation Institute of Defense Technology, Academy of Military Sciences of PLA, Beijing 100071, China

Email: xzd313@163.com

Abstract. In order to guarantee the UAV video uplink in heterogeneous networks, a novel access network selection mechanism was proposed in this paper. Due to the unknown Channel State Information(CSI), the issue was first modeled as the Multi-Armed Bandit (MAB) problem. Then, the reward function was built and the Quality of Experience (QoE) of video was considered. By means of three typical solving algorithms, decisions could be made between utilization and exploration. The simulation results show that the mechanism can effectively solve the network selection problem of video transmission in heterogeneous networks.

1. Introduction

The rapid development of Internet of Things and aviation technologies has made aerial photography through Unmanned Aerial Vehicles (UAVs) more and more popular. Video streaming captured by the UAVs could be sent to the ground by all kinds of wireless networks in real time. The overlaps of heterogeneous wireless networks indeed provide more convenience for UAV communication. But, at the same time, they also trigger the access network selection problem. Because of the particularity of video transmission service, the selection strategy needs to start from the application layer and take into account the characteristics of different wireless channels.

Due to the time-varying property of wireless channel and the mobility of UAV, the bandwidth of access network will not always keep the same. This bandwidth fluctuation adds uncertainties to the selection of access network. At the same time, due to channel fading effect and other reasons, packet loss is inevitable in video transmission, which is always random and increases the complexity of network selection. In the case of unknown CSI, the UAV needs to constantly perceive the transmission environment, learn from the network, so as to provide reference for their next decision. Therefore, the solution of this problem will be a sequential decision-making process, which can be solved by the Multi-Armed Bandit (MAB)[1] model. As a classical reinforcement learning algorithm, the model has been widely used in many aspects, such as spectrum access and networks handover[2,3]. It generally includes a decision maker and multiple optional arms. The decision maker needs to make selection among the optional arms and keep a balance between utilization and exploration.

Traditional access network selection methods often focus on Quality of Service(QoS) of the network, ignoring the needs of users. When considering the special scenario of UAV video transmission, it is necessary to start from the perspective of improving QoE for users. The main contributions of this paper are as follows:

- The network selection mechanism of UAV video uplink is modeled as a Multi-Armed Bandit problem, and the optimization objective is determined.
- Based on a QoE function according to video content classification, the reward function of the UAV video transmission is established.
- The access network selection model is solved by three different algorithms.
- The model is verified by some simulations and the differences of the three solving algorithms are analyzed.

2. Network Selection Model Based on MAB

We consider a specific service area W in the communication. It is in the overlap area of M heterogeneous access networks, as shown in the red circular area in figure 1. UAV D performs video shooting and uplink missions during flight. Assuming that UAV D is always flying in area W and T is the total transmission time slots of video, the optional access network set can be expressed as $\mathbf{S} = \{s(1), s(2), \dots, s(T)\} \subseteq \mathcal{M}$, and the reward function set is $\mathbf{U} = \{u(1), u(2), \dots, u(T)\}$.

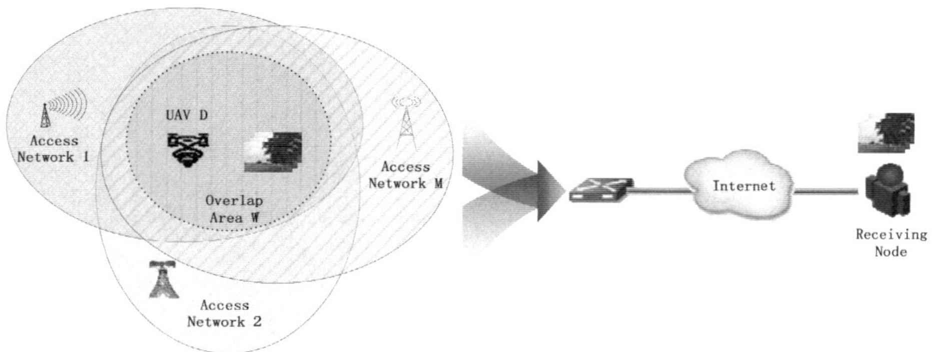

Figure 1. System Model

Because the CSI of each access network is time-varying, we will make a statistical description of its uncertainty. It is assumed that the CSI remains unchanged in each slot, and the changes between different slots follow an independent and identical distribution. First, in the slot $i \in [1, 2, ..., T]$, the transmission rate that the network v can carry, i.e. the available channel bandwidth of v, is a fixed value, $r_v(i)$. Then, in this slot, the bandwidth information vectors of M heterogeneous networks can be expressed as $\Re(i) = \{r_1(i), r_2(i), \dots r_M(i)\}$. The total mean bandwidth vector of the M networks is $\overline{\Re} = \{\overline{R_1}, \overline{R_2}, \dots, \overline{R_M}\}$. Similarly, in the slot i, the packet loss rate of the network v is set to a fixed value $l_v(i)$. Then the packet loss information vector of M channels can be expressed as $\Gamma(i) = \{l_1(i), l_2(i), \dots, l_M(i)\}$, and the mean vector of M channels is $\overline{\Gamma} = \{\overline{L_1}, \overline{L_2}, \dots, \overline{L_M}\}$.

Next, we model the network selection problem as an online learning problem based on the MAB. The heterogeneous network set $\mathcal{M} = \{1, 2, \dots, M\}$ can be regarded as a set of optional arms, each of which corresponds to a certain reward $\mathcal{Q} = \{q(1), q(2), \dots, q(M)\}$. Because of the randomness of CSI, the value of this reward will change in different time slots, and its expected value in all the time remains unchanged, which is recorded as $\boldsymbol{\mu} = \{\mu_1, \mu_2, \dots, \mu_M\}$. As a decision maker, UAV D needs to continuously select an arm from the set of optional arms $\mathcal{M}$, so as to make a selection in an access network. Generally speaking, the selection set made in different time slots can be recorded as $\mathbf{S_{MAB}} = \{s_{MAB}(1), s_{MAB}(2), \dots, s_{MAB}(T)\} \subseteq \mathcal{M}$, and the corresponding reward function set is $\mathbf{U_{MAB}} = \{u_{MAB}(1), u_{MAB}(2), \dots, u_{MAB}(T)\}$. We record the historical information of this reward, and feedback this information to the system continuously for decision-making. Therefore, the reward function is not only an evaluation method of current slot selection strategy, but also a reference basis for future decision-making. The selection and feedback process is iterated until the selection results do not change, and the algorithm converges, when the UAV will ultimately choose the arm (access network) that maximizes the cumulative reward function. The optimal selection strategy $S_{opt} = \arg\max_{s} \boldsymbol{\mu}$ is called Deterministic Stationary Policy (DSP).

Because of the unknown CSI, UAV D is not always able to make optimal judgments. Therefore, there will be a difference between the actual choice and the optimal choice, also known as regret value, which can be used to measure and judge the performance of decision-making. When the actual selection vector is $\mathbf{S_{MAB}}$ and the reward function vector is $\mathbf{U_{MAB}}$, the regret value can be expressed as $R_{\mathbf{S_{MAB}}}(i) = i \cdot \mu^* - E[\sum_{k=1}^{i} u_{MAB}(k)]$, where $\mu^* = \max \boldsymbol{\mu}$ is the optimal selection reward expectation. When UAV D transmits video in M heterogeneous access networks, we expect greater cumulative reward through continuous decision-making. This problem can be described as

$$\begin{aligned} &\mathbf{S_{MAB}} = \arg\max_{\mathbf{S}} \sum_{k=1}^{i} u_{MAB}(k) \\ &s.t.\ \mathbf{S} \subseteq \mathcal{M}, \mathbf{S_{MAB}} \subseteq \mathcal{M} \\ &i \in \{1, 2, \ldots, T\} \end{aligned} \tag{1}$$

$\mathbf{S_{MAB}}$ is the vector of selection results in different time slots in the iteration process.

3. QoE-oriented Reward Function

In order to effectively evaluate whether the access network selection algorithm can meet the needs of video users, we consider using QoE as a part of the reward function. We hope to establish a good matching relationship between CSI observations and users' real viewing experience by reasonably setting QoE-oriented reward functions in each slot. Usually, video QoE is closely related to the bit rate of video coding, packet loss rate in the network and frame rate in video playback. In this paper, we assume that the frame rate is a fixed value. This part of the reward function can be modeled as a function of bandwidth information vector, packet loss information vector and access network selection strategy of heterogeneous networks, $\mathbf{U_{QoE}} = K\{\mathbf{S}[\Re(i), \Gamma(i)], \Re(i), \Gamma(i)\}$. $\mathbf{S}[\Re(i), \Gamma(i)]$ represents the network selection vectors made by UAV D in M networks with CSI of $\Re(i)$ and $\Gamma(i)$ in each slot.

In addition to the above three factors of bit rate, packet loss rate and frame rate, the impact of video content characteristics on QoE can not be ignored. Literature [4] proposes that video can be divided into three categories: Slight Movement (SM), Gentle Walking (GW) and Rapid Movement (RM). In practical applications, the video uplinked by UAV may totally cover the three categories mentioned above. Assuming that the selected network in the slot i is $s(i) \in \mathcal{M}$, the channel rate is $r_s(i)$ and packet loss rate is $l_s(i)$, then the reward function QoE can be expressed as[5]

$$u_{QoE}(i) = K\{\phi, s(i), r_s(i), l_s(i)\} = \frac{c_1(\phi) + \ln[\frac{1 + r_s(i)}{c_2(\phi)}] + c_3(\phi) \cdot f}{c_4(\phi) + e^{c_5(\phi) \cdot l_s(i)}} \tag{2}$$

$\phi \in \{1, 2, 3\}$ represents that the video segment in current slot can be classified into one of the three categories mentioned above. $c_1(\phi), c_2(\phi), c_3(\phi), c_4(\phi), c_5(\phi)$ represents the relevant constants when video content belongs to different categories and their typical value can be found in [5]. f is frame Rate for Video.

In addition to the QoE, the reward function also needs to take into account the related cost. This paper considers the difference between QoE function and cost as the reward function, $\mathbf{U} = \mathbf{U}_{\mathbf{QoE}} - \mathbf{U}_{\mathbf{COST}}$. Usually, the cost needs to consider two factors: channel bandwidth leasing and energy consumption, which are linearly related to video transmission rate. If the correlation coefficient of the total loss of each network is set as $\boldsymbol{\lambda} = \{\lambda_1, \lambda_2, ..., \lambda_M\}$, then, the total reward function of the slot can be expressed as $u(i) = u_{QoE}(i) - u_{COST}(i) = u_{QoE}(i) - \lambda_s \cdot r_s(i)$.

In all, in order to solve the problem of UAV D's access selection in heterogeneous networks, we need to find a strategy to maximize the cumulative reward function by using the MAB model

$$\begin{aligned} \mathbf{S_{MAB}} &= \arg\max_{s(i)} \sum_{k=1}^{i} \{K\{\phi, \mathrm{s}(k), r_s(k), l_s(k)\} - \lambda_s \cdot r_s(k)\} \\ &s.t.\ \ i \in \{1, 2, \dots, T\}, s(i) \in \mathcal{M}, \phi \in \{1, 2, 3\} \end{aligned} \tag{3}$$

When using regret value to measure the performance, the expected reward of each network can be expressed as $\boldsymbol{\mu} = K\{\overline{\Re}, \overline{\Gamma}\} - \lambda \cdot \overline{\Re}$, and the regret value is $R_{\mathbf{S_{MAB}}}(i) = i \cdot \max \boldsymbol{\mu} - E[\sum_{k=1}^{i} u_{MAB}(k)]$.

4. Solution Algorithms

In Section 2, we have modeled the network selection as the MAB problem. Then, how to balance the utilization and exploration of the MAB becomes very crucial. We adopt three algorithms to solve the problem.

4.1. ε-Greedy Algorithm[6]

ε-Greedy Algorithm first sets a smaller probability value $\varepsilon \in (0,1)$. As the decision maker, UAV D will choose the arm with the largest expected reward with probability $1-\varepsilon$, and the other $M-1$ arms with small probability ε. Suppose that in slot i, the choice made by UAV D is network v. Then in slot $i+1$, the probability that D still takes the access network v as the selection arm is $p_v(i+1) = \begin{cases} 1-\varepsilon+\varepsilon/M & if \ \ v = \arg\max \mathcal{Q} \\ \varepsilon/M & otherwise \end{cases}$. Suppose the average sampling value of the reward is $\overline{\mathcal{Q}} = \{\overline{q(1)}, \overline{q(2)}, \dots, \overline{q(M)}\}$, then

$$s_{MAB}(i) = \arg\max \overline{\mathcal{Q}} \tag{4}$$

4.2. UCB Algorithm[7]

The basic idea of UCB Algorithm is to maintain confidence intervals of various expected rewards by using appropriate centralized inequalities. With the increasing number of samples, the confidence interval of sample expectations becomes more and more tight, so that the real average reward can be estimated more accurately. Let $\mathcal{A} = \{A(1), A(2), \dots, A(M)\}$ represent a cumulative value of the number of times each arm has been selected, then

$$s_{MAB}(i) = \arg\max_{s \in \mathcal{M}} [\overline{\mathcal{Q}} + \sqrt{\frac{\alpha \cdot \ln i}{2\mathcal{A}}}] \tag{5}$$

4.3. SoftMax Algorithm[8]

In SoftMax Algorithm, the probability of an arm being selected is related to both its own expected reward and all the expected reward of other arms. This probability is the ratio of the two. The improved SoftMax Algorithm based on Boltzmann exploration can be expressed as

IOP Conf. Series: Materials Science and Engineering **719** (2020) 012044 doi:10.1088/1757-899X/719/1/012044

$$s_{MAB}(i) = \arg\max_{v \in \mathcal{M}} \left[\frac{e^{\overline{q(v)}/\tau}}{\sum_{k=1}^{M} e^{\overline{q(k)}/\tau}} \right] \quad (6)$$

5. Simulations and Analyses

We conduct some simulations to evaluate the mechanism. The number of arms is set 3, which means there are three networks for selection. Firstly, we assume that the video transmitted by UAV D has fast moving speed, and it belongs to RM class. We compare the three solving algorithms under different parameters in figure 2. As the number of iterations increases, all curves tend to be smooth. This means all of the three algorithms will make the model converge to a stable state. 'Selection ratio' means the probability a network would be selected. From figure 2(a), 2(c) and 2(e) we can find SoftMax Algorithm has the fastest convergence rates. All of its choices are closer to zero or one, which means we could make a clear judgement of selection. The reward values of MOS in figure 2(b), 2(d) and 2(f) have similar trend.

The Performance of ε-Greedy Algorithm, under Different ε

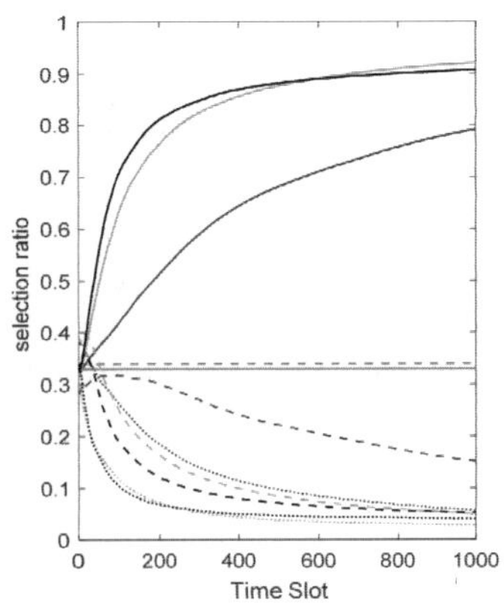

(a)

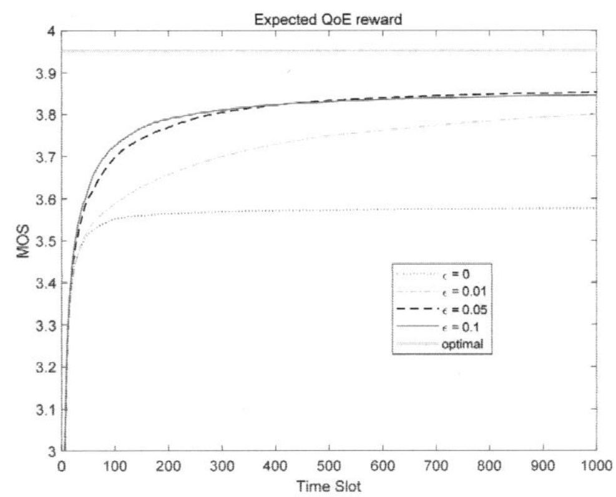

(b)

The Performance of SoftMax Algorithm, under Different τ

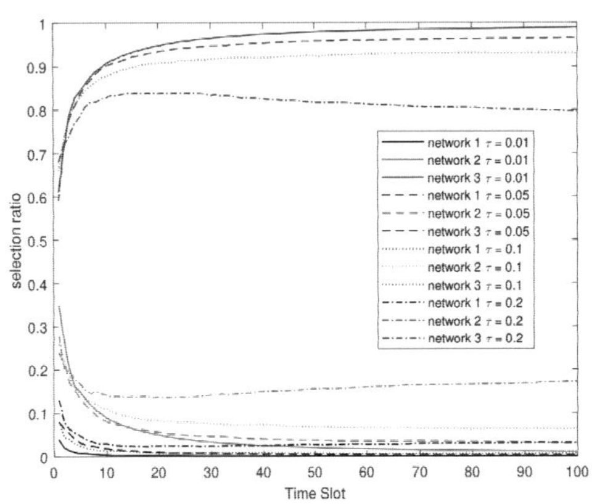

(c)

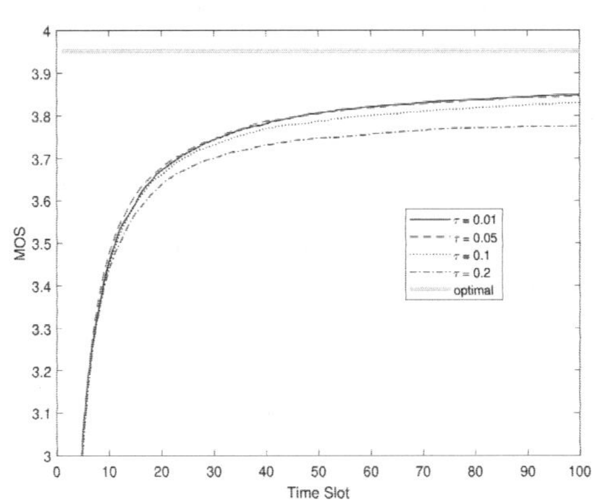

(d)

The Performance of UCB Algorithm, under Different α

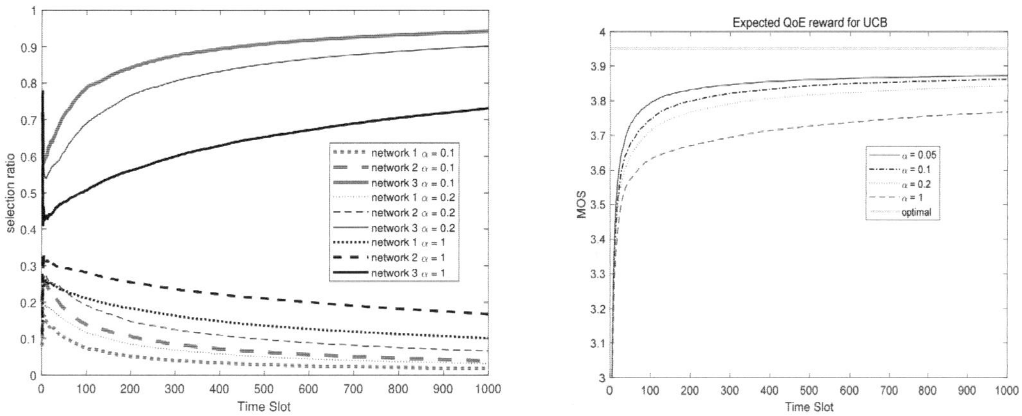

(e) (f)

Figure 2. The Results of Three Algorithms under Different Parameters

Then, we set $\varepsilon = \tau = \alpha = 0.05$ and analyze the performance of different algorithms when three different types of video, SM, GW and RM, appear randomly. Five algorithms are compared, including the ideal one which brings expection MOS and the no learning one which is completely random selection. The MOS results are shown in figure 3. We can find that the results of SoftMax Algorithm is most close to the ideal one, while the worst one is no learning algorithm.

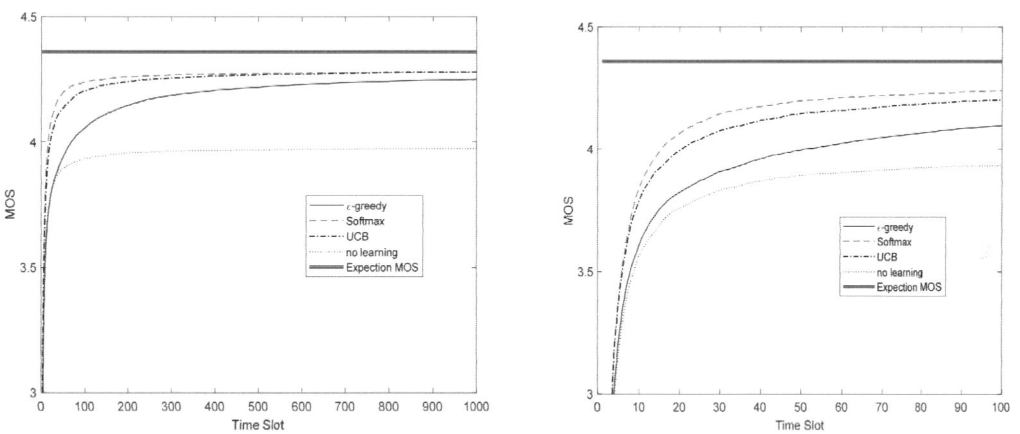

Figure 3. Comparison of Different Algorithms when Video Types are Random

6. Conclusion

By taking the video content classification and unknown CSI into consideration, this paper proposes an access network selection mechanism oriented to video QoE. The sequential decision-making problem is described as the MAB model. And three different algorithms are adopted to effectively solve it. The simulation results show that the SoftMax Algorithm outperform other solving method, and the proposed mechanism could effectively solve the problem of access network selection of UAV video transmission in heterogeneous networks.

7. Acknowledgments

This work was funded by the Project of Natural Science Foundations of China (No. 91738201 and 61401507) and China Postdoctoral Science Foundation (No.2017M613403).

8. References

[1] Robbins H. Some Aspects of the Sequential Design of Experiments[M]//Herbert Robbins Selected Papers: Springer New York, 1985:169–177.

[2] M. Bande and V. V. Veeravalli, "Multi-User Multi-Armed Bandits for Uncoordinated Spectrum Access," 2019 International Conference on Computing, Networking and Communications (ICNC), Honolulu, HI, USA, 2019, pp. 653-657.

[3] S. C. Pakhrin and D. R. Pant, "Multi-Armed Bandit Learning Approach with Entropy Measures for Effective Heterogeneous Networks Handover Scheme," 2018 International Conference on Advances in Computing, Communication Control and Networking (ICACCCN), Greater Noida (UP), India, 2018, pp. 451-455.

[4] Khan A, Sun L, Ifeachor E. Content Clustering Based Video Quality Prediction Model for MPEG4 Video Streaming over Wireless Networks[C]//2009 IEEE International Conference on Communications, 2009:1–5.

[5] Z. Deng, Y. Liu, J. Liu, X. Zhou and S. Ci, "QoE-Oriented Rate Allocation for Multipath High-Definition Video Streaming Over Heterogeneous Wireless Access Networks," in *IEEE Systems Journal*, vol. 11, no. 4, pp. 2524-2535, Dec. 2017.

[6] Vermorel J, Mohri M. Multi-armed Bandit Algorithms and Empirical Evaluation[C]//European Conference on Machine Learning.[S.l.]: Springer-Verlag, 2005.

[7] K. Saito, A. Notsu, S. Ubukata and K. Honda, "Performance Investigation of UCB Policy in Q-learning," *2015 IEEE 14th International Conference on Machine Learning and Applications (ICMLA)*, Miami, FL, 2015, pp. 777-780.

[8] Kuleshov V, Precup D. Algorithms for the multi-armed bandit problem[J]. Machine Learning Research, 2010, 4(10):1–48.

Performance Analysis of Opportunistic Relay Selection Cooperative Communication in Cascaded Fading Channels

Mingwei Wang

College of Electronic Information and Artificial Intelligence, Shaanxi University of Science & Technology, Xi'an 710021, China
E-mail: wangmingwei@sust.edu.cn

Abstract. Cooperative communication enables users to transmit information in a cooperative way, which is one of the key technologies used in wireless communication systems to improve spectral efficiency. A cascade Nakagami-*m* fading function is proposed to describe the fading characteristics of complex wireless channels, and a communication model of decoding-and-forward opportunistic relay selection (DF-ORS) cooperative communication under total power constraints is established. Theoretical analysis and simulation results show that DF-ORS cooperative communication can effectively reduce outage probability and improve spectrum efficiency under total power constraints and complex cascade Nakagami-*m* fading channels. The transmission power allocation of source node and relay node has an important impact on communication performance. Equal power allocation scheme cannot guarantee the optimal system performance, and the obtained conclusions have more general applicability.

1. Introduction

Multiple-Input and Multiple-Output (MIMO) wireless communication technology is one of the key technologies currently used to resist multipath fading and improve channel capacity. The cooperative communication technology developed in recent years makes full use of the broadcast characteristics of wireless channels, realizes "virtual MIMO" through mutual assistance among nodes, effectively overcomes the technical limitations of traditional MIMO and can retain its technical advantages to the greatest extent. Cooperative communication technology has been widely applied to wireless cognition communication, cellular networks and other fields [1,2].

The opportunistic relay selection (ORS) cooperative communication strategy proposed by Bletsas A. et al. is even better than the cooperative communication performance in which all relays participate under certain conditions[3]. ORS has many advantages, such as not requiring ideal synchronization, which greatly simplifies the physical layer design of the network. In recent years, the research on opportunistic relay selection cooperative communication is still one of the hot spots in the field of wireless cooperative communication, involving different channel environments, signal combining schemes, combining various existing specific communication technologies, showing the broad research and application space [4,5,6]. Different Channel fading characteristics are assumed to be Rayleigh, Nakagami-*m* and Rician, etc. by researchers. However, due to its simple form and few adjustable parameters, the accuracy of describing the fading characteristics of wireless signals in complex urban environments with many obstacles such as multiple diffraction, scattering, refraction and reflection is poor. In recent years, the cascade channel fading model proposed by some scholars overcomes the above problems and has been effectively applied [7,8,9,10,11].

In this paper, cascade Nakagami-*m* channel fading (N* Nakagami-*m*) model is used to describe the urban fading channel with many obstacles and the fading caused by multiple diffraction, scattering, refraction and reflection of wireless signals. The decoding-and-forward opportunistic relay selection (DF-ORS) cooperative communication model under the condition of total power is established and its communication performance is studied. on this basis, the node power allocation scheme is further studied, and a more general conclusion in line with the actual situation is obtained. Different cascade orders and fading parameters can be selected to be applied to different fading scenarios, and the conclusions have wider applicability.

2. System Model

As shown in Figure 1, a half-duplex dual-hop DF-ORS cooperative communication system model is established, where the direct path between the source and destination is blocked by an intermediate wall or obstacles, while relays are located at the periphery of the obstacle (around-the-corner). Because the most practical wireless communication situation in the urban environment is due to obstacles such as buildings, trees and the like blocking or experiencing strong signal attenuation, there is no possibility of direct communication between the source node S and the destination node D, and relay forwarding is required. In addition to the source node S and the destination node D, there are relay nodes $R_k \left(k=1,\cdots,\mathrm{K}\right)$.

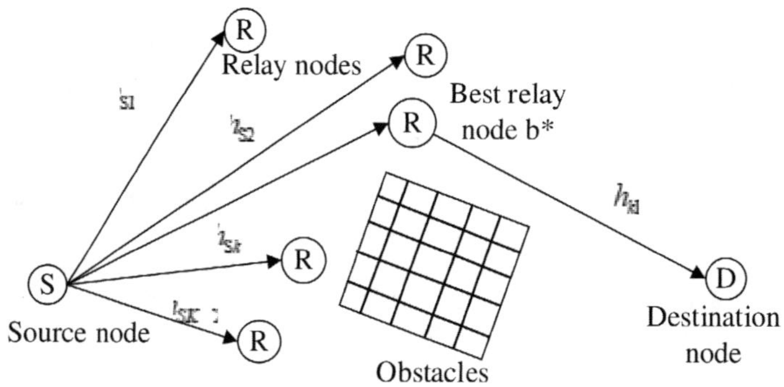

Figure 1. Schematic diagram of opportunistic relay selection cooperative communication

In Figure 1, h_{Sk} and h_{kD} $\left(k=1,\cdots,\mathrm{K}\right)$ are the channel fading coefficients from the source node to the relay node and from the relay node to the destination node respectively. Distinguishing from the influence of additive noise, channel fading will cause random non-additive signal disturbance of received signal power in wireless communication. More reasonable mathematical models and statistical functions are needed to accurately analyse and describe it. When a wireless signal is transmitted between two stations (transmitter and receiver), if there is multiple scattering, modelling the actual fading conditions with cascade Nakagami-*m* fading is more practical than commonly used fading models such as Rayleigh, Nakagami-*m* and Rician, etc. For example, in urban and forest areas, it is extremely appropriate to describe the signals using cascade channels due to the diffraction, scattering, refraction and reflection of signals by a large number of trees and building edges, street corners, moving cars, and propagation of holes. Under special conditions, the model of cascade Nakagami-*m* fading is easily simplified to commonly used channel fading models, such as Rayleigh, Nakagami-*m*, Rician, etc.

3. Theoretical Derivation

Assuming that the channel characteristics in DF-ORS cooperative communication in Figure 1 obey quasi-static flat fading, channel state information (CSI) is known and remains unchanged in one transmission. In the node-to-node link $\mathrm{A}\rightarrow\mathrm{B}$, the signal received at the node B is

$$y_{\mathrm{B}}=h_{\mathrm{AB}}x_{\mathrm{A}}+n_{\mathrm{B}} \tag{1}$$

where x_A is the transmission signal at the node A, the channel gain factor h_{AB} is between links $A \to B$, and $n_B \sim CN(0, N_0)$ is the additional complex white Gaussian noise introduced at the node B, $CN(\mu, \sigma^2)$ means that the mean value is μ and the difference is σ^2 of the additive complex white Gaussian noise. $|h_{AB}|^2$ indicates instantaneous signal power gain. If the node A is the source node, $E\{|x_A|^2\}=P_s$ is the transmission power of the source node. Similarly, if the node A is a relay node, $E\{|x_A|^2\}=P_r$ is the transmission power of the relay node.

According to the DF-ORS cooperation strategy the source node S broadcasts code word to the relay nodes, and a set of successfully decoded relay nodes is generated. In the second phase, according to the criterion of minimum interruption probability, the best relay is chosen to forward the signal to the destination node D from the decoded successful relay set. Each relay in this set has the opportunity to become a forwarding relay. During the whole cooperative transmission period, the communication process of the two-phase relay forwarding through the relay nodes participating in the cooperation can realize the transmission efficiency because the same code word is transmitted twice. If the optimal relay is selected, the maximum channel capacity achieved between source node A and destination D is

$$C_{\max} = \frac{1}{2}\log_2\left\{1 + \max\min\left(|h_{Sk}|^2 \frac{P_s}{N_0}, |h_{kD}|^2 \frac{P_r}{N_0}\right)\right\} \tag{2}$$

which means that the optimal relay node is selected according to the following principle

$$b^* = \arg\max \underbrace{\min}_{k\in S_{\text{relay}}} \left\{|h_{Sk}|^2 \frac{P_s}{N_0}, |h_{kD}|^2 \frac{P_r}{N_0}\right\} \tag{3}$$

For DF-ORS cooperative communication strategy, from the point of view of realizing complete cooperative diversity, wireless communication from source node S to destination D node can choose any path through relay node supporting the required transmission rate. The decoded successful relay set D_l formed by relays is empty, or when the link from any relay node of the decoded successful relay set D_l to destination node cannot realizing rate $2R(\text{bps/Hz})$, interrupt occurs. Where D_l is the subset formed by l decoded successful relays, $D_l \subseteq S_{\text{relay}}$. The optimal relay node according to formula (3) can be selected in the relay set D_l according to the following principle

$$b^* = \arg\underbrace{\max}_{k\in D_l} \frac{1}{2}\log_2\left\{1 + |h_{kD}|^2 \frac{P_r}{N_0}\right\} = \arg\underbrace{\max}_{k\in D_l}\left(|h_{kD}|^2\right) \tag{4}$$

The probability of forming a correctly decoded relay set in the first phase is

$$\Pr(D_l) = \prod_{k\in D_l} \Pr\left\{\frac{1}{2}\log_2\left(1+|h_{Sk}|^2 \frac{P_s}{N_0}\right) > R\right\} \prod_{k\notin D_l} \Pr\left\{\frac{1}{2}\log_2\left(1+|h_{Sk}|^2 \frac{P_s}{N_0}\right) < R\right\} \tag{5}$$

If the link from the relay of the set D_l to the destination node is interrupted, it means the DF-ORS cooperative communication is interrupted, and the interruption probability is

$$P(\text{outage}) = \Pr\left\{\frac{1}{2}\log_2\left[1+\underbrace{\max}_{k\in D}\left(|h_{kD}|^2 \frac{P_r}{N_0}\right)\right] < R \mid D_l\right\} = \sum_{l=0}^{K}\sum_{D_l} \Pr(D_l)P_{D_l} = \sum_{l=0}^{K}\sum_{D_l}\left[\prod_{i\in D_l} \Pr\left\{|h_{Si}|^2 \frac{P_s}{N_0} > 2^{2R}-1\right\}\right] \tag{6}$$

In this paper, we present single-selection opportunistic relaying with decode-and-forward strategy in cooperative communication network and analyze its outage probability under a source power constraint $P_s = \zeta P_{\text{total}}$,where the total power P_{total} of the entire cooperative network is a fixed value, and the power P_s is allocated to the source node. $\zeta \in (0,1]$ is the power allocation coefficient. An opportunistic relay power constraint $P_R = \sum_{k=1}^{K} P_k = (1-\zeta) P_{\text{total}}$, where the power $P_R = (1-\zeta) P_{\text{total}}$ allocated to the optimal relay node. The total signal-to-noise ratio of cooperative communication under a certain total power of the cooperative network is defined as $SNR = P_{\text{total}} / N_0$. The introduction of power constraints is also conducive to power allocation between source nodes and relay nodes, thus optimizing network performance.This opportunistic relay selection yields and minimizes the conditional outage probability in (6) as

$$P(\text{outage}) = \sum_{l=0}^{K} \sum_{D_l} \left[\prod_{i \in D_l} \Pr\left\{ |h_{Si}|^2 \xi SNR > 2^{2R} - 1 \right\} \right] \tag{7}$$

Next, let's consider the channel fading. If a wireless signal experiences multiple Nakagami-*m* fading (assuming N orders) during point to point transmission, the power of the received signal can be expressed as cascade products with the same probability density function independent of N orders

$$P = \prod_{k=1}^{N} P_k \tag{8}$$

where the probability density pdf of P_k satisfies Gamma distribution

$$f(p_k) = \frac{p_k^{m_k - 1}}{b_k^{m_k} \Gamma(m_k)} \exp(-\frac{p_k}{b_k}), k = 1, 2, ..., N \tag{9}$$

where m is a parameter related to the channel fading degree, the average power $E\{p\} = mb_k$ of the signal can be obtained by the moment function of Gamma function.

When these random variables satisfy the independent and identically distributed (i.i.d.). The probability density function (PDF) of signals received by the receiver via cascade fading can be obtained from the product theory of multiple random variable pdf, transformed by Moment Generating Functions (MGFs) and Laplace, expressed as MeijerG function [9].

$$f(x) = \frac{1}{x \prod_{k=1}^{N} \Gamma(m_k)} G_{0.N}^{N,0} \left(\frac{x}{\prod_{k=1}^{N} \Gamma(b_k)} \middle| \begin{matrix} - \\ m_1, m_2, ..., m_N \end{matrix} \right), \quad k = 1, 2, ..., N \tag{10}$$

where $G(\cdot)$ is MeijerG function [‡‡‡‡].

In actual communication environment, multi-order cascade N*Nakagami-*m* fading environment has good adaptability. It can be simplified as Nakagami-*m* channel fading as $N = 1$ and can be further simplified as cascaded N*Rayleigh channel fading as $m = 1$. It is simplified as $N=2$ and can be further simplified as double cascade Rayleigh fading channel when $m = 1$. Double cascade Nakagami-m channel fading model is suitable for Next-generation wireless communication systems with high user mobility, where the fading effects are more severe than classical channel models including Rayleigh and Nakagami-*m* [§§§§]. The mixed Rayleigh and double Rayleigh fading channels as an adequate multipath fading channel model for vehicle-to-vehicle communication scenarios [*****].

Defines the instantaneous signal-to-noise ratio per symbol $\gamma = pP / N_0$ of signals received through a two-point direct communication link, $\Omega = E\{p\}$ is average signal power, and $\bar{\gamma} = \Omega P / N_0$ is the average signal-to-noise ratio. By performing variable substitution and integration on equation (9) and using the integration characteristics of MeijerG function, it is obtained that the outage probability at the receiving end of point-to-point N*Nakagami-*m* cascade fading channel is

$$P_{\text{out}}\left(\gamma_{\text{th}}\right)=\int_0^{\gamma_{\text{th}}} f_\gamma\left(\gamma\right)d\gamma=\frac{1}{\Gamma^N\left(m\right)}G_{1,N+1}^{N,1}\left(m^N\frac{\gamma_{\text{th}}}{\bar{\gamma}}\middle|\begin{matrix}1\\ \underbrace{m,m,...,m,0}_{N+1}\end{matrix}\right) \tag{11}$$

where γ_{th} is the threshold of the signal-to-noise ratio.

Combining equation (7) with equation (11) and further simplifying, the outage probability of DF-ORS cooperative communication in asymmetric N*Nakagami-*m* cascade channel fading environment under total power constraint is obtained as follows

$$\begin{aligned}P\left(\text{outage}\right)&=\prod_{k=1}^{\text{K}}\left\{1-\left[1-\Pr\left(\frac{\gamma_{\text{S}k}}{\bar{\gamma}_{\text{S}k}}\le k_1\right)\right]\left[1-\Pr\left(\frac{\gamma_{k\text{D}}}{\bar{\gamma}_{k\text{D}}}\le k_2\right)\right]\right\}=\prod_{k=1}^{\text{K}}\left\{1-\left[1-\frac{1}{\Gamma^{N_{\text{S}k}}\left(m_{\text{S}k}\right)}\right]\right.\\&\left.\times G_{1,N_{\text{S}k}+1}^{N_{\text{S}k},1}\left(m_{\text{S}k}^{N_{\text{S}k}}k_1\middle|\underbrace{m_{\text{S}k},m_{\text{S}k},...,m_{\text{S}k},0}_{N+1}\right)\right\}\end{aligned} \tag{12}$$

where $k_1=\left(2^{2R}-1\right)/\zeta\,\Omega_{\text{S}k}SNR$, $k_2=\left(2^{2R}-1\right)/\left(1-\zeta\right)\Omega_{k\text{D}}SNR$ and $SNR=P_{\text{total}}/N_0$.

Note that (12) implies that the outage event happens only when all relays are in outage, and when N and *m* take different parameter values, equation (15) can equivalently describe various common and different fading functions. For example, when *N*=2 and *m*=1, it is reduced to double cascade Rayleigh channel fading.

4. Simulation Analysis

Figure 2 shows the numerical simulation results according to formula (10). Monte Carlo (MC) simulation method is used to ensure the correctness of theoretical analysis results, which counting times is in this paper. The MC simulation curve and the theoretical numerical simulation curve in the same figure have a good fit, indicating that the theoretical and practical modelling results are consistent. Overall, the curves in the figure show that the outage probability of DF-ORS cooperative communication under the total power constraint decreases with the increase of the total power SNR, that is to say, increasing the total power can reduce the outage probability and improve the cooperative communication performance.

Combined with the cascade fading channels of simulation curves, parameters $N=1$ and $m=1$ in Figure 2, the cooperative communication system model is simplified to an opportunistic relay selection cooperative communication system model in a single order Rayleigh channel fading environment. As parameter $N=2$ and $m=1$, the cooperative communication system model simplified to an DF-ORS in a two-order cascaded fading environment, written as 2*Rayleigh. The simulation results in the figure show that even if the relay node and the destination node have the same average power of the received signal, the two-order cascade Rayleigh fading makes the power of the received signal of the wireless channel have greater randomness, the communication quality becomes poor, and the outage probability increases. When the cascade fading channel with parameters $N=1$ and $m=1.5$ are adopted, the cooperative communication model is simplified to a DF-ORS cooperative communication model in a fading environment representing Nakagami-*m* channel. When a cascade fading channel with parameters $N=2$, $m=1.5$ are adopted, the cooperative communication model represents the DF-ORS cooperative communication model under two cascade 2*Nakagami-*m* fading environments. The curve in the figure shows that the two-order cascade 2*Nakagami-*m* fading has greater randomness than Nakagami-*m* channel fading, which leads to poor communication quality and increased outage probability.

Figure 2 also shows the performance analysis curves of decode-and-forward selection combination (DF-SC) cooperative communication under cascade 2*Rayleigh and cascade 2*Nakagami-*m* fading environments. DS-SC and DF-ORS both depend on fading arises from the availability of several potential paths towards the destination to select the best one. This finding reveals that the cooperation process of DF-SC is similar to DF-ORS, except that under the constraint of total power of the whole network, DF-ORS causes power to be distributed to each relay node, thus reducing the performance of the cooperative communication system. Under the same total power constraint and channel fading parameters, the performance of DF-ORS cooperative communication system is better than that of DF-ORS cooperative communication system. It can also be seen from the Figure 2 that different fading parameters m and cascade orders N mean different fading degrees of the channel, which has an important impact on the performance of DF-ORS.

Figure 3 shows the relationship between the outage probability of DF-ORS cooperative communication and the number of relay nodes under the total power constraint condition. As can be seen from the figure, with the increase of the number of relay nodes, cooperative diversity gain can be effectively realized, because more relay nodes provide more potential optimal paths for the implementation of opportunistic relay selection strategy. When DF-ORS cooperates under the condition of total power constraint, the optimal relay node selected from the K nodes participating in cooperation will use all relay power to send information in channel with the best quality, and its performance is obviously better than the traditional cooperation mode in which all relay nodes participate. The traditional cooperative mode in which all relays participate in cooperation and relay nodes evenly distribute relay power and transmit information to destination nodes in channels with different quality.

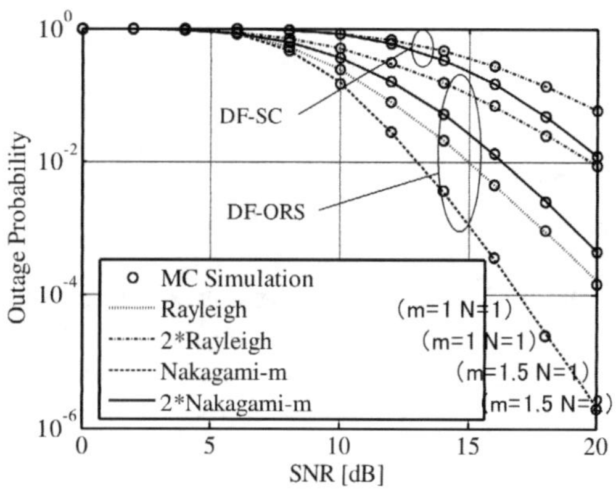

Figure 2. Relationship between outage probability and total power SNR

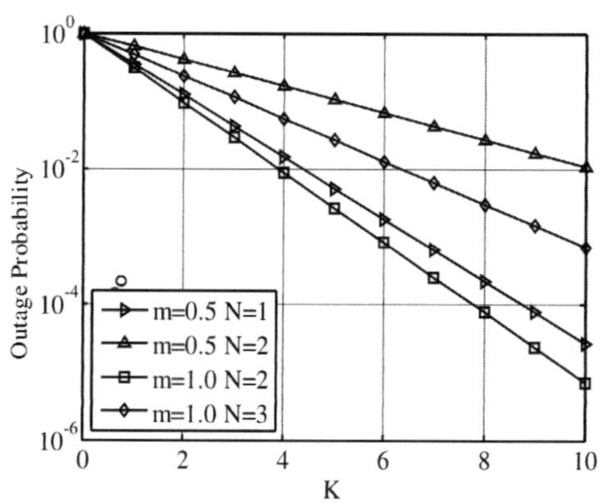

Figure 3. Relationship between outage probability and number of nodes

The simulation results in Figure 4 show the relationship between the outage probability of DF-ORS cooperative communication and the cascade order of cascade channels under total power constraint. It is shown in the figure that with the increase of cascade fading order, the outage probability of DF-ORS continues to increase and the communication performance decreases under the total power constraint of the whole network. When the channel cascade order increases, the wireless signal undergoes multiple diffraction, scattering, refraction and reflection in the path propagation process. The randomness of the signal increases and the quality of the received wireless signal decreases. It can also be seen from the Figure 4 that different fading parameters mean different fading degrees of the channel even if the cascade order is the same, which has an important impact on the performance of DF-ORS cooperative communication.

Figure 5 shows that the minimum outage probability occurs under asymmetric channel conditions is not the equal power allocation $\zeta \neq 0.5$ between source node and destination node. The value of allocation coefficient is restricted by the channel fading conditions. In the figure, the Nakagami-*m* parameter are set for channel fading as $m_{Sk} = m_{kD} = 1$, $N_{Sk} = 1$. By changing the cascade order N_{kD}, the wireless link from the relay node to the destination node meets different N*Nakagami-*m* cascade fading channel distribution conditions. When increasing the value of N_{kD}, the curve in the figure shows that the optimal power allocation coefficient ζ is decreasing while the outage probability is increasing, which means that more power needs to be allocated to relay nodes to overcome the channel quality loss caused by cascade channel fading. It can also be seen from the Figure 5 that different channel fading parameters m and cascade orders N mean different fading degrees of the Nakagami-*m* fading channel, which have an important impact on the outage probability performance of DF-ORS cooperative communication.

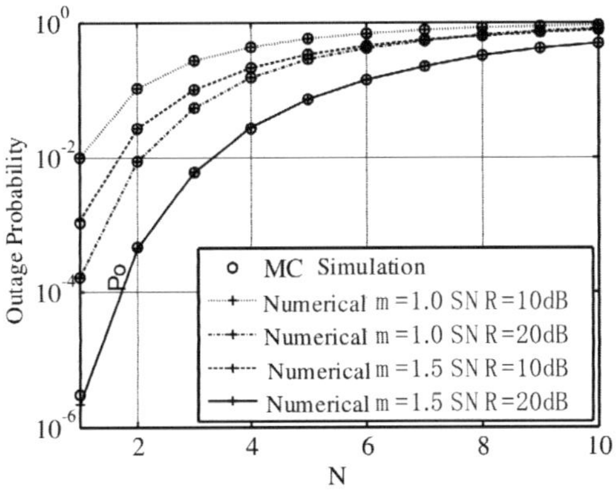

Figure 4. Relationship between outage probability and cascade channel order

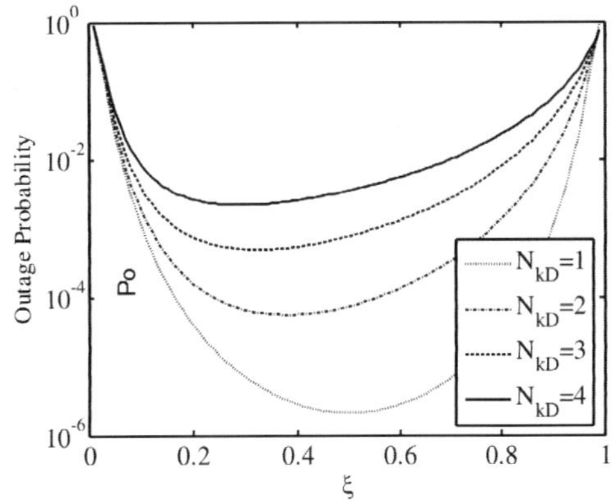

Figure 5. Relationship between outage probability and power allocation coefficient in asymmetric cascade fading channel

The main difficulty here is to have the network as a whole entity cooperate in order to rapidly discover the best path with minimal overhead. The source node and the relay node need to know the CSI of the whole network through feedback and other methods, and the best power allocation coefficient calculated will make the communication performance optimal. Although the unequal power distribution method can obtain the best performance, it will lead to increased overhead.

5. Conclusion

DF-ORS cooperative communication under total power constraint condition has been widely studied and applied due to its high diversity performance and easy physical implementation. A communication performance analysis model for DF-ORS cooperative communication strategy under total power constraint in cascade channel fading environment is proposed. The closed expression of outage probability theory is derived. The relationships between outage probability and total power signal-to-noise ratio, spectrum efficiency, number of nodes, different cascade orders of fading and power allocation coefficient are analysed, and general conclusions are obtained. It is further discussed that reasonable power allocation between source node and optimal relay node is beneficial to improve communication performance. The optimal power allocation coefficient minimizes the outage probability and optimizes the performance of DF-ORS cooperative communication under the condition of total power constraint, but it also requires some overhead.

6. Acknowledgment

This research was supported by the Doctoral Scientific Research Foundation of Shaanxi University of Science & Technology (No.2019BJ-03) and Scientific Research Project of Xianyang Municipal Science & Technology Bureau (No.2020K03-17).

7. References

[1] Let G S, Bala G J, Winston J J and Raj M D, et al. *Proc. Int. Conf. on Circuit, Power and Computing Technologies*, (Kollam: India) pp 1-4.
[2] Zhang Y and Liu F 2019 *Proc. IEEE Int. Conf. on Energy Internet* (Nanjing: China) pp 548-53.

[3] Bletsas A, Shin H, Win M Z 2007 *IEEE Tran. on Wireless Communications* **6** 3450-60.
[4] Zhou Q, Zang G Z and Gao Y Y, et al. 2017 *Proc. 2nd Int. Conf. on Advanced Information Technology, Electronic and Automation Control* (Chongqing: China) pp 2573 -77.
[5] Deng J, Tirkkonen O and Freij-Hollanti R, et al. 2017 *IEEE Communications Magazine* **55** *pp 94-101.*
[6] Xu X, Yang W, Cai Y 2017 *IET Communications* **11** pp 335-343.
[7] Karagiannidis G K, Sagias N C and Mathiopoulos P T 2015 *Frequenz* **55** pp 1453-58.
[8] Ibdah Y and Ding Y 2013 *Proc. 9th Int. Conf. on Information, Communications & Signal Processing,* (Tainan: Taiwan) pp 1-5.
[9] Shankar B P M 2012 *Fading and Shadowing in Wireless Systems* (Switzerland: Springer Nature).
[10] Sofotasios P C, Mohjazi L and Muhaidat S, et al. 2015 *IEEE Antennas & Wireless Propagation Letters* **2015** pp135-38.
[11] Benkhelifa F, Rezki Z and Alouini M 2014 *Proc. IEEE Int. Conf. on Wireless Communications and Networking* (Istanbul:Turkey) pp 154-59.
[12] Gradshteyn I S and Ryzhik I M 2007 *Table of integrals, series, and products* (USA: Academic Press).
[13] Ata S Ö 2018 *Proc. 41st Int. Conf. on Telecommunications and Signal Processing,* (Athens: Greece) pp 1-4.
[14] Pandey A and Yadav S 2018 *IEEE Trans. on Vehicular Technology* **67** pp 10615-30.

An Overview of the Wireless Key Technologies of 5G

Yang Yu

Wuhan No.2 Ship Development Design Institute, Wuhan, China
544738692@qq.com

Abstract. With the accelerate growth of mobile communication technology, 5G emerges at a historic moment, has been developing at high speed and devotes itself to open the curtain of the information age. This paper summarizes the wireless key technologies of 5G including technology trials. Finally, progress on implementation issue of these techniques are also reviewed.

1. Introduction

Driven by services and technology and to achieve a true 'Internet of everything', the 5th-generation (5G) mobile communication system has become the hottest topic and will be used in the next ten years. A great many countries, organizations, and units are keeping investing in 5G in order to bring new development opportunities and new economic growth. For example, the 5G Industry Association (5GIA) from Europe, IMT-2020(5G) [1] Promotion Group from China, and 5G Mobile Forum (5GMF) from Japan are official 5G organizations. To push forward the global unified 5G standard and to promote the 5G industry, these organizations emphasize international cooperation by initiating global 5G summits which discuss topics including spectrum collaboration, technology and standards trends for 5G. With the promotions of the summits, 5G has been moving forward rapidly. Different from the former generations of mobile communication, 5G will serve not only the telecommunication industry, but also the whole information society. The two typical kinds of services of 5G are the Internet of Things and Mobile Internet. From the standards aspect, we have a complete 5G standard in 2019 and even an approved 5G version from the ITU around 2020. The International Standards Organization of the 3rd Generation Partnership Project (3GPP) has officially released an enhanced 5G version for commercial using, which makes 5G become more mature and competitive. We can expect 5G pre-commercial and even a commercial network towards 2020.

In this paper, we give an overview of the wireless key technologies of 5G including technology trials.

2. 5G Wireless Key Technologies

To realize 5G user requirements of the system capacity, spectrum efficiency, and operational efficiency, many innovative wireless technologies have been invented and adopted, including main MIMO (massive multiple-input multiple-output) [2], NOMA (non-orthogonal multiple access), UDN (ultra dense networking), advanced coding and modulation and flexible spectrum access.

2.1. Massive MIMO

MIMO will be further enhanced in 5G and massive MIMO has been approved to be the most efficient technology for enhancing spectrum efficiency. The 'traditional' MIMO techniques have been researched extensively and deeply in 4G and the previous communication systems before 4G. There was an agreement on that there is a linear interrelationship between the MIMO capacity and the scale of antennas array. However, in recent years, by the reason of the antennas size and the terminal complexity problems, the antennas number in practice is less than or equal to 8 even with latest upgrades of 5G standard. It is obviously impractical to boost the communication capacity by increasing the antenna number. Nevertheless based on the research conclusions of [3], if the antenna number increases greatly, in other words, orders of magnitude exceeds former configuration, even the simplest MIMO system would get excellent performance, which is the so-called massive MIMO.

Because one base station can be configured dozens of antenna ports and hundreds antennas, it can be used for distributed or centralized massive MIMO. In different cases, technology needs to be improved and optimized including the design of channel estimation, reference signals, multi-user scheduling mechanism, receiving algorithms, and channel information feedback. Therefore, TDD is a more promising candidate for 5G systems with large antenna arrays.

The massive MIMO technique will bring benefit to the 5G communication systems in the following respects:

Capacity: The base stations in cellular system can be equipped with hundreds antennas but the antenna number of user equipments (UEs) is limited, e.g. less than 8. The capacity of MIMO communication system would be improved greatly in both uplink and downlink transmission.

Latency: Fading affects the transmission latency of wireless communication deeply. By adopting massive MIMO technique, MIMO signal processing and space diversity would give benefit to the receiving signal, for example precoding and beamforming, which compensate fading.

Cost and power: Although in the massive MIMO application, there is up to hundreds antennas, the energy efficiency is increased dramatically [3]. Massive MIMO could get a higher transmission gain with a lower emitting power consuming per antenna by sharping the signal in a very small region. In fact, the total power consuming of massive MIMO is much less than the traditional one.

2.2. NOMA

In former generations of mobile communication systems, orthogonal multiple access techniques in the frequency, time, and code domain have been utilized mainly. For a 5G system, connecting 1 million devices per square kilometer requires much evolved hardware capability. Non-orthogonal multiple access (NOMA) [4] has been regarded as a prospective technique to reach the objectives in terms of service latency, user connectivity, and system capacity.

Non-orthogonal multiple access, which multiplexes multiple users in the power domain, is the newest multiple access technique to meet the demand of 5G communication system for wireless broadband spectrum. By adopting NOMA technique, multiple users could be allocated at different power levels, according to their channel conditions, where it needs to be emphasized that the communication among these users is occurring at the same time, frequency and code channels.

Pilot researches in the way of system-level simulation and analysis on NOMA have been implemented in recent years [4,5]. The results of these researches prove that NOMA technique could attain higher spectral efficiency than other mainstream orthogonal multiple access techniques, for example, frequency division multiple access (FDMA), time division multiple access (TDMA), code division multiple access (CDMA) and orthogonal frequency division multiple access (OFDMA). The primarily reason why there exists such a performance improvement is because that NOMA can choose the right chance to utilize the wireless channel conditions, however, other existing mainstream orthogonal multiple access techniques, such as FDMA, TDMA, CDMA, OFDMA, could not consider channel conditions.

A more strict and comprehensive method to assess NOMA performance is to take not only the randomly distributed small scale fading gains, but also the random distances between users to the base station into consideration. Stochastic geometry can be used as an ideal mathematical method to acquire the random geospatial location of the mobile user equipments (UEs), which was adopted to NOMA in [5], which proves that even if mobile UEs are randomly distributed, NOMA could also reach higher performance than other existing conventional mainstream orthogonal multiple access techniques.

2.3. *Ultra Dense Networking (UDN)*
UDN is aimed at supplying a very high transmission rate to different users for hotspot or indoor scenarios, for example, offices, stadiums, and dense residential areas. Typical supports of UDN are for very high traffic, user, and access point (AP) density.

Key challenges of UDN include interference management, cell virtualization, integrated access, wireless backhaul, and user-centric design.

2.4. *Advanced Coding and Modulation*
In addition to traditional binary Turbo code and quadrature amplitude modulation (QAM), 5G needs to support advanced coding and modulation methods for wider services. Novel signal constellation mapping and de-mapping, faster-than-Nyquist (FTN) signaling, and low density parity check (LDPC) could be adopted to further boost spectrum efficiency in physical layer for high speed rate transmission services. In some low signal-noise-ratio (SNR) and short code-length conditions, low code rate convolution codes and polar codes could be adopted to approach the theoretical Shannon limit for small packet and low speed rate transmission services. Fast encoding and decoding schemes are chosen for low delay services. The 'error floor' effect should be removed from decoding algorithms for high reliability services. Besides, network coding could improve the backhaul system capacity in a dense network where exists a lot of wireless backhaul links.

In 3GPP, LDPC coding was accepted for the NR data channel, while polar coding was accepted for the NR control channel.

2.5. *Flexible Spectrum Access*
Different from former generations, 5G operates at not only high-frequency (HF) bands of 6–100 GHz but also low-frequency (LF) bands of below 6 GHz. The HF bands are the new supplementary bands which are used to reach high transmission rates in hotspot areas, for example, in the dense urban area. The LF bands are the core bands which are used for seamless coverage. To fully exploit the advantages of different frequencies, 5G needs to support hybrid networks with LF and HF bands to simultaneously meet the requirements of seamless coverage, ultra-high data rates, and ultrahigh capacity.

Considering the big differences between HF bands and traditional LF bands, research on 5G spectrum access focuses mainly on key technologies including channel measurement and modeling, new air interface design, networking technology, and radio frequency components for HF mobile communications.

Besides technologies discussed for HF bands and other traditional methods for spectrum efficiency, innovative spectrum utilization methods are also efficient ways to extend the usable frequency for 5G systems. Potential flexible spectrum sharing schemes include intra-operator inter-RAT, inter-operator, in secondary access, and spectrum sharing in unlicensed band.

3. 5G Technology Trial
Different from former generations, the 5G technology trial was conducted early in the same period as 5G standardization to serve as a guide to the standard. As shown in Table 1, major organizations have issued their 5G testing or commercial plan, making a huge investment to invite international vendors and operators to attend. Among them, China led the fastest and first 5G scale tests, and started 5G communication network commercialization in early 2019.

Table 1. 5G test and commercial plans from major organizations in different countries and regions.

Country/Region	Test and commercial plans
EU	Investing €1.4 billion on 5GPPP projects Conduct small scale tests to verify 5G standards in 2018
USA	A local 5G standard was led by Verizon in mid-2016 Released frequency in 2016 and started to cost more than $400 million in the next several years to establish a pilot 5G communication network
Japan	Carrying out 5G technology trials with leading operator NTT DoCoMo Planning to offer formal 5G commercial services ahead of the 2020 Summer Olympics
South Korea	Released national 5G strategy Launched a pre-commercial 5G trial at the 2018 Winter Olympics and then commercialize by the end of 2020
China	Dominated the 5G trial since the plan was issued in 2016 Started 5G network commercialization in early 2019

4. Conclusions

Thanks to a new opportunity offered by economic growth, 5G has gained wide attention and is developing very fast. As expected, 5G will bring unprecedented changes to people's daily life and even to the whole society. The 5G industrial ecosystem is at a building phase, and the involvement of and cooperation with industry need to be strengthened. 5G is finally expected to be large-scale commercial in 2020. It could be predicted that the next 10-year mobile communications will come.

5. References

[1] ITU-R WP5D TD-0591, Naming for International Mobile Telecommunications[R], 2015.

[2] Liu G, Hou X, Wang F, et al, Achieving 3D-MIMO with Massive Antennas from Theory to Practice with Evaluation and Field Trial Results, IEEE Systems Journal, 2016, 11: 62–71.

[3] Larsson E G, Edfors O, Tufvesson F, et al. Massive MIMO for next generation wireless systems. IEEE Commun Mag, 2014, 52: 186–195

[4] Ma Z, Persson D, Larsson E G, et al. Multiple symbols soft-decision metrics for coded frequency-shift keying signals. Sci China Inf Sci, 2013, 56: 022305

[5] Ashikhmin A, Marzetta T L. Pilot contamination precoding in multi-cell large scale antenna systems. In: Proceedings of IEEE International Symposium on Information Theory, Cambridge, 2012. 1137–1141

Entrepreneurship Introductory Course Designed for Students Majoring in Network Engineering

Fang Fang and XingE Luo*

Network Engineering and Research Center, South China University of Technology, Guangzhou, China, 510641
Email: ffang@scut.edu.cn; Xluo@scut.edu.cn

Abstract. This introductory course is designed to build interest for students in the field of entrepreneurship. Entrepreneurship introductory courses for students majoring in network engineering can inspire students' creativity. Innovative teaching methods link the professional knowledge of network engineering and project cases, enhance students' entrepreneurial practice ability, and strive to solve the disconnection between teaching and practice. The course provides some guidelines, insights, and advice needed to launch successful information technical ventures, helps students majoring in the network engineering develop entrepreneurial thinking. It is suitable for students who are interested in owning their own information technology business. The course design contains classroom interactive activities to make the teaching process more active.

1. Introduction
The thinking and ability of innovation and entrepreneurship are the key elements in the cultivation of new engineering talents. Innovative entrepreneurship education is the core competitiveness need to train students to cope with future changes. The development of a new economy will inevitably lead to the emergence of a number of new industries that are closely integrated with Internet technology. At the same time, accelerating the transformation and upgrading of traditional industries will inevitably promote the vigorous development of a large number of innovative enterprises, requiring students to have the ability to innovate and start businesses. [1]

2. The Importance of Learning Entrepreneurial Knowledge
As a kind of soft power, entrepreneurial culture is mainly about encouraging the spirit of entrepreneurship and the idea of dare to start a business, compared with the hard power of scientific research strength and talents. [2]
Entrepreneurship education is important because:
- It helps you think of new ways to earn money in a tough economy.
- It can help prepare you for a job.
- It can help you think of new ways to be a valued employee in any job.
- It demonstrates a variety of ways that your technical skills can be used in the workplace.
- It creates a foundation for additional learning experiences. [3]

3. Learn from the Successful Experience of Stanford University
Stanford pioneered the innovation model that combined the university, government and industry, accomplishing the optimal balance of scientific research, practical application and education. Stanford supplies the government and society with talents and research accomplishment and continuously

creates new enterprises. Almost two-third of all technical PhDs and half of engineering bachelor and master degrees holders will start a company during their careers.
Stanford's entrepreneurial resources are primarily in the curriculum. At Stanford, entrepreneurship-related courses can be roughly classified into four categories:

- Product-oriented: This type of course requires students to make prototypes of testable products within one semester. This type of course is generally offered at the College of Engineering.
- Case-oriented: This type of course focuses on the analysis and discussion of entrepreneurial cases, generally in business College. This type of class is not only a compulsory course for business College students, but also students from other colleges.
- Focused: This type of course covers one aspect of the entrepreneurial process and is spread across all majors. For example, there is a course devoted to Need finding, on how to find user needs in a systematic way.
- Lecture type: In addition to the one-time lecture course, Stanford also has a more than ten years of regular course Entrepreneur thought leader seminar. Invite an entrepreneur or investor to share an hour a week. [4]

Cultivating talents is the top priority of the university. Through the complete entrepreneurship curriculum system, the university offers various levels of entrepreneurship courses to bachelors, masters and doctors, and provides entrepreneurial extracurricular activities and practical opportunities according to the different needs of students, thus enriching students' entrepreneurial knowledge. Inspire students' entrepreneurial enthusiasm.

4. Course Modules

There are five modules in this course. Each module has case study. Each case study discusses basic business management skills.

The entrepreneur of the course case, Wang, is a university student majoring in network engineering. After evaluating his skills, Wang conducts market research and identifies the target market. Lee showed his plans to customers and began to create a small IT business that is ready to grow.

4.1. Find and Research a Business Opportunity.

Skills can be divided into two main categories: hard skills and soft skills. Hard skills are quantifiable and easy to demonstrate. For example, knowing a foreign language is easy to demonstrate by passing a test or conversing with someone. Hard skills are often associated with formal training, a certification, or degree. Soft skills show how we relate to others. These include effective communication, creativity, problem solving, flexibility, reliability, and the ability to listen. These skills are highly valued by customers and employers, but they are more difficult to demonstrate. [3]
Classroom interactive activities: Students discuss what hard skills and soft skills they have for entrepreneurship.

Wang finds out that he has extensive knowledge in computer network hardware and software installation. He has a good understanding of networking, and he has both practical and educational experience in these areas. Wang likes the idea of using his skills to make money. He also likes the flexibility that business ownership provides. The idea that Wang uses his computer network skills to conduct business is not a business opportunity unless Wang can prove that others need his skills and are willing to pay for them.

When entrepreneur realizes that his problem has no readily available solution, he considers starting a new business to solve the problem. This is entrepreneurial thinking. He recognizes that a problem may be a business opportunity. Problem solving is just one way to recognize a new business opportunity. There are other opportunities:

- Problem solving - Obstacles or challenges that the business could solve
- Changes - Laws, situations, or trends that evolve over time
- Inventions - Totally new products that are developed
- Competition - Doing the same thing more effectively than others
- Technological advances - Adapting new technology to an existing product

- Unique knowledge - Special understanding of what the people in your area want

A company can sell products or services. A product is something that exists in nature or is made by people. It is tangible, meaning that it can be touched. A service is work that provides time, skills, or expertise in exchange for money. It is intangible. [3]

Wang must decide:

- What product or service to offer?
- Where the product will be sold?
- How customers will learn of the product?
- What they should charge for the product?

4.2. Market Research

Find the people who need the skills that Wang has. Wang uses research to help him locates the people who need their product or services. These people are his target market. Communicating to the target market is called "marketing." Asking them to purchase the product or service is called "selling." Entrepreneurs sell by showing the customer the benefits they will receive when they purchase the product.

Market research is the collection of information about the industry that the entrepreneur wants to enter. In Wang's case, it is information about the need for computer services in his area. Wang needs to gather information about the industry he is entering. This information helps him determines which customers he will serve and who the competition is. Market research provides Wang with facts so that he can make good decisions. When Wang starts a new business, he will encounter many decisions. Business plans are a record of these decisions. More importantly, the business plan also includes research to support each decision.

Business plans are used to guide the management and to explain the business to people outside of the company. There are usually five sections in a business plan:

- The Executive Summary
- The Business Description
- The Environmental Analysis
- The Management Biographies
- The Financial Section

4.3. Providing Outsource Services

Outsourcing is a great way for a company to fill a temporary need or to hire expertise they may not have within their own organization. The people or companies a business hires to provide temporary labour or expertise are called contractors. Finding, communicating, and building relationships with other businesses is important to a contractor. These relationships are often defined by a contract. [3]

In the classroom case study, Wang successfully established a small computer network service business. Wang found a large Information technology company might hire a contractor for maintenance or customer support work. IT companies are not the only type of company that uses contractors. Some ways that companies in many different industries use contractors include: Cleaning services Construction work Food service Security Personnel needs.

Classroom interactive activities: Students discuss which other business is suitable for outsourcing.

Wang is hired by the large Information technology company to perform a specific task and he will receive payment for his services, but Wang isn't employees. Wang and the large Information technology company that hire him, formalize their agreement by writing and signing a contract. Common contracts Wang will encounter as he operates and expand his business include: partnership agreements, employment contracts, loan agreements, lease agreements, confidentiality agreements, and purchase agreements.

A request for proposal is a written document outlining in detail what the purchaser wants to buy. An RFP includes the following sections: introduction, requirements, bid process, selection criteria, and schedule. A bid proposal is the supplier's response to an RFP. If the bid proposal is accepted by the hiring company, it becomes a legal contract. [3]

4.4. E-Business
Wang plans to help small businesses in the community implement e-Business. He intends to use the following steps: Define a business problem; Identify, evaluate, and choose e-Business solutions; Prepare an implementation plan; Evaluate the business using business metrics.

When a company uses Internet applications as a part of business operations, they are participating in e-Business. E-Business is simply the use of Internet technology to perform any business activity.

Customer communication is facilitated through e-Business applications types: Websites; Email; Online discussion groups; Affiliate and associate programs; Blogs. Communicating with customers is a high priority for any business. Today, many companies are using e-Business applications to increase and improve their customer communications. Business owners may also use e-Business applications for operational functionality to include inventory control, utility cost management, and shipping, receiving, and order tracking. E-Business applications can also help business owners with the operational functions of their business. Businesses must research and choose which tools apply to their specific Internet goals. One of the most attractive features of e-business is the ability to customize applications to meet a business owner's specific needs. [3]

Classroom interactive activities: Students discuss how e-business can help Wang's business.

4.5. Giving Good Customer Service
Getting and keeping customers is the only way a company can continue to operate. Everyone in an organization is involved in customer service. Keep a "customer first" attitude. Handle conflicts professionally. Gain customer loyalty.

Classroom interactive activities: Ask students to think about how Wang's business can provide better customer service.

5. Summary
Based on the characteristics of network engineering, we attempt to explore the design of entrepreneurship introductory courses. The construction of entrepreneurship courses at domestic universities is still in its infancy. We need to learn from Stanford University's successful experience, actively improve the systematic curriculum, open a variety of entrepreneurial courses with rich levels, and build a combination of general education, professional education and entrepreneurial skills education; the integration of arts and sciences; comprehensive curriculum and independence Entrepreneurial courses complement each other in the entrepreneurial talent development curriculum system. The University strives to provide various forms of entrepreneurial practice opportunities, so that students can master the knowledge, skills and abilities of entrepreneurship in university, and can make objective and rational judgments on the risks of future entrepreneurship, thus laying the foundation for their future entrepreneurship. [5]

6. References
[1] Zheng qinghua, Leading with innovation and entrepreneurship education to create "new engineering" education model, China University Teaching, 2017(12), pp8-12. (In Chinese).
[2] Wu jijun, The Enlightenment of American Entrepreneurial University to the Development of Chinese Colleges and Universities, Industrial&Science Tribune, 2018(17)3, pp13-14. (In Chinese).
[3] https://www.cisco.com/, 2018
[4] Huang hai, https://www.zhihu.com/question/28833187 2015.3. (In Chinese).
[5] Zhu chunkui, Li wenjuan, Gong chen, Yang tianbo, The Experience and Enlightenment of Building an Entrepreneurial University in Stanford University, Science and technology of China, 2018(6),pp28-30. (In Chinese).

IOP Conf. Series: Materials Science and Engineering **719** (2020) 012048 doi:10.1088/1757-899X/719/1/012048

Simulation Research on Transmission Performance of UAV Data Link in Multipath Channel

Xilei Wu

College of Computer Science and Electronic Engineering, Hunan University, Tianma Student Apartment, Juzizhou Street, Yuelu District, Changsha, Hunan.
Email:1403839770@qq.com

Abstract. Low elevation Angle is the common working environment of Unmanned Aerial Vehicle (UAV) data link. Naturally, data link transmission performance will be affected by White Gaussian Noise (WGN). Besides, multipath interference is more characteristic and usually more harmful. Taking Phase Shifting Keying (PSK) modulation system commonly used in data link as the object and typical UAV data link point-to-point configuration as the prototype, this paper analyzes the transmission principle of this system and the characteristics of multipath channel transmission, and then establishes the simulated transmission model of measurement and control communication link in multipath channel on this basis. Through simulation experiment, part of the simulation data and conclusions are given. According to the paper, increasing the link level redundancy design is not the fundamental measure to solve multipath fading. And the paper suggests that in order to improve the transmission performance of data link in multipath channel, it is feasible to add adaptive equalization technology at the level of signal processing.

1. Overview

UAV data link is the nerve link between all nodes of UAV system and the key facility of integrated information platform of cross-domain unmanned systems [1-2], which has been widely used and continuously concerned in military, civil and commercial fields. PSK system is a transmission system which is commonly used in UAV systems and also a way of digital modulation with constant envelope. Due to its high power utilization and high band utilization, it is widely used to transmit command information and remote sensing information between unmanned air platform and ground command and control node. Owing to the particular working environment of UAV systems, such as aerial platform flight level is usually lower, for example ranging from several meters to thousands meters, which is too low compared with the distance ranging from tens of meters to hundreds of kilometer between UAV and ground node, so UAV usually work in a state of low elevation (e.g., 1 ° ~ 3 °). Besides, the surrounding environment is generally more complex, such as forest, mountain, near suburb or dense buildings, etc., leading to more severe multipath interference problems. This paper focuses on the PSK modulation system commonly used in UAV data link, establishes the channel transmission model of data link in multipath channel on the basis of principle analysis and analyzes the transmission performance under the condition of multipath channel through simulation, so as to provide reference for the evaluation of UAV system information transmission efficiency.

2. Principle Analysis

2.1. Introduction of Typical UAV Data Link Prototype System

Prototype system is the premise and foundation of simulation model. UAV data link has multiple configurations such as point-to-point, point-to-multi-point, multi-point to multi-point and ground-to-air network application [3-4]. In order to simplify analysis, this paper mainly establishes a data link channel transmission model based on point-to-point configuration. The typical prototype system of point-to-point UAV data link is shown in figure 1.

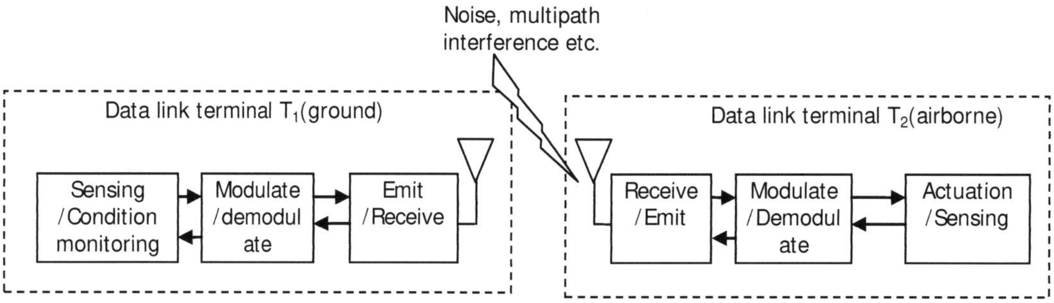

Figure 1. Typical structure of the prototype system of point-to-point UAV data link.

Figure 1 shows the prototype system of point-to-point data link configuration. The system consists of the ground data terminal (T1, Abbreviation GDT), which is generally adjacent to or unite with the ground command and control node of UAV systems, and the airborne data terminal (T2, Abbreviation ADT). The communication mode between T1 and T2 is commonly full duplex mode. Generally the T1 to T2 direction data link is called the uplink, and T2 to T1 direction data link is called the down link. Uplink usually apply PSK modulation, while down link usually apply QPSK modulation or other high order modulation and its basic communication principle is similar to uplink. Noise and multipath are the natural external interference between T1 and T2, which are different from the external artificial interference.

2.2. Principle Analysis of PSK Modulation System

The basic principle of PSK modulation system is shown in figure 2.

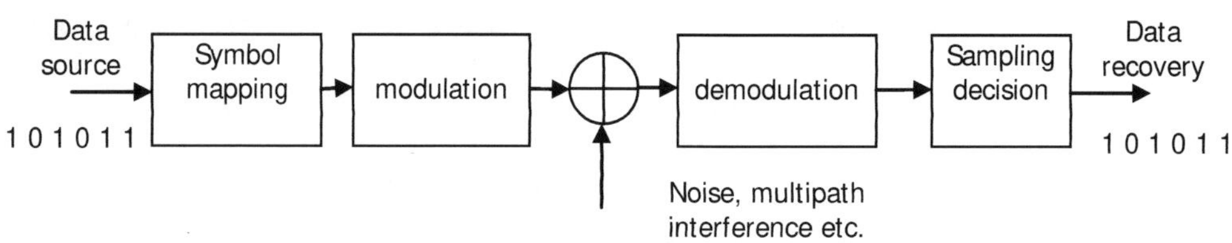

Figure 2. Basic principle of PSK modulation system.

In the figure 2, PSK modulation signal can be represented by equation (1), which is shown as

$$s(t) = A_c \cos(2\pi f_c t + \theta_c + \phi), \qquad 0 \le t \le T_b \tag{1}$$

In equation (1), A_C is the carrier amplitude, f_c is the carrier frequency, and θ_c is the initial phase of the carrier. ϕ is the modulation phase mapping of the data to be transmitted (binary symbol 0 or 1), there is the following phase mapping relationship:

$$\begin{aligned} &'0', \quad \phi = \pi \\ &'1', \quad \phi = 0 \end{aligned} \tag{2}$$

on the receiving end,

$$S_1(t) = A_c \cos(2\pi f_c t + \theta_c + 0) = \sqrt{2E_b/T_b}\cos(2\pi f_c t + \theta_c), \quad \text{symbol '1'}$$

$$S_0(t) = A_c \cos(2\pi f_c t + \theta_c + \pi) = \sqrt{2E_b/T_b}\cos(2\pi f_c t + \theta_c + \pi), \quad \text{symbol '0'} \tag{3}$$

In the equation above, T_b is the duration per bit and has $E_b = \frac{1}{2}A_c^2 T_b$.

PSK modulated signal passes through the Additive Gaussian White Noise (AWGN) channel with noise n(t), and when coherent detection is adopted, its theoretical error rate per bit is calculated as follows [5] :

$$P_b = Q\left[\sqrt{\frac{2E_b}{No}}\right] \tag{4}$$

In the equation (4), $Q(X) = \frac{1}{\sqrt{2\pi}}\int_x^{\infty} e^{-y^2/2dy}$; E_b is the energy per bit, and N_0 is the power spectral density (PSD) of the received input noise.

2.3. Analysis of Multipath Channel Transmission Characteristics

In general, the geometric relationship between airborne and ground data terminal antennas of UAVs data link is shown in figure 3.

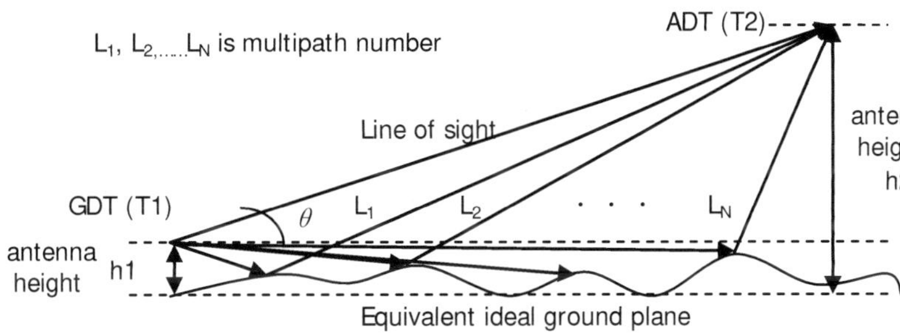

Figure 3. Geometric relation diagram of multipath channel.

In the figure, θ is the line-of-sight elevation angle between ADT and GDT. When $\theta \leq 10°$, the terminals in the data link are in the zone of low elevation angle. In addition to the line-of-sight main path component, there are more multipath components (i.e. L_1, L_2... L_N). At this point, assuming that Gaussian thermal noise n(t) is not considered, the signal at the receiving end is superposed by signal components from different paths including the main path, as shown in equation (5).

$$\begin{aligned} r(t) &= \sum_{i=0}^{N} A_{ci}(t)\cos\left[2\pi f_c\left(t-\tau_i(t)\right)+\phi_c+\phi\right] \\ &= \sum_{i=0}^{N} A_{ci}(t)\cos\left[2\pi f_c t+\phi_c+\phi+\varphi_i(t)\right] \end{aligned} \tag{5}$$

In the equation (5), $A_{ci}(t)$ is the amplitude of receiving signals in the path i and $\tau_i(t)$ is the delay of receiving signals in the path i.

Usually, $A_{ci}(t)$ and $\tau_i(t)$ change slowly with respect to f_c, so multipath signal r(t) can be expressed as

$$\begin{aligned} r(t) &= \sum_{i=0}^{N} A_{ci}(t)\cos\varphi_i(t)\cos(2\pi f_c t+\phi_c+\phi) - \sum_{i=0}^{N} A_{ci}(t)\sin\varphi_i(t)\sin(2\pi f_c t+\phi_c+\phi) \\ &= X_c(t)\cos(2\pi f_c t+\phi_c+\phi) - X_s(t)\sin(2\pi f_c t+\phi_c+\phi) \\ &= V(t)\cos\left[(2\pi f_c t+\phi_c+\phi)+\varphi(t)\right] \end{aligned} \tag{6}$$

In the equation (6),

$$\begin{aligned} X_c(t) &= \sum_{i=0}^{N} A_{ci}(t)\cos\varphi_i(t) \\ X_s(t) &= \sum_{i=0}^{N} A_{ci}(t)\sin\varphi_i(t) \\ V(t) &= \sqrt{X_c^{\,2}(t)+X_s^{\,2}(t)} \\ \varphi(t) &= \arctan\left(\frac{X_c(t)}{X_s(t)}\right) \end{aligned} \tag{7}$$

According to equations (5) ~ (7), after transmission of PSK modulated signal through multipath channel, the originally determined carrier signal becomes a narrow-band signal with both envelope and phase modulated and also causes frequency dispersion in the spectrum.

The amplitude and arrival angle of the multipath signal are both random and statistically independent. When there is a main static signal component (the main path signal) between the receiving and sending ends, the envelope of the composite signal obeys the Rice distribution. In the Rice distribution, different multipath components are superimposed on the main line-of-sight (LOS) component, while the amplitude of other multipath signals is subject to the Gaussian distribution and the phase is subject to the uniform distribution from 0 to 2 π. When the main LOS signal component is lower than the multipath components or has been submerged in the multipath components, the envelope of the superimposed signal obeys the Rayleigh distribution, the amplitude of each multipath component obeys Gaussian distribution, and the phase obeys the uniform distribution from 0 to 2 π[6-7].

Typically, the main signal component is stronger than multipath signal components in the LOS data link of UAV communication and generally the multipath channel presents the characteristics of Rice channel. However, under several special circumstances, the main signal component will be submerged in the multipath signal components. In this case, the transmission channel presents the characteristics of Rayleigh channel.

3. Establishment of Multipath Channel Transmission Simulation Model

Refer to Figure.1, UAV data link is easy influenced by modulation and demodulation procedure in the signal processing. At same time the antenna will also bring in external noise and various interference factors including multipath. Therefore, the transmission simulation analysis model under multipath channel is established as shown in figure 4.

Data source module outputs binary data flow $d_0(t)$ (0 or 1). Data flow goes through the symbol constellation mapping and converts to binary non-return to zero data flow d(t) (-1 or 1, respectively corresponding to phase π or 0). In order to reduce intersymbol interference and compress transmission bandwidth, the shaping filter is added. And then after PSK modulation, we get the modulation signal s (t). After going through multipath channel or/and AGWN channel, s(t) reaches the receiving demodulation end. When receiving signal r(t) only considers the multipath case, its signal form is expressed by equation (5). When simultaneously superposing Gaussian noise n(t), n(t) should be added to equation (5). When the influence of multipath signal is not considered, n(t) can just be added to equation (3). r(t) is demodulated in the PSK demodulation module and then delivered to the receiving filter module, going through the integral reset and sampling decision procedure and outputting d '(t) code stream. Then the code stream turns to binary data stream $d'_0(t)$ after symbol inverse mapping or constellation map inverse transformation procedure.

In figure 4, PSK modulation adopts the coherent demodulation method. However, since synchronization of the carrier at both ends of the transceiver is not the focus of this paper's simulation, we simplify the simulation model without obviously affecting the simulation result of the transmission performance. It is assumed that the carrier at both ends of the data link is already in synchronization when the link is established. Meanwhile, data link transmission delay should be considered when comparing BER between the data resource $d_0(t)$ and recovered data flow $d'_0(t)$. The original data delays forτ_d value which can be acquired through the cross-correlation of $d_0(t)$ and $d'_0(t)$).

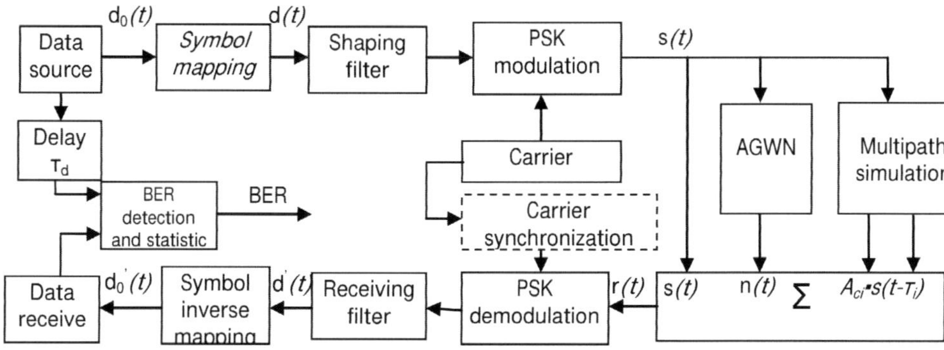

Figure 4. Multi-path channel transmission simulation model of UAV data link.

4. Simulation and Result Analysis

4.1. Simulation Environment and Related Parameters

According to the simulation model of multipath channel data link transmission shown in figure 4, the simulation circuit is built in the System Vue environment. Main parameters are set as follows

Table 1. Simulation environment and related parameters.

Type	Data rate R_b (Mbps)	Modulated carrier frequency f_c(GHz)	Modulation method	Shaping filter type	Number of multipath signal
Parameter	1.024	1.6	PSK	Raised cosine roll-off, roll-off factor: 0.5	1 path of main signal, 1 path of adjustable multipath signal, adjustable delay, random phase

There are many multipath paths in the workplace of data link, and it is impossible and meaningless to exhaust them in simulation. And usually it is the strongest multipath signal that has the greatest influence on transmission performance. In addition to the amplitude A_i (or signal Power P_i) of this multipath signal, the transmission delay τ_i of the signal and random change of carrier phase which is caused by multipath will also have an effect on the quality of the received signal. In order to simplify the complexity of simulation, this paper only includes 1 path of main signal and 1 path of multipath signal which has the maximum signal strength.

4.2. Simulation and Result Analysis

In the first step, modeling and simulation experiments are carried out to analyze the receiving and demodulation performance under the situations of several signal noise power spectral density ratios. Figure 5-7 show the baseband receiving signal eye diagram when E_b/N_0 = 5, 8 and 11dBHz, respectively.

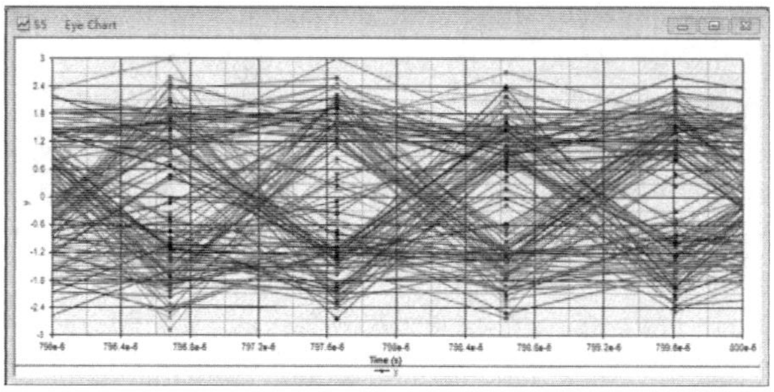

Figure 5. Baseband signal eye graph of receiving end when E_b/N_0=5dBHz.

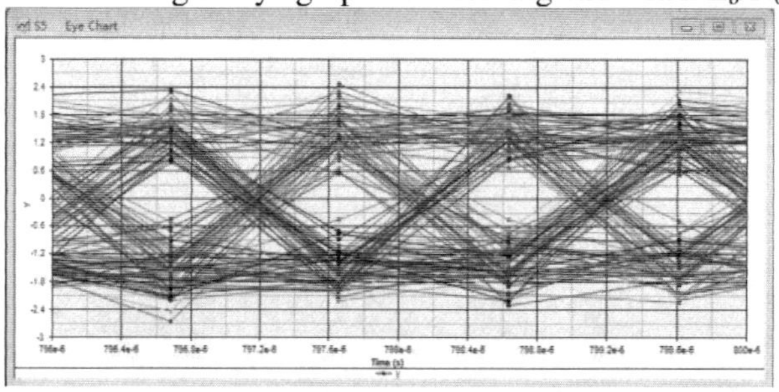

Figure 6. Baseband signal eye graph of receiving end when E_b/N_0=8dBHz.

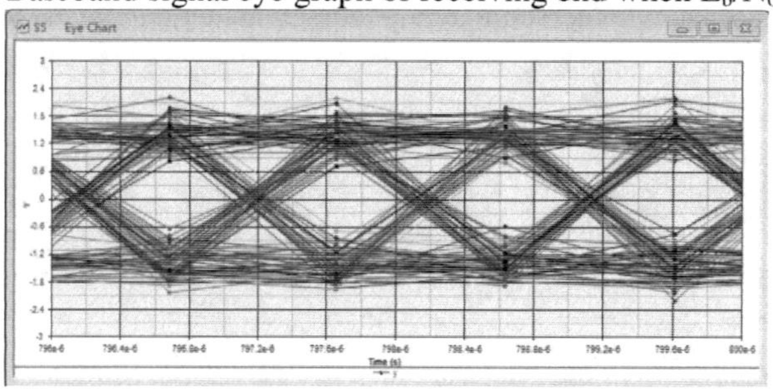

Figure 7. Baseband signal eye graph of receiving end when E_b/N_0=11dBHz.

Figure 8 shows the superposition graph of the simulated BER curve and the theoretical BER curve when Eb/N0 = 1,2...10 dBHz.

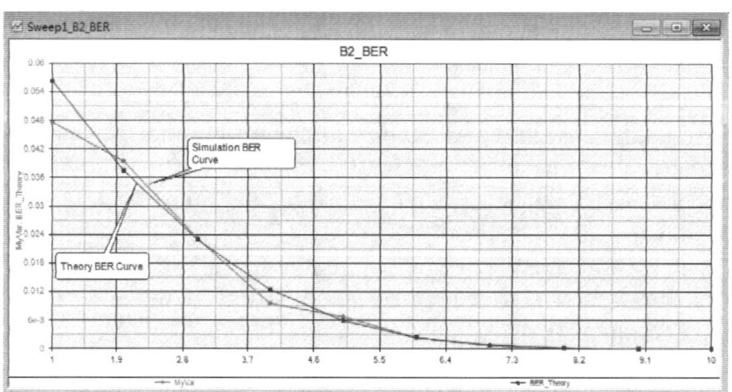

Figure 8. Superposition curve graph of simulated BER and theoretical BER.

According to the simulation results shown in figure.5-8, this paper simulates the transmission performance of the simulation model under AGWN channel, and verifies the correctness and effectiveness of the simulation model.

The second step, we change the AGWN channel to multipath transmission channel, and set the multipath fading factor G_i and multipath transmission delayτ_i. The multipath channel carrier phase obeys the uniform distribution from 0 to π and will be generated randomly by the function in the simulation model. Figure 9-13 show the comparison graph of the quality of receiving signal under different fading conditions when PSK modulation signal goes through multipath channel with transmission delay τ_1=4T_b and carrier phase φ_1 randomly changed, which is shown in the form of eye graph.

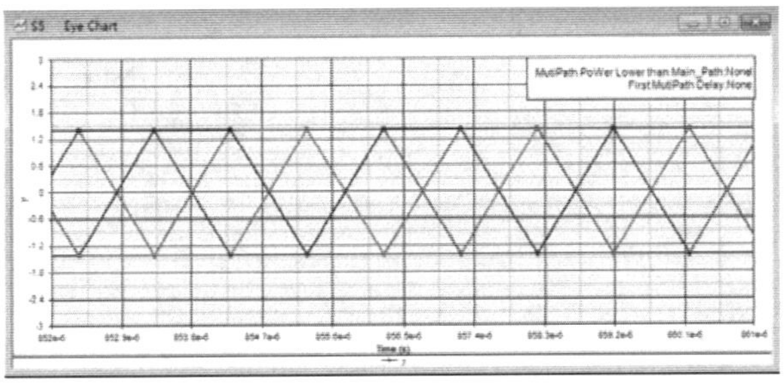

Figure 9. Eye graph of baseband signal at receiving end without multipath component.

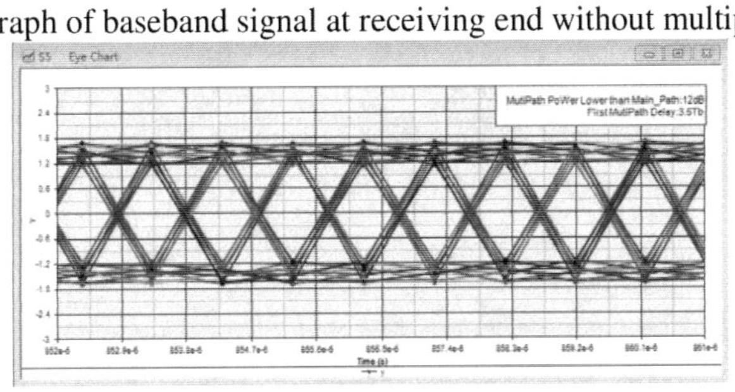

Figure 10. Eye graph of baseband signal at receiving end when the power of multipath signal is 12dB lower than the power of main signal.

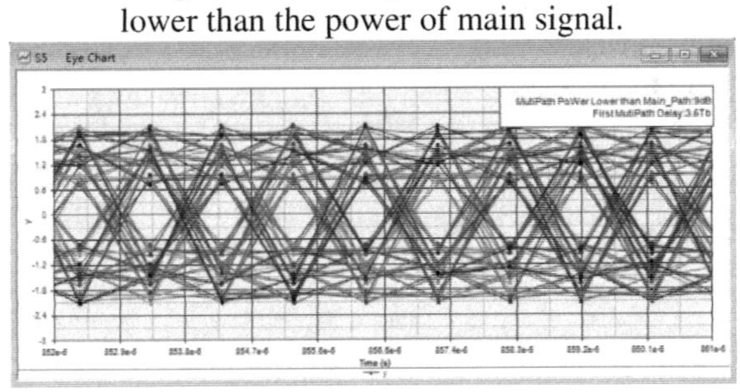

Figure 11. Eye graph of baseband signal at receiving end when the power of multipath signal is 9dB lower than the power of main signal.

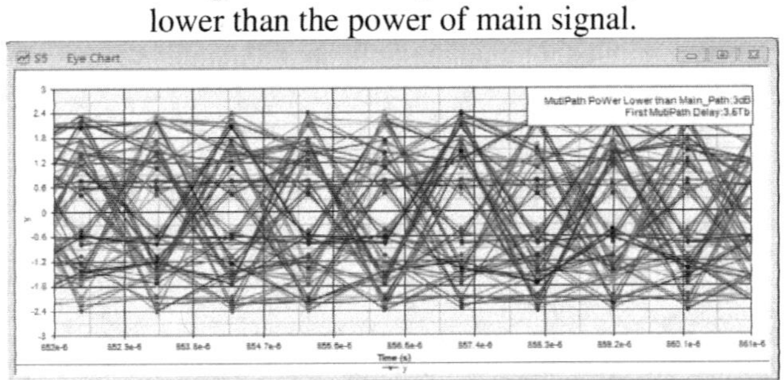

Figure 12. Eye graph of baseband signal at receiving end when the power of multipath signal is 6dB lower than the power of main signal.

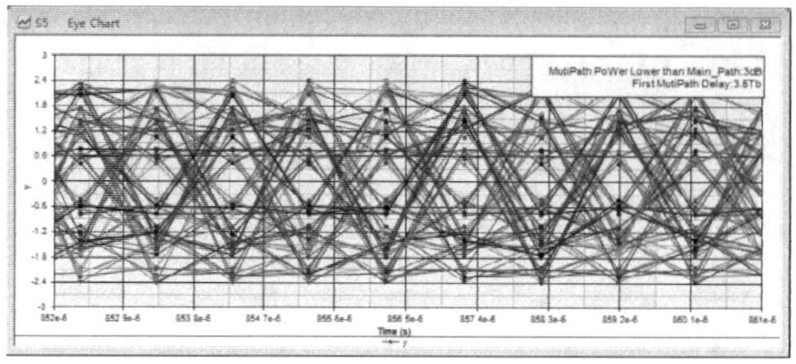

Figure 13. Eye graph of baseband signal at receiving end when the power of multipath signal is 3dB lower than the power of main signal.

As can be seen from figure 9-13, with random phase of multipath signal increasing and determined time delay, signal quality deteriorates significantly. When the power of multipath signal is 3dB lower than the power of main signal, eye diagram is nearly closed. At this time, the quality of received signal is too poor to effectively recover the data.

Figure 14-17 respectively simulate the quality of received signal when multipath reflection coefficient is fixed and the phase is random while multipath time delay changing within a certain range. Figure.18 is the spectrum graph of the transmitted signal at the modulating transmitting end. Figure.19 is the spectrum graph of the multipath superposed signal at the receiving end when the power of maximum multipath signal is 6dB lower than the power of main signal, the phase is random and the delay is $2.4T_b$.

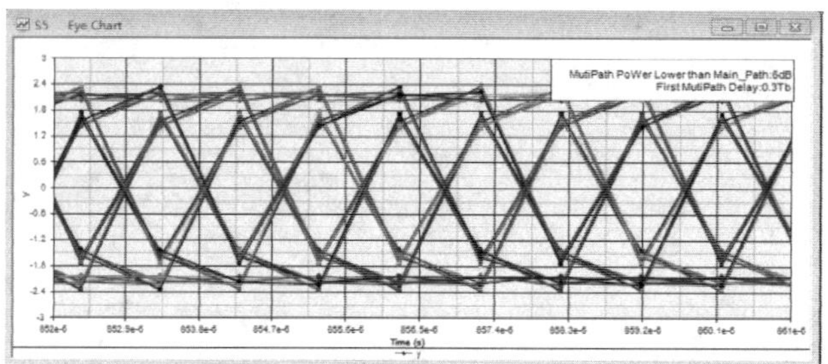

Figure 14. Eye graph of baseband signal at receiving end with multipath power lower 6dB than main path and delay of $0.3T_b$.

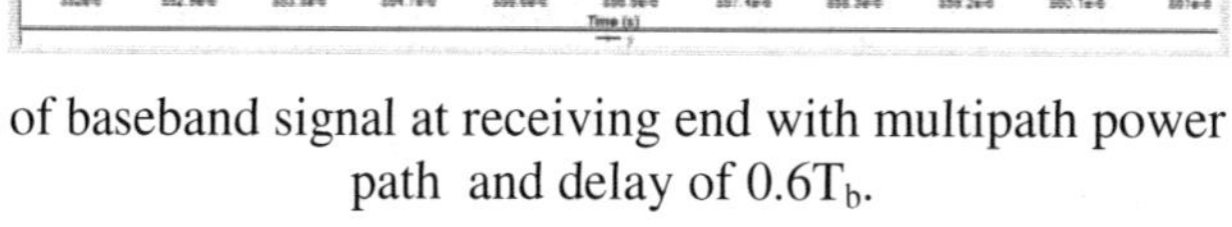

Figure 15. Eye graph of baseband signal at receiving end with multipath power lower 6dB than main path and delay of $0.6T_b$.

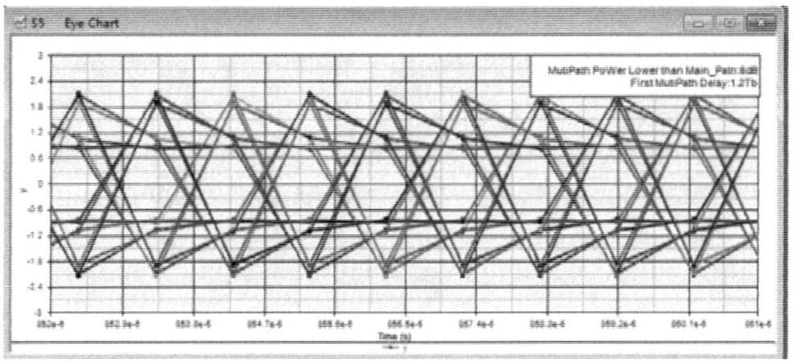

Figure 16. Eye graph of baseband signal at receiving end with multipath power lower 6dB than main path and delay of 01.2T_b.

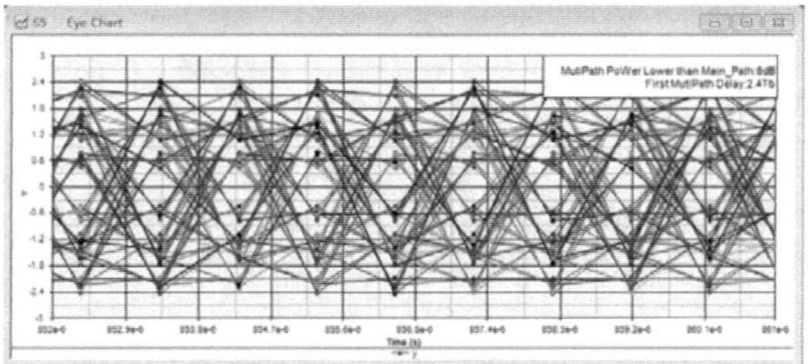

Figure 17. Eye graph of baseband signal at receiving end with multipath power lower 6dB than main path and delay of 2.4T_b.

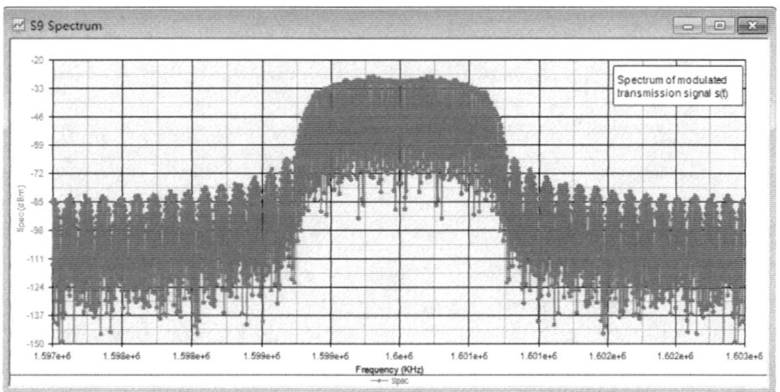

Figure 18. Spectrum of modulated and transmitted signal.

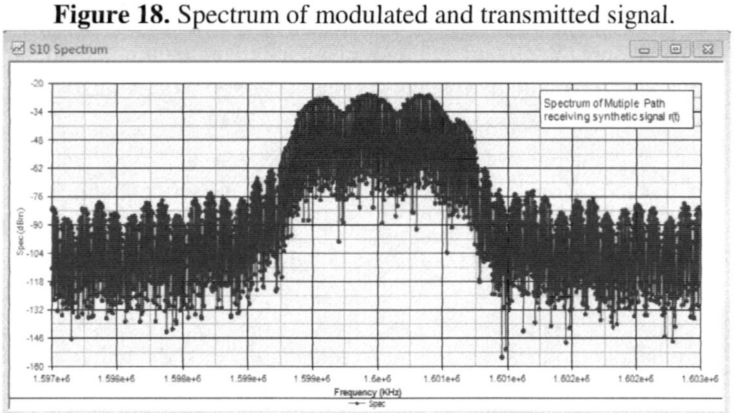

Figure 19. Spectrum of multipath superposed signal at the receiving end.

According to the simulation process and the data shown in Figure.14-19, the analysis shows that when the multipath time delay is within the range of 0-0.5T_b, the eye graph is relatively ideal, and it is generally not affected by random phase changes. In the case that the amplitude of multipath signal declines at a certain level and the multipath time delay is within the range of 0.5T_b-1T_b, the eye graph deteriorates greatly, but it is slightly affected by random phase changes. When the multipath time delay is greater than one T_b, the eye graph becomes significantly worse and is greatly affected by the random phase changes. At this time, the spectrum of the multipath superposed signal presents multi-peak state.

5. Conclusion

In this paper, the transmission performance of UAVs' data link under multipath channel is simulated by modeling the common modulation system with main working parameters, and some simulation test data are presented. According to the simulation process and result, the author thinks in UAV data link, the link level redundancy design for AGWN channel is effective. In view of multipath channel, because of the link level redundancy design, not only has the main signal been increased, but also has the multipath signal at the receiving end been increased. That is to say, the past view that the link level redundancy design is mainly aiming at multipath fading is obviously not entirely correct. In addition to the data link modulation system, spread spectrum modulation and/or adaptive equalization in signal processing should be the basic measures to improve the anti-multipath interference of UAV data link [8-9].

6. Reference

[1] Li Guihua. Development status and trend of foreign military UAV data link [J]. Telecommunications technology.2014, 54(6):851-856.
[2] Wu Huixi, Xu Peng et al. Application of UAV data link technology based on information integration system [J]. Sichuan journal of military engineering, 2013,34(2):114-116.
[3] Wu Qian. Status and development trend of UAV measurement and control system [J]. Telecommunication technology.2009(9):90-94.
[4] Headquarters U.S. Air Force Air Force Unmanned Aerial System (UAS) Flight Plan 2009-2047.2009
[5] John G Proakis. Modern communication system (Matlab edition, 2nd edition)[M]. Beijing: electronic industry press, 2008.
[6] Wang Yang, Ge Lindong. Multipath independent Rayleigh fading channel simulation model [J]. Journal of university of information engineering.2007,8(2):202-205.
[7] Li Linfeng, Sui Shaoyong, Zhao Guoqing. Research and simulation of MPSK signal transmission characteristics in multipath fading channel [J]. System simulation technology.2006,2(4):202-204.
[8] Liu Qiwei, Pu Haiyin et al. Overview of anti-jamming technology of uav data link [J]. Aircraft design.2017,37(6):13-16.
[9] Zhou Lin, Chen Jian, Kuo Yonghong. Progress and prospect of adaptive balancing technology for MIMO systems [J]. Modern electronic technology. 2007(5):7-8.

The Gas Detection Application of Deep Well Production Based on Wireless Sensor Network

Lejiang Guo, Fangxin Chen, Lei Xiao and Yahui Hu

The Department of Early Warning surveillance intelligent ,Air Force Early Warning Academy,430019, Hubei,China
Email:radar_boss@163.com, Fangxin@163.com, Shaoxiong@263.com, Yahui @263.com

Abstract. Based on wireless sensor network, a system-level solution of wireless sensor network for oil well safety information collection is constructed, including hardware and software design of data acquisition, data transmission and data analysis and processing. The system has the characteristics of low cost, wide range and flexible layout, which solves the problems of cable networking, such as wiring, location and slow troubleshooting problem. The experiment shows that the system meets the actual requirement and runs stability.

1. Introduction

Wireless sensor network (WSN) is a new technology which integrates sensor technology, embedded computing technology, network and wireless communication technology, distributed information processing technology. It has a wide range of application scenarios. The distribution of oil drilling is relatively scattered. It is difficult to establish a wired safety production monitoring network, and it requires a lot of manpower and material resources to maintain the monitoring network. Safety production is restricted by objective conditions. In the process of oil drilling and gas production. There are many factors affecting the environment and safety. Safety production has become an important issue. The gas is easily produced in oil drilling and development, such as Hydrogen sulphide, carbon monoxide and other toxic or flammable explosive gases which threaten the personal health and safety of field workers. Therefore, the monitoring and protection of these flammable and explosive toxic gases are very important in the process of oil drilling and production.

2. The Design of the Gas Detection Application Based on WSN

2.1. Characteristics and Applications of Wireless Sensor Networks

Wireless sensor networks (WSNs) are composed of self-organizing nodes, which have independent functions of data acquisition, forwarding and transmission. The nodes form a multi-hop network topology through interconnection. There is a data receiving node in the network which is responsible for fusing all collected data and then transferring it to the local server for subsequent processing and analysis. There are many data transmission paths on each node. According to the energy consumption of the path and the interference signal on the path, users can specify the superior path of data transmission on some indicators to improve the efficiency of network utilization. At the same time, the characteristics of multi-hop and multi-path transmission make the network stronger anti-jamming and anti-destructive which is superior to wired network to a large extent in Fig.1.

Sensor technology is the most widely used technology in data acquisition. At present, the application of wireless sensor networks is just emerging. Compared with traditional sensor networks, it has the following characteristics:

The scale of the network includes two meanings: one is the size of the detection area of the network, the other is the size of the network node density. Generally, the network size can be chosen freely according to the needs of users and actual conditions. If the detection conditions are single, such as measuring only a certain quantity, the network size can be reduced appropriately. If the network size can be increased appropriately in order to ensure the reliability and accuracy of the measurement.

Self-organization nodes in the network are randomly distributed in a region. Nodes have the ability of self-organization to communicate with the surrounding nodes and build the network topology. If the nodes in the network fail or communication is blocked, the new topology can be built automatically after a period of delay between the nodes.

The reliability of network topology and the accuracy of measured data can be greatly improved by self-organization of nodes and multidimensional measurement of the same variable using multiple nodes to measure variables at a certain location.

The network data is data-centric. IP address is used as the basis of access in traditional Internet. But in wireless sensor networks, users only need to submit a query command to the network to obtain data from the network.

Resource saving and adaptability of traditional data acquisition networks are wired networks, such as optical fibber sensor networks. Cable networks are not only expensive, but also taken up a lot of resources and space when the network is huge. At the same time, when a transmission line goes wrong, it will cause a point or even an area to be unable to read data. However, wireless sensor networks do not need any wires which greatly saves wiring resources and production costs. The network has the ability of self-organization and self-adaptation. When a transmission link in the network fails, the network can reconstitute new data links to ensure the normal operation of the network.

The advanced power management technology is used to effectively manage and optimize the node power supply according to the response time of the network, so as to achieve the technical target of low power consumption.

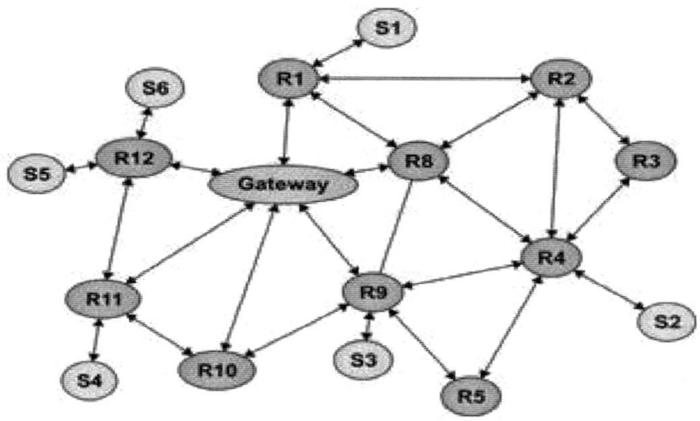

Figure 1. The structure of wireless sensor network

At present, wireless sensor networks have been applied in military detection, agricultural production, security monitoring and other fields. In the process of oil drilling and production, it is often outdoor operations, and the distribution of oil drilling is relatively scattered. It is difficult to establish a wired safety production monitoring network and require a lot of manpower and material resources to maintain the monitoring network. Safety production is restricted by objective conditions. Wireless sensor network technology can effectively solve the objective environmental constraints faced by outdoor monitoring. It provides a safe and effective monitoring network for outdoor operations and ensure the safety of personnel and equipment.

2.2. Detection Technology of Hydrogen Sulfide and Carbon Monoxide

In oil drilling and production sites, real-time detection of hydrogen sulfide and carbon monoxide produced in oil wells is needed to ensure the personal health and safety of operators. Therefore, the detection of hydrogen sulphide and carbon monoxide is the key technology affecting the efficient operation of the whole system.

According to the previous analysis, trace hydrogen sulphide and carbon monoxide will cause harm to human health in a short time. Therefore, the sensor designed is required to have high sensitivity, fast response and be able to detect the presence of toxic gases in time. Secondly, wireless monitoring network is established to detect toxic gases such as hydrogen sulphide and carbon monoxide in oil drilling and production sites. It requires sensor nodes to be small in size. It is easy to install and carry with low-power consumption, long-term continuous operation, strong selectivity. It can accurately detect a variety of gases with anti-jamming performance to work normally in complex environments. Based on the above analysis, electrochemical sensors are selected to detect hydrogen sulphide and carbon monoxide.

The electrochemical sensor has a small detection volume which is easy to install and carry. It has high sensitivity to gases and detect the presence of trace gases quickly and accurately.

3. The Realization of the Gas Detection Application based on WSN

The oil well safety production monitoring network system based on wireless sensor network is mainly divided into the following three parts, data acquisition, data transmission and data analysis. Combining with the schematic diagram of the system in Figure 2.

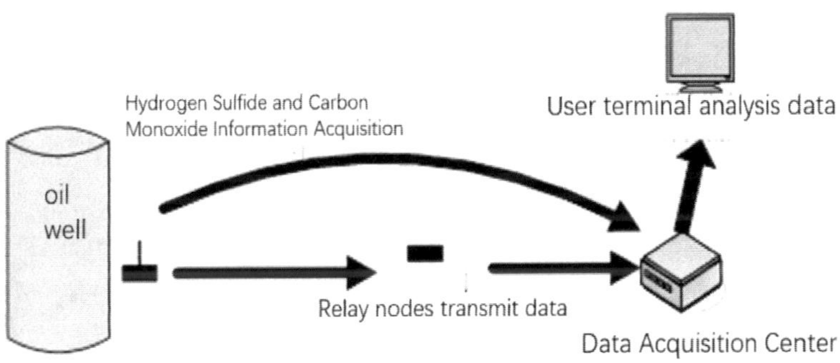

Figure 2. The System Operation Workflow

Data acquisition nodes arranged in oil wells are equipped with electrochemical sensor probes to detect the real-time concentration of hydrogen sulphide and carbon monoxide in oil wells. The detected information is transmitted to relay nodes in real time through wireless transceiver.

Data transmission refers to the process of transferring data collected by sensors from oil wells to remote user terminals. The sensor node collects data, and the data acquisition center collects the data transmitted by each sensor node, and uploads the data for the user terminal to analyse the data. If the sensor node transmits the data directly to the data acquisition canter within the effective communication distance. Otherwise, the data can be transmitted through the relay node.

Data analysis module is to analyse the collected data and display the concentration of hydrogen sulphide and carbon monoxide through the transmitted data value, and then it gives corresponding prompts with concentration value of gas.

3.1. Node Deployment

According to the need of hydrogen sulphide, carbon monoxide sensor sampling data and oil well field monitoring area. Firstly, the number of sensor nodes needed in target area and oil well, the specific location of sensor nodes is analysed. In target area, each sensor node has its own unique. According to the location of sensor nodes oil wells and sensor nodes are corresponded so that the user terminal can monitor the concentration of hydrogen sulphide and carbon monoxide in specific oil wells with the sensor data. At the same time, according to the different installation location and communication distance of different sensor nodes, relay nodes are reasonably arranged to forward data. Reasonable arrangement of nodes can not only optimize the network system, but also save the energy consumption of nodes by using the smallest number of nodes to achieve the maximum monitoring range.

The system collects the concentration of hydrogen sulphide and carbon monoxide in oil wells through sensor nodes. As shown in Figure 3, sensor nodes include gas detection, signal and data processing, data reception and transmission.

In the gas detection part, hydrogen sulphide and carbon monoxide are detected by high precision electrochemical gas sensors. The output signal of the electrochemical sensor is sent to the signal and data processing part. The signal and data processing part includes signal processing circuit and MCU module. The signal processing circuit filters and amplifies the output signal of the sensor, and then transmits the signal to MCU and converts it into digital signal. In the data receiving and transmitting part, the wireless transceiver circuit transmits the data signal to the network for the analysis and processing of the user terminal.

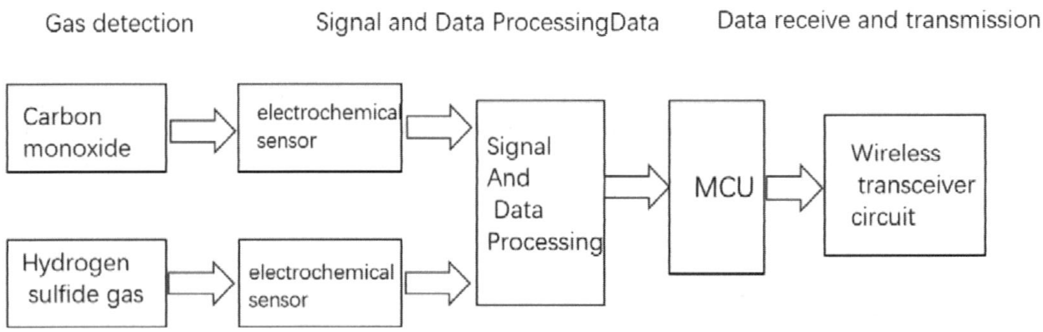

Figure 3. The Sensor Node Structure Block Diagram

3.2. Data Transfer

Data transmission refers to the process of transferring data from sensor nodes to user terminal database through network. In the process of data transmission, network topology and routing strategy are the key to high-speed and accurate data transmission.

After the nodes are deployed, the network topology can be basically confirmed according to the transmission distance of the nodes and the deployment structure of the nodes in the monitoring area. Because of the self-adaptive characteristics of the network, the nodes will automatically find the neighbouring nodes, so that they can connect with the neighbouring nodes. When all the nodes are connected with the neighbouring nodes, the topological structure diagram of all the nodes is the most original network topological structure diagram. The data of nodes can generally reach the user terminal, but with the increase of the number of nodes, data transmission between nodes will inevitably produce conflicts, affecting the real-time performance of the network. Therefore, the deployment of nodes and the construction of network topology and redundant multi-channel transmission is adopted. By applying the data forwarding function of relay nodes, the real-time performance of the network is effectively improved. The data can be sent back to the user terminal in time and accurately.

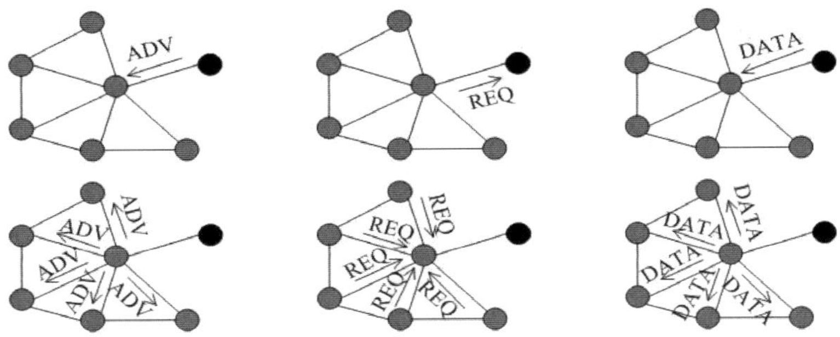

Figure 4. Data transmission

From Fig. 4,it is a schematic diagram of data transmission from node A to data sink center, where node E is a redundant relay node. Normally, the data of node A will be transmitted to the neighbor node D, node D forwards the data of node A to the data sink center, and node B and node C will also transmit the data to the data sink center through node D. when node A and node C send data to node D at the same time, data transmission conflicts will occur, resulting in transmission delay. The application of redundant multi-channel technology of relay nodes solves this problem well.

3.3. Data Analysis and Processing

When the base station receives the data from all the nodes, it stores the data in the database. Terminal application software extracts data from the database, analyses the data and extracts the concentration values of hydrogen sulphide and carbon monoxide. In the terminal software, the well number and the corresponding sensor node data are combined, so the concentration of hydrogen sulphide and carbon monoxide in each well can be dynamically displayed.

As an expert system, the user terminal can analyse the data collected according to prior knowledge, and then monitor the working environment of each oil well in real time according to the concentration of hydrogen sulphide and carbon monoxide in the oil well. The safety factor in the oil well can be judged by the staff according to the prompt information of the system.

3.4. Customer Service Software

Customer service software is the window through which the whole system interacts with users. It mainly provides users with the functions. The server software can display the concentration data of hydrogen sulphide and carbon monoxide in each oil well and store the historical data according to user needs or query. Network operation monitoring can display the network topology and data communication link status as well as the operation of each node. It is convenient for users to monitor the operation and eliminate the possible operation failure. The expert system analyses the sensor data, obtains corresponding operation tips and alarm information, and provides reference for oil well workers.

4. Conclusion

With the combination of wireless sensor technology and gas detection technology, the oil well safety monitoring system based on wireless sensor network has the following characteristics:

Wireless communication technology is used to realize the wireless transmission between data, which is convenient to work in outdoor and hard environment. Each sensor node forms a complete sensor network to facilitate users to monitor the target area comprehensively.

According to the requirements of users, the concentration of hydrogen sulfide and carbon monoxide in oil wells can be monitored in real time all day, so that users can grasp the gas situation in oil wells in time.

Wireless sensor networks have self-organizing and self-adapting functions. When a node in the network fails, new communication links can be reconstructed in time to ensure the normal flow of data in the network. Secondly, through the client software, we can real-time monitor the operation of the network, accurately find the specific location of the fault node, so as to facilitate timely troubleshooting.

The whole network is powered by advanced power management technology, single sensor node is powered by ordinary batteries, which can work continuously outdoors for a long time, and the maintenance cost of the network is greatly reduced.

According to the collected data, the client software can display the concentration of hydrogen sulfide and carbon monoxide in each oil well in real time. The expert system synthetically analyses the data and gives the corresponding prompt and alarm information.

Sensor nodes are small in size, powered by ordinary batteries, and do not need special power supply lines and communication transmission lines. It is easy to install and build networks, and easy to maintain. Compared with the traditional wired monitoring network, the cost is significantly reduced.

5. References

[1] Warunee Krudpun, Nonchanutt Chudpooti, Nattapong Duangrit, Panida Lorwongtragool, Somporn Seewattanapon, Prayoot Akkaraekthalin, "A PSE-Coated Interdigital Resonator Gas Sensor for Agricultural Applications",Antenna Measurements & Applications (CAMA) 2018 IEEE Conference on, pp. 1-3, 2018.

[2] Anuj Kumar, Abhishek Singh, Ashok Kumar, Manoj Kumar Singh, Pinakeswar Mahanta, Subhas Chandra Mukhopadhyay, "Sensing Technologies for Monitoring Intelligent Buildings: A Review", Sensors Journal IEEE, vol. 18, no. 12, pp. 4847-4860, 2018.

[3] Irfan Al-Anbagi, Melike Erol-Kantarci, Hussein T. Mouftah, "Delay Critical Smart Grid Applications and Adaptive QoS Provisioning", Access IEEE, vol. 3, pp. 1367-1378, 2015.

[4] A. Kosterev, T. Moseley, F. Tittel, "Impact of humidity on quartz-enhanced photoacoustic spectroscopy based detection of HCN", Appl. Phys. B, no. 85, pp. 295-300, 2006.

[5] J. N. Al-Karaki, A. E. Kamal, "Routing techniques in wireless sensor networks: a survey", IEEE wireless communications, vol. 11, no. 6, pp. 6-28, 2004.

[6] P. Bonnet, J. Gehrke, P. Seshadri, "Towards sensor database systems" in International Conference on Mobile Data Management, Springer, pp. 3-14, 2001.

IOP Conf. Series: Materials Science and Engineering **719** (2020) 012050 doi:10.1088/1757-899X/719/1/012050

Using Blockchain for Flow Measurement of 5G Base Station

Li Lu[1] and Donglin Wu[2]

[1]Beijing Max Technologies Co., Ltd., 2215# Building B Dongya Wangjing Center Chaoyang Beijing China 100102
[2]Rohde & Schwarz (China) Technology Co., Ltd., Rohde & Schwarz Building 1, 18 Ziyue Road, Chaoyang District, Beijing, P.R.China 100012
Email: lu.li@max-rf.com, Donglin.Wu@rohde-schwarz.com

Abstract. 5G is a hot area for its high speed data transmission by millimetre wave. Thus a huge amount of 5G base stations are required to be established, which brings heavy burden to 5G operators. In this paper, we propose a new flow measurement system based on blockchain to settle accounts for base stations belonging to different operators securely and fairly. For the base stations can verify the data flow for each other, only real data transmission can be recorded by blockchain. The proposed system can utilize existing consensus mechanism to make it stable and distributed. A concrete scheme is constructed and the scheme is proved to be secure. With this system, the base stations can be encouraged to be setup by different 5G operators and other companies to gain incentive by transmitting data service. This system can solve the problem of huge demand of base stations, and it's secure, fair, and suitable for real deployment in practice.

1. Introduction

With the development of the mobile Internet, IoT and its applications, wireless data traffic demands are expected to increase 1,000 fold by 2020 [1] and 10,000 fold by the year 2025 [2]. The fifth generation (5G) cellular networks will be used not only for communication between people, but also for communication between people and objects, as well as between things and things. At present, 5G services can be roughly divided into three scenarios: eMBB (Enhanced Mobile Broadband), mMTC (Massive Machine Communication), and uRLLC (Ultra Reliable Low Delay Communication). It provides high speed data transmission for end nodes.

With the increase of access equipment and the increase of communication bandwidth demand, higher requirements are put forward for the number of base stations and the quality of service. At the same time, due to the low penetration of millimeter wave, more micro-base stations will be built in each cell to provide stable data connection for mobile terminals, which makes infrastructure providers spend a lot of money to establish a sufficient number of base stations. To make matters worse, there are usually many 5G operators. In the construction of 5G base stations, different operators will build their own 5G base stations, resulting in a huge amount of base station construction, and excessive consumption of resources, a large number of base stations occupy a large amount of space.

If the 5G base stations built by various operators can be used in common with each other, the above problems will be effectively solved. However, this will lead operators to blame each other in the process of 5G construction, or make their base stations lazier in providing 5G data services for users of other operators. If we can provide a credible settlement mechanism in the 5G universal process, so that the services provided by the base stations built by operators can be effectively recorded, and the use of base stations can be settled between different operators within a fixed period, the above problems will be effectively solved. Therefore, there is a lack of a fair and credible technology for uniform traffic settlement between 5G base stations across operators.

Blockchain is the core supporting technology of the digital encryption currency system represented by Bitcoin [3]. Bitcoin is a peer-to-peer network of nodes that distributes records, and every node collects transactions from the broadcast network. Blockchain can use the encryption, time stamp, digital signature and distributed consensus technology to achieve decentralized architecture [4]. Data on blockchain is untampered, public and traceable, ensured by cryptography, without any help of any third-party organizations. Blockchain is an append-only bulletin board maintained in a distributed fashion by nodes in networks [5].

Similar to the block chain network architecture, 5G base stations have the characteristics of large number and point-to-point communication. Therefore, connecting 5G base stations belonging to different operators into block chain system, trustworthy recording of their transmission data and providing certain incentives for their builders can effectively solve the data settlement problem of 5G base stations. Using this technology, 5G base stations can be built by different operators to provide services together, to gain benefits according to the number of services, and to ensure the fairness and feasibility of the process. Even in the future, private users can build 5G base stations by themselves and make profits after providing data services.

The organization of this paper is as follows. In Section 2 we describe our contributions in this paper. Then in Section 3 we show some related work. Section 4 provides the system model. Section 5 provides the concrete work flow of the proposed scheme. And after this, a security analysis of this scheme is given in Section 6. Finally, Section 7 concludes this paper.

2. Our Contributions

In this paper, we propose a new flow measurement system of 5G base station based on blockchain. Our contribution can be summarized as follows.

Firstly, we design a new blockchain system model for 5G base station. The architecture of this system is described. The system is divided into five layers, data collection layer, communication layer, distributed storage layer, consensus layer and application layer. Distributed storage layer is proposed to store all data flow data, which is hash locked periodically on blockchain to ensure untampered.

Secondly, the concrete scheme is provided. In this system, we utilize the communication data between different base stations to calculate data flow for each other. In this way, the data transmitted by each base station is measured fairly.

Finally, the system can be built based on existing consensus mechanism, e.g. Proof of Work (PoW), Delegated Proof of Stake (DPoS), etc. Due to the limited source of base station, the DPoS can be achieve by the method that different 5G operators set up different nodes as miners.

The security of this system is analysed. The proposed system can provide data flow measurement of 5G base stations fairly and securely, and it can be deployed in practice.

3. Related Works

5G technology has developed for several years [6][7][8][9], and the technology of 5G base station has also been developed since then. Qian et.al. proposed the base station based centralized network architecture for 5G mobile communication systems [10]. Feng et.al. summarized the approaches and challenges for base station on-off switching in 5G wireless networks [11]. Chen et.al. proposed base station switch-off with mutual repulsion in massive MIMO networks [12]. Yang et.al. proposed relay base-station handover in 5G [13].

Blockchain is an emerging technology, which is widely used in bitcoin [3]. In the blockchain system, the last block contains the hash value of its previous one, which ensures the data untemperable [14]. Due to the valuable property of blockchain, several blockchain-based applications have been developed [15][16]. Zyskind et.al. used blockchain to protect personal data [17]. Huh et.al. proposed a blockchain platform to manage IoT devices [18]. Yuan et.al. proposed blockchain-based intelligent transportation systems [19]. And Mettler summarized the future application of blockchain in healthcare [20]. Dorri used blockchain to ensure security and privacy in IoT [21]. It also has several applications in bank industry [22] and sharing services [23]. Blockchain technology achieves secure

IOP Conf. Series: Materials Science and Engineering **719** (2020) 012050 doi:10.1088/1757-899X/719/1/012050

and fair system for decentralized architecture, and it has important influence in lots of application scenarios.

4. System Model

There are five layers in this system, as shown in Figure 1, including Data Collection Layer, Communication Layer, Distributed Storage Layer, Consensus Layer, Application Layer.

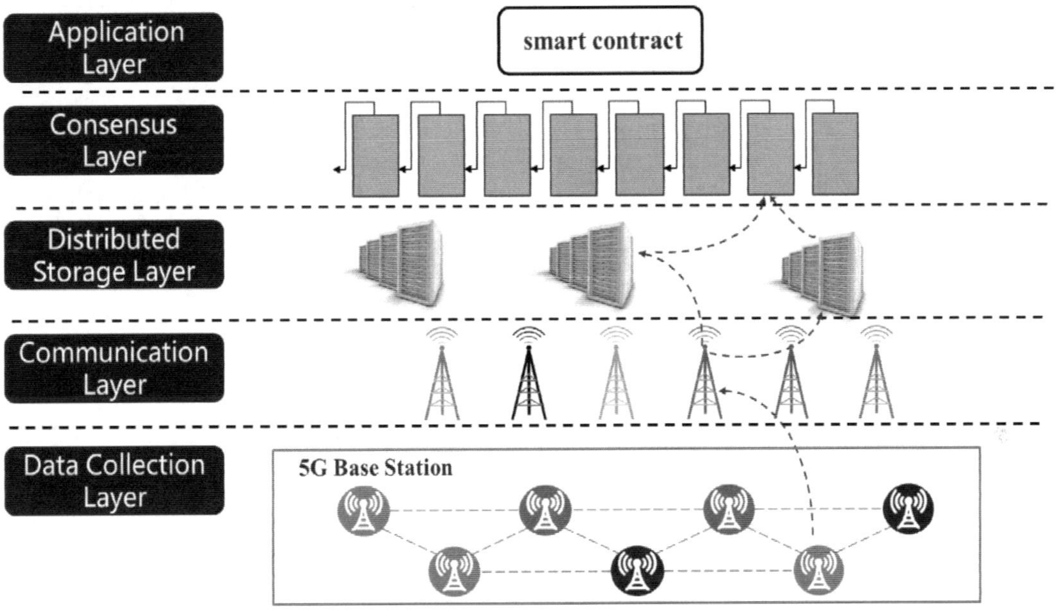

Figure 1. System Model

- The Data Collection Layer is built on the 5G base stations, and there are huge amounts of based stations in this system, which belong to different 5G operators. When the data transmitted from one base station to another, the base station would add its digital signature at the end of the data package, and the next base station would verify its validity. Due to the base stations deployed in the whole 5G network, all data is transmitted by all base stations on the route from data sender to receiver. Thus the base stations can collect all the transmitted data and identify the data route. Then all base stations would record the work of each base station that the related data it transmitted. Finally, the base station would send the result of data flow measurement to the network and the result that verified by other base stations would be recorded in distributed storage layer.
- The Communication Layer is built by the base stations and network routers. The base stations in the Data Collection Layer are linked with each other as a peer-to-peer network by wireless connection. Through this layer, the collected flow data will be spread in the whole network.
- The Distributed Storage Layer is composed of secure storage devices, and the flow date collected by the base stations can be stored in this layer securely. The system will generate a message digest for each segment of flow data which is collected by a base station over a period of time.
- The Consensus Layer is the core algorithm of the blockchain system. All base stations have the same consensus and they are combined as a blockchain system. The base stations in the system will check the correctness of the flow data. And the message digest of correctly collected data will be recorded in the blockchain. And it can assure the flow data adopted by blockchain is trusted or agreed by the most of the base stations.
- The Application Layer is carried out by the Consensus Layer. Correct flow data emerging in all stage of the system is recorded in the block and thus there can be many smart contracts

based on the blockchain to process the data according to flexible rules. For example, the smart contract can periodically calculate flow data from different base stations, and settle accounts between different 5G operators.

5. Concrete Scheme

To describe the concrete scheme in detail, we take the minimum system as an example as shown in Figure 2. Base stations $S_1,\ldots,S_n$ are deployed to transmit data and sign the data to mark its transmission work and verify the correctness of the mark from others.

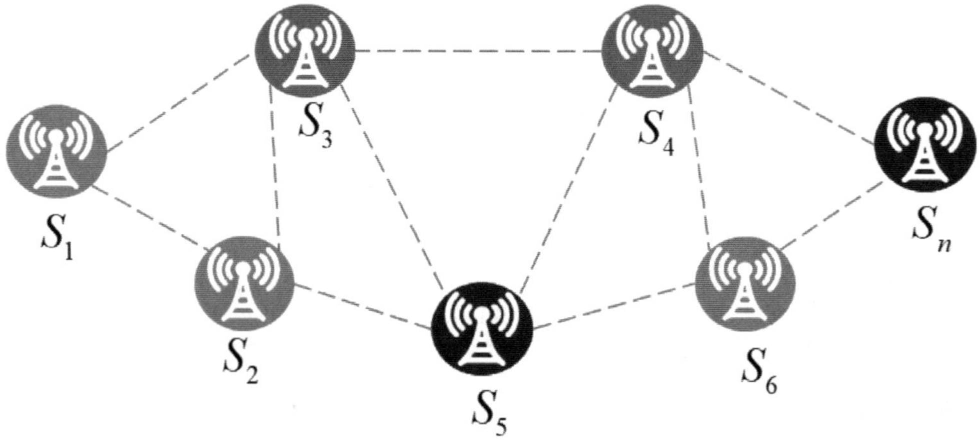

Figure 2. Concrete Scheme

The base stations in this system are also the miners of the blockchain and we adopt the existing Delegated Proof of Stake (DPoS) consensus mechanism. In DPoS, the miners are selected from the committee and generate new blocks one by one, and the current miner is named witness. While if the base stations are computing power limited, the miners can also be play by many high performance computers, which should also be produced from different manufactory or have different frameworks and run by different 5G operators. And the procedures of the system are as follows.

Step 1: Initialization. This algorithm is run at the first to setup the system at each layer of the system model. On input the security parameter λ , the system selects cryptography algorithms with relative security level, and the base stations build up the peer-to-peer network and the blockchain genies block between each other. The blockchain parameter is also initialized in this step including the number N of witness in the DPoS consensus mechanism.

Step 2: Data Collection. When a mobile sends data package from the base station S_1 and then the data goes to $S_2 \to S_5 \to S_6 \to S_n$, S_1 would mark the data package by its identity and using identity-based signature to sign the data. Then the data package with the mark and signature would be sent to S_2, other base stations and distributed storage layer. After receiving the data package, S_2 would do the same as what S_1 does. This step ensures the transmission data is broadcasted to the whole network.

Step 3: Data Verification. After receiving the data package with mark from base station, all the other base station would check the correctness, e.g., the data marked by S_1 would then be marked by S_2, which would also be broadcasted to the network. Then the other base stations would verify its correctness and if the data package is marked by S_2, then it means the data is really transmitted from S_1 to S_2. After verification, the base stations would send all the data to distributed storage layer. Periodically, the data is hash locked by generating a message digest. And thus the data flow is recorded. And this step ensures the collected data is trusted verified.

Step 4: Block Generation. When the flow data is verified to be correct, i.e. the data that is almost the same in more than half of the base stations, the witness will record the digest in the block. And then it will publish the new block to the whole system. This step ensures the data recorded in the distributed storage devices is secure and cannot be tampered, which serves for data audit.

Step 5: Smart Measurement. There can be many smart contract utilized based on the flow data. And based on data processing algorithm like Kalman Filter, the witness will calculate the difference value between each operator's base stations, and settle accounts between operators automatically. This step ensures the flow measurement is carried out honestly and fairly in the system.

Step 6: Block verification. When block published by the appointed witness of DPoS, the other miners will verify whether the data in the block is correct or the witness runs the smart measurement honestly. Once the block is not correct, each miner would reject the new block, and alert the operator. If the block is verified to be correct, the witness will add the block to the end of the existing blockchain.

6. Security Analysis

In this system, we utilize existing consensus mechanism of blockchain, which is verified to be secure in real deployment. The miners are run by different 5G operators, who must verify the flow data honestly, because they have to pay for the flow from other operators, and once some operators don't behave honestly or fairly, more than half of all operators would not record such flow data, and other operators would reject to give service to them. Besides, the signature algorithm and hash algorithm in this system are standard and proved to be secure. Above, the proposed system is secure and stable to record flow data for base station honestly and fairly.

7. Conclusion

In this paper, we proposed a new flow measurement of 5G base station system based on blockchain. The system model and concrete scheme were constructed with existing consensus mechanism and standard cryptography algorithm. It was proved that the system is secure and can provide fair flow measurement, which would have huge research and practice value in the future.

8. References

[1] Raaf B, Zirwas W, Friederichs K J, et al. Vision for beyond 4G broadband radio systems[C]//2011 IEEE 22nd International Symposium on Personal, Indoor and Mobile Radio Communications. IEEE, 2011: 2369-2373.

[2] Cudak M, Ghosh A, Kovarik T, et al. Moving towards mmwave-based beyond-4G (B-4G) technology[C]//2013 IEEE 77th Vehicular Technology Conference (VTC Spring). IEEE, 2013: 1-5.

[3] NAKAMOTO S. Bitcoin: A peer-to-peer electronic cash system[OL]. https://bitcoin.org/bitcoin.pdf, 2008.

[4] Y. Yong and W. Fei-Yue, "Blockchain: The state of the art and future trends," 2016.

[5] I. Miers, C. Garman, M. Green, and A. D. Rubin, "Zerocoin: Anonymous distributed e-cash from bitcoin," in 2013 IEEE Symposium on Security and Privacy, SP 2013, Berkeley, CA, USA, May 19-22, 2013. IEEE Computer Society, 2013, pp. 397–411.

[6] Andrews J G, Buzzi S, Choi W, et al. What will 5G be?[J]. IEEE Journal on selected areas in communications, 2014, 32(6): 1065-1082.

[7] Rappaport T S, Sun S, Mayzus R, et al. Millimeter wave mobile communications for 5G cellular: It will work![J]. IEEE access, 2013, 1: 335-349.

[8] Boccardi F, Heath Jr R W, Lozano A, et al. Five disruptive technology directions for 5G[J]. arXiv preprint arXiv:1312.0229, 2013.

[9] Roh W, Seol J Y, Park J, et al. Millimeter-wave beamforming as an enabling technology for 5G cellular communications: Theoretical feasibility and prototype results[J]. IEEE communications magazine, 2014, 52(2): 106-113.

[10] Qian M, Wang Y, Zhou Y, et al. A super base station based centralized network architecture for 5G mobile communication systems [J]. Digital Communications & Networks, 2015, 1(2):152-159.
[11] Feng M, Mao S, Tao J. Base Station ON-OFF Switching in 5G Wireless Networks: Approaches and Challenges[J]. IEEE Wireless Communications, 2017, 24(4):46-54.
[12] Chen J, Ge X, Song X, et al. Base Station Switch-off with Mutual Repulsion in 5G Massive MIMO Networks[J]. Iet Communications, 2018.
[13] Yang Z Y, Leu F Y. Relay Base-Station Handover in a 5G Environment[C]// International Conference on Broadband & Wireless Computing. 2017.
[14] Garay J, Kiayias A, Leonardos N. The bitcoin backbone protocol: Analysis and applications[C]//Annual International Conference on the Theory and Applications of Cryptographic Techniques. Springer, Berlin, Heidelberg, 2015: 281-310.
[15] Crosby M, Pattanayak P, Verma S, et al. Blockchain technology: Beyond bitcoin[J]. Applied Innovation, 2016, 2(6-10): 71.
[16] Pilkington M. 11 Blockchain technology: principles and applications[J]. Research handbook on digital transformations, 2016, 225.
[17] Zyskind G, Nathan O. Decentralizing privacy: Using blockchain to protect personal data[C]//2015 IEEE Security and Privacy Workshops. IEEE, 2015: 180-184.
[18] Huh S, Cho S, Kim S. Managing IoT devices using blockchain platform[C]//2017 19th international conference on advanced communication technology (ICACT). IEEE, 2017: 464-467.
[19] Yuan Y, Wang F Y. Towards blockchain-based intelligent transportation systems[C]//2016 IEEE 19th International Conference on Intelligent Transportation Systems (ITSC). IEEE, 2016: 2663-2668.
[20] Mettler M. Blockchain technology in healthcare: The revolution starts here[C]//2016 IEEE 18th International Conference on e-Health Networking, Applications and Services (Healthcom). IEEE, 2016: 1-3.
[21] Dorri A, Kanhere S S, Jurdak R, et al. Blockchain for IoT security and privacy: The case study of a smart home[C]//2017 IEEE international conference on pervasive computing and communications workshops (PerCom workshops). IEEE, 2017: 618-623.
[22] Guo Y, Liang C. Blockchain application and outlook in the banking industry[J]. Financial Innovation, 2016, 2(1): 24.
[23] Sun J, Yan J, Zhang K Z K. Blockchain-based sharing services: What blockchain technology can contribute to smart cities[J]. Financial Innovation, 2016, 2(1): 26.

ELF Chirp Communication based on Modulated Heating Low Ionosphere

Kaijie Zhou [1,2], Huali Wang [1] and Peipei Cao [2,3]

[1] Institute of Communications Engineering,
PLA Army Engineering University, Nanjing 210007, China.
E-mail: zhoukaijie@hytc.edu.cn, huali.wang@ieee.org
2 College of Physics and Electrical Engineering,
Huaiyin Normal University, Huai'an 223300, China
[3] School of Electronic and Optical Engineering,
Nanjing University of Science and Technology, Nanjing 210094, China.
E-mail: halscp@163.com

Abstract. ELF can be generated by HF heating the low ionosphere for submarine communication. QPSK has been used on ELF carried generated via modulated ionosphere heating to transmit digital data. However, ELF band is disturbed frequently by the power frequency harmonic interference, and QPSK modulation can't effectively suppress interference. In this paper, the Chirp-BOK(Binary Orthogonal keying) modulation method for ELF communication based on ionosphere heating is proposed. The up-chirp and the down-chirp modulation waveforms are designed to transmit the symbols "0" and "1" respectively. Simulation results show that the variation of electron temperature and conductivity in the ionosphere is basically consistent with the frequency modulation characteristics of power modulation when the ionosphere is heated by Chirp-BOK power modulation high frequency wave. The performance of Chirp-BOK and QPSK under harmonic interference is compared, and the simulation results show that Chirp-BOK has better anti-interference performance.

1. Introduction

Extremely low frequency (ELF: 0.3-3khz) communication has a wide range of applications, such as navigation, geophysical exploration, remote sensing and so on. When the ionosphere is periodically heated by the AM(amplitude modulation) modulated high frequency waves, it will lead to periodic oscillation of the currents to radiate ELF waves, and the "virtual antenna" is formed in the ionosphere[1]. Comparing to the traditional method which is directly radiating the ELF waves with ground antennas; it is small in size, flexible and not vulnerable to attack.

However, there are few reports on the scheme to realize the whole ELF communication by heating the ionosphere. In [2], Jin et al examined the experimental results of using quaternion phase shift keying (QPSK) to transmit digital data on ELF carriers generated by heating the ionosphere through modulation, and achieved the communication rates of 100bps,400bps,800bps. Whereas, the ELF band is always disturbed by the power frequency harmonic, and the BER performance of QPSK communication will be decreased sharply under harmonic interference. The chirp communication system has strong anti-jamming ability, narrowband anti-jamming ability and multipath anti-jamming ability [3]. Therefore, this paper propose a a chirp-bok waveform modulation method by combining chirp modulation method with AM(amplitude modulation) to heat the ionosphere. Finally, the feasibility of the proposed scheme is verified and its anti-interference performance is compared.

IOP Conf. Series: Materials Science and Engineering **719** (2020) 012051 doi:10.1088/1757-899X/719/1/012051

2. System Model

2.1. Theoretical Model of Heating Ionosphere
Current modulation theory is the basis of ELF generated by periodic heating of ionosphere by high frequency waves. When the high power high frequency transmitter is started, the ionosphere begins to absorb high frequency energy and the temperature rises. When the transmitter is turned off, the ionosphere cools to its ambient temperature and the conductivity returns to its ambient value. Therefore, by changing the power of the transmitter at ELF frequency, ionospheric conductivity and current radiate ELF waves at the same frequency. Therefore, the virtual antenna is formed in the low ionosphere, as shown in figure 1.

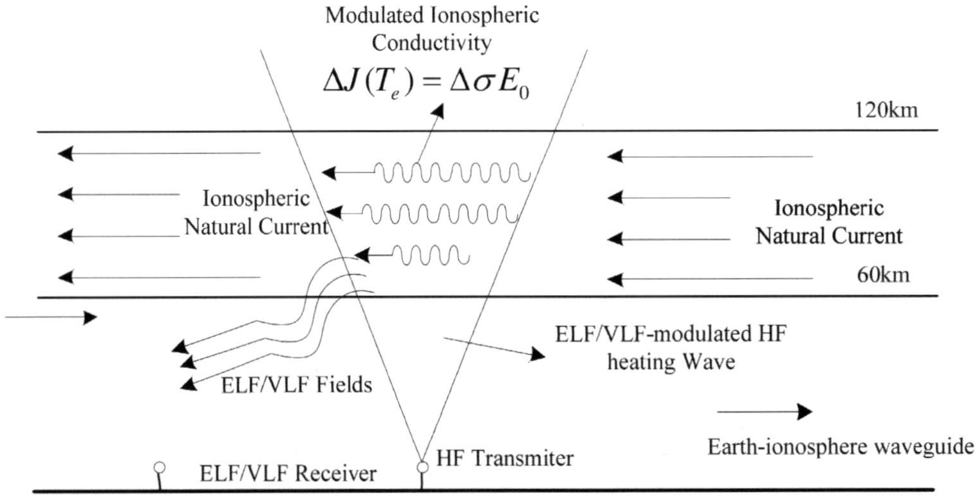

Figure 1. Cartoon depicting ELF/VLF waves generated via modulated ionospheric heating

Current modulation theory is the basis of ELF generated by periodic heating of ionosphere by high frequency waves. In the process of high frequency heating, the electron temperature distribution is controlled by the ionospheric energy balance equation, ignoring the influence of electron thermal conductivity and plasma transport. The energy balance equation is:

$$\frac{3}{2}N_e k_B \frac{dT_e}{dt} = 2k\chi(T_e)S - L(T_e, T_0) \tag{1}$$

Where, k_B, N_e, T_e and k are the Boltzmann's constant, the electron density, the heated electron temperature and the wave number respectively. S is the high frequency power density. χ is the imaginary part of the refractive index, calculated by the Appleton-Hartree equation, and L is the rate of electron energy loss, which is a function of the ambient electron temperature T_0 [4] and T_e. In [5], the formula of AM heating the low ionosphere were given.

2.2. Chirp-BOK Waveform Modulation Method
The square wave amplitude modulation efficiency of heating ionosphere is the highest, and the square wave amplitude modulation and power modulation functions are as follows:

$$\begin{aligned} M_r(t) = A_r(t) &= \frac{1}{2}\sum\nolimits_n P_{T/2}(t - \frac{T}{2} - nT_1) \\ &= \frac{1}{2}[1 + 2\sum\nolimits_{k=1} sinc(\frac{k\pi}{2}) cosk\omega_1 t] \end{aligned} \tag{2}$$

Where $P_a(x)=1$ for $|x|<a$ and 0 for $|x|>a$ is square pulse function, T_1 is the period of modulation, $\omega_1 = 2\pi / T_1$. Combining the Chirp modulation technology with square wave AM, ionospheric heating adopts Chirp-BOK waveform modulation, and its amplitude and power modulation functions are as follows:

$$\begin{aligned} M_r(t) &= A_r(t) \\ &= \frac{1}{2}[1+2\sum_{k=1} sinc(\frac{k\pi}{2})\cos k(\omega_1 t + \pi\mu t^2)] \end{aligned} \tag{3}$$

Where, ω_1 is the center frequency of the signal, μ is the rate of the linear frequency. The bandwidth of chirp signal is $B=|\mu|t$. If the symbol time T_1 is fixed, the bandwidth increases with the increase of μ. As shown in Eq.(3), when $\mu>0$, the signal of chirp is up-chirp, and when $\mu<0$, the signal of chirp is down-chirp. Due to the up-chirp and the down-chirp have good autocorrelation and cross-correlation, we can use up-chirp and the down-chirp to transmit information. The symbol '0' and '1' are denoted by up-chirp and down-chirp respectively, and the two signals are given as follows:

$$\begin{cases} S_0(t) = \frac{1}{2}[1+2\sum_{k=1} sinc(\frac{k\pi}{2})\cos k(\omega_1 t + \pi\mu t^2)], -\frac{T_1}{2} < t < \frac{T_1}{2} \\ S_1(t) = \frac{1}{2}[1+2\sum_{k=1} sinc(\frac{k\pi}{2})\cos k(\omega_1 t - \pi\mu t^2)], \frac{T_1}{2} < t < \frac{T_1}{2} \end{cases} \tag{4}$$

3. Numerical Analysis

3.1. Feasibility Test

The background of simulation was at local time 12:00 on 15 August 2016 at Nanjing, China($31.1°N$, $118.2°E$). Frequency of pump wave: 6MHz; ERP: 200MW; Modulation scheme: Chirp-BOK. Direction of high frequency heating beam: vertical; As shown in Fig.2, the chirp square wave AM is used to heat the ionosphere, and the electron temperature varies with time at an altitude of 75 km.

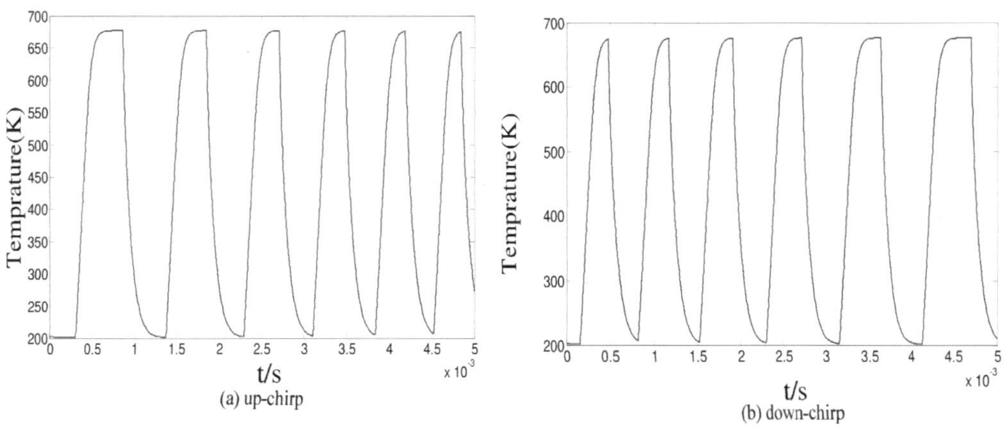

Figure 2. Electron temperature variations with time (75km height)

As can be seen from Fig.2, electron temperature frequency has the obvious frequency modulation characteristics, so does the Pedersen and Hall conductivity variations. The change of Pedersen and Hall conductivity with time at an altitude of 75 km is shown in Fig.3. The ionospheric electronic temperature rises from about 200K to 680K, with a change of about 480K. The Periodic perturbations of the electron temperature and Pedersen and Hall conductivity cause periodic perturbations of ionospheric currents. Then, ELF signals are obtained through Chirp-BOK ionospheric heating, and the up-chirp (symbol '0') and down-chirp (symbol '1') ELF/VLF signals are transmitted for practical communication.

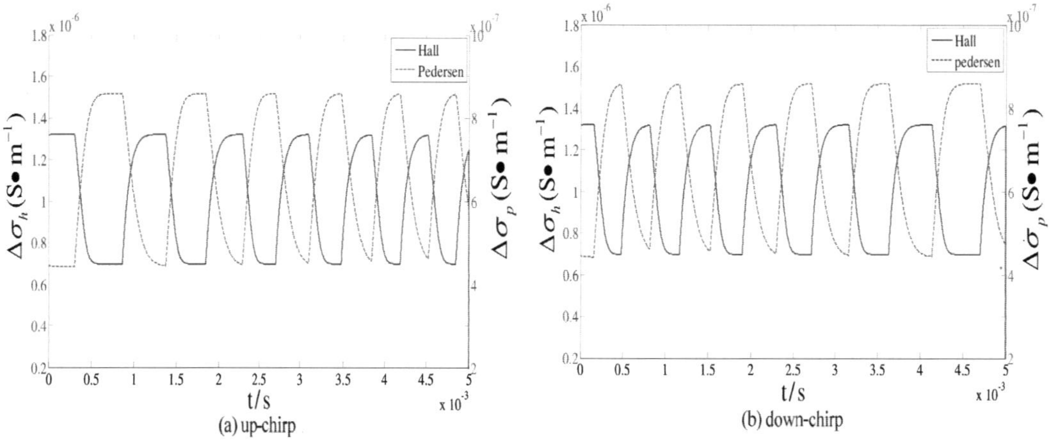

Figure 3. Pedersen and Hall conductivity variations with time (75km height)

3.2. BER Analysis

Broadband ELF channel noise is similar to gaussian noise, and the ELF band is often subject to power frequency harmonic interference, so the performance of QPSK, MSK and chirp-BOK under gaussian noise and harmonic interference are compared respectively. The simulation results are shown in Fig. 4.

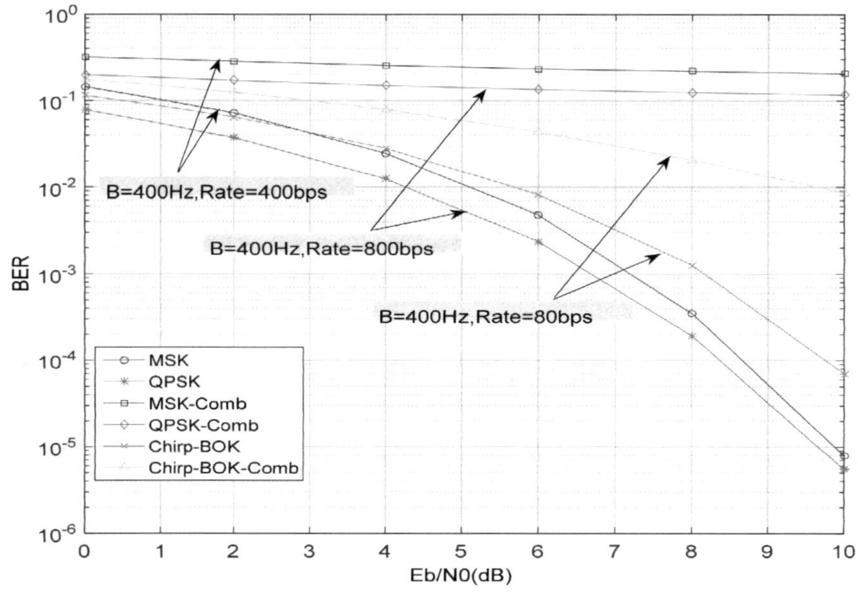

Figure 4. BER performance under gaussian noise and harmonic interference

It can be seen from the figure that MSK and QPSK are very sensitive to harmonic interference, and when using comb filter to filter the harmonic interference with bandwidth of 3Hz, the performance deteriorates sharply, while Chirp-BOK has the best BER performance under harmonic interference

4. Conclusion

If the ionosphere is periodically heated by high frequency high-energy waves, the temperature and current of the ionosphere will oscillate periodically and emit ELF/VLF waves. The up-chirp (symbol "0") and the down-chirp (symbol "1") modulation waveforms are designed and verified respectively with the ionospheric heating model. In this paper, the proposed Chirp-BOK modulation scheme can be used in ionospheric heating to generate ELF waves. Due to the use of chirp signal, Chirp-BOK has a high time-bandwidth product, so it has a strong anti-interference ability. Compared with QPSK and MSK, Chirp-BOK has better anti-interference performance under harmonic interference.

5. References

[1] S. P. Kuo, "Ionospheric very low frequency transmitter," *Physics of Plasmas*, vol. 22, p.022901, 2015.

[2] G. Jin, M. Spasojevic, M. B. Cohen, and U. S. Inan, "Utilizingnonlinear ELF generation in modulated ionospheric heating experiments for communications applications," *Radio Science*, vol.48, no. 1, pp.61–68, 2013.

[3] Xia, Li, et al, "Performance of Chirp Code-Orthogonalized Transmitted-Reference Ultra-Wideband system in presence of interference," *International Congress on Image & Signal Processing IEEE*, 2010.

[4] P. Stubbe and W. S. Varnum, "Electron energy transfer rates in theionosphere," *Journal of radio science (Chinese)*, no. 5, pp. 883-887,2008.

[5] Li Qingliang, Yang Jutao, et al, "ELF/VLF waves radiated by modulated HF heating ionospheric in mid-latitude and equatorial regions," *IEEE Microwave Magazine*, vol. 8, no. 2, pp.77-82, Apr. 2007.

Cylindrical Unidirectional Wave Transmission and Experiments Based on UWB-SWPL

Diying Wu[1,*], Yi Lu[1], Yunfeng Guo[1], Fanbo Meng[2], Dongdong Wang[2], Ran Li[1], and Xiangzhen Li[3]

Department of Material Science and Engineering, University of Bucharest, Faculty of Transport, room JC004, 313 Splaiul Independentei St., 6th sector, Bucharest, Romania.
[1]State Grid Liaoning Electric Power Co., Ltd. Shenyang Power Supply Company, China
[2]State Grid Liaoning Electric Power Co., Ltd., China
[3]DaJan Holding Group Electric Power Science & Technology Co., Ltd
Email: wudiying1974@163.com

Abstract. Current studies have found that electromagnetic waves propagate in both positive and negative directions on the surface of magneto-optical media, and even have unidirectional propagation characteristics. This unique physical phenomenon has attracted the attention of many scholars in the industry. Through the derivation and theoretical analysis of vector potential equation, scalar wave equation and dielectric wrapped conductor waveguide, the existing TM mode and TE mode are shown. Then the millimeter wave energy is effectively radiated by the horn antenna to achieve a high efficiency coupling device. The electromagnetic energy is gathered on the surface of the conductor and transmitted along the conductor. At last, the validation of surface wave transmission is realized through experiments, which proves its authenticity.

1. Introduction

At present, there are two methods to realize this kind of one-way electromagnetic surface wave. One is that Haldane and Raghu analogize the surface modes of quantum Hall effect in non-reciprocal photonic crystals to obtain this one-way effect [1]. The other is to realize the one-way effect by using the asymmetric dispersion relation of the mode of electromagnetic wave on the surface of magneto-optical medium or rotating magnetic medium.

Photonic crystal is a periodic medium which controls the propagation of electromagnetic wave in photonic forbidden band. In periodic medium, the description of photonic crystal is similar to that of electron wave in crystal, and can also be described by photonic band structure. Band gaps can also occur between photonic bands, i.e. gaps between them, but the light in the band gaps can not be transmitted in photonic crystals. In the forbidden band frequency range, the propagation of light wave can be controlled by introducing a specific defect.

Scholars have further studied unidirectional propagation [3], tunable power divider [4], broadband annulus [5], multimode unidirectional waveguide [6], which is composed of unidirectional waveguide, unidirectional coupling. In addition, the ferrite is used to realize the unidirectional propagation of electromagnetic waves on different structures of gyromagnetic photonic crystals. Slow waves in unidirectional waveguides are studied, and multi-mode unidirectional waveguides with Dachen's size are verified by experiments.

Unidirectional waves can also be realized based on the asymmetric dispersion relation of surface modes of electromagnetic waves on magneto-optic or rotating magnetic media. This asymmetric dispersion relation refers to the different interaction between left-handed and right-handed swirling waves in non-reciprocal media. When the two interactions are very different in a particular frequency band, there may be only one polarization mode. Mathematically, the asymmetric dispersion relation is derived from the asymmetric characteristic equation of surface modes with relative propagation constants. In a particular frequency band, the dispersion relation can have a single group velocity direction. In addition, there are also different methods to obtain such unidirectional surface waves, such as surface plasma in magnetic field, interface between two magnetic domains with reverse magnetic field [7], and waveguide composed of metal-dielectric-gyromagnetic medium [8]. Compared with the way of realizing unidirectional wave in photonic crystal, this way is easy to understand and can be realized through thin film material without periodic structure. However, there are many theoretical and experimental studies in this field. In this paper, through theoretical analysis, calculation and numerical simulation, a simple radiation and coupling device is designed to verify the transmission of surface wave.

2. Summary of Surface Wave Transmission

2.1. Vector Potential Equation and Scalar Wave Equation

2.1.1. Vector potential equation

According to the Maxwell equations in complex frequency domain, the following results are obtained:

$$\begin{cases} \nabla\times\overline{E} = -j\omega u\overline{H} \\ \nabla\times\overline{H} = j\omega\varepsilon\overline{E} \\ \nabla\bullet\overline{E} = 0 \\ \nabla\bullet\overline{H} = 0 \end{cases} \tag{1}$$

In the above formula, the right current source and charge source of the equation system are 0, because the analyzed region is passive (requiring that the medium characteristics are linear, isotropic and uniform, so operator ∇ can be independent of (ε,μ,σ)).

From the above formula, the vector potential equation of the passive region of the electromagnetic field is shown as follows, where $k^2 = \omega^2\varepsilon\mu$.

$$\nabla^2\overline{A} + k^2\overline{A} = 0 \tag{2}$$

$$\nabla^2\overline{F} + k^2\overline{F} = 0 \tag{3}$$

The solution of electromagnetic field in accordance with Maxwell equation (1) can be expressed as:

$$\overline{E} = -\nabla\times\overline{F} - j\omega\mu\overline{A} + \frac{1}{j\omega\varepsilon}\nabla(\nabla\bullet\overline{A}) \tag{4}$$

$$\overline{H} = \nabla\times\overline{A} - j\omega\varepsilon\overline{F} + \frac{1}{j\omega\mu}\nabla(\nabla\bullet\overline{F}) \tag{5}$$

Different from the corresponding sources $\overline{J}$ and $\overline{M}$ of $\overline{A}$ and $\overline{F}$, $\overline{A}$ and $\overline{F}$ are understood as the universal sources corresponding to electric field intensity and magnetic induction intensity. According to the superposition theorem of electromagnetic field, the solution of analytical region can be superposed by the solution of single source $\overline{A}$ and the solution of single source $\overline{F}$.

$$\begin{cases} \overline{E} = -j\omega\mu\overline{A} + \frac{1}{j\omega\varepsilon}\nabla(\nabla \bullet \overline{A}) \\ \overline{H} = \nabla \times \overline{A} \end{cases} \tag{6}$$

$$\begin{cases} \overline{H} = -j\omega\varepsilon\overline{F} + \frac{1}{j\omega\mu}\nabla(\nabla \bullet \overline{F}) \\ \overline{E} = -\nabla \times \overline{F} \end{cases} \tag{7}$$

Formulas (13) and (14) are generally applicable to passive and linear media. In the case of attenuated media, only the corresponding media constants need to be replaced by complex numbers. Such as: $\varepsilon = \varepsilon^{'} - j\varepsilon^{''}, \mu = \mu^{'} - j\mu^{''}$.

Based on $\overline{A}$ and $\overline{F}$ sources, there are two special cases as follows:

$$\overline{A} = \hat{e}_z\psi, \overline{F} = 0 \tag{8}$$

$$\overline{A} = 0,\ \ \overline{F} = \hat{e}_z\psi \tag{9}$$

These two cases correspond to TM mode and TE mode respectively. Formula (8) shows that only $\overline{A}$ is not zero. Because of $\overline{H} = \nabla \times \overline{A}$, the magnetic field is perpendicular to A. From formula (9), only $\overline{F}$ is not zero. Because of $\overline{E} = -\nabla \times \overline{F}$, the electric field and F are perpendicular to each other. By substituting formulas (8) and (9) into formulas (7) and (8), the solutions of electromagnetic fields in TM mode and TE mode can be obtained.

Formulas $\nabla^2\overline{A} + k^2\overline{A} = 0$ and $\nabla^2\overline{F} + k^2\overline{F} = 0$ can be further solved in Cartesian coordinates, and the Helmholtz equation is satisfied in each direction.

$$\nabla^2\psi + k^2\psi = 0 \tag{10}$$

Therefore, for the case of formula (8) above, it is only necessary to find ψ in the Z direction based on formula (10); the same is true for formula (9).

In passive, linear and uniform media, the electromagnetic field can be obtained by superposition of TM mode field and TE mode field. Because by properly choosing ψ, corresponding TM and TE can be constructed, such as: $\frac{\partial^2\psi^f}{\partial z^2} + k^2\psi^f = jw\varepsilon H_z$.

2.1.2. Solutions of helmholtz equation

The boundary and interface of the region studied in this paper have regular shape. After choosing the appropriate coordinate system, the equation can be simplified by the method of separating variables. Therefore, the cylindrical coordinate system should be chosen to simplify the description of boundary conditions, that is, the boundary conditions are expressed by simple functions. According to Laplacian operator: ∇^2, the concrete expression in cylindrical coordinate system is as follows:

$$\nabla^2\psi = \frac{1}{\rho}\frac{\partial}{\partial\rho}(\rho\frac{\partial\psi}{\partial\rho}) + \frac{1}{\rho^2}\frac{\partial^2\psi}{\partial\phi^2} + \frac{\partial^2\psi}{\partial z^2} \tag{11}$$

$$\psi = R(\rho)\Phi(\phi)Z(z) \tag{12}$$

The Helmholtz equation is substituted and sorted out:

$$\frac{1}{\rho R}\frac{\partial}{\partial \rho}(\rho\frac{\partial R}{\partial \rho})+\frac{1}{\rho^2 \Phi}\frac{\partial^2 \Phi}{\partial \phi^2}+\frac{1}{Z}\frac{\partial^2 Z}{\partial z^2}+k^2=0 \tag{13}$$

The general solution of the equation can be expressed as:

$$\psi=\sum_n\sum_{k_z}C_{n,k_z}\psi_{k_\rho,n,k_z}=\sum_n\sum_{k_z}C_{n,k_z}B_n(k_\rho\rho)h(n\phi)h(k_z z) \tag{14}$$

According to the boundary conditions of the solved region, the appropriate expression of the solution is selected. Then k_ρ,n,k_z is determined according to the boundary value. The constant C_{n,k_z} is determined by the source in the region. $\psi_{k_\rho,n,k_z}=B_n(k_\rho\rho)h(n\phi)h(k_z z)$ is the fundamental solution of the equation.

2.1.3. Dielectric wrapped conductor waveguide theory
The model of the dielectric wrapped conductor is shown in the following figure:

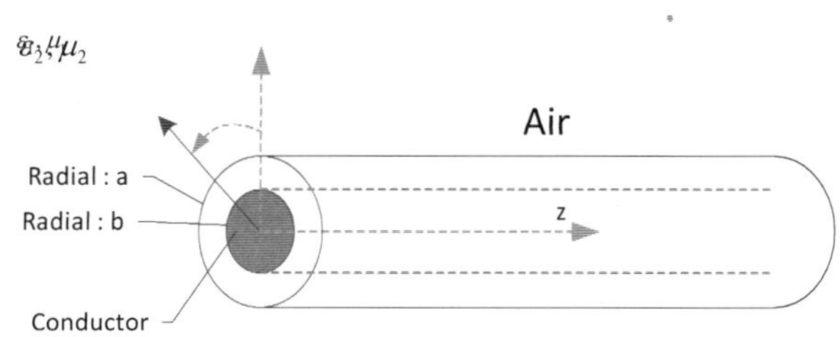

Figure 1. Diagram of dielectric wrapped conductor waveguide.

The inner layer is a conductor with a radius of b, which acts as an ideal conductor in the calculation below; the outer layer is a dielectric layer-region 1 with a radius of a and its dielectric constant is ε_1,μ_1; the outer layer is air-region 2, where the dielectric constant ε_2,μ_2 is used. TM and TE modes are analyzed respectively.

Because the innermost layer is an ideal conductor, its internal electromagnetic field is 0. Only 1/2 of the region needs to be analyzed. The basic solution of Helmholtz equation is chosen in the case of region 1:

$$\psi^{m1}=AB_n^{m1}(k_{\rho1}\rho)\cos(n\phi)e^{-jk_z z} \quad \text{(15) TM Mode}$$

$$\psi^{e1}=BB_n^{e1}(k_{\rho1}\rho)\sin(n\phi)e^{-jk_z z} \quad \text{(16) TE Mode}$$

In the case of Region 2:

$$\psi^{m2}=CB_n^{m2}(k_{\rho2}\rho)\cos(n\phi)e^{-jk_z z} \quad \text{(17) TM Mode}$$

$$\psi^{e2}=DB_n^{e2}(k_{\rho2}\rho)\sin(n\phi)e^{-jk_z z} \quad \text{(18) TE Mode}$$

Among them:

$$k_z^2+k_{\rho1}^2=k_1^2=\omega^2\varepsilon_1\mu_1 \tag{19}$$

$$k_z^2+k_{\rho2}^2=k_2^2=\omega^2\varepsilon_2\mu_2 \tag{20}$$

IOP Conf. Series: Materials Science and Engineering **719** (2020) 012052 doi:10.1088/1757-899X/719/1/012052

3. Design of Coupling Device and Dielectric Lens

3.1. Coupling Device

At present, the key technology to be tackled is the coupling device. Coupling devices are divided into tight-binding coupling device and extraction coupling device. Firstly, the tight-binding coupling device is designed and simulated in this paper.

According to the theory of electromagnetic binding, if the binding effect is to be formed in the corresponding region, the reactance boundary conditions need to be formed in this region. There are two ways: the boundary between two different dielectric constant media and the boundary between conductors.

The electromagnetic energy is bound to the power line, and the power line has a dielectric layer, which belongs to the coexistence of the conductor boundary and the dielectric boundary. However, when the incident angle of the electromagnetic wave is at a certain angle with the conductor, it is approximately mirror reflection at the boundary of the conductor; at this time, most of the energy will be reflected out.

3.2. Coupling Device

3.2.1. Design scheme

First of all, the dielectric lens is discussed, and the implementation framework is designed. The design scheme is as follows.

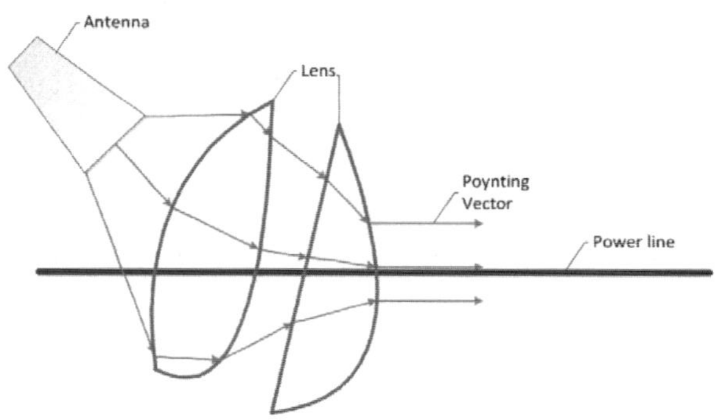

Figure 2. Working schematic diagram of lens aggregation.

The energy transmission path is shown in the red line above. A series of lenses (possibly a combination of lenses) are used to change the transmission path of the wave, eventually gathering the energy around the power line and forming a surface wave propagating along the line.

3.2.2. Lens shape selection

From the analysis of the horn antenna in front, it can be seen that after leaving the antenna aperture, the electromagnetic waveform forms a spherical wavefront, but gathers in a solid angle, and diffuses with the distance from the antenna aperture. Therefore, the incident wave of a lens can be seen as a wave propagating outward from a certain point to a certain solid angle. But using point source in simulation will not affect the result analysis.

4. Experimental Analysis

4.1. Experimental Principle

Power Line is illuminated directly by rectangular waveguide (simulated by metal rod). For the Power Line region, that is, the area between two rectangular waveguides, it is only the space area of the Power Line. After electromagnetic shielding around the waveguide, the electromagnetic field mode can not be leaked into space. Because the working frequency band is located in millimeter wave, the energy radiated from waveguide to space is concentrated in the range of its rays, so the direction of TX and RX ends is staggered, which will greatly avoid electromagnetic energy conducting RX ends directly from TX ends in the air. The signals received at the RX end can only arrive along the Power Line. According to the theoretical analysis of the above dielectric wrapped conductor, the electromagnetic energy transmitted must conform to the solution of Maxwell equation system in the area outside Coupling Cavity - the situation that dielectric wrapped conductor exists in passive, linear and uniform area, that is, the electromagnetic field forms the transmission effect of surface wave along the power line.

4.2. Experimental Result

Table 1. Test result.

Connection state	Receiving power	Attenuation	Transmission attenuation	picture	Remarks
Direct connection	-22.16	22.16	0	5,6	1.419GHz
diagonally	-87.57	87.57	65.41	7,8	1.3GHz
diagonally+PCB board+medium	-69.62	69.62	47.46	9,10	1.3GHz
horn diagonally+Metal mesh	-88.45	88.45	66.29	11,12	1.3GHz
horn diagonally+Metal mesh+PCB board+medium	-93.07	93.07	70.91	13,14	1.3GHz
horn diagonally+Metal mesh+Aluminum tube+medium	-100.98	100.98	78.82	15,16	1.3GHz

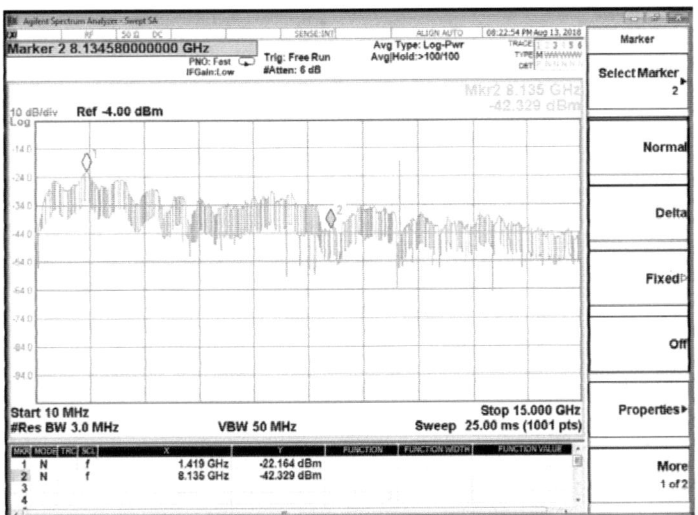

Figure 3. Experimental results.

As shown in the figure above, the higher the frequency, the greater the loss of frequency conversion, of which 61.4 GHz frequency conversion loss is the smallest, so it is recommended to use about 61 GHz frequency band for testing, which is more efficient.

As shown in the table above, when the horn antenna is obliquely aligned, the received signal of the spectrum analyzer is caused by the reflection of the test environment, i.e., the transmission loss is caused by the propagation and reflection loss of the environment; when the PCB board is placed in front of the horn antenna, the transmission loss is reduced, and it is suspected that the reflection of the PCB board to the high frequency signal increases between the up and down converters. Shielding network is added to block the reflection signal of the gap; when the shielding network is added, when PCB board or aluminium tube is placed in front of the horn antenna, the transmission loss increases, indicating that the power line surface wave has not formed, but reflects the signal of the gap, which is the increase of the transmission loss of the signal in the environment.

4.3. Analysis of Test Results

The surface wave phenomenon was not detected in the test results. The experimental principle is analyzed: (1) The energy is directly reflected by the rectangular waveguide irradiating the metal or dielectric-coated conductor at different incident angles, but not gathered on the surface of the conductor. (2) The radiation efficiency of rectangular waveguide is low, and most of the energy can not be transmitted effectively. (3) Metal rods have no dielectric layer, or the dielectric layer is too thin to bind most of the energy to the surface of conductors.

According to the theoretical and simulation analysis, the directions to be improved are: 1) using Power Line with dielectric layer thickness to test; 2) improving the energy coupling device.

5. Summary

In this paper, through theoretical analysis, calculation and numerical simulation, a simple radiation and coupling device is designed to verify the transmission of surface wave.

However, the results of this stage are based on simple experimental devices, and the transmission of real power lines needs to be further improved in terms of test configuration, excitation and coupling devices, and the performance of UWB signal transceiver needs to be further tested. The coupling device between transmitter and receiver is very important, which involves antenna technology, wave aggregation technology, dielectric construction technology and so on. Next, the direction of realization can be defined as using dielectric lens to collect and extract electromagnetic waves.

6. References

[1] Haldane, E D. M. &Raghu, S. Possible realization of directional o Vical waveguides in photonic crystals with broken time-reversal symmetry. Phys. Rev. Lett., 2008, 100(1): 13904.
[2] Ao, X, Lin, Z. & Chan, C. One-way edge mode in a magneto—optical honeycomb photonic crystal[J].Phys.Rev.B,2009,80(3):033105.
[3] Fang, K., Yu,Z. & Fan, S. Microscopic theory of photonic one-way edge modem Phys.Rev.B,2011,84(7):075477.
[4] Qiu, W., Wang, Z.& Soljabi, M. Broadband circulators based on directional coupling of one way wave guides[J].Opt.Express,2011,9(22):48-57.
[5] Hartstein, A., Burstein, E ., Maradudin, A., Brewer, R. &Wallis, R. Surface polaritons on semi—infinite gyromagnetic media[J]. J. Phys. C:Solid State Plays.6,1266-1276.
[6] Shen, L., You.Y, Wang, Z. & Deng, X. Backscattering-immune one-way surface magneto plasmons at terahertz frequencies[J].opt.Express,2015,23(2):950-962.
[7] Davoyan, A.& Engheta, N. Electrically controlled one—way photon flow in plasmonic nano structures[J].Nature Communications,2014,5:5250.
[8] Zhang, X., Li, W. & Jiang, X. Confmed one—way mode at magnetic domain wall for broadband high-efficiency one-way waveguide, splitter and bender[J].Appl. Phys. Lett., 2012,100(4):041108.

A Carrier Frequency Offset Estimation for OFDM System via Local Searching Capon Algorithm

Xi Chen[1], Gaofeng Zhao[2] and Changbo Ye[3]

1. Department of Electronic Information Engineering, Nanjing University of Aeronautics and Astronautics, Nanjing, China
2. China Mobile Group Jiangsu Company Limited, 21000, Nanjing, China
Email: [1]njxnd88@126.com, [2]zhaogaofeng@nuaa.edu.cn, [3]2714985990@qq.com

Abstract. Orthogonal frequency division multiplexing (OFDM) technology is an important modulation technology in wireless communication. One of the main disadvantages of OFDM system is that it is very sensitive to carrier frequency deviation. Therefore, estimation of carrier frequency deviation is one of the key technologies of OFDM system. In this paper, the global search Capon algorithm is improved, and a local search Capon algorithm is proposed. Simulation results show that the algorithm has low complexity and high performance. In the local search, P spectral peaks are superimposed, which can effectively overcome the problem that the estimation performance of traditional Capon algorithm is seriously reduced due to the unapparent spectral peaks or the missing spectral peaks in the case of low SNR., and it has a good application prospect in power wireless OFDM communication system. Since contemporary information-retrieval systems rely heavily on the content of titles and abstracts to identify relevant articles in literature searches, great care should be taken in constructing both.

1. Introduction

OFDM, started in the 1990s. The characteristic of this technology is to make full use of the orthogonality of subcarriers to carry out multi-carrier modulation. At present, OFDM system has been widely used in HDTV signal transmission, digital video transmission, wireless LAN and other fields, and applied in wireless wide area network and the fourth generation cellular communication network [1-2]. Compared with other systems, OFDM system has many advantages, such as high spectrum utilization and good anti-frequency selective fading ability [3]. However, OFDM system is particularly strict on the orthogonality of each subcarrier self-test [4-5]. Any point of carrier frequency offset (CFO) will affect the orthogonality between subcarriers. Therefore, the CFO in OFDM system needs to be accurately estimated and compensated to ensure the performance of the system. CFO estimation methods currently fall into two categories: non-blind CFO estimation and blind CFO estimation [6-7]. Non-blind CFO estimation methods include pilot based method and cyclic prefix based method. Blind CFO estimation adopts traditional spectral estimation methods, such as MUSIC algorithm, ESPRIT algorithm based on rotation-invariant signal parameters estimation, and CFO estimation algorithm based on kurtosis. Compared with the non-blind CFO estimation method, blind CFO estimation method has higher frequency band utilization and some algorithms have higher CFO estimation performance [8-9].

Compared with the non-blind estimation method, the global search Capon algorithm involved in this paper has better estimation performance, but has higher computational complexity. The local Capon algorithm proposed in this paper is improved on the basis of global search, and only one subcarrier is searched, so it has lower computational complexity. According to the simulation results, when the SNR is high, the local Capon algorithm proposed in this paper has the CFO estimation performance close to that of the global search Capon algorithm on the basis of low complexity. In the case of low SNR, the CFO estimation performance of global search Capon algorithm is improved.

2. System Model

Consider an uplink OFDM system containing N subcarriers, P of them are used for data transmission, and the rest N-P ones are virtual carriers. A looped prefix with length L is used to eliminate inter channel interference caused by multipath effect, where L is greater than the maximum delay of signal. After inserting the cyclic prefix, the signal is transmitted through multiple weak channels. After CP is removed, the output signal of the receiver can be denoted as

$$\boldsymbol{x}(k) = \boldsymbol{E}\boldsymbol{F}_P diag(\boldsymbol{h})\boldsymbol{s}(k)e^{j2\pi\Delta f(k-1)(N+L)} \tag{1}$$

Where $\boldsymbol{E} = diag(1, e^{j2\pi\Delta f/N}, \ldots, e^{j2\pi\Delta f(N-1)/N})$ is the CFO matrix, Δf is CFO, $\boldsymbol{F}_P$ consists of the first P columns of the inverse discrete Fourier transform Matrix. The channel vector of frequency is $\boldsymbol{h} = [\boldsymbol{H}(1), \boldsymbol{H}(2), \ldots, \boldsymbol{H}(P)]^T$, $\boldsymbol{H}(n) = \sum_{l=0}^{L_m-1} h(l)e^{-j2\pi nl/N}$. $\boldsymbol{s}(k) = [s_1(k), s_2(k), \ldots, s_P(k)]^T$ is the kth symbol of transmit signals. According to (1), it follows that

$$\boldsymbol{X} = \boldsymbol{A} diag(\boldsymbol{h})\boldsymbol{B}^T \tag{2}$$

Where $\boldsymbol{B} = diag(1, e^{j2\pi\Delta f(N+L)}, \ldots, e^{j2\pi\Delta f(K-1)(N+L)})\boldsymbol{S}$; $\boldsymbol{S} = [\boldsymbol{s}(1), \boldsymbol{s}(2), \ldots \boldsymbol{s}(K)]^T$ is all K data blocks transferred by P channels. The vandermonde Matrix **A** is

$$\boldsymbol{A} = \boldsymbol{E}\boldsymbol{F}_P = \begin{bmatrix} 1 & 1 & \cdots & 1 \\ e^{j2\pi\left(\frac{0}{N}+\Delta f\right)} & e^{j2\pi\left(\frac{1}{N}+\Delta f\right)} & \cdots & e^{j2\pi\left(\frac{P-1}{N}+\Delta f\right)} \\ \vdots & \vdots & \ddots & \vdots \\ e^{j2\pi(N-1)\left(\frac{0}{N}+\Delta f\right)} & e^{j2\pi(N-1)\left(\frac{1}{N}+\Delta f\right)} & \cdots & e^{j2\pi(N-1)\left(\frac{P-1}{N}+\Delta f\right)} \end{bmatrix} \tag{3}$$

In the case of noise, the received signal can be denoted as

$$\widetilde{\boldsymbol{X}} = \boldsymbol{A}\boldsymbol{S} + \boldsymbol{W} \tag{4}$$

Where **W** is the noise in the receiving process, $\boldsymbol{S} = diag(\boldsymbol{h})\boldsymbol{B}^T$.

3. Local Search Capon Algorithm for CFO estimation

3.1. Calculate the Estimated Value of the Signal Covariance Matrix

The estimated value of the covariance matrix is as follows

$$\widehat{\boldsymbol{R}}_x = \widetilde{\boldsymbol{X}}\widetilde{\boldsymbol{X}}^H/K \tag{5}$$

The basic algorithm of this paper is global search Capon algorithm, which is as follows:
The Capon algorithm is also known as the least variance method(MVM), the optimization problem can be expressed as

$$\varphi = arg\min_{\omega} P(\omega) \tag{6}$$

The constraint conditions are, $\omega^H \boldsymbol{a}(\varphi) = 1$, $P(\omega) = E[\boldsymbol{Y}^2] = \omega^H E[\boldsymbol{X}\boldsymbol{X}^H]\omega = \omega^H \boldsymbol{R}\omega$ is the power of the output signal.

The optimal weighted vector can be solved by Lagrange multiplier method, the result is

$$\omega_{Cap} = \frac{\hat{\boldsymbol{R}}_x^{-1}\boldsymbol{a}(\varphi)}{\boldsymbol{a}^H(\varphi)\hat{\boldsymbol{R}}_x^{-1}\boldsymbol{a}(\varphi)} \tag{7}$$

Where $\widehat{\boldsymbol{R}}_x$ is the received signal covariance matrix. For the data model in (4), its data covariance matrix can be estimated by $\hat{R}_x = \tilde{X}\tilde{X}^H/K$. Combining equation (6) and constraint conditions, the spatial spectrum of Capon algorithm can be obtained, and the CFO estimate can be obtained by searching the spectral peak of the spectral function

$$P_{Capon}(\varphi) = \frac{1}{\boldsymbol{a}^H(\varphi)\hat{\boldsymbol{R}}_x^{-1}\boldsymbol{a}(\varphi)} \tag{8}$$

Where $a(\varphi) = \left[1, e^{j2\pi\varphi}, \dots, e^{j2\pi(N-1)\varphi}\right]$. The global spectral peak search is carried out for the spectral function, search area is $[0,2\pi P/N]$, P spectral peaks can be obtained, and the corresponding frequency offset respectively is$(\frac{0}{N} + \Delta f, \frac{1}{N} + \Delta f, \dots, \frac{P-1}{N} + \Delta f)$.

The estimated true frequency deviation can be calculated from the corresponding frequency deviation values of P spectral peaks

$$\Delta f = \frac{\sum_{i=1}^{P} \phi_i - \sum_{j=1}^{P} \frac{i-1}{N}}{P} \tag{9}$$

Where ϕ_i is the frequency deviation value corresponding to the *i*th spectral peak.

3.2. CFO Estimation

In order to solve the problem of traditional global search Capon algorithm, we can use an improved local search approach for CFO estimates. The CFO corresponding to P spectral peaks can be expressed as

$$\phi_p = 2\pi\left(\Delta f + \frac{p-1}{N}\right), p = 1,2, \dots, P \tag{10}$$

So, similar to (8), we can construct the following spectral function to obtain CFO estimate through local search

$$P_{LS-Capon} = \frac{1}{\sum_{p=1}^{P} \boldsymbol{a}(\varphi + \frac{p-1}{N})^H \hat{\boldsymbol{R}}_x^{-1} \boldsymbol{a}(\varphi + \frac{p-1}{N})} \tag{11}$$

Where $\boldsymbol{a}(\varphi + \frac{p-1}{N}) = \left[1, e^{j2\pi(\varphi + \frac{p-1}{N})}, \dots, e^{j2\pi(N-1)(\varphi + \frac{p-1}{N})}\right]$. The spectrum peak search range is $[0,2\pi/N]$.

It can be seen from (10) and (11), the search range of local Capon algorithm is greatly reduced, so the computation goes down. In addition, since the spectral function of this algorithm is superimposed by P spectral peaks, even if the individual spectral peaks are not obvious or the spectral peaks are missing, the true frequency deviation of the obtained spectral function still corresponds to the unique spectral peak in the case of low SNR.

The steps for local search of Capon CFO estimation are as follows:

Step 1. Calculate the estimated value of the signal covariance matrix $\hat{R}_x = \tilde{X}\tilde{X}^H/K$.

Step 2. According to equation (10), CFO estimate can be obtained through local spectral peak search.

Advantages of the algorithm:

(1) this algorithm adopts a local spectral peak search with a sub-carrier range, while the traditional Capon algorithm adopts a global search with all P sub-carriers, and a larger search range leads to a larger operation.

(2) this algorithm inherits the high estimation performance of traditional Capon algorithm and is superior to ESPRIT algorithm.

(3) this algorithm superimposes P spectral peaks in the local search, which can effectively overcome the problem that the estimation performance of traditional MUSIC algorithm is seriously reduced due to the lack of spectral peaks or spectral peaks in the case of low SNR.

3.3. Complexity Analysis

The complexity of each algorithm is shown in table 1, where N and P are the number of subcarriers and the number of channels used by the system respectively, and K represents the number of snapshots. n_g represents the number of searches by the MUSIC algorithm or Capon global search, n_l, represents the search times of local search Capon algorithm.

Table 1. Complexity of different methods

algorithm	complexity
PM	$O(KN^2 + PN(P+N) + 3P^2(P-1) + P^3)$
ESPRIT	$O(KN^2 + N^3 + 3P^2(N-1) + P^3)$
MUSIC	$O(KN^2 + N^3 + n_g(2N^2 + N))$
Capon	$O(KN^2 + n_g N^3)$
Local search Capon	$O(KN^2 + n_l(N^3 + N^2P + P^2N)$

General Capon algorithm adopts global spectral peak search, n_g is large and complex. The Capon algorithm proposed in this letter adopts local spectral peak search, n_l is much less than n_g, greatly reducing the complexity of the algorithm.

4. Simulation Results

In order to verify the effectiveness of the algorithm and describe the Angle estimation performance of the algorithm, Monte Carlo simulation is adopted in this section to simulate the algorithm, and the simulation times are all 1000. The definition of root mean square error (RMSE) is shown in equation (1.12).

$$RMSE = \sqrt{\frac{1}{M}\sum_{m=1}^{M}\left(\frac{\Delta\hat{f}^m - \Delta f}{1/N}\right)^2} \quad (12)$$

Figure 1 is the frequency offset estimation graph corresponding to the global search Capon algorithm when SNR=0dB. Figure 2 is the frequency offset estimation graph corresponding to the global search Capon algorithm when SNR=20dB. Among them, the frequency offset Δf=0.4, the subcarrier number N = 32, number of transmission channels P=20, cyclic prefix number L=8.

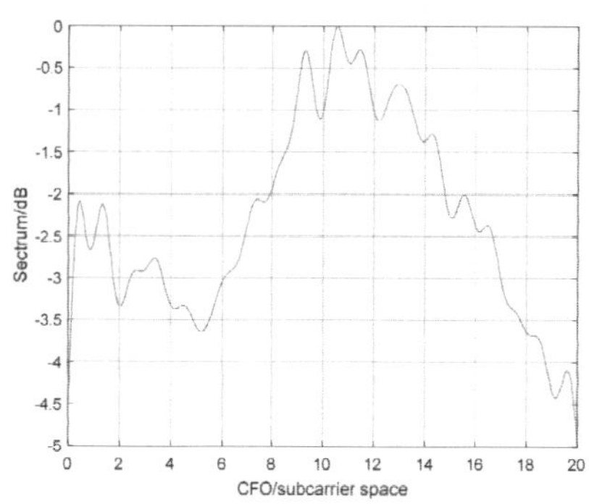

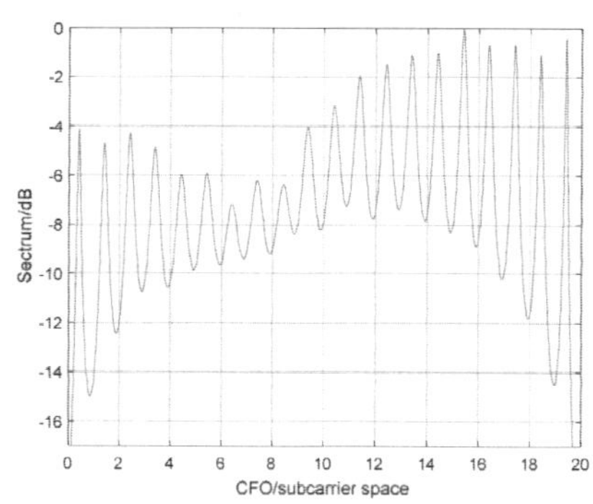

Figure 1. Global search Capon with SNR=0dB **Figure 2.** Global search Capon with SNR=20dB

Figure 3 is the frequency offset estimation diagram corresponding to the local search Capon algorithm when SNR=0dB. Among them, the frequency offset Δf=0.4, the subcarrier number N=32, number of transmission channels P=20, cyclic prefix number L=8. Figure 4 is the frequency offset estimation diagram corresponding to the local search Capon algorithm when SNR=20dB. Among them, the frequency offset Δf=0.4, the subcarrier number N=32, number of transmission channels P=20, cyclic prefix number L= 8.

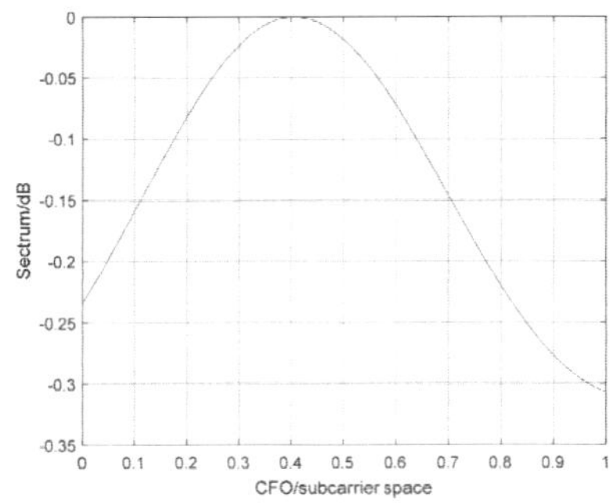

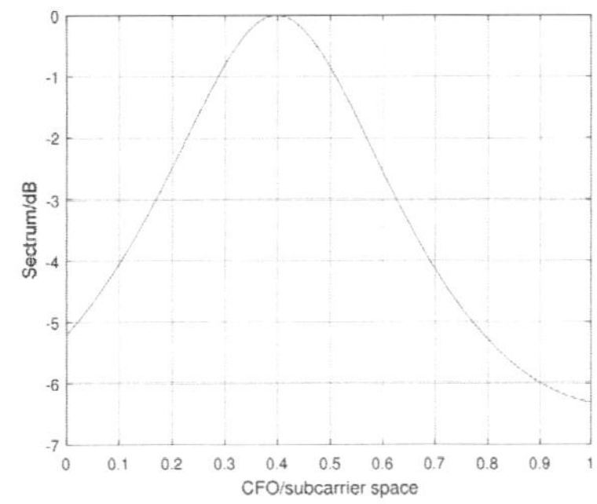

Figure 3. Local search Capon with SNR=0dB **Figure 4.** Local search Capon with SNR=20dB

Figure 5 is the performance comparison diagram of the local search Capon algorithm, ESPRIT algorithm and PM algorithm in this paper. According to the figure, the estimated performance of the algorithm in the present invention is much higher than that of the other two algorithms. Where, N=32, P=20, L=8, K=200, and the number of simulation times is 500. Figure 6 shows CFO estimation performance under different CFO scenarios. According to the figure, the local search Capon algorithm has very similar CFO estimation performance for different CFO. Where N = 32, P =20, K = 200,SNR=20dB and CFO range $[0, \omega]$, $\omega = 2\pi/N$ is normalized subcarrier space.

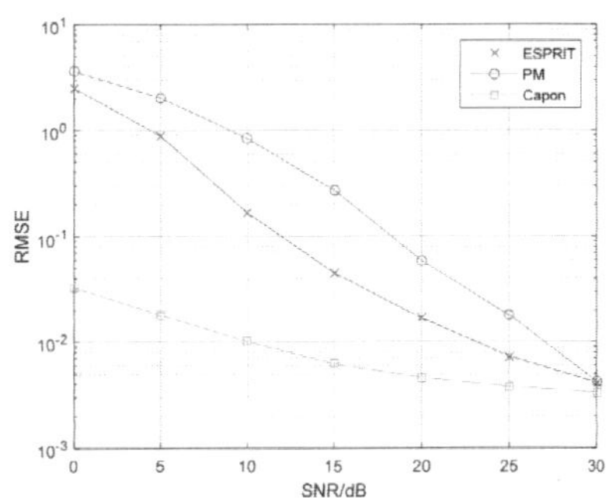

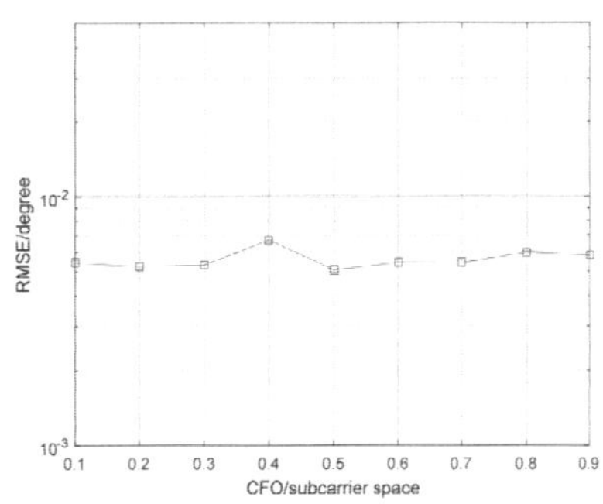

Figure 5. CFO estimation of different algorithms **Figure 6.** CFO estimation with different CFOs

Figure 7 to figure 9 are performance graphs of local search Capon algorithm under different parameters. Figure 7 and figure 8 respectively show the estimation performance of local search Capon algorithm under different subcarrier number N and different snapshot number K. The estimation performance of CFO improves with the increase of K and N. Figure 9 shows the estimated performance of the algorithm under different channel number P, and it can be seen that the performance of CFO algorithm decreases with the increase of P. As the number of channels increases, the interference between channels increases, leading to the CFO estimation of performance.

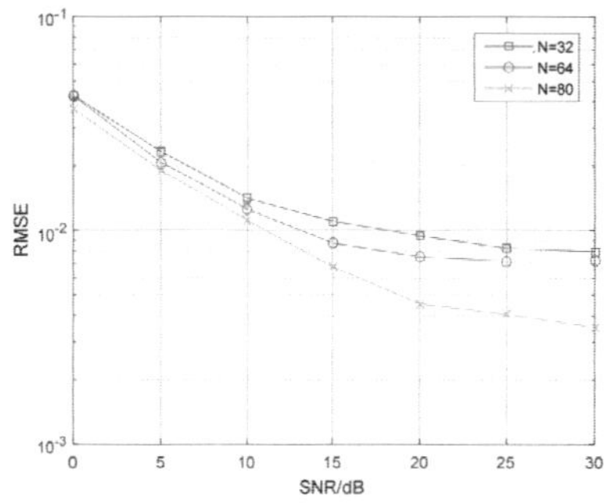

Figure 7. CFO estimation performance versus M

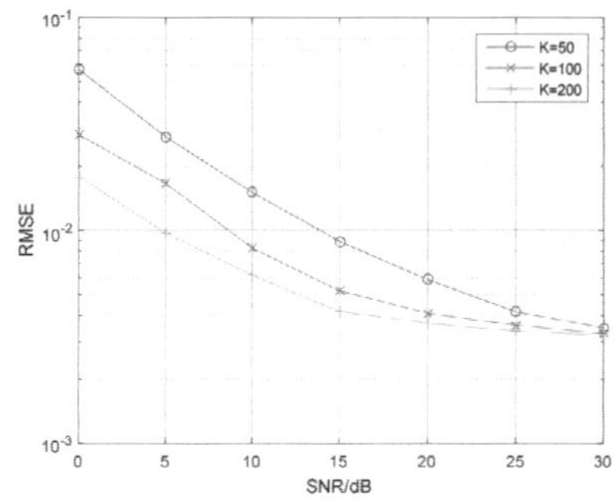

Figure 8. CFO estimation performance versus K

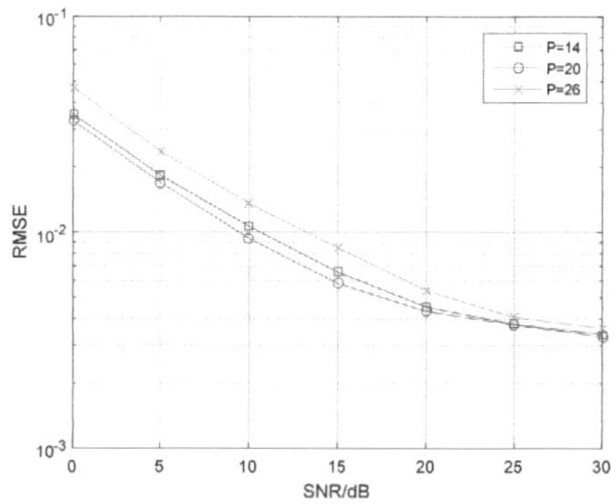

Figure 9. CFO estimation performance versus P

5. Conclusions

In this letter, we present a carrier frequency offset estimation algorithm for OFDM System by utilizing local search Capon. The algorithm adopts local spectral peak search, and the search range is a sub-carrier, which makes the search range small and the operation amount greatly reduced compared with the traditional Capon algorithm. And its estimated performance is better than ESPRIT algorithm. In addition, the superposition of P spectral peaks in local search can effectively overcome the problem that the estimation performance of traditional Capon algorithm is seriously reduced due to the lack of spectral peaks or the absence of spectral peaks in the case of low SNR. Simulation results show that the algorithm is effective.

6. References

[1] Weile Zhang, Feifei Gao, Bobin Yao. Blind CFO estimation for multiuser OFDM uplink with large number of receive antennas[J]. 2016 IEEE International Conference on Acoustics, Speech and Signal Processing (ICASSP), 2016.

[2] T. Pollet, M. Van Bladel, M. Moeneclaey. BER Sensitivity of OFDM Systems to Carrier Frequency Offset and Wiener Phase Noise[J]. IEEE Transactions on Communications, 1995, 43: 191-193.

[3] Jun-Han Oh, Jin-Goog Kim, Jong-Tae Lim, Blind Carrier Frequency Offset Estimation for OFDM Systems with Constant Modulus Constellations[J]. IEEE Communications Letters, 2011,15(9): 971-973.

[4] Xu Wang, Bo Hu. A Low-Complexity ML Estimator for Carrier and Sampling Frequency Offsets in OFDM Systems[J]. IEEE Communications Letters, 2014, 18(3): 503-506.

[5] Jaechan Lim, Daehyoung Hong. Gaussian Particle Filtering Approach for Carrier Frequency Offset Estimation in OFDM Systems[J]. IEEE Signal Processing Letters, 2013, 20(4): 367-370.

[6] L. Wu, X-D. Zhang, P-S. Li, Y-T. Su. A closed-form blind CFO estimator based on frequency analysis for OFDM systems[J]. IEEE Transactions on Wireless Commun, 2009, 57(6): 1634-

1637.
[7] Y. Yao, G. B. Giannakis. Blind Carrier Frequency Offset Estimation in SISO, MIMO, and Multiuser OFDM Systems[J]. IEEE Transactions on Communications, 2005, 53(1): 173-183.
[8] L. Wu, X.-D. Zhang, P.-S. Li, Y.-T. Su. A closed-form blind CFO estimator based on frequency analysis for OFDM systems[J]. IEEE Transactions on Communications, 2009, 57(6): 1634-1637.
[9] Xiaofei Zhang, Dazhuan Xu. Blind CFO estimation algorithm for OFDM systems by using generalized precoding and trilinear model[J]. Journal of Systems Engineering and Electronics, 2012, 23(1): 10-15.

IOP Conf. Series: Materials Science and Engineering **719** (2020) 012054 doi:10.1088/1757-899X/719/1/012054

Research on Multi-UAV Networks in Disaster Emergency Communication

Jianguo Zhou [1], Jintao Yang [1] and Lu Lu[1]

[1]The School of Electronic Information, Wuhan University, Wuhan, 430072, China
Email: zjg@whu.edu.cn

Abstract. Unmanned Aerial Vehicles (UAVs) are an emerging technology that can be used in military, public, and civilian applications. Military use of UAVs is more than 25 years old, mainly for border surveillance, reconnaissance and so on. UAVs can do the job more efficiently and cost-effectively than with a single UAV system. In disaster emergency communications system, UAVs system can be deployed rapidly and its coverage can be dynamically adjusted. It provides network support for disaster site rescue in a timely and efficient manner, and real-time information of disaster scene can be returned to the Rescue center to better rescue the affected people. This dissertation proposed SDN-based disaster emergency UAVs networks to realize the flexible deployment and management of high dynamic dis-aster area networks. Meanwhile, it has a significantly improvement in network lifetime and node switching time comparing with Optimized Link State Routing protocol (OLSR) and Ad hoc On-demand Distance Vector Routing protocol (AODV).

1. Introduction

Large-scale natural disasters often test the most basic survival instinct of human beings by causing huge and unpredictable losses of human life and property. Various types of natural disasters, such as geological disasters (earthquakes, volcanic eruptions, landslides and avalanches), hydrological disasters (floods, river dam breaks and mine intakes) and meteorological disasters (tropical storms, hurricanes, dust storms and heavy rain), result in many deaths [1]. Many people are trying to figure out how to respond to a natural disaster in a timely and effective way from the beginning, so as to quickly reduce the loss, to solve problems such as the rescue of victims and restore normal conditions as soon as possible.

Other When a disaster occurs, people tend to organize themselves to build a disaster relief command center and a real-time disaster observation platform to effectively direct disaster relief workers to carry out disaster relief work in real time according to the dis-aster situation. Thus, the communication between the disaster relief command center and the rescue personnel is particularly important in the whole process of disaster relief, and the commercial network (such as network operators) in the disaster area is often destroyed by the disaster.

When ground communications are not possible (such as earthquake recovery, forest fires, etc.), the use of UAV networks, also known as Unmanned Aeronautical Ad-hoc Network (UAANET) [2]. The UAV network could be deployed in disaster scenes and other critical scenarios to provide communications services to ground nodes that could be people carrying portable equipment such as smartphones and personal computers. In addition, a UAV network can be deployed at the object location to monitor the disaster zone, so that rescuers can search and track people or animals in need. Moreover, the UAV network can span multiple square kilometres and be able to adapt to changes on the ground. Since the UAV can move around the region on demand, the use of UAV networks in emergency situations is flexible and scalable.

Recently, more and more unmanned aerial vehicle (UAV) network display and application show that it can provide a wire-less network set up quickly and large coverage, but the survival time of UAV network nodes generally shorter, especially small rotary-wing UAV system, low prices for rapid deployment of emergency communication much lower costs UAV network has brought certain difficulties, and making it easier for the network backbone nodes drops due to run out of battery, and make the network topology changes frequently and make the network coverage unstable.

In order to enhance the network stability and stabilize the coverage of emergency networks for disaster area in the front of disaster area and the rear of the command, this paper proposes an emergency communication multi-UAV network scheme based on SDN. In our method, UAV creates a backbone network with variable topology, which has strong advantages in network lifetime time, service response time, etc., and can provide long-term stable network coverage service in disaster emergency communication network construction. To this end, the rest of this article is organized as follows. Section 2 discusses and proposes an emergency communication multi-UAV network scheme based on SDN. Section 3 presents the proposed scheme and performance evaluation of two common routing protocols. Finally, section 4 summarizes and prospect of the future work.

2. Multi-UAV Network Architecture Based on SDN

At present, the research on integrating SDN into the network of drones is still underway, and many of the most critical new network architectures have been proposed. In the control layer, application layer and data layer, the deployment of the three layers is mostly the same. The biggest difference is the key deployment in the control layer. In the early days, the east-west interface in the SDN controller has not been carefully studied, which has deployed on the ground station.

NOCC (Network Operation Control Centre) is the application layer of SDN. It is mainly responsible for route calculation and service-driven policy generation. The controller is deployed on the ground, transforming instructions from the upper NOCC into simple data structures (such Open Flow's flow table) and sending them to the underlying infrastructure layer [3]. Many drones organize the infrastructure layer of the network architecture, and have their network coverage areas, in which ground nodes and other drones can directly interact with the drone network. The UAV network architecture distributes routing policies through ISLs, or configuration instructions for network devices.

To achieve real-time and flexible central control of the UAV network, routing tables and configuration policies are uploaded to all UAVs via single-sided ISL forwarding or broadcast. At the same time, the agent on the drone continuously collects the link information of the UAV coverage area, including the remaining link bandwidth, power status and payload status, and sends it to the control layer drone above it. The control layer drone (same as the high altitude drone) then sends status information directly to the ground. All information obtained will eventually be processed on the NOCC.

For single-layer ISL forwarding, we can reduce the cost of redundant broadcast packets by using the shortest path tree and the multicast tree. These trees are included in the snapshot routing table obtained in the last update. The UAVs in the middle and low altitudes are constantly rotating and moving, and the UAV links and the UAV-to-Ground links also change. These changes require an update to the configuration, but there will be some delay in the update [4]. It is therefore difficult to decide the correct time to switch to the updated routing table and configuration of the entire network. Therefore, in this architecture, the control layer drone is used to broadcast the entire network for network update. By taking advantage of the wide coverage of the control layer drones, only three control layer drones can coverage almost 12 square kilometres of, so updates can be sent to the control layer drones. After the current control layer drone receives the update information, it sends it to the other two control layer drones and then sends them all to the covered medium/low altitude drone. We found that control layer drone broadcast updates are significantly faster and easier than ISLs forwarding. However, when we only need to update a small number of drones, the broadcast properties of the control layer drone will interfere with many other drones at the same time, because these work waste bandwidth [5]. Therefore, the control layer drone is responsible for broadcasting large-scale updates, while the small-scale configuration update is more suitable for single-layer drones' link forwarding. The hybrid control of the control layer drone and the medium/low altitude drones greatly reduces the update time and reduces the dependence on the ground nodes [6].

2.1. NOCC: Application Layer

NOCC acts as an application layer, running a large number of computing and control applications. The track of the drone is predictable. The NOCC can calculate the routing strategy of the UAV network and the configuration strategy of the network device by using the predicted data and collecting the link state information transmitted from the UAV. As the most important part, the routing method uses a simple and controllable snapshot routing method. This method divides the operating cycle of the system into a number of small time segments, and each time segment corresponds to changes in the network topology, particularly the destruction or reconstruction of the inter-UAVs link caused by the drone motion. During the divided time period, the network topology can be considered as unchangeable, and each time period becomes a snapshot. Based on the topology information, the NOCC calculates the entire network routing path using the current network traffic status information and then uploads it to the drone through the ground node.

2.2. Ground node: Control Layer

The ground node acts as a controller in the SDN framework and is separate from the NOCC for a clearer and simpler hierarchy. The configuration strategy calculated by the NOCC cannot be directly used for the underlying physical device, and needs to be converted into a data structure in the OpenFlow protocol through the ground node. The transformation is critical to the open interface design of the infrastructure layer, so drones can provide great flexibility and complex functional structures through simple implementation. However, various protocol conversions and open interface design make the controller more complex than the other two layers. The motion state of the drone can be monitored by the ground node, combined with the UAV network status information received from the drone. This information is very important for NOCC to construct a global network view and calculate the transmission rules, but the information must transform from ground nodes. The controller is deployed on multiple ground nodes and runs a coherent protocol to avoid single node fault and to ensure a consistent global network view between these nodes.

2.3. Drone: Infrastructure Layer

In traditional drone networks, drones are the most expensive to develop and the most complex facility. In the SDN-based UAV architecture, the functionality of the UAV in the architecture can be simplified. They receive user management policies, hardware configuration information and updated new routing tables from the ground station to enable NOCC deployment on the UAV network. During the operation of the network, the drone transmits the UAV data and network status information collected in real time back to the NOCC, so that the NOCC constructs the topology of the UAV network. Simplification on drones reduces drone development and production costs, saving resources. The software-defined mode simplifies management and makes the network more flexible and controllable.
Due to the development of the east-west interface of SDN control, the controller deployment is gradually enriched. It is proposed that the controller is mainly deployed in the cluster on the high-level UAV and ground nodes. On the basis of the single-controller UAV network, the controller cluster UAV network architecture is shown in Figure 1.

IOP Conf. Series: Materials Science and Engineering **719** (2020) 012054 doi:10.1088/1757-899X/719/1/012054

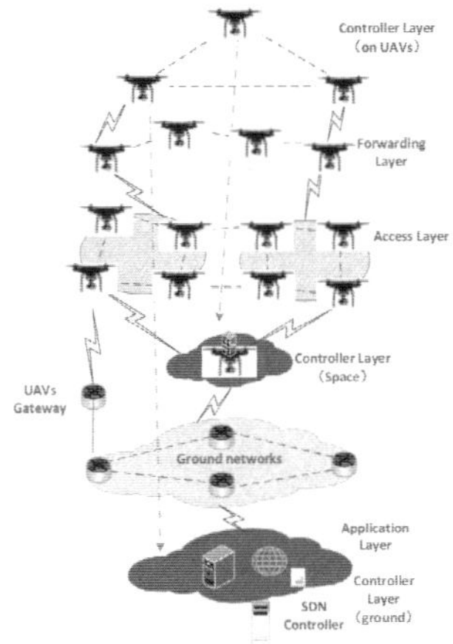

Figure 1. Multi-controller cluster drone network architecture.

2.4. *Application Layer*
Due to the limited processing power of the drone, the application layer is responsible for routing policy calculation, security and resource management strategies. In other words, it is the coordinator of the entire network. The application layer formulates policies based on the drone network conditions collected by the control layer drones and user needs. These policies are sent to the control layer drone to direct traffic between the drones.

2.5. *Control Layer*
Considering the global view of the entire network and the reliability of the link, the control layer is divided into three layers: drone, space, and ground control layer. That is, the controller cluster is deployed on the high-altitude drone, and the space drone (such as an airship), and ground stations within the coverage of high altitude drones. The high-altitude drone collects the link state information between the drones. Through the consistency protocol, the information is synchronized on the space and the ground controller, and is managed by the application layer. At the same time, the indication of the application layer is controlled by the ground controller, which is synchronized to the control layer drone and then sent to other drones. When the application layer instructions interact with the ground controller, they are converted to OpenFlow flow tables and distributed to medium altitude drones. Therefore, the transmission of drone data is controlled by a high altitude drone.

2.6. *Forwarding Layer*
The control decision part and the forwarding hardware are separated, and only the forwarding function is reserved on the medium altitude drone. The flow table of the medium altitude drone is configured by the dynamic controller. When data from a medium altitude drone/low altitude drone arrives, the mid-elevation drone searches the flow table to match forwarding information. These packets are then forwarded to the next drone and the flow table is used to reserve network resources for flow table. This can greatly reduce the processing burden of the mid-elevation drone.

2.7. *Access Layer*
Compared with the high-altitude drone/middle-altitude drone, the processing burden on the low-altitude drone is not too great, and the low-altitude drone is less difficult to develop and cheaper [7]. Only the basic access function and partial forwarding function are left on the low altitude drone. If the transmission distance is short, the data is forwarded between the low altitude drones via the ISL.

3. Disaster Emergency Multi-UAV network simulation
This simulation experiment is aimed at emergency UAVs communication network based on common task scenario building simulation model, with KVM virtual machine + entity SDN switches in the

form of the semi-physical simulation platform of real business simulation, and build SDN-based network architecture and traditional routing based respectively, on the basis of this to the entire communication process to make a detailed performance test and evaluation, and compared the performance differences between different implementations.

3.1. Simulation Environment Construction

The structure diagram of the simulation system is shown in Figure 2. The following describes the hardware and software environment required by the system according to different layers.

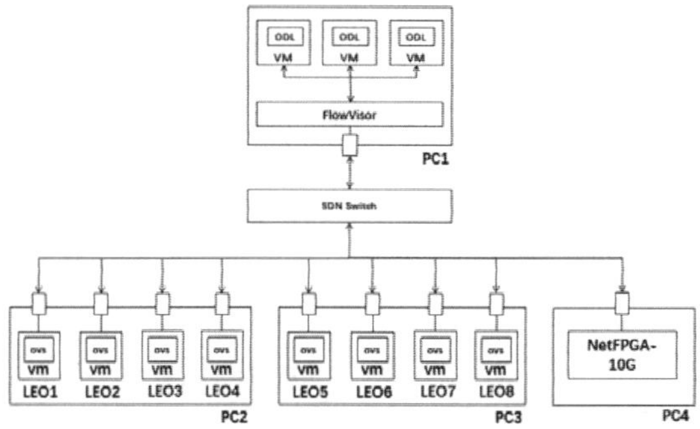

Figure 2. Simulation system block diagram.

Table 1. Computing Environment Description Table.

Computing environment description				
	CPU	RAM	hard disk	Network card
Server 1	Intel Core i7 quad-core and above	24 g-DDR3	240GSSD	Intel E1G44ET2
Server 2	Intel Core i7 quad-core and above	24 g-DDR3	240G SSD	4 x1000base PCI - E - T
Server 3	Intel Core i7 quad-core and above	24 g-DDR3	240G SSD	4 x1000base PCI - E - T
Server 4	Intel ® Pentinum ® G620	8 g - DDR3	240G SSD	Realtek PCIe GBE Family Controller

Table 2. Network Environment Description Table.

Network environment description		
	Introduction	Performance parameter
SDN switches	Shengke V350 cost-effective high-performance SDN /OpenFlow switch	240Gbps (8x1GE+12x10GE) forwarding capability with rich Open Flow features
Ordinary switch	TP-LINK TL - WR842N	4 10/100M adaptive LAN ports, 1 10/100M adaptive WAN port, auto flip (Auto MDI/MDIX)
NETFPGA	Xilinx XC5VTX240TFFG1759 Virtex - 5	Support SFP+ 10Gbps interface

Table 3. Software Environment Description Table.

Software environment description		
	Introduction	Version
OpenVswitch	Distributed high quality multi-layer virtual switch supporting multiple physical machines	2.5.0
Flowvisor	Network virtualization tool based on OpenFlow	2.1
Opendaylight	JAVA-based open source SDN controller	Be 0.4.1
Iperf	Mainstream network performance testing tools	0.2
VLC player	Is an open source cross-platform multimedia player and framework for playing most multimedia files	1.1.9
KVM	Open source system virtual machine based on Linux platform	0.12
Ubuntu Kylin	Ubuntu-based Chinese customized operating system	14.04

3.2. Simulation Scene Description

This scenario simulates a wireless communication network composed of multiple airship nodes covering a large area, and using an instance of remote video on demand service to realize uninterrupted data flow. The diagram of simulation scenario is shown in Figure 3.

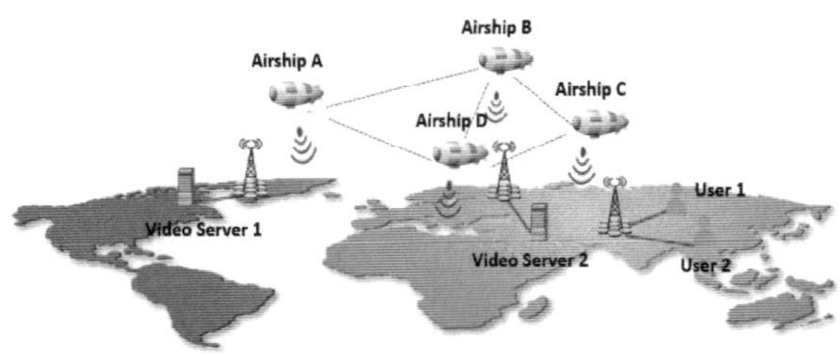

Figure 3. Schematic diagram of simulation scene.

User 1 initiates video server 1 to the video on demand service request, smooth playback of video to start a business after a successful build, followed by user 2 initiate a video-on-demand to video service 2, we use the schemes of traditional distributed routing and SDN-based centralized control to test separately. And evaluate the performance of the two schemes in network traffic scheduling.

3.3. Contrast Test Plan

3.3.1. Traditional routing scheme.

The network topology of this test solution is shown in Figure 4. Two traffic flows 1 and 2 are constructed in the figure, from server 1 to user 1 and server 2 to user 2.

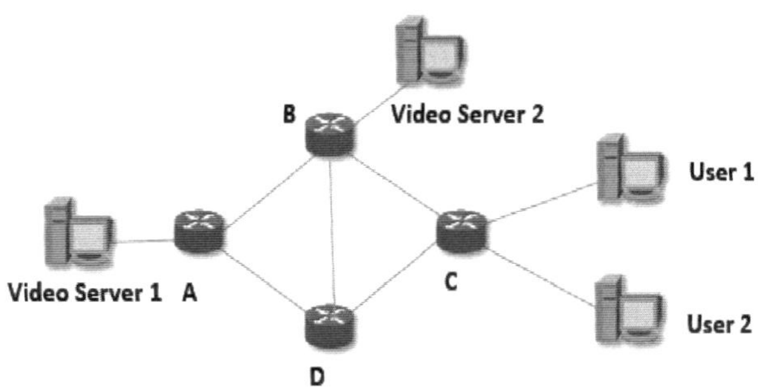

Figure 4. Traditional routing test solution topology.

Now the AODV routing protocol and OLSR routing protocol are respectively used to test the traditional routing solution:

First of all, AODV(on-demand distance vector routing protocol), as a kind of optional driven routing protocol, is an algorithm based on distance vector, and only needs to maintain the required route, without requiring the node to maintain the route of the inactive destination node during the communication process. Therefore, in this scenario, two traffic flows are always kept using the two paths with the shortest distance. (Even if the link is severely congested.)

For OLSR proactive table driven routing protocol, it is different from AODV routing protocol in that it periodically broadcasts routing information grouping, exchanges routing information and actively discovers routing. Its routing standard is to analyse the link status, including the delay, hop count and other indicators are finally weighted, and the cost of cutting the route is large. In the transmission of the second service flow, the test is the same as the AODV routing protocol. In the OLSR protocol test, B-->C is also selected as the transmission path.

3.3.2. SDN solution

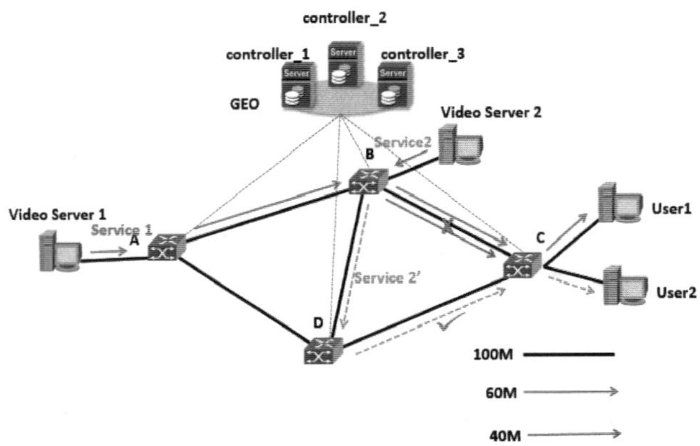

Figure 5. SDN test plan topology.

SDN solution test topology as shown in Figure 5, A, B, C, D for four SDN switch nodes, both belong to the control domain of the ODL controller cluster. First, we assume that the communication link bandwidth between the airships is 100M. User 1 initiates a video on demand service to the video server 1, assuming that the service needs to occupy 60 M bandwidth resources. According to the traditional routing protocols, this data flow will be along A --> B --> C flow path to video server 1, due to the whole link resources are more abundant, after the success of the service building, the video playback is clear and smooth.

Then the user 1 to the video server 2 launched a second video on demand service, under the assumption that the service need to 40M bandwidth resources, in the traditional network, according to the traditional routing protocol choice B - -> C for the shortest path, and set up along the path B --> C the business flow, at this point, the link B --> C because carries two business flow at the same time, the bandwidth load of close to 100%, the link quality serious decline, a more obvious delay and packet loss, reflected in service QOS is that the user's viewing experience of the video is degraded, and the video has obvious flower screen and stagnation. The main reason for this situation is that the traditional routing protocol is distributed and lacks the control of the whole network view, so it is difficult to optimize the whole forwarding path from a global perspective. As a result, some links are overloaded, while some links are idle for a long time.

However, in the SDN- based spatial information network architecture, we can send flow table to the SDN switch deployed on the airship to perform traffic scheduling by deploying an SDN controller deployed on the airship .See Figure 4,switch the data stream of service 2 to the relatively idle path of B -->D --> C to relieve the bandwidth load on the path of B --> C, so as to ensure the smooth establishment of both services and enable both user 1 and user 2 to watch the clear and smooth video program.

3.4. Test Result Analysis

Table 4. UAV resource consumption comparison table.

	Average CPU usage	Averagememory consumption
SDN solution	9.32%	162 Bps
Traditional routing scheme	14.52%	63 Bps
Contrast curve	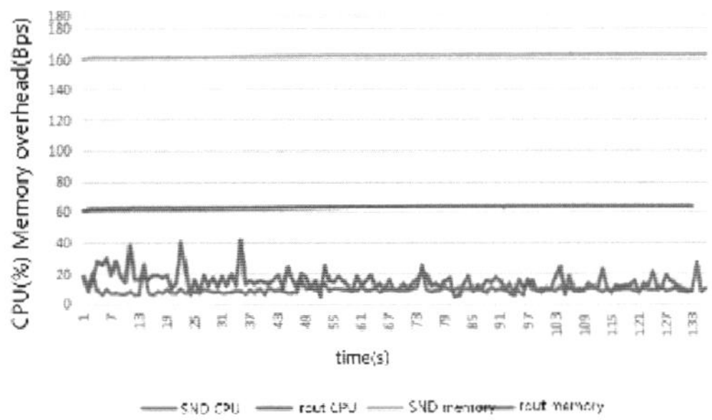	
conclusion	The SDN solution reduces the CPU consumption by about 35% compared to the traditional routing scheme.	

Table 5. Comparison table of signaling overhead.

	Mean signaling overhead
SDN solution	810.9416
Traditional routing scheme	1655.91
Contrast curve	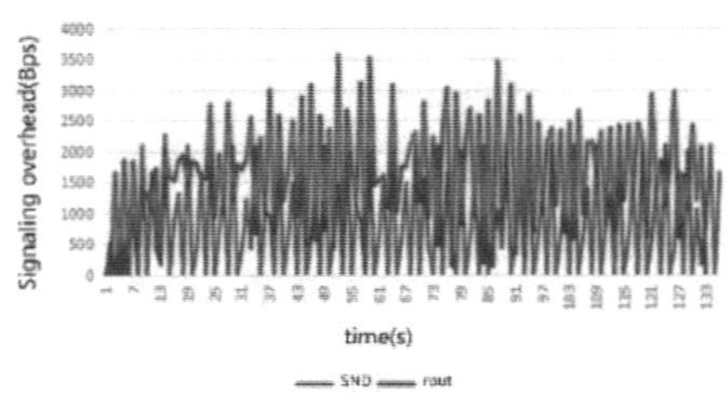
Conclusion	Compared with the routing scenario, the signaling overhead of SDN is reduced by about 51.03%
Controller-Switch Signal overhead	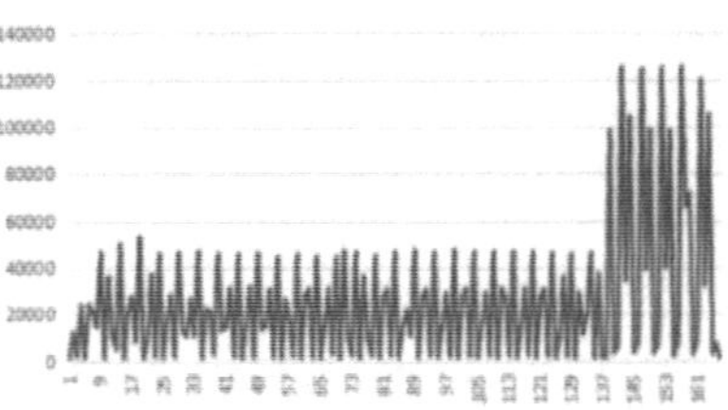
Controller- Controller Signal overhead	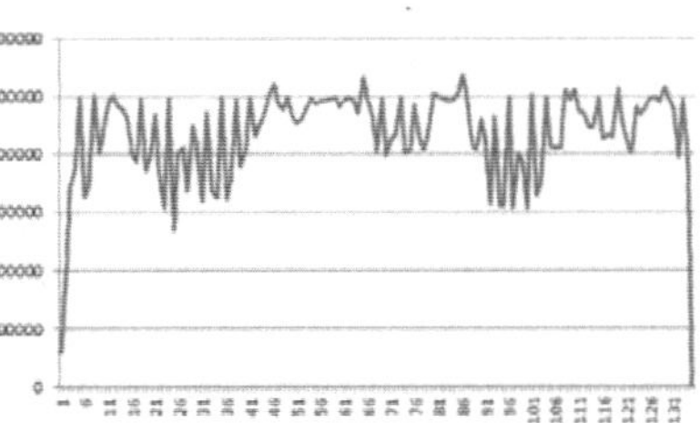
Conclusion	Signaling between controller clusters is expensive and requires a lot of communication resources

IOP Conf. Series: Materials Science and Engineering **719** (2020) 012054 doi:10.1088/1757-899X/719/1/012054

Table 6. Comparison table of business responsiveness.

	Business response speed	
	Business flow 1	Business flow 2
SDN solution	2.53 seconds	2.48 seconds
Traditional routing scheme	2.027 seconds	2.07 seconds
Contrast curve	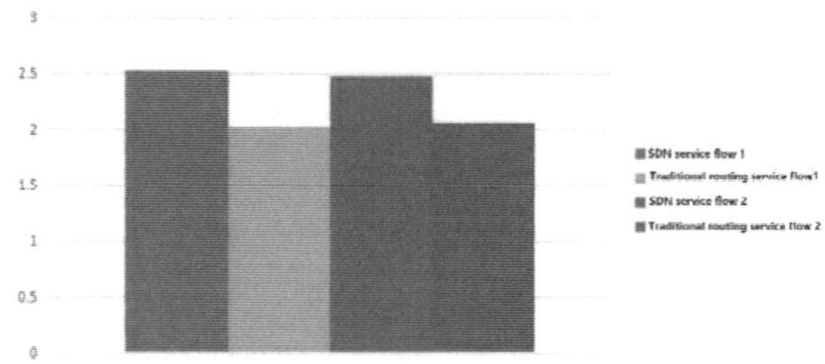	
Conclusion	Compared with the routing scenario, SDN service response speed is slightly slower	

4. Conclusion

In this paper, we focus on the performance analysis of multi-UAV system based on software-defined network in disaster emergency communication. Simulation results show that, compared with AODV and OLSR, the method proposed in this paper can save 30%-50% energy consumption for the entire network, at the same time, it can also provide a relatively stable business responsiveness, and significantly improve the network lifetime on the basis of meeting business requirements. The following work intends to start from reducing the signalling overhead between controllers to further save system resources and improve the lifetime of the emergency disaster rescue UAV network. Our goal is to conceive the future of emergency communications systems, as well as solve emergency communication system can quickly build, long and stable operation of the problems and challenges, and thus a step forward in communications disaster rescue command information systems.

5. Acknowledgment

This work was supported by the National Key Research and Development Program of China under Grants 2017YFB0504103.

6. References

[1] Tang Z, Zhao B, Yu W, Feng Z and Wu C 2014 *IEEE Com. Com. Ap.* pp 127-132.

[2] Mohammad M ,Walid S,Mehdi B and Merouane D 2016 *IEEE Commun. Conf.* pp 1–6.

[3] Sgambelluri A, Giorgetti A, Cugini F, Paolucci F and Castoldiet P 2013 *IEEE/OSA J. Opt. Commun. Net.* **51** pp 066-1075.

[4] Galkin B, Kibiłda J and DaSilva L A 2017 *IEEE Globe Commun. Conf* . pp 4-8.

[5] Tuncer D, Charalambides M, Clayman S and G. Pavlou 2015 *IEEE Trans.Netw.Serv.Mange.* **12** pp 18-33.

[6] Matolak D W and Sun R 2015 *IEEE Veh. Tech. Mange.* **10** pp 79–85.

[7] Motlagh N H, Taleb T, and Arouk O 2016 *IEEE Internet Things J.* **3** pp 899–922.

IOP Conf. Series: Materials Science and Engineering **719** (2020) 012055 doi:10.1088/1757-899X/719/1/012055

Research on Indoor and Outdoor Positioning System for Special Population

Jun Zhong[1], Shasha Zhao[1], Xuefeng Han[1,2], Yongfeng Liu[1] and Kai Guo[1,*]

1.Suzhou Institute of Biomedical Engineering and Technology, Chinese Academy of Sciences, Suzhou 215163, China.
2. Changchun Guoke Medical Technology Development Corporation, Changchun 130000,China.
*Corresponding author: Kai GUO
Email:guok@sibet.ac.cn

Abstract. Special populations, such as empty-nesters, people with a mental health condition, active children, prison inmates, etc., have a considerable number. The WHO noted that there were 47.5 million people with dementia worldwide in 2016. As the number of people with dementia in low- and middle-income countries continue to increase, it is predicted that the total number of people with dementia will reach 75.6 million in 2030 and reach 135.5 million in 2050. The National Health and Family Planning Commission released the "China Family Development Report (2015)", which is half of the total number of elderly people. The number of severely mentally disabled patients registered in the country reached 4.297 million, and there were 1,650 national mental health professional institutions and more than 20,000 psychiatrists. As of 2012, there were 681 prisons in the country, 300,000 people in prison, and 1.64 million prisoners. More than 80% of the prisons have established video surveillance systems. The Ministry of Public Security's "Emergency Release Platform for Children's Missing Information" is about 100 people abducted every year. The number of missing children who are about one-millionth of a year is not found, and the missing rate is about three-tenths of a million. The real-time positioning of the above-mentioned special population has become the key to solving related problems. Relevant nursing homes, community institutions, mental health institutions, and prisons have become the main demand side.

1. Introduction

At present, indoor and outdoor positioning technologies have their characteristics, and the scope of application and positioning accuracy are different. The satellite positioning technology realizes wide-area outdoor coverage, and the positioning accuracy reaches the meter level, the decimeter level, or even the higher precision, but there is no indoor positioning capability.

In terms of indoor positioning technology, Wi-Fi/WSN/RFID/UWB can achieve local meter-level indoor positioning, and Wi-Fi local area communication network has been established in many public places, such as smartphones and handheld computers. Wi-Fi or Bluetooth communication capabilities provide support for indoor location services. The existing 2G/3G/4G mobile communication network forms the communication network with the largest coverage on the ground and has billions of users. Therefore, the navigation satellite cooperates with the ground network, the coordination of multiple ground positioning technologies, and the high precision seamless of indoor and outdoor coordination. Positioning technology has become a trend in the future.

IOP Conf. Series: Materials Science and Engineering **719** (2020) 012055 doi:10.1088/1757-899X/719/1/012055

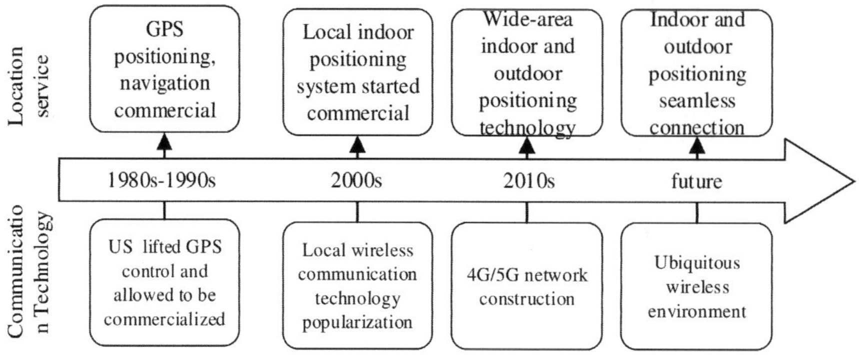

Figure 1. Development of wireless positioning technology.

2. Characteristics of Different Positioning Technologies

With the development of wireless communication technologies, emerging wireless network technologies, such as infrared, ultrasonic, radio frequency identification (RFID), Wi-Fi, ZigBee, Bluetooth, and ultra-wideband (UWB), have been widely used in offices, homes, factories, etc. application.

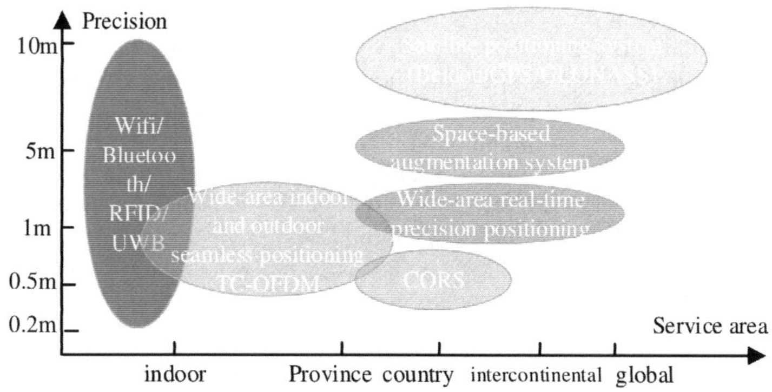

Figure 2. Different positioning system accuracy.

Infrared is only suitable for short-distance propagation, and is easily interfered by fluorescent lights or lights in the room, and has limitations in precise positioning.

Ultrasonic positioning has higher overall positioning accuracy and simple structure, but ultrasonic waves are greatly affected by multipath effects and non-line-of-sight propagation. At the same time, a large amount of underlying hardware facilities investment is required, and the cost is too high.

The shortcoming of Bluetooth is that the price of Bluetooth devices and devices is relatively expensive, and for complex space environments, the stability of the Bluetooth system is slightly poor, and it is greatly interfered by noise signals; the advantage of RFID is that the volume of the logo is relatively small, and the cost is relatively low, but the effect is Close distance, no communication capability, and inconvenient integration into other systems.

Compared with the traditional narrow-band system, the ultra-wideband system has the advantages of strong penetrating power, low power consumption, good multi-path resistance, high safety, low system complexity, and accurate positioning accuracy.

Wi-Fi positioning is a positioning solution for the IEEE 802.11 standard of the wireless local area network. The system uses a combination of empirical testing and signal propagation models, is easy to install, requires few base stations, can use the same underlying wireless network structure, and has a high total system accuracy.

Wi-Fi, Zigbee, pseudo-satellite, etc., achieve localized indoor 1-3m precision positioning but need to arrange a large number of nodes, and the cost is high. Wi-Fi, Zigbee, and other systems cannot be used in the event of a fire.

Table 1. Advantages and disadvantages of different positioning technologies

	advantage	Disadvantage
Infrared positioning	Low cost and high positioning accuracy	Small launch angle, easy to be blocked, small positioning range
RFID identification	Small size and low cost	Small coverage, no communication capability, and poor compatibility with other systems
Ultrasonic positioning	Close range distance measurement, high cost	Vulnerable to non-line-of-sight propagation and multipath effects
Bluetooth positioning	Wide application range and high positioning accuracy	High cost, easy to be disturbed by noise, slightly less stable, and high construction cost
UWB positioning	Low complexity, high positioning accuracy, and safety	Low real-time positioning
Sensor positioning	High positioning accuracy	Vulnerable to environmental interference, positioning accuracy depends on the resolution of the camera
AGPS positioning	Low cost and high positioning accuracy	Not able to locate quickly, positioning is not real-time
WLAN positioning	Low cost, wide coverage and wide use	Influenced by obstructions when propagating in non-line-of-sight environments, interference is present in different channels

The coverage of mobile base stations is wide, and the high-precision indoor positioning of mobile base stations is 1-3 meters, which is of great significance for the development of indoor location services. The accuracy of different positioning systems is shown in Figure 2.

3. The Technical Route Adopted by the Positioning System

Our special crowd positioning system uses the universal GPRS combined with GPS and WIFI positioning method to realize the outdoor positioning function. For the indoor positioning scene, it adopts the form of a Wi-Fi label.

Its function features are as follows: Firstly, the positioning method is seamlessly connected, and the most suitable positioning means is automatically adopted for different scenarios; secondly, the indoor positioning system is very popular for wireless LAN based on Wi-Fi, and the price is cheap, easy to set up and manage; Ultra-low power consumption, the label can be replaced or charged, and it is safe to the human body and has no radiation damage.

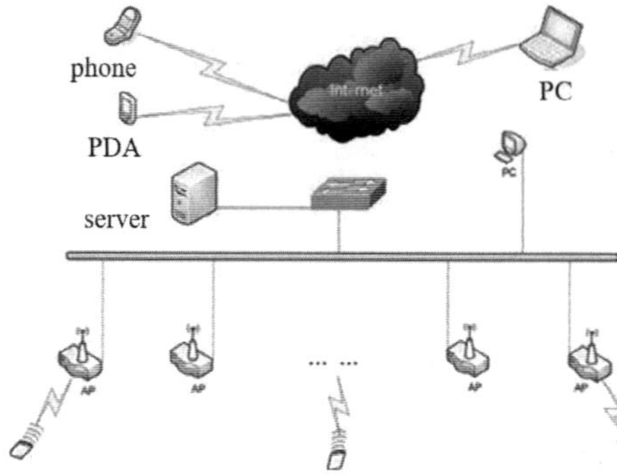

Figure 3. Positioning system software and hardware framework.

With the rapid development of smart terminals, smart children's watches, smart shoes, smart jewelry, and other terminals with outdoor positioning functions and indoor wireless positioning systems are constantly being introduced, partially satisfying people's needs.

However, these terminals have the following disadvantages: First, the concealment is poor, especially in children's watches, when encountering accidents such as kidnapping, trafficking, etc., they are often removed at the first time and cannot play the role; then It is easy to lose. If it is improperly worn or the product is accidentally lost, it is easy for the guardian to fall into the search for difficulties. Finally, the indoor and outdoor wireless positioning functions are not available at the same time, and the application scenario is limited. Once the environment changes, the specific location cannot be obtained in time.

Given the above defects, our advantages are: First, the concealment is good, through the miniaturization, flexibility, and flattening of the terminal, the combination of positioning interruption and clothing is achieved, thereby achieving the purpose of strong concealment; then, it is not easy to lose due to clothes. The specialty is greatly reduced compared with other wearable terminals. Finally, through the combination of GPRS+GPS, it has the commonly used wireless positioning function. On this basis, WIFI is actively sent to transmit tag information to realize indoor wireless. Positioning function, which expands the application surface of the smart terminal.

4. Special Crowd Positioning Software System

We built the MYSQL database storage positioning data on the Linux cloud server, and at the same time, established the positioning display webpage by using the node.js framework, and the positioning device transmits the positioning data in real-time through the network. The website can display the device track in real-time, as shown in Figure 4.

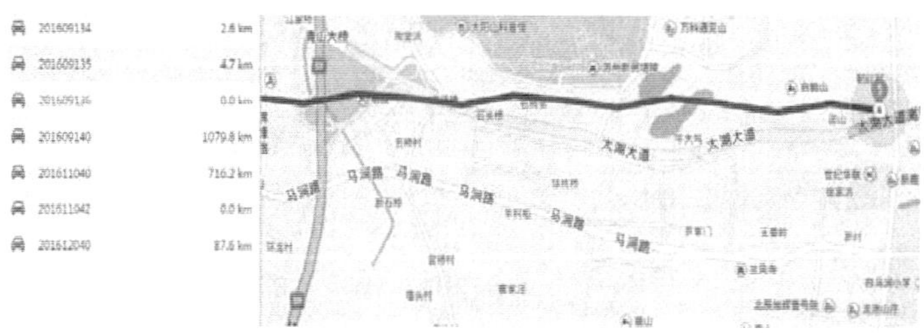

Figure 4. Positioning display webpage based on node.js framework.

We have developed mobile display software based on the ionic mobile app framework. It accesses the cloud real-time location data through the API connected to the EXPRESS framework and displays it in real-time through the front end. The mobile APP display software of the special crowd positioning system is shown in Figure 5.

Figure 5. Special crowd positioning APP.

5. Conclusion

In this paper, by using different positioning technologies, the cloud server stores the positioning data, and the webpage and the APP access the API to display the real-time positioning of the special crowd. In summary, through the combination of different positioning technologies, indoor and outdoor positioning can be used to a certain degree of accuracy, which is suitable for different special population positioning.

6. Acknowledgments

This work was supported by "Jilin Province Science and Technology Development Plan Project:"(No20191102004YY), "Jiangsu Science and Technology Plan Project" (SBE2017740113 "Jiangsu Province Science and Technology Plan Key Project" (No. BE2017007-2).

7. References

[1] Ma C, Yang J, Chen J, et al. Indoor and outdoor positioning system based on navigation signal simulator and pseudolites[J]. Advances in Space Research, 2018:S0273117718305684-.

[2] Shi Z, Li H, Lin H, et al. A Nano-Quadcopter Formation Flight System Based on UWB Indoor Positioning Technology[J]. 2018:1-4.

[3] Hutabarat D P, Hendry H, Pranoto J A, et al. Human tracking in certain indoor and outdoor area by combining the use of RFID and GPS[C]// Wireless & Mobile. 2017.

[4] Angelats E, Navarro J A. Towards a Fast, Low-Cost Indoor Mapping and Positioning System for Civil Protection and Emergency Teams[J]. ISPRS - International Archives of the Photogrammetry, Remote Sensing and Spatial Information Sciences, 2017, XLII-2/W8:9-15.

[5] Lee B, Lim C, Lee K. Classification of indoor-outdoor location using combined global positioning system (GPS) and temperature data for personal exposure assessment[J]. Environmental Health & Preventive Medicine, 2017, 22(1):29.

[6] Alsina-Pagès R M, Navarro J, Alías F, et al. homeSound: Real-Time Audio Event Detection Based on High Performance Computing for Behaviour and Surveillance Remote Monitoring:[J]. Sensors, 2017, 17(4):854-.

[7] Yang Z, Qiang L, Lin H, et al. Research on visual autonomous navigation indoor for unmanned aerial vehicle[J]. Journal of Shanghai Jiaotong University (Science), 2017, 22(2):252-256.
[8] Chao M, Wang J, Chen J. Beidou compatible indoor positioning system architecture design and research on geometry of pseudolite[C]// 2016 Fourth International Conference on Ubiquitous Positioning, Indoor Navigation and Location Based Services (UPINLBS). 2017.

Smart Grid Monitoring Systems based on Advanced Encryption Standard and Wireless Local Area Network

Chuankai Yang[1], Jingfeng Wu[1], Longyang Wang[2], Xiaolan Zhang[1], Liangshu Li[1], Shan Liu[2]

[1] State Grid Shaanxi Electric Power Research Institute, Xi'an, Shaanxi Province, China
[2] School of Electrical Engineering and Automation, Wuhan University, Wuhan, Hubei Province, China
Email: wanglongyang1994@163.com

Abstract. China's power grid is developing in the direction of informatization and increasingly large-scale at present. In view of the problems of optical cable interference and poor data security in the existing power grid detection system, a grid detection system based on wireless local area network (WLAN) and improved advanced encryption standard (AES) is proposed. Firstly, wireless local area network technology is introduced on the basis of the original power grid detection system to realize wireless transmission of power data. Then the newly-proposed system improves the S-BOX in the traditional AES encryption algorithm to improve the security of the algorithm. The improved AES encryption algorithm is applied to the grid detection system to realize the secure transmission of power data. Finally, the practical value of the grid detection system is verified by simulation.

1. Introduction

In recent years, the power grid equipment has potential safety hazards under heavy load due to the rapid growth of national electricity consumption. In order to ensure the normal operation of the grid equipment, the establishment of a safe and reliable high-efficiency smart grid is the development trend of the existing grid system [1]. Among them, the grid monitoring system is used to monitor the safety situation of the grid equipment as a problem finder in the power grid. It is an important component of the smart grid, and its normal operation directly affects the working condition of the entire grid. Therefore, the design of the grid monitoring system is of great significance for the construction of the grid. In order to ensure the safe and reliable operation of the grid monitoring system, actual and reliable monitoring of the information transmission process is one of the cores focuses when designing the system.

At present, some scholars have conducted relevant research on the power grid monitoring system. Chen Gang [2] designed the information transmission of the monitoring system which mainly used optical cable for wired transmission, which greatly improves the capacity of information transmission meaning it can transmit a large amount of information at the same time. However, the cable needs to be pre-laid and is affected by the transmission distance and the external environment. Li Min [3] analysed the characteristics of information networks in smart grids in order to overcome the information security issues that are increasingly prominent in smart grids, and established the importance of information network security to ensure the safe and efficient operation of smart grids. In order to pursue efficient and fast data transmission, Meikang Qiu et al. [4] used symmetric encryption algorithms such as DES algorithm and RC5 algorithm for data encryption. The encryption speed is accelerated and the energy loss is reduced, but the both sides of the communication use the same decryption key, resulting in reduction of the security level in data transmissions. Zeng Fanyu [5] proposed combining AES encryption algorithm with ECC algorithm, combining the advantages of symmetric encryption algorithms and asymmetric encryption algorithms, and improving the security of encryption, but the encryption process is cumbersome and cannot achieve fast and efficient encryption.

Aiming at the limitations of wired transmission in the power grid, low data transmission efficiency and poor security, this paper proposes an improved power grid monitoring system. The system uses WLAN for data transmission and uses the improved AES encryption algorithm to improve the security of data transmission. Finally, the paper verifies the correctness and feasibility of the system through verification.

2. System Design

The main purpose of the grid monitoring system is to monitor the security situation of each station and send corresponding commands to the station according to the monitoring situation [6]. According to the above requirements, this paper divides the grid data monitoring system into separate layers: the sensing layer, the network layer and the application layer. Each layer has its own function. The overall framework of the grid monitoring system is shown in Figure 1.

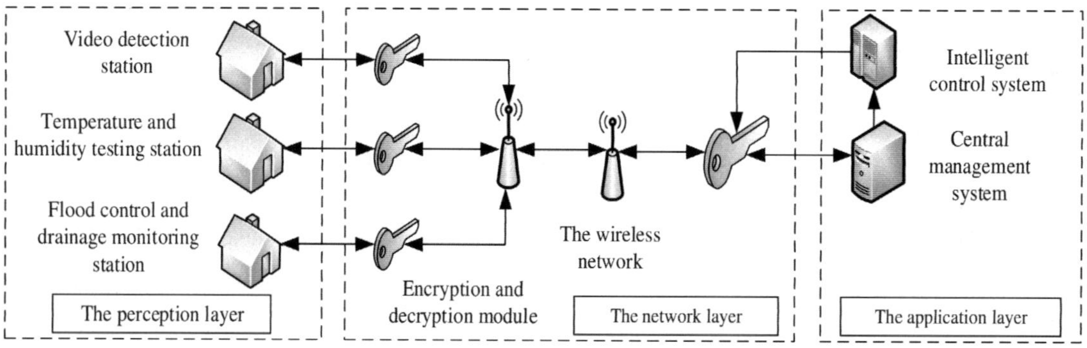

Figure 1. Grid monitoring system.

- Perceptual layer: mainly collects device data, including information in the video subsystem, temperature and humidity monitoring subsystem, flood control and drainage monitoring subsystem. They collect device data, and transfers the information to the central management system through the network transport layer;
- Network layer: mainly based on AES encryption algorithm to encrypt and decrypt data in WLAN. The collected data is encrypted by an improved encryption algorithm to improve the security and efficiency of data transmission;
- Application layer: mainly includes the central management system and intelligent control system. After receiving the encrypted data, the central management system decrypts, analyses and stores the encrypted data, and the intelligent control system uses the decrypted data to convey specific instructions to the power grid equipment to achieve control;

In the data transmission process of the power grid, the traditional wired communication needs to pre-lay the optical cable, the construction process is complicated, and has certain requirements concerning the transmission distance [7]. The system proposed in this paper uses wireless local area network for communication [8], which is not limited by distance. The WLAN established by WIFI technology is a combination of computer network and wireless communication technology. It mainly uses the open radio frequency (RF) technology of 2.4 GHz and 5.8 GHz bands instead of wired technology to form a local area network, which has all the functions of traditional wired LAN. The user can realize the broadband network access at anytime, anywhere and at random.

A schematic diagram of a grid monitoring system using a wireless local area network is shown in Figure 2 below.

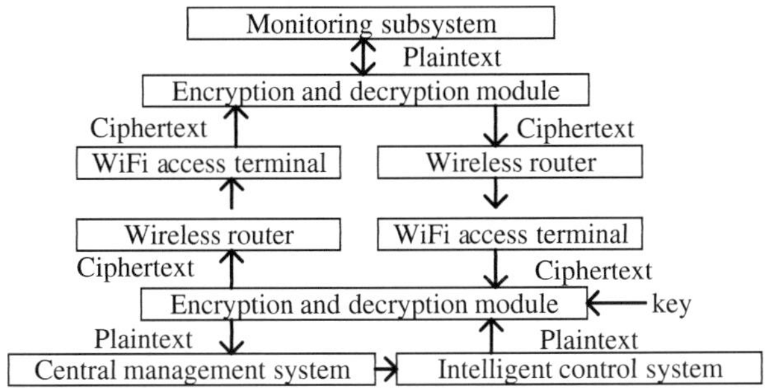

Figure 2. Schematic diagram of the grid monitoring system

The system information transmission is not limited by distance, and the information is encrypted before being sent out, which can effectively prevent the monitoring information from being stolen. After receiving the information, the receiving end decrypts the decrypted text, and the decrypted plaintext can directly perform subsequent processing, which can realize the high efficiency of monitoring information transmission.

3. Improved AES Encryption Algorithm

3.1. Analysis of AES Encryption Algorithm

In the process of data transmission, in order to protect the transmitted information, it needs to be encrypted. The commonly used encryption algorithms are RC4, RC5, etc. The system uses AES encryption algorithm. Advanced Encryption Standard (AES), also known as Rijndael encryption in cryptography, is a block encryption standard adopted by the US federal government. AES is the most common symmetric encryption algorithm, using the same key for encryption and decryption. In the encryption process, the key and the plaintext are operated in bytes [9].

The operation in the AES encryption process needs to be performed in the S-box. The iteration period of the traditional S-box on the GF(2^8) domain is too short, which affects the algebraic nature of the S-box and the anti-attack capability of the algorithm. Since the packet keys of AES is generated by initial key expansion, there is inevitably a certain degree of correlation between each packet key. Correlation can lead to two bad results. One is to increase the possibility of the initial key being compromised, affecting the security of encryption; the second is to make the errors in the transmission spread ferociously once occurred, reducing the accuracy of encryption and decryption. Therefore, the use of improved S-box and elimination of correlation between packet keys is crucial to improving the security and accuracy of the AES algorithm.

3.2. Improvement of S-box

In the AES encryption algorithm, the SubByte operation is its only non-linear part. It is implemented by the pre-calculating table S-box, so the quality of the S-box has a significant impact on the performance of the algorithm [10]. The complexity of the S-box algebraic equation is one of the important indicators reflecting its security, which is related to the ability of the S-box to resist attacks. The traditional S-box algebra equation only has 9 items [11], so the traditional S-box algebraic equation is not complex enough and is lacking in security.

In this paper, the two operations of modifying the affine transformation pair and adjusting the S-box calculation order are applied to the encryption algorithm of the grid monitoring system, which can significantly increase the iteration period of the S-box, enhance the algebraic nature of the S-box, and improve the security of the information transmission of the grid monitoring system[12]. A new affine transform pair ('A7', '6F') is used, and the following three steps are used to calculate a new S-box. (1) An affine transformation is performed on affine transformation pair ('A7', '6F'). (2) Find the multiplicative inverse element. (3) The affine transformation pair ('A7', '6F') is subjected to affine transformation again, and achieving output result. The operation of the inverse S-box is the same as that of the S-box, and the affine transformation pair is ('D0', '35'). The new S-box iterative output period and affine transformation pair period are longer, and the number of algebraic equation items of S-box and inverse S-box is also increased from 9 to 255, which greatly improves the complexity of S-box.

3.3. Improvement of Key

In the traditional AES encryption algorithm, each round of encrypted key is obtained by a key expansion algorithm, and the 4-word initial key is expanded into a 44-word round key for ten rounds of encryption operation [13]. Due to the reversibility of the extended algorithm, the round key and the initial key can be calculated from each other, and therefore, the correlation between the traditional round keys is strong. The attacker can easily obtain all the key content after obtaining a partial key, and the security is low.

This paper proposes to apply the stream cipher idea to the information transmission process of the grid monitoring system, and generate a random sequence by using a pseudo-random algorithm to generate a random key. A pseudo-random sequence can satisfy the non-speculative needs in cryptographic operations under certain conditions. Commonly used pseudo-random number column generation algorithms include square-to-square algorithm and BBS algorithm. In this paper, the pseudo-random algorithm used is the linear congruence algorithm [14]. The flow of generating pseudo-random numbers is shown in Figure 3 below.

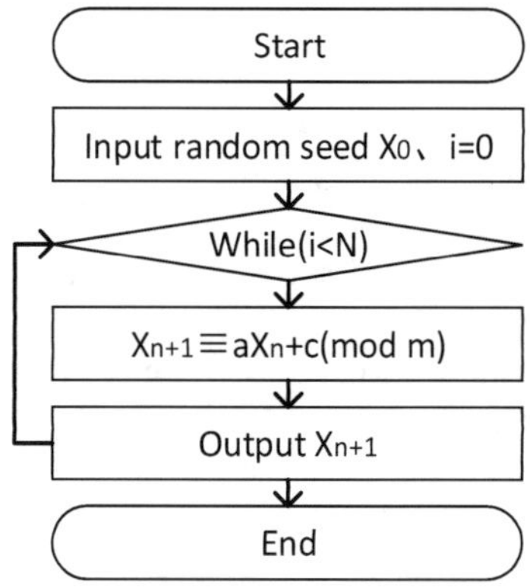

Figure 3. Pseudo-random number generation process

This is a widely used pseudo-random number generation algorithm whose recursion formula is $X_{n+1} = (aX_n + c) mod\ m,\ n=1、2、3...n$. he linear congruence algorithm has the advantages of fast generation speed and long sequence period of generation, making it an ideal pseudo-random number generation algorithm.

The specific process of encrypting by using a pseudo-random number is as follows: a random sequence is generated by a pseudo-random number generation algorithm, then the random sequence produces a key stream sequence, and the plaintext and the key stream are packet-encrypted according to a certain length. The byte replacement operation in the encryption process uses the improved S-box described above. For the cryptanalyst, even if a sufficient number of plaintext and ciphertext pairs are intercepted, the entire key sequence cannot be inferred because the encryption keys of each packet are not correlated.

4. Simulation and Analysis of Improved AES Encryption Algorithm

4.1. Encryption and Decryption Simulation of Improved AES Encryption Algorithm

The improved AES encryption algorithm proposed in this paper is based on C++ programming, and a palindrome is selected as plaintext for independent encryption simulation. The X_0 of the generated random sequence is 167, and the seed key is randomly selected in the pseudo-random number column, and the packet length is 128 bits. The number of encryption and decryption rounds is 10, and the generated key stream sequence is 1408 bits in total. The encryption and decryption simulations were performed twice using the same plaintext, and the results are shown in Tables 1 below.

Table 1. Data and results of encryption and decryption simulation.

FIRST	**Plaintext**	123456abccba654321
	Ciphertext	de,99,1d,20,68,90,85,44,f3,ed,77,a2,d5,8,h7,e0,3b,53
	Decryption result	123456abccba654321
SECOND	**Plaintext**	123456abccba654321
	Ciphertext	2a,4e,35,7a,19,h6,31,d0,55,fc,c6,ad,4e,3h,5c,ea,1,40
	Decryption result	123456abccba654321

The traditional AES encryption algorithm obtains the same ciphertext by independently encrypting the same plaintext multiple times, and since the packet key is expanded by the original seed key, the correlation between the ciphertexts of each packet is stronger. In this paper, the simulation is performed independently using the same plaintext. The decrypted result is the same as the plaintext, and the correctness of the improved AES encryption algorithm is verified. However, in these two simulations, the ciphertext is completely different. This is because the key in the improved AES algorithm is a random sequence obtained by the linear congruence algorithm using the stream cipher idea. Therefore, the keys used for each encryption are different. Even if the same plaintext is independently encrypted multiple times, the keys are different each time, and the ciphertext obtained each time changes.

4.2. Security Analysis of Improved AES Encryption Algorithm

In the process of improving the AES encryption algorithm, the affine transformation pair of the S-box is firstly modified, the calculation order of the S-box is adjusted, the iteration period of the S-box is increased, and the algebraic property of the S-box is enhanced. After changing the key generation method, instead of using the key expansion of the traditional AES algorithm, a pseudo-random algorithm is selected to obtain a random key stream sequence. Therefore, the improved AES encryption algorithm is a hybrid encryption algorithm combining packet cipher and sequence cipher, which is more secure than the algorithm using packet cipher or sequence cipher alone.

In the improved key, a random number sequence is generated using the function $X_{n+1} = (aX_n + c) mod\ m,\ n=1、2、3...n$. This function is a one-way function [15], which is a function that is easy to solve in the forward direction and is difficult or impossible to solve in the reverse direction. Therefore, when one or more random numbers in a random sequence are known, it is extremely difficult to detonate the entire sequence or seed key. Each time a different seed key is used for encryption, the random sequence generated each time is different, so that each encryption and decryption use a different key stream. Since the key stream sequence is not generated by key expansion, there is no correlation between the packet keys. The cryptanalyst cannot infer the key stream sequence even if a certain number of plaintext and ciphertext pairs are intercepted. The unpredictability of the key greatly increases the difficulty of the algorithm being compromised, so it can be considered that the improved AES encryption algorithm is more secure.

5. Conclusion

This paper optimizes the data transmission process in the grid detection system, by improving the traditional AES encryption algorithm, S-box and keys, it also simulates and analyses the improved algorithm. The conclusions are as follows:

- Data transmission using wireless local area network solves the problem that the optical cable in the traditional wired communication is limited by the distance and the pre-laying process is complicated. Wireless transmission is not limited by distance and has a higher degree of transmission liberty, which provides great convenience for power grid monitoring;

- In the process of data transmission, the widely used AES encryption algorithm is adopted. In the improvement process of the S-box and the key, the complexity of the S-box and the unpredictability of the key are significantly improved, and the anti-attack capability of the AES encryption algorithm is improved.
- After the simulation, it is found that the plaintext before encryption is the same as the plaintext after decryption, but the ciphertext is different. The correctness and the feasibility of the improved AES encryption algorithm is verified.

The power grid monitoring system designed in this paper can carry out long-distance transmission of monitoring data safely and efficiently, and realize the protection of the power grid.

6. Acknowledgments
This work was supported by the Science and Technology Project of State Grid Corporation of China under Project of “Research on the third generation substation auxiliary system” with NO. 520530180015.

7. References
[1] Farhangi, and H. "The path of the smart grid." IEEE Power & Energy Magazine 8.1(2010):18-28.
[2] Chen Gang, and Wang Xiaoshan. "Design and implementation of optical cable intelligent distribution monitoring system for regional electric power communication network." Modern Electronics Technique 09(2017):174-176+180.
[3] Li Min, et al. "The Information Security Risk Analysis of Smart Grid." NORTH CHINA ELECTRIC POWER1(2017):62-65.
[4] Qiu Meikang, et al. "Energy Efficient Security Algorithm for Power Grid Wide Area Monitoring System." IEEE Transactions on Smart Grid 2.4(2011):715-723.
[5] Zeng Fanjing, and Liu Xiaodong. "The research of AES algorithm based on the WLAN encryption." Electronic Design Engineering 05(2017):119-122.
[6] Zhai Hucheng. "Research and Analyze the Operating Condition Monitoring Technology of Transmission Line under Smart Grid." Science and Technology & Innovation 24(2015):96-96.
[7] Fang Yuesheng, Chen Hao, and Zeng Fanxing. "Development of Optical Fiber Technology and Its Application in Electric Power Communication." ELECTRIC POWER ICT 12.8(2014):20-26.
[8] Fan Yulin. "Review of Wireless Local Area Networks." Modern Information Technology v.03; No.9(2019):60-61+64.
[9] Mohammad, Omer K. Jasim, et al. "Innovative Method for Enhancing Key Generation and Management in the AES-Algorithm." Computer Science 4.4(2015):14-20.
[10] Shreenivas, Pai N, et al. "Logic optimization of AES S-Box." International Conference on Automatic Control & Dynamic Optimization Techniques IEEE, 2017.
[11] Das, Suman, U. Z. J. K. M. Sadique, and R. Ghosh. "Study of Randomness in AES Ciphertexts Produced by Randomly Generated S-Boxes and S-Boxes with Various Modulus and Additive Constant Polynomials." Journal of the Institution of Engineers (India): Series B 97.2(2016):193-208.
[12] Shen Xiaochen, Han Meng. "Improved S-box Based on Strict Avalanche Distance Criterion." Microelectronics & Computer v.35; No.6(2018):92-96.
[13] Fan Lichen, and Wang Guihai. "Research and Improvement on the Key Expansion of Encryption Algorithm AES." Information Technology and Information 03(2015):82-84.
[14] Zhang Dawei, Shao Yinghai, and Zuo Lei. "Pseudo random number generation algorithm based on linear congruence method." Journal of Eastern Liaoning University (Natural Science Edition)v.25; No.99.03(2018):55-60.
[15] Zhang, Xiaomin. "Verifiable multi-secret sharing scheme based on linear one-way function." Journal of Computer Applications 33.5(2013):1391-1393.

IOP Conf. Series: Materials Science and Engineering **719** (2020) 012057 doi:10.1088/1757-899X/719/1/012057

Design and Implementation of EMC System Layout for Rail Trains

Gang Song[1], Tong Wu[2], Dafa Jiang[3], Xiaotong Mu[4], Baichen Huo[5], Jilei Song[6], Baoying Shang[7], Liwen Hu[8] and Kai Liu[4]

[1] CRRC Changchun Railway Vehicles Co., Ltd., Changchun, 130062, China
[2] China National Institute of Metrology, Beijing, 100029, China
[3] CRRC Zhuzhou Locomotive Co., Ltd., Zhuzhou, 412001, China
[4] Beihang University, Beijing, 100191, China
[5] Shenzhen Beihang Testing Co., Ltd., Shenzhen, 518000, China
[6] Beihang Engineering Technology Center (Shenzhen) Co., Ltd., Shenzhen, 518016, China
[7] Foshan Shunde Aerial Survey Electromagnetic Compatibility Technology Co., Ltd., Foshan, 528305, China
[8] Suzhou Smart Testing and Certification Co., Ltd., Suzhou, 215001, China
Email: songgang@cccar.com.cn

Abstract. Nowadays, China's urbanization process is rapid, and the rail transit system has become a good solution for the traffic congestion problem. The electromagnetic compatibility (EMC) of rail transit system is an important guarantee for the safe and reliable operation of rail trains. To this end, this paper proposes an EMC layout scheme for rail trains, and designs their electrical equipment layout and electrical wiring layout. In addition, according to the distribution of the train body equipment, the rail trains are divided into three parts, i.e. the roof, the interior and the bottom, and their associated EMC partition layouts are given. Finally, the layout scheme is tested by taking the Shanghai Metro Line 6 as a test rail train. The test results show that the proposed scheme can greatly improve the EMC performance of the rail trains and achieve the expected design goals.

1. Introduction

As a travel mode with less land occupation, large traffic volume, less environmental pollution, safety and comfort, urban rail trains have sprung up in China's major cities with rapid economic development since the opening of the first metro line in Beijing in 1956 [1]. It is the primary task to ensure that the rail train design meets the standards in the factory and is safe during its operation. There are two major factors affecting the safe operation of rail trains, namely the electromagnetic interference (EMI) from their own equipment and the electromagnetic radiation formed throughout the city. Most of electrical and electronic systems in rail trains, such as signal transmission systems (including track circuits and other train detection systems), train traffic control systems and railway side control systems, use high-power inverters, rectifiers and various advanced signal controller. This makes the rail trains become huge EMI sources in the city. Therefore, EMI occurs between the equipment inside the rail trains, between the rail trains and the railway infrastructure, and between the rail trains and their external environment [2].

Literature [3] presents a theoretical basis for the electromagnetic compatibility (EMC) design of high-speed multiple unit trains. Literature [4] adopts data mining (DM) technique to analyze the electromagnetic emission measurement data of high-speed trains, and can predict the EMC data of the equipment, thus giving the key equipment for anti-EMI. In [5], a wiring scheme based on high frequency structure simulator (HFSS) is designed by SOLIDWORKS modeling and simulation. In [6], the EMC problem in the subway station is studied, and a method to determine the minimum space for the coexistence of the strong and weak electric systems is proposed. Based on this, the layout of the electromechanical equipment is reasonably planned.

The EMC problem of rail trains should be analyzed from three aspects, namely the interference source, EMI propagation path and sensitive equipment. The EMI sources of rail transit trains include train interference sources and trackside system interference sources. Firstly, the train interference sources include the traction motor, high speed circuit breaker, main switch box, variable voltage and variable frequency (VVVF) main inverters, filter reactor, auxiliary inverters, transformers, and bus bar breakers, etc. Secondly, the EMI sources of the track systems include the power supply equipment of overhead lines and the signal system equipment in the track. The sensitive equipment of rail trains includes lighting systems, air conditioning systems, communication systems, passenger information systems, train control systems, door operating systems, train signal systems and auxiliary systems.

Once the interior of the rail train is subject to EMI, the video in the rail train may appear to be jittered, or the communication quality of the passengers' mobile phone in the train is affected. More seriously, the command signals may be changed to give an error indication, thereby inducing a safety accident. Because the EMC problem seriously affect the safety and reliability of rail trains, a reliable and comprehensive EMC design scheme is particularly important.

For this reason, this paper proposes an EMC layout scheme for rail train systems from the general viewpoint. The different parts of train body, which are divided into three modules: the roof, the bottom and the inside of the train, are different in EMC design. Therefore, the detailed EMC partition layout scheme is then carried out for them. Finally, some test scenarios for testing the EMC performance of the rail train are set based on Shanghai Metro Line 6, and the test results are given.

2. EMC System Layout of Rail Trains

The overall EMC layout scheme for the railway trains is determined by some factors, such as the overall train structure, train operating environment, performance requirements, and maintenance. This involves in the coordination and cooperation of various aspects of the systems. It is very necessary to make a certain electromagnetic evaluation before the EMC design scheme. Then, especially the layout of the electrical wiring and the design of the wiring harness should be considered, which is closely related to the train body, pipeline structure arrangement, internal structure, and interface design of the electrical equipment. Therefore, in the design process, it is also necessary to comprehensively consider from various aspects to ensure that the EMC system layout is optimal. From the viewpoint of the rail train structure, it mainly consists of two parts, namely, the equipment that plays various roles and the cables that connect the equipment. Therefore, the overall layout of the equipment inside the rail trains and the overall layout of the electrical wiring are the basis of the EMC system layout scheme, and also the principle of the EMC design scheme for each subsystem. This paper firstly designs the overall EMC system layout scheme for the rail trains, and then designs the EMC partition layout scheme based on their three parts, i.e. the roof, the bottom and the inside.

2.1. Overall Layout of Electrical Equipment

The overall layout of the electrical equipment should make the entire system concentrated and compact. The specific layout scheme is listed as follows.

First, fully consider the train characteristics, modularize the roof, interior and bottom, and functionally combine similar equipment. For example, the main equipment of main circuit, large equipment, and electrical components with large noise and vibration are placed under the train as much as possible or isolated with other equipment or components. For functional or comfortable electrical equipment, such as switchboards, and water and sanitation systems, are placed at both ends of rail trains for easy access, maintenance and passenger use. Passenger information systems (PISs), such as broadcast and video systems, and commonly-used switch buttons, should be placed in convenient and obvious locations.

Second, focus on the design of wiring path. On the one hand, trunking, conduit and wire bracket should be set up. Except the special design structure requirements, weld slag, riveting and screws should be avoided in the cable passing area. Otherwise, there will be hidden dangers for breaking lines. On the other hand, according to the wiring and installation area, the protective area of the wiring in the structural drawing should be identified. Technical requirements are presented in the structural assembly or component drawings. That is to say, the structure with cable passing and the fitting surface need to be removed from the protrusions and corners. The wiring path is designed to avoid cluttering of cable connections between the devices, or the appearance of bare wires. These cases provide the conditions for unnecessary EMI. Therefore, the EMC design must be carried out in strict accordance with the proposed scheme.

Third, design the electrical wiring connection method. When planning and resolving the connections between electrical equipment, between electrical connection boxes, and between electrical connection boxes and controlled devices, three principles should also be fully considered:

- The connection terminals with standard specifications according to the current size and the number of input and output lines should be selected.
- A connection terminal or an industrial connector is used between the electrical cabinets or their controlled boxes, in order to facilitate the correction and withstand voltage test of the lines after the train assembly is completed.
- For the selection of connection terminals and connectors, the accessory interchangeability of different train types should be selected on the premise of achieving the technical performance.

2.2. Overall Layout of Electrical Wire

The electric field effect generated by cables of different voltage levels in rail trains will become the EMI interference source of the whole system. In order to suppress EMI and improve the stability of the overall wires, the electrical wiring design scheme should follow the principles of top-down and module design. The specific layout design principles are as follows:

- The driving safety equipment line, the control wire and the radio line are pre-wired, and after the pre-wiring under the rail train is completed, the rail train can be lifted and erected. The wires are located in the wiring grooves on both sides of the train body, and the wire grooves should be grounded nearby at multiple points.
- The EMC design of traction motor cable is very important [7]. The pre-wired method is adopted for the traction motor wires and auxiliary circuit wires. The auxiliary circuit wires should be as far as possible from the traction motor wires and protected by electromagnetic shielding with metal protection.
- The metal casing of equipment must be grounded nearby through a braided wire, and the grounding wires should be coated with electrical contact grease to ensure good grounding.
- Light wires are arranged separately from control wires and under the top cover.
- The control wires entering into the operator cab also adopt the form of pre-wiring, and enter into the operator cab from the top left side of the train body. The auxiliary wires enter into the operator cab from the bottom of the middle corridor floor.

The above scheme scientifically and rationally arranges the main cables of the rail trains. Meanwhile, the electromagnetic protection measures for the wires are indispensable. Next, this paper designs a scheme to suppress EMI between wires for rail trains from the viewpoint of electromagnetic coupling. The electromagnetic coupling ways in rail trains mainly belong to non-conducting coupling, i.e. electric field coupling, magnetic field coupling and hybrid coupling of electric field and magnetic field. According to their characteristics, the following EMI suppression methods are designed.

- Methods of suppressing electric field coupling: reduce the distributed capacitance between wires by increasing their distances, shortening the wire length or increasing the ground plane, and use a shielding layer on the wires.
- Methods of suppressing magnetic field coupling: reduce the mutual inductance between circuits by making the wires as close to the ground plane as possible and making the magnetic field directions perpendicular to each other, and use a shielding layer on the wires.

- Method suppression hybrid coupling: add shielding layer and ground.

3. EMC Partition Layout Scheme for Rail Trains

In order to meet the EMC requirements, the layout design of rail trains should consider its three elements. Firstly, it is necessary to control the emission of the interference sources. Secondly, from the perspective of EMI propagation, the coupling impedance should be increased in the electromagnetic coupling. Thirdly, measures are taken to improve the anti-jamming capability of sensitive equipment.

Completing any of these three links can interrupt the closed loop of "EMI source-coupling path-sensitive components" in the EMC problem. And in order to reduce a large part of EMI, it is important to partition the rail trains because not only high-voltage, high-frequency and high-power interference source equipment, but also a lot of sensitive equipment that is susceptible to interference concentrate within the limited space of rail trains, which cause severe mutual EMI interference.

Based on this, this paper divides the rail trains into three modules, i.e. the roof, the inside and the bottom. High-voltage power supply interference source of the pantograph mainly concentrates on the roof. The main communication system, PIS, train control system, door system, train signal system and sensitive equipment are distributed inside the rail trains. At the bottom of the rail train, system interference sources, such as traction motors, high-speed circuit breakers and main switch boxes mainly concentrates on. Next, this paper expands the design from the above three modules.

3.1. Roof Layout of Rail Trails

The high-voltage equipment on the roof of the rail trains needs to have a certain margin in electromagnetic shielding and insulation performance on the basis of satisfying their electrical performance [8], in case it is eroded by harsh natural conditions, such as wind, sand, rain and snow. Figure 1 shows the equipment layout plan on the roof of rail trains.

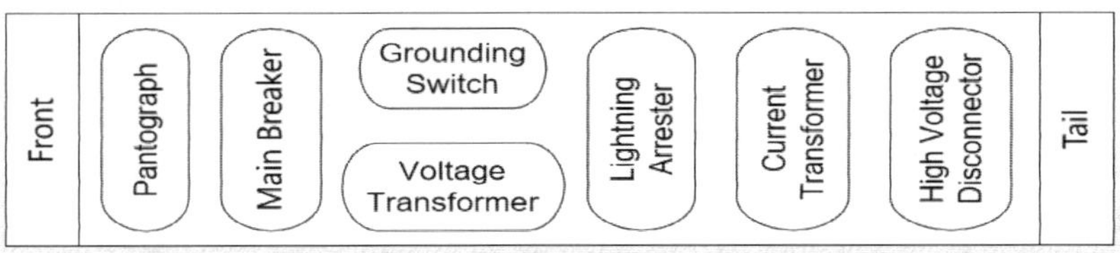

Figure 1. Equipment layout on the roof of rail trains.

- Pantograph: It is a device that obtains electric energy from an overhead contact system by a rail train and is installed on its roof. When the pantograph is raised, the air will be compressed, and the pantograph slide will contact the overhead lines to introduce 25 kV single-phase AC into the rail train from the overhead lines. When the pantograph is fallen, the compressed air will be discharged [9]. Its EMC design mainly considers the overvoltage caused by the pantograph as the power input terminal. It is arranged close to the front of rail train.
- Main circuit breaker: As the main switch and total protection of the railway train power, when the rail train has a serious fault, its total power supply can be quickly cut off to protect it. The main circuit breaker consists of a combination of high-voltage and low-voltage parts, which requires appropriate electromagnetic shielding and is placed close to the pantograph.
- High-voltage voltage transformer: After detecting the voltage of the overhead contact system, the voltage signal is supplied to the network control system of the rail train to control and protect the corresponding system.
- Grounding switch: It is installed next to the main breaker and is used in conjunction with the main breaker to ensure the safety grounding of the rail train during maintenance and repairing.
- Lightning arrester: It used to protect the electrical system on the rail trains to prevent lightning overvoltage and overpressure.
- High-voltage current transformer: Similar to the function of high-voltage voltage transformer, the only difference is that its transmitted signal is the current signal.

- High-voltage isolating switch: It automatically connects the two carriage roofs when they are connected. It is generally installed at the rear of the rail train.

3.2. Bottom Layout of Rail Trail

The EMC layout of the rail train bottom should follow the principle that the wiring and electrical equipment are closely combined, and their weight balance calculation is reasonably combined with the train boundary. Minimizing the electromagnetic radiation of equipment and cables within a controlled range is the goal of EMC design. In the specific design process, firstly, the wire trunking is used reasonably for wiring. In this way, not only the limited train space is saved, and the number of wires and tubes is reduced, but also the standardized design of EMC and electromagnetic radiation calculation can be facilitated. Meanwhile, from an overall perspective, the efficient combination of electrical equipment and cables can improve EMC performance. In addition, in order to ensure the safety and reliability of the rail trains, some factors, such as safety, fire prevention, overhaul, and cold resistance, should be fully considered when designing the bottom layout.

3.2.1. Bottom equipment layout of rail trains.

Large-size equipment at the bottom of rail trains may become the interference source. Therefore, its proper configuration not only meets EMC requirements, but also makes the overall layout reasonable.

- The larger equipment must at first meet the weight balance requirements, and then is arranged in accordance with electrical principles, EMC standards and wiring sequences.
- For equipment requiring ventilation cooling, such as braking resistors and motors, sufficient space must be provided to prevent other equipment from blocking its ventilation passages while avoiding accidental EMI [10].
- In order to avoid EMI to other equipment, the arcing direction of the equipment with outward arcing function, such as busbar high-speed circuit breaker, should face the carriage outside.
- Heating equipment should follow the principle of staying away from heat sensitive equipment and near non-sensitive equipment. If necessary, heat insulation should be added to prevent heat from being transmitted to the passenger compartment, and accelerating the aging of the wires to cause unexpected EMI.

Based on these principles, figure 2 shows the layout of the main equipment for the rail trains. The specific design ideas are as follows. The main inverter is placed in the middle of the rail train bottom to provide high voltage for each tractor. The auxiliary inverter is arranged at the front and rear parts of the rail trains, and the auxiliary system is powered by the high-voltage isolating switch. Its position is mainly determined by the action and position of the high-voltage isolating switch. Both the traction fans and the compressors belong to the auxiliary motors on the rail train. The compressor provides the wind source for the locomotive. The traction fan forcibly cools the rectifiers, compressors and traction motors. It needs to be arranged in multiple places at the bottom of the rail train to reduce the temperature of the electronic equipment. As a cooling and exhausting device in an air conditioning system, the cooling tower needs to be avoided in a high temperature environment. Therefore, the cooling tower is arranged around the traction fans. These devices are distributed in different positions according to their functions and EMC features, and are supplemented with different electromagnetic shielding methods to improve the EMC performance.

3.2.2. Bottom cable layout of rail trains

The wiring carrier of the undercarriage of the rail trains is mainly divided into three types: wire trough, electric conduit and coupling box. The layout design of the wire trough is as follows:

- The material of the current wire trough is divided into aluminum and steel. Both types of wire troughs have their own advantages and disadvantages. For aluminum wire troughs, their flatness is good, but it is not easy to be curved. It is generally used when there is a large space at the bottom of train. Steel wire troughs are easier to bend than aluminum, and are generally used when other electrical equipment needs to be properly avoided. In terms of EMC, the aluminum wire trough has better magnetic field shielding effect, and is usually selected.

- If the cable laid inside the wire trough has different voltage level inclusions and DC/AC inclusions, the internal division of the wire trough is inevitable. In order to ensure stable operation between the subsystems, the cables with different natures need to be isolated.
- In the special cases, such as the maintenance space of the reserved equipment, the cross section of the wire trough should be deformed on the basis of ensuring the basic cross section.

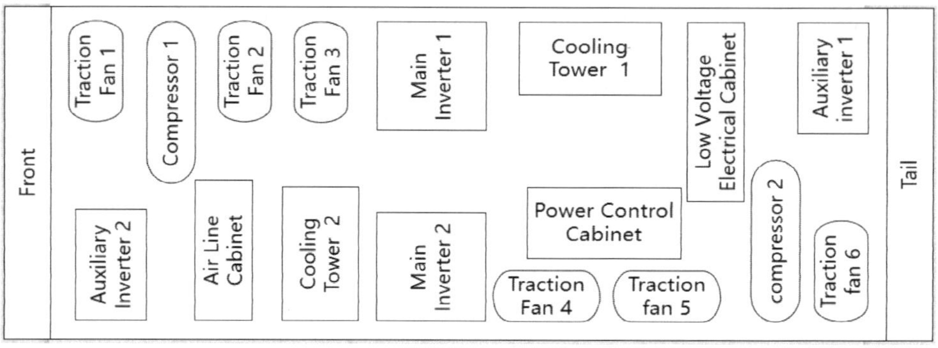

Figure 2. Equipment layout at the bottom of rail trains.

The layout design of the electric conduits is as follows:

- Similar to the current wire trough, the material of the wire conduit is generally divided into steel and aluminum. Most rail trains use internally coated aluminum tubes to reduce both body weight and tube deformation, and enhance shielding.
- The electric conduit must be close to its associated equipment, and its relative position should be selected according to the principle of connection between the various devices, so that the external interface can be better connected to avoid EMI generated by improper connection.
- Overall, electric conduits distribute in an obvious area division with the brake pipe to ensure the lowest degree of line coupling between them. In their shared area, it is strictly necessary to ensure that they do not interfere with each other.

The layout design of the coupling box is as follows:

- Similar to the electric conduit, the coupling box is also made of steel and aluminum. The aluminum is generally selected in the rail trains, because the aluminum coupling box can better weaken the EMI of the wires to the electrical system.
- The size of the coupling box is determined according to the actual situation. The following factors are mainly considered, i.e. the number of wires in the coupling box, the height difference between it and the external interface of other equipment, the space size at the coupling box, and whether it interferes with the surrounding electrical equipment.
- From the viewpoint of appearance, according to the actual situation, the coupling box can be locally widened and partially narrowed on the basis of the standard rectangular parallelepiped.

3.2.3. Bottom interface layout of rail train

The rail train system integrates a traction system, a braking system, and an auxiliary power supply system. These systems need to install equipment supporting their functions. There are mutual and complex EMI between these various systems. It is very important to clarify the layout design of the bottom interface of rail trains to improve the EMC performance. The interface layout scheme for the bottom of the rail trains is as follows:

- Based on different actual conditions, the external interface of the electric equipment under the rail trains can be selected from bare wire type, joint type, splint type and connection type.
- The protection and grounding of equipment is the most important. When the electrical equipment, wire trough, electric wire conduit and coupling box are made of metal, they must be protected and grounded. Grounding protection should be considered for electrical equipment, such as combiner boxes, switchboards and low-voltage circuit coupling boxes.

- The isolation of equipment from lines can also reduce EMI. For the devices or lines that are close to each other, it is necessary to consider their signal coupling or signal interference, and if necessary, use an isolator. For the devices with high noise, use noise isolators to reduce noise interference. For high voltage circuits, ensure their insulation and safety distance.
- To avoid aging and damage of the interface under extreme weather or improper operation, its fire protection, and heat and cold resistance are considered for better EMC performance.

3.3. Interior Layout of Rail Trains

In rail train system, most of control components are centrally installed in the electrical cabinets, which have become the control hub of rail trains. Therefore, EMC system layout of the rail trains is mainly reflected in the electrical cabinet layout. It is of great significance to make reasonable EMC analysis and find practical and feasible EMC measures to ensure the safety and stability of each system.

Each system host and its control unit are installed in the electrical cabinet. As shown in figure 3, its main parts include broadcast host, PIS play host, network system host, pyrotechnic alarm host and monitoring host. Only by making reasonable shielding design and grounding treatment design according to the operation principle and work frequency of these devices can the EMC standard be met. Prior to this, cable shielding and grounding design are critical to the EMC design of rail trains.

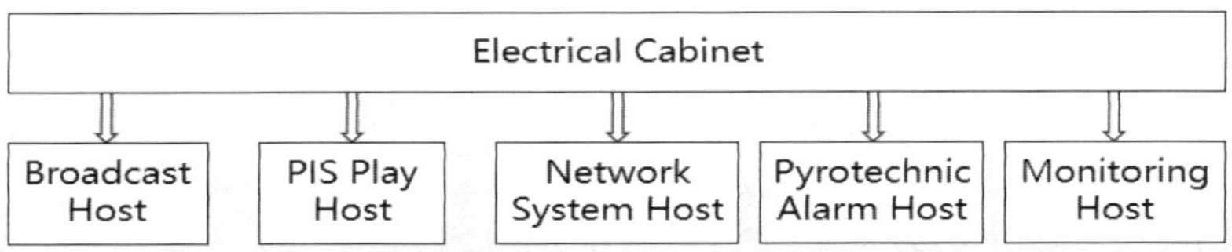

Figure 3. Host equipment inside electrical cabinet.

The cable shielding is to wrap a shielding metal mesh around the cable. A small part of the electromagnetic energy is converted into thermal energy through the eddy current loss inside the metal shielding body, and most of the electromagnetic energy is introduced into the earth through the grounding point. Therefore, the cable shielding must be matched with the grounding. Depending on the signal frequency, the grounding of the shielding layer is also different. For high-frequency signals, double-ended grounding is required to shield the electric field, and for low-frequency signals, only single-ended grounding is needed to achieve the purpose. When the signal frequency is in the low frequency range ($f<1$ MHz), if both ends of the cable are grounded, a ground loop will be formed, and a magnetic field will be formed in the shield layer, and noise interference will occur between the cables. When the signal frequency is in the high frequency ($f>1$ MHz), the impedance of the shielded cable cannot be ignored. If there is only single-ended grounding, the noise current will easily cause a voltage drop on the shield layer to make the electric potential different and affect the shielding effect. The EMC design of each host is described below.

3.3.1. Broadcast host

The broadcast host acts as an important device for transmitting voice signals inside the rail trains. It broadcasts the stations after reading the station information or the train running status information.

The broadcast host uses the data buses and audio buses of local operation network (LON) to connect with other main devices. It connects the internal auxiliary equipment through RS485 data buses and audio buses, and connects the external devices through the multifunctional vehicle bus (MVB), RS485, RS232 and standard audio buses. All its internal and external audio buses and data buses use shielded twisted lines to resist EMI. Its data bus transmissions adopt the form of differential signals. Therefore, they must be grounded at both ends. The audio bus transmissions adopt the form of analog level transmission, which can be single-ended grounding according to the equipment installation situation.

3.3.2. PIS play host

The PIS player host is mainly used to provide passengers with some riding information, train TV advertisements and emergency information. It is connected to external devices through VGA interface, RJ45 interface, RS232 interface, RS485 interface, standard audio interface, aviation plug and other interfaces to realize communication display function. Its inner and outer cables are shielded and twisted, and are single-strand multi-core cables. Its internal data use digital signal transmission. Therefore, each device is grounded at one end.

3.3.3. Network system host

The network system host includes four main modules: central control unit (CCU), remote input and output module (RIOM), RS485/MVB gateway, and data recorder. Their functions and their cable EMC design are listed as follows.

- CCU is the bus management host of the train network monitoring system, and its functions include bus management, process data transceiving, message transceiving and bus status monitoring. More importantly, CCU is responsible for the control, monitoring and protection of the entire rail trains, completing a series of control logic and fault diagnosis. Its external power cable uses a single-strand multi-core cable.
- RIOM collects the digital and analog signals in the rail trains, and then transmits the collected signals to the MVB network. Its external power cord uses a single-strand multi-core cable.
- RS485/MVB gateway connects MVB device and RS485 device, and processes the data from one to another. After such steps, the data interaction between the MVB device and RS485 device is completed. Its external RS485 bus cable uses a shielded three-core cable, and its external power cable uses a single-strand multi-core cable.
- The data recorder obtains status information of various devices on the rail train through the MVB bus and records information. Its power cable uses a single-strand multi-core one.

All MVB buses, RS485 buses and control local area network (CAN) buses transmit differential signals. Therefore, these three types of buses must be grounded at both ends.

3.3.4. Pyrotechnic alarm host

Pyrotechnic alarm host amplifies and filters the sensed smoke signals in the detector and suppresses their interference to perform a pyrotechnic alarm, and the alarm data is transmitted through a CAN bus or an RS485 bus. It uses low-pass filters, surge suppression devices, electrical transient suppression devices, radiated electromagnetic fields, and conducted disturbance suppression devices to suppress many outside interfering signals, while using filter capacitors and opto-isolation technology to reduce the internal interference.

3.3.5. Monitoring host

The monitoring host displays and stores the video data collected by the train driver's cab and the guest room camera, and simultaneously uploads the real-time monitoring video to the monitoring center in response to its instructions. Its inner and outer cables are shielded twisted wire and single-strand multi-core wire. Each device adopts single-ended grounding because of digital signal transmissions.

4. EMC Test Result

In order to evaluate the reliability of the EMC system layout scheme for rail trains, this paper selects the Shanghai Metro Line 6 for EMC testing.

4.1. Internal Anti-Interference Test

The internal anti-interference test is to verify the EMC performance between any electronic equipment, systems and subsystems inside the rail trains. Three items are set to test separately, as shown in table 1. According to the test results, the interference between the electronic devices inside the rail train is within the standard range, and does not cause EMI between the electronic devices, thus affecting the safe operation of the entire rail train.

4.2. Magnetic Field Strength Test of Train Body

This experiment is mainly to test whether the distribution of the magnetic field strength of the train body and the magnetic field emission intensity meet the requirements of the relevant standards and the owner's contract. Figure 4 shows the test method. There is a high-voltage cable between the pantograph and the high-voltage equipment box under the rail train. The cable has a rated operating current of 200 A and a frequency of 50 Hz. Firstly, test the magnetic field strength generated at different positions of the high-voltage cable. The four measuring points are respectively located under the top plate of the ultra-high-voltage (UHV) cable, the top of the anti-magnetic plate of the side-wall special pressure cable, the center position of the anti-magnetic plate, and the UHV line under the rail train. Secondly, three test points are arranged in the lower compartment of the pantograph to test the maximum electromagnetic induction intensity at 30 cm, 90 cm and 150 cm from the measuring point.

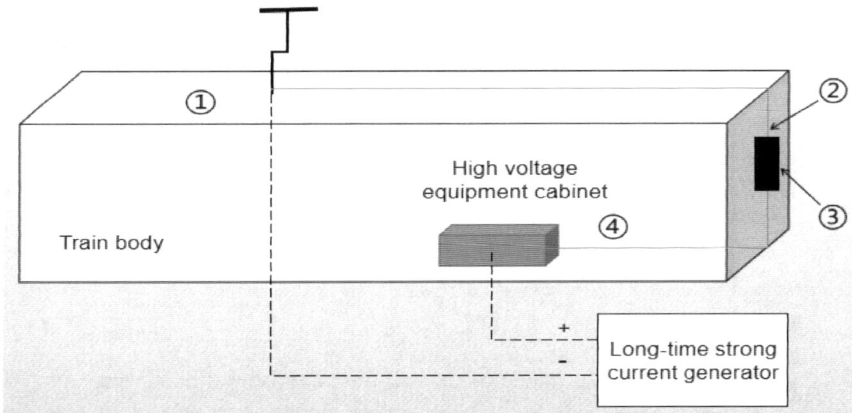

Figure 4. Magnetic field strength test of train body.

Table 1. Internal anti-interference test of rail trains.

Project name	Test train status	Test procedure	Experimental results
Interference of main and auxiliary converters to control circuits	Static Full acceleration Maximum braking force deceleration	1) Observe brake control unit (BCU) of electronic control circuits, traction/ auxiliary/ battery unit, door control unit (DCU), HVAC system, PIS, train CCU. 2) Read the information from the event recorder.	Each electronic control circuit operates in accordance with its respective specifications and there is no significant error in the event recorder of fault diagnostic circuit.
Interference between main and auxiliary converters	Static Full acceleration Maximum braking force deceleration	1) Observing the electronic control circuits of main inverters and auxiliary inverters. 2) Read the information from the event recorder.	The individual traction inverters and auxiliary inverters operate according to their respective specifications and there is no significant error in the event recorder.
Transient impulse interference when opening and closing of switches, relays, and contactors	Static Full acceleration Maximum braking force deceleration	1) Turn on and off the switches, relays and contactors in the main and auxiliary circuits. 2) Read data information from the event recorder.	There is no important error in the event recorder.

Table 2 shows the test results. It can be seen from the table that according to the magnetic induction intensity at 30 cm of the measuring point, it exceeds 100 μT (1 GS=100 μT), and the farther away from the interference source, the smaller the electromagnetic induction intensity. Therefore, if a passenger is exposed to a distance of 30 cm from the interference source for a long time, it may cause physical discomfort. At the same time, it will cause strong interference to the electronic equipment carried by passengers. In the position where the anti-magnetic plate is provided, the magnetic field strength is reduced by about 40% compared with the measuring point without the anti-magnetic plate, which indicates that the anti-magnetic plate has a certain magnetic separation effect. In summary, the stainless steel anti-magnetic plate with different thicknesses should be added to the roof panel directly under the pantograph, and no passenger seat should be provided directly under the pantograph. In addition, the magnetic induction at all other test points meets the magnetic field strength standard of the rail trains.

Table 2. Test results of magnetic field strength test of train body.

Distance (cm)	Current (A)	Measuring point 1 (GS)	Measuring point 2 (GS)	Measuring point 3 (GS)	Measuring point 4 (GS)
30	200	1.21	1.92	1.14	1.06
90	200	0.46	0.80	0.51	0.22
150	200	0.33	0.43	0.32	0.18

4.3. Immunity Test

4.3.1. Conducted Immunity Test

Test purpose: Test the EMC performance of the rail trains and the infrastructure next to the track. From a quantitative point of view, whether the harmonic current generated at a specific frequency point exceeds the required limit during train operation.

Test train status:

- Turn on all on-board equipment that may affect the interference current, including but not limited to auxiliary AC module (ACM), battery charger (BC), air conditioning unit (ACU), and lighting devices of rail train.

Train operating conditions:

- Continuous high speed, intermittent high speed, simulation operation.

Test equipment:

- Current probe, data acquisition card, acquisition signal computer.

4.3.2. Radiation immunity test

Test purpose: Verify whether the train radiation emission meets the compatibility requirements by testing the electromagnetic radiation emission of the whole rail train.

Test train status:

- Static state: The traction circuit is powered on but does not work, the auxiliary converter remains in operation, and all electrical systems on the rail train capable of generating radiation emissions are turned on.
- Slow state: The working state of the train electrical system is exactly the same as the static state, and accelerates or decelerates by 1/3 of the maximum traction force when passing through the antenna.

Test environment:

- In order to minimize the influence of weather conditions on the measured values, the weather conditions during the test should be: the temperature is not lower than 5 °C, the wind speed is not more than 10 m/s, and the humidity is kept from being condensed on the power supply.
- Environmental noise: Trees, walls, bridges or tunnels should be avoided within 10 m of the test point. There should be no substations or transformers nearby. Other rail trains cannot be driven within 20 m of the same power supply section.

4.3.3. Immunity test results
After testing according to the above test plan, the conducted immunity and radiated immunity of rail trains are within the reference range, as shown in table 3. In other words, the EMC layout scheme for the rail trains designed in this paper fully meets the requirements of EMC performance.

Table 3. Immunity test results.

Criterion	Project	Reference	Measurement	Qualified or not
IEC 61000-4-6	Conducted immunity	150 kHz-80 MHz	200 kHz	Qualified
IEC 61000-4-3	Radiation immunity	Peak is less than 20 V/m	18V/m	Qualified

5. Conclusion
In this paper, an EMC system layout scheme for the rail trains is proposed and then tested on Shanghai Metro Line 6. Firstly, this paper designs the equipment layout plan and the electric wiring layout for the rail trains from the whole. Secondly, according to the difference of equipment distribution, a rail train is divided into three parts, i.e. its roof, its interior and its bottom, and based on this partition, an EMC partition layout scheme is proposed. The test results show that the proposed EMC layout scheme fully meets the EMC requirements of rail trains, minimizes the risk of EMI before the operation of the rail trains, and provides a favorable guarantee for the safety and reliability of urban rail train operation.

6. Acknowledgments
This work was supported by the National Key R&D Program of China under Grant No. 2016YFB1200100.

7. Reference
[1] Fang J 2010 Research on EMC of urban subway *Dalian Jiaotong University*
[2] Lu J 2018 EMC test and analysis of subway trains *Automation Application* pp 30–31
[3] Ma Y 2013 Research on key technologies of EMC of a new generation of EMU *Beijing Jiaotong University*
[4] Sun X 2017 Electromagnetic emission measurement and data analysis of high-speed trains *Beijing Jiaotong University*
[5] Li Y, Zhang C and Wang Z 2019 Simulation study on EMC of metro line slot based on HFSS *Inverter World* pp 67–69
[6] Jiao J 2013 Research on EMC of co-constructed co-location of urban rail transit strong and weak electric system *Beijing University of Posts and Telecommunications*
[7] Sun J, Yin G, Gao Z and Zhao Y 2018 Design of EMC test scheme for a metro traction converter *Mechanical and Electrical Engineering* **35** pp 1232–1236
[8] Zhang S 2018 EMC of high-voltage train equipment *Dalian Jiaotong University*
[9] Wang G 2019 Research on the design of a roof train roof structure *Rail Transit Equipment and Technology* pp 1–2&8
[10] Wu D 2009 Study on EMC of linear motor metro trains on Guangzhou Metro Line 4 *Central South University*

IOP Conf. Series: Materials Science and Engineering **719** (2020) 012058 doi:10.1088/1757-899X/719/1/012058

Cut-Node Tree and Loops of LDPCs

Huanming Zhang

Institute of Electronics and Information Engineering
Foshan University, Guangdong, China, 528000

Abstract. Bipartite graph is used to illustrate Low-density parity-check (LDPC) code by Tanner [1]. In this thesis, we introduce a new algorithm to depict LDPC codes. Cut-node tree(CNT), although appearing to be a tree defined in graph theory, can figure out all cycles of an LDPC. And this method can apply to decoding, help to prevent superfluous iteration in belief propagation (BP).

1. Introduction

In coding and decoding domain, Tanner graph(TG) is suitable for indicating constraints or equations in LDPC, and iterative decoding of forward error correcting codes [2]. See Fig.1. It can commendably compose longer codes, [3][4].

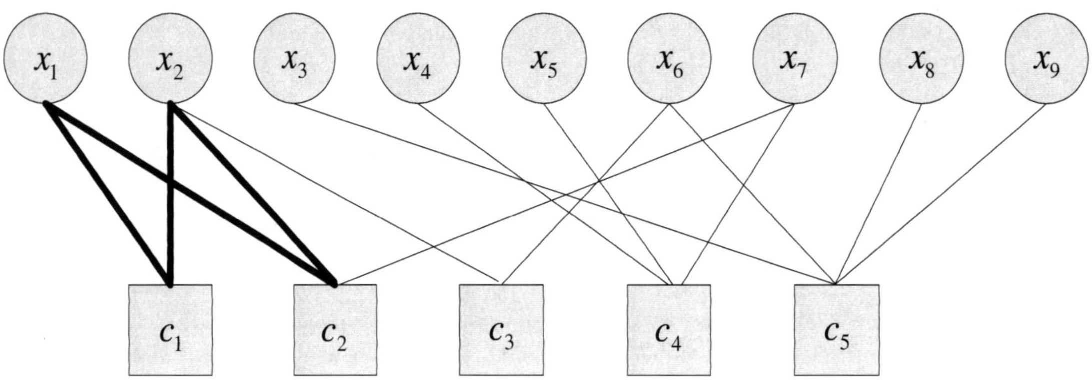

Figure 1. TG of an LDPC code

For a codeword of FEC, N is the length of codeword. the transmitted variable, supposing, is $x=\{x_1,\cdots,x_N\}$ and unknown; instead of received value $y=\{y_1,\cdots,y_N\}$, M Parity-check equations can be got, $c=\{c_1,c_2,\cdots,c_M\}$. $f=\{f_1^a,\cdots,f_N^a\}$ is initial value of transmitted codeword(variable). Where x_i is ith bit for a binary system of variable.

Belief propagation (BP) algorithm of FEC or particularly LDPC code can be elaborated by TG [5]-[7], See Fig.2.

R_{ji}^a represents check value from check c_j to variable v_i, and Q_{ij}^a variable value from v_i to check c_j.

IOP Conf. Series: Materials Science and Engineering **719** (2020) 012058 doi:10.1088/1757-899X/719/1/012058

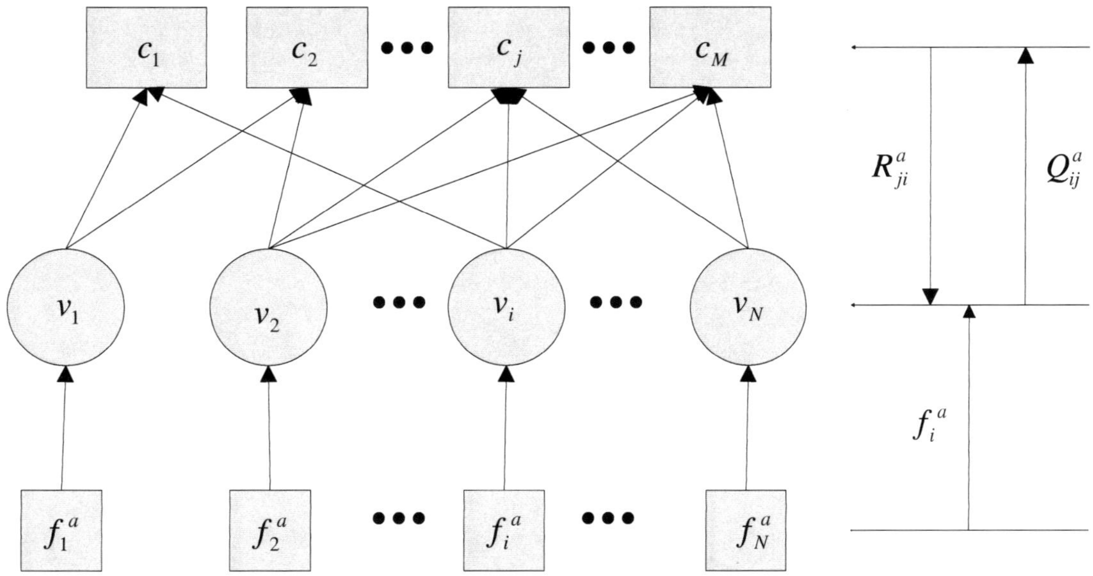

Figure 2. BP decoding Algorithm of LDPC

$$R_{ji}^{0}(t+1)=\frac{1}{2}\left(1+\prod_{j'\in N(i)\backslash j}\left(1-2Q_{ij'}^{1}(t)\right)\right) \tag{1}$$

$$R_{ji}^{1}(t+1)=1-R_{ji}^{0}(t+1) \tag{2}$$

$$Q_{ij}^{0}(t+1)=\alpha_{ij}(1-P_i)\prod_{i'\in M(j)\backslash i}R_{ji}^{0}(t+1) \tag{3}$$

$$Q_{ij}^{1}(t+1)=\alpha_{ij}P_i\prod_{i'\in M(j)\backslash i}R_{ji}^{1}(t+1) \tag{4}$$

Where α_{ij} is normalization value, satisfy $Q_{ij}^{0}(t+1)+Q_{ij}^{1}(t+1)=1$.

2. Theory Of CNT

In order to figure out all cycles of code. TG is re-drew to a tree, cut-node tree:

(1)selecting a value '1' in H, its variable node considered as root-node;

(2)correspondingly, all check nodes connected to the root-node as first-order child nodes;

(3)when growing, a node, it is a leaf originally or it appeared in previous node, will be forbid to grow and become an end-node;

(4)repeating(2) (3), at the end, a CNT can be got. If all nodes are connected, a single CNT can be got. Repeating this process until all end nodes are either cut-nodes or leaf nodes. See Fig.3:

IOP Conf. Series: Materials Science and Engineering **719** (2020) 012058 doi:10.1088/1757-899X/719/1/012058

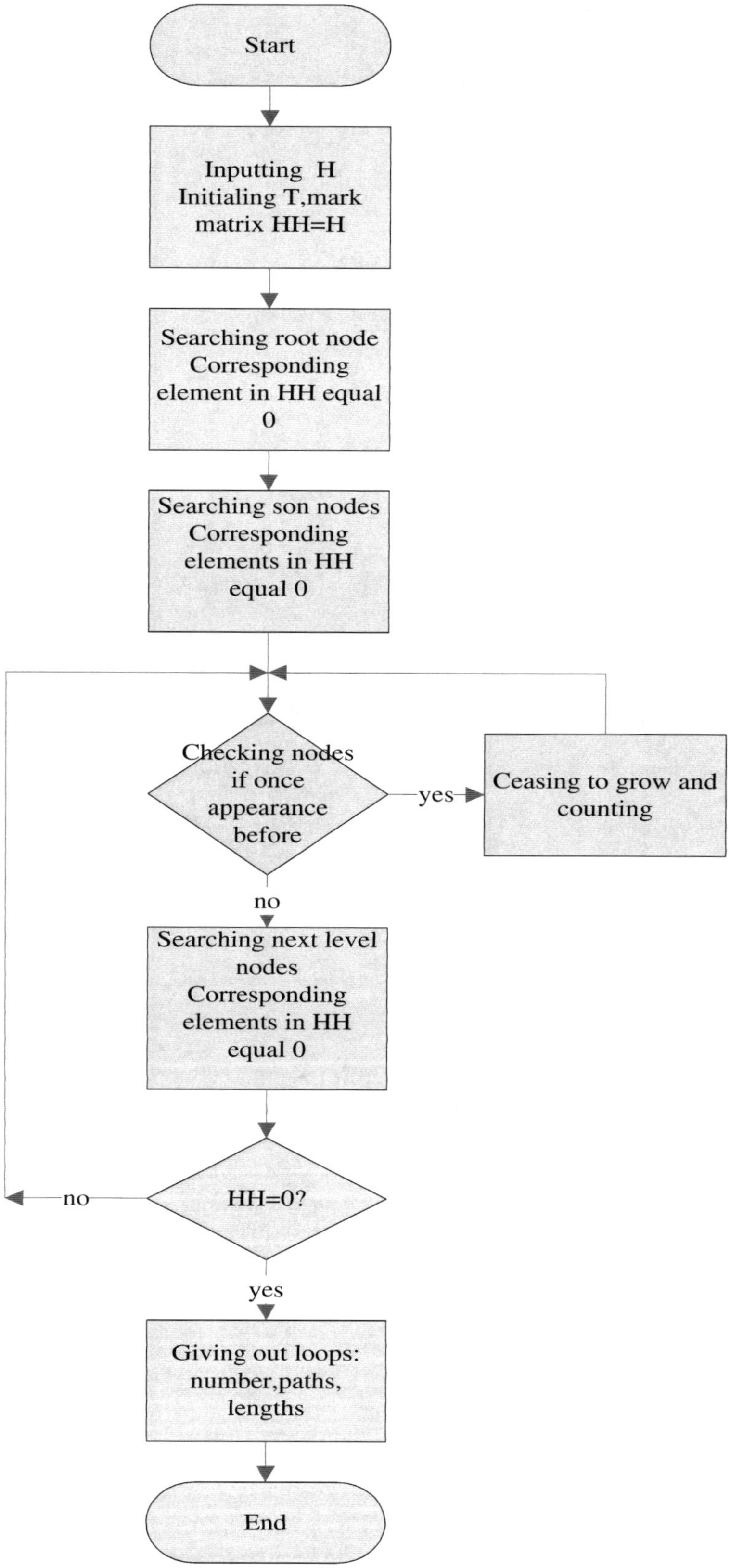

Figure 3. Diagram of CNT algorithm

The initial value of matrix HH is H, when a node appeared in CNT , the corresponding element in HH replaced by zero; all elements of HH change to zero at the end of algorithm.

3. Implement of Principle and Results

Here for one example:

(1)first, choosing a variable node $v_i, h_{ij} = 1$, as root-node, the son nodes of this root-node are those check nodes which connect to it, $v_i, h_{ij} = 1$;

(2)according to this principle, all child-nodes can be obtained. [6].

Such as an LDPC code with check *H*:

$$H = \begin{bmatrix} 1 & 1 & 1 & 0 & 1 & 0 & 0 \\ 1 & 1 & 0 & 1 & 0 & 1 & 0 \\ 1 & 0 & 1 & 1 & 0 & 0 & 1 \end{bmatrix} \tag{5}$$

CNT matrix used to express *H* by matlab. In fact, here tree matrix matches TG by element [i j k l m], [i j] states current node, [k l] states father node, m states times being cut, m=-1inherited, m=0 leaf.

Root-node: [1 1 0 0 -1]

Son-node Level1: [2 1 1 1 -1] [3 1 1 1 -1]

Son-node Level2: [2 2 2 1 -1] [2 4 2 1 -1] [2 6 2 1 0] [3 3 3 1 -1] [3 4 3 1 1] [3 7 3 1 0]

Son-node Level3: [1 2 2 2 -1] [3 4 2 4 2] [1 3 3 3 -1]

Son-node Level4: [1 1 1 2 1] [1 3 1 2 1] [1 5 1 2 0]

So, searched by a computer, all cycles of Eq.5 can be get by this algorithm. See Table 1.

Cycle 1: [1 1]-[1 2]-[2 2]-[2 1]-[1 1]

Cycle 2: [1 3]-[1 2]-[2 2]-[2 1]-[1 1]-[3 1]-[3 3]-[1 3]

Cycle 3: [3 4]-[2 4]-[2 1]-[1 1]-[3 1]-[3 4]

Table 1. Features of cycles about above example

Parameters	Values
Sparsity	0.43
Cycles	3
Total length of cycles	14
Average length	4.67
Girth	4
Maximum length	6
Relativity	1.5

Here, *H* is a single CNT, with 5 cut-nodes, 3 leaf node and 3 cycles.

In a graph without cycle, CNT just a tree defined by graph theory, but with cycles CNT can be got by cutting all cycles.

CNT has all characters which TG contains.

4. Conclusion

In this paper, we give out CNT algorithm. If matrix *H* is large, such as *H* in DVB-S2, cycles entangle with each other, it is hard to solve all cycles. LDPC's performances have certain relationship with characters of cycles. By CNT and matlab's cell array, cycles can be figure out relatively easily.

On the other hand, CNT can be applied in decoding algorithm to prevent redundant iteration in belief propagation algorithm.

5. Reference

[1] R. G. Gallager, Low-Density Parity Check Codes, MIT Press, Cambridge, MA, 1963.
[2] D. J. C. Mackay, "Good Error-Correcting Codes Based on Very Sparse Matrices", IEEE Trans.

Inform. Theory, vol. 45, 399-431, Mar. 1999.

[3] R. M. Tanner, "A recursive approach to low complexity codes," IEEE Trans. Inf Theory, pp. 533-547, Sept. 1981

[4] R.Kschischang, B.J.Frey, and H.A.Loeliger, “Factor graphs and the sum-product algorithm” IEEE Trans. Inform. Theory, vol.47, no.2, pp.498–519, Feb 2001.

[5] N.Wiberg, Codes and decoding on general graphs, Linkoping Studies in Science and Technology, Ph.D. dissertation No. 440, Univ. Link oping, Sweden, 1996.

[6] Reinhard D, Graph Theory[M]. New York: Springer-Verlag, 1997

[7] Martin J. Wainwright, Sparse Graph Codes for Side Information and Binning, IEEE SIGNAL PROCESSING MAGAZINE [47],47-57, SEPTEMBER 2007

[8] Martin Wainwright, Tommi Jaakkola, and Alan Willsky, Tree-based reparameterization analysis of belief propagation and related algorithms for approximate inference on graphs with cycles, ISIT2002, Lausanne, Switzerland, June 30-July 5, 2002

[9] Gholami Mohammad, Nassaj Akram, LDPC codes based on Mobius transformations, IET Communications,v13,n11,p1615-1624;ISSN:17518628,DOL:10.1049/iet-com.2018.6192

[10] Xu, Hengzhou, Zhu, Hai,Xu, Mengmeng, Zhang, Bo, Zhu, Sifeng,Girth analysis of tanner (5,11) quasi-cyclic LDPC codes, Proceedings - 14th International Conference on Computational Intelligence and Security, CIS 2018, p 210-214, December 5, 2018, Proceedings - 14th International Conference on Computational Intelligence and Security, CIS2018; ISBN-13: 9781728101699; DOI: 10.1109/CIS2018.2018.00053; Article number:8564291; Conference: 14th International Conference on Computational Intelligence and Security, CIS 2018, November 16, 2018 - November 19, 2018;

Joint Power Allocation and Relay Beam-forming in Orbital Angular Momentum Amplify-and-Forward Relay Networks

Shuai Li

School of Electronics and Information Technology, Sun Yat-sen University, Guangzhou 510006, Guangdong, China.
Email: lish66@mail2.sysu.edu.cn

Abstract. Orbital Angular Momentum (OAM) is a promising technology that can effectively improve the spectral efficiency of wireless communications. However, it is important for OAM to keep the antenna alignment of transmitter and receiver. The alignment requirements greatly increases the difficulty of the practical application of OAM, especially when there is a barrier or a large turning angle between uniform circular arrays (UCAs). In this letter, we studied the application of amplify-and-forward (AF) relay in multiple-input multiple-output (MIMO) OAM system to solve the problem caused by the existence of UCAs corners. In this work, our goal is to achieve the maximum achievable rate by jointly optimizing the source power allocation matrix and the relay pre-coding matrix, constrained by the power of the source and the relay. The problem is expressed as a convex problem by mathematical operations, and then based on the Lagrange dual method, we solved it by proposing an iterative algorithm. Numerical simulations show the superiority of the algorithm.

1. Introduction

With the rapid development of 5G communication technology and the universal application of terminal equipment, it is urgent to find a way to efficiently use spectrum resources. As a relatively new wireless communication mode, OAM is receiving increasing attention due to its great potential in improving the spectral efficiency (SE) of wireless communications. In short-range line of sight (LOS) applications, OAM is playing a more and more important role.

In order to ensure the performance of the OAM system, the source and destination antenna arrays are required to be fully aligned. If this requirement cannot be strictly enforced, the OAM system cannot be easily commercialized. In [1], the author models the UAM-based OAM channel and analyses the effect of the tilt angle on the OAM channel capacity.

A relay communication network is a technology that utilizes one or more relay terminals to obtain spatial diversity to improve transmission performance. In the AF relay strategy, the relay amplifies the received signal and then forwards it to the destination, which has lower complexity [2]. As far as we know, the application of AF technology to the OAM system to solve the problem of UCAs misalignment has not been analysed yet.

In this paper, a three node MIMO AF relay system, including a source, an AF relay and a destination is investigated to maximize the information rate by jointly optimizing the source power allocation matrix and the relay pre-coding matrix, constrained by the power of the source and the relay. The problem is expressed as a convex problem by mathematical operations, and then solved by an iterative algorithm based on the Lagrange dual method. The rest of the paper is organized as follows. In the second section, we introduce the system model and analyse and simplify the problem. The third section proposes an iterative algorithm for power allocation. The simulation results are given in the fourth section, and the article is summarized in the fifth section.

Notations: Boldface lowercase and uppercase letters denote vectors and matrices, respectively. The transpose, conjugate transpose and Frobenius norm of matrix $\mathbf{A}$ are denoted as $\mathbf{A}^T$, $\mathbf{A}^\dagger$ and $\|\mathbf{A}\|$, respectively. $\mathcal{CN}(0,\mathbf{I})$ denotes the distribution of a circularly symmetric complex Gaussian vector with mean 0 and covariance $\mathbf{I}$.

2. System Model and Problem Formulation

In this section, we present the considered OAM MIMO AF relay system mode and formulate the power allocation problem to maximize the achievable rate of the destination.

2.1. OAM MIMO AF Relay System

We consider a three node MIMO AF relay system, including a source, a relay and a destination. Assuming that all the nodes are equipped UCAs for signal transmitting, but the source's UCAs can't aligned with and the destination's because of the obstacle or the deployment restrictions of antenna installation sites.

To solve this problem, we can introduce an AF relay to realize the function of turn the signal in OAM system with UCAs used. The source's transmitting UCAs are aligned with the receiving UCAs of the AF relay, and the distance between the centres of them is d_1. As well as the transmitting UCAs of the AF relay are aligned with the receiving UCAs of the destination, and the distance between the centres of them is d_2. All the UCAs are equipped with N elements and the radii of UCAs for source, transmitter of relay, receiver of relay and destination are given by r_s, r_1, r_2 and r_d, respectively. There can generate N different OAM modes, counting the model number with l, where $l \in \{0,1,\ldots,N-1\}$ [3].

By inputting the same signal x_l to all the antenna unit of the UCA and performing phase shift sequentially, we can get number l OAM wave. Denote the input signal as $\boldsymbol{x} \in \mathbb{C}^{N\times 1}$, where $\boldsymbol{x} = [0,1,\ldots,N-1]^T$. So the transmission vector the OAM wave with all the modes activated can be expressed as

$$\boldsymbol{s} = \mathbf{WP}\boldsymbol{x} \tag{1}$$

where $\boldsymbol{x} \sim \mathcal{CN}(0,\mathbf{I})$, $\mathbf{P} = \mathrm{diag}([p_0, p_1, \ldots, p_{N-1}])$ is the power allocation matrix with p_l as the power allocation factor of mode l, $\mathbf{W} \in \mathbb{C}^{N\times N}$ is the discrete Fourier transform (DFT) matrix with w_{nl} being the element in the n-th row and l-th column of $\mathbf{W}$, where

$$w_{nl} = \frac{1}{\sqrt{N}}\exp\left(-\mathrm{j}2\pi\frac{nl}{N}\right) \tag{2}$$

In this paper, we use the free-space propagation model to simulate the channel as

$$h(d_{ij}) = \beta\frac{\lambda}{4\pi d_{ij}}\exp\left(-\mathrm{j}2\pi\frac{d_{ij}}{\lambda}\right) \tag{3}$$

where β and λ representing attenuation and phase rotation, which are complex constant associated with the carrier frequency, respectively. Given the complete alignment of the transmit and receive UCAs, the distance between the antenna elements denoted as $d_{ij}, i \in \{0,1,\ldots,N\text{-}1\}, j \in \{0,1,\ldots,N\text{-}1\}$ is [4]:

$$d_{ij}^{sr} = (d_1^2 + r_s^2 + r_{r1}^2 - 2r_s r_{r1}\cos\phi_{ij})^{1/2} \tag{4}$$

$$d_{\mathrm{ij}}^{\mathrm{rd}}=(d_2^2+r_{\mathrm{d}}^2+r_{\mathrm{r2}}^2-2r_{\mathrm{d}}r_{\mathrm{r2}}\cos\phi_{\mathrm{ij}})^{\frac{1}{2}} \tag{5}$$

where $\phi_{\mathrm{ij}}=\frac{2\pi(i-j)}{N}$.

Then the channel matrix between source and relay , and between relay and destination, denoted as $\mathbf{H}_{\mathrm{sr}}\in\mathbb{C}^{N\times N}$, $\mathbf{H}_{\mathrm{rd}}\in\mathbb{C}^{N\times N}$, respectively, can be expressed as $\mathbf{H}_{\mathrm{sr}}=\left[\left(h(d_{\mathrm{ij}}^{\mathrm{sr}})\right)\right]$, $\mathbf{H}_{\mathrm{rd}}=\left[\left(h(d_{\mathrm{ij}}^{\mathrm{rd}})\right)\right]$, where $\mathbf{H}=\left[\left(h(d_{\mathrm{ij}})\right)\right]$ means that $h(d_{ij})$ is the element of the i-th row and j-th column of **H**. We can find that $\mathbf{H}_{\mathrm{sr}}$, $\mathbf{H}_{\mathrm{rd}}$ are circulant matrixes because that the UCAs are used on both the source and the destination and there exists a precise symmetry.

The AF transmission scheme is employed. During the first slot, the source transmits the signal vector $\boldsymbol{s}$. Then the received signal at the relay is

$$\boldsymbol{y}_{\mathrm{r}}=\mathbf{H}_{\mathrm{sr}}\boldsymbol{s}+\boldsymbol{n}_{\mathrm{r}}=\mathbf{H}_{\mathrm{sr}}\mathbf{WP}\boldsymbol{x}+\boldsymbol{n}_{\mathrm{r}} \tag{6}$$

where $\boldsymbol{n}_{\mathrm{r}}\sim\mathcal{CN}\left(0,\mathbf{I}\right)$ is the additive Gaussian noise at the relay. The transmit power of the source is constrained to P_S as

$$P_{\mathrm{s}}\triangleq\|\mathbf{WP}\|^2=\|\mathbf{P}\|^2=\sum_{i=0}^{N-1}p_{\mathrm{i}}^2\le P_{\mathrm{S}} \tag{7}$$

During the second slot, the relay process the received signal by multiplying a pre-coding matrix $\mathbf{F}\in\mathbb{C}^{N\times N}$, $\mathbf{F}=\mathrm{diag}\left(\left[f_0,f_1,\ldots,f_{\mathrm{N-1}}\right]\right)$ and forwards it to the destination. But before the process of pre-coding, we realize the OAM mode de-multiplexing by the multiplying the received signal $\boldsymbol{y}_{\mathrm{r}}$ with the inverse DFT (IDFT) matrix $\mathbf{W}^{\mathrm{H}}$ [5] as

$$\tilde{\boldsymbol{y}}_{\mathrm{r}}=\mathbf{W}^{\mathrm{H}}\boldsymbol{y}_{\mathrm{r}}=\mathbf{W}^{\mathrm{H}}\mathbf{H}_{\mathrm{sr}}\mathbf{WP}\boldsymbol{x}+\mathbf{W}^{\mathrm{H}}\boldsymbol{n}_{\mathrm{r}}=\boldsymbol{\Lambda}_{\mathrm{sr}}\mathbf{P}\boldsymbol{x}+\mathbf{W}^{\mathrm{H}}\boldsymbol{n}_{\mathrm{r}} \tag{8}$$

where $\boldsymbol{\Lambda}_{\mathrm{sr}}\in\mathbb{C}^{N\times N}$ is a diagonal matrix. We can get it by the diagonalization of matrix $\mathbf{H}_{\mathrm{sr}}$ through the DFT matrix,

$$\boldsymbol{\Lambda}_{\mathrm{sr}}=\mathrm{diag}(\mathrm{eig}(\mathbf{H}_{\mathrm{sr}}))=\sqrt{N}\mathrm{diag}(\mathbf{W}\boldsymbol{h}_{\mathrm{sr1}}) \tag{9}$$

where $\boldsymbol{h}_{\mathrm{sr1}}$ is the first column of $\mathbf{H}_{\mathrm{sr}}$. $\boldsymbol{\Lambda}_{\mathrm{sr}}$ contains the eigenvalues of $\mathbf{H}_{\mathrm{sr}}\left(\xi_0,\xi_1,\ldots,\xi_{\mathrm{N-1}}\right)$. We denote $\boldsymbol{\Lambda}_{\mathrm{sr}}=\mathrm{diag}\left(\left[\lambda_{\mathrm{sr0}},\lambda_{\mathrm{sr1}},\ldots,\lambda_{\mathrm{srN-1}}\right]\right)$. Then we multiples the signal $\tilde{\boldsymbol{y}}_{\mathrm{r}}$ with the pre-coding matrix **F**, and after that we take the processing of OAM multiplexing by the DFT matrix **W**. So the received signal at the destination is

$$\boldsymbol{y}_{\mathrm{d}}=\mathbf{H}_{\mathrm{rd}}\mathbf{WF}\boldsymbol{\Lambda}_{\mathrm{sr}}\mathbf{P}\boldsymbol{x}+\mathbf{H}_{\mathrm{rd}}\mathbf{WFW}^{\mathrm{H}}\boldsymbol{n}_{\mathrm{r}}+\boldsymbol{n}_{\mathrm{d}} \tag{10}$$

where $\boldsymbol{n}_{\mathrm{d}}\sim\mathcal{CN}\left(0,\mathbf{I}\right)$ is the additive Gaussian noise at the destination. The transmit power of the relay is constrained to P_{R} as

$$P_{\mathrm{r}}\triangleq\|\mathbf{WF}\boldsymbol{\Lambda}_{\mathrm{sr}}\mathbf{P}\|^2+\|\mathbf{WF}\|^2=\sum_{i=0}^{N-1}(\lambda_{\mathrm{sri}}^2p_{\mathrm{i}}^2+1)f_{\mathrm{i}}^2\le P_{\mathrm{R}} \tag{11}$$

Just like the processing at the relay, at the destination, we realize the OAM mode de-multiplexing by multiplying the received signal $\boldsymbol{y}_{\mathrm{d}}$ with the IDFT matrix $\mathbf{W}^{\mathrm{H}}$ as

$$\begin{aligned}\tilde{\boldsymbol{y}}_{\rm d}&=\mathbf{W}^{\rm H}\boldsymbol{y}_{\rm d}=\mathbf{W}^{\rm H}\mathbf{H}_{\rm rd}\mathbf{W}\mathbf{F}\boldsymbol{\Lambda}_{\rm sr}\mathbf{P}\boldsymbol{x}+\mathbf{W}^{\rm H}\mathbf{H}_{\rm rd}\mathbf{W}\mathbf{F}\mathbf{W}^{\rm H}\boldsymbol{n}_{\rm r}+\mathbf{W}^{\rm H}\boldsymbol{n}_{\rm d}\\&=\boldsymbol{\Lambda}_{\rm rd}\mathbf{F}\boldsymbol{\Lambda}_{\rm sr}\mathbf{P}\boldsymbol{x}+\boldsymbol{\Lambda}_{\rm rd}\mathbf{F}\mathbf{W}^{\rm H}\boldsymbol{n}_{\rm r}+\mathbf{W}^{\rm H}\boldsymbol{n}_{\rm d}\\&=\boldsymbol{\Lambda}_{\rm rd}\mathbf{F}\boldsymbol{\Lambda}_{\rm sr}\mathbf{P}\boldsymbol{x}+\boldsymbol{\Lambda}_{\rm rd}\mathbf{F}\tilde{\boldsymbol{n}}_{\rm r}+\tilde{\boldsymbol{n}}_{\rm d}\end{aligned} \tag{12}$$

where $\boldsymbol{\Lambda}_{\rm rd}\in\mathbb{C}^{N\times N}$ is a diagonal matrix, like before

$$\boldsymbol{\Lambda}_{\rm rd}={\rm diag}({\rm eig}(\mathbf{H}_{\rm rd}))=\sqrt{N}{\rm diag}(\mathbf{W}\boldsymbol{h}_{\rm rd1}) \tag{13}$$

where $\boldsymbol{h}_{\rm rd1}$ is the first column of $\mathbf{H}_{\rm rd}$. We denote $\boldsymbol{\Lambda}_{\rm rd}={\rm diag}\left(\left[\lambda_{\rm rd0},\lambda_{\rm rd1},...,\lambda_{\rm rdN\text{-}1}\right]\right)$.

Based on (12), the achievable rate at the destination is

$$\begin{aligned}R&=0.5\log_{10}\left|\mathbf{I}+(\boldsymbol{\Lambda}_{\rm rd}\mathbf{F}\boldsymbol{\Lambda}_{\rm sr}\mathbf{P})^{\dagger}\mathbf{Q}^{-1}\boldsymbol{\Lambda}_{\rm rd}\mathbf{F}\boldsymbol{\Lambda}_{\rm sr}\mathbf{P}\right|\\&=0.5\log_{10}\left|\mathbf{I}+\mathbf{P}^{\dagger}\boldsymbol{\Lambda}_{\rm sr}{}^{\dagger}\mathbf{F}^{\dagger}\boldsymbol{\Lambda}_{\rm rd}{}^{\dagger}\mathbf{Q}^{-1}\boldsymbol{\Lambda}_{\rm rd}\mathbf{F}\boldsymbol{\Lambda}_{\rm sr}\mathbf{P}\right|\end{aligned} \tag{14}$$

where $\mathbf{Q}=\boldsymbol{\Lambda}_{\rm rd}\mathbf{F}(\boldsymbol{\Lambda}_{\rm rd}\mathbf{F})^{\dagger}+\mathbf{I}={\rm diag}[\lambda_{\rm rd0}\lambda_{\rm rd0}{}^{\dagger}f_0{}^2+1,\ \lambda_{\rm rd1}\lambda_{\rm rd1}{}^{\dagger}f_1{}^2+1,...,\ \lambda_{\rm rdN\text{-}1}\lambda_{\rm rdN\text{-}1}{}^{\dagger}f_{\rm N\text{-}1}{}^2+1]$. It's obviously that the matrixes in (14) are diagonal matrix, so we can rewrite the equation as

$$\begin{aligned}R&=0.5\log_{10}\left|{\rm diag}\left[\left(SNR_{\rm i}\right)\right]\right|\\&=0.5\log_{10}\prod_{i=0}^{N\text{-}1}SNR_{\rm i}\\&=0.5\sum_{i=0}^{N\text{-}1}\log_{10}SNR_{\rm i}\end{aligned} \tag{15}$$

where

$$SNR_{\rm i}=1+\frac{\tilde{p}_{\rm i}\tilde{f}_{\rm i}\tilde{\lambda}_{\rm sri}\tilde{\lambda}_{\rm rdi}}{\tilde{f}_{\rm i}\tilde{\lambda}_{\rm rdi}+1} \tag{16}$$

where $\tilde{p}_{\rm i}=p_{\rm i}^2,\tilde{f}_{\rm i}=f_{\rm i}^2,\tilde{\lambda}_{rdi}=\lambda_{\rm rdi}\lambda_{\rm rdi}^{\dagger},\ \tilde{\lambda}_{\rm sri}=\lambda_{\rm sri}\lambda_{\rm sri}^{\dagger}$. And in (13), ${\rm diag}\left[\left(SNR_{\rm i}\right)\right]$ means that the diagonal matrix takes $SNR_{\rm i}$ as its i-th element in the diagonal.

2.2. Problem Formulation

In this work, we aim to maximize the achievable rate of the destination, subject to the power constraint of the source and relay, by jointly optimizing the source power allocation matrix **P** and the relay pre-coding matrix **F**. So the problem of the OAM AF system can be formulated as (**P1**)

$$\begin{aligned}(\mathbf{P1})\ \ &\max_{\mathbf{P},\mathbf{F}} R=0.5\sum_{i=0}^{N-1}\log_{10}SNR_{\rm i}\\ &{\rm s.t.}\ \ P_{\rm s}\le P_{\rm S}\\ &\qquad P_{\rm r}\le P_{\rm R}\\ &\qquad p_{\rm i}\ge0,f_{\rm i}\ge0,\forall i\in\{0,1,...,N\text{-}1\}\end{aligned} \tag{17}$$

The problem (**P1**) is convex, because the objective function in (17) is jointly convex with respect to $p_{\rm i}$ and $f_{\rm i}$, and the constraint are linear. We can solve the problem by the tool of CVX.

3. Iterative Algorithm

To solve this problem, we propose an iterative algorithm based on the Lagrange dual method.

3.1. When Pre-Coding Matrix F is Given

When **F** is given, then f_i can be determined, and we can rewrite (**P1**) as (**P2**):

$$\begin{aligned}(\mathbf{P2})\ \max_{\mathbf{P}}\ & 2R=\sum_{i=0}^{N-1}\log_{10}(1+\alpha_i\widetilde{p}_i)\\ \text{s.t.}\ & \sum_{i=0}^{N-1}\widetilde{p}_i\le P_S\\ & \sum_{i=0}^{N-1}\gamma_i\widetilde{p}_i\le\widetilde{P}_R\\ & \widetilde{p}_i\ge 0,\forall i\in\{0,1,\ldots,N-1\}\end{aligned}\tag{18}$$

where $\alpha_i=\frac{\widetilde{f}_i\widetilde{\lambda}_{sri}\widetilde{\lambda}_{rdi}}{\widetilde{f}_i\widetilde{\lambda}_{rdi}+1}$, $\gamma_i=\widetilde{f}_i\widetilde{\lambda}_{sri}$, and $\widetilde{P}_R=P_R-\sum_{i=0}^{N-1}\widetilde{f}_i$.

We can find that (**P2**) is a convex problem, so we can use the Lagrange dual method to find the closed solution. The Lagrangian of (**P2**) with the Lagrange multiplier $\eta_1\ge 0,\eta_2\ge 0$ is given as

$$L\ (\mathbf{P},\eta_1,\eta_2)=\sum_{i=0}^{N-1}\log_{10}(1+\alpha_i\widetilde{p}_i)+\eta_1(P_S-\sum_{i=0}^{N-1}\widetilde{p}_i)+\eta_2(\widetilde{P}_R-\sum_{i=0}^{N-1}\gamma_i\widetilde{p}_i)\tag{19}$$

The derivative for the transmit power $\widetilde{p}_i$ is

$$\frac{\partial L}{\partial\widetilde{p}_i}=\frac{\alpha_i}{1+\alpha_i\widetilde{p}_i}-\eta_1-\eta_2\gamma_i\tag{20}$$

Setting this derivative to zero, we can get that

$$\widetilde{p}_i^{\,*}=\left(\frac{1}{\eta_1+\eta_2\gamma_i}-\frac{1}{\alpha_i}\right)^+\tag{21}$$

Equation (21) is a max-operation $(x)^+=\max(0,x)$ to ensure that the result is non-negative. The Karush-Kuhn-Tucker conditions guarantee the optimality of the solution even when equation (21) has been used.

The dual function is given by

$$f(\eta_1,\eta_2)=\max_{\mathbf{P}}L\ (\mathbf{P},\eta_1,\eta_2)\ \ \text{s.t.}\ \widetilde{p}_i\ge 0,\forall i\in\{0,1,\ldots,N-1\}\tag{22}$$

The dual problem is

$$(\mathbf{D2}):\min_{\eta_1,\eta_2}=f(\eta_1,\eta_2)\ \ \text{s.t.}\ \eta_1\ge 0,\eta_2\ge 0\tag{23}$$

Since the dual problem is always convex, but is generally non-differentiable, we can use the sub-gradient-based method, such as the ellipsoid method, to obtain the optimal dual solution as η_1^*,η_2^*.

3.2. When Source Power Matrix P is Given

When **P** is given, then p_i can be determined, and we can rewrite (**P1**) as (**P3**).

$$\begin{aligned}(\mathbf{P3})\ \max_{\mathbf{F}}\ & 2R=\sum_{i=0}^{N-1}\log_{10}\left(1+\frac{\theta_i\widetilde{f}_i}{\widetilde{\lambda}_{rdi}\widetilde{f}_i+1}\right)\\ \text{s.t.}\ & \sum_{i=0}^{N-1}\mu_i\widetilde{f}_i\le P_R\\ & \widetilde{f}_i\ge 0,\forall i\in\{0,1,\ldots,N-1\}\end{aligned}\tag{24}$$

where $\theta_i = \tilde{p}_i \tilde{\lambda}_{sri} \tilde{\lambda}_{rdi}, \mu_i = \tilde{p}_i \tilde{\lambda}_{sri} + 1$.

We can find the objective function is of the form $f(x) = \log_{10}\left(1 + \frac{\theta_i x}{1+\tilde{\lambda}_{rdi} x}\right)$, which can be proved to be convex because that the outer function is convex and non-decreasing, the inner function is convex. So (**P3**) is also a convex problem, we can solve the problem as before. The Lagrangian of (**P3**) with the Lagrange multiplier $v_1 \geq 0$ is given as

$$L\left(\mathbf{F}, v_1\right) = \sum_{i=0}^{N-1} \log_{10}\left(1 + \frac{\theta_i \tilde{f}_i}{\tilde{\lambda}_{rdi} \tilde{f}_i + 1}\right) + v_1(P_R - \mu_i \tilde{f}_i) \tag{25}$$

The derivative for the transmit power $\tilde{f}_i$ is

$$\frac{\partial L}{\partial \tilde{f}_i} = \frac{\theta_i + \tilde{\lambda}_{rdi}}{1 + (\theta_i + \tilde{\lambda}_{rdi})\tilde{f}_i} - \frac{\tilde{\lambda}_{rdi}}{1 + \tilde{\lambda}_{rdi}} - v_1 \mu_i \tag{26}$$

Setting this derivative to zero, we can get that

$$\tilde{f}_i^* = \left(\frac{-(\theta_i + 2\tilde{\lambda}_{rdi}) + \sqrt{\Delta}}{2(\theta_i + \tilde{\lambda}_{rdi})\tilde{\lambda}_{rdi}}\right)^+ \tag{27}$$

where $\Delta = \frac{4(\theta_i + \tilde{\lambda}_{rdi})\tilde{\lambda}_{rdi}\theta_i}{v_1 \mu_i} + \theta_i^2$

Also, we can use the same method to solve the dual problem (**D3**) of (**P3**) to obtain the optimal dual solution as v_1^*.

We can conclude the alterative algorithm as shown in table 1:

Table 1. The proposed iterative algorithm for OAM AF system.

1: Initialize	n = 0, $\mathbf{F}^{(0)}$
2:Repeat	Solve Problem (**P2**) and (**D2**) to obtain $\mathbf{P}^{(n)}$ and η_1, η_2
	n:=n+1;
	Solve Problem (**P3**) and (**D3**) to obtain $\mathbf{F}^{(n)}$ and v_1
3:Until	Convergence.

4. Simulation Results

The simulation results verify that our algorithm can improve the achievable rate when using AF relay.

In the simulation, we set the operation frequency f_c = 10 GHz, the wavelength λ = 0.03m, the UCA radius of the source, the relay, and the destination as $r_s = r_{r1} = r_{r2} = r_d = 20\lambda$ = 0.6m, the UCA configurations as N = 4, the transmission distance $d_1 = 500\lambda$ = 15m, $d_2 = 500\lambda$ = 15m, the average transmission power constraint $P_S = P_R$ are settled from 50dB to 90dB, and the antenna parameter β = 1.

We present the average achievable rates achieved by the proposed scheme, and we compare it with two other situation. Situation 1: the aligned situation, in which the source and the destination are aligned, and they will communicate directly with each other; Situation 2: the non-aligned situation, in which the source and the destination will have an angle offset, but they will ignore the offset and communicate directly with each other, too.

IOP Conf. Series: Materials Science and Engineering **719** (2020) 012059 doi:10.1088/1757-899X/719/1/012059

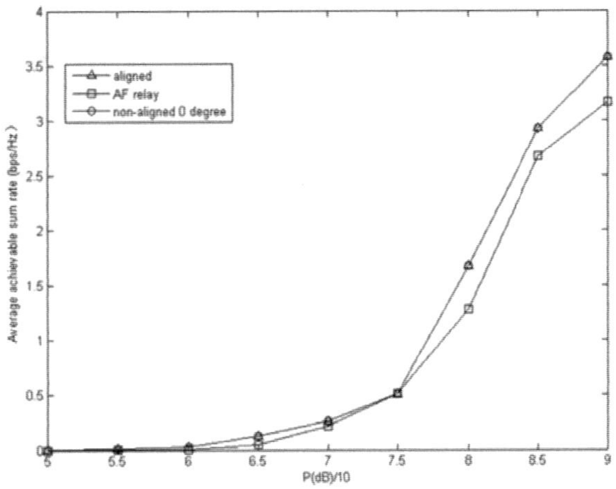

Figure 1. The achievable sum rate versus power, obtained by "aligned", "The AF relay" and "0 degree offset".

In Fig. 1, we settle the angle offset as 0 degree. Obviously, the curves of the aligned situation is coincided with the situation 2's curves. And we can find that, situation 1 is better than the AF relay situation, that is caused by the energy cost at the relay.

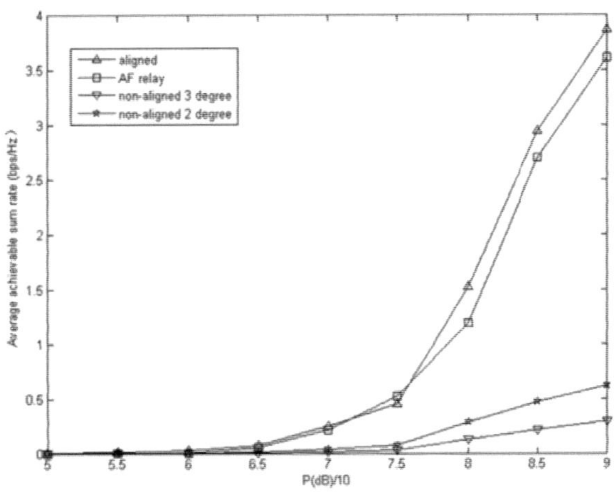

Figure 2. The achievable sum rate versus power, obtained by "aligned", "The AF relay" and "3 degree offset", and "2 degree offset".

In Fig. 2, we settle the angle offset as 2 and 3 degree, respectively. Obviously, the curves of the aligned situation is better than all other curves. And we can find that, situation 2 is worse than the AF relay situation, that is because the non-aligned of the UCAs. And, we can find that, with the degree increase of angle offset, the system performance is getting worse without the AF relay.

5. Conclusion

In this paper, an OAM MIMO AF relay system is investigated to maximize the information rate by jointly optimizing the source power allocation matrix and the relay pre-coding matrix, constrained by the power of the source and the relay. The problem is expressed as a convex problem by mathematical operations, and then solved by an iterative algorithm based on the Lagrange dual method. The simulation results demonstrate the superiority of our proposed scheme.

6. References

[1] Zhang Y, Feng W and Ge N 2013 *Proc. 8th Int. Conf. Commun. Netw. China (CHINACOM)* pp 271–5
[2] Xue C, Zhang Q, Li Q and Qin J 2017 *IEEE Trans. Veh. Technol.* vol 66 no 8 pp 7558–62
[3] Yang Y, Cheng W, Zhang W and Zhang H 2018 *IEEE Trans. Veh. Technol.* vol 67 no 11 pp 10704–14
[4] Chen R, Xu H, Marco M and Li J 2018 *IEEE Wireless Commun. Lett.* vol 7 no 4 pp 582–5
[5] Ertugrul B 2018 *IEEE Trans. Wireless Commun.* vol 17 no 3 pp 2029–37

Design and Test of Anti-EMI for Rail Trains

Tong Wu[1], Xiaotong Mu[2], Baichen Huo[3], Jilei Song[4], Baoying Shang[5], Liwen Hu[6] and Kai Liu[2]

[1] China National Institute of Metrology, Beijing, 100029, China
[2] School of Electronics and Information Engineering, Beihang University, Beijing, 100191, China
[3] Shenzhen Beihang Testing Co., Ltd., Shenzhen, 518000, China
[4] Beihang Engineering Technology Center (Shenzhen) Co., Ltd., Shenzhen, 518016, China
[5] Foshan Shunde Aerial Survey Electromagnetic Compatibility Technology Co., Ltd., Foshan, 528305, China
[6] Suzhou Smart Testing and Certification Co., Ltd., Suzhou, 215001, China
Email: mxt@buaaec.cn

Abstract. With the continuous improvement of the intelligence level of the rail transit system, the installation density of electrical and electronic equipment in rail trains becomes larger and larger, which causes the increasingly prominent electromagnetic interference (EMI) problem among any equipment, systems and subsystems. This paper proposes an anti-EMI design scheme for rail trains, which improves the anti-EMI capability of the entire train system from four aspects, i.e. equipment grounding, filtering, shielding and wiring. By taking Shanghai Metro Lines 4 and 6 as an example, we carried out the system implementation and testing. The test results show that the proposed scheme is verified to be effective and reliable to improve the electromagnetic compatibility (EMC) performance.

1. Introduction

With the continuous improvement of the intelligence level of the rail transit system, the installation density of electrical and electronic equipment in the rail trains becomes larger and larger, which causes the increasingly prominent electromagnetic interference (EMI) problem among any equipment, systems and subsystems [1]. The interference signals generated by various devices are superimposed on each other, and make the electromagnetic environment in the rail trains quite worse, which may cause malfunction of train equipment, frequent failure of communication networks, network degradation operation, and damage to electronic components, etc. [2]. In order to make the whole rail train system work stably, it is necessary to design an efficient anti-EMI scheme when assembling the rail trains to improve the anti-EMI capability of the entire train system.

In this paper, we at first analyze the EMI generation mechanism, and based on this, propose the basic anti-EMI method. Then according to the corresponding national standards and engineering practices, we propose an efficient anti-EMI design scheme.

2. General Rules of Anti-EMI for Rail Trains

EMI have three elements, i.e. interference source, coupling path, and interfered object. There are two ways for interference sources to transfer electromagnetic energy to the interfered objects, i.e. conduction coupling and radiation coupling. Conduction coupling includes direct conduction coupling, common impedance coupling and transfer impedance coupling. Radiation coupling is divided into field/antenna-to-antenna coupling, field-to-cable coupling and cable-to-cable coupling [3].

To realize conduction coupling, there must be a complete circuit connection between an interference source and its interfered object, and the interference signal is transmitted to the sensitive equipment along this connection circuit, causing EMI phenomenon. Radiation coupling means that electromagnetic energy spreads through the medium in the form of radiation electromagnetic wave, and the interference signal is emitted to the surrounding space in the form of electromagnetic wave, thus affecting the interfered object. Therefore, there are mainly three basic anti-EMI methods, and the implementation of each basic anti-EMI method has several different technologies, as shown in table 1.

Table 1. Anti-electromagnetic interference technology.

Basic method	Specific technology
Eliminate the interference source or reduce the intensity of EMI that it outputs to the outside	Reasonable grounding, circuit filtering, and electromagnetic shielding
Destroy the path of interference propagation or reduce the interference coupling degree	Reasonable wiring, and electromagnetic shielding
Special protection against sensitive equipment to improve its resistance to EMI	Reasonable grounding, circuit filtering, and electromagnetic shielding

It can be seen from table 1 that there are four main technologies for anti-EMI: reasonable grounding, electromagnetic shielding, circuit filtering, and reasonable wiring. Designing a proper grounding, suitable shielding, appropriate filtering and reasonable wiring method for the characteristics of electrical and electronic equipment in a rail train can improve the anti-EMI capability of entire train system. They are introduced below.

2.1. Reasonable Grounding

As a system with relative motion to the ground, in order to better realize the current loop, train body and its interior subsystems should be grounded in a proper way under normal operating environment. Grounding is usually divided into protective grounding, working grounding and shielding grounding.

Protective grounding is the grounding of the exposed conductive parts of the equipment for the purpose of ensuring personal safety and preventing direct or indirect contact with electricity.

Working grounding is a necessary condition for the normal operation of the entire electrical system of a rail train. On the one hand, it provides a channel for discharging charge or establishing a reference voltage. On the other hand, in the process of electromagnetic compatibility (EMC) design, grounding can introduce noise voltage into the earth and reduce noise interference. Correct working grounding is one of the important ways to suppress electromagnetic noise and prevent EMI.

Shield grounding refers to the grounding of cable shield. A cable shield is a layer of shielded metal mesh wrapped around a protected wire. When the EMI signal occurs, a small part of the electromagnetic energy is converted into thermal energy through the internal eddy current loss of the shielding layer, and most of them are introduced into the earth through the grounding point of the shielding layer. Therefore, the cable shielding layer must be grounded to play the role of shielding.

2.2. Electromagnetic Shielding

Electromagnetic shielding refers to the isolation of two spatial regions by a shield in order to control the induction and radiation of an electric field, a magnetic field and electromagnetic waves in one region to another region. For both external and internal electromagnetic waves from wires, cables, components, circuits or systems, a suitable shield absorbs their energy by eddy current loss, reflected their energy by interfacial reflection of electromagnetic waves on the shield, and cancels their energy by a reverse electromagnetic field on the shielding layer generated by electromagnetic induction [4]. Therefore, the electromagnetic shielding body has the function of reducing interference.

A rail train vehicle is a system integrating high voltage, frequency conversion, network communication and computer control. Its interior adopts high power supply equipment with rich harmonics, such as AC drive system and semiconductor static inverter, a large number of microprocessors for real-time control and diagnosis of traction and braking systems, and the installed systems for vehicle signals, wireless communications and passenger information. In order to guarantee the normal operation of electrical equipment on the vehicle and avoid the interference with each other, effective electromagnetic shielding measures can be taken to enclose the interference source to prevent the electromagnetic field from propagating outward through the space, and the electromagnetic shielding technique is also used to encapsulate the sensitive objects and protect them from any external electromagnetic fields.

2.3. Circuit Filtering

The circuit filtering technology is the main means to suppress the EMI of electrical and electronic equipment, especially conducted interference. It is also an important measure to protect the overall or partial shielding effectiveness of equipment. Practice has shown that even a well-designed product with the correct shielding and grounding measures still has electromagnetic disturbances entering into the device through the cable. Filters can be used to filter out related disturbances. Their role is to greatly attenuate the non-working signals, i.e. EMI.

According to different viewpoints, many types of filters are classified as follows:

(1) Filters can be divided into absorption filters and reflection filters based on the filtering principle.

(2) Filters can be divided into active filters and passive filters according to their working conditions.

(3) Filters can be classified into low-pass, high-pass, band-pass, and band-stop filters according to their frequency characteristics.

(4) Depending on the application, filters can be classified into power supply filters, signal filters, control line filters, anti-electromagnetic pulse filters, etc.

2.4. Reasonable Wiring

Due to the particularity of rail trains, there are various levels of voltage cables in the train, such as DC 1500 V/750 V/110 V and AC 380 V/220 V. There may be capacitive coupling and inductive coupling between the cables in these different circuits and functional units. The capacitive coupling between the cables occurs when the cables are laid in parallel. Its coupling degree depends on the length of the cables laid in parallel, the distance between the cables, and the height of the cable from the ground. Inductive coupling occurs in the loop conductor, and the strength of its inductive interference depends mainly on the area of the loop conductor [5].

In the process of wiring for rail trains, its coupling mechanism is mostly non-conducting coupling, which includes three ways, i.e. electric field, magnetic field and mixed mode (i.e. electric field and magnetic field work together). From the actual situation of train wiring, the most effective way to suppress electric field coupling is to reduce the distributed capacitance between the wires as much as possible, and to use the shielding layer. Herein, reducing the distributed capacitance can be achieved by increasing the distance between the wires, shortening the length of the wires, or adding a ground plane under the conductor. The most effective way to suppress magnetic field coupling is to reduce the mutual inductance between the circuits and to use a shielding layer. Herein, reducing the mutual inductance between the loops can be achieved by widening the conductor spacing, shortening the length of the conductors, making the conductors as close to the ground plane as possible and making the magnetic field directions perpendicular to each other. The most effective way to suppress the mixed mode is shielding and good grounding.

3. Anti-EMI Design for Rail Trains

From the above four aspects, an anti-EMI design scheme is proposed. Different anti-EMI measures are adopted to improve the anti-EMI performance of the train vehicle for vehicle operating environment and the different EMC characteristics of vehicle system equipment.

3.1. Grounding

For rail trains, all cabinets, brackets, and assembly framework of on-board electronic control units need to be grounded. The following grounding design is proposed based on their characteristics:

(1) The brackets of electronic control units ground the assembly architecture through the possible large surface area to ensure low impedance, and if necessary, a toothed spring washer should be used.

(2) The assembly architecture should be connected to the vehicle framework by the possible large surface area and as many points as possible. If grounding is not possible by bolting or welding, an additional strap will be used to connect the assembly framework to the vehicle body.

(3) Doors, floors, screen cabinets, connectors, etc. should also form a highly interconnected, large-area grounding design to ensure continuity of electrical conduction.

(4) The surface treated by spraying, painting or anodizing should use the tooth-shape elastic pad to make it contact reliably.

(5) For an electrical device that may be charged in the event of a fault, any its conductive part shall be internally connected to the equipotential ground. For any electrically conductive component not belonging to the electrical equipment, it shall also be internally connected to the equipotential ground.

In addition, for all electronic control units, they are equipped with DC/DC converters with potential isolation. The reference potential of the battery should be connected to the grounding point of the chassis, and the reference potential of the electronic control unit should be connected to the assembly framework with low impedance and low inductance, and these connections should be as short and wide as possible.

3.2. Shielding

Four main parts of rail trains need to be shielded, as shown below:

3.2.1. Shielding design of train body

(1) Highly conductive materials (i.e. copper, aluminum and steel), or metal plating and conductive coatings should be selected as shielding materials.

(2) Minimize the electrical discontinuity of the structure in order to control the leakage radiation through the casing and cabinet.

(3) Vents are designed to use perforated metal plates, as long as the diameter of the holes is small enough to achieve the required shielding effectiveness. If the requirements for ventilation are high, the honeycomb panel must be used; otherwise the shielding and ventilation requirements cannot be balanced.

(4) Measures to improve the shielding effectiveness of the gap include increasing its depth, reducing its length, adding an electromagnetic sealing gasket on the joint surface, and applying a conductive coating on the joint.

(5) When lap jointing with screws or by riveting, ensure that the fastening method has sufficient pressure to minimize the distance between the screws so as to maintain good surface contact when there is deformation stress, impact and vibration.

(6) Shielded pipe joints are used at the entrance and exit holes of the box, and metal pipes or metal braided net sleeves are connected.

3.2.2. Shielding design of electronic control unit

(1) If the assembly structure or casing is used for shielding, they shall be composed of a conductive material, and all the outer casing members shall be connected with low impedance, low inductance, large area, and ensure continuity of electrical conduction.

(2) If there is equipment in the cabinet that generates strong interference radiation, shielding should be provided between these equipment and sensitive equipment. For example, using metal plates, these boards and assembly areas are supported for multi-point low-impedance connection.

(3) The power supply filters and signal filters should have a reference ground plane and be connected to the casing or assembly structure. The filters should be placed at the entrance of the circuit.

(4) Doors, covers, etc. should be electrically connected to the framework and its side at multiple points. The connection can be pulled at the hinge of the door, connected by fasteners at the cover, and there should be other contact faces at the edge of the door.

(5) Unshielded signal cables need to be shielded when entering into the shielded area or filtered at the entrance to the shielded area.

3.2.3. Shielding design of signal and data line

(1) Both ends of the shield of the signal and the data line are grounded, because only the shield with both ends grounded can provide effective shielding to the interference field, and the filtering technology is not suitable for the signal and data lines.

(2) The shielding layer should be in contact with the possible widest surface area, and the low impedance and low inductance should be guaranteed to be connected.

(3) The cable shielding layer should be directly connected to the casing entrance to prevent interference current from flowing into the interior of the casing through the shielding layer.

(4) If it is not possible to maintain a minimum space between conductors that cause considerable interference, and the conductors are kept parallel over long distances, particularly sensitive conductors, such as data buses, must be placed separately in a metal conduit, and connected to the vehicle ground.

3.2.4. Shielding design of motor equal power unit

(1) The electronic power device is placed in a closed metal box to achieve shielding. The ventilation hole should be a punching plate, a metal grid, etc., and the cover, the cover plate and the ventilation hole should be sealed together with the motor casing;

(2) All outlets of the motor casing shall be shielded, filtered or electrically isolated.

3.3. Filtering

On the rail train, a low-pass filter is generally used to filter its power supply, and a band-pass filter is used to filter its signal devices. In order to maximize the filtering efficiency, the following filter design is proposed:

(1) The filter should be placed directly at or near the shield entrance. The filter shell or filter reference ground is connected to the bracket, shielded sheath or cabinet with low impedance and inductance.

(2) The input and output wires should be arranged separately. If a bypass capacitor or bypass filter is used, it should be bolted to the shielding ground.

(3) The filter leakage current cannot flow to the internal reference potential, and should be connected to the shielding or vehicle ground through a low impedance and low inductance connection.

(4) If the assembly area is not a shielded area or the filter is outside the shield, the connection from the filter to the electronic devices should be shielded.

(5) If the filter is located in the shielded area, the wiring from the shield to the filter should be shielded. All battery-powered electronic control units should be equipped with suitable power filters.

3.4. Wiring

The motor wires, inverter wires and auxiliary equipment wires of the rail trains are particularly prone to generate the interference, while the signal lines, data transmission lines, antenna leads, and audio and video wires are particularly sensitive to field inductive coupling.

Table 2. Cable type.

Cable category	Cable model
A1	Power supply cable
A2	Motor cable, brake resistor cable, heating cable, grid side filter (converter side) cable, power supply cable of auxiliary equipment, starter cable of internal combustion engine, transformer and four quadrant inverter leads, auxiliary equipment leads, reactor leads on the side of the reverse transformer
B	110V battery lead, control cable, unshielded signal cable
C	Shielded cables such as data buses, signal transmission wires (e.g. PT100, speed sensors, etc.), antenna cables, speaker cables, audio and video transmission cables

The wiring should comply with the regulations on EMC in TB/T 3153—2007 [6]. In order to ensure that the rail vehicle cabling meets the standard requirements, the cables are divided into three categories, as shown in table 2.

Different types of cables should be routed separately. To ensure adequate decoupling, the minimum spacing between different types of cables should be in accordance with table 3.

Table 3. Minimum spacing between different types of cables.

Cable category	A1	A2	B	C
A1		0.1 m	0.1 m	0.2 m
A2	0.1 m		0.1 m	0.2 m
B	0.1 m	0.1 m		0.1 m
C	0.2 m	0.2 m	0.1 m	

To achieve the EMC performance and improve the system ability to resist EMI, the following design is proposed:

(1) Separate the different categories of cables and try to maintain the spacing recommended in table 3.

(2) If the minimum spacing of different cable types is not guaranteed, especially for Class C cables, they can be separated by metal conduits and metal trunking, which are connected to the ground.

(3) If different types of cables cross at right angles, there is no need to consider the minimum spacing.

(4) In order to utilize the attenuation effect, the cables should be placed as close as possible to the vehicle ground, such as metal cable tubes, conduits, etc.

4. Test Results

The anti-EMI test is carried out on Lines 4 and 6 of Shanghai Metro to verify the stability and reliability of the proposed anti-EMI design for the rail trains.

4.1. Internal Interference Test

In order to verify the EMC performance between the electronic equipment inside the rail train and its various subsystems, we carry out its internal interference test [7]. The reference standard is GB/T 28806–2012/IEC 61133: 2006 [8]. This test consisted of three subtests as shown below.

4.1.1. Interference test of main and auxiliary converters on control circuits

All vehicle-mounted electrical appliances are powered on, and then the rail trains are in three operating states: stationary, full acceleration, and maximum braking force deceleration. Observe the three work states of the brake control units of electronic control circuits, traction/auxiliary/battery units, door control units, heating ventilation and air conditioning (HVAC) system, passenger information system (PIS), train central control units, and then read information from event recorder.

4.1.2. Interference subtest between main and auxiliary converters

All vehicle-mounted electrical appliances are powered on, and then the rail trains are in three operating states: stationary, full acceleration, and maximum braking force deceleration. Observe the operating conditions of the main inverter electronic control circuits and auxiliary inverter electronic control circuits in three states, and then read the information from the event recorder.

4.1.3. Transient pulse interference subtest when opening/closing switches, relays and contactors

All vehicle-mounted electrical appliances are powered on, and then the rail trains are in three operating states: stationary, full acceleration, and maximum braking force deceleration. In the three states, the switches, relays and contactors in the main and auxiliary circuits are turned on and off, and their operating conditions are observed and the data information is read from the event recorder.

4.1.4. Results

Table 4 shows the test results. In the three subtests, all the observed objects do not have over-voltage, under-voltage and under-current problems in three operating states of the rail trains, and the control circuits corresponding to each object work properly. Meanwhile, the event recorder does not report an error. It can be concluded that the EMC design between the internal electronic equipment and each subsystem can meet the requirements of the standard, and it has good anti-EMI capability.

Table 4. Internal interference test results of rail trains.

Three working states: Static/full acceleration/maximum brake deceleration			
Subtest name	Over-voltage, under-voltage,or under-current	Control circuit operation	Errors in event recorder
Interference of main and auxiliary converters to control circuits	No	Normal	No
Interference between main converters and auxiliary converters	No	Normal	No
Transient impulse interference when opening/closing switches, relays, and contactors	No	Normal	No

4.2. Radiation Immunity Test

In view of the fact that the distance between the antennas of mobile communication devices and the sensitive devices on the rail train is less than 1 m, the radiation emission intensity generated by the former is sufficient to interfere with the normal working intensity of in-vehicle electronic devices, which may lead to their performance degradation or even the loss of some important functions. Therefore, the test was set up to verify the immunity performance of the rail train to radiation interference sources, such as portable radio equipment.

The interference sources selected in this experiment mainly include the following two parts:

(1) Public communication equipment used by passengers or train employees.

(2) Hand-held walkie-talkie or wireless communication system used by train employees.

According to the standard requirements and daily operation, the selection of the equipment to be tested is based on the location where the mobile communication device is easily accessible. Therefore, the test equipment of this test includes train monitoring display, driver console, driver room door control, driver room machine control, passenger room display, guest room dynamic map, guest room door control, and guest room video camera.

The rail train used for the equipment test is stationary and should be in normal operating conditions without any addition or reduction of equipment. Devices that may form interference coupling paths are not allowed to appear because these interference coupling paths may affect its electromagnetic vulnerability. All equipment should be operated in accordance with established normal procedures. Use a variety of frequency bands to test the impact of interference sources on the devices under test.

The specific steps of the test are as follows:

(1) Check if the devices under test are installed correctly.

(2) Calibrate the interference intensity of the interference sources (i.e. mobile phone/interphone on the corresponding frequency band) and determine the distance from the equipment under test.

(3) Place the selected mobile phone/interphone next to the in-vehicle devices to be tested, keep the current position of the phone (just bell, and do not answer the call), observe the working status of the devices under test or check the interfered intensity of each device through the train control and communication units, and record the results truthfully.

(4) If there is a sensitive phenomenon and the performance level does not meet the requirements, adjust the distance between the mobile phone and the devices under test, and record the disturbed distance as the basis for subsequent analysis.

The test results are shown in table 5. It can be seen from table 5 that all the tested equipment work normally under various interference sources and have high radiation immunity.

5. Conclusion

This paper proposes an anti-EMI design scheme for rail trains. From the four aspects of grounding, shielding, filtering and wiring, the scheme integrates the characteristics of each part of the rail trains. The assembly tests are carried out on the Shanghai Metro Line 4 and Line 6 to verify its effectiveness and reliability. The test results show that the proposed scheme can provide useful reference for the subsequent anti-EMI design of railway trains, improve the reliability of train systems, ensure the safety train operation, passenger safety and reduce the operation and maintenance cost of rail trains.

Table 5. Test results.

Test site	Working status	Test site	Working status
Monitoring display	Normal	SIV controller	Normal
In-vehicle radio display	Normal	Driver room light controller	Normal
Broadcast control	Normal	Headlight controller	Normal
In-vehicle radio keyboard	Normal	Brake mode controller	Normal
Broadcast control keyboard	Normal	Drivers cab camera	Normal
HMI screen	Normal	Flute controller	Normal
HMI screen indicator	Normal	Dashboard and surrounding	Normal
Mode control display	Normal	Right door switch controller	Normal
MMI display	Normal	Left door switch controller	Normal
Speed control device	Normal	LCD screen	Normal
Door operation console	Normal	Guide screen	Normal
Lighting console	Normal	Surveillance video cameras	Normal

6. Acknowledgments

This work was supported by the National Key R&D Program of China under Grant No. 2016YFB1200100.

7. References

[1] Sun P 2015 Electromagnetic compatibility standards for trains and their applications *Railway Quality Control* **43** pp 7–10

[2] Cao H, Zhang A and Jiang F 2008 Treatment of EMI in the subway trains *Railway Locomotive and Cars* **28** pp 130–134

[3] Zhang H, Wu D and Zhang A 2011 EMC designing technology for rail vehicles *Electric Drive for Locomotives* pp 61–64

[4] Wang J 2002 Mechanism and development of electromagnetic shielding materials *New Chemical Materials* **30** pp 16–18

[5] Shan Q 2013 Research on key techniques of electromagnetic compatibility for China high-speed EMU *Beijing Jiaotong University*

[6] Railway application-rolling stock: rules for installation of cabling 2007 TB/T 3153–2007

[7] Yan Z 2007 Research on electromagnetic compatibility test method of subway screen door system *South China University of Technology*

[8] Railway application-rolling stock: Testing of rolling stock on completion of construction and before entry into service: GB/T 28806–2012/IEC 61133 2012 General Administration of Quality Supervision, Inspection and Quarantine of the People's Republic of China/China National Standardization Administration Committee

Travel Route Prediction Using Travel Habits and Real-Time Traffic Condition

Wanguo Mu

Lianyungang Jari Electronics CO.,LTD
18036670164@163.com

Abstract. The traffic problems of cities at home and abroad, especially major cities, have become focus issues. To solve this problem, traffic guidance is one of the most important solutions. For traffic guidance, precisely prediction of travel route of the users becomes one more and more urgent issue. To find out which factor affects the prediction mostly or to propose one efficient route prediction algorithm is attracting more researchers' attentions. In this paper, we proposed one route prediction model to deal with the problem by considering both travel habits and real-time traffic condition. By embedding LSTM model, the proposed approach has the ability of predict sequence data to predict the travel route. Also the logistical model is introduced to embedding the real-time traffic condition data into the proposed model. Experiment results have evaluated the performance of the proposed approach.

1. Introduction

With the rapid development of social economy and the improvement of people's living standards, the number of urban residents' car and resident trips have been increased constantly. As a result, the traffic problems of cities at home and abroad, especially major cities, have become focus issues. Problems such as poor road network, insufficient facilities, crowded traffic, etc, have been more and more prominent. In many cities, blocked traffic, parking difficulties, disordered traffic, etc have have been increasingly emerging, resulting in great impact and pressure of transportation system management[1].

In addition to paying attention to urban transportation facilities, traffic guidance is also important for solving the problem of heavy traffic. Efficient traffic guidance provides the means to improve road resources utilization, improve traffic speed and ease traffic congestion[2].

The research results of travel route prediction and guidance methods at home and abroad show that travelers' response speed can be accelerated to adjust routes in time and avoid traffic congestion by certain methods applied to accurately predict the travel routes in theory[3].

Therefore, this paper aims to analyze routes selection by effective survey methods and determine to introduce new considered variables in the process of travel routes prediction, so as to make the prediction more practical[4, 5].

The LSTM and Binary Logistical model can be used to predict road segments and overall travel route selection, and to guide car travel, which can help residents avoid traffic congestion, reduce travel time, make travel convenient and improve travel efficiency.

2. Related Work

2.1. ADVANCE System of America

ADVANCE (Advanced Driver and Vehicle Advisory Navigation Concept) is a typical distributed traffic guidance system[6]. Traffic Information Center acquires real-time traffic data by collecting the data of floating cars and fixed detectors on the road, and forms synthetic impedance function of road segments by traffic state discrimination and travel time estimation. Thetn the impedance function will be sent to related vehicle-mounted device, which could provide the driver with optimal driving path and guide the car to the starting point according to driver's individual requirement.

2.2. Ali-Scout System of Europe

Ali-Scout System is a centralized traffic route guidance system in Europe[7]. Its central processing platform is responsible for data maintenance, travel time calculation and optimal route calculation. In addition, it can connect with vehicles in the road network through infrared beacons installed on both sides of the intersection. When a vehicle pass through the beacon, the beacon will send information, such as local map, shortest pate, etc, to the vehicle and the vehicle will transmit data, such as road operation condition, to the data center.

2.3. VICS System of Japan

Japan's Vehicle Information and Communication System, established on July 1, 1995, was set up and managed by Japan Road Traffic Information Center[8]. This system aims to improve road traffic safety, road traffic liquidity and road environment. It is regarded as the most successful system in the world for providing traffic information. The VICS system has been spread throughout Japan. In actual operation, it consists of four parts: information collection, information processing, information publishing and information application. VICS can provide traffic information for travelers through radio beacon, light beacon and FM multicast.

The internal traffic guidance system is currently in the research stage of positioning system, electronic map, both-way communication, etc. Those traffic guidance means that have already put into use are based on radio broadcasting and variable message sign (VMS). With relatively simple functions, for them, there are many issues needed to be improved further. The guidance information generation and management system with perfect functions and architecture has not been formed yet. (1) System-based optimal route prediction

Fu Dongmian proposed a minimum transfer algorithm based on the breadth search for sets of several nodes so as to calculate travel route with minimum transfer times[9].

Zhang Shujian analyzed features of urban traffic network and summarized problems existing in the traffic network to apply GIS-T System with function of spatial analysis into travel route algorithm. He proposed a dynamic route guidance algorithm on the basis of GIS- and genetic algorithm[10].

Kong Huixin studied a route recommendation method based on real-time passenger flow distribution. She proposed the calculation method of its passenger carrying probability and measurement standard - PVC (Potential Vacant Cost) equation - of route recommendation through K-means clustering algorithm to divide, so as to design the minimum cost recommendation algorithm for taxi travel routes[11].

3. Route Prediction Based on Travel Habits and Real-time Traffic

A traveler's route selection depends to a great extent on the traveler's habits. The traveler's route selection can be regarded as the accumulation process of habitual routes, that is, acquiring habitual routes from historical routes selection and choosing the next route according to habitual routes. Each route selection has a certain probability, as a result, it can be described and predicted by LSTM. According to the investigation and modeling analysis of the travel routes selection mentioned above, it can be seen that a traveler's selective behaviour of route not only depends on his/her traveling habits, but also affected by the actual road conditions.

The information input the model includes three aspects. The first is traveler's basic information, including traveler's social, economic and travel attribute. The second is historical route selection data, that is, the actual routes selection data of all kinds of travel in several days, including travel routes data and actual routes selection for several days. The third is real-time traffic conditions. After obtaining the relevant data, the model is applied to predict the rout selection[12].

IOP Conf. Series: Materials Science and Engineering **719** (2020) 012061 doi:10.1088/1757-899X/719/1/012061

3.1. LSTM
LSTM is one of the most common approaches for sequence data prediction[13]. As shown in Figure 1, the repeat cell of LSTM can be embedded into a neural net model to predict the sequence information. As described in equation (1), the sigmoid layer called the "input gate layer" decides updated values is introduced.

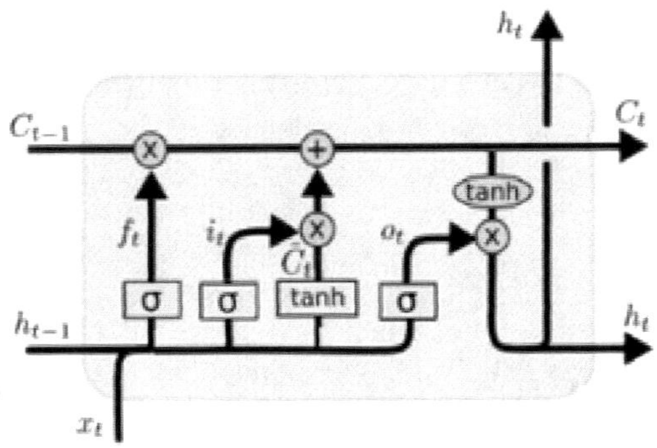

Figure 1. The structure of LSTM

$$f_t = \sigma\left(W_f \cdot [h_{t-1}, x_t] + b_f\right) \tag{1}$$

To calculate the input of time t, equation (2) is introduced:

$$i_t = \sigma\left(W_i \cdot [h_{t-1}, x_t] + b_i\right) \tag{2}$$

Then a tanh layer creates a vector of new candidate values is added into the state as show in equation (3).

$$\tilde{C}_t = \tanh\left(W_C[h_{t-1}, x_t] + b_C\right) \tag{3}$$

Using equation (4), the old cell state C_{t-1} is added into the new cell state C_t.

$$C_t = f_t * C_{t-1} + i_t * \tilde{C}_t \tag{4}$$

Then a sigmoid layer which decides what parts of the cell state we're going to output as shown in equation(5).

$$o_t = \sigma\left(W_o[h_{t-1}, x_t] + b_o\right) \tag{5}$$

Finally LSTM puts the cell state through tanh (to push the values to be between −1 and 1) and multiply it by the output of the sigmoid gate, so that we only output the parts we decided to.

$$h_t = o_t * \tanh\left(C_t\right) \tag{6}$$

3.2. Logistical Prediction Model
Logistical prediction model is one of the most famous prediction models in route prediction[14]. It assumes utility maximization hypothesis on the observed data. The corresponding utility function of it is defined as follows:

$$U_{ip} = o_{ip} + e_{ip} \tag{7}$$

As shown in equation (7), i is signal of choice. p is choice of user. o_{ip} means the observed influence factor and e_{ip} is the influence factor of unobserved. When i equals 1, it means that user chooses the customary section.

$$o_{ip} = \sum_{k=1}^{K} \alpha_k x_{ik} \tag{8}$$

x_{ik} means that the user choose the k-th influenced factor of route i.

To calculate model, optimization technique should be used. Model parameters optimality is usually defined with a loss function and the objective task is to minimize the sum of losses on the observed data.

3.3. The Proposed Model

In the following, we present our algorithm which is generalized from LSTM and logistical. If the collected ratings are entering sequentially, we could adjust the model by considering the coming data only. Dual-average method for the two model which absorbs previous information in an approximate average gradient of the loss is one of the most widely used methods to solve optimization problems[15]. The proposed algorithm is detailed as follows:

ALGORITHM1: *Proposed Model*

Step 1: Prediction of habits-based routes selection. Markov Prediction Model is used for habits-based route selection. Based on historical data of route selection, the model can predict selective probability of traveler's alternative routes under the condition of considering habits only and set the route with maximum probability as the habitual route. The road segments consisted all the alternative routes can be regarded as the habitual segments in the route.

Step 2: If there are more than one optional road segment in the road segment selection, the selective probability of habitual routes and non-habitual routes should be calculated repeatedly under different road conditions and ratiometric conversion should be carried out to determine the selective probability of each segment in one route selection. If there are multiple segment selections, above steps should be repeated until the end of the route.

Step 3: Utility calculation of route selection based on habits and road conditions. On the basis of the selective probability of each route based on habits in the first step and selective probability of each road segment in the third step, the selection utility can be obtained by amending the selection probability of each route according to the segment composition of each route.

Step 4: The selection probability calculation of the travel route. The selection probability of route selected by the model can be calculated according to the selection utility of each route calculated in the third step.

return *prediction result*

4. Experiments

To evaluate the performance of our algorithm, several experiments are conducted on a dataset that is described in section III. The questions we want to deal with include:

1) How is the performance of the proposed algorithm?
2) In which way do parameters affect the method's performance?
3) What is the time overhead of the proposed method?

4.1. Evaluation Metrics

We carried out several experiments to evaluate the performance of the proposed algorithm. The experiments are conducted on a dataset.

For the purpose of measuring the prediction quality of our method in comparison with other approaches, hit-ratio metrics which is one of the widely used methods in the field of prediction system is embodied, and it is defined as:

$$hit - ratio = \frac{hit}{testlists} \tag{9}$$

where hit is the actual hits of selected results assigned by user, testLists is estimated value. The higher the hit-ratio, the better the performance[16].

4.2. Performance Evaluation

Table 1 presents the results of prediction accuracy performance of the approaches mentioned above. According to these experiments, we have several important observations. The total prediction hit-ratio is 82.5. When the observation sample is the customary routine, the prediction performance is high and the hit-ratio is 91.6%. When the observation sample is the uncustomary routine, the prediction performance is low and the hit-ratio is 68.9%. At this point, the model overestimates the probability of choosing the habits without considering the road conditions. In fact, road conditions have a certain influence on the choice of road sections, and some of the samples will choose unfamiliar road sections because of the road conditions.

Table 1. Prediction results

observations		*prediction results*		
		route choice		*hit-ratio*
		customary	*uncustomary*	
routine choice	*customary*	522	82	91.6%
	uncustomary	220	321	68.9%
hit-ratio		--	--	82.4%

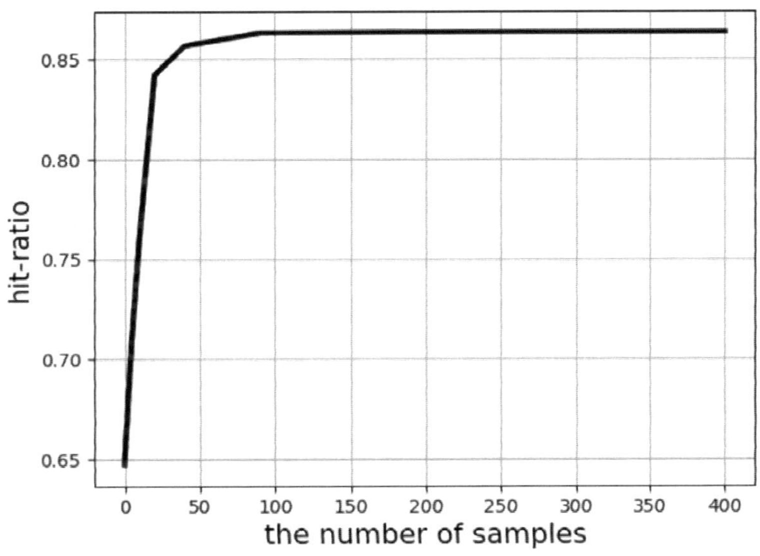

Figure 2. The hit-ratio under different number of samples

A respondent whose number is 18 was performed a one-sample analysis. The respondent #18 belongs to age group 18-30 years old, with a driving age of less than one year, the occupation is a student, the monthly income is below 4000, and the education is bachelor. The purpose of the trip is to go to school, and the departure time is the evening peak. Bringing these independent variable information into the model, predicting the probability that the respondent #18 will select the custom road segment and the uncustomary road segment (adjust the road segment according to the road condition) under different road conditions. The prediction result is shown in Table 2.

Table 2. Prediction Results

C	*Probability of choice*	
	customary route(%)	*uncustomary route(%)*
C1	77.3	22.7
C2	27.4	72.6
C3	36.5	63.6
C4	92.3	7.7
C5	76.8	23.2
C6	53.2	46.8
C7	92.1	7.9
C8	63.7	36.3
C9	72.1	28.9

The results show that, in general, the habit of influence is greater than the road conditions. Specifically, (1) when the road conditions of the two road sections are the same, the respondent tends to follow the habit; (2) when the customary road section is better than the unfamiliar road section, the weight of the habit is considered to be greater than the weight of the road condition. (3) When the road condition of the unfamiliar road section is better than the custom road section, the traveler will prefer to consider the road condition. When the customary road section is normal and the unfamiliar road section is unblocked, although the uncomfortable road section is better than the custom road section, because the customary road sections can pass normally, the weight of the habit is still greater than the weight of the road conditions[17].

5. Time Overhead

The proposed approach takes more than 40 seconds to finish 500 iterations using an implementation in a Win8 PC with Core 2.8 GHz processor and 8GB memory. The time overhead becomes much more significant along with the growth of the data.

As shown in Figure 3, the different algorithm has different time overhead.

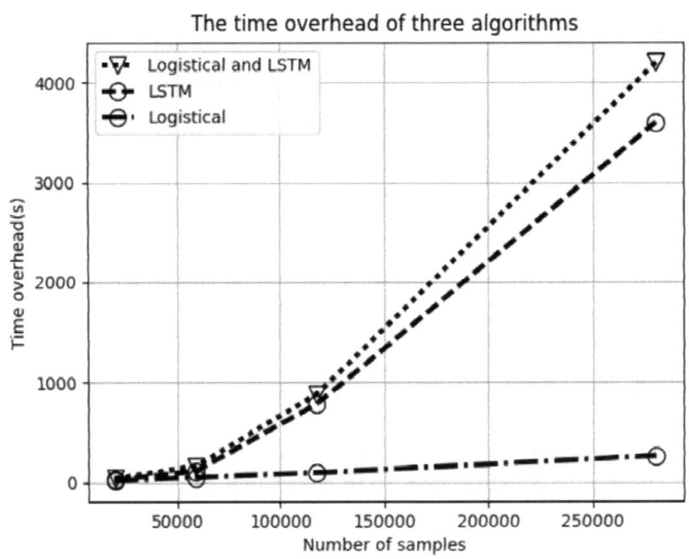

Figure 3. The time overhead under different number of samples

6. Conclusions and Future Work

In this paper, we proposed an routine prediction algorithm based on travel habits and real-time traffic condition. We first give account of LSTM for series data prediction. For the purpose of dealing with real-time traffic condition data route prediction, we then embody "logistical model" and propose a combined model of LSTM and logistical model. Experiments show the propose model can get better performance and has less time overhead.

Our future work includes conducting more experiments to improve the performance of our proposed method. We also plan to explore how to solve more complex traffic condition problems.

7. References

[1] M. C. Lin, J. Sewall, D. Wilkie. Transforming GIS Data into Functional Road Models for Large-Scale Traffic Simulation[J], IEEE Transactions on Visualization & Computer Graphics, 2012, 18(6):890-901.

[2] N. M. Garcia, P. Lenkiewicz, M. M. Freire, P. P. Monteiro. On the Performance of Shortest Path Routing Algorithms for Modeling and Simulation of Static Source Routed Networks -- an Extension to the Dijkstra Algorithm[C], In Proc of International Conference on Systems & Networks Communications, 2007:

[3] M. Z. Al-Ani, F. M. M. Al-Naima, S. Y. Amin. Simulation of a New Constraint Based Routing Algorithm for Multi-Protocol Label Switching Networks[C], In Proc of International Conference on Information & Communication Technologies: from Theory to Applications, 2008:

[4] D. Devajyoti, R. A., P. K. D., G. J. P., S. Vivek. Simulation of a heavy rainfall event during southwest monsoon using high-resolution NCUM-modeling system: a case study[J], Meteorology & Atmospheric Physics, 2018, 3):1-20.

[5] A. Schollmeyer, B. Froehlich. Efficient and Anti-aliased Trimming for Rendering Large NURBS Models[J], IEEE Transactions on Visualization & Computer Graphics, 2018, PP(99):1-1.

[6] M. Morning. ADVANCE, ADVANCED DRIVER AND VEHICLE ADVISORY NAVIGATION CONCEPT[J], April, 1991,

[7] R. V. Tomkewitsch. DYNAMIC ROUTE GUIDANCE AND INTERACTIVE TRANSPORT MANAGEMENT WITH ALI- SCOUT[J], IEEE Transactions on Vehicular Technology, 2002, 40(1):45-50.

[8] K. TAMURA. Toward Realization of VICS-Vehicle Information and Communications System[C], In Proc of 1993:

[9] Fu Dongmian. Algorithm of Least Transfer in Traffic System and Its Implementation[J], Journal of Huaqiao University(Natural Science), 2001, 22(4):348-350.

[10] Zhang Shuijian. Studies on Optimal Route Guidance Algorithm of Urban Transportation Based on GIS-T[D], 2010, Southwest Jiaotong University.

[11] Kong Huixin. Research on City Traffic Distribution and Taxi Travel Route Recommendation Algorithm[D], 2015, Beijing University of Posts and Telecommunications.

[12] B. K. Rout, G. Brooks, M. A. Rhamdhani, Z. Li, F. N. H. Schrama, J. Sun. Dynamic Model of Basic Oxygen Steelmaking Process Based on Multi-zone Reaction Kinetics: Model Derivation and Validation[J], Metallurgical & Materials Transactions B, 2018, 49(2):537-557.

[13] K. Greff, R. K. Srivastava, J. Koutník, B. R. Steunebrink, J. Schmidhuber. LSTM: A Search Space Odyssey[J], IEEE Transactions on Neural Networks & Learning Systems, 2016, 28(10):2222-2232.

[14] D. J. Bowersox. Logistical Management[J], McGraw-Hill international editions, 1974,

[15] J. P. Epperlein, J. Monteil, M. Liu, Y. Gu, R. Shorten. Bayesian classifier for Route prediction with Markov chains[J], 2018,

[16] H. Gomaa, G. G. Messier, C. Williamson, R. Davies. Estimating Instantaneous Cache Hit Ratio Using Markov Chain Analysis[J], IEEE/ACM Transactions on Networking, 2013, 21(5):1472-1483.

[17] N. K. Nariani, I. K. G. Bendesa, M. Budiarsa. Development of Bedulu Village Towards Future Tourism[J], Management Research: English, 2019, 3):247-256.

IOP Conf. Series: Materials Science and Engineering **719** (2020) 012062 doi:10.1088/1757-899X/719/1/012062

Research on Assisted Driving Technology based on Improved YOLOv3

Long Zhao[1,2,*], Xiaoye Liu[3], Qiang Wang[1] and Honglong Chen[1]

[1]Big Data Institute, Heilongjiang Oriental University, Harbin, Heilongjiang 150066, China
[2]College of Information and Computer Engineering, Northeast Forestry University, Harbin, Heilongjiang, 150040, China
[3] Harbin Far East Institute of Technology, Harbin, Heilongjiang, 150016, China
*Correspondence should be addressed to Long Zhao; zl_oriental@163.com

Abstract. In the era of automobile popularization, China and the developed countries such as Europe and the United States are facing the same problem of high car accidents. This paper uses deep learning technology to detect and identify lane lines, traffic lights, vehicles and pedestrians during driver driving. Improve the efficiency and accuracy of convolutional neural network training through migration learning and data enhancement techniques. Based on the current advanced YOLOv3 network, we have improved the network structure and loss function. The KITTI dataset has achieved the highest 2D target recognition accuracy, and the target recognition speed is higher than 36 frames/sec. The price of the assisted driving equipment we developed does not exceed RMB 5,000, which has a good market prospect.

1. Introduction

Traffic accidents caused by dangerous behaviors such as fatigue driving, distracted driving, drinking and driving, and speeding have become "the world's number one" [1]. According to data released by the World Health Organization, 1.2 million people worldwide die in car accidents every year, and one person is killed in traffic accidents every 25 seconds. In China, with the increase in car ownership, the number of people killed in car accidents has also risen sharply. At present, China has become one of the countries with the highest death rate from road traffic accidents.

In the existing domestic and international automatic driving research, target recognition is one of the core tasks, including road and road edge identification, lane line detection, vehicle identification, vehicle type identification, non-motor vehicle identification, pedestrian identification, traffic sign recognition. , obstacle identification and avoidance, etc. [2-3]. The target recognition system uses computer vision to observe the traffic environment, automatically recognizes the target from the real-time video signal, and provides a basis for real-time automatic driving, such as start, stop, steering, acceleration, and deceleration. Due to the extremely complicated actual road conditions, the performance of assisted driving technology based on traditional target detection is difficult to be greatly improved. The existing automatic driving technology generally relies on advanced radar systems to compensate, which significantly increases the cost of system implementation.

The video sequence captured by our proposed method in the road traffic scene contains various video targets, such as pedestrians, cars, roads, obstacles, various objects in the background, etc., and even identifies in the test image. The target object of the category of interest. At the same time, the program can also feed the test results to the driver in the form of voice announcements, provide the basis for decision-making of the vehicle control system, and remind the driver to drive safely. Since we have improved the network structure and loss function of YOLOv3, our method achieves the best performance in the KITTI dataset.

2. Related Work

In recent years, due to the rise of deep neural networks, some significant advances have been made in the research of vehicle detection and tracking algorithms at home and abroad, mainly in vehicle detection algorithms based on deep convolutional neural networks [4-7] and correlation filtering. Vehicle tracking algorithm [8-10]. These target detection studies are distinguished from target frame selection methods and can be divided into 2D target detection and 3D target detection. From the application medium, they can be divided into LIDAR data for deep learning and image data for deep learning. Most of the initial research relied on expensive lidar systems such as Velodyne and hand-labeled environmental maps. In contrast, recent efforts have attempted to replace laser radar with cheap car cameras that are readily available on most modern cars. Although some algorithms have a significant improvement in accuracy, this comes at the cost of increased system complexity and computational cost. This is difficult to apply in the field of assisted driving with extremely high real-time requirements.

3. Approach

3.1. Enhance Image Datasets Using "Polar Coordinate Transformation"

The so-called polar coordinate transformation is that the pixel is represented by the original (x, y) by polar coordinate transformation (r, θ), and then represented as a two-dimensional image. Involved in mathematical Equations 1 and 2.

$$\theta_u = 2\pi \cdot u/U \qquad \forall u e\{0, \cdots, U-1\} \tag{1}$$

$$x = \|v \cdot \cos(\theta_u)\| \; and \; y = \|y = \sin(\theta_u)\| \tag{2}$$

Enhancing the image dataset can significantly improve the generalization ability of the model and reduce the possibility of overfitting. Use the diagram to indicate as shown in Figure 1.

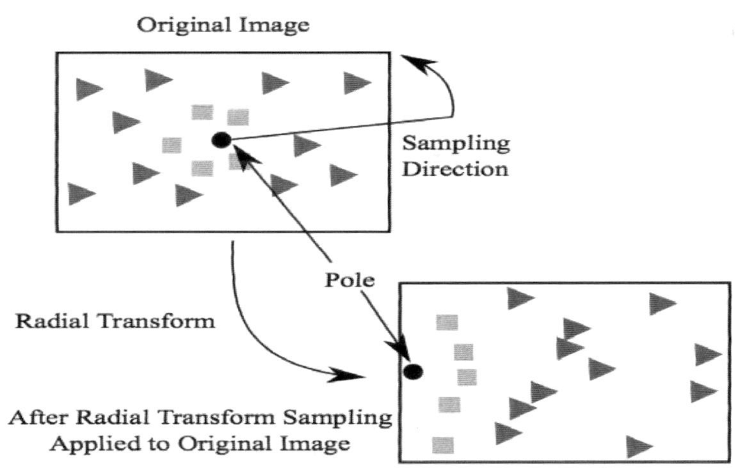

Figure 1. Polar coordinate transformation diagram

3.2. Using Migration Learning to Speed up the Training of the Model

In traditional classification learning, in order to ensure the accuracy and high reliability of the classification model obtained by training, there are two basic assumptions: (1) The training samples used for learning and the new test samples satisfy the independent and identical distribution. (2) There must be enough training samples available to learn a good classification model. However, in practical

IOP Conf. Series: Materials Science and Engineering **719** (2020) 012062 doi:10.1088/1757-899X/719/1/012062

applications we find that these two conditions are often not met. First, over time, the previously available tagged sample data may become unavailable, with semantics and distribution gaps in the distribution of new test samples. In addition, tagged sample data is often scarce and difficult to obtain. This raises another important issue in machine learning. How to use a small number of tagged training samples or source domain data to build a reliable model to predict target areas with different data distributions. In recent years, migration learning has caused widespread concern and research. Migration learning is a new machine learning method that uses existing knowledge to solve different but related domain problems. It relaxes two basic assumptions in traditional machine learning, with the goal of migrating existing knowledge to solve learning problems in the target domain with only a small number of tagged sample data or even no. Migration learning is widely existed in human activities. The more factors shared by two different fields, the easier it is to migrate learning. Otherwise, the more difficult it is, even the "negative migration", which has side effects.

3.3. Target Recognition Network Design and Loss Function Design

YOLOv3 is the representative of the advanced one-stage target detection model [11]. YOLOv3 uses Darknet-53 as its backbone network. The existing CNN model learns the characteristics of objects by stacking multiple convolution and pooling layers, but the YOLOv3 network is a full-convolution network that uses a lot of residual hopping connections. The advantages of using residual structure: (1) A key point of the depth model is whether it can converge normally. The residual structure can ensure that the network structure can still converge under deep conditions, and the model can be trained. (2) The deeper the network, the better the characteristics of the expression, and the effect of classification and detection will increase. (3) 1*1 convolution in the residual, using the idea of network in network, greatly reducing the channel of each convolution, on the one hand reducing the amount of parameters (the larger the parameter, the larger the saved model), On the other hand, the amount of calculation is reduced to some extent. The structure of YOLOv3 is shown in Figure 2.

	Type	Filters	Size	Output
	Convolutional	32	3 × 3	256 × 256
	Convolutional	64	3 × 3 / 2	128 × 128
1×	Convolutional	32	1 × 1	
	Convolutional	64	3 × 3	
	Residual			128 × 128
	Convolutional	128	3 × 3 / 2	64 × 64
2×	Convolutional	64	1 × 1	
	Convolutional	128	3 × 3	
	Residual			64 × 64
	Convolutional	256	3 × 3 / 2	32 × 32
8×	Convolutional	128	1 × 1	
	Convolutional	256	3 × 3	
	Residual			32 × 32
	Convolutional	512	3 × 3 / 2	16 × 16
8×	Convolutional	256	1 × 1	
	Convolutional	512	3 × 3	
	Residual			16 × 16
	Convolutional	1024	3 × 3 / 2	8 × 8
4×	Convolutional	512	1 × 1	
	Convolutional	1024	3 × 3	
	Residual			8 × 8
	Avgpool		Global	
	Connected		1000	
	Softmax			

Figure 2. YOLOv3 Structure

We borrowed from the literature [12] to use the spatial attention module on low-level features, and used the idea of channel attention module on advanced features to improve the network structure of YOLOv3 and improve the accuracy of target recognition. At the same time, the loss function of YOLOv3 is improved, and the target positioning accuracy of the image edge is improved. We recognize that certain levels of inaccurate information can cause performance degradation or even mispredictions. It's important to filter these and more valuable features. According to the characteristics of different levels of features, the channel is used to focus on high-level features, and spatial attention is used to select effective features for low-level features. In addition, spatial attention is not used for advanced features, as advanced features that contain high abstract semantics do not require filtering of spatial information. Because there are few semantic differences between the different channels of the low-level function, channel attention is not used for low-level functions.

In order to quickly deal with emergencies and minimize the probability of traffic accidents, the driver hopes that the field of view is large enough and the goal is clear enough. YOLOv3 is a typical 1-stage target detection algorithm with a huge speed advantage, but the positioning accuracy is not very high. We have improved the loss function of YOLOv3 as shown in Equation 3.

$$L_s = -\sum_{i=0}^{size(Y)}(a_s Y_i \log(P_i) + (1-a_s)(1-Y_i)\log(1-P_i)) \quad (3)$$

Where Y is the ground truth and P is the salient map of the network output, indicating the balance parameters of the positive and negative samples, set a_s= 0.536, which is calculated from the groundtruth of the training set. However, the loss function simply provides a general guide to generating a saliency map. A simpler strategy is used to emphasize the generation of significant object boundary details. First, the Laplacian operator is used to obtain the boundary of the ground truth and saliency map of the network output, and then the cross entropy loss is used to supervise the generation of significant object boundaries. Figure 3 and Figure 4 below show the YOLOv3 test results before improvement and the improved YOLOv3 test results. By comparison, we found that the improved network small target detection ability is stronger.

Figure 3. YOLOv3 target detection results

Figure 4. Improved target detection results for YOLOv3

3.4. Image Partition Recognition and Voice Prompts

In order to make it easier for car drivers to deal with various types of events in driving, we have identified and collected the images captured by the camera. The way we partition images is to perform horizontal dynamic partitioning according to specific needs. The height and width of the images we collect are H and W, respectively. The number of partitions to be partitioned is K (if K_Y is used for average partitions, and K_N is not used for average partitions), the area of each area after average partitioning. The calculation formula is as shown in Equation 4.

$$S_Y=H*W/K_Y \tag{4}$$

In order to give the driver the most direct voice prompt, we called Baidu's API.

4. Experiment

We evaluate the proposed network on the challenging KITTI dataset [13], which contains 7481 training images and7518 testing images with calibrated camera parameters. The KITTI data set is one of the most authoritative data sets for evaluating the performance of algorithms in the autopilot domain. We made detailed comparisons with other mainstream algorithms on the KITTI dataset. The comparison results are shown in Table 1. From Table 1, we can see that our target recognition accuracy is the highest, especially when it is difficult to achieve other goals.

Table 1. Performance comparison on KITTI dataset

	Cars			Pedestrians			Cyclists		
	Easy	Moderate	Hard	Easy	Moderate	Hard	Easy	Moderate	Hard
LSVM-MD PM-sv	68.02	56.48	44.18	47.74	39.36	35.95	35.04	27.50	26.21
DPM-VOC +VP	74.95	64.71	48.76	59.48	44.86	40.37	42.43	31.08	28.23
Regionlets	84.75	76.45	59.70	73.14	61.15	55.21	70.41	58.72	51.83
Faster R-CNN	86.71	81.84	71.12	78.86	65.90	61.18	72.26	63.35	55.90
M3DOD	92.33	88.66	78.96	80.35	66.68	63.44	76.04	66.36	58.87
Ours	93.28	90.01	80.25	80.67	67.26	64.13	76.67	67.23	60.56

5. Conclusions

In this paper, we propose a new target recognition algorithm based on YOLOv3. We used the Spatial Attention Module on low-level features, the Channel Attention Module on advanced features, and we improved the loss function. Our method has been tested on the KITTI dataset to achieve the best results, especially when it is difficult to achieve other goals.

6. Acknowledgments

All the works were conducted at Forestry Intelligent Equipment Engineering Research Center.

The work is supported by Heilongjiang Oriental university key scientific research project in 2019(HDFKY190101) and Heilongjiang province education science "13th five-year plan" 2018 key project (GBB1318041)

7. References

[1] Mukhtar Amir, Xia Likun, Tang Tongboon. Vehicle detection techniques for collision avoidance systems: A review[J]. IEEE Transactions on Intelligent Transportation Systems, 2015, 16(5): 2318-2338.

[2] X. Chen, K. Kundu, Z. Zhang, H. Ma, S. Fidler, and R. Urtasun. Monocular 3d object detection for autonomous driving. In Conference on Computer Vision and Pattern Recognition(CVPR), pages 2147–2156, 2016.

[3] Song Wenjie, Yang Yi, Fu Mengyin, et al. Real-time obstacles detection and status classification for collision warning in a vehicle active safety system[J]. IEEE Transactions on Intelligent Transportation Systems, 2018, 19(3): 758-773.

[4] X. Chen, H. Ma, J. Wan, B. Li, and T. Xia. Multi-view 3d object detection network for autonomous driving. In IEEECVPR, volume 1, page 3, 2017.

[5] X. Cheng, P.Wang, and R. Yang. Depth estimation via affinity learned with convolutional spatial propagation network. In European Conference on Computer Vision, pages 108–125, 2018.

[6] S. Ren, K. He, R. Girshick, and J. Sun. Faster R-CNN: Towards Real-Time Object Detection with Region Proposal Networks. NIPS, 2015.

[7] T.-y. Lin, F. Ai, and P. Doll. Focal Loss for Dense Object Detection. ICCV, 2017.

[8] W. Luo, B. Yang, and R. Urtasun. Fast and Furious: Real Time End-to-End 3D Detection,Tracking and Motion Forecasting with a Single Convolutional Net. CVPR, 2018.

[9] S. Wang, D. Jia, and X. Weng. Deep Reinforcement Learning for Autonomous Driving. arXiv:1811.11329, 2018. URL http://arxiv.org/abs/1811.11329.

[10] S. Casas, W. Luo, and R. Urtasun. IntentNet: Learning to Predict Intention from Raw Sensor Data. CoRL, 2018.

[11] J. Redmon and A. Farhadi, "Yolov3: An incremental improvement," *arXiv preprint arXiv:1804.02767,* 2018.

[12] Ting Zhao,Xiangqian Wu.Pyramid Feature Attention Network for Saliency detection. CVPR, 2019.

[13] A. Geiger, P. Lenz, and R. Urtasun. Are we ready for autonomousdriving? the kitti vision benchmark suite. In ComputerVision and Pattern Recognition (CVPR), pages 3354–3361. IEEE, 2012.

IOP Conf. Series: Materials Science and Engineering **719** (2020) 012063 doi:10.1088/1757-899X/719/1/012063

Deepbot: A Deep Neural Network based approach for Detecting Twitter Bots

Linhao Luo[1], Xiaofeng Zhang[1], Xiaofei Yang[1] and Weihuang Yang[1]

Department of Computer Science Harbin Institute of Technology Shenzhen, China
luolinhao@stu.hit.edu.cn, zhangxiaofeng@hit.edu.cn, yangxiaofei@stu.hit.edu.cn, yangweihuang@stu.hit.edu.cn

Abstract. Social networks have played a very critical role in very aspect of our daily life. However, a wide variety of bots have been found which are designed for some malicious purposes such as spreading spam mes- sages and faking news. Although various techniques have been proposed, this task is still challenging if we want to judge whether the tweets are posted by a bot or not merely based on the textual information. For this challenge, the Deepbot is designed which adopts the Bi-LSTM model to analyze tweets and a Web interface is provided for public access which is developed using Web service. From our empirical studies, this system can achieve better classification accuracy.

1. Introduction

Social networks, e.g., Twitter and Facebook, have played a more and more important role in our daily life. Users prefer to share their information via Social networks and are susceptible to the messages posted by other users. This naturally results in the pervasive of social media bots which keep on forwarding messages or faking news. This phenomenon is particularly serious for Twitter. Generally, Twitter users could post millions of *tweets* including textual messages and other rich format messages such as images and videos per day. According to [10], it is estimated that around 48 million registered Twitter users are bots. Some of these bots simply forward news and automatically update its status. On the contrary, a majority of these bots keep on spreading spam messages or fake news which may cause serious consequence such as misleading the political election.

To detect such bots, various techniques have been proposed in the literature. However, this task is still challenging. Actually, bots with fixed patterns such as regularly updating status or forwarding messages could be easily detected. However, some bots try to mimic the behaviour of human beings and thus are hard to detect. This attracts more and more research efforts from both the industry and the academic. Some commercial websites like Botcheck[1 https://botcheck.me/] detects account related features and tries to discover the outliers from these features such as Join date, follower count, tweeting rate, retweeting rate, and tweet text. Bot Sentienl[https://botsentinel.com] identifies bot based on the inappropriate activities of bots. [11] Extracts several graph-based features as well as some content-based features to detect the bots sending spam messages.

Apparently, this task is challenging as the latest bots are designed to mimic human beings to post tweets. Thus, the semantic content of tweets posted should be analyzed. In this paper, we propose a Bi-LSTM based Web application which can quickly detect bots based on only one piece of tweet. Based on the analysis results, the system outputs the probability that whether the input tweet is posted by a bot or not.

IOP Conf. Series: Materials Science and Engineering **719** (2020) 012063 doi:10.1088/1757-899X/719/1/012063

2. Related Works

Social network bot detection has long been studied in the literature [1, 2, 4, 9]. Most of existing works focus on the account level classification. In these approaches, the abnormal accounts are detected as they may share similar features like email address and account creation datetime [6]. Some works found that bots usually post tweet using automatic devices with hijacked IP address [1]. Similarly, their social ties as well as their topological network structure could be used to detect abnormal accounts.

On the contrary, some researchers treat the bot detection issue as a classification task. In these works, a number of features are designed for the classification task which are the number of followers or friends, the ratio of the friends to followers, the percentage of bi-directional friends, and the standard deviation of unique numerical IDs of followers and friends [7]. Authors in [11] proposed a social graph using trending topics, replies and mentions to build the classifier. In, the authors utilized the synchronization features of bots for detection. And [5] adopted the deep learning approach to detect bots. In their work, a basic version of LSTM model is built and both the account information and the tweets' contents were used as the model input. Although this work could achieve comparably good results, they need a large amount of labeled tweets to train model which involves a high manual labeling cost. Alternatively, we propose this new technique which is Bi-LSTM based approach and it only requires a small number of training data. This proposed approach can save the labeling cost and can achieve better classification results.

3. The Proposed Deepbot

The proposed Deepbot contains two components: a trained Twitter bot classifier and a Web interface developed using Web service for public access.

3.1. Twitter Bot Classifier

The Twitter bot classifier is proposed based on a deep neural network model to determine whether the input tweet is posted by a bot or not. To represent the textual features of tweets, First, we embed the tweets into vectors using the Global Vectors for Word Representation (GloVE) [8]. This pre-trained word embedding matrix is denoted as $E \in R^{|e| \times |V|}$, where $|e|$ denote the length of each word after embedding and $|V|$ is the total number of vocabulary V . Let D denote the number of words in the i-th tweet (S_i). Then, each tweet S_i could be embedded as a matrix $S\ R^{D \times |e|}$ by replacing all the words with the corresponding word vector v in$E$.

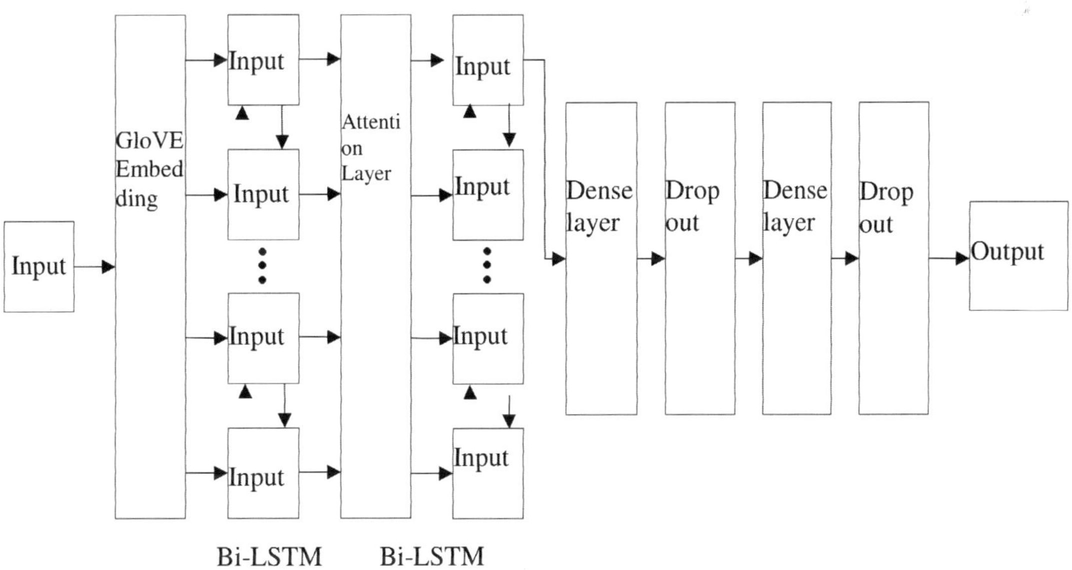

Figure1. The architecture of the proposed Deepbot.

The proposed Deepbot customizes the Bi-directional Long Short Term Memory (Bi-LSTM) [12] to analyze the input tweets and automatically extracts the important textual features. By doing so, it can largely save the manual cost for feature selection and is helpful to build a robust feature space for the learning of a more accurate classifier. For each tweet matrix *S*, it consists of *D* word vectors and we have $S = [v1, v2, \ldots, vD]$. As shown in Figure 1, *S* is treated as a sequence and will be fed into the Bi-LSTM as the input. Then, the Bi-LSTM will store both the long term useful historical textual features but also the short term textual features in the cell unit. The first layer of the adopted Bi-LSTM outputs the hidden state $h_{l1} \in R^{D\times 2u}$, *u* is the unit number of the single Bi-LSTM cells. And the $hl1$ is then sent to an attention layer which can calculate the importance probability *P* for each word. This attention mechanism can help the network focus on more important words. Then, the $h_{l1} \times P_i$ is sent to the second Bi-LSTM layer. The second layer only outputs the results of the last cell, i.e., $h_{l2} \in R^{1\times 2u}$. The h_{l2} is then inputted to two fully connected layers with the Relu as the active function. To avoid the over fitting issue, two drop layers are placed between the fully connected layer and the final output layer.

To train the Deepbot, the public dataset is adopted and its link is provided [https://pan.webis.de/clef19/pan19-web/author-profiling.html]. This dataset contains 412,000 annotated tweets posted by 2,060 bots and 2,060 humans, respectively. Its training and testing data are already prepared. The training set includes 144,000 tweets posted by bots and 144,000 tweets by human and the testing set includes 62,000 bots' tweets and 62,000 humans' tweets. In the model training process, we choose "Binary Cross Entropy" as the loss function. The optimizer is "Adam" [3] and the batch size is set to 512. The Deepbot is trained on RTX2080. Finally, we got 79.64 accuracy in test set and the ROC is 87.04.

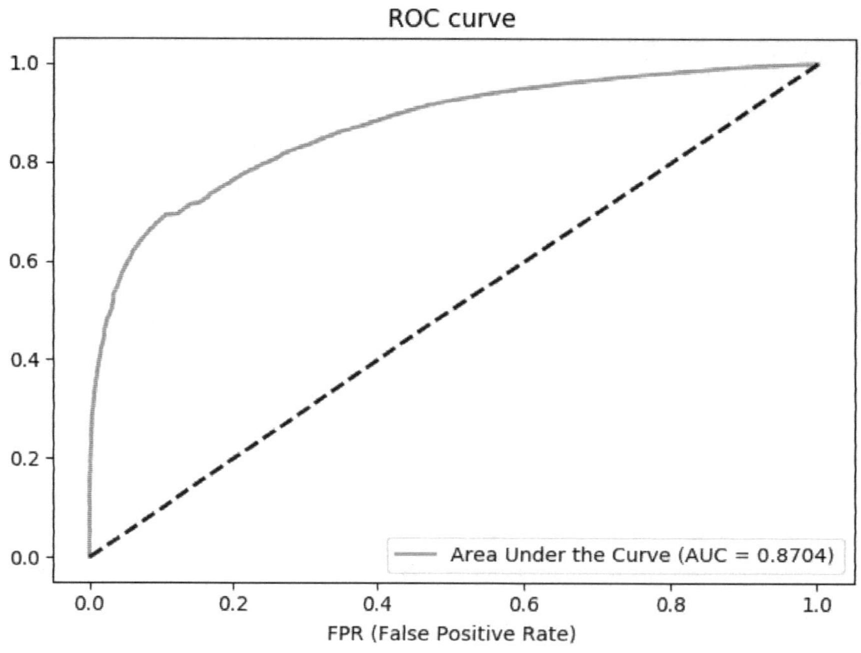

Figure 2. The ROC curve.

3.2. Web Interface

To allow users to access this Twitter bot classifier through the Internet, we provide a public accessible interface developed using Web service. This interface allows a user to upload the tweet and returns the classification result (probability that the tweet is posted by a bot or not) to that user. For the server side, we use Flask, a micro framework developed in Python, to return JSON data generated by the Deepbot.

The working procedure is as follows. After the Web service starts, then the classifier is loaded into the main memory and wait for the message. For the front-end side, we adopt Vue.js to control the data flow, such as receiving user's message, posting the tweets to **the** server, and displaying the returned results.

4. Conclusion

The proposed Deepbot consists of two components which is the bot classifier and the Web interface. In the near future, we will further enhance the model classification ability by designing a more sophisticated deep neural network structure and try to make the Web system of Deepbot to support the high concurrency access.

5. Acknowledgments

This paper is partially supported by Shenzhen Science and Technology Program under Grant No.JCYJ20170811153507788, and the Guangdong Province Science and Technology Department Project under Grant NO.2017B090901022. This work is also partially supported by the National Science Foundation of China under grant No.61872108.

6. References

[1] Cook, D.M., Waugh, B., Abdipanah, M., Hashemi, O., Rahman, S.A.: Twitter deception and influence: Issues of identity, slacktivism, and puppetry. Journal of Information Warfare 13(1), 58–71 (2014)

[2] John, J.P., Moshchuk, A., Gribble, S.D., Krishnamurthy, A., et al.: Studying spam- ming botnets using botlab. In: NSDI. vol. 9 (2009)

[3] Kingma, D.P., Ba, J.: Adam: A method for stochastic optimization. arXiv preprint arXiv:1412.6980 (2014)

[4] Krebs, B.: Twitter bots drown out anti-kremlin tweets. Krebs on Security 11 (2011)

[5] Kudugunta, S., Ferrara, E.: Deep neural networks for bot detection. Information Sciences 467, 312–322 (2018)

[6] Lee, K., Eoff, B.D., Caverlee, J.: Seven months with the devils: A long-term study of content polluters on twitter. In: Fifth International AAAI Conference on Weblogs and Social Media (2011)

[7] Morstatter, F., Wu, L., Nazer, T.H., Carley, K.M., Liu, H.: A new approach to bot detection: striking the balance between precision and recall. In: 2016 IEEE/ACM International Conference on Advances in Social Networks Analysis and Mining (ASONAM). pp. 533–540. IEEE (2016)

[8] Pennington, J., Socher, R., Manning, C.: Glove: Global vectors for word representation. In: Proceedings of the 2014 conference on empirical methods in natural language processing (EMNLP). pp. 1532–1543 (2014)

[9] Stone-Gross, B., Holz, T., Stringhini, G., Vigna, G.: The underground economy of spam: A botmaster's perspective of coordinating large-scale spam campaigns. LEET 11, 4–4 (2011)

[10] Varol, O., Ferrara, E., Davis, C.A., Menczer, F., Flammini, A.: Online human- bot interactions: Detection, estimation, and characterization. In: Eleventh inter- national AAAI conference on web and social media (2017)

[11] Wang, A.H.: Machine learning for the detection of spam in twitter networks. In: International Conference on E-Business and Telecommunications. pp. 319–333. Springer (2010)

[12] Wilson, T., Wiebe, J., Hoffmann, P.: Recognizing contextual polarity in phrase- level sentiment analysis. In: Proceedings of Human Language Technology Conference and Conference on Empirical Methods in Natural Language Processing (2005)

IOP Conf. Series: Materials Science and Engineering **719** (2020) 012064 doi:10.1088/1757-899X/719/1/012064

Robust Vehicle Detection on Multi- Resolution Aerial Images

Ziao Wang[1], Xiaofeng Zhang[1], Xiaofei Yang[1] and Wu Xia[1]

[1]Department of Computer Science, Harbin Institute of Technology, Shenzhen, China
Corresponding author: Xiaofeng Zhang (e-mail: zhangxiaofeng@hit.edu.cn).

Abstract. Detecting vehicles in aerial images is one of the core components for intelligent transportation system. This task is challenging due to the comparably small size of the target objects, the complex background and multi-perspective views. It is particularly difficult for the real-time detection cases where only several tens of milliseconds delays are allowed. In this paper, we propose an approach to detect vehicles from aerial images with different resolutions and perspectives in an approximate real-time manner. The proposed model is robust and can detect vehicles in various detection scenarios. It can detect vehicles from an image within 47 milliseconds. We evaluate our method on a challenging data set of original aerial images over Munich and our data set collected using an unmanned aerial vehicle (UAV). The experimental results have demonstrated that our proposed method is superior to the state-of-the- art algorithms with respect to accuracy, recall and detection time.

1. Introduction

With the rapid development of UAV [1], it has been widely adopted in many application domains, such as intelligent transportation systems. UAV is equipped with automatic positioning and stability system of small high-altitude operating platform. UAV can easily collect video or images which are generally used for the object detections.

For vehicle detection problem using UAV, the traditional detection methods are not appropriate as they are trained on the car camera images. Compared with the traditional perspective of car cameras, UAV images generally cover a larger range of objects which is considered to be able to facilitate the object detections. There are a lot of problems for approaches which are based on aerial images [2], such as low detection accuracy, slow detection speed and unstable detection results when cope with multi-resolution and multi-perspective images [3]. This work is motivated by aforementioned challenging difficulties. Specifically, we propose a fast vehicle detection approach which is robust to the practical application scenes. It is also compatible with a variety of resolutions and multiple perspectives. The detection speed is very fast and can reach 21 FPS with high detection accuracy.

In order to rapidly detect vehicles, a fully convolutional neural networks (FCNN) [4] is chosen as a fast feature extractor. Then, a unified detection for labeling and localization is performed on the feature set extracted by FCNN. Multi-resolution vehicle detection scheme is then introduced to the basic vehicle detector, which makes it compatible with various resolution aerial images. Moreover, a multi-perspective vehicle detection scheme is also integrated into the proposed vehicle detectors making it to be able to work on aerial images collected from a variety of different shooting perspectives.

We evaluate the proposed approaches on two data sets to demonstrate the efficacy of the proposed vehicle detector. One data set is Munich Images adopted in [2], another data set is collected by us which contain multi-resolution and multi- perspective UAV images. In this paper, the detection accuracy rate, recall rate and detection time are chosen as the evaluation criteria of performance.

Our main contributions can be summarized as follows: 1) This approach adopts FCNN to extract the features of images, which makes the detection speed very fast. The anchor box extracted in FCNN is utilized to localize the objects which is particularly suitable for small object detection. 2) A traffic flow detection method based on weak supervised learning is proposed to solve the problem of annotation data of aerial traffic image. 3) We designed a joint training model using the aerial image of multi-resolution and multi-perspective which can achieve a good model robustness. 4) We have collected a set of aerial images of multi-resolution and multi-perspective as shown in Fig. 1, in order to simulate the practical detection scene.

(a) (b) (c) (d)

Figure 1. Aerial images with four shooting perspectives. (a) height = 90m and θ = 90o. (b) height = 120m and θ = 90o. (c) height = 90m and θ = 45o. (d) height = 120m and θ = 45o

2. Related Work

In recent years, the deep learning algorithm has achieved great success in the field of image detection and recognition. [5, 6] are state-of-the-art methods based on region proposal, the advantage of those algorithms is detection accuracy. There are some state-of-the-art methods not based on region proposal, such as [7, 8], the advantage of those algorithms is detection speed. Whether it is based on region proposal, those algorithms have achieved good results from high resolution situations, but they acquire poor results when target object is small or images are low pixel. In the literatures, Kaiming et al. proposed deep residual network of extracting highly descriptive features, [9] and it can improve the accuracy of small objects detection and recognition, but it is time consuming. It is difficult that the detection method has high detection accuracy and high detection speed simultaneously. After successful in face recognition, deep learning has been rapidly applied to many fields. However, many fields do not have public data sets as many as face recognition. In some fields, annotation data is very scarce, for example, medical domain [10] and aviation[11] domain. Therefore, more and more people study the effect of using semi-supervised [12] or weak-supervised [13,14,15] deep learning method instead of strong supervised deep learning method to train with less annotation data and achieve a large amount of annotation data to train the model results. Changyu Jiang et al. proposed a weak supervised vehicle learning algorithm for deep learning [16]. This algorithm uses image-level annotation data to train the first few layers of convolutional layers in the detection network and then uses a small amount of annotation data fine-tune the model to obtain an efficient detection model.

There exist a number of literatures about vehicle detection from aerial images. Rodney LaLonde *et al.* proposed a multi-frame multi-object detection method based on FCNN [17], the performance is superior when compared with many state-of-the-art methods. Nassim Ammour *et al.* proposed deep learning approaches to extract highly descriptive features of Vehicle Detection in UAV Images [18], and them method outperformed state-of-the-art methods in terms of accuracy and computational cost.

Yongzheng Xu et al. proposed a combination of the Viola–Jones + SVM algorithm and an HOG + SVM algorithm for vehicle detection in UAV Images [19], which uses a detector switching strategy based the different descending trends of detection speed of both algorithms to improve detection efficiency. Kang Liu *et al.* applied a fast binary detector using integral channel features in a soft-cascade structure for vehicle detection in UAV Images, and estimated the orientation and type of the vehicles [2]. All of the above algorithms are based on aerial images of the same shooting perspective, and they seldom consider the practical issues on vehicle detection using UAV images with different perspectives.

3. Our Approach

This section describes the whole process the proposed vehicle detection network. The design idea of a unified detection network is described.

3.1. Unified Detection Network

In this paper, we propose a fast vehicle detection network based on FCNN detection framework which unifies bounding box prediction and object prediction. As shown in Fig. 2, FCNN is used to directly convert image from the pixel space to the feature space, then the feature map is divided into N×N grids, each grid is responsible for predicting object of the center point falling within it. This detection network predicts whether a grid contains an object or not and predicts the possible location of objects simultaneously.

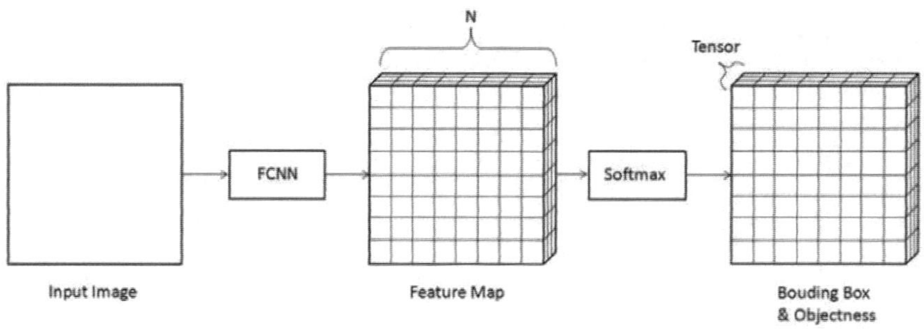

Figure 2. A Unified vehicle detection network based on FCNN. FCNN is used to extract the feature of the input image. Feature map is divided into N×N grids. Feature map is encoded by softmax. A tensor is used for object annotation and object localization of box

Let g_o and g_t denote prediction value and truth value whether grid g contains an object or not, and the range of its value is {0, 1}. The loss of g at grid g is calculated as (1).

$$\mathcal{L}_{ob}(g) = (g_o - g_t)^2 \quad (1)$$

Let $\mathcal{L}_{de}(g)$ denote detection loss of grid g. The overall loss of the network is calculated as (2).

$$\mathcal{L} = \sum_{g\epsilon G} \mathcal{L}_{ob}(g) + \sum_{g\epsilon G} g_{ot} * \mathcal{L}_{det}(g) \quad (2)$$

Where $\boldsymbol{g_{ot}} = \boldsymbol{g_o} * \boldsymbol{g}$, G denotes all grids in the feature map. The first term in (2) represents the cumulative loss when $\boldsymbol{g_o}$ is different from $\boldsymbol{g_t}$. The latter term in (2) represents the cumulative loss when $\boldsymbol{g_o}$ is same as $\boldsymbol{g_t}$.

Due to the easy-to-collect but not easily annotated nature of the aerial data of the subject, we also employ weak supervised learning as illustrated in Fig. 2 with the target of improving the vehicle detection accuracy. As shown in Fig. 3.

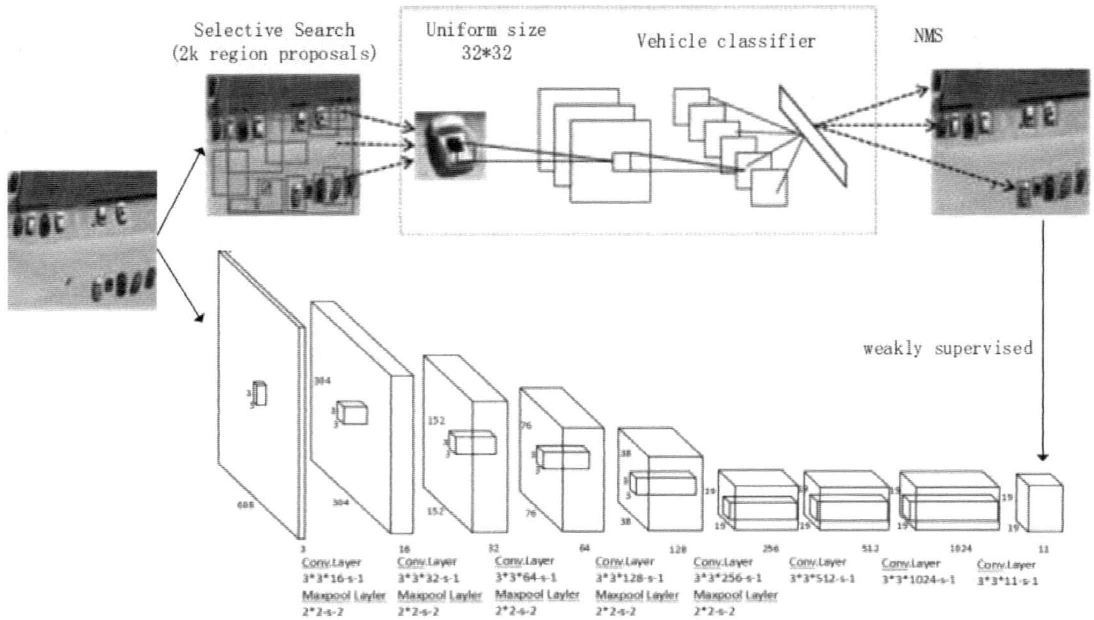

Figure 3. A vehicle detection network based on weakly supervised. The top half is used to generate weakly labeled data. The lower half will use weakly labeled data for training

Each unlabeled image will first extract several candidate boxes using the selective search [20] algorithm, and then classify these candidate boxes using a network similar to LeNet. Finally, the result of the rough annotation is filtered by the non-maximal suppression algorithm. In the rough annotation process, a network similar to LeNet is used to classify the candidate boxes which contains two convolution layers, two maximum pooling layers, two fully connected layers and a lost layer.

The class label $y^{(i)} \in \{0,1\}$ indicates that the problem is a dichotomous problem, and the loss layer uses a logistic regression to calculate the loss. Logistic regression function is given as,

$$h_\theta(x) = \frac{1}{1+\exp(-\theta^T x)} \tag{3}$$

where θ is the model parameter and x is the input eigenvector. To train the model parameter θ, L_{cls} is denoted as,

$$L_{cls} = -\sum_{i=1}^{m}\{y^{(i)}\log h_\theta(x^{(i)}) + (1-y^{(i)})\log\left(1-h_\theta(x^{(i)})\right)\} \tag{4}$$

where L_{cls} is classification loss and m is the number of pictures annotated in the training set.

Our network is a two-step one. After training a rough flow detection model using weak supervision, the algorithm uses a traffic detection network based on FCNN. When calculating the prediction loss, the image is treated as annotation data. Because there is some noise in the data learned by the weak supervisor used in the training process of the network, we can reduce the impact of noise on the model training by modifying the loss function. The improved network loss function is given as,

$$\lambda_{coord} \sum_{i=0}^{S^2} \sum_{j=0}^{B} \Gamma_{ij}^{obj} \bar{P}_{ij}((X_i - \bar{X}_i)^2 + (Y_i - \bar{Y}_i)^2)$$

$$+ \lambda_{coord} \sum_{i=0}^{S^2} \sum_{j=0}^{B} \Gamma_{ij}^{obj} \bar{P}_{ij}((\sqrt{\omega_i} - \sqrt{\bar{\omega}_i})^2$$

$$+\left(\sqrt{h_i} - \sqrt{\bar{h}_i})^2\right)$$

$$+ \sum_{i=0}^{S^2} \sum_{j=0}^{B} \Gamma_{ij}^{obj} \bar{P}_{ij}(C_i - \bar{C}_i)^2$$

$$+ \lambda_{noobj} \sum_{i=0}^{S^2} \sum_{j=0}^{B} \Gamma_{ij}^{noobj} (C_i - \bar{C}_i)^2$$

$$+ \sum_{i=0}^{S^2} \Gamma_{i}^{obj} \sum_{c \in classes}(P_i(c) - \bar{P}_i(c))^2 \tag{5}$$

Where $\bar{P}_{ij}$ represents the probability that the true box corresponding to the j-th target detection box in the i-th grid containing a car.

The loss function is mainly divided into three parts. The 1-3 items of (5) are the first part, indicating that the prediction box has the classification loss corresponding to the real position, the fourth items is the second part, indicating that the predicted box does not find the loss corresponding to the real box, and the fifth item is the third section, which indicates the classification loss of each grid. After the weak supervision and training, the first part is mainly modified, that is, when the prediction box can find the corresponding box in the annotation data, $\bar{P}_{ij}$ is the weight of the first part of the loss. This weakly supervised training method can integrate the model into the characteristics of more target vehicles, and finally use the strong supervised training method to make the model converge.

4. Experiment

We evaluate our approach on two different data sets, one is Munich Images adopted form [2], and another one is UAV image data sets collected by us including images of different heights and different perspectives. The evaluation criteria are detection time and AP (Average Precision). Rigorous experiments are performed on these two data sets to verify the efficacy of the proposed approach.

Our data set is collected from UAV Images, we use UAV to take video on five different roads with different flying height and shooting angle, a total of 40 videos are collected, accumulating up to 400 minutes. A total of 3000 images were captured from the video, which evenly includes four shooting scenarios (1.height = 90m and θ = 90o; 2.height = 120m and θ = 90o; 3.height = 90m and θ = 45o; 4.height = 120m and θ = 45o). We randomly choose 2000 images as training data set which contain 32646 cars, and the rest 1000 images as testing data set which contain 17548 cars. The image size is 800×450 pixels. The results on our UAV Images are reported in Table 1.

Table 1. Results on Our UAV Images

Method	Ground Truth	AP (%)	Time (ms)
Liu-Mattyu's	17548	76.1	76
Faster RCNN	17548	57.4	257
SSD	17548	63.5	39
yolo v2	17548	60.5	**29**
Ours	17548	**88.8**	47

Our method achieves the best detection accuracy on our UAV images with multi-resolutions and multi-perspectives images, which reaching high detection accuracy of 88.8% AP. Yolo v2 has the fastest detection speed with 28 milliseconds per frame, but the performance is not good in terms of detection accuracy. Our method detects an image of 800×450 pixels only took 47 milliseconds which are an approximate real-time speed. Resolution of cars is lower in our UAV Images than Munich Images, it difficult to detect vehicle on our UAV Images. And the result of other methods proved that vehicle detection is more difficult on our UAV Images. As shown in Fig. 4, our detection results are compared with Liu-Mattyu's to verify the robustness of the proposed approach.

5. Conclusion

We proposed a robust model that can detect vehicles on aerial images of different resolutions and perspectives at approximate real-time speed. This model is very suitable for vehicles detection in aerial images with different detection scene. The method of multi-resolution and multi-perspective training is effective, it can improve the robustness of the detector. Our method outperformed state-of-the-art methods for vehicle detection on multi-resolution and multi-perspective aerial images. It is a good idea to detect the road [23] and then detect the vehicle. As future work, the performance could be further improved by using information which extracts from video frames [24] by LSTM [25], the speed of the detector would still keep fast for vehicles detection in aerial videos.

(a) (b)

Figure 4. Comparison of detection results on four different perspectives. (a) Results of Liu-Mattyu's method. (b) Results of our method. When θ close to 90o, the results of our method and Liu-Mattyu's method are almost the same. However, when θ close to 45o, the detection result of Liu-Mattyu's method is not good, and there are many cars that are not detected. But our detection results are still great on four different perspectives. It indicates that our method is more robust than Liu-Mattyu's method on multi-resolution and multi-perspective aerial images

6. Acknowledgments

This work was supported by the Guangdong Province Science and Technology Department Project under Grant NO. 2017B090901022. This paper is partially supported by the National Science Foundation of China under grant No.61872108, and Shenzhen Science and Technology Program under Grant No.JCYJ20170811153507788.

7. References

[1] Pérez A, Chamoso P, Parra V, and Sánchez A. J. Ground vehicle detection through aerial images taken by a UAV. Information Fusion (FUSION), 2014 17th International Conference on. IEEE, 2014:1-6.

[2] Liu K, Mattyus G. Fast multiclass vehicle detection on aerial images. IEEE Geoscience and Remote Sensing Letters, 2015, 12(9):1938-1942.

[3] Khoshelham K, Nardinocchi C, Frontoni E, et al. Performance Evaluation of Automated Approaches to Building Detection in Multi-Source Aerial Data. Isprs Journal of Photogrammetry and Remote Sensing, 2010, 65(1):123-133.

[4] Dai J, Li Y, He K, et al. R-FCN: Object Detection via Region-based Fully Convolutional Networks. Advances in neural information processing systems, 2016, 26(1):397-389.
[5] Girshick R. Fast r-cnn. Proceedings of the IEEE International Conference on Computer Vision, 2015:1440-1448.
[6] Ren S, He K, Girshick R, Sun J. Faster r-cnn: Towards real-time object detection with region proposal networks. Advances in neural information processing systems, 2015:91-99.
[7] Redmon J, Farhadi A. YOLO9000: Better, Faster, Stronger. arXiv preprint arXiv:1612.08242, 2016.
[8] Liu W, Anguelov D, Erhan D, Szegedy C, Reed S, Fu C Y, and Berg A C. SSD: Single shot multibox detector. European Conference on Computer Vision. Springer International Publishing, 2016:21-37.
[9] He K, Zhang X, Ren S, and Sun J. Deep residual learning for image recognition. Proceedings of the IEEE Conference on Computer Vision and Pattern Recognition, 2016:770-778.
[10] Shen D, Wu G, Suk H I. Deep Learning in Medical Image Analysis. Annual Review of Biomedical Engineering, 2017, 19(1):221-248.
[11] Tao Y, Wang X, Yao B, et al. Small Moving Vehicle Detection in a Satellite Video of an Urban Area. Sensors, 2016, 16(9):1528-1542.
[12] Zhang J, Han Y, Tang J, et al. Semi-Supervised Image-to-Video Adaptation for Video Action Recognition. IEEE Transactions on Cybernetics, 2016, 47(4):960-973.
[13] Jia Z, Huang X, Chang E I, et al. Constrained Deep Weak Supervision for Histopathology Image Segmentation. IEEE Transactions on Medical Imaging, 2017, 27(7):1-16.
[14] Pathak D, Krahenbuhl P, Darrell T. Constrained Convolutional Neural Networks for Weakly Supervised Segmentation. Proceedings of the IEEE International Conference on Computer Vision, 2015:1796-1804.
[15] Croft W B, Croft W B, Croft W B, et al. Neural Ranking Models with Weak Supervision. arXiv preprint arXiv:1704.08803, 2017.
[16] Jiang C, Zhang B. Weakly-Supervised Vehicle Detection and Classification by Convolutional Neural Network. Proceedings of the International Congress on Image and Signal Processing, 2017:570-575.
[17] LaLonde R, Zhang D, Shah M. Fully Convolutional Deep Neural Networks for Persistent Multi-Frame Multi-Object Detection in Wide Area Aerial Videos. arXiv preprint arXiv:1704.02694, 2017.
[18] Ammour N, Alhichri H, Bazi Y, Benjdira B, Alajlan N, and Zuair M. Deep Learning Approach for Car Detection in UAV Imagery. Remote Sensing, 2017, 9(4): 312-322.
[19] Xu Y, Yu G, Wang Y, Wu X, and Ma Y. A hybrid vehicle detection method based on Viola-Jones and HOG+ SVM from UAV images. Sensors, 2016, 16(8): 1325-1348.
[20] Uijlings J R, Sande K E, Gevers T, et al. Selective Search for Object Recognition. International Journal of Computer Vision, 2013, 104(2):154-171.
[21] Jia Y, Shelhamer E, Donahue J, Karayev S, Long J, Girshick R, and Darrell T. Caffe: Convolutional architecture for fast feature embedding. Proceedings of the 22nd ACM international conference on Multimedia. ACM, 2014:675-678.
[22] Srivastava N, Hinton G, Krizhevsky A, et al. Dropout: A Simple Way to Prevent Neural Networks from Overfitting. Journal of Machine Learning Research, 2014, 15(1):1929-1958.
[23] Cheng G, Wang Y, Xu S, et al. Automatic Road Detection and Centerline Extraction via Cascaded End-to-End Convolutional Neural Network. IEEE Transactions on Geoscience and Remote Sensing, 2017, 55(6):3322-3337.
[24] Pu F, Xie W, Cheng Y, et al. Implementation of Real-Time Vehicle Tracking in City-Scale Video Network. Cybernetics and Systems, 2016, 47(4):249-260.
[25] Greff K, Srivastava R K, Koutník J, Steunebrink B R, and Schmidhuber J. LSTM: A search space odyssey. IEEE transactions on neural networks and learning systems, 2016: 2222-2232.

IOP Conf. Series: Materials Science and Engineering **719** (2020) 012065 doi:10.1088/1757-899X/719/1/012065

A Prediction Method of Node Attack Path based on Optimized-AG

Kunfu Wang, Wei Feng and Wanfeng Mao

System Engineering Research Institute of China State Shipbuilding Corporation, 1 Fengxiann East Road, Haidian District, Beijing P.R. China
Email:wangkunfu1166@163.com

Abstract. An optimization algorithm for reducing the scale of attack graphs and an attack path prediction method are presented herein to cover the difficulty of predicting the attack path in traditional attack graphs. Firstly, a new distance cost attack graph model is generated and converted into a marked attack graph via the algorithm. Secondly, the original attack graph is optimized by introducing an attack distance weight value and a path redundancy coefficient in order to determine the possible attack paths and calculate the cost of path attack behavior. Finally, the preferred attack path is determined based on an evaluation function. According to the experimental results, the above scheme eliminated the path redundancy of attack graphs and achieved comprehensive prediction of attack paths.

1. Introduction

As the internet industry advances rapidly, IT plays an important role in enterprises and the enterprise intranets are facing increasing threats. According to CSI/FBI's Computer Crime and Security Survey [1] for consecutive five years, more than 85% of the damages resulting from computer security accidents are incurred by internal security vulnerabilities instead of viruses. The Development Status of China Internet Sites and Security Report in 2018 [2] illustrates a well-known case of the virus WannaCry specific to enterprises in May 2017. The attacker spread the virus via dangerous vulnerabilities. On this basis, future network attacks are likely to be more violent and have stronger impact on enterprises. In the Review of China's Internet Network Security Situation in 2017 [3], the number of security vulnerabilities listed in China National Vulnerability Database (CNVD) rises continuously in recent years. Since 2013, the number of vulnerabilities listed in CNVD has grown by an average of 21.6% per year.

In recent years, researchers have built various attack graph models by analyzing vulnerability information of intranets [4] and created attack paths using the resource information and attack behavior in the network [5].

2. Related Research

Kaynar K[6] proposed an attack graph modeling method to simulate potential attack paths in the network. However, the method could only be used to obtain data through simulation experiments and was incapable of changing with the real environment. Muñozgonzález L[7] et al. used the Loopy Belief Propagation algorithm to perform static and dynamic analysis and drew a conclusion that the number of attack nodes expanded linearly. Even through this method could be applied to evaluate the risks of system damages in case of vulnerability interconnection, it was impossible to reduce the size of an attack and failed to provide any reference for analyzing the internets of larger enterprises.

Poolsappasit N[8] et al. were based on the Bayesian network and CVSS system and proposed to quantify the node confidence to costs and benefits. Then, they compared the cost and benefit values of different attack paths to optimize the Bayesian attack graph and select a preferred attack path. However, they failed to take into account the path redundancy in attack graphs, thus the path filtering in the attack graph was affected.

Man D[9] et al. developed a global attack graph generation algorithm based on the breadth-first search, which reduced the scale of an attack graph by limiting the attack steps and the probability of success of attack paths. Although they achieved some effects, attack costs and other factors were not included in the optimized attack graph and the predication of preferred attack paths was incomplete.

Wu Hongrun [10] et al. proposed an attack path prediction method based on attack costs in the analysis of network attacks. However, they ignored the fact that the attacker acted as an intelligent agent while launching a network attack and would integrate the factors of attack distance and cost of an attack path. Moreover, they only took the attack cost into account and could not predict the attack path accurately.

3. Establishment of Network Attack Graph Model NETAG

The existence of network resources (vulnerabilities) is the reason why an attacker launches attacks. Attackers gain permissions by exploiting the vulnerability information on the network and implement basic attacks to change the status of network resources, thereby occupying the desired resources. Then, they will attack the network again until they gets what they want. Therefore, the following definitions need to be considered while creating an attack graph model:

Definition 1: Attack vulnerabilities in the network system are defined as vulnerabilities by which an attacker obtains appropriate permissions.

Definition 2: The complexity of an attack is an measure of an attacker's ability to obtain appropriate permissions successfully by virtue of these vulnerabilities. Due to a variety of factors, the attack complexity cannot be calculated precisely for each of these vulnerabilities, but some quantitative indexes can be used to illustrate the difference of attack complexity between these vulnerabilities approximately. The authors herein referred to the study of Chen Xiaojun[12] et al. on the complexity of attacking vulnerabilities and developed a quantitative standard of complexity in percentages upon changes according to the actual situation. The vulnerability is less likely to be attacked successfully when the complexity is closer to 100. The quantitative standard value are shown in Table 1.

Table 1. Quantitative Standard of Attack Complexity

Grade	Complexity	Description
1	10	No attack tools required and detailed attack methods
2	30	Available attack tools and detailed attack methods
3	50	Without attack tools but with relatively detailed attack methods
4	70	Report vulnerabilities publicly and refer to attack methods roughly
5	90	Report vulnerabilities publicly but not refer to attack methods

3.1. Definition of Network Attack Graph

A network attack is a complex multi-step process that involves a series of closing related basic attack behavior. An attack graph is a directed graph of possible attacks from the perspective of the attacker as well as a dependency graph composed by the attacker implementing attacks with a vulnerable host in the target network. Based on the above analysis, the following definitions are given:

Definition 3: The network attack graph $NETAG = (V, A, E, D, W)$ represents a five-tuple directed acyclic graph with one or more OR nodes.

- $V = \{v_i \mid i = 1,2,3,\ldots N\}$ represents the set of resource nodes: The system has open network services (i.e., attack vulnerabilities), and these attack vulnerabilities are also defined as a set of resource nodes.
- $A = \{a_j \mid j = 0,1,2,3,\ldots N\}$ represents the set of attack nodes: The attacker occupies the next resource

node by initiating attacks via the attack nodes, and this is a nonempty finite set of OR nodes.

- $E=\{E_A \cup E_V\}$ represents the set of directed edges connecting nodes: The set $E_A \subseteq A\times V$ is a collection of attack behavior edges from attack nodes to resource nodes, in which $e_{ji} =< a_j, v_i >$ and $e_{ji} \in E_A$. The set $E_V \subseteq V\times A$ is a collection of edges that resource nodes triggers attack nodes, in which $e_{ji} =< a_j, v_i >$ and $e_{ji} \in E_A$.
- D represents the set of attack distance weight: $D=\left\{d_{ji} \mid j=0,1,2,...,N, i=1,2,...,N\right\}$, and $a_0 \to v_1 \to a_2 \ldots a_j$ for the sequence of sequence points, in which d_{ji} is an attack distance weight value assigned to the attack behavior edges $<a_j,v_i>$ given the difficulty of attacking the resource nodes when the attack nodes a_j attacks against the resource nodes v_i.
- W represents the set of attack permissions: Refer to Definition 10 for details.

Definition 4: OR node means that all father nodes Pre(n) of node n are in an OR relationship, i.e., the node n is triggered if any node in the Pre(n) set is satisfactory.

3.2. Generation and Conversion of Network Attack Graph

3.2.1. Generation of network attack graph

From the attacker's initial access to attack permissions, the attack graph model NETAG is required to contain the status of all accessible networks of the attacker in order to analyze the status of each node in the network accurately.

For this purpose, the authors herein proposed an attack graph generation algorithm for forward search[13] and the specific steps are described as follows:

Step 1: collect service information in the network, i.e., vulnerabilities (resource nodes), attack nodes and network topologies and other information.

Step 2: construct a sequence of attack sequence points based on the network vulnerability information and the vulnerability exploit rules.

Step 3: find the network vulnerabilities v_i corresponding to the attack nodes a_j.

Step 4: create an attack behavior edge e_{ji} that a_j attacks against and occupies v_i with the exploit rules and store it to E_A; create a edge e_{ji} that a_j can be triggered through v_i and store it to E_V.

Step 5: construct a network attack graph utilizing the combination of E_A and E_V according to the association between network statuses.

In order to detail the above steps of generating an attack graph, the authors herein put forward an attack graph generation algorithm FRG-Alg for forward search as follows:

Algorithm 1: An attack graph generation algorithm FRG-Alg (a_0, A and V) for forward search

Description: An attack graph is generated by virtue of the directed relationship between attack nodes and resource nodes.

Input: the initial attack node a_0, the attack node set A and the resource node set V.

Output: NETAG..

1) For (a sequence of attack sequence points)
2) While ($A \neq \phi, V \neq \phi$)
3) For each $a_j \in A$
4) Search $a_j \to v_i$ //find the open service information v_i of a_j
5) Relationship a_j and v_i to Edge e_{ji} //compose the edge e_{ji} based on the
10) $E_V \leftarrow e_{ij}$
11) $\forall e_{ij}$ the nodes are in an OR relationship
12) End For
13) End For
14) $E \leftarrow (E_A \cup E_V)$ //store elements in E_A and E_V to the edge set E
15) Relationship e_{ij} and e_{ji} //the edge e_{ji}

relationship between a_j and v_i

6) $E_A \leftarrow e_{ji}$ //store e_{ji} to the edge set E_A

7) For each $v_i \in V$

8) Search $v_i \rightarrow a_j$

9) Relationship v_i and a_j to Edge e_{ij}

related to e_{ij}

16) Rank e_{ij} and e_{ji} //arrange the sequence of e_{ij} and e_{ji}

17) NETAG$\leftarrow e_{ij}$, e_{ji};

18) Return NETAG

The attack graph NETAG generated by the above algorithm is as shown in Figure 1:

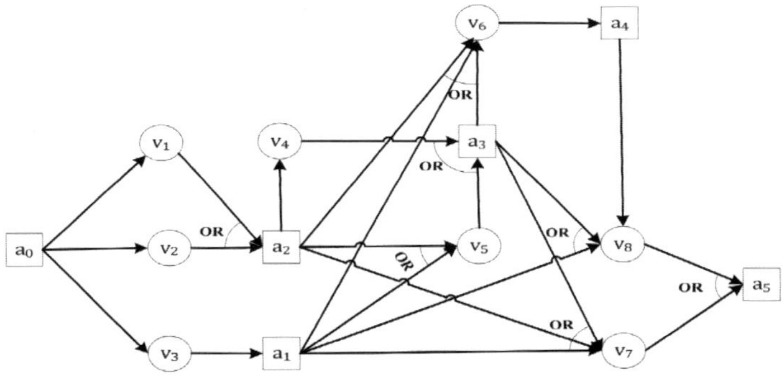

Figure 1. Initial Attack Graph NETAG

3.2.2. Description of attack path based on network attack graph

When an attacker attacks a network target, he/she attacks the resource node of the target first to change its status, and then attacks other resource nodes until the target node is occupied finally. The attack track of the attacker in this process is called an attack path. As shown in Figure 1, the attacker starts at the starting node a_0 and arrives at the destination node a_5 after passing through the nodes v_2, a_1, v_7, and the sequence composed by nodes a_0, v_2, a_1, v_7, a_5 in order is an attack path. Accordingly, an attack path is defined as follows:

Definition 5: If there is an ordered sequence of nodes $a_0, v_1, a_1, v_2, a_2 \dots v_i, a_j$ from the starting node a_0 and any two neighbor nodes in the sequence meet $<v_i, a_j> \in E_V$ or $<a_j, v_i> \in E_A\left(0<i<n, 0 \le j<n\right)$, the sequence of nodes $a_0, v_1, a_1, v_2, a_2 \dots v_i, a_j$ is called an attack path of the network attack graph and expressed in $AStep$. All attack paths are set to $AStep_i \in AS$.

3.2.3. Conversion of network attack graph

In order to conduct further studies on the relationship between attack nodes and resource nodes in the attack graph and on the complexity of attack behavior (i.e., the assignment of distance weight of attack behavior edges), the authors herein converted the original attack graph and marked the converted attack behavior, thus different marks indicates different attack distance weight.

As v_i is a resource node and represents the open services (i.e., vulnerability information) of the host a_j, v_i is marked with a Bugtraq ID from the appropriate vulnerability database. d_{ji} refers to the attack distance weight value of the attack behavior edge $<a_j, v_i>$, i.e., the attack complexity of v_i. In this paper, the ID of vulnerability information v_i is assigned to the attack behavior edge $<a_j, v_i>$, and the ID corresponds to different attack distance weight d_{ji} by reference to the quantitative standard value of vulnerability information of the host. To describe the conversion process explicitly, a conversion algorithm AGC-Alg is designed for the original attack graph as detailed below:

Algorithm 2: A attack graph conversion algorithm **AGC-Alg** (NETAG, A, V, E_A and AS)

Input: NETAG, the attack node set A, and the resource node set V, the edge set E_A and the path set AS.

Output: the marked edge set $M\text{-}E_A$, the attack permission set $D=\{d_{ji} \mid j=1,2,...,N, i=1,2,...,N\}$ and the marked attack graph M-NETAG.

1) $M\text{-}E_A \leftarrow \emptyset$;
2) $D \leftarrow \emptyset$;
3) While(true) do
4) For (each $AStep_i$ in AS)
5) Search ID$\leftarrow v_i$; //Find the vulnerability ID corresponding to v_i
6) Mark ID on $<a_j,v_i>$; //Mark edges with this ID
7) $<a_j,v_i>' \leftarrow <a_j,v_i>$;
8) Search Weight $\leftarrow$ID; //Find the attack distance weight corresponding to this ID
9) $d_{ji} \leftarrow$ Weight;
10) $D \leftarrow d_{ji}$; //Store the weight to the set D
11) $<a_j,v_i>' \leftarrow d_{ji}$;
12) End For
13) $M\text{-}E_A \leftarrow <a_j,v_i>'$; //Store the marked behavior edges to $M\text{-}E_A$
14) M-NETAG$\leftarrow M\text{-}E_A$;
15) End While
16) Return M-NETAG

Upon the above algorithms, the converted attack graph M-NETAG is as shown below:

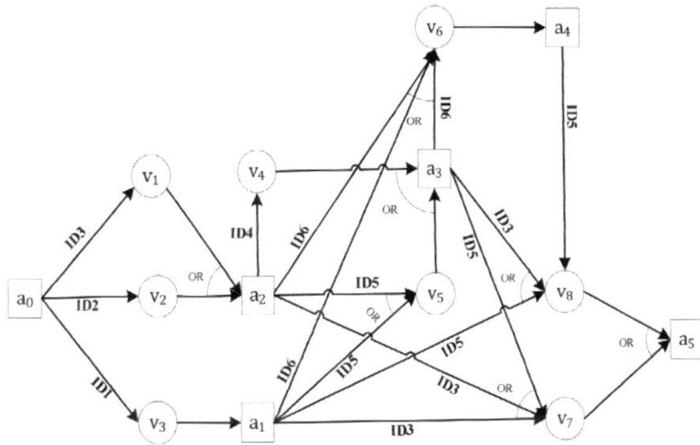

Figure 2. Marked Attack Graph M-NETAG

According to the converted attack graph, each attack behavior edge is assigned an ID to reflect the relationship between nodes. The value corresponding to the ID represents the distance weight from an attack node to a resource node.

3.3. Optimization based on M-NETAG

According to the experiment environment, attackers as an intelligent group prefer to select an easier attack path. But in the attack graph generated via the above algorithm FRG-Alg, there might be a large number of redundant attack paths. The authors herein analyzed the attack distance weight of accessible attack paths and introduced an attack path redundancy coefficient to optimize the above attack graph and reduce the scale of attack graph and the redundancy of attack paths. In this way, it is possible to analyze the optimized attack graph easily and prepare for the predication of the preferred attack path of attackers.

Definition 6: If d_{ji} is the attack distance weight of an attack behavior edge $<a_j,v_i>$ in the attack graph M-NETAG, the attack distance weight L_d represents the sum of weight of attack behavior edges in a single accessible path $AStep_i$ and is calculated as follows:

$$L_d = \sum d_{ji} \quad (1)$$

Definition 7: The maximum attack distance weight $L_{\max}$ represents the maximum attack distance weight among all attack paths $AStep_i$ from a_j to a_n if nodes a_j and a_n is accessible in the attack graph M-NETAG.

Definition 8: The minimum attack distance weight L_{min} represents the minimum attack distance weight among all attack paths $AStep_i$ from a_j to a_n if nodes a_j and a_n is accessible in the attack graph M-NETAG.

Definition 9: According to the Definitions 7 and 8, the attack path redundancy coefficient ξ represents the ratio of the maximum attack distance weight to the minimum attack distance weight among all attack paths $AStep_i$ from a_j to a_n and is calculated as follows:

$$\xi = \left(L_{\max} / L_{min}\right) \quad (2)$$

If larger ξ is obtained, there are more redundant paths in the attack graph, thus an attack distance weight threshold value γ is set to be divided by the redundant attack paths having an attack distance L_d greater than the threshold. The threshold γ is calculated as follows:

$$\gamma = \xi * \Delta + L_{min} \quad (3)$$

Δ represents the quantitative standard value of the path redundancy coefficient and is defined by an experienced expert.

According to the above definitions, the authors proposed an optimization algorithm AGO based on the attack graph M-NETAG. The optimized attack graph retains some effective attack paths for the reference of administrators. The attack graph optimization algorithm is detailed as follows:

Algorithm 3 An optimization algorithm **AGO** (d_{ji} and AS) for network attack graphs.

Description: The attack distance L_d is calculated for $AStep_i$ in the set AS of the attack graph M-NETAG and the attack paths with the L_d less than the threshold is sorted out. The L_d value of each path is marked using the vector $U = (U_1, U_2, \cdots U_i)$ and stored in the set ϖ in sequence.

Input: The distance weight d_{ji} of $<a_j, v_i>$ in the set D and the accessible attack path set AS

Output: the optimized attack path $OAStep_i$ and the path set OAS

1) $OAS \leftarrow \emptyset$
2) $\varpi \leftarrow \emptyset$
3) While (true) do
4) For (AS each $AStep_i$)
5) $L_d = d_{ji}$++
6) $U_i \leftarrow L_d$
7) $\varpi \leftarrow L_d$
8) End For
9) Select $L_{\max}$, L_{min} in ϖ
10) Calculate $\xi = (L_{\max} / L_{min})$, γ
11) Find $U_i \leftarrow ((L_d < \gamma) \text{ in } \varpi)$ //find the marks with weight less than the threshold
12) $AStep_i \leftarrow U_i$ //find the corresponding path by the marks
13) Reserve v_i and a_j in $AStep_i$ //reserve the attack and resource nodes in the attack path
14) Store $V' \leftarrow v_i$, $A' \leftarrow a_j$
15) Edge a_j and v_i to e_{ij} and e_{ji} //Compose an edge according to the relationship between a_j and v_i
16) Rank e_{ij} and e_{ji}
17) $OAStep_i \leftarrow (e_{ij}, e_{ji})$
18) $OAS \leftarrow OAStep_i$
19) End While
20) Return OAS

According to the Algorithm 3 and all accessible attack paths from the initial attack node a_0 to the destination node a_5 in Figure 2, the minimum attack distance weight, maximum attack distance weight, path redundancy coefficient ξ and attack distance weight threshold are calculated. After the attack paths with the attack distance weight greater than the threshold is removed in M-NETAG, the optimized attack graph Optimized-AG reserves the effective attack paths for the reference of administrators, as shown in Figure 3.

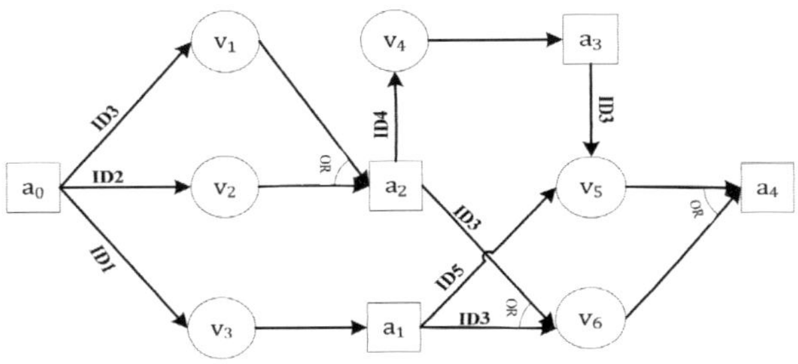

Figure 3. Optimized Attack Graph

4. Prediction of Attack Path Based on Optimized-AG.

According to the Optimized-AG, if there are attack paths with equal attack distance L_d, such effective attack path is defined as equivalent. Since launching a network attack is an economic behavior, attackers have to pay appropriate attack costs when they occupy the desired resources. Therefore, attack behavior cost is introduced in the paper and the preferred attack path of attackers among various effective attack paths is determined by calculating the total cost of the accessible paths and combining the analysis of L_d.

Definition 10: W means the set of attack permissions and $\forall w \in W$ is related to each node and expressed in two-tuple (d_{ji}, c). d_{ji} represents the attack distance weight of edges $<a_j, v_i>$ and c represents the attack costs paid when the attack behavior edges $<a_j, v_i>$ occurs.

Based on the above definition, each attack behavior edge $<a_j, v_i>$ is assigned a weight w and the current attack graph is shown as follows:

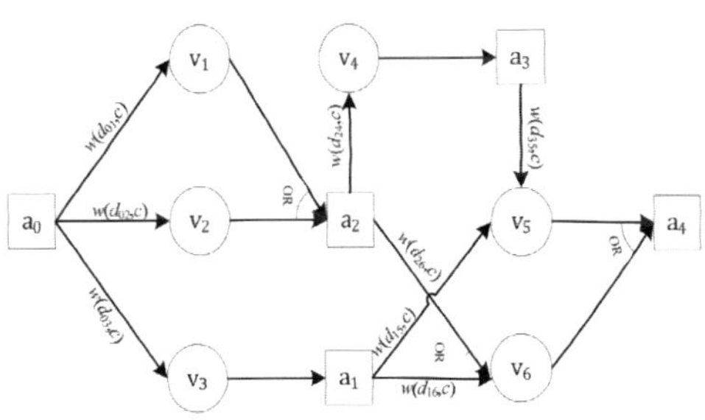

Figure 4. Optimized-AG with Weight

4.1. *Cost Analysis of Node Attack Path Based on Optimized-AG*

In this paper, the cost required to implement a single-step attack in the attack path of Optimized-AG is expressed as $Ecostep(e_{ji})$, i.e., the parameter c mentioned in Definition 10 above. Ignoring other aspects, the total cost of an attack sequence or an attack path is calculated as follows:

$$TCost(e_{ji}) = \sum_{e_{ji} \in E_A} Ecostep(e_{ji}) \tag{4}$$

Where, $Ecostep(e_{ji})$ consists of two parts, namely the operation cost $OCost(e_{ji})$ and the risk cost $RiskCost$. In the literature[14], a mathematical model of operation cost is developed in the process of analyzing the cost of attacks.

$$OCost\left(e_{ji}\right) = \mu * cost(Meta-operations) + \eta * cost(Sequence) \tag{5}$$

Where, $e_{ji} \in E_A$; and $cost(Meta-operations)$ is the meta-operation cost of a network node attack, so that the result depends on the meta-operations and the use of resources; $cost(Sequence)$ is the operation sequence cost of a network node attack; μ is the probability of a network node attack and η is path redundancy with low correlation in network node attack paths.

Moreover, the attacker has to take the risks of being discovered by the network administrators when he/she launches an attack, that is, he/she has to pay appropriate risk costs $RiskCost$. Therefore, the risk cost factor needs to be included in the cost analysis of an attack and $RiskCost$ is related to network security risk coefficient, attacker's experience and operation costs.

Network security risk coefficient is a measure of the likelihood of an attack being discovered and is calculated with the following formula:

$$Risk(e_{ji}) = \Psi(v_i) * \mathrm{I}(a_j) \tag{6}$$

In the formula (6), $Risk(e_{ji})$ is determined by the influence coefficient $\Psi(v_i)$ of a resource node on attack behavior as well as the influence coefficient $\mathrm{I}(a_j)$ of the attack node itself.

As the attack node proceeds and the attack distance L_d increases, the attacker needs to spend more time and takes higher risks of being exposed while attacking a resource node is increasingly complex. In this way, it is understood that the attacker will prefer a shorter attack distance to avoid risks as the cost of each further step grows exponentially. Therefore, attack distance is treated as one of the factors in the evaluation of risk cost in the paper. With the introduction of the attack path redundancy coefficient mentioned above, the risk cost is calculated as follows:

$$RiskCost = OCost(e_{ji}) * Risk(e_{ji}) * Ex(e_{ji})^{count-1} * \xi \tag{7}$$

Where, $OCost(e_{ji})$ is the operation cost; $Risk(e_{ji})$ $\left(Risk(e_{ji}) > 1\right)$ is security risk coefficient and is defined by an experience expert; $\left(0 < Ex(e_{ji}) < 1\right)$ is the attacker's empirical dependence coefficient; ξ is the attack path redundancy coefficient and $count$ is the number of attacks.

According to the definitions above, the cost $Ecostep(e_{ji})$ of attack behavior between attack nodes is calculated as follows:

$$Ecostep(e_{ji}) = \partial * OCost(e_{ji}) + \ell * RiskCost \tag{8}$$

Where, ∂、ℓ represent the attack cost weight coefficient. According to the definitions above, the total cost of a single attack path is calculated as follows:

$$TCost(e_{ji}) = \sum_{e_{ji} \in E_A} \left(\partial * OCost(e_{ji}) + \ell * RiskCost\right) \tag{9}$$

4.2. Prediction Method of Node Attack Path Based on Optimized-AG

The most possible effective attack paths are reserved in the Optimized-AG for the reference of administrators, and the paths might have different attack distance and attack cost, same attack distance and attack cost, same attack difference but different attack cost, or same attack cost but different attack distance. If only the attack path distance weight or attack cost is considered, locating the preferred attack path is not comprehensive and time-consuming.

To balance the role of the two factors in the process that the attacker prioritizes an attack path among all effective attack paths, a reasonable parameter-evaluation function needs to be designed as the criteria for determining the path selection. A path with a higher probability of being selected requires least attack path distance weight and attack cost, i.e., less evaluated value. Thus, the evaluation function is calculated as follows:

$$\oint(OAStep_i) = \lambda * L_d + \gamma * TCost(e_{ji}) = \lambda * \sum d_{ji} + \gamma * \sum_{e_{ji} \in E_A} Ecostep(e_{ji}) \tag{10}$$

Where λ and γ represents attack distance and attack cost weight coefficient respectively ($0<\lambda<1$, $0<\gamma<1$, $\lambda+\gamma=1$ and $\lambda<\gamma$); L_d is the distance weight of an effective attack path and $TCost(e_{ji})$ is the attack behavior cost of an effective attack path. Based on the path prediction method above, a path prediction algorithm OASJ is designed. The preferred attack path of an attacker is sorted out by calculating the evaluated value of a path $OAStep_i$ for the reference of administrators. The algorithm is implemented as follows:

Algorithm 4 Attack path determination algorithm OASJ (OAS, d_{ji} and $Ecostep(e_{ji})$)

Description: The evaluated value of each path $OAStep_i$ is marked with a vector $\Theta=(\Theta_1,\Theta_2...\Theta_i)$, and the values are stored in the set Ω in sequence. The preferred attack path can be located by comparing the evaluated values in the set Ω.

Input: the set OAS in Optimized-AG, the attack distance weight d_{ji} of edges $<a_j,v_i>'$ and the attack behavior cost $Ecostep(e_{ji})$

Output: the evaluated value of a path $\oint(OAStep_i)$ and the preferred attack path $U\text{-}OAStep_i$

1) $\Theta \leftarrow \emptyset$, $\oint(OAStep_i) \leftarrow \emptyset$
2) While (true) do
3) For (OAS each $OAStep_i$)
4) $L_d = d_{ji}{++}$
5) $TCost(e_{ji}) = Ecostep(e_{ji}){++}$
6) $c \leftarrow TCost(e_{ji})$
7) $W \leftarrow w(d_{ji},c)$ //Store two-tuple values in the attack permission set
8) Calculate $\oint(OAStep_i)=\lambda * L_d + \gamma * TCost(e_{ji})$
9) $\Theta_i \leftarrow \oint(OAStep_i)$ //Mark the evaluated value with Θ_i
10) $\Omega \leftarrow \oint(OAStep_i)$
11) End For
12) Select $\oint(OAStep_i)_{\min}$ in Ω
13) $\Theta_i \leftarrow \oint(OAStep_i)_{\min}$ //find the corresponding marks by the evaluated value
14) Through Θ_i finding $OAStep_i$
15) $U\text{-}OAStep_i \leftarrow OAStep_i$
16) End While
17) Return $U\text{-}OAStep_i$

5. Experimental Verification

5.1. Experimental Network Configuration

To verify the feasibility of the attack path prediction method in this paper, a small Windows-based enterprise intranet is designed and has been subject to corresponding simulation experiment. The network typology applied in the experiment is as shown in Figure 5:

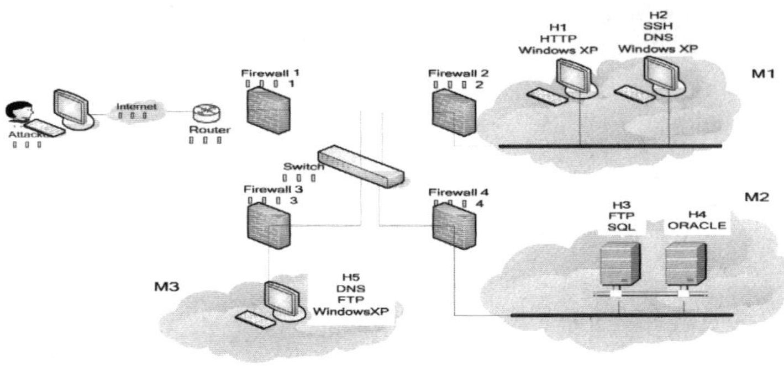

Figure 5. Topological Graph of Local Networks

In the experiment network, the safe zones M1, M2 and M3 in the Internet and LAN are separated from each other by firewalls, thus the attack host Attack (i.e. H0) assesses the LAN via Internet. In the zone M1 of LAN, the host H1 providing open HTTP services and the host H2 providing SSH and DNS services are separated from other networks by Firewall 2. In M2, a SQL server and an ORACLE server serves as the hosts H3 and H4 and are separated from other networks by Firewall 4. Moreover, H3 provides FTP services for hosts in M1 and M3. In M3, H5 provides DNS and FTP services and serves as a target host that stores important data. H5 is separated from other networks by Firewall 3. In the typological graph of local networks above, the firewalls will only allow the external host H0 to access the hosts H1 and H2, and any other external accesses are blocked. Moreover, the internal hosts H1 and H2 cannot access each other, but other hosts can access the open services of each host.

5.2. Experimental Data and Analysis

The experimental network is scanned with a scanner and the vulnerabilities and related information of each host are obtained as shown in Table 2. The quantitative value of the difficulty level of these vulnerabilities being exploited can be gained from querying the quantitative criteria of attack complexity of vulnerabilities in Table 1, and the results is shown in Table 3.

Table 2. Vulnerability information of each Host

Safe Zone	Host	Service	Bugtraq ID
M1	Host1	HTTP	ID1=CVE-2017-2666
	Host2	SSH/DNS	ID2=CVE-2018-0497
			ID3=CVE-2018-5538
M2	Host3	SQL/FTP	ID4=CVE-2018-0607
			ID5=CVE-2017-3329
	Host4	ORACLE	ID6=CVE-2018-3091
M3	Host5	DNS/FTP	ID3=CVE-2018-5538
			ID5=CVE-2017-3329

Table 3. Quantitative Value of Vulnerability Information in Each Host

Bugtraq ID	Distance Weight
CVE-2017-2666	10
CVE-2018-0497	10
CVE-2018-0607	50
CVE-2018-3091	70
CVE-2018-5538	30
CVE-2017-3329	90

According to the design of experiment and the information of hosts and vulnerabilities in the network, the generated network attack graph is shown in Figure 1. Based on the information in Table 2, the attack graph in Figure 1 is transformed into the attack graph marked with a vulnerability ID in Figure 2 by applying the conversion algorithm AGC-Alg, and the quantitative value of attack distance weight corresponding to the vulnerability ID is shown in Table 3. All accessible attack paths from the attack host H0 to the target host H5 are displayed in Figure 1, and the attack distance weight of all attack paths in Figure 1 is calculated with the attack graph optimization algorithm AGO mentioned above, thus the attack distance weight of each path goes through a gradual change process as shown in Figure 6 below:

IOP Conf. Series: Materials Science and Engineering **719** (2020) 012065 doi:10.1088/1757-899X/719/1/012065

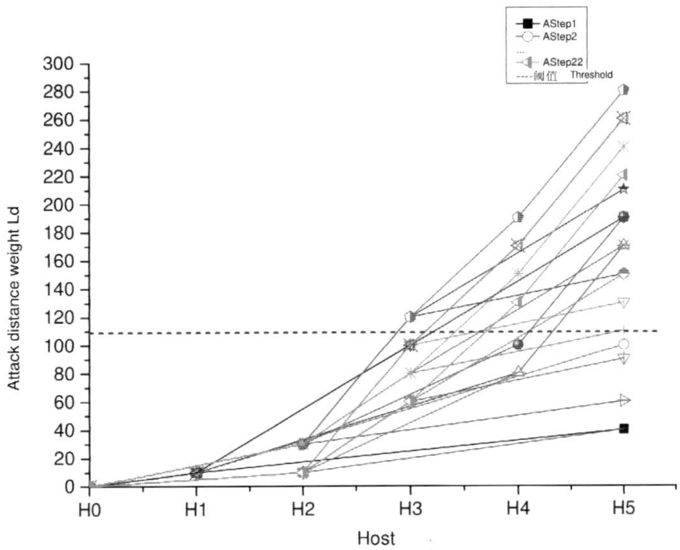

Figure 6. Weight graph of attack path

In Figure 6, each polygonal line represents an accessible attack path, and the fluctuation of a corresponding value indicates a process in which the distance weight of the attack path H0→…→ H5 overlaps. Among all attack paths in Figure 6, the minimum attack distance weight is 40 and the maximum attack distance weight is 280, thus the path redundancy coefficient ξ of the attack graph is 7. The quantitative standard value of the path redundancy coefficient is set to 10 by an experienced expert, and the calculated attack distance weight threshold γ is 110, as shown by the dotted line in the Figure. Therefore, the attack paths with the path weight over 110 are removed in Figure 6. The optimized attack paths are displayed in Figure 3 and the corresponding experimental network topological graph is shown in Figure 7:

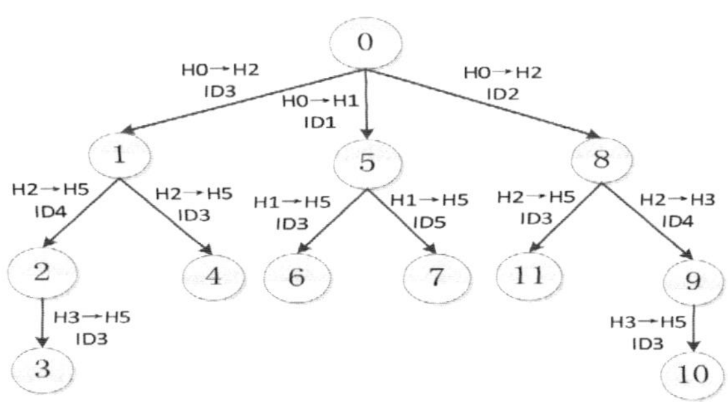

Figure 7. Optimized Experiment Topological Graph

Six accessible attack paths to the target host H5 are retained in Figure 7 and the attack behavior cost of each path is analyzed on this basis. According to the definitions of attack path behavior cost $TCost(e_{ji})$ above and based on the experiment of an expert, the appropriate parameters are set as follows: ∂、ℓ are set to 0.3 and 0.7 individually; $Risk(e_{ji})$ is set to 2 and indicates that the cost doubles for each further step; $Ex(e_{ji})$ is set to 0.5 and indicates that the second similar attack has a cost half of the original cost and ξ is set to 7 calculated above. The cost result analysis of attack paths is as shown in Table 4:

IOP Conf. Series: Materials Science and Engineering **719** (2020) 012065 doi:10.1088/1757-899X/719/1/012065

Table 4. Cost Information of Effective Attack Paths

$OAstep_i$	$Ecostep$(c)	$TCost(e_{ji})$
$OAstep_1$	$<a_0,v_1>$(2.1), $<a_2,v_6>$(4.6)	6.7
$OAstep_2$	$<a_0,v_1>$(0.7),$<a_2,v_4>$(1.9),$<a_3,v_5>$(4.0)	6.6
$OAstep_3$	$<a_0,v_2>$(3.2), $<a_2,v_6>$(6.7)	9.9
$OAstep_4$	$<a_0,v_2>$(0.3),$<a_2,v_4>$(0.7),$<a_3,v_5>$(1.4)	2.4
$OAstep_5$	$<a_0,v_3>$(2.1), $<a_1,v_5>$(4.5)	6.6
$OAstep_6$	$<a_0,v_3>$(1.2), $<a_1,v_6>$(2.5)	3.7

The preferred attack path is sorted out by analyzing the attack distance weight and cost and computing the evaluated value of each path in Figure 7 and Table 3. According to the experiment of an expert, λ is set to 0.45 and γ is set to 0.55 in $\oint(OAStep_i)$. The attack distance weight is calculated in a centesimal system and the attack cost is obtained on a score of ten, thus the path distance weigh need to be divided by 10 for uniformity and the final value $L_d{}'$ is substituted into the formula for L_d. The analysis of attack paths is shown in Figure 5.

Table 5. Effective Attack Path in Experimental Topological Graph

$OAstep_i$	Step Description	L_d	L_d'	$TCost$	$\oint$
$OAstep_1$	H0→(H2,DNS)→(H5,DNS)	60	6	6.7	6.385
$OAstep_2$	H0→(H2,HTTP)→(H4,SQL)→(H5,DNS)	110	11	6.6	8.58
$OAstep_3$	H0→(H2,SSH)→(H5,DNS)	40	4	9.9	7.245
$OAstep_4$	H0→(H2,SSH)→(H4,SQL)→(H5,DNS)	90	9	2.4	5.37
$OAstep_5$	H0→(H1,HTTP)→(H5,FTP)	40	4	6.6	5.43
$OAstep_6$	H0→(H1,HTTP)→(H5,DNS)	100	10	3.7	6.535

According to the simulation experiment, it can be seen from Table 5 that the path $OAStep_4$ is calculated to have a minimum evaluated value among all effective attack paths that may have same distance weight or same distance cost. If the attacker takes into account the attack distance weight and cost factors of the path and can access the path with less effort, he/she would select the path H0→(H2,SSH)→(H4,SQL)→(H5,DNS) as the attack path. The network administrators can prioritize the attacked path for reference and prepare security defense strategies in advance.

6. Conclusion

A new distance cost attack graph model NETAG is presented herein to cover the shortage of considerations in the existing attack path prediction methods of attack graphs. Firstly, the original model is transformed into a marked attack graph via the AGC-Alg algorithm and the attack distance weight and attack path redundancy coefficient are introduced to optimize the attack graph. Secondly, the Optimized-AG is subject to a cost calculation of attacking a path and the evaluation function is utilized to calculate the evaluated value of an attack path. Finally, the preferred attack path is located. The experiment suggests that the attack path prediction method of attack graphs presented herein eliminates path redundancy and the preferred attack path is determined based on the path distance weight and cost filtering. However, there still some deficiencies in the study. Although the authors provide the quantitative criteria of attack distance, the distance weight value is not detailed enough and needs to be assigned by an expert. Besides, the parameters are too complex in the mathematical model of attack cost and can only be obtained by an expert empirically. Therefore, further studies will focus on how to gain specific parameter values of distance weight and attack cost by conducting a large number of simulation experiments and referring to the experience of experts.

7. References

[1] *2017 Computer Crime and Security Survey* [EB/OL]. http://www.fbi.gov
[2] Development Status of China's Internet Sites and Safety Report (2018)[EB/OL]. http://www.cnii.com.cn/mobileinternet/2018-07/13/ 2082449.html
[3] National Computer Network Emergency Response Technical Team/Coordination Center of China [EB/OL]. http://www.cert.org.cn/publish/main/upload/File/situation.pdf
[4] Ni G, Ling G, Yiyue H E, et al. Dynamic Security Risk Assessment Model Based on Bayesian Attack Graph[J]. Journal of Sichuan University, 2016.
[5] Lee P P C, Misra V, Dan R. Distributed algorithms for secure multipath routing in attack-resistant networks[J]. IEEE/ACM Transactions on Networking, 2007, 15(6):1490-1501.
[6] Kaynar K. A taxonomy for attack graph generation and usage in network security[J]. Journal of Information Security & Applications, 2016, 29(C):27-56.
[7] Muñozgonzález L, Sgandurra D, Paudice A, et al. Efficient Attack Graph Analysis through Approximate Inference[J]. 2016, 20(3).
[8] Poolsappasit N, Dewri R, Ray I. Dynamic Security Risk Management Using Bayesian Attack Graphs[J]. IEEE Transactions on Dependable & Secure Computing, 2011, 9(1):61-74.
[9] Man D, Zhang B, Yang W, et al. A Method for Global Attack Graph Generation[C]// IEEE International Conference on Networking, Sensing and Control. IEEE, 2008:236-241.
[10] Wu Hongrun, Qin Jun, Zheng Bojin. Anti-attack Ability Based on Costs in Complex Networks[J]. Computer Science, 2012, 39(8):224-227.
[11] Jian S I, Chen P, Ningping G U, et al. Network attack graph backward depth-first building algorithm[J]. Computer Engineering & Applications, 2017.
[12] Chen Xiaojun, Shi Jinqiao, Xu Fei et al. Algorithm of Optimal Security Hardening Measures Against Insider Threat[J].Journal of Computer Research and Development, 2014,51(7):1565-1577.
[13] Qing Dapeng, Zhang Bing, Zhou Yuan et al. Depth-first Method for Attack Graph Generation [J].Journal of Jilin University (Engineering and Technology Edition), 2009, 39(2):446-452.
[14] Wang Hui, Liu Shufen. A Scalable Predicting Model of Insider Threats[J].Chinese Journal of Computers, 2006, 29(8):1346-1355.

IOP Conf. Series: Materials Science and Engineering **719** (2020) 012066 doi:10.1088/1757-899X/719/1/012066

Design of an Access Control System Based on Face Recognition

Qiong Wu, Qiang Wang, Yuhao Zhang, Bin Jiang, Yongcheng Cao and Fang Liu

Department of Computer Science and Electrical Engineering
Heilongjiang Oriental College Harbin, China
E-mail: 282977435@qq.com

Abstract. This paper studies design of an access control system based on face recognition for the major purpose of programming face recognition and access systems for embedded equipment. The access control system studied in this paper has functions regarding security, access control, personnel management, travel logs and time management. In this paper, face recognition algorithms are theoretically analyzed and derived. Based on face recognition algorithms, access control systems, particularly including face recognition algorithms, image processing and software, are designed and realized. Overall systematic debugging practices prove that aforementioned algorithms and software designs are correct and effective.

1. Introduction

Since the ancient times, people have used various door locks for protecting their properties or other articles, in order that people couldn't enter their rooms without permission. This simple measure has been taken for quite a long period. In general, door locks are always inseparable from keys, and a key is needed for each lock. As a result, more and more areas shall be locked. Under this situation, people have to carry many keys. However, it is inconvenient to take numerous keys at the same time. In addition, keys are easily lost and stolen. All these phenomena are rather unsafe. With the rapid development of today's sciences and technologies, traditional door locks can no longer accommodate people's needs for security [1].To thoroughly reverse this situation, different smart door control systems have been developed. These developments have been performed based on technologies of electronics, machinery, optics, biology, computers and communications. Mart door control systems are superior products of a new era.

2. Image Processing and Face Recognition Algorithms

2.1. Image Processing

The main purpose of image preprocessing is to eliminate irrelevant information and other information with insignificant impacts from images, recover useful information, retain or highlight primary information, increase detectability of related information, and simplify data to the greatest extent, in order to extrct, separate, match and recognize image features more reliably. A range of interfaces available from Open Cv are helpful for image processing [3].

The filtration process is performed for two purposes: On one hand, features of objects are extracted for image recognition; on the other hand, noises mixed at the time of image digitalization are eliminated [4], so as to satisfy requirements for computer processing. However, there are two requirements for filtration. Firstly, important information such as image contours and images can't be destroyed; secondly, images are clear and highly visible, or else it will deviate from original purposes of filtration [5]. as shown in Fig 1:

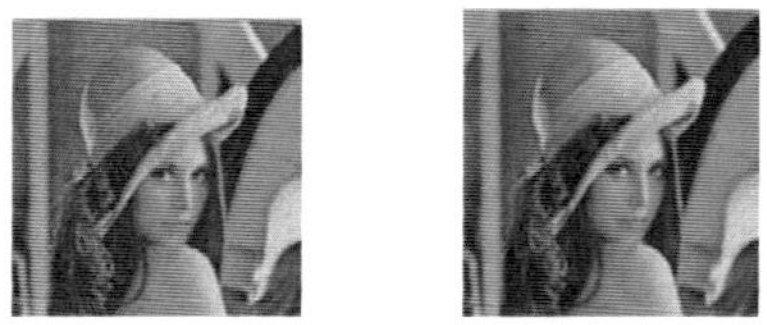

Figure 1. Results of Median Filtering

As shown in the figure, median filters blurred images and destroyed their details[6]. Therefore, different filtering results will be gained by adjusting the median filters, namely parameters of thee medianBlur (srcImage, dstImage, ksize).

By adjusting value of ksize, different results can be gained[7]. In case of varying aperture, the filter results are as shown in Fig 2:

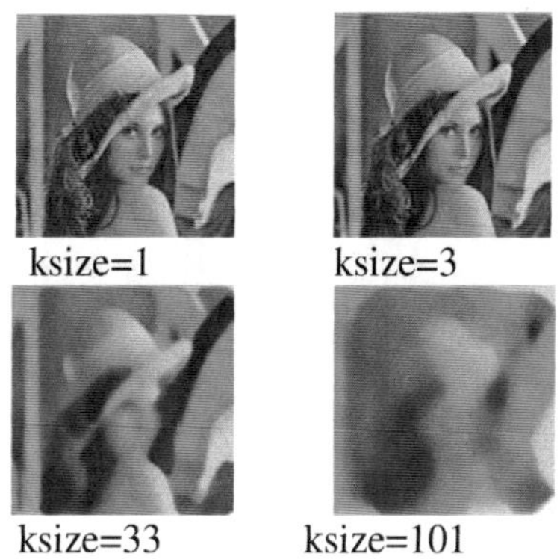

Figure 2. Median Filtering Results for Different Aperture

2.2. Traditional PCA-based Face Recognition Algorithms

To apply the PCA-based algorithm is applied in face recognition, all faces are represented in low-dimension linear spaces, and faces are separable within the space. When the algorithm is used, a new group of orthogonal bases is obtained from the high-dimensional image space after linear transformation. These orthogonal bases are chosen and ranked in an order of priority, while some higher-rank orthogonal bases are retained to generate a low dimensional space for human faces, and dimensionality of human faces. is reduced to obtain a characteristic subspace[8].

The steps are as follows:

1. Face images are preprocessed.
2. The images are read and saved in a database of human faces, trained to obtain characteristic subspaces.
3. Trained and tested images are projected into characteristic subspaces obtained at the preceding step.
4. Judgments are made according to certain rules.

Images of human faces are preprocessed, denoised via median filtering. They are converted into H*W pixels through size transformation and by converting grayscale images.

3. Software Programming

Programs of this system are designed based on VMware10.01, Ubuntu12.04 Linux, OpenCv2.4.9 and CMake (compiling tool). The basic functions of this system include personnel management, time management, face recognition and travel records. Generally, there are three basic requirements for an access control system: who can enter, how and when to enter.

IOP Conf. Series: Materials Science and Engineering **719** (2020) 012066 doi:10.1088/1757-899X/719/1/012066

3.1. Reading of Configuration Files and Assignments to Parameters

According to the analysis results, many parameters such as total number of people in the training database, each person's name, port number and IP address of IPV4 necessary for networking as well as image resolution (width+height), match ratio, start/end time of authorized transit time, number of each person' face images in the training database, and the sole identification number of the access controller used for remote control are initially assigned in the running process of the program. In this design, above parameters are included in configuration files and read during running. It is unnecessary to re-translate or alter source codes. The program operations can be impacted by altering content of the configuration files anytime. On this part, the flowchart is shown in Fig 3 as follows:

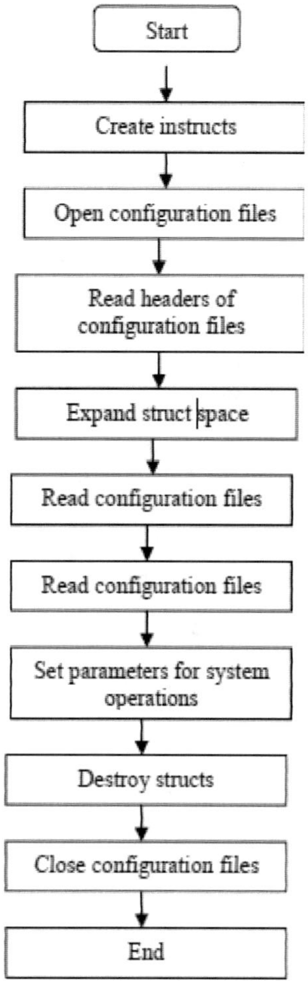

Figure 3. Flowchart of the Module of Configuration Files

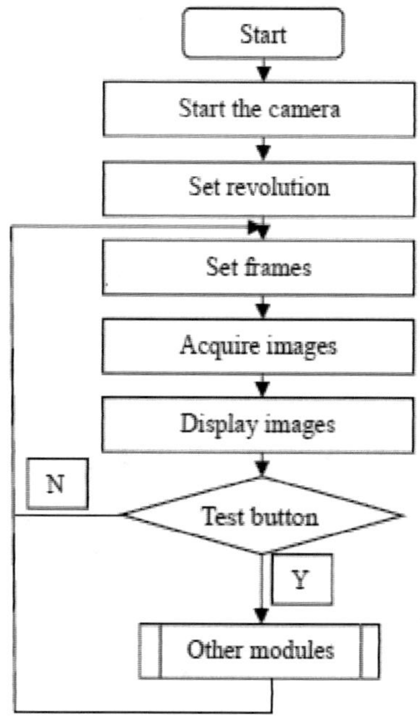

Figure 4. Flowchart of Camera Operation

3.2. Camera Operation

In this design, USB cameras are used. They are easy to use and cheap. They are called through related interfaces of OpenCv2.4.9. Most USB cameras available in markets mostly use USB2.0 interfaces. For these cameras, signal transmission and power supply share the same wires. Unstable power supply or too low voltage causes mutual interference, as a result of which cameras of controllers don't work. Under this circumstance, select timeout errors in program operation. On this part, the flowchart is shown in Fig 4 as follows:

3.3. Module of Mode Detection

In the course of mode detection, durations and working state are tested. To test the working state, STATE is set as a global variable in the system. There are three states, namely STATE_NORMAL, STATE_CLOSE and STATE_OPEN. These states mean normal working, normal close and normalopen states respectively. When the system is working normally, subsequent tests are performed; if it is normal close, it will deny all accesses and show that access is banned at the current time, while the access will be kept normal close; provided that the system is normal open, all accesses are allowed and all people are permitted to access, while the access is normal open.

3.4. Module of Logs

For an access control system, it is fairly necessary to keep travel logs. In important occasions, people stepping in and out shall be registered as proofs for future query. the flowchart is shown in Fig 5 as follows:

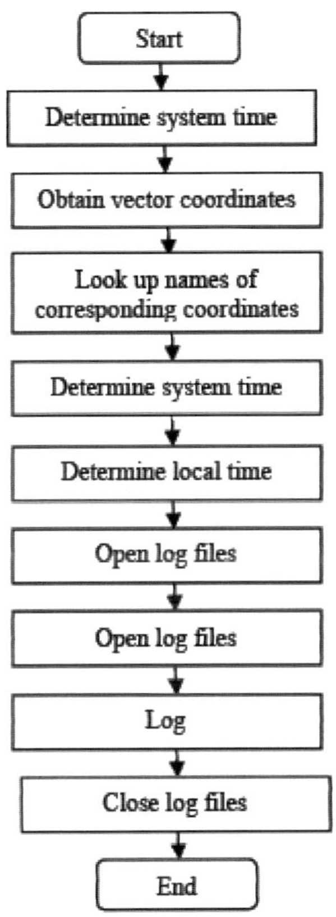

Figure 5. Flowchart of the Module of Mode Detection

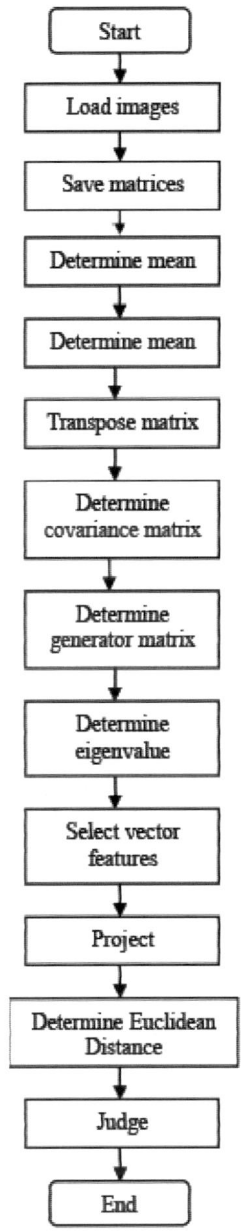

Figure 6. Module of Face Recognition

IOP Conf. Series: Materials Science and Engineering **719** (2020) 012066 doi:10.1088/1757-899X/719/1/012066

3.5. Face Recognition
Programmed design for face recognition involves many functions, including matrix multiplication, matrix transposition, mean calculation, covariance matrix, eigenvalue determination, feature vectors, projection and recognition. The specific algorithms have been introduced in the foregoing section. The flowchart is shown in Fig 6.

3.6. Server Design
The Linux server of this design is useful for received files uploaded at clients, indicating states of doors to clients and reading standard inputs. All these functions can be realized at the same time by setting unobstructed socks and unobstructed reading modes.

The server serves clients through a thread pool, and preliminary requests are made for thread resources, in order not to make repeated requests for resources and cause poor efficiency. The thread pool makes synchronization possible between threads by its mutex and cond. Function pointers are utilized to have the thread pool widely used and highly scalable.

4. Overall Debugging

4.1. Reading Configuration Files and Setting Running Parameters
Parameters shall be written in a configuration file altogether. After it is read, the configuration file shall be closed. Parameters shall be assigned to corresponding variables. Some running results are shown in Fig 7.

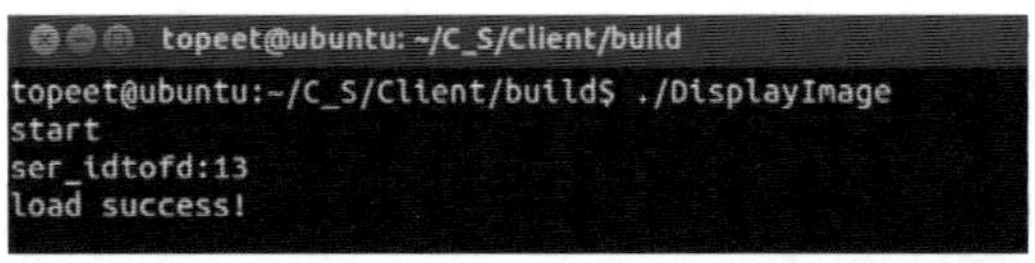

Figure 7. Running Results of the Module of Configuration Files

4.2. Network Connection
Through the last step, the IP address of the server and corresponding port number are identified. The networking sockets are determined in accordance with parameters. Try connecting network and set operation mode of the system based on return results.

In case of successful networking, running results are as follows (Fig 8):

a) Server; b) Clients.

Figure 8. Running Results on Networking

In case of failed networking, running results are as follows (Fig 9):

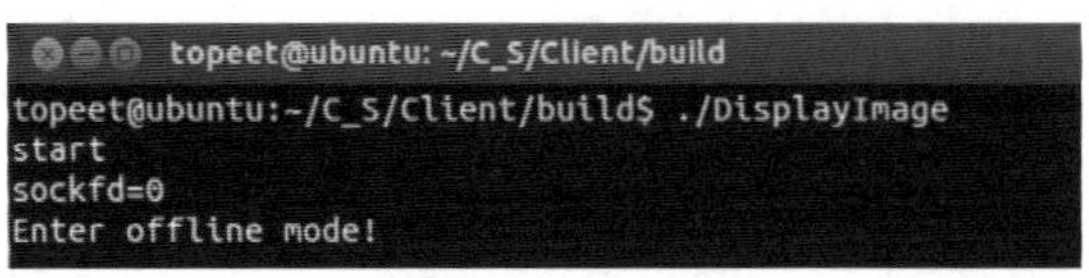

Figure 9. Running Results on Networking - Clients

4.3. Reading Face Databases and Face Modelling
Create corresponding matrix spaces and read the images of each face according to the grayscale

images. Save grayscales of each image pixel inside matrices. Determine mean of matrices and calculate covariance matrices. Calculate eigenvalues and feature vectors. Then, project them into characteristic subspaces and generate egenfaces. The running results are shown in Fig 10.

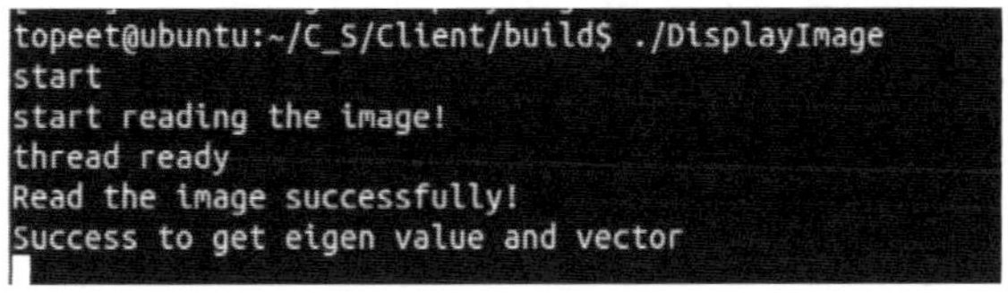

Figure 10. Running State of the Module for Face Modelling

When the module runs successfully, it will give a prompt that images have been read, while eigenvalues and feature vectors have been determined.

4.4. Opening Cameras and Setting Parameters

Set frames, window name and resolution for the camera. Start the camera. Set 33 frames for the camera and name the window as "capture". The resolution is 640*480. After the camera is started, the running results are shown in Fig 11.

Figure 11. Operation State of Camera

4.5. Grayscale Conversion

The results of grayscale conversion are shown in Fig 12:

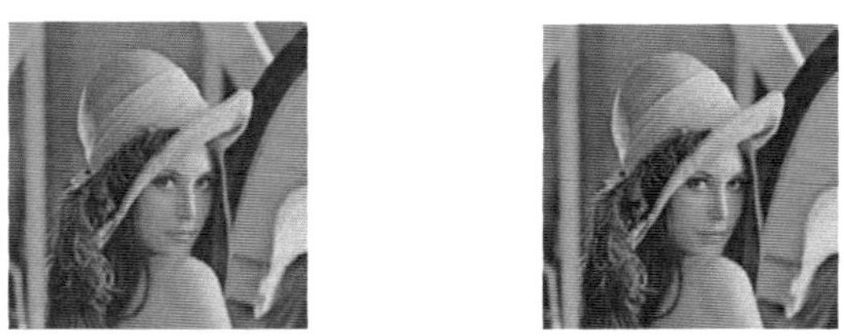

a) Raw Images for Grayscale Conversion b) Results of Grayscale Conversion

Figure 12. Grayscale Conversion

5. Conclusions

This access control system is designed based on face recognition technologies. The face recognition algorithm is used based on the traditional PCA algorithm, and the optimized algorithm is useful for general face recognition. By this algorithm, faces are recognized via grayscale conversion, filtering and denoising. However, it is impossible to guarantee the accuracy when there are very few samples in the training database.

The system designed in this paper is useful for identity recognition, travel recording, control and execution as well as remote control, log uploading and time management. Furthermore, travel records are locally kept. All logs are uploaded together every 24 hours, and local files are cleared. The server names and saves received logs pursuant to unified naming rules for the convenience of future checks. The test results suggest that the design scheme for access control system proposed in this paper is effective.

6. Acknowledgment

This research was supported by the Scientific Research Project of Heilongjiang Oriental University in 2018 (Numbering: HDFHX180112).

7. References

[1] Ruan Kui. Research and Application of Face Recognition Technologies [D]. National University of Defense Technology, 2006.
[2] Sui Yushun, Qi Sumin. Research on Template Matching and Face Tracking Algorithms Based on Face Detection [J]. Electronic Technology, 2016, 45(12):26-29.
[3] Wang Fei. Research on Face Recognition Algorithms Based on Deep Learning [D]. Lanzhou Jiaotong University, 2017.
[4] Geng C, Jiang X. Face recognition using sift features[C]// IEEE International Conference on Image Processing. IEEE, 2010:3313-3316.
[5] Geoffrey E, Hinton, Simon Osindero, Yee-Whye Teh. A Fast Learning Algorithm for Deep Belief Nets [J]. Neural Computation. 2006 (07):1527- 1544.
[6] Li Yuemin. Face Detection: A Survey [A]. Computer Application Federation of China Instrument and Control Society. The 16th CACIS Collection of Papers [C]/ Computer Application Federation of China Instrument and Control Society: April 2004.
[7] Shen Jie. Research and Comparison of Face Recognition Algorithms [D]. Nanjing University, 2017.
[8] Kong An. Design and Realization of Face Recognition and Management Systems Based on PCA and LDA [D]. Hunan University, 2016.

IOP Conf. Series: Materials Science and Engineering **719** (2020) 012067 doi:10.1088/1757-899X/719/1/012067

Design of Embedded Medical Automatic Transfusion Supervision System

Qiang Wang, Qiong Wu, Long Zhao, Hongfei Sun, Hejing Wu , Hui Wang, Qiang Wang and Yue Cao

Department of Computer Science and Electrical Engineering
Heilongjiang Oriental College, Harbin, China
E-mail: 123818364@qq.com

Abstract. This paper puts forward a new type of automatic transfusion supervision system to solve the common problem that manual monitoring and manual replacement of medicine bottles take up more time and energy of family or nurses in the hospital, and devises the real-time monitoring and management function of the whole process of venous transfusion. The upper computer uses vb as the control interface, and the lower computer uses STM32 as the core controller, and the ZigBee self-networking technology is adopted to realize the long-distance wireless communication function. The controlled object is the stepping motor and transducer by using ZigBee Ad-Hoc Network technology to realize the function of remote wireless communication. Among them, in order to stabilize the process of replacing the bottle,the encoder is used to detect, and the PID control algorithm is designed. The experiment shows that this design can effectively reduce the flow of the infusion process, improve the working efficiency, so as to promote the digital development of the medical level in domestic.

1. Introduction

It is well-know that infusion and replacement of liquid bottles are the most common tasks for clinical medical staff. However, because of the patient is relatively more, infusion quantity and varieties are different, manual monitoring and manual replacement of medicine bottles take up more time and energy of family or nurses, high power of work, low efficiency, as well as the accidents caused by artificial forgetting, which bring a lot of inconvenience to patients and medical staff[1]. A kind of peristaltic pump automatic control infusion equipment has been developed previously in foreign countries, but the loss is large and the material requirement of infusion equipment is relatively high. So a variety of infusion equipment similar to peristaltic pump have been developed[2]. Some peristaltic pump infusion equipment has also been introduced in China, but its cost is quite expensive, and it can only be used in some critical patients and special patients, so it can not be popularized. Therefore, the hospital urgently needs equipment that can simplify the infusion process, reduce the manpower expenditure, effectively improve the safety of intravenous infusion, and reduce the working intensity of medical and nursing staff[3]. In order to meet the needs of the hospital and promote the development of the national health care industry, we have carefully designed the equipment.

The system takes ARM-STM32 as the control core, detects the liquid level by using the capacitance sensor[4], and detect the droplet speed by using the infrared photoelectric sensor. By the high-precision stepping motor, the medicine bottle conversion and pin-pulling function can be realized[5], the nurse and families can know the infusion condition in time, and the infusion information can be displayed in real time and the sound and light alarm can be carried out[6]. In view of the characteristics such as large number of monitoring points, the variety of sensing quantities and the complexity of wiring in hospital wards, the Ad-Hoc Network and wireless transmission of ZigBee technology are used for network control[7], and long distance wireless control can be carried out, as well as multiple lower computers can communicate with the upper computer system, and the multi-computer system can simultaneously communicate with the upper computer system[8]. If the system encounters a sudden and abnormal condition, there will be a reminder to protect the site and stop the work Compared with the problem of manual control and monitoring of traditional infusion, it has the following advantages:

(1) Remote control: the design adopts two 42-step motors as the power source of the control mechanism, can be automatically plugged and changed, and can carry out on-site display and remote monitoring through STM32 controller and ZigBee self-networking communication.

(2) PID angle and speed double closed-loop control algorithm is used to control the changing of the stepping motor. The device has the characteristics of stability, accuracy and high speed.

(3) Centralized control: the nurse can control the infusion situation of the whole department in the computer of the duty room, and can also see the abnormal infusion at a certain terminal, medication situation, remaining time and so on. The nurse can monitor the current data of each drop in real time to judge the occurrence of the emergency.

2. System Hardware Block Diagram Design

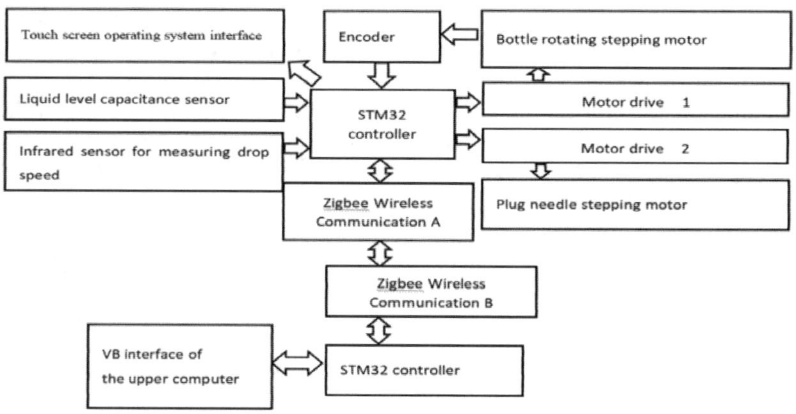

Figure 1. System hardware block diagram design

3. Subsystem Design

3.1. The Design of the Interactive Touch Screen

The system on the patient infusion rack is the lower end of the machine, and its operating screen is touch type, which can display the working state and set the number of bottles of infusion. Touch display screen are shown in Fig 2.

Figure 2. Touch Display Screen Interface

The interactive capacitance touch screen forms a capacitance at the intersection of the transverse and longitudinal ITO electrodes on the glass surface. The scanning method of interactive capacitance

IOP Conf. Series: Materials Science and Engineering **719** (2020) 012067 doi:10.1088/1757-899X/719/1/012067

can detect the capacitance value of each intersection and the change of capacitance after touch. The touch screen is connected with the upper computer, and the number of working bottles of the drop is selected, and the lower computer and the upper computer perform the "exchange" display window. The core is written based on ARM-STM32, the response is sensitive, the acquisition is accurate, and the safety and the university can be ensured.

3.2. Operating System Interface of the Upper Computer

3.2.1. Login interface. The upper computer supervision system is installed in the medical studio. You need to enter the correct user name and password to log on to the system. The system login interface is shown in Fig 3.

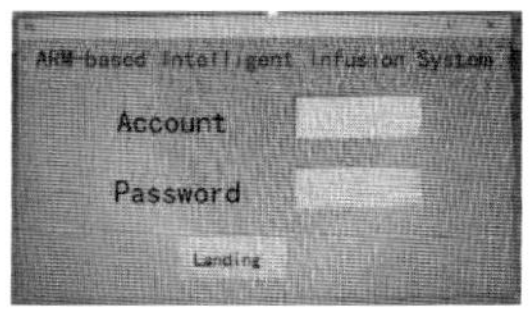

Figure 3. System login interface

Staff can set passwords according to their own conditions, can prevent work leakage and prevent non-operators from logging into the operating system to ensure the rigour of the work.

3.2.2. Control main interface of the upper computer. Enter the main interface after logging into the system. Fig 4 is an operation main interface of the upper computer, and the worker can monitor the working state and control the lower computer.

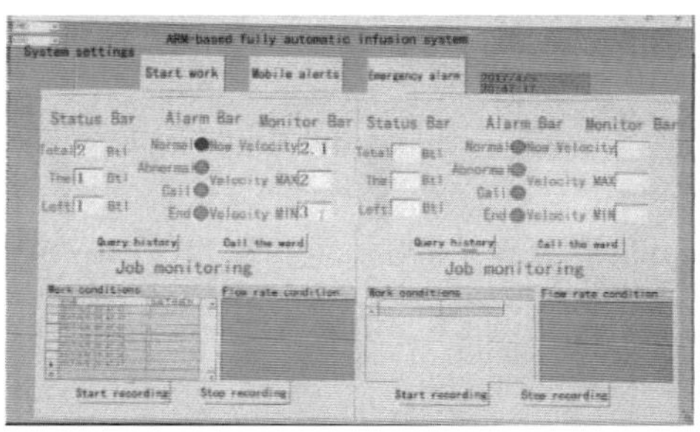

Figure 4. Operation main interface

3.3. Query History Interface

The historical data is stored in the database, and the staff can query the relevant data through the query history interface. The query interface is shown in Fig 5.

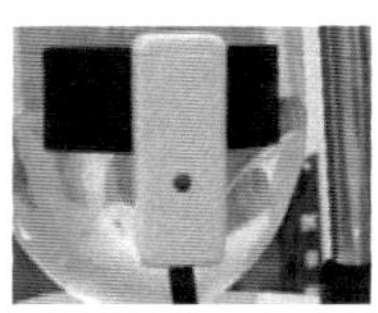

Figure 5. Query interface

Operators can query the critical time period of patient data in the database page.

3.4. Capacitive Sensor

The liquid level of the medicine bottle is detected by capacitive sensor, and Fig. 6 shows the schematic diagram of the capacitance sensor attached to the medicine bottle. The capacitive sensor is small in size, sensitive and less space occupied. The liquid level can be detected. Once the drug runs out, the signal will be sent to ARMSTM32 for the next automatic drug exchange operation.

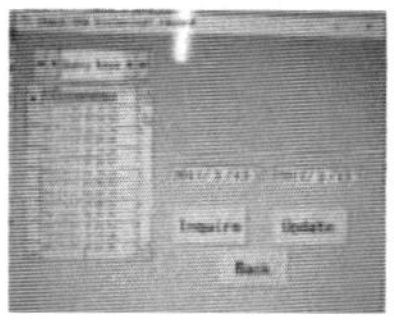

Figure 6. Capacitance transducer

3.5. Design of Straight Blade Stepping Guideway

The needle-pulling and changing medicine of the system adopts a straight blade stepping guideway, as shown in Fig 7.

Figure 7. Straight blade stepping guideway

The rotating force of the two-phase stepping motor is converted into the vertical force by using the principle of the mechanical arm. The stepper motor operates stably, which can control the distance, speed and precise implementation of needle pulling and drug exchange function.

3.6. Embedded Needle Slot

The fixed needle is realized by embedded needle slot. Fig 8 shows the card slot physical diagram and three-dimensional model diagram.

The embedded needle slot is made according to the size of the needle can effectively fix the needle. When changing dressing, it can make the needle bear better force, the card slot design is exquisite and reasonable, and it is convenient for nurses to operate.

a) Physical diagram of card slot

b) Side diagram of card slot

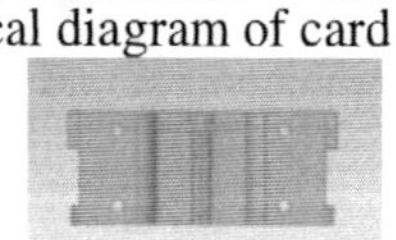

c) Overlooking view of card slot

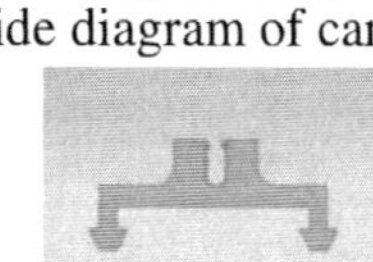

d) Left view of card slot

Figure 8. Embedded needle slot

3.7. Fixed Bottle Flange Design

The rotating dressing changing device of the medicine bottle adopts flange. Fig 9 a) is a three-dimensional model diagram of the flange plate, and Fig. 9 b) is a real drawing of the flange plate.

IOP Conf. Series: Materials Science and Engineering **719** (2020) 012067 doi:10.1088/1757-899X/719/1/012067

a) Flange for fixing medicine bottle b) Flange solid diagram

Figure 9. Flange model

Flange can be stuck in different sizes of drop bottle mouth, so that the drop bottle better force, convenient needle extraction, dressing change. The flange plate can be moved up and down, and is suitable for drip containers of different sizes.

3.8. Three-Dimensional Model Diagram of the Changing Area of Dripping Rod

Fig 10 is a main structure of a drip bar, which is convenient for fixing and replacing a medicine bottle, and the flange plate can adjust the distance up and down on the rod.

a) Side vertical view of the drop bar b) Side upward view of the drop bar

Figure 10. Main structure of a drip bar

4. System Software Design

4.1. System Program Flow Chart

Fig 11 is the upper computer program flow chart, and Fig 12 is the lower computer flow chart, from which the work flow can be easily understood.

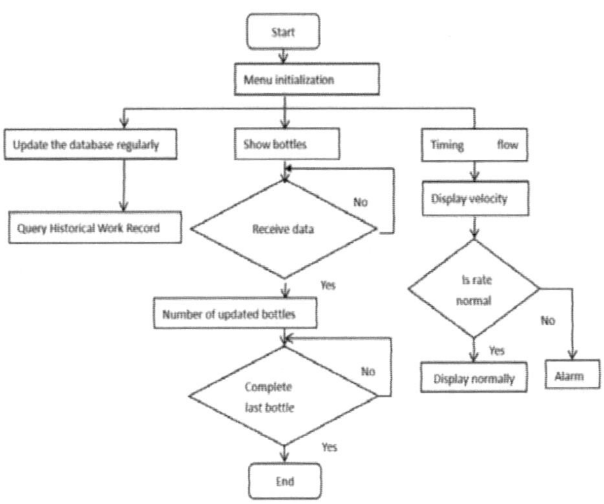

Figure 11. Flow chart of the upper computer

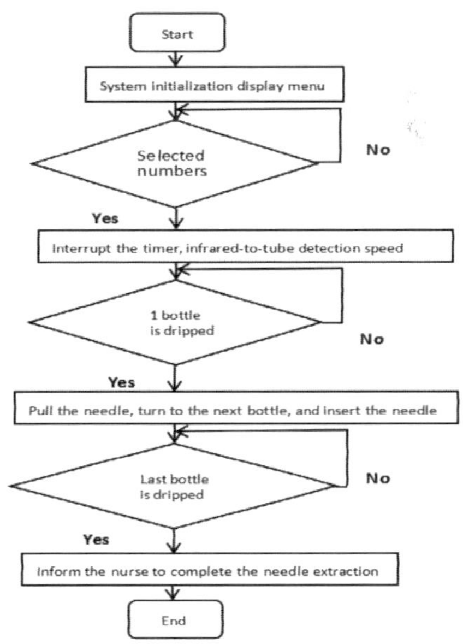

Figure 12. Flow chart of the lower computer

4.2. Real-time Communication between Upper and Lower Computer

As shown in Fig 13, the data acquisition protocol adopts opc way to ensure the real-time, precision and stability of its data.

Instantaneity: The transmission speed must be guaranteed. If it can't be too slow, the upper and lower computer needs to be combined. Otherwise, the problem will be solved.

Precision: The signal can't be lost, and skipped. It can't have too much error and can't bring too much interference and noise.

Stability: Failure, such as communication disconnection, crash and so on, can be automatically reconnected, stable enough to not cause crash, automatic restart, fault-tolerant mechanism and error log.

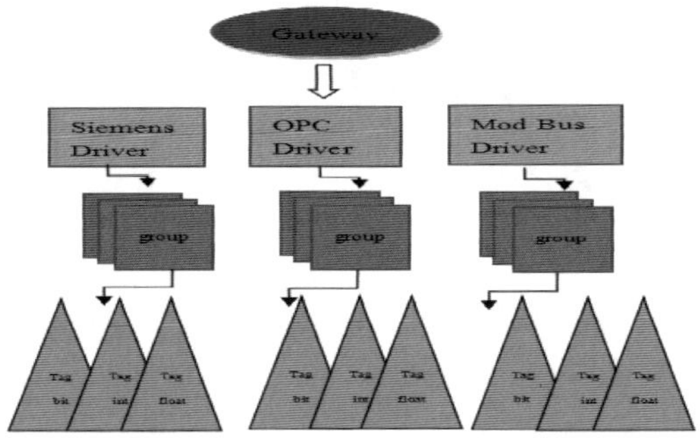

Figure 13. Schematic diagram of opc data acquisition

4.3. The Accurate Alignment of the Change-of-drop Bottle is Realized by Using the PID Control Algorithm

In order to achieve the control precision, the position and speed double closed-loop PID control method is adopted for the rotation control of the stepping motor used for changing the dropping bottle. The step angle of the stepping motor is 1.8 degrees, the measuring angle is 512 encoders, and the accuracy can reach the requirement by the detection of the pulse. When position control is carried out, the corresponding expression of position PID, is as follows:

$$u_k = K_p e_k + K_I \sum_{j=0}^{k} e_j + K_D (e_k - e_{k-1}) + u_o \tag{3-1}$$

In the formula, u_k represents the k corresponding output value, e_k represents the k input deviation of sampling. It is more suitable to adopt incremental PID in speed control, and its expression is as follows:

$$\Delta u_k = K_p (e_k - e_{k-1}) + K_I e_k + K_D (e_k - 2e_{k-1} + e_{k-2}) \tag{3-2}$$

As shown in Fig 14, the control block diagram of the position and speed double closed loop can slow down the speed and finally smooth the position corresponding to the needle. Fig 15 and 16 are the waveform diagram corresponding to the different PID parameters of the outer ring and the inner ring when the upper computer software is used to test the outer ring and the inner ring. It can be seen from the diagram that the stability, accuracy and rapidity of the system with the parameters of figure 16 are better.

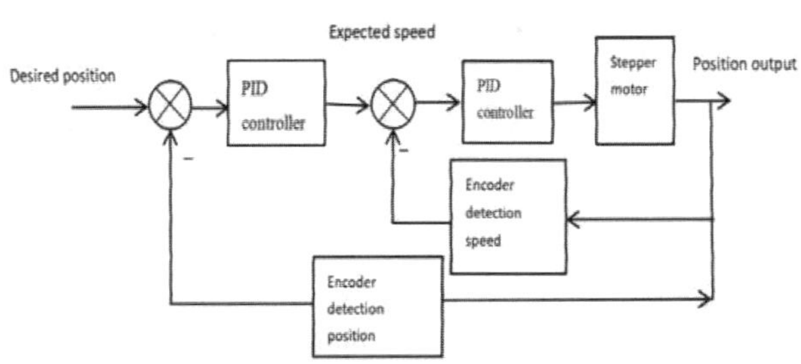

Figure 14. Control block diagram of position and speed double closed loop

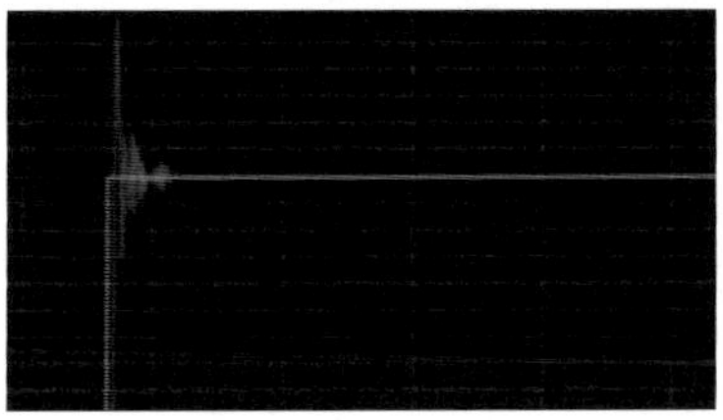

Figure 15. Oscillograph of Kop=400, Koi=0, Kod=0, Kip=20, Kii=0, Kid=0

Figure 16. Oscillograph of Kop=100, Koi=0.2, Kod=400, Kip=25, Kii=0, Kid=30

5. Product Physical Diagram
Fig 17 is the front, back and side of the design.

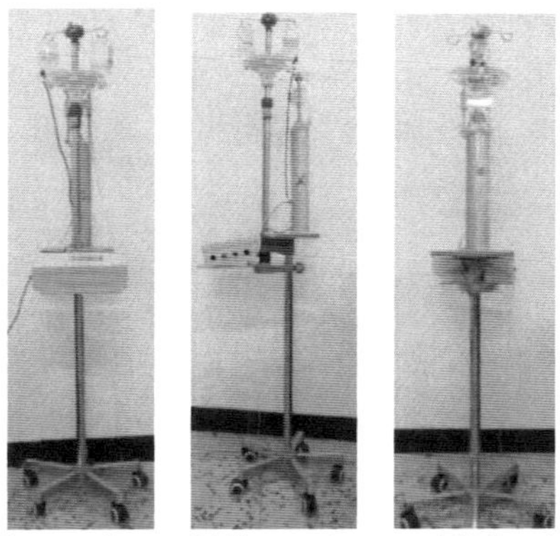

a)Front view b)Rear view c)Side view

Figure 17. System Physical Map

6. Conclusions
This paper introduces an improved medical automatic infusion supervision system, puts forward the solutions based on embedded technology, gives the 3D printing technology UG model of the whole system, designs the VB upper computer program and the lower computer program, carries on the PID double closed loop control of the stepping motor, and carries on the ZigBee Ad-Hoc Network. It can realize the functions of automatic dressing change and remote wireless operation, so as to improve the work efficiency, reduce the work pressure of family and nurses, and save the human, financial and material expenses of the hospital. And the cost of this design is low, so that different levels of patients can be used, which is convenient for the popularization of medical and health places. The experimental results show that work efficiency can be increased by about 40%.

7. Acknowledgment
This research was supported by the Scientific Research Project of Heilongjiang Oriental University in 2017(Numbering: HDFHX170114).

8. References
[1] Yang Lixiao, Wang Shuyi, Shang Changhao. Development of Automatic Liquid Exchange Intravenous Infusion Device[J]. Chinese medical equipment, 2015(04): 35-37.
[2] Chen Junlong, Shi Longzhao. Research on the Non-contact Liquid Level Linear Measurement System of Coplanar Capacitor [J]. Journal of Wuyi College, 2015,34(12):74-78.
[3] Walter H.Dzik.New technology for transfusion safety[J]. British Journal of Haematology, 2007 ,136(2):90-181.

[4] McIntosh, Robert B., Mauger, Philip E., Patterson, Steven R. Capacitive transducers with curved electrodes.[C]//IEEE Sensors Journal. 2006:125-138.

[5] Lin Min, Rong Bixian, Cheng Yinan. Automatic Bottle-changing and Withdrawing Integrated Infusion Machine[J]. Chinese Technology Information, 2012(18):98.

[6] Tong Yaoying. Development of Automatic Infusion Alarm[J]. Medical and Health Equipment, 2008,29(12):59-61.

[7] Liu Jian. Design and Research of Telemedicine Monitoring Node based on ZigBee Technology [D].Xi'an University of Science and Technology, 2011.

[8] Wang Yiwang, Lu Jun, Zhang Chengcheng, Chen Hai. Design of Intelligent Infusion Monitoring System based on Wireless Sensor Network[J]. Observation and Control Technology,2015,34(11):64-66.

Steganalysis of Intra Prediction Mode and Motion Vector-based Steganography by Noise Residual Convolutional Neural Network

Peng Liu and Songbin Li

Haikou Laboratory, Institute of Acoustics, Chinese Academy of Sciences, Haikou 570105, China.
Email: liup@dsp.ac.cn

Abstract. In this paper, we present a universal steganalysis method for both intra prediction mode and motion vector-based steganography based on deep learning. Since the embedding process is eventually reflected in the modification of pixel values in decoded frames, we design a Noise Residual Convolutional Neural Network (NR-CNN) from the perspective of the spatial domain, which is the first CNN-based approach for this subject. In NR-CNN, feature extraction and classification modules are integrated into a unified and trainable network framework. It automatically learns features and implements classification in a data-driven manner, which effectively solves the existing problems. Experimental results show that NR-CNN has better performance of steganalysis than the related method.

1. Introduction

Steganography is the art and science of data hiding, which realizes covert communication by embedding secret data into an innocent-looking cover media, such as digital image, audio, video, et al., without arousing any suspicion. In contrast, steganalysis aims to expose the presence of hidden data. In this paper, we are focusing on the intra prediction mode [1-3] and motion vector-based [4-6] steganography.

To detect the intra prediction mode-based steganography, Li et al. [7] design a series of features based on Markov chain to quantify this correlation property. Zhao et al. [8] conduct the steganalysis based on intra prediction mode calibration; In the aspect of steganalysis for motion vector-based steganography, some feature-based methods have been presented in recent years. They can be divided into three major categories. The first category designs features based on neighboring motion vector difference [9-10] The second category uses calibrations to enhance the features [11-12]. The third category utilizes the statistics of Sum of Absolute Differences (SADs) to construct features [13-15].

In this paper, we propose a universal steganalysis method for both intra prediction mode and motion vector-based steganography based on deep learning. Since the process of intra prediction mode and motion vector-based steganography is eventually reflected in the modification of pixel values in decoded frames, we design a Noise Residual Convolutional Neural Network (NR-CNN) from the perspective of the spatial domain. Feature extraction and classification modules are integrated into a unified and trainable network framework. It automatically learns features and implements classification in a data-driven manner, which effectively solves the existing problems.

2. Network Architecture

The embedding process can be considered as adding low-amplitude noise in the cover image, so the secret information has an extremely low SNR compared to the image content. The low SNR mainly leads to two types of problems. The first is that commonly used activation functions such as ReLU do not fully apply to this type of task. Since the proportion of the useful signal in the input signal itself is already very low, forcibly losing half of the signal each time will cause a large number of invalid filters in the training process; the second is the problem of parameter initialization. The network usually cannot converge when using the most commonly used weights initialization method in the field of computer vision.

Recently, Ye et al. [16] give some ideas to solve these problems. They introduce an Image Steganalysis Network (IS-Net) based on CNN which uses 30 high-pass filters as weights to initialize the first convolutional layer. In addition, they introduce a new activation function called Truncated Linear Unit (TLU). The idea of IS-Net can help us to solve the problems in video steganalysis. But IS-Net is designed specifically for image steganalysis. In this paper, we present a Noise Residual Convolutional Neural Network (NR-CNN) based on the ideas of IS-Net. The main idea of the NR-CNN is to extract features from noise residuals, which are used for classification. The network structure is shown in figure 1. Improvements are made from three aspects: First, we add four global filters in the residual computation part to obtain more feature maps of steganographic noise residual signal; second, a new activation function called Parametric Truncated Linear Unit (PTLU) is proposed to better capture the structure of embedding signals; third, a steganalysis residual block structure is presented, which can improve the learning ability of steganographic noise residual signal. As we can see from figure 1, the network is divided into three main parts, which are used for residual computation, feature extraction, and binary classification.

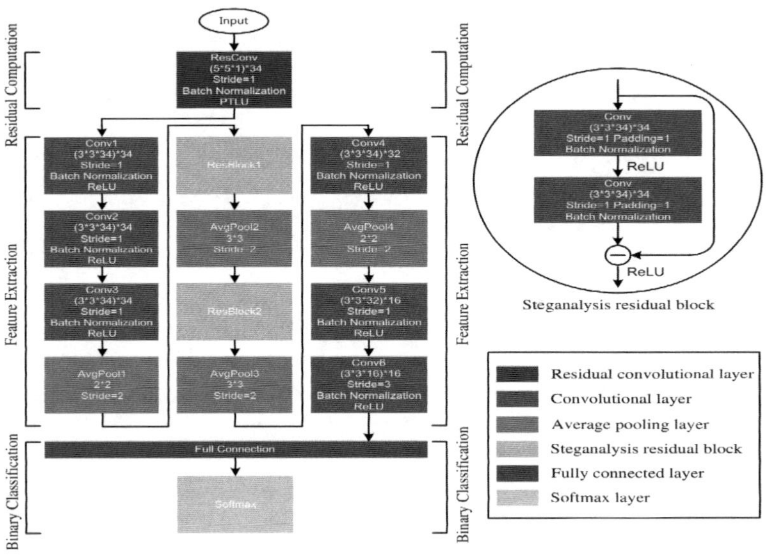

Figure 1. The architecture of the proposed convolutional neural network.

2.1. Residual Convolutional Layer

The residual convolutional layer "ResConv" is used to compute the steganographic noise residual. Current convolutional neural networks tend to learn features from the image content. However, the embedded secret information is independent from the image content. Thus, the residual convolutional layer is very important, whose role is to obtain steganographic noise residual features that are independent of the image content.

Figure 2 shows the visualizations of high-pass filters used in the first convolutional layers of IS-Net. Each filter corresponds to a noise residual feature, and we can see that only the filter in the fifth row and the second column is a global filter that covers every pixel in the 5×5 area. In the video steganography, information hiding is conducted in units of one block. Modification of the intra prediction mode or motion vector will changes the pixel value of one block. So, we consider that more global filters are needed to represent the steganographic noise residual signals. Figure 3 shows the new global filters introduced in this paper. Therefore, a total of 34 convolution kernels are used in the residual convolutional layer of NR-CNN. Although initialization using these 34 convolution kernel parameters is better than random initialization, these parameters are not the best. These parameters will be optimized by global optimization during training process.

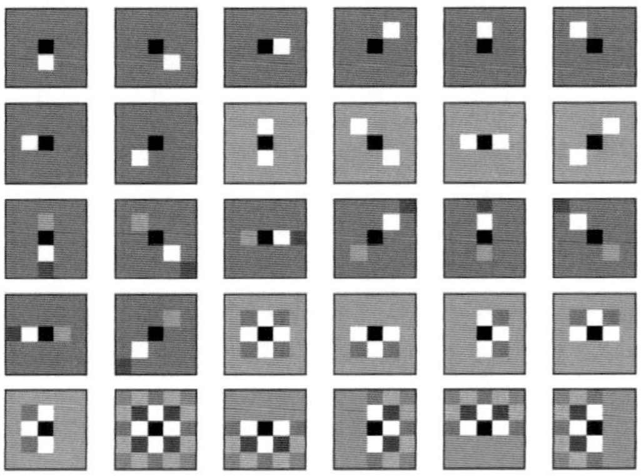

Figure 2. Visualization of high-pass filters used in the first convolutional layer of IS-Net.

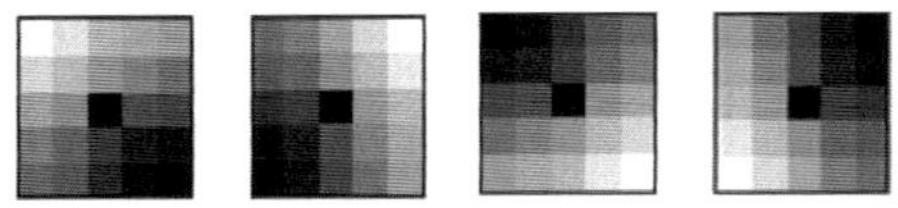

Figure 3. Visualizations of new global filters.

The difference between the residual convolutional layer and the common convolutional layer is that the parameters of residual convolutional layer are initialized with fixed values. The input data for this layer is 256×256 single-channel image data. This layer contains 34 filters of size 5×5×1, where 1 represents the number of channels. The convolution step size is 1 and the output of this layer is 34 feature maps with a size of 252×252.

2.2. Convolutional Layer

The feature extraction part contains six convolutional layers. The convolutional layers "Conv1", "Conv2", and "Conv3" all contain 34 filters of size 3x3x34 with a step size of 1, and use ReLu as the activation function. Among them, the output of the convolutional layer "Conv1" is 34 feature maps with a size of 250×250. The output of the convolutional layer "Conv2" is 34 feature maps with a size of 248×248, and the output of the convolutional layer "Conv3" is 34 feature maps with a size of 246×246. The convolutional layer "Conv4" contains 32 filters of 3x3x34, and uses ReLU as an activation function. "Conv4" outputs 32 feature maps with a size of 28x28. The convolutional layer "Conv5" contains 16 filters of 3x3x32 with a step size of 1, and uses ReLU as the activation function. The output is 16 feature maps with a size of 12x12. The convolutional layer "Conv6" contains 16 filters of 3×3×16 with a step size of 3. It uses ReLU as the activation function, and outputs 16 feature maps with a size of 4×4. It should be noted that we use the batch normalization operation to normalize the data before the activation function in each convolutional layer.

2.3. Activation Function

Ye el. [16] present a new activation function for image steganalysis called Truncated Linear Unit (TLU). They prove that TLU is better than ReLU for image steganalysis. We extend TLU and propose a new activation function called Parametric Truncated Linear Unit (PTLU), which is defined as

$$f(x)=\begin{cases} T, & x>T \\ x, & 0\le x\le T \\ \alpha x, & -T/\alpha\le x<0 \\ -T, & x<-T/\alpha \end{cases} \tag{1}$$

For PTLU, the coefficient of the negative part is not constant and can be learned adaptively. When $\alpha=1$, PTLU is equivalent to TLU.

2.4. Steganalysis Residual Block

Since the depth of CNN has a great influence on the classification performance, it is usually considered that deeper network result in the better result. However, a deeper network will lead to higher training error than the shallow network. This can be understood as data disappearing through too many layers of the network, which will lead to a worse result. To this end, He et al. [17] propose the ResNet, which contains several residual blocks as shown in figure 4(a).

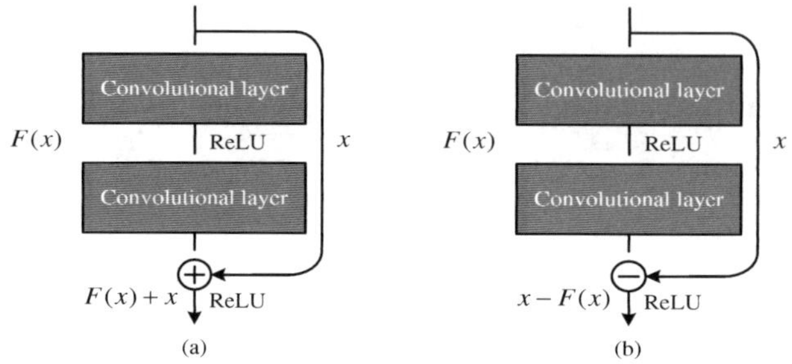

Figure 4. The comparison of residual blocks.
(a)The structure of traditional residual block. (b) The structure of steganalysis residual block.

In this paper, an improved structure of the residual block is proposed for steganalysis. As shown in figure 4(b), the improved structure is called steganographic residual block. The main improvement is to change the final mapping function from $F(\boldsymbol{x})+\boldsymbol{x}$ to $\boldsymbol{x}-F(\boldsymbol{x})$. In the steganalysis problem, the input data $\boldsymbol{x}$ can be regarded as the sum of the carrier image content $\boldsymbol{c}$ and the steganographic residual signal $\boldsymbol{m}$. Ideally, the content of the carrier image in the input data $\boldsymbol{x}$ has been already filtered out in the previous processing, and at this time $\boldsymbol{c}$=0. However, in the actual situation, the carrier image content cannot be filtered completely, which means $\boldsymbol{c} \neq 0$. The purpose of the steganographic residual block is to further suppress the carrier image content so as to reduce $\boldsymbol{c}$. In the steganographic residual block, $F(\boldsymbol{x})$ is used to filter out the steganographic residual signal $\boldsymbol{m}$ so as to only retain the carrier image content $\boldsymbol{c}$. Then the steganographic residual signal $\boldsymbol{m}$ can be retained as much as possible by $\boldsymbol{x}-F(\boldsymbol{x})$. Therefore, the steganographic residual block is very suitable for learning the steganographic residual signal $\boldsymbol{m}$. In this paper, two steganographic residual blocks are used for steganalysis.

Each steganographic residual block contains two convolutional layers. Each convolutional layer contains 34 filters with a size of 3×3×34 and uses zero padding. The step size is 1 and ReLU is used as the activation function. Batch Normalization is conducted before ReLU. Due to zero padding, the size of output feature maps is the same as that of input. The output of the steganographic residual block "ResBlock1" is 34 feature maps with a size of 123×123. The output of "ResBlock2" is 34 feature maps of a size of 61×61.

2.5. Pooling Layer

The main role of the pooling layer is to reduce the dimension of input feature, thereby reducing the parameters and computation of the entire network, and suppressing overfitting.

The feature extraction part contains four pooling layers, all using average pooling. The pooling layer "AvgPool1" has a kernel size of 2x2 and a step size of 2, which outputs 34 feature maps with a size of 123×123. The pooling layer "AvgPool2" has a kernel size of 3x3 and a step size of 2, which outputs 34 feature maps with a size of 61x61. The pooled layer "AvgPool3" has a kernel size of 3×3 and a step size of 2, which outputs 34 feature maps with a size of 30×30. The pooling layer "AvgPool4" has a kernel size of 2x2 and a step size of 2, which outputs 32 feature maps with a size of 14x14.

3. Experimental Results and Discussion

In this section, we will evaluate the availability and effectiveness of the proposed NR-CNN. Since

NR-CNN is inspired by IS-Net, we will compare with it in this section.

3.1. Experimental Settings

The dataset used in this paper contains three parts: training set, verification set and test set. There are 200,000 frames in the training set, 20,000 frames in the verification set, and 200,000 frames in the test set. The training set and the verification set are used to train the network. The test set is used to evaluate the information hiding detection accuracy.

In the experiment, we use the methods in [1] and [6] for intra prediction mode and motion vector-based steganography, respectively. For the training set and verification set, we use 100% embedding rate for intra prediction mode-based steganography and 20% embedding rate for motion vector-based steganography. For the test set, we use five embedding rates of 20%, 40%, 60%, 80%, and 100% for intra prediction mode-based steganography, and use 5%, 8%, 10%, 15%, and 20% for motion vector-based steganography. IS-Net and NR-CNN are implemented on the deep learning framework PyTorch, and the batch size is 32. The network optimizer is AdaDelta with a learning rate of 0.4, a momentum value of 0.95, a weight decay of 5×10^{-4}, and a "delta" parameter of 1×10^{-8}. The number of training iterations epoch is 150, and the goal of the training process is to minimize the cross-entropy cost function.

3.2. Comparison of NR-CNN and IS-Net

Previous experiments show that both NR-CNN and IS-Net have the best detection performance when T=7. Thus, we will compare the two networks under this threshold. Table 1 shows the detection accuracy for intra prediction mode-based steganography under different embedding rates. It can be seen from the table that NR-CNN performs better than IS-Net under each embedding rate. Table 2 shows the detection accuracy for motion vector-based steganography under different embedding rates. It can be seen from the table that NR-CNN performs better than IS-Net under all embedding rates.

Table 1. The detection accuracy for intra prediction mode-based steganography under different embedding rates.

Method	Embedding rates				
	20%	40%	60%	80%	100%
NR-CNN	59.82%	85.67%	98.2%	98.95%	99.74%
IS-Net	58.07%	83.41%	96.25%	97.69%	98.7%

Table 2. The detection accuracy for motion vector-based steganography under different embedding rates.

Method	Embedding rates				
	5%	8%	10%	15%	20%
NR-CNN	62.53%	73.82%	82.37%	91.48%	95.39%
IS-Net	60.39%	70.86%	77.69%	90.28%	93.45%

4. Conclusion

In this paper, a universal steganalysis method is proposed for both intra prediction mode and motion vector-based steganography based on deep learning. Since the embedding process is eventually reflected in the modification of pixel values in decoded frames, we design a Noise Residual Convolutional Neural Network (NR-CNN) from the perspective of the spatial domain. Feature extraction and classification modules are integrated into a unified and trainable network framework. It automatically learns features and implements classification in a data-driven manner, which effectively solves the existing problems in the framework of “feature extraction-feature classification”. Experimental results show that NR-CNN has better performance of steganalysis than IS-Net.

5. Acknowledgments

This work is supported partly by the Important Science & Technology Project of Hainan Province under grant ZDKJ201807, part by the National Natural Science Foundation of China under grant U1636113, part by the Hainan Provincial Natural Science Foundation of China under grant 618QN309, part by the Scientific Research Foundation Project of Haikou Laboratory, Institute of Acoustics, Chinese Academy of Sciences, and part by the IACAS Young Elite Researcher Project QNYC201829 and QNYC201747.

6. References

[1] HU Y, ZHANG C, SU Y. Information hiding based on intra prediction modes for H.264/AVC[C]. Proceedings of the 2007 IEEE International Conference on Multimedia and Expo. Beijing, China, July 2007: 1231-1234.

[2] YANG G, LI J, HE Y, et al. An information hiding algorithm based on intra-prediction modes and matrix coding for H.264/AVC video stream[J]. AEU-International Journal of Electronics and Communications, 2011, 65(4): 331-337.

[3] XU D, WANG R, WANG J. Prediction mode modulated data-hiding algorithm for H.264/AVC[J]. Journal of Real-Time Image Processing, 2012, 7(4): 205-214.

[4] ZHU H, WANG R, XU D. Information hiding algorithm for H.264 based on the motion estimation of quarter-pixel[C]. International Conference on Future Computer and Communication, IEEE. 2010:V1-423-V1-427.

[5] Yao Y, Zhang W, Yu N, et al. Defining embedding distortion for motion vector-based video steganography[J]. Multimedia Tools and Applications, 2015, 74(24), 11163-11186.

[6] ZHANG H, CAO Y, ZHAO X. Motion vector-based video steganography with preserved local optimality[J]. Multimedia Tools and Applications , 2016 , 75 (21):1-17.

[7] LI S, DENG H, TIAN H, et al. Steganalysis of prediction mode modulated data-hiding algorithms in H.264/AVC video stream[J]. annals of telecommunications - annales des télécommunications, 2014, 69(7-8):461-473.

[8] ZHAO Y, ZHANG H, CAO Y, et al. Video steganalysis based on intra prediction mode calibration[C]. International Workshop on Digital Watermarking. Springer, Cham, 2015: 119-133.

[9] SU Y, ZHANG C, ZHANG C. A video steganalytic algorithm against motion-vector-based stegangraphy[J]. Signal Processing, 2011, 91(8): 1901-1909.

[10] Hao-Tian Wu, Yuan Liu, Jiwu Huang, and Xin-Yu Yang. 2014. Improved steganalysis algorithm against motion vector based video steganography. In Proc. IEEE Int. Conf. Image Processing (ICIP). 5512–5516.

[11] CAO Y, ZHAO X, FENG D. Video steganalysis exploiting motion vector reversion-based features[J]. IEEE signal processing letters, 2012, 19(1): 35-38.

[12] Yu Deng, Yunjie Wu, and Linna Zhou. 2012. Digital video steganalysis using motion vector recovery-based features. Appl. Opt. 51, 20 (Jul. 2012), 4667–4677.

[13] WANG K, ZHAO H, WANG H. Video steganalysis against motion vector-based steganography by adding or subtracting one motion vector value[J]. IEEE Transactions on Information Forensics and Security, 2014, 9(5): 741-751.

[14] Yanzhen Ren, Liming Zhai, Lina Wang, and Tingting Zhu. 2014. Video steganalysis based on subtractive probability of optimal matching feature. In Proc. 2nd ACM Workshop Inf. Hiding Multimedia Security. 83–90.

[15] ZHANG H, CAO Y, ZHAO X. A steganalytic approach to detect motion vector modification using near-perfect estimation for local optimality[J]. IEEE Transactions on Information Forensics and Security, 2017, 12(2): 465-478.

[16] YE J, NI J, YI Y. Deep Learning Hierarchical Representations for Image Steganalysis[J]. IEEE Transactions on Information Forensics & Security, 2017, 12(11):2545-2557.

[17] HE K, ZHANG X, REN S, et al. Deep Residual Learning for Image Recognition[C]. Proceedings of the IEEE conference on computer vision and pattern recognition, 2016:770-778.

IOP Conf. Series: Materials Science and Engineering **719** (2020) 012069 doi:10.1088/1757-899X/719/1/012069

Measurement and Evaluation of Marine Intelligent Transportation PNT Data Based on BDS and DGNSS

Xiang Wang[1, 2] Jingxian Liu[1], Zhao Liu[1] and Zhi Yuan[1]

[1.] Hubei Key Laboratory of Inland Shipping Technology, School of Navigation Wuhan University of Technology, Wuhan, Hubei, China

[2.] Donghai Navigation Safety Administration (DNSA) of the Ministry of Transport (MOT), Shanghai, China

Email: 2444984719@qq.com

Abstract. This paper proposes the acquisition of PNT data based on Beidou navigation satellite system and the joint BDS/GPS joint pseudorange difference algorithm. This paper relies on the DGNSS infrastructure of the Dajishan lighthouse management station of the Donghai Navigation Safety Administration (DNSA) of the ministry of transport of china. BDS and DGNSS were measured and evaluated using static measurement and evaluation methods. The results show that in the scene with strong interference, the static positioning level error of BDS exceeds 20 meters, DGNSS, static positioning level error is less than 2 meters. Therefore, the DGNSS algorithm is used to acquire the PNT data method, which has the characteristics of availability, accuracy, reliability, continuity and robustness, and can be used in the marine intelligent transportation system.

1. Introduction

Differential Global Navigation Satellite System (DGNSS) is a shore-based radio-directed broadcast system that uses DGNSS differential data and is broadcast to the sailing vessel in Radio Technical Commission for Maritime services - Special Committee (RTCM-SC) 104 format [1-3]. It is a precision positioning navigation aid system [4]. The DGNSS provides voyage users with differential global positioning data with availability, accuracy, reliability, continuity and robustness [5]. The International Maritime Organization (IMO) and the International Association of Marine Aids to Navigation and Lighthouse Authorities (IALA) have indicated that DGNSS is recommended for PNT data for navigation users [6,7]. This paper relies on the DGNSS base station of the Dajishan Lighthouse Management Station of the Donghai Navigation Safety Administration (DNSA) of the Ministry of Transport (MOT) [8]. We conducted experiments and evaluations of BDS and DGNSS in Shanghai, China.

This paper is organized as follows. In Section 2, this paper describes the DGNSS data broadcast method based on BDS. In Section 3, experimental design. In Section 4, the results of the experiment and discussion. In Section 5, conclusions and work to be carried out in the next step.

2. Method

2.1. The Basic Principle of Pseudorange Difference

According to the literature, at t_0, the distance from the base station to the satellite i that is accurately calculated by the known coordinates of the base station and the satellite ephemeris is t_0, and the rate of change of the pseudorange correction number and the pseudorange correction number are respectively:

$$\Delta P_b^i = R_b^i - P_b^i = c\delta\tau_b - c\delta\tau^i + d_{b,orb}^i + d_{b,trop}^i + d_{b,ion}^i$$
$$\Delta \dot{P}_b^i = \frac{\Delta P_b^i}{\Delta t} \tag{1}$$

The base station sends ΔP_b^i and $\Delta \dot{P}_b^i$ to the user in a timely manner, and the corrected pseudorange observation value is:

$$P_{u,corr}^i = P_u^i + \Delta P_b^i + \Delta \dot{P}_b^i (t - t_0) = R_u^i - c(\delta\tau_u - \delta\tau_b) - \left(d_{u,orb}^i - d_{b,orb}^i\right) - \left(d_{u,ion}^i - d_{b,ion}^i\right) - \left(d_{u,trop}^i - d_{b,trop}^i\right) + \Delta \dot{P}_b^i (t - t_0) \tag{2}$$

Among them, $d_{*,orb}^i$ is the distance deviation caused by satellite ephemeris error, $\delta\tau_*$ is the receiver clock difference, $d_{*,ion}^i$, $d_{*,trop}^i$ are the ionosphere and tropospheric delay respectively ($*=$ b, u,b is the base station, u is the user rover); $\delta\tau^i$ is the satellite clock difference; c for the speed of light; Δt is the correction time calculation interval.

When the distance between the user and the base station is relatively close, the spatial correlation of the error is strong. At this time, it can be considered that the tropospheric delay, the ionospheric delay, and the distance deviation caused by the satellite ephemeris of the base station and the user are approximately equal, then:

$$P_{u,corr}^i = R_u^i - c(\delta\tau_u - \delta\tau_b) + \Delta \dot{P}_b^i (t - t_0) \tag{3}$$

In order to avoid solving the whole-circumference ambiguity, in practical applications, the pseudo-range observations after solving the following formula are:

$$\begin{cases} \breve{P}(k) = \frac{P(k)}{k} + \frac{k-1}{k}\left(\breve{P}(k-1)\right) + \Delta L(k, k-1) - \frac{2\Delta\left(dL_{1,2}\right)(k,k-1)}{\gamma - 1} \\ \breve{P}(1) = P(1) \end{cases} \tag{4}$$

Among them, $\breve{P}(k)$ is the phase smoothing pseudorange $(k = 2, \cdots, n)$ of the k epoch, $P(k)$ is the original pseudorange observation value of the k epoch, $\Delta L(k, k-1)$ is the difference of the phase observation values of the two epochs before and after, $\gamma = f_1^2/f_2^2$, $\Delta\left(dL_{1,2}\right)(k, k-1)$ are the front and rear the difference between the phase observations of the different frequency of the epochs and the difference between the epochs, the formula is as follows:

$$\Delta\left(dL_{1,2}\right)(k, k-1) = \left(L_2(k) - L_1(k)\right) - \left(L_2(k-1) - L_1(k-1)\right) \tag{5}$$

2.2. Beidou Pseudorange Difference Algorithm

The Beidou pseudorange difference algorithm is the same as GPS, but since the issue of data ephemeris (IODE) is the same in the big dipper calendar, the GPS number method is used to correct the number broadcast, and the rover cannot guarantee that the same ephemeris is used for the correction number calculated by the base station. This paper proposes to use the reference time in the ephemeris to generate a new IODE (0-252) in the following way to solve the problem.

The time of ephemeris (TOE) range is (0~ 604/ 800), and the TOE is different at different times in the same week. The IODE can be uniquely determined by a mapping function that uses a range of TOE in chronological order:

$$\text{IODE} = \begin{cases} int(TOE/1\,200), 0 \le TOE \le 302\,400 \\ int(TOE - 302\,400), 302\,400 < TOE < 604\,800 \end{cases} \tag{6}$$

Where int is the rounding symbols. After obtaining the new IODE, the Beidou pseudorange differential positioning can be performed according to the conventional pseudorange difference algorithm.

2.3. BDS/GPS Combined Pseudorange Difference Algorithm

Firstly, the above algorithm is used to calculate the Beidou and GPS carrier phase smoothing pseudoranges after the correction of the rover. Then, according to the BDS/GPS joint single-point positioning mode, the corrected carrier phase smoothing pseudorange is used to solve the rover coordinates. The error equation as follows:

$$BX = L \tag{7}$$

which is:

$$\begin{bmatrix} b_0^1 & \cdots & 0 \\ \vdots & \ddots & \vdots \\ b_0^{n+m} & \cdots & 1 \end{bmatrix} \begin{bmatrix} \Delta x \\ \Delta y \\ \Delta z \\ c\delta\tau_{u,GPS} \\ c\delta_{u,BDS} \end{bmatrix} = \begin{bmatrix} l^1 \\ \vdots \\ l^{n+m} \end{bmatrix} \tag{8}$$

Among them, b represents the direction cosine of the satellite, and the satellite is represented above, the first n is a GPS satellite, the last m is a Beidou satellite, and $n + m \geq 5$, and the subscripts 0, 1, and 2 respectively represent x, y, z axes. $\Delta x, \Delta y, \Delta z$ represents the receiver clock difference between GPS and BD, respectively, $c\delta\tau_{u,GPS}$ and $c\delta\tau_{u,BDS}$ represent the receiver clock difference between GPS and Beidou, respectively, and l represents a constant term. Can get $X = (A^T PA)^{-1}(A^T PL)$, where P is the weight matrix, expressed as follows:

$$p = \begin{bmatrix} P_{GPS} & 0 \\ 0 & P_{BDS} \end{bmatrix} = \begin{bmatrix} P_1 & & \\ & \ddots & \\ & & P_{n+m} \end{bmatrix} \tag{9}$$

The weight P is selected by the height angle, namely:

$$P = \begin{cases} \sin(ele), \frac{\pi}{6} \leq ele \leq \frac{\pi}{2} \\ \sin^2(ele), 0 \leq ele \leq \frac{\pi}{6} \end{cases} \tag{10}$$

3. Experiment

Relying on the infrastructure of the DNSA, it can help the precise navigation of vessels and provide PNT data for marine intelligent transportation systems. On august 8, 2019, we conducted experiments in the waters of the Yangtze River estuary in Shanghai, China.

The base station, the DGNSS base station is located at the DNGSS base station at the Dajishan lighthouse management station in Shanghai, as shown in Figure 1.

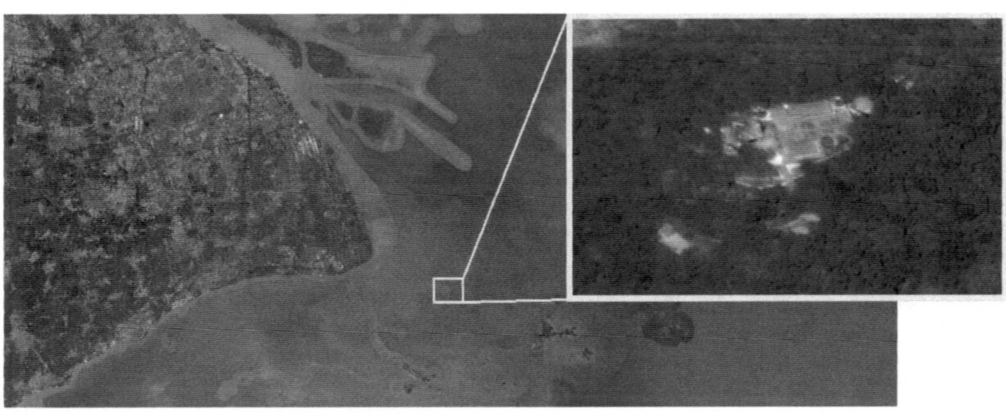

Figure 1. DGNSS infrastructure

The receiver was set up in Pudong New Area, Shanghai. The experimental scene setup of the receiver is located between Shanghai Tower, Shanghai World Financial Tower and Shanghai Jinmao Tower, as shown in Figure 2.

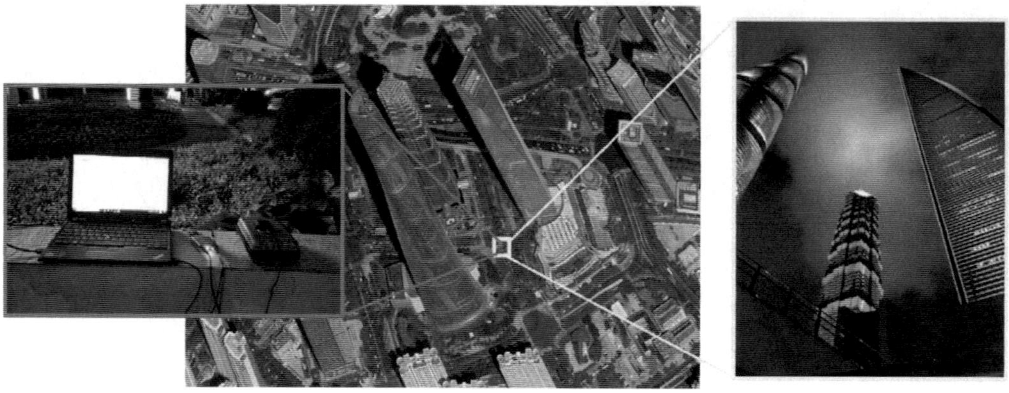

Figure 2. Experimental scene

4. Results and Discussion

4.1. Results

We use PNT data similar to that required for intelligent ship navigation as a base scene. The specific conclusions are as follows:

Result 1: Through the measurement, we set the receiver in a high-rise scene. The use of BDS to obtain PNT data has a large position error, which is more than 20 meters than the actual position of the receiver and the location of the PNT data, as shown in Figure 3.

Result 2: Through the measurement, we set the receiver in a high-rise scene. Using DGNSS to obtain PNT data, there is an error compared to the actual position of the receiver, only within 2 meters, as shown in Figure 3.

Result 3: Through measurement, it shows that the DGNSS method can greatly improve the receiver's ability to acquire PNT data. Compared with the single-frequency BDS receiver, the ability to acquire PNT data can be improved to less than 2 meters. Vessels navigation, narrow waterways, PNT data within 2 meters, can support the ability of intelligent ship positioning, navigation, timing and data acquisition.

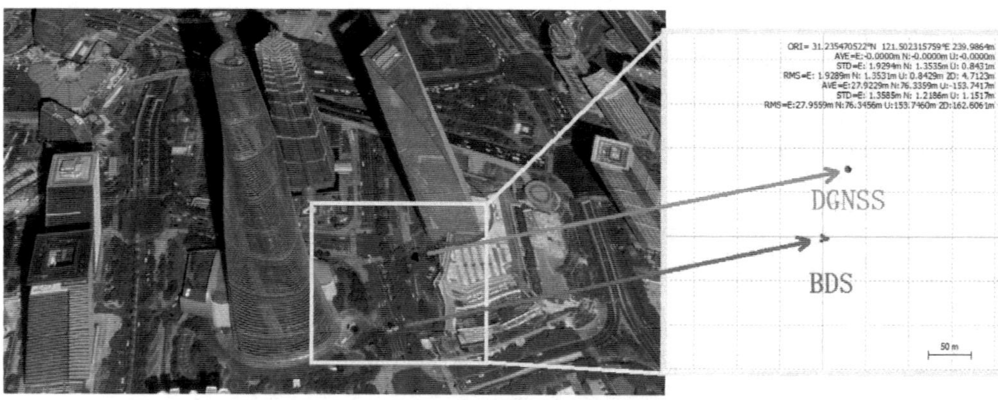

Figure 3. Experimental results

4.2. Discussion

The core of the marine intelligent transportation system at sea is to obtain low-latency PNT data. In all kinds of environments, obtaining PNT data with high reliability is the most important when the ship sails to a narrow waterway, taking into account the rich water depth, the navigation capacity of the channel, the collision avoidance of the ship, and the accurate PNT data. In this paper, the BDS-based DGNSS joint positioning differential positioning method can improve the ability of the marine intelligent transportation system to acquire PNT data. The most demanding marine traffic conditions provide PNT data with availability, accuracy, reliability, continuity and robustness for marine vessel navigation.

Through experiments, it is shown that the DGNSS joint differential method can increase the PNT data of the marine intelligent transportation system to about 2 meters, which can effectively improve the navigation capability of the seaborne vessels.

5. Conclusion

In this paper, the algorithm of obtaining PNT data based on BDS-based DGNSS is used to measure and evaluate in the waters of the Yangtze River estuary. Experiments show that this method can greatly improve the ability of multi-system and multi-source navigation satellites to obtain PNT data. Compared with navigation satellite systems, the availability, accuracy, reliability, continuity and robustness of PNT data can be greatly improved. The experimental scenes of high-rise buildings show that the static positioning data of PNT to about 1 meter. On the one hand, the vessels can rely on DGNSS to obtain reliable PNT data, which can effectively manipulate the ship and carry out precise navigation. On the other hand, marine intelligent transportation systems require PNT data with availability, accuracy, reliability, continuity and robustness. Therefore, DGNSS can be used as an important PNT data acquisition method. Next, we will conduct data fusion researchs on more systems and more data sources for marine intelligent transportation.

6. Acknowledgment

This work was supported in part by the National Natural Science Foundation of China (No. 51809207), and a grant from the Fundamental Research Funds for the Central Universities (No. 2019IVB040).

7. References

[1] Yijun T, Dong Q, Zhao Q, Zhou C and Sai W, BDS ambiguity resolution with the modified TCAR method for medium-long baseline 2017 *J. Advances in Space Research* **59** 670-681.

[2] Chuang S, Fu Z, Yidong L, Shengfeng G, Weixing Z, Xiaolei D, Xianjie L, Hailin G and Xiaopeng G, National BDS Augmentation Service System (NBASS) of China: Progress and Assessment 2017 *J. Remote Sensors* **9(8)** 837.

[3] Ying X, Chen W, Lei L, Lizi Y, Min L and Shengli W, GPS/BDS Medium/Long-Range RTK Constrained with Tropospheric Delay Parameters from NWP Model 2018 *J. Remote Sensors* **10(7)** 1113.

[4] Xin X, Tao G, Qile Z, Jingnan L and Bin W, Performance of BDS-3: Measurement Quality Analysis, Precise Orbit and Clock Determination, 2017 *J. Sensors* **17(6)** 1233.

[5] Wujiao D, Qiang S and Changsheng C, Characteristics of the BDS Carrier Phase Multipath and Its Mitigation Methods in Relative Positioning 2017 *J. Sensors* **17** 796.

[6] Gangyuan Y, Dongdong L, Ying-Chang Z, Rui L and Victor C. M., Optimal Resource Allocation in Full-Duplex Ambient Backscatter Communication Networks for Wireless-Powered IoT, 2018 *J. IEEE Internet of Things* **25**.

[7] Rui T, Rui Z, Pengfei Z, Jinhai L and Xiaochun L, An approach for real-time fast point positioning of the BeiDou Navigation Satellite System using augmentation information, 2017 *J. Measurement Science and Technology* **29**.

[8] Xuan Z, Zongnan L, Min L, Weiming T, Chenlong D, Liang C, Chen W and Chuang S, Modeling BDS pseudorange variations and models assessment, 2017 *J. GPS Solutions* **21** 1661.

Generating Test Data for Path Coverage Based on Genetic Algorithm

Shuping Fan[1], Baoying Ma[2], Nianmin Yao[3], Yan Zhang[1*], Chunyan Xia[1] and Dan Zhang[1]

[1] School of Computer and Information Technology, Mudanjiang Normal University, Mudanjiang 157011, China
[2] School of Health Management, Mudanjiang Medical University, Mudanjiang 157011, China
[3] School of Computer Science and Technology, Dalian University of Technology, Dalian 116024, China
Corresponding author: Yan Zhang (E-mail: zhangyan@mdjnu.cn)

Abstract. Test data generation is an important part of software testing. The imbalance of data crossing program branches is often ignored in generating test data. As a result, there is much data crossing some branches while little data crossing other branches. To solve the phenomenon so as to generate test data effectively, we introduce branch balance and program balance in the evolutionary generation of test data. First, the number of individuals crossing the true and false branch of each branch node on the target path are computed. Then, the calculation methods of branch balance and program balance are given. Finally, the fitness value function which considers the change of program balance before and after an individual joining is presented. And an individual that can improve the balance will be retained in the evolution process. Experiments show that our method is better than the other method in running time and success rate.

1. Introduction

Software testing is a time-consuming and labor-consuming work, whose goal is to find as many errors as possible with less test data to reduce the cost of software development. An efficient and feasible method to generate test data automatically is of great significance to test process [1]. Path coverage is a test criteria with the highest coverage in white-box testing, which refers to selecting enough test data so that every possible path of the program can be executed at least once. In this paper, we consider the balance of individuals crossing program branches, which is used to generate test data in Genetic Algorithm (GA). So an individual crossing new branch or improving the program balance will obtain higher fitness value and it will be preserved in the evolution process to enhance the efficiency of test data generation.

2. Related Work

In recent years, many scholars have studied the application of evolutionary theory to generate test data satisfying path coverage, and they have developed many new methods for automatically generating test data covering target paths. Ahmed et al. [2] proposed a method to generate multi-path test data for the first time by transforming the test data generation problem into a multi-objective optimization problem, which enables the application of GA to generate multiple test data crossing multiple target paths at one time. Gong et al. [3] presented to generate regression test data based on the existing test data. Dang et al. [4] proposed to generate executable paths for weak mutation testing by examining the correlation between the true branches of mutation statements, so that test data covering these paths can kill all mutants. Ding et al. [5] used a complete model for fast generation of test data from the perspective of engineering practice of software testing, which realized the path representation for key point. It improved the efficiency for test data generation. In addition, some scholars have discussed the application of other technologies in test data generation. For example, Yao et al. [6] proposed to integrate with neural network, which can effectively reduce the time consumption of running programs comparing with previous methods. Reference [7] applied the capture technology of rare data, which protected these data traversing nodes that are difficult to cover, and then they calculated the individual contribution degree to the generation test data to adjust the individual fitness value.

The goal of software testing is to realize the sufficiency and comprehensiveness of program testing. Though the above techniques effectively generate test data covering target path by applying evolutionary algorithm, they do not consider the balance of test data traversing program branches. If a balanced mechanism is introduced during the evolution of test data so that the generated data can traverse multiple branches in a balanced way, thus the generation efficiency of test data covering target path will undoubtedly be improved.

Therefore, we introduced the balanced mechanism in the evolution of test data. After test data running the program, the number of individuals traversing each branch node on target path is counted, and the fitness function is designed according to it so individuals that can improve the program balance are retained to improve the generation efficiency of test data.

3. Basic Concepts

For convenience, the concept of Control Flow Diagram is given.

(1) Control Flow Diagram

Control Flow Diagram called CFD is a graphical representation of the program control structure, which is a directed graph G (*N*, *E*, s, e), where *N* is called the set of nodes in G and each node corresponds to a program statement, and *E* is a set of edges. (n_i,n_j) is called a edge of G, and it indicates that there is control flow from n_i to n_j. Generally, the CFD of each program contains a unique entry node s and exit node e. The Figure 1(b) shows the CFD of the program, and the source program is in Figure 1(a) [8].

```
void Function( )
{
1  if(a>2)
2     c=a;
   else
3     c=b;
4  if(a+b-c>1)
5     w=a-b;
6  if(w >5)
7     printf("It is OK");
   else
8     printf("error");
}
```

(a) Source program

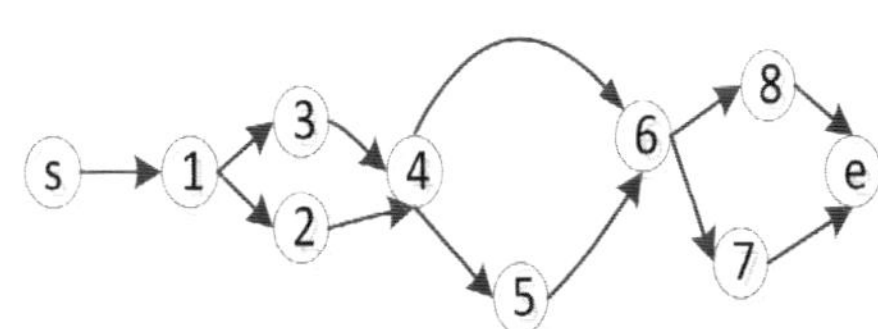

(b) Control Flow Diagram

Figure 1. Source program and its CFD

(2) Branch nodes

In the CFD, nodes with an out-degree greater than or equal to two are called branch nodes. $node_1, node_4, node_6$ in Figure 1(b) are all branch nodes. As the switch statement can be expressed as a double-branch selection structure, and according to Z-path coverage [9], the loop structure can also be converted into a double-branch selection structure basing on the number of times the loop body executing zero and at least once. Therefore, switch nodes and loop nodes are all regarded as branch nodes in this paper.

(3) True and false branches of branch nodes

The two edges starting from the branch node are the true and false branches of the branch node respectively. When the predicate of a branch node is true, the true branch of the branch node is executed, otherwise the false branch is executed.

4. Calculation of the Program Balance

From the CFD in Figure 1(b), it can be seen that the number of test data crossing the sequence node and its direct successor node must be equal. Therefore, the sequence nodes are not considered in the calculation of the program balance to reduce calculation cost. After all the corresponding data of individuals run the program, the program balance is calculated. According to individuals crossing the true and false branches of each branch node, first we calculate the branch balance of each branch node, then we take the sum of all the branch balance as the program balance.

4.1. Calculation of Branch Balance

The branch balance is used to indicate the balance of test data crossing the true and false branches of a branch node. Supposing the number of branch nodes on target path is n. After the corresponding data of all individuals run the program, the number of individuals crossing the true and false branch of the kth branch node in the tth generation population is counted. Here denote it as $Num_{kT}(t)$ and $Num_{kF}(t)$ respectively, so the branch balance of the kth ($k = 1,2 \cdots, n$) branch node can be expressed as:

$$bb_k(t) = \begin{cases} 0 & , Num_{kT}(t) = Num_{kF}(t) = 0 \\ |Num_{kT}(t) - Num_{kF}(t)|/max(Num_{kT}(t), Num_{kF}(t)), & \text{else} \end{cases} \tag{1}$$

Where $max(Num_{kT}(t), Num_{kF}(t))$ is to get the maximum value of $Num_{kT}(t)$ and $Num_{kF}(t)$. From Eq. (1), we can see that the defined branch balance actually reflects the balance of all individuals crossing the true and false branches of the kth branch node. If the difference in the number of individuals crossing the true and false branches is small, the value is small, which indicates that the individuals crossing the true and false branches of the node are more evenly. It is not difficult to obtain that the smaller of the branch balance, the better.

4.2. Calculation of Program Balance

According to Eq. (1), the branch balance of all branch nodes on the target path is calculated, and the sum of the branch balance of all branch nodes is taken as the program balance, as is seen in Eq. (2). According to the definition of the branch balance, the smaller the program balance is, the better of the program balance is. The smaller value reflects data crossing the program in a balanced way.

$$pb(t) = \sum_{k=1}^{n} bb_k(t) \tag{2}$$

Taking the program in Figure 1 as an example. Supposing the chosen target path is "s,1,3,4,6,8, e", which means that there are three branch nodes on the path, they are $node_1$, $node_4, node_6$. Assuming there are four individuals $x_1 \sim x_4$ in the population, and the number of individuals crossing the true branch and the false branch of the branch node of $node_1$ in the 6th generation are $Num_{1T}(6) = 3$ and $Num_{1F}(6) = 1$ respectively. The branch balance of $node_1$ is further calculated according to Eq. (1):$bb_1(6) = |Num_{1T}(6) - Num_{1F}(6)|/max(Num_{1T}(6), Num_{1F}(6)) = |3 - 1|/\max(3,1) \approx 0.67$.

Assuming that the number of individuals crossing the true branch of node_4 and node_6 are 2, 4 respectively, crossing the false branch of these two nodes are 2 and 0 respectively. Then the branch balance of the kth ($k = 2,3$) branch node can be calculated in the same way. The values are $bb_2(6)$=0 and $bb_3(6)$=1. Finally, according to the calculation method in Eq. (2), the program balance is achieved. $pb(6) = \sum_{k=1}^{6} bb_k(6) \approx 0.67 + 0 + 1 \approx 1.67$.

5. Evolutionary Generation of Test Data

To check whether an individual can improve the program balance, the method used is: after deleting an individual, recalculate the branch balance of each branch node on the target path, then calculate the program balance. So the effect of the individual basing on the change of the program balance before and after deleting the individual can be computed, and finally the effect is taken as fitness value of the individual.

5.1. *Calculation of Program Balance after an Individual Deletion*

To calculate the fitness value of an individual, this section gives the calculation method of program balance after deleting an individual. Supposing the number of individuals crossing the true branch and false branch of the kth branch node after deleting the individual x_w is $Num'_{kT}(x_w,t)$ and $Num'_{kF}(x_w,t)$ respectively, the branch balance of the kth branch node can be expressed as $bb'_k(x_w,t)$:

$$bb'_k(x_w,t)=\begin{cases}0 \ , Num'_{kT}(x_w,t)=Num'_{kF}(x_w,t)=0\\ |Num'_{kT}(x_w,t)-Num'_{kF}(x_w,t)|/max(Num'_{kT}(x_w,t),Num'_{kF}(x_w,t))) \ ,\text{else}\end{cases} \quad (3)$$

So we get the program balance after x_w is deleted, which can be expressed as:

$$pb'(x_w,t)=\sum_{k=1}^{n} bb'_k(x_w,t) \quad (4)$$

5.2. *Calculation of Individual Fitness Value*

Generally speaking, individuals with higher fitness value in current population will be copied to the next generation with large probability. In the paper, the calculation of an individual fitness value consider the program balance obtained before and after deleting an individual x_w. And according to the change on the program balance, we get the fitness value of x_w, which can be expressed as:

$$f(x_w,t)=\begin{cases}pb'_w(t)-pb(t), pb'_w(t)>pb(t)\\ 0, \qquad pb'_w(t)\le pb(t)\end{cases} \quad (5)$$

As is seen from Eq. (5) that the change of program balance before and after deleting x_w is considered when calculating the individual fitness value. If the deletion of x_w increases the program balance, that is, the existence of x_w can effectively improve the program balance, then x_w is preferentially retained in the evolution process. If there are multiple individuals like this at the same time, it can be ensured that the greater improvement on the program balance, the larger fitness value of an individual gets by Eq. (5). On the other hand, if the deletion of x_w leaves the program balance unchanged or reduced, that is, x_w cannot improve the program balance, then the individual should be deleted by setting its fitness value to 0.

Taking the program in Figure 1 as an example. In the 6th generation, when deleting x_1, assuming that the number of individuals crossing the true and false branch of $node_1$ is $Num'_{1T}(x_1,6)$=3, $Num'_{1F}(x_1,6)=0$ respectively, thus the branch balance of $node_1$ after deleting $x_1$can be calculated, $bb'_1(x_1,6)=|Num'_{1T}(x_1,6)-Num'_{1F}(x_1,6)|/max(Num'_{1T}(x_1,6),Num'_{1F}(x_1,6))=|3-0|/\max(3,0)=1$. In the same way, we get the branch balance of $node_4$ and $node_6$. So the program balance in this case can be calculated. Supposing the program balance after deleting x_1 by Eq. (4) is : $pb'(x_1,6)=\sum_{k=1}^{n} bb'_k(x_1,t)=1$. And the program balance after deleting other individuals can be calculated in the same way, supposing the program balance after deleting $x_2{\sim}x_4$ are $pb'(x_2,6)=3$, $pb'(x_3,6)=2$, and $pb'(x_4,6)=2.5$ respectively. As the program balance calculated before deleting x_1 is $\mathrm{pb}(6)\approx 1.67$, and the value after deleting it is $pb'(x_1,6)=1$. Since $pb'(x_1,6)<pb(6)$, the fitness value of x_1 can be calculated according to Eq. (5), $f(x_1,6)=0$. Similarly, we get $f(x_2,6)=1.33$, $f(x_3,6)=0.33$,$f(x_4,6)=0.83$. As $f(x_2,6)>f(x_i,6)(i=1,2,3)$, it shows that x_2 can improve the original program balance to the greatest extent in all individuals. Besides, both $f(x_3,6)$ and $f(x_4,6)$ are greater than 0, which indicates that both x_3 and $x_4$can effectively improve the balance of program coverage. These three individuals will have a greater probability to be preserved in evolution process. The fitness value of x_1 is 0 and it will be deleted in the evolution process. It can be seen that the proposed method can effectively distinguish the effect of different individuals on the program balance.

6. Experiments

To validate the effectiveness of our method, we apply it to triangle classifier program in [7], a benchmark program commonly used in software testing. The program is written in C language. The selected comparison method is described in reference [10]. The two methods adopt the same experimental parameters. Comparing the evaluation times, running time and success rate to find the test data covering the target path, which are used in [7]. The evaluation times means sum of evolutionary generations all individuals in one experiment. The smaller the evaluation times and the shorter the running time, the better the performance of the algorithm, and the success rate refers to the ratio of the number of experiments that successfully generate the test data to the total number of experiments within the maximum running generation. All results are the average values by many experiments. The experimental results are shown in Table 1.

Table 1. Experimental Results of Triangle Classification Program

Experimental settings			The proposed method			The method in [10]		
Data range	Population size	Maximum generation	Evaluation times	Time (ms)	Success Rate(%)	Evaluation times	Time (ms)	Success rate(%)
[1,128]	30	3000	4966.0	19.7	100	41974.0	84.9	100
[1,256]	50	8000	7750.0	20.5	100	337880.0	628.3	40
[1,512]	80	10000	11226.7	32.7	100	670917.3	1285.1	33.3
[1,1024]	150	30000	28690.0	74.4	100	3415370.0	7217.0	46.7
[1,2048]	300	80000	65360.0	256.7	100	8003560.0	19422.8	66.7

As can be seen from Table 1:

(1) When the data range is [1,128], the average value of evaluation times in [10] and our method are 41974.0 and 4966.0 respectively, which is more than our method. When the data range is [1,2048], the value in [10] is 8003560, which is also more than ours of 65360. For different experimental conditions in Table 1, the evaluation times of our method are obviously less than that of the other method. This shows that comparing with the similar method in [10], ours can successfully generate test data with less evaluation times. That is because less generations of GA is used to get the terminal conditions of GA by our method.

(2) The average running time required by our method is less than that of [10] for different experimental conditions. For example, When the data range is [1,2048], the average running time of [10] is 19422.8ms and 256.7ms of our method, which is less than the value in [10]. Besides, under different data ranges, population size and maximum evolution generations in Table 1, our method all consumes less time. This is because the method in [10] considers the layer proximity and branch distance when calculating the fitness value, which needs large amount of computation and consumes more time. Our method mainly needs to calculate the branch balance and program balance before and after an individual joining, which needs less time.

(3) From the average success rate, the value of the proposed method can all reach 100% in different conditions, while the method in [10] can't. For example, When the data range is [1,1024], the value of [10] is 46.7%, which is obvious lower than our method. It demonstrates the effectiveness of the proposed method in generating test data.

We can conclude from the experimental results above that compared with the method in [10] , our method is advantageous in evaluation times, running time and success rate for triangle classifier program.

7. Conclusion

The paper presents a test data generation method based on GA. After all the corresponding data of individuals run the program, the number of test data traversing each branch node on the target path is counted. Based on this, a calculation method for the effect of individuals on the program balance is designed. The individual traversing the new path is preferentially retained. If all the generated data do not traverse the new path, the effect of each individual on the program balance is calculated. Furthermore, the fitness value of individuals which can improve the program balance and the test data generation efficiency is raised. So these individuals will be retained in evolution process. Experimental results show that the test data generation efficiency of our method is higher than that of the comparison method. However, the experimental results are just on the triangle classification program, and we will further study the experiments on other programs in the future.

8. Acknowledgement

This study was jointly funded by the Research Projects of Basic Scientific Research Business Expenses in Institutions of Higher Learning of Heilongjiang Province(Grant Nos.1353ZD003 and 2018-KYYWFMY-0104); the Scientific and Technological Plan Project of Mudanjiang City (Grant Nos.Z2018g023, Z2016s0027 and Z2018g022); the Science and technology research project of Mudanjiang Normal University(Grant No.YB2019003);the National Natural Science Foundation of China (Grant No. 61573362); the Natural Science Foundation of Heilongjiang(Grant No. F2016039); and the Innovation Foundation of Science and Technology of Dalian (Grant No. 2018J12GX045).

9. References

[1] MIAO X X, ZENG P J. Research on an Automatic Generation Algorithm of Test Data[J]. Measurement and control technology. 2018, 37, 36-38.
[2] Ahmed M A, Hermadi I. GA-based multiple paths test data generator [J]. Computers and Operations Research. 2008, 35(10): 3107-3124.
[3] Gong D W, Ren L. Evolutionary Generation of Regression Test Data [J]. Chinese Journal of Computers. 2014, 3: 489-499.
[4] Dang X Y, Gong D W, Yao X J. Feasible path generation of weak mutation testing based on statistical analysis [J]. Chinese Journal of Computers. 2016, 39(19): 1-17.
[5] Ding R, Dong H B, Zhang Y. Fast Automatic Generation Method for Software Testing Data Based on Key-Point Path[J]. Journal of Software, 2016, 27(4): 814-827.
[6] Yao X J, Gong D W, Li B. Evolutional Test Data Generation for Path Coverage by Integrating Neural Network[J]. Journal of Software. 2016, 27(4): 828-838.
[7] Zhang Y, Gong D W. Evolutionary Generation of Test Data for Paths Coverage Based on Scarce Data Capturing[J]. Chinese Journal of Computers. 2013, 36(12): 2429-2439.
[8] Xia C Y, Zhang Y, Song L. Evolutionary generation of test data for paths coverage based on node probability[J]. Journal of Software, 2016,27(4):802-813
[9] Xia H, Song X, Wang L. Research of test case auto generating based on Z path coverage[J]. Modem Electronics Technique. 2006, 6: 92-94.
[10] McMinn P. Evolutionary search for test data in the presence of state behavior [D]. University of Sheffield, England, 2005

IOP Conf. Series: Materials Science and Engineering **719** (2020) 012071 doi:10.1088/1757-899X/719/1/012071

Study on the Prediction of Driving Braking Behaviour Based on FPNN

Jinjian Xiao[1], Yubo Weng[2] and Yingna Xie [1]

[1] School of Automobile and Transportation, Tianjin University of Technology and Education, No.1310, Dagu South Road, Hexi District, Tianjin, China
[2] Tianjin Transportation Group, Tianjin, China
Email: Xjj119@126.com

Abstract. The driving braking recognized and predicted was complicated. The driving braking samples of various braking actions were gained, by means of 3 setting of driving braking including braking of automobile in front, roadblocks cutting off in other lane were planed and simulated in the real road environment. Experimental data of driving actions including automobile velocity, braking velocity and sustained action time were acquired by the driving action collecting system. Taking advantage of fuzzy aggregation analysis, the driving braking action sample was unified for the fuzzy probabilistic neural network (FPNN). Under different action groups of braking action information selected from braking action samples, the FPNN network was constructed and trained. The analytical results show that the hit rate is 95.3 % when the training braking action data number is 46. Meanwhile, the results show that the relative fuzzy unified and FPNN is useful with sufficient driving braking action data. The driving FPNN are a valid way for prediction of driving action time.

1. Introduction

Domestic and international study and analysis show that with using of normal braking assist system in general automobiles, the driving braking time is decreased, and following distance is reduced effectively with 3% or more, then the number of road accidents is less. At same time, by means of prediction of driving braking actions, controlling braking lights time, braking action would be judged, would gain the more braking reaction time. The driving braking reaction time would be decreased with 0.8 s, and the automobile rear-end collision risk can be lower[1,2]. For sake of controlling automobile braking action and braking lights working time, braking actions, especially driving operating time, should be predicted correctly.

Because of nonlinearity, fuzziness of driving judgments and operations [3,4], driving braking behaviour action would be identified by neural network and fuzzy aggregation. Braking operating parameters were gathered in different automobile braking velocity in driving action settings with different following automobiles or roadblocks settings. By means of lots of braking action factors were involved, accurately reflect relatively reactions volatility of each driving factor. The fuzzy membership was used to unify driving action data, hence the accurate and effective operation data for braking model and driving braking training were provided. The consequence show that the FPPN model has precise mapping and action recognition ability [5,6,7], so the FPNN model of driving actions is set up, based on PNN and fuzzy aggregation, to realize braking actions identification and tracking. Finally, tracking ability and accuracy of braking action identification are analysed with driving braking action data.

2. Probability Mapping Relations

So as to depict driving action model precisely, action factors should be considered when automobile is braking. The automobiles drivers would take controlling operations according to following distance and distance to roadblocks [8,9]. The driving operation shows that distances are the leading factors to effect and describe the drivers' operation condition. The changing rate of braking pedal velocity is far over distinct factor in normal driving actions [10, 11]. So, braking velocity fluctuation is the key factor of driving action factors[12]. Under the variant braking velocity, different braking actions would be used according to following distance, velocity and time. Automobile braking velocity of driving action would be considered as the key factor. At last, factors *V* of braking actions are gained.

$$V = (V1, V2, V3, V4, V5) \quad (1)$$

$V1, V2, V3, V4$ *and* $V5$ are the automobile velocity, following distance, braking velocity and action time severally in formula (1). The Impact factor of driving and normal driving actions are set as the same number. The driving action probability mathematic model z_V is gained.

$$\text{If } z_A p_A > z_B p_B,\ then\ z_V \in A;\ \text{if } z_A p_A < z_B p_B,\ then\ z_V \in B \quad (2)$$

In formula (2), z_A, z_B are driving probability according to the non-normal driving actions *A* and the normal driving *B*, $z_A = L_A/L$, $z_B = L_B/L$, z_A, z_B are driving training sampling size of action *A* and action *B* , p_A, p_B are PDF (probability density function) of driving action *A* and action *B*, z_V is the driving action model *A* or *B* . Braking model mapping space is foundation of probability neural network model. The event 0 and event 1 correspond to braking action space *A* and *B* respectively.

3. Braking Data Unified Disposal

Based on above analyses, four estimated sub parameters are used in the normal driving action model: automobile velocity, following distance, braking velocity and action time. If there are *n* groups of action sample data, then the driving action sample numerical matrix *M* is gained.

$$M = (m_{ij})_{n\times 4} \quad (3)$$

The m_{ij} is no. *j* sub parameter of no. *i* braking action sample, assumed to be no. *i* braking action vector in formula (3). The raw normal driving action samples is not unified. The data change range is much bigger. The driving action FPPN tracking accuracy would be affected. Then the action samples dada are pre-processed with 3 steps.

3.1. Braking Sample Mini-Max Uniformization

The bigger quantitative parameters in clustering rank directly to original braking action sample data process may be focused. The non-quantitative method is applied in braking action samples. The action sample data will be changed into the range [0, 1] with a certain rate. The original braking action samples are processed using mini-max uniformization method.

$$m_{ij} = (m_{ij} - m_{jmin})/(m_{jmax} - m_{jmin}) \quad (4)$$

The m_{jmax} and m_{jmin} in formula(4)are mini and max braking action data of no. *j* parameter respectively.

3.2. Braking Sample Fuzzy Clustering

The fuzzy braking action relationship matrix *S* with the uniformization method is built. The similar parameter s_{ij} is constructed using the cosine angle method.

$$s_{ij} = \sum_{j=1}^{4} m_{ij}m_{kj} \Big/ \sqrt{\sum_{j=1}^{4} m_{ij}^2 \sum_{j=1}^{4} m_{kj}^2} \quad (5)$$

The property of transitivity doesn't in the braking action relationship matrix *S*. The *S* has to be calculated. When $S^k = S^{2k}$, the matrix *S* calculation will be stopped. If $S' = S^{2k}$, then the fuzzy relationship matrix is S'. $\beta \in [0,1]$ of the braking action fuzzy relationship matrix S' is built up. If the sample quantitative value is bigger, then β is set 1. If the sample quantitative value is smaller, then β is set 0. The cut matrix of β is gained. Based on this way, the braking action samples will be divided into the smaller cluster sample.

3.3. Braking Action Sample Standardization

The average value $\overline{m_m}$ of braking action samples using with the clustering ranking method can be obtained.

$$\overline{m_m} = \sum_{l=1}^{h} m_l / h \tag{6}$$

The clustering rank m'_m of no. *m* is taken as the driving action sample *h* in formula (6). The m_l is taken as the braking action sample. When setting $m_i \in m'$, the braking action sample is regrouped. The v_l is the braking action original sample according to m'_m (l = 1,2, ..., h). The original braking action sample m_i, that is v_l , is standardized with fuzzy clustering rank of no. *m*.

$$r_m = [(v_l - \bar{m}_m)^2/(h-1)]^{0.5} \tag{7}$$

$$v_{lj} = v_{lj} - \overline{m_{mj}}/r_m \tag{8}$$

The β is bigger, the driving action sample classification is better. When β is 0.925, the processed braking action samples simplified with above methods are applied in the normal braking action recognition.

4. Braking Action Neural Network

There are total 4 layers in the braking action neural network structure based on Bayes' rule of PNN. The in-put layer has 4 nodes standing as automobile velocity, following distance, braking velocity and time based on above analyses about normal braking action parameters and probability function. The braking action neural network is shown as Fig. 1.

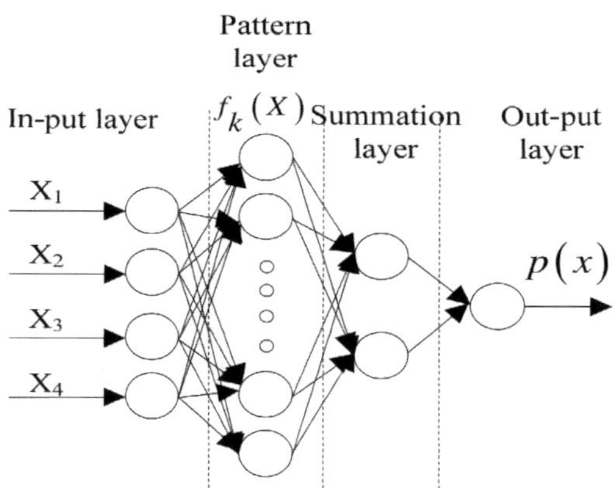

Figure 1. Driving action neural network

The no. 2 layer of braking action neural network adopted the radial basis function as the algorithm. The Parzen method is used Probability density function $p_k(V)$ to build the probability density function from braking action samples.

$$p_k(V) = 1/(2\pi)^{p/2}\alpha^p\, N_k \sum_{j=1}^{N_k} [\|V - W_{ij}\|^2 / 2\alpha^2] \tag{9}$$

V is braking action sample vector. N_k is number of braking action samples, *k* is *A* or *B*, α is the smoothing factor. Weight vector W_{ij} is made by w_{ij} connecting no.1 layer, no.2 layer and no.3 layers. The *i* is the in-put layer's nodes. The *j* is the hidden layer's nodes. A competitive nerve unit is used in the out-put layer as the out-put of braking action neural network. The out-put unit adopts the equation z(v) shown as the formula (10).

$$\mathrm{z(v)} = \begin{cases} 1, & \max_N(z(v)_{N_A}, z(v)_{N_B}) \\ 0, & others \end{cases} \tag{10}$$

If the input data is normal driving samples, then the out-put results of FPNN would be 1. The FPNN provides possibility to recognize braking actions.

5. Driving Action Sample Acquisition

Driving action samples were acquired in the professional testing ground. The automobile different automobile velocity and following distance were simulated under normal and non-normal braking actions. The professional drivers were selected with the age range is 26 to 50 years. The driving years range is 5 to 20. And the experimental automobile was a car. The experimental car followed the car in front under different testing velocity. The velocity range was 50 and 90 km/h. The following distance range was 30 and 80m. Driving action settings included the car in front braking action, sudden reducing velocity action, and human form blocks suddenly crossing the road. The driving action times in one setting traffic environment was be limited to 2 times. These experimental requirements prevented from being familiar with traffic settings. Driving action samples were recorded with driving action acquisition system shown in Fig. 2.

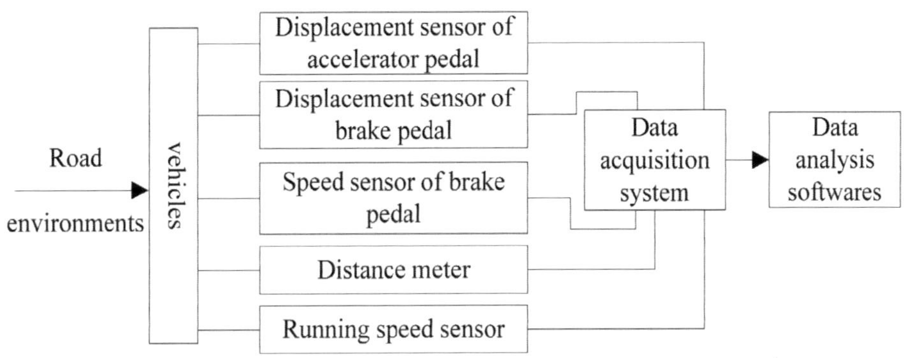

Figure 2. Driving action acquisition system

The braking action sample of simulation normal actions experiment is processed. The processed braking action sample in Table 1. The formula (5) is used to process the driving action samples in table1 and in table 2 with unification algorithm. In table 3, the action samples is translated into cut-off matrix at $\beta = 0.925$. Five classes ($m_{1,2,3,4,5}$, $m_{6,7,8}$, $m_{9,10,11,12}$, m_{13}, $m_{14,15,16}$) are gained from 16 groups braking action samples. Then, in table 4 the braking action training sample class and the trained class are gained.

Table 1. Driving action samples

Car velocity	Following distance	Braking velocity	Action time
76.21	35.90	54.62	0.98
78.70	32.46	55.69	0.86
76.52	36.21	56.60	0.77
79.03	34.97	55.30	0.72
69.27	30.78	46.05	0.77
69.36	45.63	39.86	0.76
70.95	44.32	34.65	0.78
69.58	42.68	33.04	0.79

Table 2. Driving action sample with fuzzy unification algorithm

Car velocity	Following distance	Action velocity	Action time
0.9433	0.0447	0.7404	1.0026
1.0010	0.0013	0.9729	0.7947
0.9610	0.0233	1.0012	0.9374
0.9735	0.0148	0.8990	0.6742
0.7234	0.0398	0.6194	0.4111
0.6456	0.3594	0.1597	0.3594
0.6847	0.3484	0.2182	0.4337
0.7490	0.2916	0.2148	0.5193

Table 3. Cutting matrix of action sample

	m1	m2	m3	m4	m5	m6	m7
m1	1	1	1	1	1	0	0
m2	1	1	1	1	1	0	0
m3	1	1	1	1	1	0	0
m4	1	1	1	1	1	0	0
m5	1	1	1	1	1	0	0
m6	0	0	0	0	0	1	1
m7	0	0	0	0	0	1	1

Table 4. Driving action sample with standardization

Car velocity	Following distance	Action velocity	Action time
0.0215	0.0168	0.2144	0.0012
0.0386	-0.0221	0.4924	0.0049
0.0192	-0.0023	0.5997	0.0059
0.0619	-0.0070	0.4047	-0.0048
-0.2912	0.0121	-1.7511	-0.0016
-0.0538	0.0494	-1.1306	-0.0019
0.1507	0.0125	0.6588	0.00038
-0.1469	-0.0638	0.4618	0.00162

6. Applying Analyses

Driving action samples are normalized firstly. Using Matlab neural networking tools, the driving action neural network is constructed and verified. The tolerance of feedback calculation and connecting weights in FPNN are needless. The classifying ability and tracking accuracy of driving action FPNN is simulated and analysed. The driving action hit rate is to analyse accuracy of FPNN.

Identification precision of the driving action neural network at $\alpha \in [0.233,0.675]$ are analysed in fig. 1. The weights is 1.0 in the no.2 layer and the no. 3 layer. The fuzzy unification algorithm is used in the no.3 layer to the no.4 layer. By means of braking action sample training, each layer weight in the input layer to pattern layer is gained. Layer weights of training samples are unified. Because the relevantly sample astringency is not involved any more, the action training process is simplified. According to action sample size and hitting rate changing tendency, the maximum hitting rate 95.3 % of trained braking action network is got at the sample number 190 in Fig. 3. The fuzzy unified samples can accurately map braking actions. The braking action neural model can reach recognition accuracy needed. The rising sample number does not affect identify accuracy at the number of braking training samples 250 and 350 respectively. When the sample number reaches 260, the identify accuracy is stabilized at 95.3%. From above analyses, the braking action neural network can correctly discern the rest of driving action samples.

7. Conclusions

Through the fuzzy unifying algorithm, braking action samples of different units are transform into the unified space. The characters of different action parameters are clearly changed into the same space providing basis for the driving action neural networking training and building.

By means of braking action neural network, samples of normal braking actions are trained and are discerned. The braking sample number impacting the identifying accuracy is achieved according to the tracking accuracy of driving action network.

Results show that while training samples are sufficient, the highest accuracy 95.3% of the driving action neural network could be reached. FPNN based on the Fuzzy cluster ranking can meet requirement of the accuracy of the normal braking action identification and prediction. The FPNN provides a valid technology method for advanced driver assistance systems.

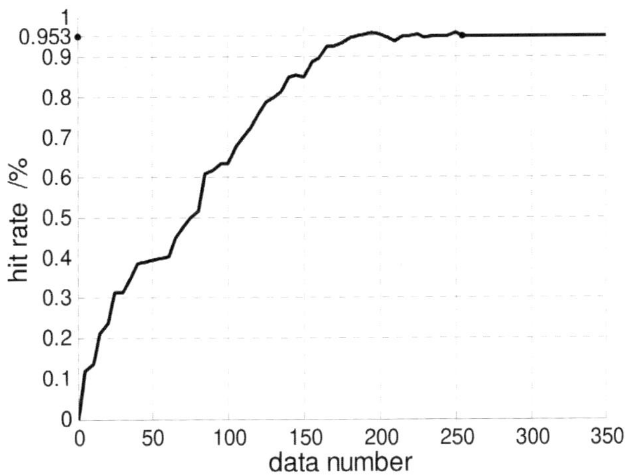

Figure 3. Results of braking behaviour neural network model

8. References

[1] Seto Yoji, Minegishi Kouki and Yang Zhengrong 2004 Research on detection of braking behaviours in normal situations Automobile System Dynamics vol 41 pp 784-790.

[2] Chang Wang, Zhen Li, Rui Fu, Mingfang Zhang and Qinyu Sun 2018 Lane change safety assessment of coaches in naturalistic driving state Safety Science vol 10.

[3] Ricardo Maia, Marco Silva, Rui Araújo and Urbano Nunes 2015 Electrical vehicle modeling: A fuzzy logic model for regenerative braking Expert Systems with Applications vol 42(22) pp 8504-8519.

[4] Jinjian Xiao and Li GAO 2007 Inference model of driving's curve safety cognition Journal of Beijing University of Technology vol 8(33) pp 813-819.

[5] Arribas Juan Ignacio and Cid-Sueiro J. 2005 A model selection algorithm for a posteriori probability estimation with neural networks IEEE Transactions on Neural Networks vol 4(16) pp 799-809.

[6] Xue-mei CHEN, Zhong-hua WEI and Li GAO 2007 Study on drivers' behaviour under normal situations Journal of Beijing University of Technology vol 4(35) pp 449-454.

[7] Reshma Rastogi and Sweta Sharma 2018 Fast laplacian twin support vector machine with active learning for pattern classification Applied Soft Computing Journal vol 5.

[8] Jozsef Suto and Stefan Oniga 2018 Efficiency investigation from shallow to deep neural network techniques in human activity recognition Cognitive Systems Research.

[9] Osama A. Osman, Mustafa Hajij, Sogand Karbalaieali and Sherif Ishak 2019 A hierarchical machine learning classification approach for secondary task identification from observed driving behaviour data Accident Analysis and Prevention vol 123.

[10] Mahdieh Allahviranloo and Leila Aissaoui 2019 A comparison of time-use behaviour in metropolitan areas using pattern recognition techniques Transportation Research Part A vol 129.
[11] Jingang Yi, Luis Alvarez and Roberto Horowitz 2002 Adaptive normal braking control with underestimation of friction coefficient IEEE: Transaetion Systems Technology vol 10 pp 381-392.
[12] Jun LIANG, Zhiqiang and SHA Long CHEN 2012 Dynamic neural network-based integrated learning algorithm for driver behaviour Journal of Transportation Systems Engineering and Information Technology vol 04.

Improving k-Nearest Neighbors Algorithm for Imbalanced Data Classification

Zhan Shi

Power Dispatch and Control Center, Guangdong Power Grid Co., Ltd., China
Email: shiz@foxmail.com

Abstract. The k-Nearest Neighbors (k-NN) algorithm is a classic non-parametric method that has wide applications in data classification and prediction. Like many other machine learning schemes, the performance of k-NN classifiers will be significantly impacted by the imbalanced class distributions of data. That is, the data instances in the majority class tend to dominate the prediction of the test instances. In this paper, we look into the data pre-processing techniques that can be used to rebalance the training data and enhance the performance of k-NN classifiers in imbalanced data sets. We conduct extensive experiments on 14 real-world data sets collected from different application domains. We also perform statistical tests to verify the significance of different data pre-processing techniques in terms of boosting k-NN classification precision.

1. Introduction

In the big data era, data mining and machine learning techniques become a fundamental building block in the application systems across a variety of domains. Researchers and practitioners take advantage of the ever-growing and diverse data from different sources for knowledge extraction, in order to obtain meaningful predications.

The quality of data is a critical challenge in real-world data mining applications. In many domains, the nature of data is imbalanced. That is, there is a majority class that dominates the data set, and data from the minority class is rare. Usually, the minority class is of great concern in these applications. For example, for social network spam detection in cybersecurity domain, most of the messages are benign and only around 5% are spam [1]. Lots of traditional machine learning algorithms have difficulties to handle imbalanced data [2-4].

The k-Nearest Neighbors (k-NN) algorithm [5] is a classic non-parametric method that is widely used for classification. For training, it only needs to store the feature vectors and class labels of the training data. In the testing phase, a test instance is given the label that is most frequent among the k training instances nearest to itself. The two-class k-NN algorithm has a nice statistical property. That is, it is guaranteed to yield an error rate no worse than twice the Bayes error rate, as the data size approaches infinity. k-NN suffers from the problem of class imbalance. This is because during the majority voting in the classification phase, instances from the majority class tend to dominate the prediction of the test instance. One way to overcome this drawback is to give different weights to the training instances. The other way is to perform data resampling, including oversampling the minority class and undersampling the majority class.

In this paper, we investigate a variety of data resampling methods for improving the classification performance of k-NN in imbalanced data sets. Specifically, we consider nine sampling schemes, such as random sampling, synthetic minority class oversampling (SMOTE), Wilsons editing, cluster-based sampling, and ensemble data sampling. The goal is to identify which schemes are more suitable for k-NN in real-world applications. For the purpose of evaluation, we conduct extensive experiments based on 14 publicly available real-world data sets that are collected from different areas, including web-usage records and medical data. The precision results of the minority class are presented, along with a hypothesis testing using one-factor analysis of variance (ANOVA) models.

The remainder of this paper is organized as follows. Section 2 presents a brief literature review of related work. Section 3 discusses the methodology of this work. Section 4 introduces the experimental design and results. Section 5 concludes this work.

2. Related Work

Class imbalance occurs in a classification problem when the classes exhibit a skewed distribution. It is ubiquitous and prevalent in many real-world applications. Generally, there are two types of methods to solve the problem. The first type is data pre-processing, which attempts to re-balance the training data by generating artificial data instances and using data sampling techniques. The second type is to adjust the weight of instances or the cost of objective functions in the classification models, which is usually model specific. In this work, we focus on the data pre-processing approaches, which are more general and applicable to different classification schemes.

In order to re-balance the training data, one can choose to remove some data in the majority class or generate some data (e.g., by duplicating instances and generating artificial instances) to compliment the minority class. For example, random oversampling selects instances in the minority class randomly and then duplicates them, and random undersampling selects data instances from the majority class in random and then removes them [6]. The drawbacks of random oversampling and undersampling are the potential overfitting of classifiers and the possibility of losing useful information. Chawla et al. [7] propose the Synthetic Minority Over-sampling Technique (SMOTE), which is extended by Han et al. [8] to the Borderline-SMOTE algorithm. Barandela et al. [9] introduce the Wilsons editing algorithm that downsizes the majority class using k-NN classifiers. The cluster-based oversampling [10] and cluster-based undersampling [11] algorithms use clustering techniques to generate balanced training sets. Kubat et al. [12] propose the one-sided selection method that creates a training set consisting of safe data by removing the instances that are considered either redundant or noisy in the majority class. An ensemble method is proposed by Wang et al. [13], which combines the outputs of various sampling methods by using the majority voting scheme.

A series of researches have investigated the approaches to perform imbalanced data mining in big data environments. Fernández et al. [2] provided an insight into the existing approaches and challenges of imbalanced classification in big data applications. The key questions towards solving the problem are how to generate artificial data instances and how to balance the trade-offs for the ratio between classes. Abdel-Hamid et al. [14] propose an imbalanced data mining framework based on Spark that consists of two key modules for border handling and selective border instance sampling. Triguero et al. [15] propose a parallel model of evolutionary undersampling for imbalanced big data. Rastogi et al. [16] take advantage of locality sensitivity hashing to implement a distributed version of SMOTE based on Spark. In [17], Jedrzejowicz et al. introduce a novel approach to parallelize computations in imbalanced data classification using MapReduce.

3. Methodology

3.1. k-Nearest Neighbors Algorithm

Machine learning algorithms can be generally divided into two categories, that is, parametric and non-parametric. The k-Nearest Neighbors (k-NN) algorithm [5] is the most popular representative of the non-parametric machine learning algorithms.

The k-NN algorithm basically requires no training, just simply stores the feature vectors and class labels of the training data. Given a new instance for testing, its class is predicted as majority class label from its k nearest neighbors.

In particular, we use Euclidean distance to define the distance between instances. Besides, we set $k = 1$ in the experiment. This is the most intuitive nearest neighbor classifier, which simply assigns an instance to the class of its closest neighbor in the feature space. As the size of training set approaches infinity, the 1-NN classifier also guarantees an error rate of no worse than twice the Bayes error rate.

3.2. Data Pre-processing Techniques

In this work, we focus on improving the imbalanced data classification performance of k-NN by using data pre-processing techniques. In particular, we explore a variety of data resampling methods, which are introduced as follows.

Oversampling techniques are used to increase the number of data instances in the minority class, by either duplicating some of the minority data instances or generating some artificial data instances. We adopt five oversampling techniques in this study. In specific, we use the classic random oversampling, SMOTE [7] and its variant Borderline-SMOTE [8], as well as the cluster-based oversampling [10]. In addition, we adopt the ensemble oversampling algorithm [13] based on information decomposition, cluster-based oversampling and random oversampling.

Undersampling techniques are used to decrease the number of data instances in the majority class. Specifically, we implement four undersampling techniques in this study. These algorithms are random undersampling, Wilsons editing undersampling [9], cluster-based undersampling [11] and one-sided selection undersampling [12].

The data pre-processing techniques are used to re-balance the classes in the training data, in order to provide the k-NN model with a relatively balanced training set. The testing data set are left intact in our study. In other words, the testing set is imbalanced as in the state they are in the real world. This ensures that our results reflect the actual performance in real-world applications.

3.3. Evaluation Methods

In the experiments, the minority class is treated as the positive class and the majority class is treated as the negative class. For imbalanced data sets, the overall accuracy is not a good indicator of the actual classification performance, especially for the positive (minority) class. Therefore, we use the precision metric in our study, which is widely used in data mining and statistical test. In particular, we count the number of true positives, false positives, true negatives and false negatives. Precision is the proportion of positive results in classification that are true positive, as showed in the following equation.

$$Precision = \frac{true\ positives}{true\ positives + false\ positives} \tag{1}$$

To provide some insight into the experimental results, we conduct a statistical test using one-factor analysis of variance (ANOVA) models [18] for different data sampling methods. We consider the null hypothesis is that there is no significant difference among different sampling techniques, while against it the alternative hypothesis is that at least one is significantly different. Additionally, we perform the Tukey's HSD (honestly significant difference) test, which indicates the performance levels of different sampling techniques.

4. Experiment and Results

4.1. Experimental Design

Table 1 shows the data sets used in this work. In specific, we make use of 14 public data sets that are collected from different areas [19-21]. As can been seen in the table, the size of data sets (i.e., the number of instances) varies a lot, ranging from the smallest of 253 (i.e., mw1) to the largest of 17186 (i.e., pc5). The number of attributes is between 9 and 74. In addition, Table 1 presents the size of minority class and the imbalance rate in each data set. The imbalance rate is defined as the number of instances in the majority class divided by the number of instances in the minority class. The most imbalanced data set is mv1 that has the highest imbalance rate of 4556%, while the diabetic data set with 113% imbalance rate is roughly balanced. In short, the data sets we use in the experiments cover a variety of sizes and imbalance rates.

Table 1. Data Sets.

Data Sets	Size	Size of Minority Class	Imbalance Rate	Number of Attributes
diabetic	1151	540	113%	20
pima	768	268	187%	9
ilpd	583	167	249%	11
etoys	270	28	864%	41
onehr	1832	57	3114%	74
bmg	2417	547	342%	41
qvc	2152	386	458%	41
mw1	253	27	837%	38
mc1	1988	46	4222%	39
pc1	705	61	1056%	38
pc2	745	16	4556%	37
pc3	1077	134	704%	38
pc4	1458	178	719%	38
pc5	17186	516	3231%	39

For the purpose of evaluation, we split each of the data sets into two parts in random. The first part consists of 60% of instances, which are used as training data. The other part consists the rest instances that are used to test the classification performance.

In order to find the best parameters in the data sampling step, we apply each sampling algorithm to the training data with a range of parameters. For undersampling on the majority class, we use sampling rates of 20%, 50%, 70%, 90% and EVEN (i.e., the number of instances in both classes are even). For oversampling on the minority class, we test sampling rates of 200%, 500%, 700%, 900%, and EVEN. We also build the baseline classifier from the original imbalanced data set.

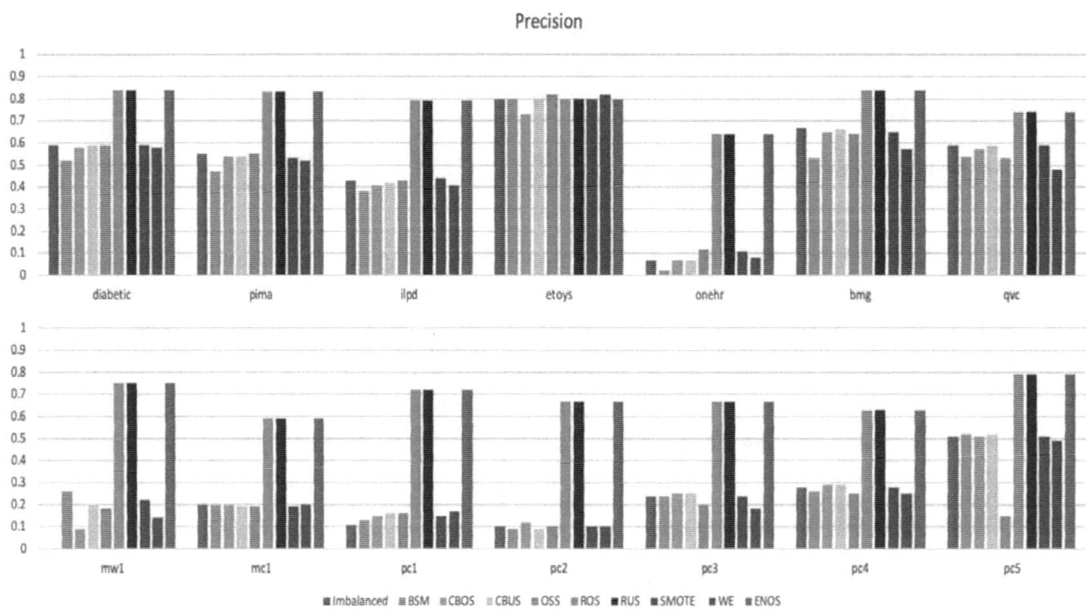

Figure 1. Precision Results.

4.2. Results

The precision results are illustrated in Figure 1. First of all, among the data sampling schemes, random oversampling (ROS), random undersampling (RUS) and ensemble oversampling (ENOS) are able to improve the precision of k-NN classifiers in most data sets. In mw1, mc1, pc1, pc2, and pc3 data sets, the improvements of precision are over 40%, while in diabetic, pima, ilpd, onehr, bmg, qvc, pc4, pc5 data sets, the improvements are between 15% and 39%. The only exception is the etoys data set, where there is no obvious improvement and the precision level is around 80% for all algorithms.

Table 2. ANOVA Models.

Data Sets	Degrees of Freedom	Sum of Squares	Mean Squares	F-statistics	P-value
diabetic	9	40.28	4.48	18.51	<<0.05
pima	9	20.44	2.27	9.81	<<0.05
ilpd	9	5	0.56	2.52	<0.05
etoys	9	7.23	0.8	7.61	<<0.05
onehr	9	67.69	7.52	162.07	<<0.05
bmg	9	4.31	0.48	2.43	<0.05
qvc	9	7.25	0.81	4.26	<<0.05
mw1	9	3.87	0.43	5.79	<<0.05
mc1	9	5.81	0.65	14.79	<<0.05
pc1	9	2.47	0.27	2.68	<0.05
pc2	9	0.35	0.04	0.98	<0.05
pc3	9	5.35	0.59	5.11	<<0.05
pc4	9	3.59	0.4	3.31	<<0.05
pc5	9	64.79	7.2	190.78	<<0.05

Table 2 and Table 3 depict ANOVA models and the HSD test results of the classification precision for various methods across data sets. We can see that the selection of data sampling techniques has a significant impact on prediction label at the 5% level (i.e., $P - \text{value} < 0.05$). In other words, at least one sampling technique is statistical significantly different from others. The HSD test shows that ROS, RUS and ENOS have outstanding performance (HSD marked as "A").

Table 3. HSD Test Results.

Methods	Mean	HSD
Imbalanced	0.37	B
Borderline-SMOTE (BSM)	0.35	B
Cluster-based Oversampling (CBOS)	0.37	B
Cluster-based Undersampling (CBUS)	0.38	B
One-sided Selection (OSS)	0.35	B
Random Oversampling (ROS)	0.74	A
Random Undersampling (RUS)	0.74	A
SMOTE (SM)	0.39	B
Wilsons Editing (WE)	0.36	B
Ensemble Oversampling (ENOS)	0.74	A

5. Conclusion

The class imbalance problem is a great challenge for most machine learning algorithms. In this work, we focus on k-NN, a popular traditional algorithm, and explore the impact of imbalanced data set. To address this problem, we use a variety of data sampling algorithms to rebalance the training set for k-NN. Experiment results obtained based on 14 real-world data sets from different areas indicate that the precision performance of k-NN classifiers in imbalanced data could be greatly improved by applying random oversampling, random undersampling and ensemble oversampling to the training data.

6. Acknowledgments

This work is supported by the Guangdong Power Grid Project (GDKJXM20162186).

7. References

[1] Liu S, Wang Y, Zhang J, Chen C and Xiang Y. Addressing the Class Imbalance Problem in Twitter Spam Detection Using Ensemble Learning. Computers & Security, 69:35-49, 2017.
[2] Fernández A, del Río S, Chawla NV and Herrera F. An insight into imbalanced Big Data classification: outcomes and challenges. Complex Intell. Syst. (2017) 3: 105.
[3] Chen C, Wang Y, Zhang J, Xiang Y, Zhou W, and Min G. Statistical Features Based Real-time Detection of Drifted Twitter Spam. IEEE TIFS, vol.12, no. 4, pp. 914-925, Apr 2017.
[4] Wang Y, Zhang J, Xiang Y, and Zhou W. Internet Traffic Clustering with Side Information. Journal of Computer and System Sciences, vol. 80, no. 5, pp. 1021-1036, Aug 2014.
[5] Fix E and Hodges JL. Discriminatory Analysis. Nonparametric Discrimination: Consistency Properties. International Statistical Review, (1989)57(3), 238-247.
[6] Sui Y, Zhang X, Huan J, and Hong H. Exploring data sampling techniques for imbalanced classification problems. Proc. SPIE 11198, Fourth International Workshop on Pattern Recognition, 1119813 (31 July 2019)
[7] Chawla NV, Bowyer KW, Hall LO and Kegelmeyer WP. SMOTE: synthetic minority over-sampling technique. Journal of artificial intelligence research, vol. 16, pp. 321-357, 2002.
[8] [8]Han H, Wang WY, and Mao BH. Borderline-SMOTE: a new over-sampling method in imbalanced data sets learning. Advances in intelligent computing, 2005, pp. 878-887.
[9] Barandela R, Valdovinos RM, Sánchez JS, and Ferri FJ. The imbalanced training sample problem: Under or over sampling? in SSPR /SPR 2004, pp. 806-814.
[10] Jo T and Japkowicz N. Class imbalances versus small disjuncts. ACM SIGKDD Explorations Newsletter, vol. 6, pp. 40-49, 2004.
[11] Rahman MM and DavisD. Cluster based under-sampling for unbalanced cardiovascular data. in Proceedings of the World Congress on Engineering, 2013.
[12] Kubat M and Matwin S. Addressing the curse of imbalanced training sets: one-sided selection. ICML, 1997, pp. 179-186.
[13] Wang C, Hu L, Guo M, Liu X, and Zou Q. imDC: an ensemble learning method for imbalanced classification with miRNA data. Genetics and molecular research, vol. 14, pp.123, 2015.
[14] Abdel-Hamid NB, ElGhamrawy S, Desouky AE, Arafat H. A Dynamic Spark-based Classification Framework for Imbalanced Big Data. J Grid Computing (2018) 16: 607.
[15] Triguero I, Galar M, Vluymans S, Cornelis C, Bustince H, Herrera F, Saeys Y. Evolutionary undersampling for imbalanced big data classification. in CEC 2015, Sendai, pp.715-722.
[16] Rastogi AK, Narang N, and Siddiqui ZA. Imbalanced big data classification: a distributed implementation of SMOTE. In Proceedings of the ICDCN '18 Workshops. ACM, NY, USA.
[17] Jedrzejowicz J, Kostrzewski R, Neumann J and Zakrzewska M. (2018) Imbalanced data classification using MapReduce and relief. J. of Info. and Tele., 2:2, 217-230.
[18] Berenson ML, Levine DM, and Goldstein M. Intermediate statiscal methods and applications: a computer package approach. 1983.
[19] Asuncion A and Newman D. UCI machine learning repository.
[20] Saar-Tsechansky M and Provost F. Handling missing values when applying classification models. 2007.
[21] Shirabad JS and Menzies TJ. The PROMISE repository of software engineering databases.

Trajectory Design and Mode Selection for a Cellular-based UAV Traffic Monitoring System

Kai Wang[2], Fanyi Wu[1], Kaiming Xu[1], Jianjun Wu[1]

[1]School of Electronics Engineering and Computer Science, Peking University, Beijing, China.
[2]Institute of War Studies, Academy of Military Science, Beijing, China.
E-mail: 1801213629@pku.edu.cn

Abstract. In this paper, we propose to enable cellular networks to support a UAV traffic monitoring system, in which multiple UAVs are required to record videos from the target roads, and then transmit their video data back to the ground stations over the cellular spectrum. In this system, the video data can be transmitted directly to the ground stations bypassing the base station (BS), or via the traditional cellular links. Each UAV should select its transmission mode from the direct mode and the cellular mode as it preforms traffic monitoring tasks. Since the mode selections are influenced by the trajectories of UAVs, we study the trajectory design problem for UAVs in consideration of their transmission modes under a reinforcement learning (RL) framework. Then, we propose a trajectory design algorithm to solve this problem. Simulation results show that our proposed algorithm outperforms the single-agent one.

1. Introduction

With high mobility and low operational cost, unmanned aerial vehicle (UAV) has been regarded as an emerging facility to provide real-time monitoring for road traffic [1]. When performing traffic monitoring tasks, UAVs should send the recorded video data to the ground stations effectively. However, in the current unmanned aerial systems for traffic monitoring, the video data is transmitted over the unlicensed spectrum, which is difficult to ensure satisfactory quality-of-service (QoS). To tackle this challenge, the cellular-enabled UAV network is proposed as a promising solution to support UAV communications [2]. In this network, UAVs can send the video data to the base station (BS), after which the BS transmits the received data to the corresponding ground stations. Moreover, when UAVs are close to their ground stations, they can directly transmit their sensory data to the ground stations bypassing the BS, which can further improve the throughput in the system.

In this paper, we investigate a cellular-based UAV traffic monitoring system, in which UAV can transmit their video data through the direct mode and the cellular mode. We consider an orthogonal frequency division multiple access (OFDMA) network consisting of multiple UAVs performing traffic monitoring tasks, where the direct mode work as an overlay to the cellular mode. Since the transmission mode selection of UAVs is influenced by their trajectories of the UAV, we aim to maximize the total utility in the system by designing their trajectories in consideration of their transmission mode.

Nevertheless, since the trajectories of UAVs are coupled with their transmission modes, it is challenging to investigate the trajectory design problem for UAVs. To tackle with this challenge, we adopt a sense-and-send protocol to schedule multiple UAVs, which can be analysed by nested bi-level Markov chains. By this means, the transmission modes of UAVs are determined simultaneously as we design their trajectories. Under this condition, the trajectory design problem can be regarded as a

IOP Conf. Series: Materials Science and Engineering **719** (2020) 012073 doi:10.1088/1757-899X/719/1/012073

Markov decision problem (MDP), and thus, we can solve the trajectory design problem using reinforcement learning (RL). Then, we propose a Trajectory design algorithm based on deep Q-network (DQN) to optimize policies for UAVs.

In the literature, several works have investigated the UAV communications over cellular networks. In [3], the authors jointly optimized trajectories, sensing locations and scheduling for UAVs performing sensing tasks in the cellular-enabled UAV network, where the BS was required to collect sensory data from UAVs. Authors in [4] jointly optimized subchannel allocation and flying speed for UAVs in a UAV-to-X cellular network, where UAVs could set up communication links with the BS as well as other UAVs However, as an important practical scenario for traffic monitoring, the direct communications between UAVs and ground stations are lack of consideration. Therefore, different from these works, we consider a cellular-based UAV traffic monitoring system, in which UAV transmissions can be conducted over both the direct mode and the cellular mode.

2. System Model

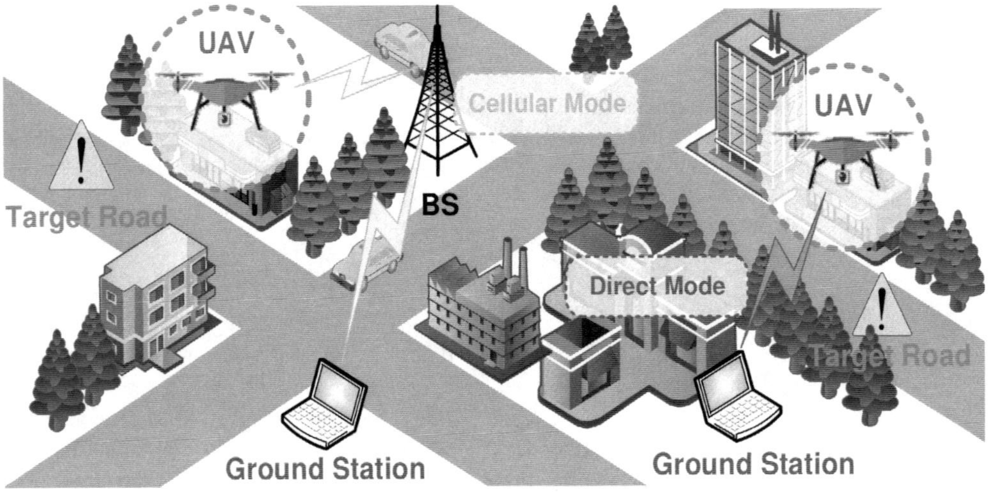

Figure 1. System model.

As illustrated in Fig.1, we consider general city road scenario, which is an OFDMA network with single BS. This network consists of N UAVs performing traffic monitoring tasks. Each UAV is required to record video from its target road and then transmit the video data to the ground stations. The data transmissions can be supported by two modes:

1) Direct Mode: The UAV transmits its video data to its ground station directly;
2) Cellular Mode: The UAV sends the video data to the BS, after which the BS transmits the received data to the ground station.

The time unit for a transmission in both two modes is a frame. To be specific, a UAV in the direct mode takes a whole frame to send the sensory data to its ground station directly, while the two phases of transmissions in the cellular mode take half frame for each.

The network owns C orthogonal subchannels to support the OFDMA transmissions among UAVs and ground station. The BS will allocate orthogonal subchannels to different UAVs, for avoiding the mutual interference among UAVs.

We adopt the probabilistic sensing model in [5] to evaluate UAV's sensing qualities. Specifically, the successful sensing probability for a UAV is an exponential function of the distance between the UAV and its target road. Besides, we adopt the air-to-ground channel model in [6] to evaluate the air-to-ground and ground-to-ground transmissions. Given a threshold, we can express the successful transmission probabilities over the direct mode and the cellular mode from their signal-to-noise ratio (SNR), as referred in [7].

In our paper, we adopt a sense-and-send protocol to coordinate multiple UAVs performing traffic monitoring tasks simultaneously. This protocol can be formulated by the nested bi-level Markov chains, in which UAV sensing and transmission are formulated as the state transitions in the outer and the inner Markov chains, respectively. More details about the protocol are introduced in [7].

3. Algorithm Design

We introduce a lattice model in [7] to describe UAVs' trajectories, in which the space is divided into a finite set of discrete spatial points. The locations of the BS, UAVs, ground stations and target roads are specified by 3D cartesian coordinates. Thus, the trajectory of a UAV can be characterized by a series of discrete spatial points, which is denoted by a set $\boldsymbol{T}$. We adopt the concept of utility in [7] to evaluate the performance of UAVs, and UAVs aim to maximize their utility by designing their trajectory. To be specific, we define the reward of UAV in the n-th frame is r_n. Besides, we introduce the discounting valuation to evaluate the timeliness of traffic monitoring, in which the discount factor is ρ. Therefore, the trajectory design problem for a UAV can be formulated as follow

$$\max_{\boldsymbol{T}} \sum_{n} \rho^{n-1} r_n$$

Since the utility of a UAV is determined by all UAVs' trajectories, the trajectory design problem is difficult to solve. Note that this problem is a MDP, we can analyze it using RL [8]. Under the RL framework, each UAV is regarded as an agent. The states and the actions of each agent are defined as the locations and the movements of the corresponding UAV, respectively.

Denote the state of UAVs in a frame as $\boldsymbol{s}$, the a UAV taking action a can receive reward r. At the beginning of each frame, a UAV observes state $\boldsymbol{s}$, and then it takes an action a based on its policy, denoted by π. As action $\boldsymbol{a}$ taken by the UAV, it receives a reward r and observes a new state $\boldsymbol{s}'$. Therefore, the trajectory design problem can be solved by maximizing the total expected discounted rewards of all UAVs by optimizing their policies. Under the RL framework, we propose a Trajectory design algorithm based on DQN to optimize policies for UAVs, which is referred in [7].

4. Simulation Result

We present the simulation results on the Trajectory design problem using MATLAB.

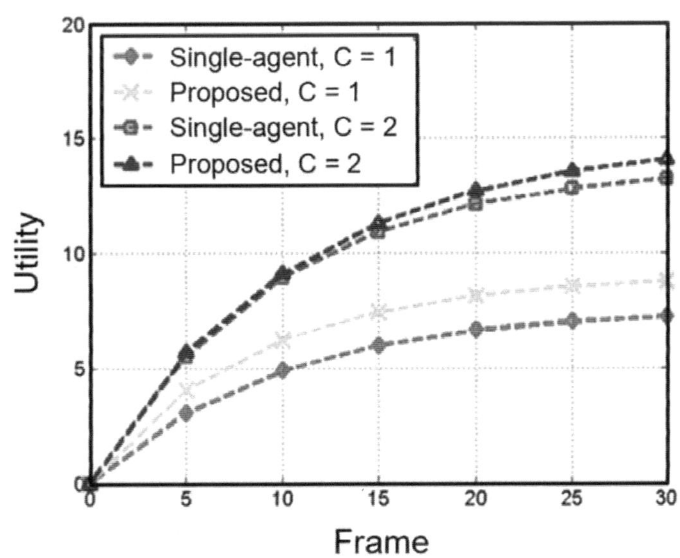

Figure 2. Performance comparison

In Fig. 2, we compare our proposed algorithm with the single-agent Q-learning algorithm [9], in which each UAV updates its policy considering other UAVs as the environment. We can observe that our proposed algorithm achieves a higher utility than the single-agent algorithm. Besides, for either algorithm, the utility increases with the number of subchannels C, since more frequency resources are utilized.

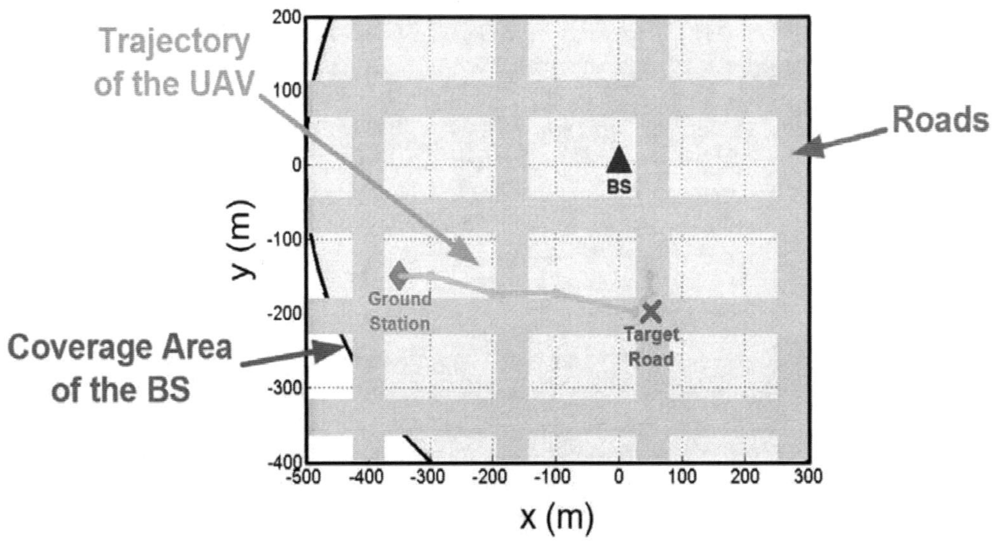

Figure 3. The trajectory of a UAV.

Besides, in Fig. 3, we present the trajectory of a UAV. We can observed that the UAV first flies towards its target road, and then hovers next to its target road rather than on its target road. This is due to the trade-off between UAV sensing and transmission.

5. Conclusion

In this paper, we investigate the trajectory design problem for a cellular-based UAV traffic monitoring system, in which multiple UAVs record videos from the target roads, and then transmit their video data back to the ground stations. We have adopted a sense-and-send protocol to schedule UAV sensing and transmission, and analysed it using nested bi-level Markov chains. As the trajectory design problem is a MDP with large state-action space, we have reformulated this problem using RL, and proposed a DQN-based Trajectory design algorithm to solve it. Simulation results have shown that our proposed algorithm outperforms the single-agent Q-learning method.

6. Acknowledgment

This study was supported by the National Natural Science Foundation of China under Grant No. 61371073.

7. Reference

[1] Y. Zeng, R. Zhang, and T. J. Lim, “Wireless communications with unmanned aerial vehicles: opportunities and challenges,” IEEE Commun. Mag., vol. 54, no. 5, pp. 36-42, May 2016.

[2] H. Zhang, L. Song, Z. Han, and H. V. Poor, “Cooperation techniques for a cellular internet of unmanned aerial vehicles,” IEEE Wireless Commun., to be published.

[3] S. Zhang, H. Zhang, B. Di, and L. Song, “Joint trajectory and power optimization for UAV sensing over cellular networks,” IEEE Commun. Lett., vol. 22, no. 11, pp. 2382-2385, Nov. 2018.

[4] S. Zhang, H. Zhang, B. Di, and L. Song, "Cellular UAV-to-X communications: Design and optimization for multi-UAV networks," IEEE Trans. Wireless Commun., vol. 18, no. 2, pp. 1346-1359, Feb. 2019.
[5] V. V. Shakhov, and I. Koo, "Experiment design for parameter estimation in probabilistic sensing models," IEEE Sensors J., vol. 17, no. 24, pp.8431-8437, Oct. 2017.
[6] 3GPP TR 36.777, "Enhanced LTE support for aerial vehicles," Release 15, Dec. 2017.
[7] F. Wu, H. Zhang, J. Wu, L. Song, ''Trajectory Design for Overlay UAV-to-Device Communications by Deep Reinforcement Learning'', in Proc. IEEE GLOBECOM, Waikoloa, HI, USA, Dec. 2019.
[8] H. Ye, and G. Y. Li, "Deep reinforcement learning based distributed resource allocation for V2V broadcasting," in Proc. IEEE IWCMC, Limassol, Cyprus, Jun. 2018
[9] J. Hu, H. Zhang, and L. Song, "Reinforcement learning for decentralized trajectory design in cellular UAV networks with sense-and-send protocol," IEEE Internet Things J., vol. 6, no. 4, pp. 6177-6189, Aug. 2019.

IOP Conf. Series: Materials Science and Engineering **719** (2020) 012074 doi:10.1088/1757-899X/719/1/012074

Face Recognition Method Based on Probabilistic Neural Network Optimizing Two-Dimensional Subspace Analysis

Haiyan Zhang[1,a] and Fenqi Qiao [2,b]
[1] University of Chinese Academy of Sciences, Beijing 100049, China
[2] School of Mechanical Engineering and Automation, Beihang University, No.37 Xueyuan Road, Haidian District, Beijing 10091,China
[a] 42054035@qq.com, [b] 15071005@buaa.edu.cn

Abstract. If there is noise in the original image of face recognition, the efficiency of face recognition will be affected. In this paper, a face recognition method based on probabilistic neural network optimizing two-dimensional subspace analysis was proposed. Firstly, discrete wavelet variation was used to preprocess the image, and then two-dimensional linear discriminant analysis was used for feature extraction. Finally, the probabilistic neural network was used to complete the face classification. According to the results of experiments conducted on ORL and Fei general face database and the database collected independently, the recognition rate can also be as high as 98.9% when noise is added, and compared with several new identification methods, this method can achieve better identification performance.

1. Introduction

Based on the obtained face sample database, face recognition recognizes or verify one or more faces from static or dynamic scenes with image processing and pattern recognition technology. Face recognition technology is extensively and widely used in important state organs and social security field. Face recognition technology can also be applied to video retrieval of multimedia databases and multimedia production. In recent years, facial expression information extraction based on face recognition technology became possible. It can be used to improve the human-computer interaction mode, thereby providing more humanized experience. Intelligent computer system face detection is the basis of automatic face recognition technology, affecting the accuracy and speed of subsequent face recognition, therefore, it is an important issue that must be solved in practical use. Only by detecting the face quickly and effectively can the recognition be instant and accurate.

As early as 1965, Chan and Bledsoe [1] published a technical report on Automated Face Recognition (AFR) on Panoramic Research Inc., which is the earliest academic paper on face recognition. It has been a history of more than 60 years from the discovery of face recognition technology to the application of face recognition technology. In recent years, more and more people have begun to study face recognition technology, therefore, the technology and algorithms are becoming more and more mature and the success rate of identifying faces and matching faces is getting higher and higher. At present, many Internet enterprises are studying face recognition technology and have developed many products. Researchers have proposed a number of face recognition algorithms to improve face recognition performance, such as support vector machine (SVM) [2], Hidden Markov Model (HMM) [3], Probability Method (Bayesian Network) [4] and Artificial Neural Network [5]. Among them, the artificial neural network is based on shape detection and classification. Among them, artificial neural network has attracted extensive attention because of its high operability in shape detection and classification.

Face recognition methods based on 2D images mainly consists of three categories: global [6], local [7] and hybrid methods [8].

Local methods recognize faces mainly through the specific features of faces (corners, mouth, nose, etc.), so this method needs prior knowledge of image [9].

The global method uses the entire face surface as a source of information without considering local features [10].

The hybrid method combines the advantages of global and local methods and recognizes faces [11] by combining geometric feature detection (or structure) with local appearance feature extraction.

Compared with the one-dimensional method that one-dimensional array (vector) represents a surface, this paper focuses on a two-dimensional feature extraction method (matrix) while preserving the original shape of the face, which is mainly based on the loss of matrix vectors. The loss is geometric and temporal information related to the image pixel.

2. System Design

2.1. Preprocessing with DWT

Discrete Wavelet Transformation [12] is a signal processing tool, which is widely used for feature extraction, the compression and denoising, etc. At present, Discrete Wavelet Transformation has been applied to all kinds of face recognition studies. The main advantage of the Discrete Wavelet Transformation lies in the time scale position of the Fourier transform, according to the literature [12], DWT can be implemented with a filter group, which includes a low-pass filter and one high-pass filter.

2.2. Extracting features with 2DLDA

A two-dimensional linear discriminant analysis (2DLDA) method was proposed in literature [13]. First, the image is projected as a vector of m-dimension features.

$$Y_i = A_j X \tag{1}$$

Where, A_j is an m×n matrix, and Y is an m-dimension feature vector of the projected image A.

Suppose L denotes the class number, M denotes the total number of training images, and the training images are represented by a matrix m×n A_j (j=1 , ….M), $\overline{A_j}(i=1,....M)$ denotes the average of all classifications, and N_i denotes the samples of each class.

The optimal vector projection is orthogonal series, which can maximize the ratio of the dispersive matrix determinants in projected images [14].

$$J_{FLD}(W_{opt}) = \arg\max_{W} \frac{|W^T S_b W|}{|W^T S_W W|} \tag{2}$$

Suppose P_b=*trace* (S_b). P_W=*trace* (S_W), where, S_b is between the classes of scattered matrices, S_w is within the class of scattered matrices, and:

$$S_b = \sum_{i=1}^{N} N_i(\overline{Y_i} - \overline{Y})((\overline{Y_i} - \overline{Y}))^T = \sum_{i=1}^{L} N_i \left[\overline{A_i} - \overline{A}X\right]\left[(\overline{A_i} - \overline{A}X)\right]^T \tag{3}$$

$$S_W = \sum_{i=1}^{L} \sum_{yk \in P_i} (\overline{Y_k} - \overline{Y})((\overline{Y_k} - \overline{Y}))^T = \sum_{i=1}^{L} \sum_{yk \in P_i} \left[\overline{A_k} - \overline{A}X\right]\left[\overline{A_k} - \overline{A}X\right]^T \tag{4}$$

Then:

$$trace(S_b) = X^T \left[\sum_{i=1}^{L} N_i \left(\overline{A_i} - \overline{A}\right)^T (\overline{A_i} - \overline{A})\right] X = X^T S_b X \tag{5}$$

$$trace(S_W) = X^T\left[\sum_{i=1}^{L}\sum_{yk\in P_i}\left(\overline{A_i}-\overline{A}\right)^T(\overline{A_i}-\overline{A})\right]X = X^T S_b X \tag{6}$$

The criterion can be expressed as

$$J(X) = \frac{X^T S_W X}{X^T S_b X} \tag{7}$$

In the formula, X is a column vector unit.

The unit vector X maximization $J(X)$ is called the projection axis. When the X_{OPT} is maximized, the optimal projection can be obtained by the following equation:

$$X_{OPT} = \arg\max_X J(X) \tag{8}$$

If S_W is reversible, the optimal solution is suitable for solving generalized characteristic number problems.

$$S_b X_{opt} = \lambda S_W X_{opt} \tag{9}$$

Where, λ is the maximum eigenvalue of $S_\omega^{-1} S_b$

In general, it is not enough to have only one optimal projection axis. Usually, it is necessary to follow the following constraints to select a set of projections. x_1, x_2, ……x_d, $\{x_1, x_2, \ldots\ldots x_d\} = \arg\max_X J(X)$ where, $X_i^T Xj = 0, i \neq j, i, j = 1, 2, \ldots\ldots d$.

In fact, the optimal projection axis $x_1, x_2, \ldots\ldots x_d$ is corresponding to the best eigenvalue "d", allowing the creation of a new eigenvector of the projected orthogonal matrix X $S_w^{-1} S_b$, which is an nxd matrix: $X=[x_1, x_2, \ldots\ldots x_d]$

Optimum projection 2DLDA x_1, x_2 ,……x_d is used to extract image features, and the following formula can be used:

$$y_k = AX_k, k = 1, 2, \ldots. d \tag{10}$$

A set of eigenvectors can be obtained from the formula y_1, y_2……y_d, which is represented by y= [y1, y2...yd] with a size of M×D of the image feature matrix A.

2.3. PNN Classification

The research shows that compared with classification based on Euclidean Distance, probabilistic neural network classification can improve the face recognition system. The theoretical basis of solving classification problems by network is based on Bayesian Decision Theory, and is realized in feed-forward network architecture. Bayesian decision theory is the principle concept of classification, which uses statistical inference to find the maximum expectation of minimum risk.

The raw data is classified into class C, and each class has m-dimensional observations, such as $X=[x_1, x_2, \ldots\ldots x_d]$. The following Bayesian decision rules classify raw data into class C: $h_i c_i P_i(X) > h_j c_j P_j(X) \forall j \neq i$, where h_c is the prior probability of class C; The loss function is the definite value; P_c is the probability density function of Class C.

The probability density function of each category can be expressed as follows:

$$P_i(x) = \frac{1}{(2\pi)^{\frac{d}{2}}\sigma^d}\frac{1}{N_i}\sum_{j=1}^{N}\exp\left[\frac{-(x-x_{ij})^t(x-x_{ij})}{2\delta^2}\right] \tag{11}$$

Where, $P_i(x)$ denotes the dimension of training vectors, σ is a smoothing parameter, d is the dimension of training vectors, N_i denotes the total number of training vectors of class i, while x_{ij} is a neuron vector and x is a test vector.

PNN mathematical expressions are as follows:

$$a = rad\ bas(\|IW - x\| b) \tag{12}$$

$$y = compet(LW\alpha) \tag{13}$$

Where Q represents the number of input targets/pairs;

R represents the number of types of input data; IW represents the input weight.

The PNN structure consists of two layers [15]:

Layer One: Calculating the distance from the input vector to the input weight (IW) and producing a vector whose elements indicate the distance from input to IW.

Layer Two: Comparing various net output probability vectors, and finally selecting the maximum value of these probabilities in the competition transfer function of the second layer output, and display the number 1 for the class of the maximum class and the number 0 of the other class. The structure of the system is shown in Figure 1.

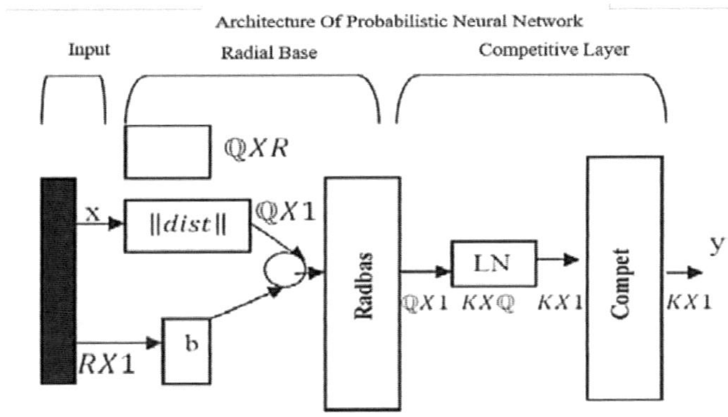

Figure 1. Architecture of probabilistic neural networks

Back-propagation neural network is used, and the probability of each hidden unit in the network can approximate any continuous nonlinear function, with a Gaussian function as the activation function:

$$rad\ bas = \exp\left[(-n^2)\right] \tag{14}$$

Finally, one or more larger values are selected as output unit, and the data points used for indication in the same category are contended to convert the output end of the summation unit into a function, namely

$$compet(n) = e_i\left[0000_1 0......0_i\right], n(I) = MAX(n) \tag{15}$$

For the task of face recognition, the subspace of the face is obtained through 2DLDA, which is defined as the orthogonal basis containing the most relevant information of the face. The vector is the feature vector of the distributed covariance matrix. The recognition process is realized by forwarding the feature matrix to the constructed classifier and discriminator, where the feature proof must be converted into a vector before it is provided to the PNN classifier.

3. Experiment

In order to evaluate and test the described in the face recognition system discussed in this paper, three databases were selected: ORL face database [16], FEI [17] and self-constructed database. All experiments are carried out in MATLAB, which is installed on personal computers with the dual-processor T5870 of 2.03GHz and 2GBRAM.

3.1. Data Set

a) ORL Face Database: The changes of light on faces generated by the environment are uniform, and facial expressions may be different. It includes 40 subjects and each has 10 different images, with an angle from 20 degrees to the top of the head, tilting left or right. The image is a gray image with 92*112 pixels. An image example is shown in Figure 2.

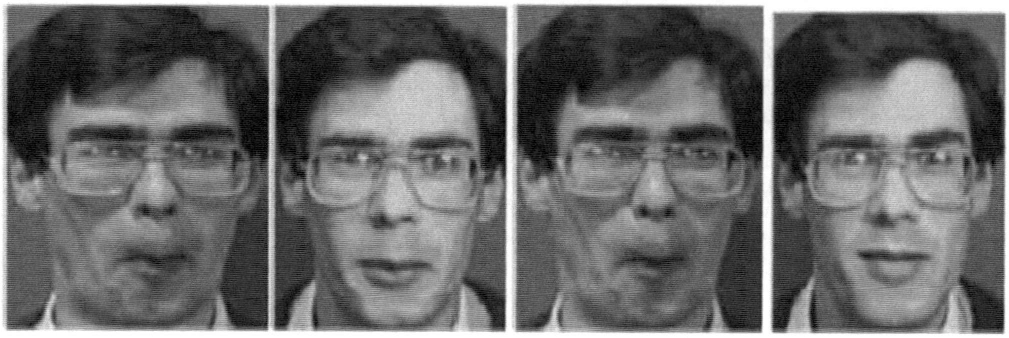

Figure 2. Four images of the same person in ORL face database

b) FEI database: The database contains 200 objects, 14 different images for each person, and the head is not tilted. The image is a gray scale image with 200*180 pixels. An image example is shown in Figure 3.

Figure 3. An example of someone's image in the FEI database

c) Self-constructed database: the capture device (camera) is used to collect face images at different times to form a database. The database selects 100 different objects, each with N=10 images, and each image is shot from a different angle, namely, 1000 face images in JPG format, an image example is shown in Figure 4.

Figure 4. Example image in self-constructed database

3.2. Effect of Parameters on Recognition Performance

ORL database was used in the experiment to study the effect change of characteristic number on recognition rate. In this experiment, Euclidean distance was used, and the recognition result is shown in figure 5.

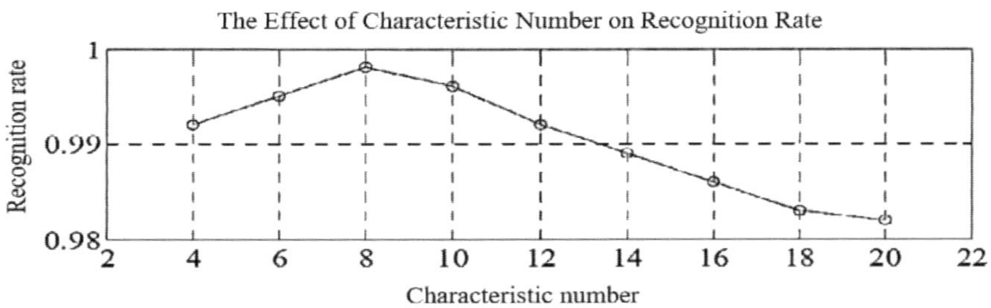

Figure 5. The Effect of Characteristic Number on Recognition Rate

It can be seen from Figure 5 that the method mentioned in this paper has the best recognition rate based on the seventh clean face (K=8) on the ORL, which can be as high as 99.8%. It can be seen from the fourteenth feature vector that the recognition rate fell to 98.9%. There is a similar trend in the FEI database that if the characteristic number is too large, it will cause over-fitting, resulting in the reduction of recognition rate.

To get the best recognition rate, the λ variance should be as small as possible; however, the characteristic number should not be too large. Therefore, it is necessary to find an optimal parameter setting. The influence curve of characteristic number changes on λ variance is shown in Figure 6.

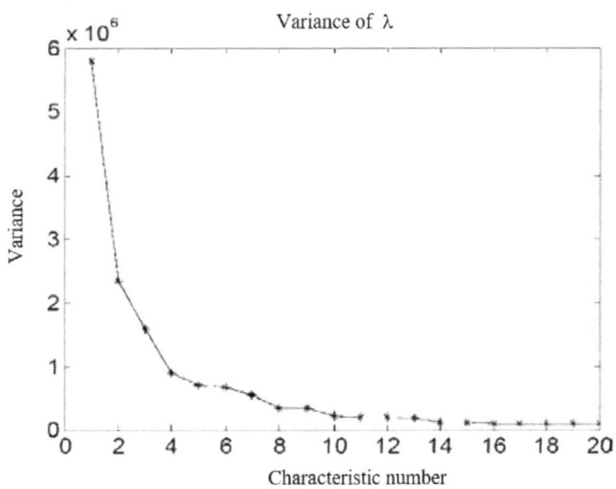

Figure 6. The effect of characteristic number on λ variance

It can be seen from Fig. 6 that as the characteristic number increases, the λ variance decreases. When the characteristic number is 8, the λ variance tends to be stable. It is also found that when the characteristic number is 8, the best recognition rate can be realized with the method discussed in this paper. Therefore, characteristic number in all subsequent experiments was set to 8.

The best recognition results of this method in various database are shown in Table 1.

Table 1. The best face recognition rate with the method in this paper (%)

Database	ORL	FEI	Self-constructed database
Recognition rate	99.8	99.5	98.5

It can be seen from Table 1 that the recognition rate of this method in ORL and FEI can be as high as 99.8% and 99.5%, respectively, and the recognition rate can be as high as 98.5% in the self-constructed database, which verifies the effectiveness of the proposed method.

3.3. Comparison and Analysis

In order to better show the excellence of this method, an experiment is conducted to test the noise and noise-free images separately. Three kinds of noise were used in the experiment: 1) salt and pepper noise with noise density a=0.06; 2) Gaussian noise, noise average m=0, variance v=0.04; 3) Gaussian noise, noise average m=0, variance v=0.06. The image example of three noises are shown in Fig. 7.

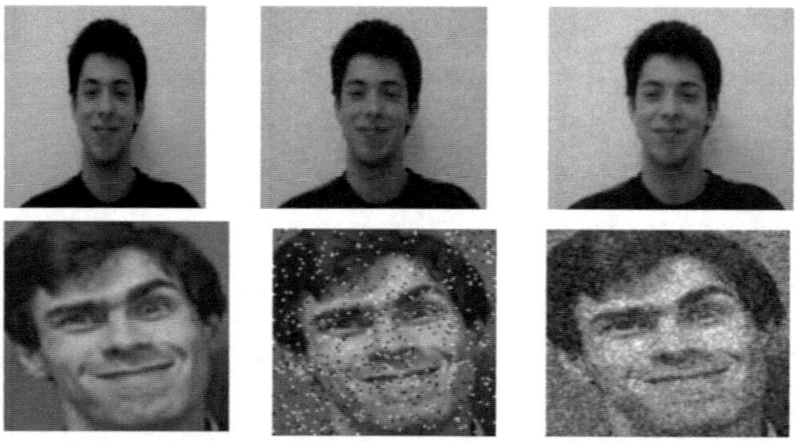

Figure 7. Image example with noise

The first half of the images of each object in the database were selected for training and the rest were used for testing. And the recognition result of the method in this paper is compared with that of several new methods, including the method based on HMM and SVD coefficients (HMM-SVDC) [3], tensor local Fisher discriminant analysis (TL-FLDA) [7], kernel fusion multiple descriptors-based multi-scale local phase quantization method (KFMD-MLPQ) [16], 2DPCA method based on wavelet transformation and SVM classification (WT-SVM-2DPCA) [10]. The recognition rates of each method with noise and without noise respectively are shown in Table 2 and Table 3. Table 4 shows the execution time of each method.

Table 2. Face recognition rate of each method without noise (%)

Method	ORL	FEI	Self-constructed database
HMM-SVDC	98.6	98.7	97.4
TL-FLDA	99.2	99.0	97.9
KFMD-MLPQ	99.4	99.2	99.2
WT-SVM-2DPCA	99.6	99.2	98.3
Method in this paper	99.8	99.5	98.5

Table 3. Face recognition rate of each method with noise (%)

Method	ORL	FEI	Self-constructed database
HMM-SVDC	96.5	96.1	94.5
TL-FLDA	97.0	96.5	95.1
KFMD-MLPQ	97.5	97.2	95.5
WT-SVM-2DPCA	97.8	97.6	96.4
Method in this paper	98.9	98.3	97.6

Table 4. Total time of each method (seconds)

Method	ORL	FEI	Self-constructed database
HMM-SVDC	25.593	30.459	14.582
TL-FLDA	21.322	28.347	12.495
KFMD-MLPQ	19.549	27.190	11.587
WT-SVM-2DPCA	17.480	25.039	9.733
Method in this paper	18.292	26.520	10.304

It can be seen from Table 2 that in most cases, the recognition rate of the method in this paper in three databases is higher than that of any other method. However, the recognition rate is not obviously improved. It can be seen from Table 3 that for images with noise, the recognition rate of this method is obviously higher than that of any other methods, indicating that the proposed method has better performance when processing images with noise. It can be seen from Table 4 that the execution time of this method is slightly higher than that of WT-SVM-2DPCA method. It is mainly because the WT-SVM-2DPCA method extract features with 2DPCA, which is quicker than the 2DLDA of this method. Compared with other methods, the execution time of this method is the shortest, indicating that this method excels in computational cost.

In summary, the following conclusion can be drawn that the method used in this paper reduces the computation time of the two-dimensional subspace method training and improves the recognition rate, indicating the excellence of this proposed method.

4. Conclusion

In order to solve the problem of face recognition with strong noise image, this paper puts forward a grid method based on DWT-2DLDA and probabilistic neural classifier. According to the experimental results, the method used in this paper has better recognition performance than other methods. It has been verified through experiments that DWT can be used as a preprocessing technology. With the help of this technology, the system mentioned in this paper has advantages of better recognition rate, high computing speed and reducing PNN memory calculation. Since this method can show better robust performance in this kind of uncontrolled environment, it will be applied to video monitoring in the future

It has been 60 years since the face recognition technology was first put forward. The face recognition technology becomes more and more mature and various algorithms come out continuously. Although great achievements have been gained, since different algorithms require different implementation environments, there are still so many difficulties in facial recognition technology that need to be solved, including face detection and key point location under complex conditions, illumination change, facial gestures, facial expressions, facial shade, changes in a person's age, large-scale face recognition problem, lack of samples, a large amount of data learning problems, collection equipment for face information and so on.

With the continuous development of technology, face recognition technology will develop in the following directions in the future: face recognition research based on multi-data and multi-method integration; feature acquisition research with dynamic tracking; face nonlinear modeling research; 3D modeling research; automatic face recognition research. It is believed that with the continuous advancement of related disciplines such as pattern recognition, computer vision, image processing and machine learning, the face recognition technology will be continuously improved and optimized. Face recognition technology will be more widely applied to daily life more conveniently and more safely.

5. References

[1] BLEDSOE W W. The model method in facial recognition [J]. Panoramic Research Inc Palo Alto, 1964(1): 55-64.
[2] Wang Yang, Yan Yunyang, and Wang Hongyuan. Face Recognition Based on Residual Space and SVM with Bidirectional Two Dimensions PCA [J]. Computer Science, 2012, 39(12): 268-271.

[3] Anand C, Lawrance R. Algorithm for Face Recognition Using HIM and SVD Coefficients[J]. Artificial Intelligent Systems and Machine Learning, 2013, 5(3):125-130.

[4] Zhang Ruyan, Wang Shitong. Multi-classification Recogition Method Applied to Facial Image on Distribution Characteristic of Heavy-tailed Noise [J]. Journal of Electronics and Information Technology, 2012, 34 (12): 523-528.

[5] Whither Biometrics Committee. Biometric recognition: challenges and opportunities [M]. National Academies Press, 2010.

[6] Chan C H, Tahir M A, Kittler J, et al. Multiscale local phase quantization for robust component-based face recognition using kernel fusion of multiple descriptors[J].Pattern Analysis and Machine Intelligence, IEEE Transactions on, 2013, 35(5): 1164-1177.

[7] Zheng Jianwei, Wang Wanliang, Yao Xiaomin et al. Face Recognition Using Tensor Local Fisher Discriminant Analysis [J]. Acta Automatica Sinica, 2012, 38 (9): 1485-1495.

[8] Su Yu, Shan Shiguang, Chen Xilin et al. Integration of Global and Local Feature for Face Recognition [J]. Journal of Software, 2010, 21 (8): 1849-1862.

[9] Mashhoori A, Zolghadri Jahromi M. Block-wise two-directional 2DPCA with ensemble leaming for face recognition [J]. Neurocomputing, 2013, 10(8):111-117.

[10] Xu W, Lee E J. Face Recognition Using Wavelets Transform and 2D PCA by SVM Classifier [J]. International Journal of Multimedia & Ubiquitous Engineering, 2014, 9(3): 243-249.

[11] Zhu Yulian, Peng Xing. Two-dimensional PCA Based on Random Sampling for Face Recogniton [J]. Journal of Chinese Mini-Micro Computer Systems, 2011, 32 (12): 2461-2465.

[12] Li D, Tang X, Pedrycz W. Face recognition using decimated redundant discrete wavelet transforms[J].Machine Vision and Applications, 2012, 23(2): 391-401.

[13] Zhang H, Jonathan Wu Q M, Chow T W S, et al. A two-dimensional Neighborhood Preserving Proiection for appearance-based face recognition [J]. Patterm Recognition, 2012, 45(5): 1866-1876.

[14] Guo Jinxin, Chen Wei. Face Recognition Based on HOG Multi-feature Fusion and Random Forest [J]. Computer Science, 2013, 40 (10): 279-282.

[15] Eslamloueyan R. Designing a hierarchical neural network based on fuzzy clustering for fault diagnosis of the Tennessee-Eastman process [J].Applied Soft Computing, 2011, 11(1): 1407-1415.

[16] Chan CH, Tahir M A, Kittler J, et al. Multiscale local phase quantization for robust component-based face recognition using kernel fusion of multiple descriptors[J].Pattern Analysis and Machine Intelligence, IEEE Transactions on, 2013; 35(5):1164-1177.

[17] Pandya J M, Rathod D, Jadav J J.A survey of face recognition approach [J].International Journal of Engineering Research and Applications (IJERA), 2013, 3(1):632-635.

The Application and Discussion of Virtual Simulation Platform in the Teaching of Ship Handling and Collision Avoidance Course

Jie Chen[1, a], Mincang Liang[2,b] and Hu Liu[3]

1Marine College of Zhejiang Ocean University, Zhoushan 316000, P. R. China
2Marine College of Zhejiang Ocean University, Zhoushan 316000, P. R. China
3Marine College of Zhejiang Ocean University, Zhoushan 316000, P. R. China
aEmail:chenjiejie2021@163.com,bEmail:liangmincang2016@163.com

Abstract. With the development of the computer, virtual reality technology and "Internet+" concept, the virtual training system was used by all walks of life to carry out pre-job training, virtual simulation teaching has become an inevitable trend of the education industry. Through the analysis of the drawbacks of professional course on ship handling and collision avoidance, combining the advantages of virtual reality technology "Internet+" practical application in simulation implementation, the application superiority and significance of the application in the ship maneuvering collision course teaching is to be given in order to promote the development of virtual simulation teaching.

1. Introduction

As early as the middle of the 20th century, the research on ship handling and collision avoidance simulation was paid attention to by some shipping developed countries in the west. Abstract concepts of ship handling and collision avoidance were quantified and calculated by these countries to conduct in-depth research on decision generation and visualization of ship handling and collision avoidance. In the 1970s and 1980s, simulation research on ship collision avoidance system was first carried out at the Liverpool University in UK and Tokyo Merchant Shipping University in Japan. A ship collision avoidance expert system developed by Liverpool University and the research of collision avoidance decision system on the experimental ship called shihirokuaru in Japan was a pioneer in the simulation of ship handling and collision avoidance. With the development of China's shipping industry, it was not until the 1990s that the research on ship handling and collision avoidance simulation system was gradually paid attention to. At the same time, the ship handling simulator required by the training of seafarers was the only ship handling simulator of Dalian maritime University in China that was certified by Norwegian classification society class A. The application of large ship handling simulator in crew training has the disadvantages of high cost, few trainees accommodation and limited practice time and venue. If a pure virtual platform can be built to simulate typical scenes of ship collision avoidance rules, the above problems can be better solved. And as the system functions are upgraded and improved, the web-based online operation training platform can be realized, enabling students to practice anytime and anywhere.

Today, with the rapid development of information technology, teaching methods based on Internet virtual platform are becoming popular. When some boring knowledge is presented to students through virtual platform, it becomes easy to understand. The application of virtual reality technology in the field of education should be further developed to create a virtual learning platform for students to change the traditional thinking mode of stuffy teaching. As a result, the teaching effect can be strengthened and teachers' teaching pressure can be relieved.

2. The Research Conditions at Home and Abroad

Virtual simulation technology has developed rapidly in foreign countries and has been widely used in college teaching, as well as in medical, military, aerospace and other fields. In China, although virtual simulation technology has been widely involved in some 3D modeling design or numerical control model teaching, in the application of driving professional training in ships, and other fields, only the large ship manipulation simulator has been currently recognized and widely used. The use of the device has many limitations, and the application of virtual simulation teaching is relatively simple. At present, when college students learn about ship collision avoidance, the use of large-scale ship manipulation collision avoidance simulator for teaching and training requires wider site and higher cost, more strict safety management regulations of the laboratory, which have many limitations and do not meet the actual conditions of most schools. Therefore, students' learning is still in the stage of theory learning in class and relatively weak practice.

3. The Existence of Problem

3.1. Uniformity of Teaching Models and Methods

Nowadays, the students of the domestic universities' navigation technology major are basically taught through the traditional teaching mode when learning the ship's manipulation and collision avoidance. The teacher explains the textbook knowledge on the podium while the practical simulation classes are not enough. What's more, exams in the internet have been used in the majority school which lead to a duck-stuffing type of learning. It not only reduces the quality of classroom teaching, but also affects students' learning enthusiasm and real practical application after boarding.

3.2. The Lack of Practical Training Positions

It demands higher requirements of Marine navigator and more power to break the forces, so the real operation risk factor is too large. In the face of the hull, the safety of the site operation is relatively high, and the cost of use and maintenance is relatively high. When the operation is touched, the operation for a long time and many times does not conform to the actual situation. It is unrealistic to actually carry out the training related to the operation and collision avoidance on the ship.

3.3. The High Cost of Training

At present, college students are mainly at the theoretical level in learning to avoid collisions, which is difficult to understand and accept at a deeper level, which is not conducive to students' mastery and understanding of knowledge. If training is carried out with a large simulator, the site requirements of the equipment and equipment are high, the cost is too high, and the safety management regulations of the laboratory, etc., have strong limitations and do not meet the actual conditions of most schools.

4. The Advantages of Platform in the Teaching of Ship Maneuvering Collision Avoidence

4.1. Make Up for the Shortage of Teaching and Training Equipment

With the rapid development of shipping market, the demand for crew members is increasing and at the same time, the related technologies of crew training are in advancing. In the process of training of related professional colleges and schools and training institutions, it is necessary to combine shipping development characteristics and the crew types, constantly updating teaching training, equipping with professional equipment for teaching and training, and then cultivating more professional talents. However, for some smaller navigation colleges and some training institutions, due to lack of space, insufficient funds and other factors, the training equipment cannot be updated immediately and meet the teaching needs. As a result, students or crew members have great problems in mastering new knowledge in the training process. However, the targeted introduction of ship maneuvering collision simulation platform can build a good learning environment for students, through the standalone version or web page version of the online virtual simulation platform, making students understand how to conduct collision operation when the ship sailing on the sea, and how to solve various problems encountered in handling collision avoidance. These knowledge can be trained on the platform, helping them acquire knowledge and skills in the virtual practice and improving the teaching effect.

4.2. Optimizing Classroom Teaching Effect and Stimulating Learning Interest

At present, there are a lot of schools or training institutions offering navigation majors, but there are only 7 colleges and universities on a large scale, such as dalian maritime university, Shanghai maritime university, jimei university and wuhan institute of technology. Among them, only dalian maritime university is affiliated to the ministry of communications, other institutions and institutions have relatively low reputation and are less attractive to students, which makes it difficult to attract more students. Besides, aging equipment and other factors make students less interested in learning. After using the virtual simulation platform for teaching, a large number of ship collision avoidance knowledge in the textbook can be digitized, visualized and video, and the three dimensional model can make the rigid knowledge points in the book more concrete. Deepen students' understanding and mastery, improve student' learning efficiency. In addition, it can also be used as a publicity point to attract more students *to* the school and increase the competitiveness of students.

4.3. To Improve the Teaching Quality through the combination of Theory and Pratice

Due to the lack of equipment in some smaller universities and training institutions, there is a delay in the implementation of courses on ship handling and collision avoidance. Such being the case, students can only finish the theoretical knowledge, and there is little chance for them to use the largescale ship handling and collision avoidance simulator for practical operation training, so they cannot master relevant knowledge well. After boarding the ship, most of the knowledge learned in the school was forgotten, which greatly reduced the enthusiasm and efficiency of learning.

The teaching of the application of virtual simulation platform can effectively avoid the occurrence of these problems, in the process of classroom teaching theory. Through the direct using of the virtual simulation platform of ship maneuvering collision process and the specific operation for intuitive demonstration, the possible problems in a timely manner can be presented in front of students. Students, through the observation and started training operation, can directly grasp the relevant knowledge of ship collision avoidance, achieving the objectives of the synchronization of classroom teaching.

4.4. Enhance the Enthusiasm of Participation and Meet the Overall Teaching Requirements

In the process of training crew, students can not carry out the experiment of independent ship handling and collision avoidance due to the limitation of equipment conditions, so they can only drill in the large-scale operation simulation room in the form of groups, leading to the fact that the actual operation ability is not satisfied, which affects students' understanding and mastering of collision avoidance knowledge. However, using virtual simulation platform can directly control computer to conduct ship collision avoidance training operation, improve the practical ability of students, strengthen their mastery of knowledge level, and meet the overall requirements of teaching. The specific operation measures are as follows:

Firstly, create a good learning atmosphere. The virtual simulation platform will not be widely promoted and used without following supports. On the one hand, it needs the government's policy support, which can be widely promoted in a short period of time. On the other hand, the school should increase its support for the simulation teaching, constantly carry out the teaching renewal reform, make most of the teachers familiar with the virtual platform, try best to stimulate the enthusiasm of the teachers, and make the virtual platform virtual. Then the use of simulation platform in teaching will be much faster.

Secondly, expand channels and resources. Resources are the basis of virtual simulation platform teaching. If we want to fully carry out virtual simulation education, we need to arrange and utilize resources reasonably. For example, by increasing cooperation with the same type of colleges and universities and with shipping companies, we can find a practical operation platform for students, so that students' practical ability can be exercised, nevertheless, the virtual simulation platform can play a greater role to improve the overall quality of students. On the basis of the virtual simulation platform, combined with Internet, we can make a web version of the ship collision avoidance related professional control system, which can make students more convenient in practice.

Thirdly, driven by radiation from point to area. In essence, at the technical level, the teaching of virtual simulation platform belongs to computer science. While vigorously promoting the teaching of virtual simulation platform in navigation related colleges and universities, it also needs the support of computer science in technical maintenance, so the computer related majors will also get corresponding development. After the success of the experiment in our own college, we can also introduce and promote it to other same colleges, so as to popularize the navigation training related industries and adhere to the teaching method of virtual simulation technology in the way of driven by radiation from point to area.

5. The Significance of Virtual Simulation Teaching in Ship Maneuvering Collision Avoidence Course

The development of operating software for the typical meeting scenarios in the international maritime collision avoidance rules can make the knowledge of collision avoidance rules easilier learnt by the students, and make the pilots of the pilots more proficient in the collision avoidance operation. The purpose of the dangerous probability of voyage is to ensure the safety of the ship. Therefore, the virtual simulation platform that which has been introduced the stand-alone version or the online version of the ship in the ship collision avoidance course has high application value.

(1) Development of ship collision avoidance simulation software can reduce ship maintenance costs and ship damage probability. In the actual collision avoidance operation, the hull navigation consumption is too large, the cost too high, and the operation mistakes of the hull will require a high maintenance cost.

(2) It can be learnt by students more easily. Only through personal operation can correct abstraction and generalization be made. It is difficult to master such abstract knowledge without the experience of one's own operatio.

(3) It is a good operation and learning opportunity for the social crew. The social crew can only conduct training on the ship when learning to avoid collisions, having few opportunities but frequent accidents. After the several-time practices, the grasp of collision avoidance is more firm and accurate, and it also makes the future navigation safer.

(4) Security is guaranteed. In actual operation, if the ship fails to avoid accidents, it may cause damage to the hull and cause damage to the personnel on board. This is a serious hazard. However, when practicing with the ship collision avoidance simulation system, the error will be promptly prompted or judged. If you fail, there will be no loss. You can have the students or the social crew perform multiple safety trainings until they have completely mastered.

6. Conclusion

Rmarkable effect will be achieved when the single or online virtual simulation platform is applied to the teaching of the ship maneuvering collision course, which is developed according to the international regulations for preventing collisions at sea in a typical encounter scenario. For instance, the teaching field and the teaching equipment investment cost will be reduced. The platform is also easy for maintenance and content updates, which will be beneficial to student's autonomous learning, practice of ship collision avoidance operation demonstration study and the improvement of efficiency of teaching resources. Moreover, a new teaching practice platform will be established when the application has been used to the professional teaching of ship handling and collision avoidance. The comprehensive teaching level of the platform also will be improved thoroughly, which is of great help to train high-quality and professional ship driving talents.

7. References

[1] Lu Zhan. Application of virtual simulation software in the teaching of new energy vehicle application and maintenance [J]Time automobile, 2019(12):42-43

[2] Xu Zheng. Research and Application of Simulation Platform for Ship Collision Avoidance Decision-making [D]. Dalian Maritime University, 2015
[3] Yang Jinbao. Maritime Engineering College Students' Values Education Research [D]. Dalian Maritime University, 2015
[4] Chang Yuan, Li Ying. Research on the application of virtual simulation technology in Clothing Technology Teaching [J]. Technology Wind, 2014(05):209
[5] Ma Xuetiao, Cheng Zhiqun, Zheng Xuefeng, Wang Yonghui. The construction and practice of virtual simulation experiment teaching platform of electronic information technology [J]. Experimental Technology and Management, 2018, 35(11):130-133.
[6] Zheng Rujing. Research on Intelligent Navigation Collision Avoidance Algorithm Based on Internet of Things [D]. Tianjin University of Technology, 2017
[7] Bao Hongyang. Research on the mathematical model and Simulation of ship automatic intelligent collision avoidance [J]. Ship Science and Technology, 2017, 39(19):164-169.
[8] Xie Hongyan. Construction and Practice of Virtual Simulation Experimental Teaching Platform for Economic Management [A]. Institute of Management Science and Industrial Engineering. Proceedings of 2018 5th International Conference on Education, Management, Arts, Economics and Social Science (ICEMAESS 2018) [C]. Institute of Management Science and Industrial Engineering: Computer Science and Electronic Technology International Society, 2018:5.
[9] Gao Jianquan, Fang Xinguo. Research on virtual simulation teaching platform of product manufacturing process [J]. The Design, 2019, 32(11):80-82.
[10] Huang Feng, Wang Wei, Luo Zuchun. Research on the application of virtual simulation teaching platform in the practical teaching of cardiopulmonary resuscitation Technology [J]. Journal of Snake, 2019, 31(02):288-289+310.

Road Segmentation based on Deep Learning with Post-Processing Probability Layer

Weibin Chen

Network center, Jimei University, 183 Yinjiang Road, Jimei district, Xiamen, China.
Email: wbchen@jmu.edu.cn

Abstract. In order to solve the problem of road image semantics segmentation, a novel semantic segmentation method based on deep learning is proposed in this paper, and it is verified that this method can optimize the segmentation effectively comaring with traditional Full Convolution Neural Network (FCN) model. Firstly, the traditional Full Convolution Neural Network model is constructed. And then according to the principle of the post-processing probability layer method proposed in this paper, the label of all road image training sets is used to compute and transform it to form a two-dimensional array which can represent the classification probability of each pixel, and it is combined with the Full Convolution Neural Network model to be used for road image semantics segmentation. Secondly, the tensorflow neural network framework is used to simulate the above two models. Finally, the experimental results show that the CNN model with the proposed post-processing probabilistic layer is able to get better results in road semantics segmentation in KITTI data sets. The pixel accuracy is improved from 88.8% to 91.3%, and the mean pixel accuracy is increased from 82.9% to 85.7%. The mean intersection over union increased from 72.5% to 77.9%.

1. Introduction

Autonomous driving technology is a hot topic in these years. Among them, accurate semantic segmentation is one of the key technologies in the process of automatic driving. For example, fuzzy c-means algorithm based on objective function is widely used[1]. In recent years, with the progress of computer hardware, the expansion of information and data, and the rapid improvement of Deep Learning technology, Convolutional Neural Networks [2-4] are proposed and applied in image classification and semantic segmentation. In order to make convolutional network better applied to image semantic segmentation, many researchers have improved and optimized it and proposed a variety of neural networks for semantic segmentation. In 2014, a fully convolutional networks (Jonathan Long et al.) [5] was proposed. By replacing the final fully connected layer with the convolution layer, the network uses deconvolution to carry out up-sampling, so that the size of the output image can be restored to the same size as the input image, thus achieving end-to-end image semantic segmentation at the pixel level. In 2015, researchers proposed SegNet neural network (Vijay Badrinarayanan et al.) [6]. In which, the symmetric structure of encoder and decoder is used, and the maximum pooling index is applied in the decoder, so that the segmentation resolution can be improved.

In order to solve the problem of image information loss caused by pooling, DeepLab neural network (liang-chieh Chen et al.) [7] was proposed. In which, a pyramid-shaped dilated pool structure is proposed by using dilated convolution and combining it with fully connected condition random field.

However, dilated convolution structure has the disadvantages of high computational cost and large footprint, so RefineNet neural network (Guosheng Lin et al.) [8] was proposed. The method solves the disadvantages of dilated convolution structure. The segmentation precision of the network is high, but the network is complex. At present, the research focus in the field of road image semantic segmentation has gradually transformed intotreatment after joining the deep learning technology, and on this basis, new research achievement constantly results, even if the application and the promotion of these results of this study is still limited to some aspects, but with the continuous development and improvement of deep learning technology, pushes the road image semantic segmentation technology into a new era.

The networks mentioned above are proposed to solve the problem of image semantic segmentation, but these networks have a wide range of applications and are not designed specifically for a certain kind of segmentation problem. Therefore, in order to make the neural network model is more suitable for solving the problem of road detection. This paper proposes a post-processing probability layer of the new model, and the model is used for road-image semantic segmentation. Comparatived with traditional fully convolution neural network algorithms, the proposed model has a better ability for road-image semantic segmentation.

2. The Design of Post-Processing Probability Layer

The roads in data set which is used to train the model are almost in the middle and lower of the image. Therefore, this paper proposes an algorithm of post-processing probability layer after CNN based on this characteristic of road image.

The method of post-processing probability layer is shown as follows.

First, transform RGB image to binary image in training data sets. The value in the result image is 1 or 0, as shown in Figure 1. A value of 1 means that the pixel is classified as a road, while a value of 0 means that the pixel is classified as a non-road (background).

Then, the all labels array sum averaging (normalized), get a the size of h*w matrix, the matrix is called a probability layer (as shown in Figure 2, 3), in which h, was the input label images high and width, the range of each value is 0 to 1 (including 0 and 1), the closer the value 1 indicates that the pixels is the more likely for road, the closer the value 0 indicates the pixel points is more likely for background.

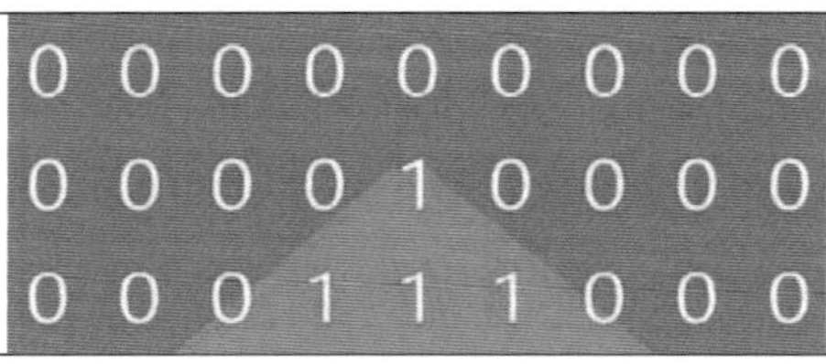

Figure 1. Schematic of the converted data label.

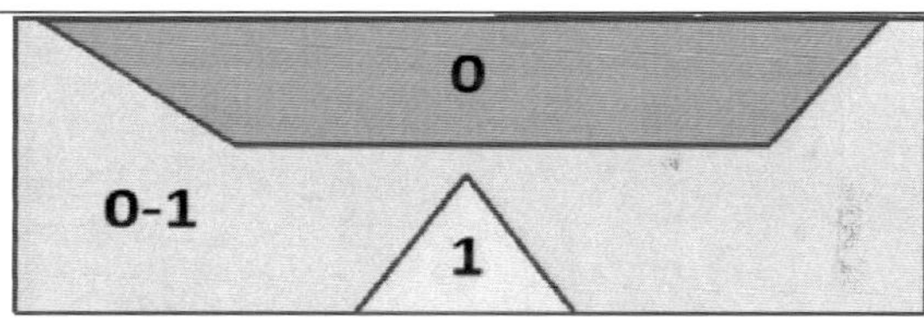

Figure 2. Probability layer concept map.

In order to enable the probability layer to act on the final semantic segmentation results, it is necessary to transform the probability layer. Subtract 0.5 from each value of the original probability layer to get the converted probability layer, as shown in Figure 4.

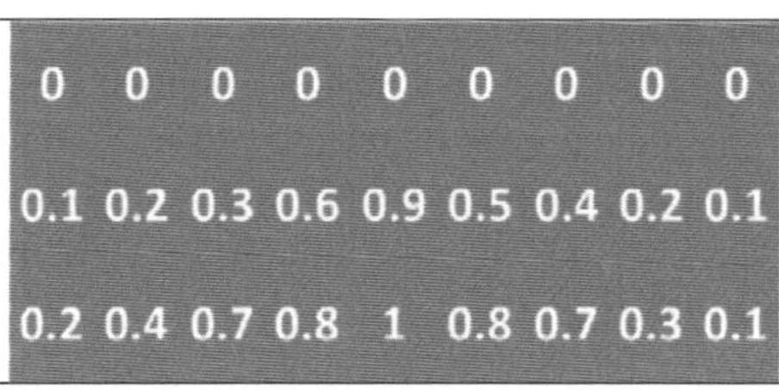

Figure 3. Original probability layer diagram.

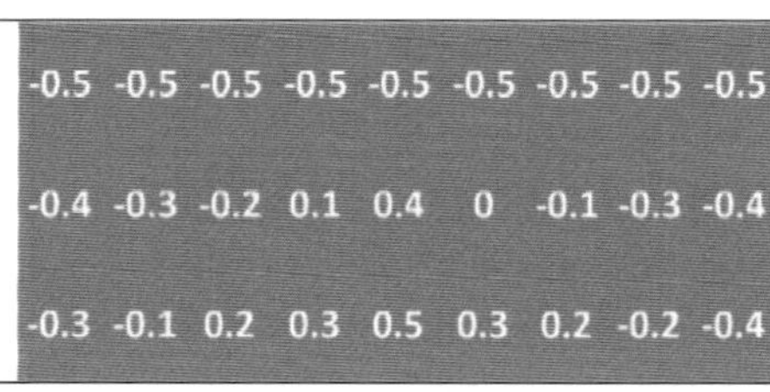

Figure 4. Probability layer schematic diagram after conversion.

Finally, when the model predicts the test set image, add the probability layer in the last layer of the model and assign an impact factor to this layer as the post-processing method of the prediction result, so as to correct some misclassified pixels.

After the road image has been trained by the neural network model, a two-dimensional array label matrix will be generated, as shown in Figure 5, with the value of 0 to 1 for each pixel. Therefore, the value of each pixel is compared with 0.5. A value greater than 0.5 indicates that the pixel is classified as a road, while a value less than 0.5 indicates that the pixel is classified as a background. As can be seen from the Figure 5, two pixel values were misclassified, and the background was predicted to be the road, and the road was predicted to be the background.

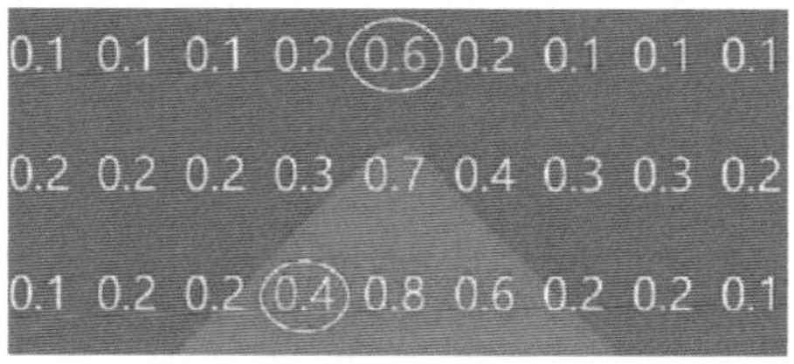

Figure 5. Prediction error label matrix

In this case, the post-processing probability layer can influence the results. By adding the predicted value of each pixel in the prediction result to the corresponding pixel value of the probability layer, as shown in Figure 6, the above prediction errors can be eliminated, thus improving the segmentation accuracy of road image semantic segmentation to a certain extent. As can be seen from the figure, two originally misclassified pixels were finally classified correctly after being processed by probability layer, and the classification results of other pixels were not changed.

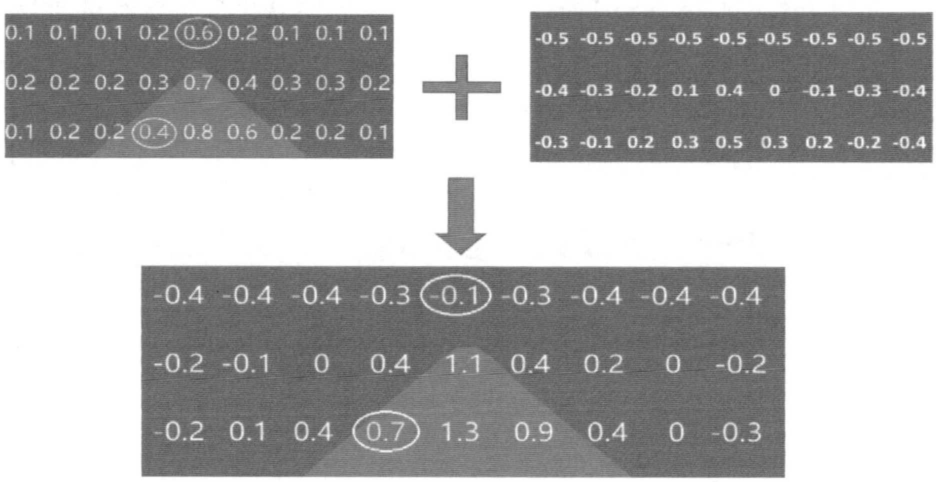

Figure 6. Combining predictional and probability layer.

Because of the particularity of road image, the road is always at a specific position in the road image. Therefore, it is possible to attach a special value to each pixel value in the image to improve the semantic segmentation result of the road image. This is the main idea of post-processing probability layer.

3. Experimental Verification and Analysis

3.1. Experimental Data

All experiments in this paper are based on road images in the data set KITTI[10] for model training. Data set KITTI was created in 2012 by Toyota's Chicago institute of technology and Karlsruhe institute of technology in Germany. It is one of the most famous data set in the field of autonomous driving. The KITTI data set mainly includes images of daily life scenes such as highways, rural areas and cities, which are mainly captured by a color camera mounted on a car. In these dataset, the road is marked purple (RGB value [255, 0,255]) and the background is marked red (RGB value [255, 0, 0]). This paper uses 289 images, and their sizes were cropped to 576*160, among which 214 images were used as the training model and 75 images were used as the result verification.

3.2. The Experimental Process

(1) Pretreatment: at the beginning of the experiment, it is necessary to preprocess the trained images and labels. First, cut the image and label; Then the label image is transformed into a vector array encoded by one-hot code to represent the classification of corresponding pixels. Finally, the training picture and the corresponding label were divided into batch of batch_size for the input of the training model.

(2)Build FCN network model

(3)Training model: the output of the neural network was normalized by softmax function, and the processed results and label results were averaged through cross entropy function to form the cost function of model training. The Adam Optimizer is then used to optimize the model in the direction of the cost function drop and to calculate the training accuracy for each iteration. In the process of model training, the learning rate is set as 0.00001.

(4)Generate probability layer: according to the probability layer principle, generate a two-dimensional array that can represent the classification probability of each pixel with all the training pictures.

(5)Model prediction: after completion of the training model, the model parameters are saved, and then input the test pictures to already trained model, and use probability layer makes these generated prediction labels, then compared prediction labels with real labels, and find out the pixel accuracy (PA), average pixel accuracy (MPA) and the average intersection over union ratio (MIoU).

3.3. Compare Experimental Results and Analysis

In order to verify the effect of this algorithm in road image segmentation, the following work is mainly done on the data set:

(1)Fully convolutional neural network (FCN) model is used to segment road image semantically.

(2)A fully convolutional neural network model based on probability layer is used for road semantic segmentation.

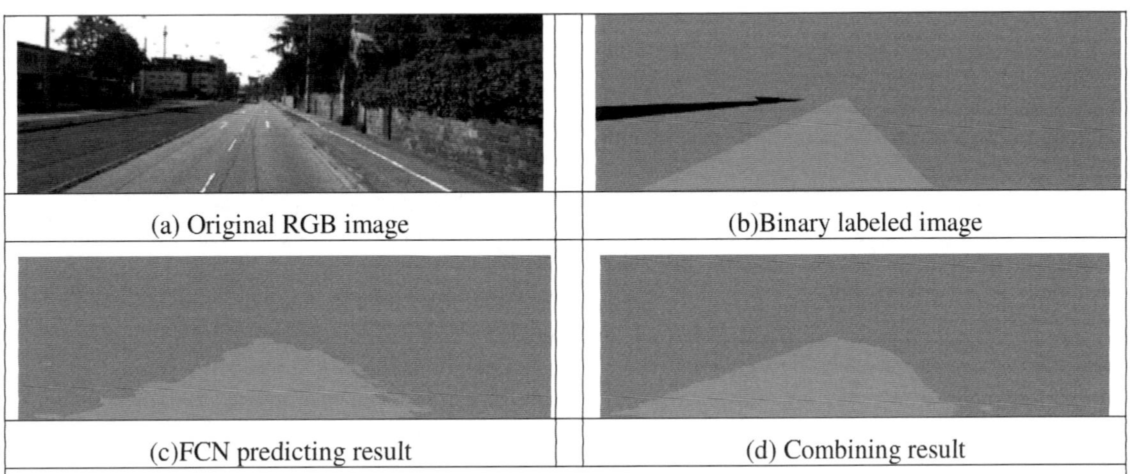

(a) Original RGB image (b)Binary labeled image
(c)FCN predicting result (d) Combining result

Figure 7. The results of road segmentation.

It can be seen from the above figures that the FCN model algorithm based on probability layer proposed in this paper has a better segmentation effect on road images than the traditional FCN model algorithm. After adding probability layer, FCN model can make the segmented road outline smoother and make the road boundary more coherent and complete. At the same time, after adding the probability layer, some unreasonable discrete points can be removed to make the prediction region as connected as possible, which is also consistent with the real road scenario.

Then, the prediction results of these two models were quantitatively analyzed and their pixel accuracy (PA), average pixel accuracy (MPA) and average intersection over union ratio (MIoU) were calculated respectively. The results are shown in table 1:

The results shown in Table 1. Compared with the traditional FCN model algorithm, the proposed algorithm in this paper has improved the segmentation accuracy. Among them, pixel accuracy (PA) improved by 2.5 percentage points to over 90%.The average pixel accuracy (MPA) increased by 2.8 percentage points to over 85%.The average intersection over union ratio (MIoU) increased by 5.4 percentage points.

Table 1. Formatting sections, subsections and subsubsections.

Metric	FCN	FCN+Probability Layer
PA	88.8%	91.3%
MPA	82.9%	85.7%
MIoU	72.5%	77.9%

According to the above results, the application of probability layer can improve the semantic segmentation effect of road image. The probability layer records the special location of the road in the road image, and makes use of the characteristic of the road image to carry on the later intervention to each pixel value in the image, so as to improve the semantic segmentation result of the road image.

4. Conclusion

According to the latest research results of FCN model in the field of road image semantic segmentation algorithm in the world, considering the particularity of road image, this paper carried out the research of road image semantic segmentation method based on the idea of transformation and calculation. The main work is as follows:

(1) In view of the traditional algorithm FCN model segmentation defect with low accuracy, according to the particularity of road image: road always in a certain location in the road image, this paper proposes a new idea, each pixel of image attach a particular value, indicates the probability of classification. That can improve the road image semantic segmentation results, and improve each evaluation index through the model of semantic segmentation;

(2) The tensor flow open source neural network framework was used to train and learn the neural network framework in this paper based on the KITTI data set. Parameters and training model were determined via multiple experiments.

(3) This paper applies the method to the real road image recognition experiment, and verifies the effect and optimization performance of post-processing probability layer model by analyzing the results.

Through comparison and analysis with experimental results of various deep learning algorithms, it is shown that compared with traditional fully convolutional neural network algorithm, the algorithm proposed in this paper can better segment road images, with higher accuracy and better reliability, and can be used for semantic segmentation of road images. The final experimental results show that the post-processing probability layer can improve the pixel accuracy from 88.8% to 91.3%, the average pixel accuracy from 82.9% to 85.7%, and the average intersection ratio from 72.5% to 77.9%.

Because of the existing open source road image data set stock rarely, this algorithm could not carried out in a large number of road image segmentation experiments, the subsequent will continue to carry out the research work on establishing road image sample libraries and label data set, further verify the algorithm effect that is applied to a large number of different scenarios of road image segmentation. At the same time, other mathematical changes can be attempted in the steps of generating the probability layer, such as nonlinear changes such as tanh function and sigmoid function. The difference between linear change and nonlinear change and their advantages and disadvantages will be explored to make them suitable for road images in various situations.

5. References

[1] Bezdek, James C. Pattern Recognition with Fuzzy Objective Function Algorithms. Advanced Applications in Pattern Recognition, 1981, 22(1171):203-239.
[2] Shi J, Malik J. Normalized cuts and image segmentation. IEEE Transactions on Pattern Analysis and Machine Intelligence, 2000, 22(8):888-905.
[3] Rother C, Kolmogorov V, Blake A. Grabcut: Interactive foreground extraction using iterated graph cuts. ACM, 2004, 23(3):309-314.
[4] Krizhevsky A, Sutskever I, Hinton G. ImageNet Classification with Deep Convolutional Neural Networks. Advances in neural information processing systems, 2012, 25(2).
[5] Long J, Shelhamer E, Darrell T. Fully Convolutional Networks for Semantic Segmentation. IEEE Transactions on Pattern Analysis & Machine Intelligence, 2014, 39(4):640-651.
[6] Badrinarayanan V, Kendall A, Cipolla R. Segnet: A deep convolutional encoder-decoder architecture for image segmentation. IEEE transactions on pattern analysis and machine intelligence, 2017, 39(12):2481-2495.
[7] Chen L C, Papandreou G, Kokkinos I, et al. Semantic image segmentation with deep convolutional nets and fully connected crfs. arXiv preprint arXiv: 1412.7062, 2014.
[8] Lin G, Milan A, Shen C, et al. Refinenet: Multi-path refinement networks for high-resolution semantic segmentation. Proceedings of the IEEE conference on computer vision and pattern recognition. 2017:1925-1934.
[9] Garcia-Garcia A, Orts-Escolano S, Oprea S, et al. A review on deep learning techniques applied to semantic segmentation. arXiv preprint arXiv: 1704. 06857, 2017.
[10] Geiger A, Lenz P , Stiller C , et al. Vision meets robotics: The KITTI dataset. The International Journal of Robotics Research, 2013, 32(11):1231-1237.
[11] Panhai Zheng, Ling Guo, Libing Ding. Research and realization of convolutional neural network based on Tensor Flow. Electronic technology and software engineering, 2018(18):20-22.
[12] Lihua Wu. Google. Application of Tensor flow framework in automatic vehicle recognition function. Automobile practical technology, 2018(20):38-40.
[13] Feng Xiao, Ting Rui, Tongwei Ren, Dong Wang. Fully convolution semantic segmentation and object detection network. Chinese journal of image and graphics, 2019, 24(03):474-482.
[14] Yanfang Xing, Hongxiu Duan, Guangwei He. Application of Tensor Flow in image recognition system[J/OL].Computer technology and development, 2019(05):1-5
[15] Zongsheng Wu, Weiping Fu, Gaining Han. Road scene understanding based on deep convolutional neural network. Computer engineering and applications, 2017, 53(22):8-15.
[16] Juting Dai, Xinyi Tang, Peng Liu, Baotai Shao. Scene semantic segmentation network based on color - depth image and deep learning. Science, technology and engineering, 2018, 18(20):286-291.

Research on UAV Point Landing based on Visual Navigation

Boyong He, Chen Meng and Liaoni Wu

School of Aerospace Engineering, Xiamen University, Xiamen City, Fujian Province, 361005, China

Email: heboyong@qq.com

Abstract. In this paper, the point landing technology based on visual navigation is studied on a rotor UAV flight platform. By contrasting different flight datum, several common image recognition and tracking algorithms are compared such as blob analysis, CamShift and KCF. Thus, automatic and manual methods are respectively designed for identifying ground targets, with rapidity, accuracy and robustness. After testing on UAV onboard equipment, the point landing technology based on visual navigation is finally achieved.

1. Introduction

With the rapid development of UAV industry, intelligent UVA has been put on the agenda. An important aspect of UAV intelligence is the development of visual navigation, so researches on UAV visual navigation have been gradually emphasized. UAV landing technology is one of the key technologies of UAV whose visual system can be used as a complement to the existing sensor system. Therefore, the research of vision based on UAV landing technology has a positive significance[1-3]. This paper mainly introduces the principle of image recognition and target tracking and the implement in opencv(Open Source Computer Vision Library)[4]. Rapidity, accuracy and robustness of visual programs are compared among four groups flight datum of different light and background. After testing on UAV onboard equipment, the point landing technology is eventually realized.

2. Navigation Approach

2.1. The Design of Image Pre-Processing

The main purpose of image pre-processing is to eliminate irrelevant information in the image, recover useful information, enhance the detectability of related information and simplify the data to the utmost, thereby improving the reliability of feature extraction, image segmentation, matching and recognition. Image pre-processing covers the following approaches: color-space conversion, gray-scale image transformation, smoothing, thresholding, region features and so on. Different image pre-processing is usually required in different computer vision fields. The following will introduce the image pre-processing algorithms involved in this paper.

In image processing, the most common color space is the RGB model, as shown in Fig.1. However, The HSV color space is insensitive to light conditions and can reflect color changes better, as shown in Fig.2. Therefore, the HSV color space can be employed to perform color segmentation of a specified color.

Second, it is necessary to transform the gray-scale. Performing a gray-scale transformation on a color image can significantly reduce the amount of image processing calculation and greatly increase the speed of image processing. According to theoretical calculations, it can increase the calculation speed by two times. Binarization is to convert the gray value of the point on the image into 0 or 255. The collection property of the image is only related to the position of the pixel value of 0 or 255, and no longer involves the multi-level value of the pixel, so that the processing becomes simple, and the data processing and compression amount small. Finally, we use the method of erosion and dilation to eliminate noise and divide the individual elements[5]. Fig.3 depicts the results of image pre-processing.

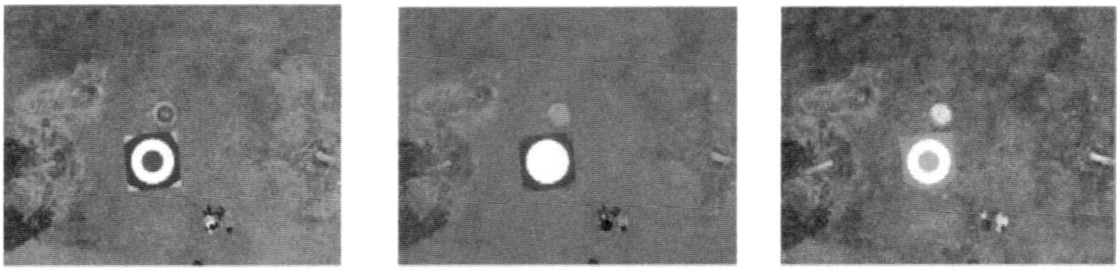

(a) (b) (c)

Figure 1. RGB single channel image

(a) (b) (c)

Figure 2. HSV single channel image

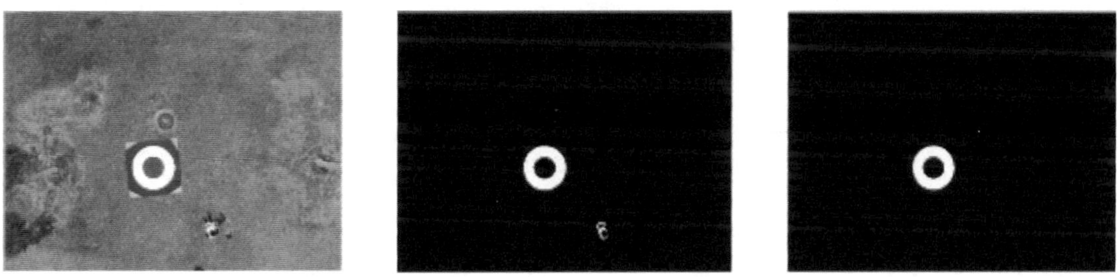

(a) (b) (c)

Figure 3. Image pre-processing. (a): gray scale transformation, (b): binarization, (c): erosion and dilation

2.2. The Design of Image Recognition

There are many algorithms for image recognition, which are generally divided into two categories. One of them bases on gray matching, and the other bases on feature matching. Gray matching is mainly used in space of one-dimensional or two-dimensional sliding template image matching, and the differences among them mainly lie in the template and the selection of related guidelines. Generally, this kind of method has a high efficiency, but the large amount of calculation and slow speed; Feature-based matching extracts features such as points, lines and regions from the original image as matching resources, and then it is performed. Generally, matching speed is faster, but matching accuracy is not high. Comparing common image recognition algorithms, this paper choose the recognition based on color and blob analysis to test[6].

After color-space conversion, there are six color channels, R(red), G(green), B(blue), H(hue), S(saturation), V(value), respectively. The main of recognition based on color is selecting one color channel with enough contrast. The result of image pre-processing is a binary image, and then use find-Contours, a function in opencv, to extracting outlines. These outlines are screened by shapes and areas. At last the target is identified by the outline. Fig.4 depicts the results of image recognition based on color .

The key of blob detection is the analysis of blob features[7]. In computer vision, Blob refers to a connected area with similar colors, textures, and other features in an image. In practice, change the threshold value at first, and run binarization continuously with threshold step. Next, use find-Contours to Extract connected domains and calculate the center of each connected domain. Some close points are classified as a group corresponding to a blob feature. Then, estimate the blob feature and radius of the target. Eventually, run Simple-Blob-Detector, a function in opencv, to detect the target, as shown in Fig.5.

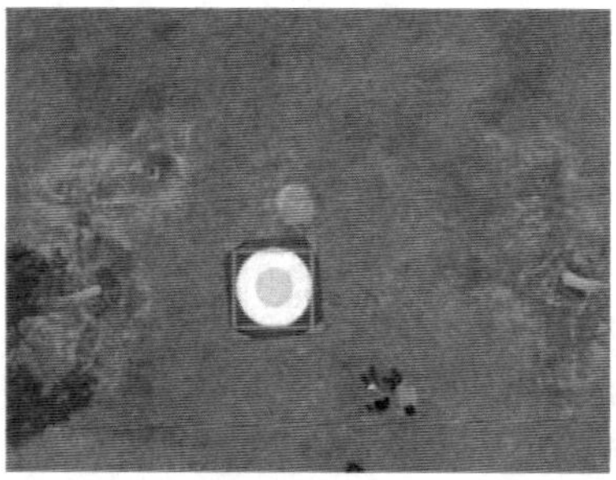

Figure 4. Image recognition based on colour

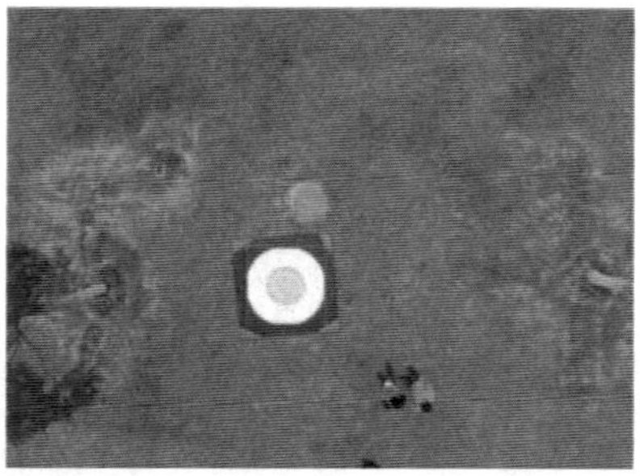

Figure 5. Image recognition with blob detect

2.3. The Design of Target Tracking

The target recognition algorithm needs to process and compare the whole image, while the target tracking algorithm only needs to recognize and calculate the initialized target, which is relatively faster and more accurate. According to the target tracking method, the tracking algorithm can be divided into two categories: generation class method and discriminant method. The generation method refers to the modeling of the target area in the current frame, and the next frame is looking for the most similar area of the model, that is, the prediction position, like Kalman-filter, particle filter, mean-shift method; The discriminant method, also called tracking detection (tracking-by-detection), refers to training classifier based on the current frame target area as the positive sample and the background area as the negative sample. The next frame uses a trained classifier to find the best region, and the typical method is TLD (tracking-learn-detect) algorithm. In recent years, the correlation filtering, such as the KCF/DCF algorithm, is also widely used. After the comparison of real time and accuracy, finally, three kinds of tracking methods, CamShift and KCF are selected to test in this paper[8-9].

CamShift(Continuously Adaptive Mean-Shift)is an improved algorithm for meanshift. Meanshift is a non-parametric method for density function gradient estimation, which locates the target by iterative optimization to find the extremum of the probability distribution[10-11]. Based on opencv, we calculate the H-component histogram of target in HSV color space at first, and use the target's color histogram model to convert the image into a color probability distribution, and initialize the size and position of a search window. Next, call the command of meanshift in opencv, and it can automatically track and adjust the center and size of the target window eventually. CamShift can effectively solve the problem of target deformation and occlusion. Furthermore, it has low requirements for system resource, fast processing speed and is insensitive to the changes in the shape and size of the target. Especially, it can automatically adjust the size of the search box and apply under a simple background. While the background is extremely complicated or there are more interferences in similar target colors, the tracking will fail[12]. Fig.6 depicts the results of CamShift.

As mentioned, KCF (Kernelized Correlation Filter) is a tracking method belonging to tracking-by-detection[13]. In opencv, it trains a target detector during the tracking process. The target detector is employed to judge whether the next frame predicted position is the target or no. Then, the latest target detector is employed to update the training set and update the target detector. When train the target detector, the target region is generally selected as a positive sample, and the target surrounding region is a negative sample. Especially, it collects samples by the circulant matrix in the region around the target, and uses the method of ridge regression to train the target detector. As consequence, it reduces the amount of calculation, increase the speed of operation and meets the real-time requirement, as shown in Fig.7.

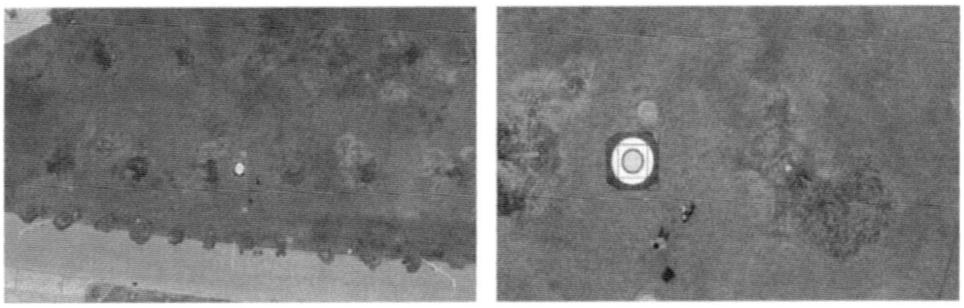

Figure 6. Target tracking with CamShift in different sclaes

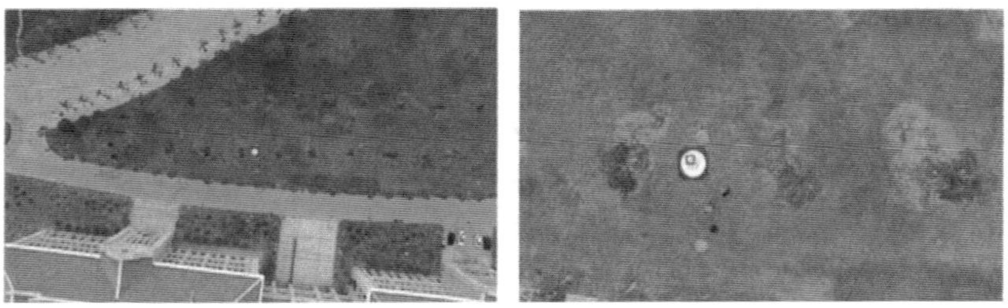

Figure 7. Target tracking with KCF in different sclaes

2.4. The Design of Position Estimation

The position estimation needs to understand the transformation relationship between the pixel plane coordinate system, the image plane coordinate system, the camera coordinate system and the world coordinate system. Besides, several significant parameters contain the followings: internal reference of camera (focal length and pixel size), a rotation matrix between the camera coordinate system and the world coordinate system(angles of pitch, roll and yaw), and the height of the UAV relative to the ground. More specifically, internal reference of camera is supplied by the manufacture, and the rotation matrix is measured by the attitude measurement of the UAV, and GPS offers the relative height at the same time[14-15].

According to these known parameters, the relative position between the UAV and the ground target can be obtained. No more explanations here as this section is not the focus of our work.

2.5. The Process of Visual Navigation

According to the above recognition algorithms, we design a visual navigation strategy that automatically recognizes ground targets. Besides, through the above tracking algorithms, we can identify manually by select the target on the ground station in real time. Due to the powerful computing power of our onboard equipment, these two image recognition algorithms or these two target tracking algorithms are able to run simultaneously. Furthermore, there is a filter to combine the results between these two image recognition algorithms or these two target tracking algorithms. After the process by filter, the image processing equipment is able to output the center of target's coordinate in the image coordinates in 10 hz, as shown in Fig.8.

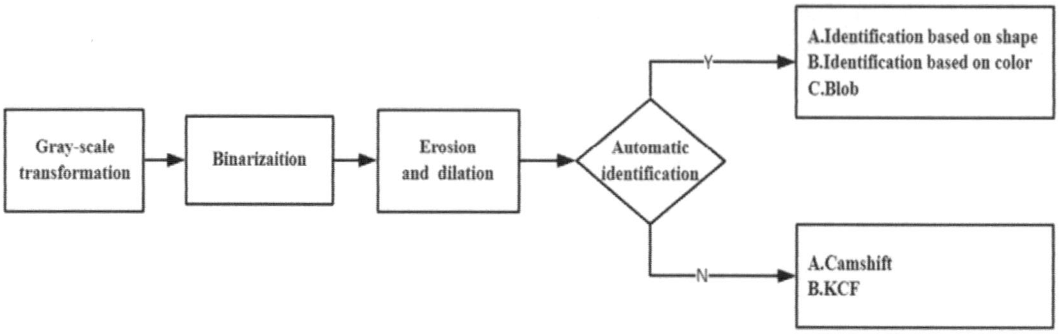

Figure 8. Flow chart of visual navigation

3. Hardware Platform and Software Platform

DJI Matrice 100 flight platform, as shown in Fig.9, is employed as a carrier which allows developers to customize and tailor the system by using the DJI SDK(Software Development Kit)and has flexible access to external devices for control and monitor applications. A NUC computer is equipped to operate GCS(Ground Control Station)and visual programs. An USB camera is added to capture videos. Fig.10 shows the structure of hardware platform[16].

In visual aspect, opencv 3.0 version is utilized to implement multiple universal algorithms on image processing and computer vision in Visual Studio 2015 environment on windows 7 system.

Figure 9. DJI Matrice 100

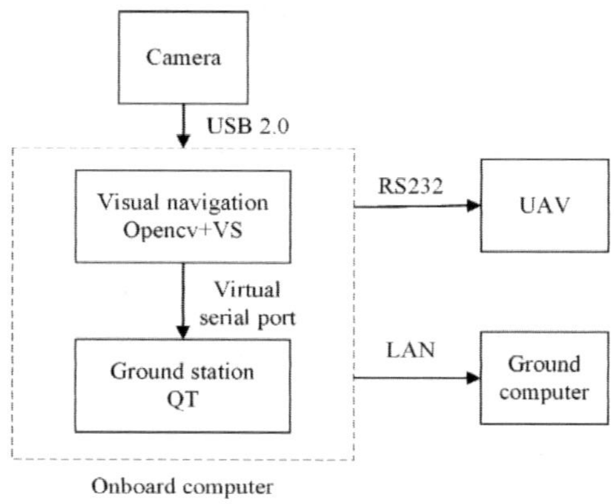

Figure 10. Structure of hardware platform

In flight control aspect, the programs of ground control station is operated in QT development environment on the on-board computer, communicating with the aircraft through a RS232 serial port. Virtual serial ports enable the communication between visual system and ground station, through which the ground station receives relative position information sent by vision program at a frequency of 10 hz, converted desired position information to speed information through PID(proportion-integral-derivative) control and then sent the speed to flight control system after softening speeds.

The ground computer connects with the on-board one via LAN(local area network).It remotely controls the programs of the on-board computer to monitor the health status of the aircraft and the progress of visual navigation in real-time on the ground.

4. Result and Discussion

4.1. The Comparison between Visual Algorithms

We designed six sets of flight scenes, and run these six sets of visual algorithms, namely the recognition based on color, blob analysis , combined recognition algorithm , CamShift and KCF, combined tracking algorithm. Fig.11.12.13.14 show that the comparison of pixel coordinate in image plane coordinate system among combined recognition algorithm combined tracking algorithm and the real pixel coordinate. From these line charts, three line almost coincide, showing that the error of these two combined algorithms are close to zero. Especially, some incorrect datum equivalent to zero means that the error between these two recognition algorithms or tracking algorithms is larger than the desired value of error, and through the filter, the output of visual system is zero.

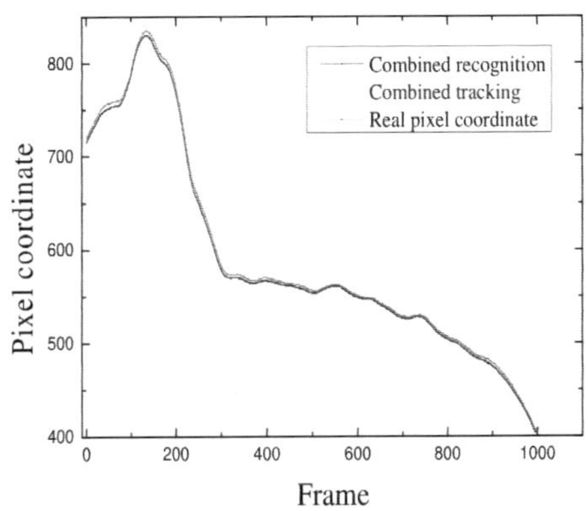

Figure 11. Flight under strong light

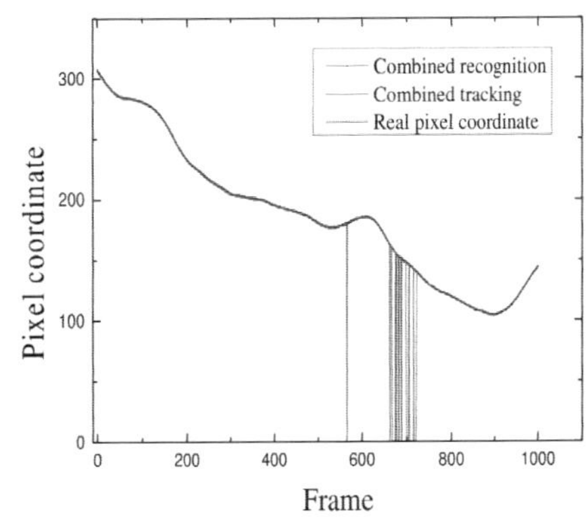

Figure 12. Flight under weak light

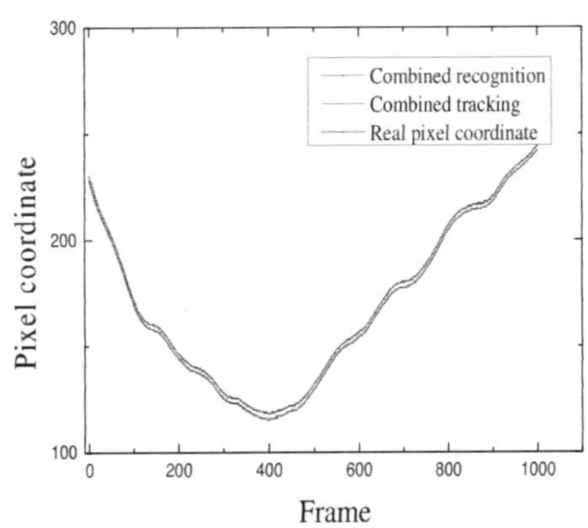

Figure 13. Flight under complex underground

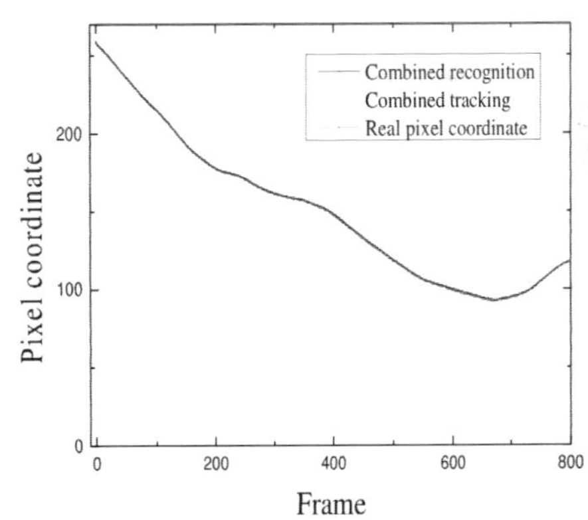

Figure 14. Flight under single underground

Table 1 shows the characteristics among these algorithms, and the computing time is what time it costs to calculate a picture (960*560 dpi) of a frame. In brief, combined recognition algorithm and combined tracking algorithm have higher accuracy and processing speed, which can meet the accuracy and real-time requirements of autonomous landing. Through selecting the color space and changing value of the thresholding visual system is able to accomplish recognition and tracking in most of the scenes. To sum up, this visual system meets the requirement of rapidity, accuracy and robustness, with simple configuration.

Table 1. Comparison between visual algorithms

Algorithm	Time	Characteristic
Image recognition based on color	2.33 ms	Simple, high processing speed and accuracy, especially on target in similar color.
Blob detect	14.56 ms	Robust, high processing speed and accuracy, but affected by complex background.
KCF	9.66ms	High processing speed and accuracy, but without self-adaptability aimed at varied target, without loss message.
CamShift	12.87 ms	High processing speed and accuracy, with self-adaptability aimed at target in brightly color, able to change target at any times.

4.2. Flight Experiment

The main process of point landing contains the following steps:

- Manipulate manually the UAV to the height of 10 meters. Make sure that the camera has been collected with the ground target and communicate normally with the ground station.
- Select ground targets manually or recognize them automatically, then visual procedure executes image pre-processing, target tracking or image recognition and the solution of relative position orderly at a frequency of 10 Hz. At the same time, the real distance will be sent to the ground station program at a frequency of 10Hz. The relative position received by the station will be converted into the desired speed with PID control. After softening, the forward desired speed and the lateral one will be sent to flight control system at the vertical speed of 0. These mentioned above are the first phase of visual navigation.
- Whether the relative position reaches the setting value will be determined by visual program loop. If so, the aircraft will hover for one second and begin to decline. Otherwise repeat the second step.
- The second phase of visual navigation will start. The ground station will send the forward and lateral desired speed to flight control system, and keep the vertical speed at 0.5m/s.
- When the height decline to 2 meters, the visual navigation will terminate. At the same time, the forward and lateral desired speed will reduce to 0, and keep the vertical speed at 0.5m/s.
- The experimental result shows that the visual system run well under different light or backgrounds. Due to the altitude error measured by GPS, and the simple position estimation, the error between the landing point and the center of target exist within 3 meters.

5. Conclusions

Image recognition and target tracking, the most essential technologies of visual navigation on UAVs were studied in this paper. The flight datum on various scenes demonstrated contrasts in rapidity, accuracy and robustness among six visual navigation algorithms. Thus, automatic and manual methods were respectively designed for identifying ground targets, with rapidity, accuracy and robustness. Onboard experiments suggested the feasibility of point landing technology based on visual navigation. in practice, the ground targets were effectively identified while the landing point had large data errors. Therefore, some improvement opinions were put forward as below for further work.

- GPS has relatively large errors when approaching the ground. Thus, ultrasonic or laser radar is more applicable for measuring the height above ground.
- The relative position estimation algorithm should be improved to accurately solve the true relative position between UAVs and the ground target.
- The cooperative target lacks the indication of forward direction towards the head of UAVs while this direction after landing need to be clarified in the actual landing scenario. Therefore, a ground target with strong directivity should be designed.

IOP Conf. Series: Materials Science and Engineering **719** (2020) 012077 doi:10.1088/1757-899X/719/1/012077

6. References

[1] Yang S, Scherer S A, Schauwecker K, et al. Autonomous Landing of MAVs on an Arbitrarily Textured Landing Site Using Onboard Monocular Vision[J]. Journal of Intelligent & Robotic Systems, 2014, 74(1-2):27-43.

[2] Maravall D, Lope J D, Fuentes J P. Vision-based anticipatory controller for the autonomous navigation of an UAV using artificial neural networks[J]. Neurocomputing, 2015, 151:101-107.

[3] Saripalli S, Montgomery J F, Sukhatme G S. Visually guided landing of an unmanned aerial vehicle[J]. IEEE Transactions on Robotics & Automation, 2003, 19(3):371-380.

[4] Kalal Z, Mikolajczyk K, Matas J. Forward-Backward Error: Automatic Detection of Tracking Failures[C]// International Conference on Pattern Recognition. IEEE, 2010:2756-2759.

[5] Kong W, Zhang D, Wang X, et al. Autonomous landing of an UAV with a ground-based actuated infrared stereo vision system[C]// Ieee/rsj International Conference on Intelligent Robots and Systems. IEEE, 2013:2963-2970.

[6] Bay H, Ess A, Tuytelaars T, et al. Speeded-Up Robust Features (SURF)[J]. Computer Vision & Image Understanding, 2008, 110(3):346-359.

[7] Wang L, Ju H. A Robust Blob Detection and Delineation Method[C]// International Workshop on Education Technology and Training & 2008 International Workshop on Geoscience and Remote Sensing. IEEE Computer Society, 2008:827-830.

[8] Henriques J F, Rui C, Martins P, et al. High-Speed Tracking with Kernelized Correlation Filters[J]. IEEE Transactions on Pattern Analysis & Machine Intelligence, 2015, 37(3):583-596.

[9] Weiss S, Siegwart R. Real-time metric state estimation for modular vision-inertial systems[C]// IEEE International Conference on Robotics and Automation. IEEE, 2011:4531-4537.

[10] Ranganatha S, Gowramma Y P. An integrated robust approach for fast face tracking in noisy real-world videos with visual constraints[C]// International Conference on Intelligent Computing and Control. IEEE, 2018:772-776.

[11] Yen S H, Wang C H, Chien J C. Accurate and robust ROI localization in a CamShift tracking application[J]. Multimedia Tools & Applications, 2015, 74(23):10291-10312.

[12] Wang C, Zhan S U. CamShift Target Tracking Based on Kalman and Surf[J]. Software Guide, 2018.

[13] Lehtola V, Huttunen H, Christophe F, et al. Evaluation of Visual Tracking Algorithms for Embedded Devices[M]// Image Analysis. 2017:88-97.

[14] Li L, Liu Y H, Jiang T, et al. Adaptive Trajectory Tracking of Nonholonomic Mobile Robots Using Vision-Based Position and Velocity Estimation[J]. IEEE Trans Cybern, 2018, 48(2):571-582.

[15] Chenavier F, Crowley J L. Position estimation for a mobile robot using vision and odometry[C]// IEEE International Conference on Robotics and Automation, 1992. Proceedings. IEEE, 2002:2588-2593 vol.3.

[16] Dey V, Pudi V, Chattopadhyay A, et al. Security Vulnerabilities of Unmanned Aerial Vehicles and Countermeasures: An Experimental Study[C]// International Conference on Vlsi Design and 2018, International Conference on Embedded Systems. IEEE Computer Society, 2018:398-403.

Public Opinion Prediction Model of Food Safety Events Network Based on BP Neural Network

Chen Zheng[1], Yinghua Song[1] and Yaping Ma[1]

[1]School of Safety Science and Emergency Management, Wuhan University of Technology, Wuhan Hubei 430070, China.
Email: 445396142@qq.com

Abstract. The food safety incident network public opinion has the characteristics of wide audience, complex and changeable, and bad influence. It is of great theoretical and practical significance to study the behavioral and influencing factors of microblog public opinion forwarding in this kind of event. This paper summarizes and enriches the index factors affecting the network's public opinion forwarding volume. Combined with BP neural network algorithm, this paper constructs a network public opinion forwarding behavior prediction model, and applies and verifies it by crawling Sina Weibo food safety event microblog data. The results show that the introduction of fan activity has a certain weight ratio, which has a greater impact on the forwarding of public opinion in food safety events network with identity authentication, hot search, hypertext and other indicators, and the prediction model combined with BP neural network has a better prediction effect.

1. Research Survey

The rapid development of new media information technology provides conditions for the expansion of public opinion dissemination of food safety incidents. The network public opinion of food safety emergencies has the characteristics of wide audience, fast propagation, long duration and great social response. If the guidance is not appropriate, it is easy to cause serious adverse effects on social. At the same time, as a representative of the new media social platform, microblog users and their interaction behaviors, such as forwarding and praising, play an important role in the dissemination and development of public opinion. In recent years, the research on influencing factors of public opinion communication behavior on Twitter and Sina Weibo platform has been widely concerned by scholars at home and abroad. Through the study of twitter data, Suh [1] found that whether the microblog contains links and topic tags has a direct impact on the forwarding rate of microblog, while the number of followers, the number of followers and the length of registration of microblog authors have an indirect impact on the forwarding rate of microblog. Rudat [2] classifies the influencing factors into three aspects: user guidance, Weibo theme and information content, and determines the different influences of various factors. Zhang [3] uses classification algorithm to verify that mentioning other people, identity authentication and other factors in Twitter has a great positive impact on Weibo forwarding. Guan [4] introduces new content factors such as inserting pictures and links. Deng [5] proposed and proved that social influence index and user active time have a high influence on microblog forwarding. Although relevant scholars have carried out a lot of research on the influencing factors of Internet public opinion users' forwarding behavior, and studied the promoting effect of many factors on the public opinion communication, all kinds of emergencies have heterogeneity characteristics, and the influence of various factors on different types of emergencies may be different, which needs further research.

BP neural network and its optimization algorithm are more suitable for dealing with public opinion events with strong outbreak and rapid diffusion because of its excellent approximation effect, faster calculation speed and strong nonlinear fitting ability. Therefore, this paper takes Sina Weibo as the research platform, through analyzing the research results of food safety incidents and public opinion prediction, establishes an index system of influencing factors of online public opinion forwarding for such emergencies as food safety, and combines with BP neural network to build a prediction model to realize the accurate prediction of online public opinion forwarding behavior of food safety.

2. Influence Index Selection

In this paper, we select the index system which has certain influence on microblog forwarding and is easy to obtain from the existing literature, and verify the factor index which affects the online public opinion forwarding volume of food safety events in the new model. User influence index: identity authentication enables users to have different identification advantages in information dissemination. According to the difficulty of authentication, they are represented by 0, 1, 2 and 3. The number of fans, number of concerns, and registration period are more intuitive to reflect the accumulation of users' fans and influence, which can be represented by their real value. High fan activity can maintain a high exposure rate of information in the microblog platform. This paper uses the weighted sum of the average number of forwarding, comments and likes of each microblog, and its calculation is shown in formula (1), where w1, w2, w3, are weight coefficients, which are represented by the calculated actual value; user activity is directly expressed by the accumulated real number of microblogs every month.

$$F = \frac{\sum(w_1 A + w_2 B + w_3 C)}{D} \tag{1}$$

F is fan activity, A is the number of forwarding, B is the number of comments, C is the number of likes.

Table 1. Impact indicators and their characterization

impact index		indicator state	characterization value
user influence index	identity authentication	No, orange v, gold v, blue v	0, 1, 2, 3
	number of fans	M	m
	attention number	M	m
	age of registration	M	m
	Weibo volume	M	m
	user activity	M	m
	original proportion	[0,1]	m
	fan activity	M	m
content influence index	emotional tendency	Negative, neutral (no comments), positive	-1, 0, 1
	visual content	Text only, pictures, video	0, 1, 2
	hot search or not	no, yes	0, 1
	super topic or not	no, yes	0, 1
	is there a web link	no, yes	0, 1
	@ others people or not	no, yes	0, 1
	time interval	2: 00-8: 00, 8: 00-14: 00, 14: 00-20: 00, 20: 00-2: 00	0, 1, 2, 3
	Weibo length	M	m

In terms of microblog content, the selected indicators include whether there are pictures, whether there is visual content in video, whether there is hot search, whether there is super topic, whether there is web link, whether there is @ others. These additional information increases the diversity of microblog content, increases the possibility and way of information acquisition for other users, and improves the amount of information contained in the content. Table 1 shows the indicators of each influencing factor and their specific characterization values, and the indicator characterization values in the table are the real values shown in M.

3. BP Neural Network

BP neural network is a kind of neural network which can learn and store a large number of input-output mode mapping relationships without establishing mathematical equations in advance. It has a strong nonlinear fitting ability. Its structure is shown in Figure 1, including input layer, hidden layer and output layer. The data is imported into the network at the input layer, and then transmitted forward. Each hidden layer will generate connection weights to build the model by itself. The fitting results will be compared with the actual values, and the backward feedback error will be fed back. The hidden layer will continue to adjust the weight self-learning combined with the feedback error, and this process will continue to cycle until the preset fitting effect is achieved. When the fitting effect is achieved, the network will have the minimum feedback error, and the connection weight in the hidden layer can be used to judge the important proportion of each input layer neuron in the network.

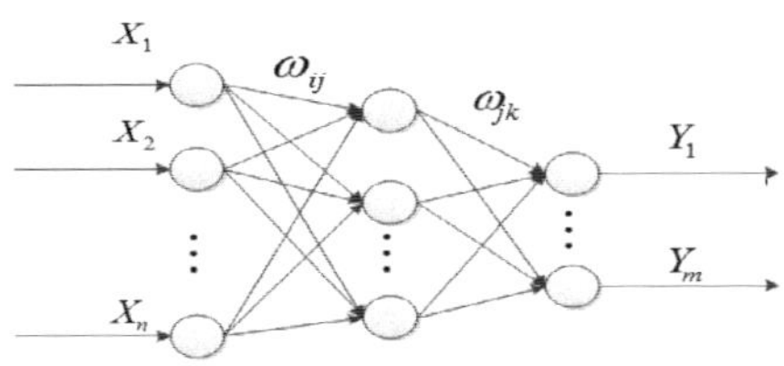

Figure 1. BP neural network structure

The number of neurons in the input layer and the output layer is determined according to the index number. The hidden layer is an important link between the preceding and the following in the BP neural network, but there is no definite criterion for the number of neurons in the hidden layer, so the number can only be adjusted according to the empirical formula and the actual test. The empirical formula is as follows:

$$H = \sqrt{in + out} + a \tag{2}$$

"in" is the number of input layer cells, "out" is the number of output layer cells, "h" is the number of hidden layer cells, and "a" is an integer between 1 and 10.

According to formula (2), the hidden layer is set from 5, and each group of experiments is carried out 20 times. The results show that when the hidden layer neuron is 6, there is the minimum training error mean. Therefore, this paper constructs a three-layer BP neural network of 16-6-1.

4. Experiment and Result Analysis.

4.1. Data Sources and Processing

According to the hot search events and event database provided by Sina Weibo, this paper crawls and collates the Weibo data of 12 typical food safety emergencies in 2019, including "Tongrentang expired honey", "Muji good carcinogenic food". 600 of the 1500 processed valid data are randomly selected as training samples and verification samples, and the remaining data are selected as prediction samples. At the same time, the micro blog forwarding volume is divided into different popularity criteria. In this paper, the forwarding volume less than 100 is set as low popularity, 100-1000 as medium popularity, and 1000 above as high popularity, with corresponding indicator values of 0, 1, 2.

4.2. Analysis of the Impact of Indicators

This paper uses MATLAB 2016 as a simulation tool. Through the learning and training of sample data, the weights of each layer in the network are obtained. The normalized result of the sum of the absolute values of the connection weights from the input index to all hidden layer nodes is taken as the influence weight of the index on the micro blog forwarding volume. The calculation formula is as follows:

$$\omega_i = \frac{\sum_{j=1}^{n} |\omega_{ij}|}{\sum_{i=1}^{m}\sum_{j=1}^{n} |\omega_{ij}|} \tag{3}$$

$$i = 1,2\cdots,m \quad j = 1,2,\cdots,n \quad ,$$

m is the number of input indicators; n is the number of nodes in the hidden layer.

Table 2. Weight of each indicator

index	connection weight						influence weight
identity	-0.301	-0.202	0.736	-0.838	0.405	-0.277	**0.071**
number of fans	-0.326	0.665	0.56	-0.118	-1	-0.348	**0.0776**
attention number	0.026	-0.748	-0.105	0.779	0.057	-0.133	0.0476
age of registration	0.5	0.301	0.426	0.54	-0.001	-0.449	0.057
Weibo volume	-0.11	0.308	-0.153	0.301	0.376	-0.283	0.0394
user activity	-0.414	-0.52	-0.965	-0.229	0.862	0.408	**0.0874**
original proportion	-0.165	-0.23	-0.333	0.804	0.004	0.043	0.0406
fan activity	0.39	-0.535	-1	0.112	-0.995	-0.154	**0.082**
emotional tendency	0.112	-0.511	0.848	0.063	0.209	-0.389	0.0549
visual content	0.634	-0.486	-0.303	0.478	-0.2	-0.091	0.0564
hot search or not	0.175	0.859	-0.532	-0.579	-0.284	-0.32	**0.0707**
super topic or not	0.579	-0.647	-1	-0.117	-0.659	-0.814	**0.0982**
is there a web link	-0.049	-0.419	-0.679	0.038	0.69	-0.162	0.0524
@ others people or not	-0.518	-0.229	-0.723	-0.45	-0.228	0.15	0.0591
time interval	-0.864	0.353	-0.153	0.25	-0.133	-0.285	0.0524
Weibo length	0.092	0.634	-0.783	0.053	-0.333	-0.179	0.0533

Table 2 shows the connection weight and influence weight of each index. The average influence weight of each index is 0.0625, among which, the six indexes such as user's identity authentication, number of fans, monthly average activity, fan activity, whether microblog has hot search, whether microblog has hypertext and so on all exceed the average, indicating that the high activity of users and their fans has a greater role in promoting the dissemination of public opinion information. Whether there is a super topic in the content of microblog has the highest weight of 0.0982, while the super topic students have the functions of settling related high-quality topics and gathering fans who are interested in such topics, etc., and have the conditions to create high popularity public opinion, so once such microblog appears, it needs to focus on monitoring. In addition, this paper introduces a fan activity weight of 0.082, which shows that high-quality fans will greatly promote the spread of public opinion, which is also an important factor in public opinion monitoring. However, the weight of the indicators such as blog volume, number of concerns and proportion of original works is low, which shows that the impact on the development of microblog public opinion is limited, which is consistent with the results of some literatures. Therefore, we can consider not as an impact factor as appropriate to simplify the model structure.

4.3. Prediction Result Analysis

In order to compare the influence of fans' activity on prediction accuracy, the index system including and excluding this index is tested. 30 sample data are randomly selected from the processed prediction data, and the popularity of its forwarding volume is simulated and predicted, and compared with the actual popularity of the forwarding volume. After repeated experiments, randomly select three test results, and make interval statistics for errors, the results are shown in Table 3. It can be found that in the popularity prediction experiment which is divided into three levels, the model with fan activity in the index system has better prediction accuracy.

Table 3. Statistical for the prediction of the error

error group model	Prediction model without fan activity				Prediction model including fans activity			
	1	2	3	probability	1	2	3	probability
[0, 0.1]	15	16	15	51.1%	16	20	18	60.0%
[0, 0.2]	21	22	22	72.2%	23	24	25	80.0%
(0.2, 1]	9	8	8	27.8%	7	6	5	20.0%

It can be seen from the above statistical results that, compared with the original index system, the index system with fan activity improves the accuracy of the prediction model to a certain extent and has a better matching effect.

5. Conclusion

This paper constructs a prediction model of online public opinion of food safety events, verifies the prediction index system of 16 indexes including identity authentication, user activity and fan activity, and confirms the high influence of identity authentication, number of fans, hot search and super talk and other factors on micro blog forwarding in the existing system. At the same time, it verifies the importance of the proposed fan activity in such micro blog forwarding, enriches and rationalizes the existing prediction index system to a certain extent, and proves that the appropriate index can improve the accuracy of BP neural network public opinion prediction.

6. Reference

[1] Suh B, Hong L, Pirolli P, et al. Want to be retweeted? Large scale analytics on factors impacting retweet in twitter network[C]//2010 IEEE 2nd International Conference on Social Computing, Minneapolis, USA:IEEE, 2010
[2] Rudat A, Buder J, Hesse F W. Audience design in Twittter: Retweting behavior between informational value and fowllower' interests[J].Computers in Human Behavior, 2014, 35:132-139.
[3] Zhang Yang, Lu Rong, Yang Qing. A study on the prediction of forwarding behavior in Weibo customers [J]. Journal of Chinese Information, 2012, 26 (4): 109 -114.
[4] GUAN Wanqiu, GAO Haoyu, YANG Mingmin. Analyzing user behavior of the micro-blogging website Sina Weibo during hot social events[J].Physica A, 2014, 395: 340-3.
[5] Deng Qing, Ma Yefeng, Liu Yi. Prediction of Weibo forwarding volume based on BP neural network [J]. Journal of Tsinghua University (Natural Science Edition), 2015, 55 (12): 1342-1347.

[6] Zhang Yuliang. Research on the Construction of risk Evaluation Index system of Public opinion Information flow in Emergency Network based on UML method [J]. Books and Information, 2016, (3): 1: 10, 18.

IOP Conf. Series: Materials Science and Engineering **719** (2020) 012079 doi:10.1088/1757-899X/719/1/012079

Research on Application of Emergency Text Classification Based on Combination Model

Long Lyu[1], Yinghua Song[1] and Dan Liu[1]

[1]School of Safety Science and Emergency Management, Wuhan University of Technology, Wuhan Hubei 430070, China.
Email:18381305006@163.com

Abstract. Based on the powerful learning ability of the deep learning model, the text uses convolutional memory network (CLSTM) to process the text information of the emergency. Firstly, the convolutional neural network is used to learn the local spatial feature information of the text, and then the text is extracted by the long and short memory neural network. The time feature information is used to scale the feature information by using the Softmax layer to finally obtain the event text category. Through experimental comparison, the precision, recall and comprehensive values of CLSTM model reached 0.879, 0.877 and 0.848, respectively, which were significantly higher than MLP and CNN models.

1. Introduction

In the past three decades, the rapid development of the Internet has made the network an important carrier of information dissemination. Every day, a large amount of network information is generated around the world, and it is transmitted on the Internet in the form of sound, text and images. Faced with such a huge amount of information, it is very important for users to get the information they care about in a short time. At the same time, the network platform will generate a large amount of news information, including the information content of emergencies such as natural disasters, social security incidents, public health incidents and accident disasters[1]. Compared with ordinary events, emergencies have obvious characteristics: sudden Sex, destructive, complex, limited information, uncertainty and variability, persistence and publicity. Because of its sudden occurrence, it will have a greater impact in society in a short period of time. At the same time, the information held by the public is very limited. It may develop in an uncontrollable direction in a short period of time and cause negative emotions to the society. Therefore, the relevant departments need to monitor the situation in real time. The emergence of incidents, as well as the media's attention to emergencies, to avoid serious impacts on society and the public.

Text classification can be roughly divided into three periods from the technical development process: traditional text classification, shallow learning text classification and deep learning text classification. The traditional text classification technology is based on the mathematical statistics knowledge. The machine learning method is used to calculate the probability that the text belongs to various categories. The most probable one is the final classification. The typical method is the naive Bayesian method[2], decision tree. Method[3] and KNN method[4]. The shallow learning text classification technology is also calculated by using mathematical statistics. The typical methods are support vector machine (SVM)[5], BP neural network[6], integrated learning method[7]. As a hot field, the deep learning method mainly constructs a neural network model by simulating the information transfer between neurons in organic organisms, extracts text information, and finally realizes classification function. The commonly used model has deep neural network (DNN)[8], Convolutional

Neural Network (CNN)[9], Recurrent Neural Network (RNN)[10], and Deep Confidence Network (DBN)[11].

In this paper, the Convolutional Memory Network (CLSTM) depth combination model is used. The model expresses the incident text information in word vector mode. First, the word vector is input to the convolutional neural network (CNN), and the convolution kernel is used to extract the local spatial features of the text. Then, the feature information is input to the long-term memory neural network (LSTM) to learn the text time feature information, the second is connected to the Softmax layer scaling feature information, and finally the full-connection layer is connected to the burst event text category.

2. Introduction to The Classification Process of Emergency Texts

The classification of emergency texts consists of two main steps: sample collection and processing and combined model training. The process is shown in Figure 1:

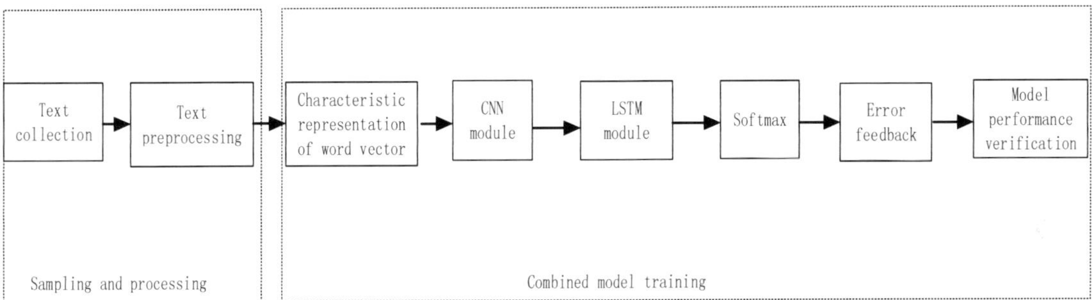

Figure 1. Emergency file text classification flow chart

2.1.Emergency Text Collection

Incidents fall into four categories: public health events, social security events, accident disaster events, and natural disaster events. Through reptile technology, this article searches for 23 categories of secondary category events in Baidu News in the form of keywords: explosions, virus infections, shipwrecks, earthquakes, vicious killings, child trafficking, aircraft crashes, tsunami disasters, traffic accidents, Financial fraud, terrorist attacks, mine accidents, storms, mass incidents, forest fires, sandstone disasters, food poisoning, urban catastrophic fires, hazardous chemicals accidents, plutonium incidents, juvenile crimes, vaccine incidents and rainwater disasters, each One class crawled 150 articles for a total of 3,450 articles. Through manual proofreading, the irrelevant text is eliminated, and 667 valid texts are finally saved, and saved in the txt file in the form of title, time, and body.

2.2.Text Preprocessing

At this point, the data text belongs to the original sample, and the text needs to be pre-processed. The main operations include: filtering non-Chinese information, eliminating the information that is insignificant to the missing text; Chinese word segmentation, so that the text represents the text feature in the form of words; Eliminate the useless words such as connectives, relational words, and modal words in the words; dataset segmentation, randomly divide the text according to the 'training set: test set=4:1' ratio; text length processing, keep the text information fixed length, redundant Intercepting, insufficient zero padding.

2.3.Text Feature Representation

There are three commonly used feature representation methods: one-hot method, TF-IDF method and word vector method.

The One-hot method only considers whether words appear, ignores the weight of words in the text, and when the amount of text data is large, there will be a phenomenon of dimensional explosion and feature stagnation, which is a big performance for computer memory and model. The challenge.

Although the TF-IDF method considers the importance of words in the text category, it ignores the semantic relationship between words. When the amount of text data is large, there will also be a phenomenon of dimensional explosion and feature stagnation.

The word vector method maps the features of each word into a vector of several dimensions. All words can be represented by vectors of fixed length. At the same time, the word vector method also takes into account the semantic relationship before and after the words, meaning or function similar. The vector space distance of words is also small, in other words, the similarity is high. The word vector mainly implements feature extraction functions by two kinds of neural networks: CBOW (continuous bag-of-word) and Skip-gram (continuous skip-gram gram). The CBOW model predicts intermediate words based on several words before and after the sentence; the Skip-gram model predicts several words in the future based on one word. The objective function of the two models is:

$$\text{loss} = \sum_{w \in c} logp(w|Context(w)) \tag{1}$$

In this paper, the gensim module based on the Skip-gram model is used to train the word vector. If the word vector dimension is too small, the word vector feature cannot be fully represented. If the dimension is too large, the resource is wasted. Therefore, it is necessary to select the appropriate dimension.

2.4. CNN Module

The CNN network is a feedforward neural network whose artificial neurons can respond to a surrounding area of a part of the coverage, which includes a convolutional layer and a pooling layer. The core advantage of this network lies in its shared convolution kernel. The convolution kernel can extract information of several words at a time. The convolution kernel can be shared in one operation, reducing model parameters. The principle is:

The corresponding word vector of the word x_i in the convolutional layer after k-dimensional embadding is $£^k$, $x_i \in £^h$, a sentence $X_{1:n}$ can be formalized as:

$$X_{1:n} = x_1 \oplus x_2 \oplus \cdots x_n \tag{2}$$

A window with h words is represented as:

$$X_{i:i+h-1} \in £^{hk} \tag{3}$$

A filter matrix of size h*k is converted into a one-dimensional vector of length h*k, expressed as:

$$\mathrm{W} \in £^{hk} \tag{4}$$

$$c_j = f(W \cdot X_{i:i+h-1} + b) \tag{5}$$

The eigenvalues extracted by a filter are:

$$\hat{c} = \max(c) \tag{6}$$

Filters can get the feature vector:

$$\mathrm{z} = [\widehat{c_1}, \widehat{c_2}, \widehat{c_3} \ldots \widehat{c_m}] \tag{7}$$

2.5. LSTM Module

The LSTM unit structure has an input gate, an output gate, and a forget gate. W_f, W_i, and W_c, and W_o are neural network weight parameters of the forgetting gate, the input gate, and the output gate, respectively; b_f , b_i and b_c, b_o represent the neural network offsets of the forgetting gate, the input gate and the output gate, respectively; $\widetilde{C_t}$ and h_t represent the state values of the neural unit at time t. Then there are the following calculations:

A, forget gate:

$$f_t = \sigma(W_f \cdot [h_{t-1}, \mathrm{P(t:)}] + b_f) \tag{8}$$

B, input gate:

$$i_t = \sigma(W_i \cdot [h_{t-1}, \mathrm{P(t:)}] + b_i) \tag{9}$$

$$\widetilde{C_t} = \tanh(W_c \cdot [h_{t-1}, \mathrm{P(t:)}] + b_c) \tag{10}$$

C, output gate:

$$o_t = \sigma(W_o \cdot [h_{t-1}, \mathrm{P(t:)}] + b_o) \tag{11}$$

$$h_t = o_t * \tanh(f_t * C_{t-1} + i_t * \widetilde{C_t}) \tag{12}$$

2.6.Model Evaluation
In the classification problem, the common prediction results are shown in Table 1:

Table 1. Classification result possibility

		actual results	
		positive	Negative
forecast result	positive	True positive(TP)	False positive(FP)
	Negative	False Negative(FN)	True Negative(TN)

The model is evaluated using precision, recall, and composite value (F1) as follows:

$$\text{precision} = \frac{TP}{TP+FP}, recall = \frac{TP}{TP+FN}, \frac{2}{\text{F1}} = \frac{1}{accuracy} + \frac{1}{recall} \tag{13}$$

3. Analysis of The Results

3.1.Experimental Tools
Dataset: 667 news reports of emergencies (including 23 categories)
System: Windows 10 64-bit operating system
Processor: AMD Ryzen7 3700U
Programming tool: anaconda3
Toolkit: Jieba module, gensim module, pandas module, sklearn module, numpy module, keras module

3.2.Model Assistant
The parameters of the emergency text classification model based on deep learning have great influence on the performance of the model. In order to optimize the performance of the model, based on the same sample set, the accuracy is obtained by repeating the simulation three times and changing the parameters of a single model each time. The average value of the recall rate and the comprehensive value is the best parameter when the comprehensive value is the largest. In order to analyze the performance of the CLSTM combined model, the MLP and CNN models are trained and compared with the same method. The optimal parameters of the model are shown in Table 2-4.

Table 2. MLP model parameters and optimal values

Parameter	Best Value	Parameter	Best Value	Parameter	Best Value
Number of hidden layer units	640	Hidden layer activation function	Relu	Dropout	0.6

Table 3. CNN model parameters and optimal values

Parameter	Best Value	Parameter	Best Value	Parameter	Best Value
Sentence length	400	Number of word dimensions	260	Number of kernel	768
Convolution kernel activation function	Tanh	Dropout	0.7	Number of hidden layer units	128
Hidden layer activation function	Elu				

Table 4. CLSTM model parameters and optimal values

Parameter	Best Value	Parameter	Best Value	Parameter	Best Value
Sentence length	400	Number of word dimensions	260	Number of the first kernel	768
First layer convolution kernel activation function	Tanh	Number of the next kernel	1024	Next layer convolution kernel activation function	Softplus
LSTM unit number	256	LSTM Dropout	0.8	Full connection layer dropout	0.7
Number of fully connected layer units	512	Full connection layer activation function	Rlu		

3.3.Model Comparison Analysis

Compare the models according to the optimal state values of all the tunable parameters in each model obtained in the previous step. All model uniform parameters have batch training size (batch_size) and sample training times (epochs), set to: batch_size =16, epochs=20. The results of each model are shown in the following table and figure:

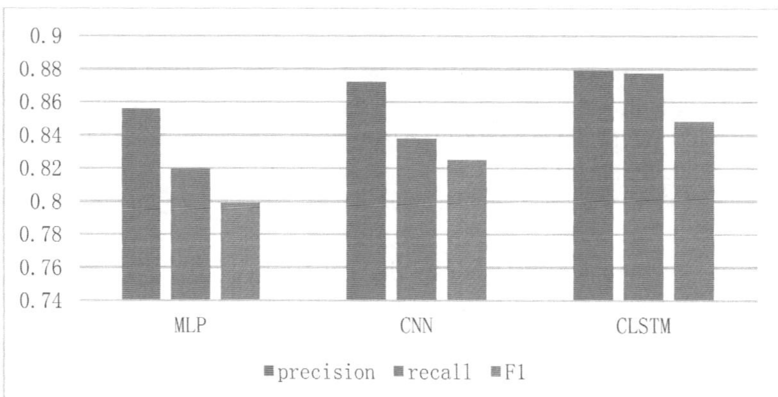

Figure 2. Model performance comparison

Through experimental comparison and analysis, we can know:

(1) Comparison of model complexity: MLP<CNN<CLSTM; accuracy ratio comparison: MLP<CNN<CLSTM; recall ratio comparison: MLP<CNN<CLSTM; comprehensive value comparison: MLP<CNN<CLSTM. It can be seen that the more complex the model structure, the more layers there are, the more the number of neural units, the stronger the learning ability.

(2) The CLSTM model accuracy rate is 0.879, which is significantly higher than MLP (0.856) and

CNN (0.872); its recall rate is 0.877, which is 5.7 percentage points higher than MLP (0.82) and CNN (0.838), respectively, and 3.9 percentage points; The comprehensive value takes into account the balance between accuracy and recall. The overall performance of the model can be considered. The integrated value of the CLSTM model (0.848) is significantly higher than MLP (0.799) and CNN (0.825).The data shows that the model has higher accuracy than the other two models.

4.Conclusion

In this paper, the CRNN model is used to learn text information. Firstly, the CNNN local information extraction capability is used to obtain the text space information, and then the LSTM model is used to obtain the sequence information to extract the text time feature information. Through experiments, the CLSTM model has a better classification effect in the emergency news classification.

5.References

[1] Yinghua Song. Introduction to emergency management of emergencies [M]. Beijing: China Economic Publishing House, 2009: 15-17, 21-25, 32-33.
[2] Zhibin Xiong, Jianfeng Zhu, Chengguo Yin et al. Automatic classification of tourism emergencies based on Weka[J].Software Guide, 2016, 15(04): 154-156.
[3] Hothorn T, Lausen B, Benner A, et al. Bagging survival trees, Statistics in Medicine, 2004, 23: 77-91.
[4] Xu Wei, Lingyu Liu, Mingming Wang. Research on Incident Detection and Trend Research Based on Cross-media Analysis[J]. Systems Engineering - Theory & Practice, 2015, 35(10): 2550-2556.
[5] Zhang D, Xu H, Su Z, et al. Chinese comments sentiment classification based on word2vec and SVM perf [J]. Expert Systems with Applications, 2015, 42(4): 1857-1863.
[6] Xinsheng Liu, Li Wei. Design and Implementation of a Text Classification System for Tourism Emergencies Based on BP Neural Network[J]. Computer and Modernization, 2011(07): 192-194+198.
[7] Hothorn T, Lausen B. Double 2bagging: Combining calssifiers by boot strap aggregation, Pattern Recognition, 2003, 36: 1303-1309.
[8] Rao A, Spasojevic N. Actionable and political text classification using word embedding and lstm [J]. Compute Science, arXiv: 1607.02501, 2016.
[9] Kim Y. Convolutional neural networks for sentence classification [J]. Eprint Arxiv, 2014, 1181.
[10] Socher R, Lin C.Y, Ng A.Y, et al. Parsing natural scenes and natural language with recursive neural networks [C]. International Conference on International Conference on Machine Learning. Omnipress, 2011: 129-136.
[11] Shubao LHU, Mingyue Wang, Pei Xiang et al. An information text classification algorithm for deep learning[J]. Journal of Harbin University of Science and Technology, 2017, 22(02): 105-111.

Dynamic Expression Recognition Based on Hybrid Features and Optimized Extreme Learning Machine Model

Xiujuan Sui, Lei Xue and Dunyu Li

Department of Communication and Information System, School of Communication and Information Engineering, Shanghai University, 99 Shangda Road Baoshan District, Shanghai, 201900, China.
Email: suixiujuanas@163.com

Abstract. Facial expression recognition is an academic subject, and it can be applied not only in the field of computer vision but also in psychology and sociology. This paper proposed a robust dynamic facial expression recognition method which using hybrid features and the PSO-ELM (Particle Swarm Optimization-Extreme Learning Machine) model. At first, six PSO-ELM models were trained on the training set of six basic expressions, respectively. Then by extracting the hybrid features which fusing geometric features and texture features, the dynamic expressions were recognized by PSO-ELM models. These experiments were performed on the Cohn-Kanade+ face database. The results show that this method has a better recognition effect on dynamic facial expression recognition.

1. Introduction

In recent years, with the improvements of computer computing speed and performance, the relationship between humans and computers has become more closely, and intelligence has penetrated into every corner of life. With the development of artificial intelligence, facial expression recognition is receiving more and more attention. Facial expression recognition involves many researching fields such as pattern recognition, image processing, computer vision, cognitive psychology, etc. Facial expressions can present a lot of emotional and psychological information, and are the main carrier of human emotions. American psychologist Mehrabian [1] believes that emotions are composed of 7% language, 38% voice and 55% facial expression.

Expression recognition generally consists of three steps [2]: face detection, facial expression feature extraction and expression classification. And the facial expression feature extraction is the most important part of the entire recognition process.

Face detection technology, mainly detecting whether the images input include faces, has matured in recent years. Face detection methods are roughly divided into two categories: statistics-based methods and structure-based methods. Specifically, these methods include binary wavelet transforms, template matching, local features matching, etc.

Expression feature extraction is to extract useful facial expression information in face images. At present, the main facial expression feature extraction methods can be roughly divided into two categories: geometric features and texture features. The method about geometric features is to extract the face contour points and the local feature points which mainly around the eyes, eyebrow, nose and mouth. Then using the geometric relationship of feature points to describe facial expression features. The main geometric feature extraction algorithms are based on ASM (the Active Shape Model) [3] or AAM (the Active Appearance Model) [4]. The Local Binary Features (LBF) method was proposed by Shaoqing Ren, Jian Sun et al. [5] in 2014. It uses the cascade and shape regression to locate and track the feature points of human faces. Texture features can represent the subtle changes in facial parts such as eyes, eyebrows, corners of mouth, etc. Common algorithms of texture features include Gabor wavelet transform [6], local binary pattern (LBP)[7] and gray level co-occurrence matrix (GLCM). Overall, the geometric features can reflect the dynamic changes of facial expressions, while the texture features focus on representing local detail changes.

The current methods of expression classification mainly use machine learning methods, such as support vector machine (SVM) [8], dynamic Bayesian network (DBN) [9], artificial neural network [10], BP neural network [11] and so on.

In this paper, we performed experiments on the Cohn-Kanade+ facial expression database. Fusing geometric features and texture features, a dynamic expression recognition method based on PSO-ELM is proposed. The geometric features are the coordinate difference of facial feature points in the normalized image sequence of the same group. The texture features are the difference of the local texture information corresponding to the feature points. Given six different expression sequences, six PSO-ELM models are obtained through training. By comparing the similarity between the predicted sequence and the real sequence, the expression category of the given sequence can be discriminated. The complete facial expression recognition process is showed in figure 1.

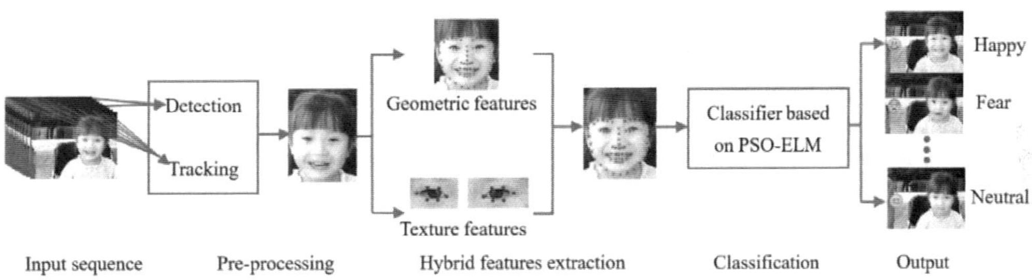

Figure 1. Expression recognition flowchart.

2. Feature Extraction

In this paper, LBF method is used to track and locate feature points of facial expression images in image sequences. Then between the normalized facial expression image and the neutral image, the geometric feature is obtained by the coordinate difference of the feature points, and the texture feature is obtained by the feature difference of the texture region corresponding the feature points.

2.1. Geometric Feature Extraction

As mentioned earlier, expression recognition generally consists of three steps, and the first is face detection. Face detection is a very active direction. Skin-based face detection [12] and Adaboost-based face detection [13] have achieved good results. In recent years, some methods based on deep learning also perform well, especially in complex scenes. These methods have high detection speed and accuracy. In this paper, we use the deep learning algorithm MTCNN [14] to detect human faces, then extract and track facial feature points by LBF method in the detected face frame and entire image sequence. The feature point positioning results in the sequence are shown in Figure 2.

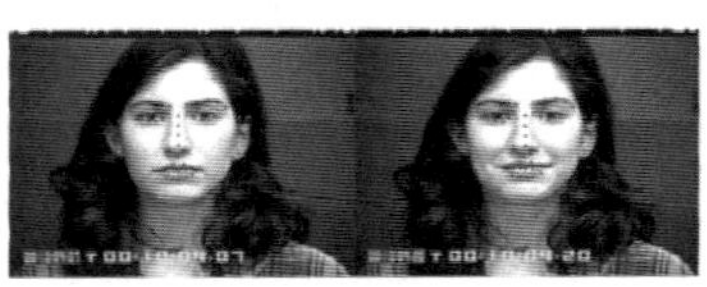

Figure 2. Geometric feature points positioning results in expression sequence.

Figure 3. Texture features area corresponding to geometric feature points.

After finding the position of the feature points, the baseline connecting the centres of the two eyes is converted to 0° orientation, and its length is standardized to 1 unit. Then all the faces are normalized by conversion and standard. The x and y coordinates of 68 feature points are calculated as a 68×2-dimensional feature vector.

2.2. Texture Feature Extraction

The texture feature is the facial detail feature of the local area corresponding to the feature point. Here, the texture difference between the neutral expression image and six other expression images is taken as texture features. The calculation of texture features is greatly affected by environmental factors, like uneven or too strong illumination. Gray level co-occurrence matrix is defined as the joint probability distribution of pixel pairs. It is a symmetric matrix. It not only reflects the comprehensive information of image grayscale in directions, adjacent intervals, and amplitude, but also reflects the position distribution characteristics between the same gray level pixels. It is not sensitive to light. The texture analysis method based on GLCM studies the spatial dependence of gray level in image texture with low computational complexity. Due to the robustness of the normalized cross-correlation coefficients, the gradient value difference between the neighborhood of the same feature point corresponding to the neutral image and the expression image can be defined as correlation:

$$\text{Corr} = {\left[\sum_{f_a}\sum_{f_b}\big((a,b)P(\text{a},b)\big) - \mu_x\mu_y\right]}\Big/{\sigma_x\sigma_y} \tag{1}$$

where μ_x, μ_y are variances, σ_x, σ_y are standard deviations. The size of the correlation value reflects the local gray correlation in the image. When the values of matrix elements are evenly equal, the correlation value is large. Corresponding to the 68 feature points extracted by the geometric features, 68 texture features can be obtained for each image. The 68x2-dimensional geometric features and the 68-dimensional texture features are fusing to obtain a total of 204-dimensional hybrid features. Each image can be represented by the 204-dimensional features. Figure 3 shows the texture feature area corresponding to geometric feature point.

2.3. PCA Dimensionality Reduction

Hybrid features have a total of 204 dimensions. The higher the dimension, the more the feature quantity. With large amount of calculation, it is easy to fall into "curse of dimensionality". In addition, as the dimension increases, the sparseness of the data becomes higher and higher. It is more difficult to explore the same data set in a high-dimensional vector space than to explore in the same sparse data set. So that we need educe the redundant information in the original data by PCA dimension reduction, and retain the most representative feature vector. The feature vector after dimension reduction can better represent the facial expression image.

3. Expression Classification

In this section we use a PSO-ELM (Particle Swarm Optimization-Extreme Learning Machine) model to perform the final expression classification.

3.1. PSO-ELM model

The Extreme Learning Machine (ELM) [15] is an algorithm proposed by Guangbin Huang for solving a single-layer neural network. The algorithm will randomly generate the deviation b_j of the hidden neuron and the linking weight ω_j between the input layer and the hidden layer. And the number of neurons in the hidden layer needs to be set. The output of ELM is:

$$\text{Output} = \sum_{j=1}^{L}\beta_j g\big(w_j.x_i + b_j\big) = O_j \tag{2}$$

where, β_j is the output weight of the jth hidden layer neuron; g ( ) is the output function of the jth implicit neuron; L is the number of implicit nodes. Learn the extreme learning machine by minimizing $\sum_j^N ||O_j - y_j||$. The traditional ELM method has fast learning speed and good generalization performance, but the input weights and deviation are set randomly. This may cause the parameter allocation of the hidden nodes to be uneven, which leads to poor robustness on classification problems.

To solve the problem of uneven allocation of hidden node parameters in the input layer matrix and the implicit layer matrix, we use Particle Swarm Optimization algorithm to optimize them. Then we can obtain the optimal network structure. The learning process of PSO-ELM model is as follows:

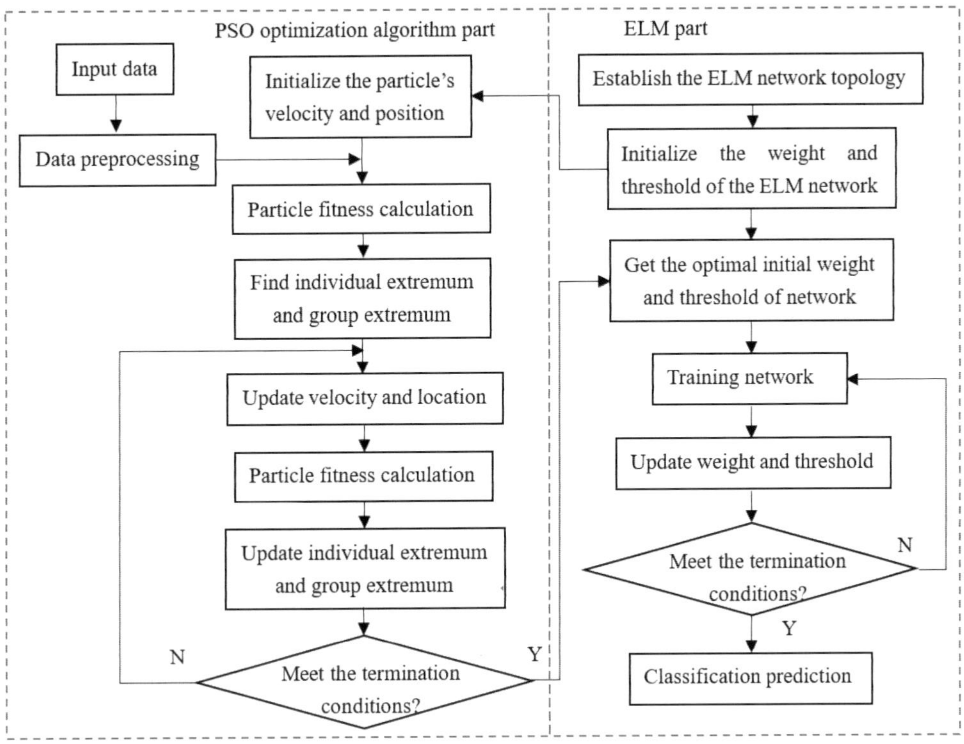

Figure 4. PSO-ELM model flowchart.

3.2. Dynamic Expression Recognition

Facial expression is a dynamic process, all expressions are from the first calm state to the most expressive state and then returning to calm state. This process sometimes is fast, sometimes slow. The expression at each moment is related to the facial muscle movement at the adjacent moment. Six PSO-ELM models correspond to six basic facial expressions. Given a set of facial expression sequences, N frames of images are selected and normalized to make up a Sequence I. Sequence I includes vectors $V_1, V_2, \dots, V_N$, where V_i, $i = 1,2, \dots, N$ and V_i is a feature vector of each frame image.

4. Experimental Results

These experiments were performed on the Cohn-Kanade+ (CK+) database [16], which is one of famous expression databases that are widely used to verify facial expression algorithms. The database consists of 593 image sequences and 123 human objects. Each sequence contains a series of dynamic facial expression images. In this database, we randomly select three-fourths of the expression sequence samples as the training set, and the remaining one-fourth as the test set. Using the cross-validation to evaluate the proposed algorithm. Figure 5 shows some samples of CK+ database.

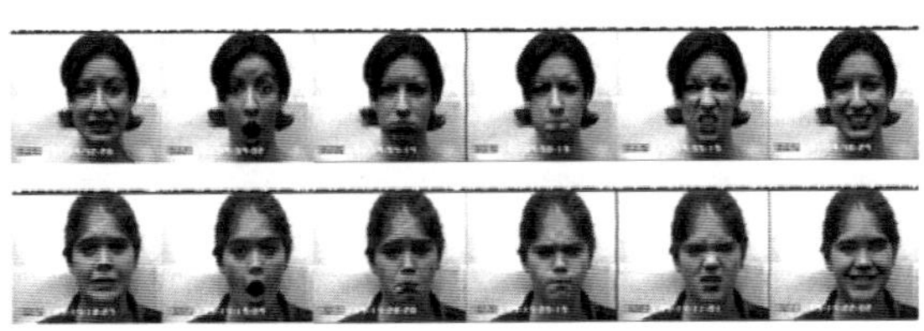

Figure 5. Samples of CK+ database.

After numerous experiments we have found that when the number of hidden nodes is between 50 and 300, the average recognition rate is constantly increasing; from 300 to 400, the average recognition rate shows a downward trend; between 400 and 500, the average recognition rate rises and tends to be stabilized, but the recognition accuracy did not reach the highest. If we continue to increase the number of nodes in the hidden layer, the recognition rate might not increase much, but the structure of the hidden layer will become more complicated, and the time consumption of the algorithm will increase. The results are shown in Figure 6. Therefore, in this paper we have set hidden node number as H=300.

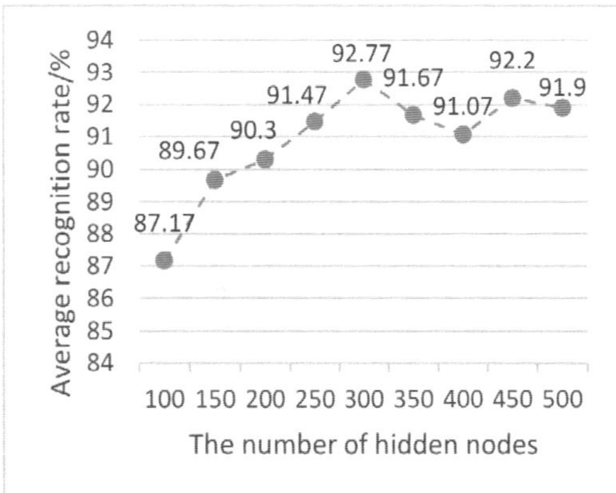

Figure 6. Relationship between average recognition rate and hidden layer nodes.

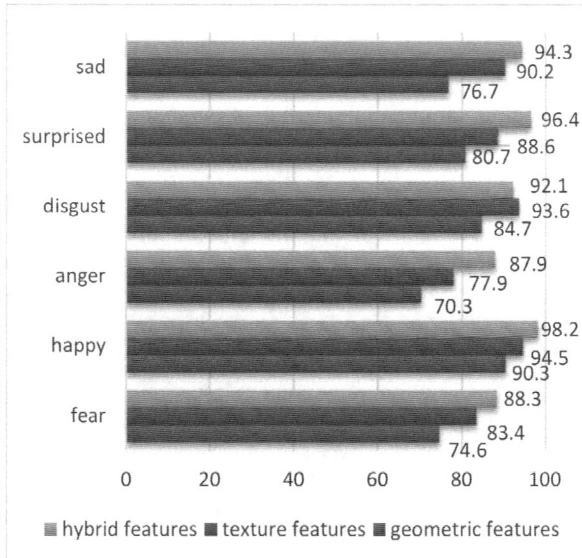

Figure 7. Experiment results using different features based on PSO-ELM (%).

Under the PSO-ELM model with hidden layer node number H=300, we performed the expression recognition experiments by extracting the hybrid features. At the same time, the experiments using other feature extraction methods were also carried out. The recognition results are shown in Table 1. Comparing the three groups of experiments: geometric features, texture features, and hybrid features, the results are shown in Table 2. And Figure7 intuitively reflects the different results of different feature vectors in expression classification.

Table 1. Test accuracy confusion matrix of our PSO-ELM model (%).

	Fear	Happy	Anger	Disgust	Surprised	Sad
Fear	88.3	1.0	2.0	2.0	4.3	2.4
Happy	0.0	98.2	0.0	0.0	1.4	0.4
Anger	2.5	1.1	87.9	3.2	3.0	2.3
Disgust	1.3	0.0	3.2	92.1	1.8	1.4
Surprise	0.3	1.2	1.7	0.4	96.4	0.0
Sad	2.3	0.4	1.5	1.5	0.0	94.3

Table 2. Experiment results using different features based on PSO-ELM (%).

	Fear	Happy	Anger	Disgust	Surprised	Sad	Average
Geometric features	74.6	90.3	70.3	84.7	80.7	76.7	79.5
Texture features	83.4	94.5	77.9	93.6	88.6	90.2	88.0
Hybrid features	88.3	98.2	87.9	92.1	96.4	94.3	92.8

According to Table 1 and Table 2, we can find that only using geometric features, the highest expression recognition rate is 84.7%, and the average recognition rate is 79.5%; only using of texture features, the highest expression recognition rate is 94.5%, and the average recognition rate is 88.0%. However, based on hybrid features, the highest expression recognition rate is up to 98.2%, and the average recognition rate is 92.8%, which is better than the other two. The experimental results according to Table 2 show that the recognition effect of the hybrid features is better than that using only geometric features or texture features.

5. Conclusion

This paper proposes a robust dynamic expression recognition method based on the hybrid features of geometric features and texture features and the PSO-ELM model. In this paper, hybrid features are obtained by fusing the geometric information and the texture information of the feature points. Six PSO-ELM classifiers are trained using six different expression sequences. The dynamic expression sequence can provide more expression information and can eliminate the influence of appearance differences. Through the trained six PSO-ELM models, comparing the similarity between the given sequence and the actual sequence, the expression of the given image sequence can be determined. The method of this paper is tested on CK+ database. The experimental results show that the proposed method has higher average recognition rate of 92.8%, and its highest expression recognition rate is up to 98.2%.

6. References

[1] Argyriou A, Evgeniou T and Pontil M 2008 *Machine Learning* **73(3)** pp 243-272

[2] Abate A F, Nappi M, Riccio D and Gabriele S 2007 *Pattern Recognition Letters* **28(14)** pp 1885-1906

[3] Cootes T F, Taylor C J, Cooper D H and Graham J 1995 *Computer Vision and Image Understanding* **61(1)** pp 38-59

[4] Jizheng Y, Xia M and Yuli X 2013 *Journal of Electronics & Information Technology* **35(10)** pp 2403-2410

[5] Ren S, Cao X, Wei Y and Sun J 2014 *Proc. Int. Conf. Computer Vision and Pattern Recognition (Columbus)* vol 25 (USA: Institute of Electrical and Electronics Engineers) pp 1685-1692

[6] Yuan L, Ye C and Yan, W 2014 *Semiconductor Optoelectronics* pp 330-333

[7] Ying Z, Tang J, Li J and Zhang Y 2008 *Acta Electronica Sinica* **36(4)** pp 725-730

[8] Tang J, Zhang Y and Ying Z 2008 *Computer Engineering and Applications* **44(8)** pp 220-222

[9] Saudagare P V and Chaudhari D S 2012 *International Journal of Soft Computing and Engineering* **2 (1)** pp 238-241

[10] Rumelhart D E and Hinton G E 1986 *Nature* **323(6088)** pp 533-536

[11] Kanade T, Cohn J F, Tian and Y L 2000 *Proc. Int. Conf. Automatic Face and Gesture Recognition* (Los Alamitos) pp 46-53

[12] Wang Z and Li S 2011 *Advanced Materials Research* **271-273** pp 165-170

[13] Viola P and Jones M J 2004 *International Journal of Computer Vision* **57(2)** pp 137-154

[14] Zhang K, Zhang Z, Li Z and Qiao Y 2016 *IEEE Signal Processing Letters* **23(10)** pp 1499-1503

[15] Huang G, Zhu Q and Siew C 2004 *Proc. Int. Joint Conf. Neural Networks* (Budapest) vol 2 (USA: Institute of Electrical and Electronics Engineers) pp 985–990

[16] Lucey P, Cohn J F, Kanade T and Saragih J 2010 *Proc. Int. Conf. Computer Vision and Pattern Recognition Workshops* (Los Alamitos) pp 94-1017

IOP Conf. Series: Materials Science and Engineering **719** (2020) 012081 doi:10.1088/1757-899X/719/1/012081

Research and Design of Integrated System of Headphone Amplifier Based on Bluetooth Technology

Hong Wang, Lina Wang and Xuehao Zhang

Shandong Xiehe University, Jinan, China
Email:82087341@qq.com

Abstract. Traditional earphones and speakers need to use audio cables to connect mobile devices and sounding units when using audio files of mobile devices such as mobile phones as audio input sources. This paper proposes a solution to this problem. This paper studies and designs a power headphone amplifier integrated system based on Bluetooth technology with STM32F407 as the main control core. The system uses CSR8670 Bluetooth audio SoC chip to realize Bluetooth audio transmission, uses TPA6112 headphone audio amplifier to realize amp function, uses HT8696 class D power amplifier to realize speaker power amplifier function, uses 3.2 inch resistive touch screen to realize human-computer interaction function, and the system introduces STemWin graphics library Implement human-machine UI interface. The system is more practical in the independent output of the amp and the speaker power amplifier circuit, the control method is more humanized, and the human-machine experience is better, which plays a driving role for the popularization of the smart home.

1. Introduction

Audio equipment is one of the important carriers for spreading music. There are many problems in the use of traditional audio equipment, for example, the wiring of equipment such as headphones and speakers is cumbersome, the degree of intelligence is low, the sound source is difficult to switch easily, and the sound source needs to be connected through the wire, and the distance is limited by the length of the connection line. Now, phone has a TRS terminal design, but the audio output is more cumbersome and not friendly to users with high sound quality requirements.[1-2]In order to solve the problems existing in traditional audio equipment and improve the intelligent portability of audio equipment, using Bluetooth headphone amplifier as an audio device can bring intelligence and portability to users.

2. Research and Design Analysis

The completion of the design is inseparable from the hardware module design and overall system design.

2.1. Hardware Module Design

2.1.1.Master chip. The master chip is the main control core of the Bluetooth Headphone Amplifier, which realizes the control of the touch screen. For the later expansion of the system, so the selected master chip is STM32F407VET6.[3-5]

2.1.2.Touch screen. The system adopts 8080 interface, the resolution is 320×240, and the resistive touch TFT LCD screen. The FSMC interface of the MCU uses the LCD screen as an external RAM to drive the LCD screen by writing data.

2.1.3.Bluetooth Audio Transmission. The Bluetooth audio transmission chip is the main function implementation chip of the Bluetooth Headphone Amplifier, and the CSR8670 is selected as the firmware for outputting the analog signal. This chip is a Bluetooth audio SoC with its own powerful DSP processor. The system performs secondary function development according to the IDE development environment provided by CSR, or burns different firmware provided by the official to realize different outputs such as I2S, SPDIF and analog signals.CSR8670 directly outputs two differential analog audio signals for left and right channels after decoding. Differential analog signal has strong anti-interference ability and simple signal processing, but the output signal is weak, and the load capacity is poor. It is difficult to directly drive the earphone. Therefore, the TPA6112 is connected as the headphone driver, and then the classifier amplifiers such as HT8696 and TPA3116D2 are used to realize the power amplifier function.[6-8]

2.2. Overall System Design
The system adopts STM32F407VET6 as the main control core, adopts 3.2-inch TFT touch LCD screen to realize human-computer interaction, adopts CSR8670 Bluetooth audio transmission module to realize Bluetooth audio transmission and decoding, and adopts NXP NTAG216 NFC tag to realize Bluetooth shortcut matching with front-end playback device. The system uses the TPA6112 headphone audio amplifier to convert the analog differential audio signal output by the CSR8670 into a balanced signal and power amplification to drive the headphones, uses the HT8696 Class D power amplifier to implement the power amplifier function to directly drive the speaker. The overall design of the Bluetooth amp amplifier(refer with figure 1).

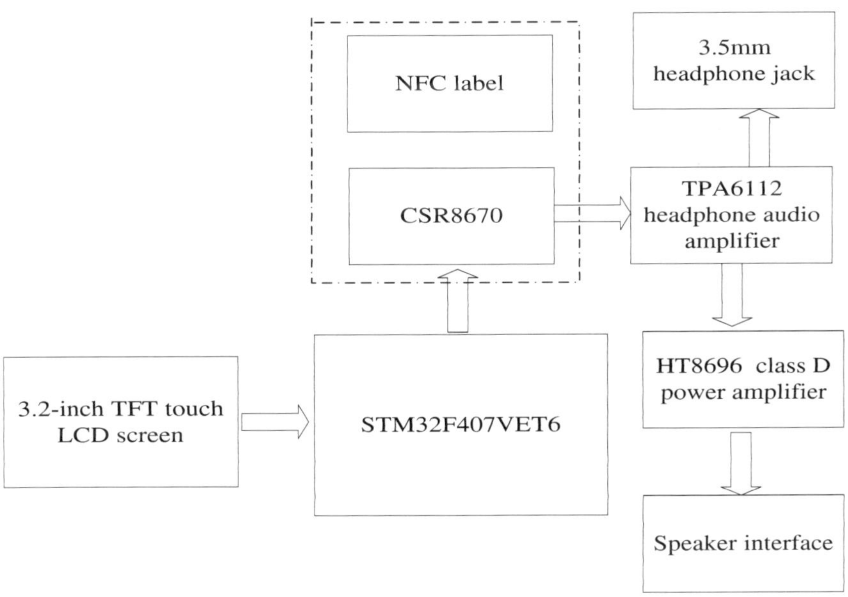

Figure 1. System overall design.

3. System Software Design

3.1. Graphical Interface Design
The system uses the SemWin graphics library of version 5.40 and uses emWin GUIBuilder V5.40a to build the framework of the display interface. The content displayed in the window is the content displayed

on the LCD screen, and the control of different functions is realized by multiple button controls. After the layout is set, save it and you can generate a C language file for the corresponding interface. Controls can call various events, such as song switching, sound adjustment. CSR8670 detects the change of the IO port level caused by the click event of the Button, thus switching the working mode.

STemWin Graphics Library Porting

Graphic display interface adopts the porting function of STemWin graphics library. The code to be written is GUI_X_UCOSII.c, GUI_X_Touch_Analog.c and LCDConf_FlexColor_Template.c. The specific implementation of GUI_X_InitOS, GUI_X_Lock, GUI_X_Unlock, GUI_X_GetTaskId, GUI_X_WaitEvent, GUI_X_SignalEvent, CheckInit, GUI_X_Init, GUI_X_GetKey, GUI_X_WaitKey, GUI_X_StoreKey functions is required in GUI_X_UCOSII.c.

GUI_X_Lock is the lock function and GUI_X_Unlock is the unlock function. They all call the lock and unlock function of the uCOS III operating system. The specific implementation code is as follows:

```
void GUI_X_Lock(void){
    INT8U err;
    OSSemPend(DispSem,0,&err);        // Request Semaphore }
void GUI_X_Unlock(void){
    OSSemPost(DispSem);            // Send Semaphore }
```

GUI_X_GetTaskId is to get the task priority function, the function returns the OSTCBCur->OSTCBPrio task priority, the specific code is as follows:

```
U32 GUI_X_GetTaskId(void){
    return ((U32)(OSTCBCur->OSTCBPrio)); // Get the task priority, which is the task ID
}
// The GUI_X_WaitEvent function is used to wait for an event to occur, using the request message mailbox
void GUI_X_WaitEvent(void){
    INT8U err;
    (void)OSMboxPend(EventMbox,0,&err); //Request message mailbox
}
// GUI_X_SignalEvent is implemented for signal events by sending a message mailbox
void GUI_X_SignalEvent(void){
    (void)OSMboxPost(EventMbox,(void*)1);  //Send a message mailbox
}
```

Touch screen driver function framework included in GUI_X_Touch_Analog.c file. The coordinate information of the X and Y axes of the touch position is read by implementing the GUI_TOUCH_X_MeasureX, GUI_TOUCH_X_MeasureY function and calling the OTT2001A_RD_Reg interface function in the emWin function.

The LCDConf_FlexColor_Template.c file includes the LCD display driver functions LCD_X_Config and LCD_X_DisplayDriver. The LCD display driver function is used to write graphic data to the LCD screen. LCD_X_Config is used to create display driver devices, set color conversion programs and display sizes. LCD_X_DisplayDriver mainly implements LCD initialization.

3.2. MCU Main Block Diagram

The main program design is realized by the uCOS III operating system and STemWin graphics library. The main program block diagram of the system(refer with figure 2).

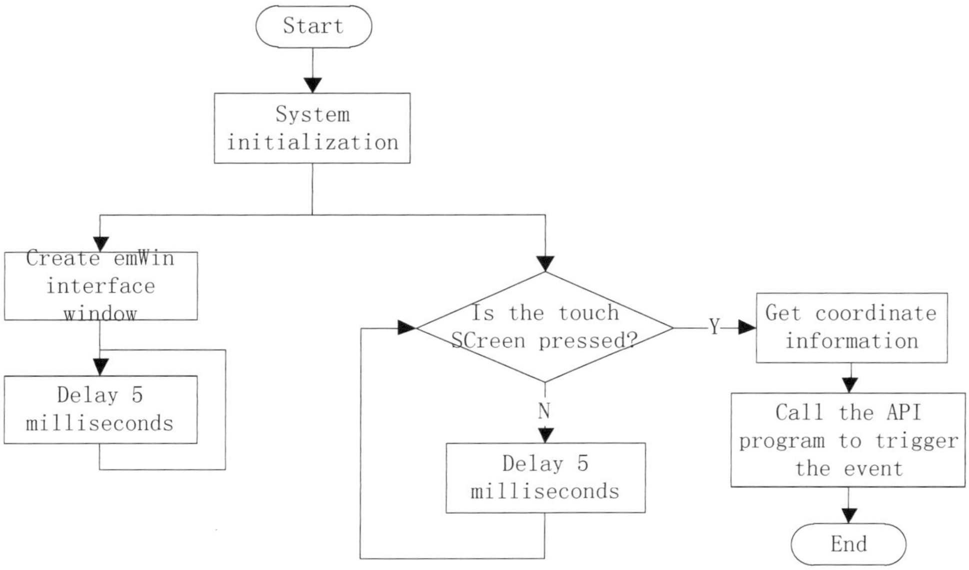

Figure 2. System main program flow chart.

3.3. NFC Fast Pairing Implementation of Bluetooth Headphone Amplifier

Enable fast pairing of Bluetooth headphone amplifier devices by NFC tag, which is implemented by installing TagWritter software on NFC-enabled mobile phones.

Creating a Bluetooth tag and writing the MAC address and device name of the Bluetooth headphone amplifier device to the tag, and selecting the category as a Bluetooth headset. NFC tags are placed close to the phone to implement NFC information write tags. Scaning this tag to pair with a Bluetooth headphone amplifier device to speed up pairing.[9-10] In view of the advantages of long life, strong anti-interference ability, and large storage capacity, NFC tags of NTAG216 are used for pairing.

After a series of debugging, the system can run normally. The physical operation effect(refer with Fig 3).

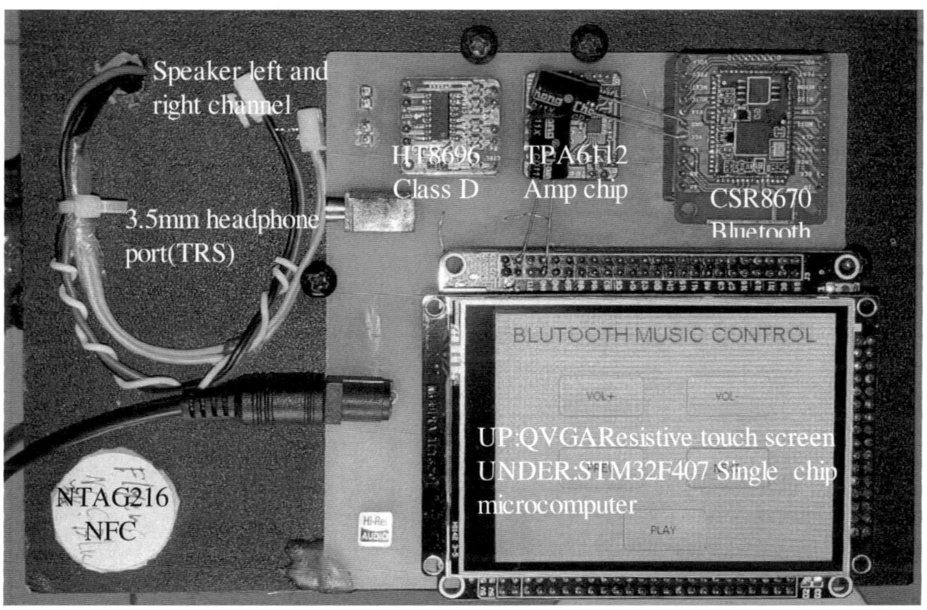

Figure 3. Physical operation diagram

4. Summary

The main research content of this paper is the display of touch interface on uCOS III RTOS system, the design of Bluetooth 5.0 audio module based on CSR8670, and how to call STemWin graphics library to realize the research of human-computer interaction.Research has enabled mobile devices to tap NFC tags for fast Bluetooth binding to connect to audio devices to play music. The touch screen controls the volume, switches songs, and amplifies the output power. But there are also shortcomings， For example: selection of filter power supply, research on audio signal transmission of single crystal copper or mixed wire, adoption of LDAC Bluetooth audio protocol, etc.

5. Acknowledgement

This research was financially supported by Shandong Xiehe University College-level scientific research project: Prediction and Diagnosis of Crop Pests and Diseases Based on Machine Learning - A Case Study of Southwestern Shandong Province(XHXY201830).

6. References

[1] Chaopu Shu 2016 *Foreign* LeTV mobile phone accelerated running: IP + Eco 3.0 *Investment in China* **19** 70-71
[2] Quan Chen 2018 Design and fabrication of power amplifier based on Bluetooth technology *Electronic Production* **9** 111-5
[3] Guancong Ding 2018 Class D audio power amplifier design *Shandong Industrial Technology* **4** 155-5
[4] Hao Zhen,Qiuyan Wang and Yanfei Liu 2017 Mobile device application low THD+N high PSRRD class power amplifier *Solid State Electronics Research and Progress* **2** 134-9
[5] Xuefang,Chen,Jishui Pan and Yanfei Liu 2018 Audio player design with Bluetooth *Journal of Dongguan University of Technology* **1** 101-3
[6] Min Qiu 2017 Millet Bluetooth Receiver Lets your speaker headset change to Bluetooth device *Computer and Network* **17** 33
[7] Zhibin Tang. Wireless digital hearing aid system based on low power consumption Bluetooth[D]. Xiamen:Jimei University,2017(**5**):13-14
[8] Cheng Wang 2017 Bluetooth NFC reader design *Electronic Production* **21** 21-22.
[9] Sriramulu Bojjagani and V.N. Sastry 2019 A Secure End-to-End Proximity NFC-based Mobile Payment Protocol *Computer Standards* 6-11
[10] NXP B.V 2018 Patent Issued for Method And Apparatus For Using An Over Sampling Architecture To Achieve A Software Defined Radio For NFC Integrated Circuits(USPTO 10,127,487) *Electronics Business Journal* 23-34.

Optimization of Waste Smoke Recovery Scheduling Strategy Based on Multi AGV

Shunkai Sun[1], Jie Li[1], Jiule Zhu[2], Haitao Chen[1], Qi Xu[1], Jingjie Liu[1], Haobo Cui[1], Guorui Li[1] and Wangdi Hu[1]

1 Ningbo Cigarette Factory, China Tobacco Zhejiang Industrial Co., Ltd., Ningbo 315504, China;
2 College of computer and software, Nanjing University of Information Science and Technology, Nanjing 210044, Jiangsu, China;
Email: shunkaisun@126.com

Abstract. AGV is an important material handling equipment for flexible, lean and intelligent production of smart factory. Because of the high cost of AGV car and the congestion of vehicles in the system, this chapter studies the waste smoke recovery path planning and material handling task scheduling strategy of multi AGV system to improve the waste smoke recovery efficiency of AGV.

1. Problem Description and Modelling of Multi AGV Waste Smoke Recovery

In the smart factory, materials are transported according to the waste smoke recovery demand of each work area, and the AGV scheduling strategy meets the production of each production area. The planning path of AGV raw and auxiliary materials handling in smart factory needs to consider multiple interference factors. When building the handling model, it needs to consider the conflict of small workshops with multiple AGVs, the connection between AGV tasks, the AGV transportation distance and the interference between AGV and production environment. The handling strategy and path planning of AGV are not in line with the production situation, which may lead to material interruption in the production area, thus affecting the production interruption of the plant. Considering the complexity of multi AGV path planning in the plant workshop, the following assumptions are made:

(1) The initial position of AGV can be in any area of the smart factory, without considering factors such as avoidance, straight-line acceleration and deceleration, turning and deceleration, and walking at a constant speed.

(2) The number and working environment of AGV are known in multi AGV system, and the reaction time of AGV stop and start and the problem of AGV failure are ignored.

(3) A single AGV is only allowed to perform one handling task. AGV can handle multiple production materials, multiple handling modes and multiple handling businesses. This paper describes the planning of AGV path for the key logistics equipment in the smart factory workshop: V vehicles are equipped in the smart factory workshop to carry AGV by fork. There are n production areas in the handling task flow to complete the material transportation. A single AGV system produces P material transportation tasks. After the material loading, AGV carries the material from one production area to another. Build a mathematical model that AGV can perform the material handling task in the minimum cycle, as shown in formula (1) and formula (2):

$$f_{ij} = \min \max_{1 \le i \le V} \left\{ \sum_{j=1}^{P} X_{ij} T_{ij} \right\} \tag{1}$$

$$X_{ij} = \begin{cases} 1 & \text{The i-th AGV completes the j-th task} \\ 0 & \text{Other} \end{cases} \tag{2}$$

Equations (1) and (2) are constraints:

$$\sum_{i=1}^{V} X_{ij} = 1 \quad \forall j \tag{3}$$

$$\sum_{j=1}^{P} \sum_{i}^{V} X_{ij} = P \tag{4}$$

Where: f_{ij} is the time for completing all the handling tasks; T_{ij} is the time required for the i-th AGV to complete the j-th task; X_{ij} is the decision coefficient for the i-th AGV to perform the j-th task; formula (1) indicates the minimum total AGV operation time for performing all the material handling tasks; formula (3) indicates that a single AGV performs a single material handling task; formula (4) indicates that multiple AGVs perform all the tasks.

2. Waste Smoke Recovery Path Time Window

The reliability and effectiveness of the algorithm structure are reflected in the time window algorithm, which is very important for the accuracy of the access path of AGV. Therefore, we will discuss different definitions of time windows for various path situations.

A shown in Figure 1, the body length of AGV is set as s, and the uniform running speed is a. the handling tasks of a single AGV in a certain road section are divided into three types:

(1) AGV in straight line section

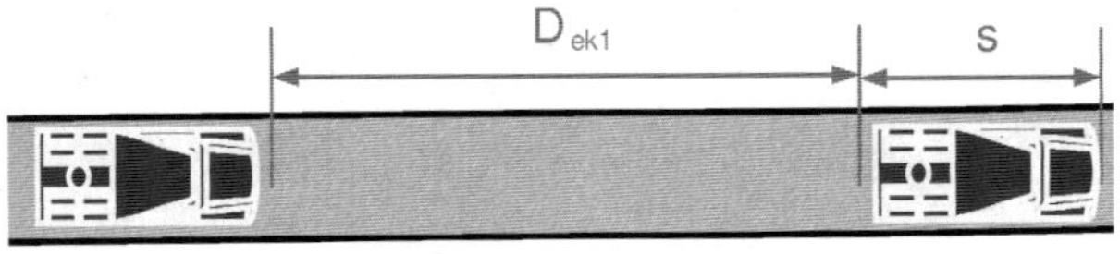

Figure 1. Straight Line Section

The calculation formula of AGV travel time in straight line section is as follows:

$$t_{straight} = \frac{D_{e_{k1}} + s}{a} \tag{5}$$

(2) AGV in curve section

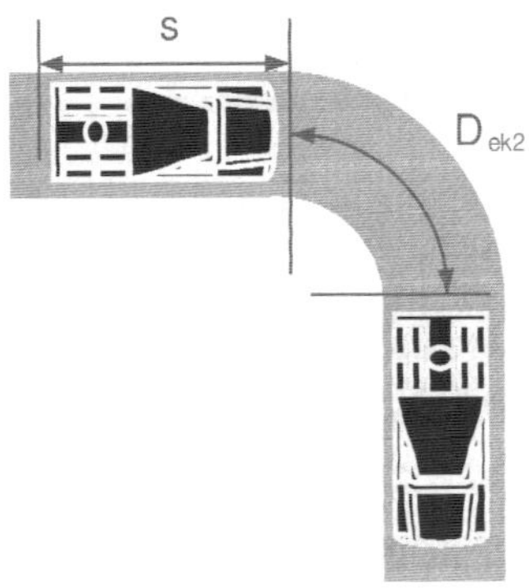

Figure 2. Curve Road Section

As shown in Figure 2, the calculation formula of AGV travel time in curve section is:

$$t_{curve} = \frac{\frac{\pi D_{e_{k1}}}{2} + s}{a} = \frac{\pi D_{e_{k1}} + 2s}{2a} \tag{6}$$

(3) AGV in mixed section

There are two AGVs competing for resources in the same road section. The relevant time window of the road section can be determined by confirming the time point they enter. As shown in Figure 3, AGV_2 on the road e_{k3} and AGV_1 on the road e_{k1} compete for the road e_{k4} section. 1#AGV passes the intersection first. Suppose that the time t_{2k4}^{in} point for entering the road section is e_{k4}, the time t_{2k3}^{out} point for 2#AGV to leave the road section is e_{k3}, and the time t_{1k4}^{out} point for 1#AGV to leave the road section is e_{k4}, then:

$$t_{2k4}^{in} = \max\left\{t_{1k4}^{out}, t_{2k3}^{out}\right\} \tag{7}$$

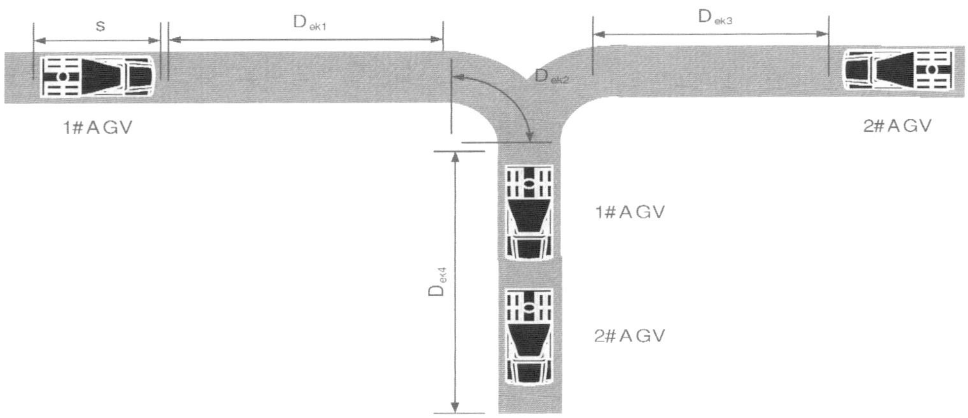

Figure 3. Two AGVs Compete for the Same Section

3. Optimization of Particle Swarm Optimization Algorithm for AGV Scheduling

With $P_i=(p_{i1},p_{i2},\cdots,p_{iD})$, $i=1,2,\cdots,n$, the i-th particle is searched for the current optimal spatial position, and the individual particle p_{best} of the basic particle swarm algorithm is searched until now. The optimal position is represented by $P_g= (p_{g1}, p_{g2}\cdots, p_{gD})$, which is the g_{best} global extremum, and each particle updates the spatial position and velocity in real time with reference to P_g and P_i.

$$\mathrm{v}_{id}(k+1)=w\mathrm{v}_{id}(k)+c_1r_1(p_{id}(k)-x_{id}(k))+c_2r_2(p_{gd}(k)-x_{gd}(k)) \tag{8}$$

$$x_{id}(k+1) = x_{id}(k) + \mathrm{v}_{id}(k+1) \tag{9}$$

$$i = 1,2,\cdots,\mathrm{n} \quad d = 1,2,\cdots,\mathrm{D}$$

In the formula, $w\geq 0$ is inertia. r_1 and r_2 are arbitrary numbers between [0, 1], The normal number learning factors c_1 and c_2. It is difficult to find the local optimal solution, thus losing the opportunity to find the global optimal solution. If the whole particle swarm will converge to P_g quickly, then P_g is a local optimal value.

This paper improves the original particle swarm optimization algorithm:

(1) V_{id} adjustment

It is necessary to adjust the particle swarm of the individual extreme P_{id}:

$$P_r = (p_{r1}, p_{r2},\cdots, p_{rD}) \qquad r = 1,2,\cdots,\mathrm{n} \tag{10}$$

Among them $P_{rj} = (p_{1j} + p_{2j} + \cdots + p_{nj}) / n \quad j = 1, 2, \cdots, D$

The updated particle speed is updated to:

$$v_{id}(k+1) = wv_{id}(k) + c_1 r_1 (p_{rd}(k) - x_{id}(k)) + c_2 r_2 (p_{gd}(k) - x_{id}(k)) \tag{11}$$

(2) P_{id} adjustment

In the evolution of the species of the t-th generation, a random number r in the range [0, 1] is generated, and R_i is obtained according to $R_i = t/G_{max}$. When $r > R_i$, randomly generate a single particle, mark the optimal spatial position of this particle as P_{rnd}, replace P_{gd} in equation (11), adjust the particle velocity in real time according to formula (12); Otherwise, according to formula (11)) to update.

$$v_{id}(k+1) = wv_{id}(k) + c_1 r_1 (p_{rd}(k) - x_{id}(k)) + c_2 r_2 (p_{rnd}(k) - x_{id}(k)) \tag{12}$$

In the formula: P_{rnd} is randomly generated from P_i of non-local particles and is not P_g, G_{max} is the largest algebra of evolution, and t is the current evolutionary algebra.

4. Implementation of AGV Scheduling Particle Swarm Optimization Algorithm

The scheduling of the tasks received by each AGV and the scheduling of each AGV handling task are two key points for the AGV system to optimize the handling tasks. The AGV scheduling strategy is a complex process of multiple interference factors. In this paper, the particle swarm algorithm of AGV scheduling strategy is designed by multi-bit spatial representation of vector particles. The flow is shown in Figure 4.

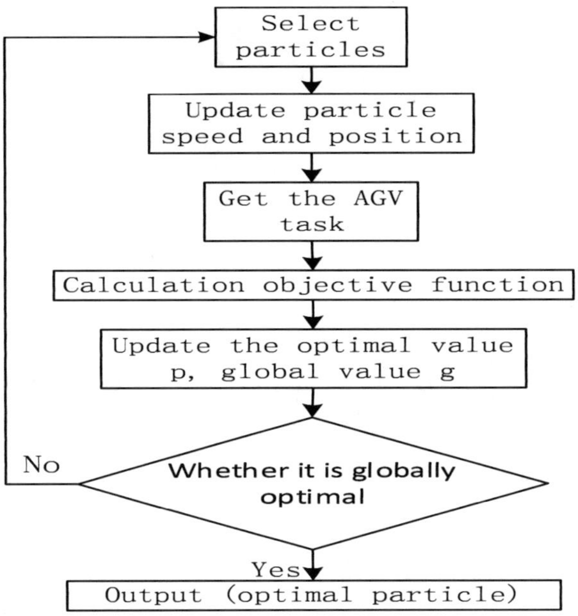

Figure 4. Implementing Flow Chart of Improved Particle Swarm Optimization Algorithm

4.1. Encoding and Decoding of Particle Swarm Optimization

The handling tasks of AGV in this paper include assignment and ordering problems. Therefore, the paper uses 3D vector representation when completing particle swarm coding. See Table 5.1:

Table 1. Representation of the First Particle

Task Number	1	2	3	4	5	…	Z
Task Assignment	z_{i1}	z_{i1}	z_{i1}	z_{i1}	z_{i1}	…	z_{iN}
Task Order	d_{i1}	d_{i1}	d_{i1}	d_{i1}	d_{i1}	…	d_{iN}

In the Table, D is the number of particle groups, $i\in[1,M]$, M is the number of AGVs in the smart factory site, Z is the number of material handling tasks; the first dimension is the AGV handling task, Z is the number of handling tasks; the second dimension It is the sorting of material handling tasks. The size of the $D_{i(1\text{-}Z)}$ value is unlocked, and the tasks belonging to the same path are queued to complete the whole column of the particle vector. The third dimension is automatically received by the upper system of the multi-AGV material handling task. The particle group is given upper and lower limits: $Z_{i(1\text{-}N)}\in[1,M+1]$, and the complete $INT(z_{ij})$ is obtained in the $[1,M]$ constant, which corresponds to the ordering of the AGV. Therefore, when the algorithm is initialized, the range of values of each dimension of the particle vector is known, and the range of feasible solutions to some extent has been given in the particle swarm optimization algorithm.

4.2. Constraint Conditions and Objective Function Settings of Particle Swarm Optimization

The AGV system of this paper determines that the total number of tasks performed by the number of tasks is at least the objective function. The handling cycle of each task z consists of two parts. Each AGV takes a fixed delivery period, and the AGV assignment task is more uniform, so it arrives from the current position. The time of the starting station and the time at which the task start station reaches the target station determine the period of the AGV handling task.

The start is based on the single-particle coding. In each group of the execution task queue, the displacement and velocity calculations are completed, and the obtained particle space position is analyzed. The first dimension is the execution task number of the AGV application; the second dimension is the task order obtained by decoding; the third dimension is the acquisition task number and the corresponding AGV. The total running time of all AGV tasks is the total operating cycle of the AGV system. Finally, the total time of all AGV operations is compared, and the shortest solution of the AGV system is selected, and the AGV time is relatively average as the objective function of the iteration.

4.3. Analysis of AGV Waste Smoke Recycling Business

Assume that the AGV of the smart factory waste smoke recovery system has 12 sets of AGVs of the same specification. Selecting a certain cycle, the AGV system tasks include 10 empty waste pipe recycling tasks, 10 machine waste recycling tasks, and AGV system tasks total 20 the task of recycling AGV waste smoke at a certain time is shown in Table 2:

Table 2. Task Data Table for AGV Jobs at a Time

No.	Start address	Aims address	Issued time	Start time	End Time	Task status
1	YPP08	8345	17:27:00	17:29:00	17:37:03	End
2	YPP07	8345	17:27:40	17:29:35	17:37:16	End
3	ZBE04	8345	17:32:30	17:37:21	17:42:19	End
4	8314	YPP07	17:39:50	17:41:41		Doing
5	8323	ZBE04	17:45:10	17:46:54		Doing
6	8334	YPP08	17:46:40	17:48:59		Doing
7	ZBE09	8345	17:52:00	17:57:02		Doing
8	8333	ZBE09	18:04:30	18:06:11		Doing
9	ZJG04	8345	18:06:45	18:08:41		Doing
10	8323	YPP02	18:09:45	18:12:20		Doing
11	ZBG03	8345	18:14:30	18:15:34		Doing
12	8333	ZBE11	18:20:00	18:23:39		Doing
13	8314	ZBG08	18:20:40	18:23:20		Doing
14	ZJG04	8345	18:21:15	18:23:19		Doing
15	ZJE06	8345	18:26:15	18:27:25		New
16	8333	YPP11	18:30:10	18:31:19		New
17	8323	YPP03	18:31:40	18:32:00		New
18	ZBF02	8345	18:32:40			New
19	8334	ZJF07	18:36:10			New
20	ZBG08	8345	19:40:10			New

Set the particle swarm initial condition: the operation precision of 0.5, the population number of particles D=40, the total number of AGVs M=12, the total number of task schedules Z=22, the maximum number of iterations G_{max}=200, the inertia weight w=0.6, the learning factor c_1=2, c_2=2, particle coding uses a three-dimensional vector group.

5. Conclusions

The total waste smoke recovery task performs an improved particle swarm optimization algorithm with a minimum running time as an objective function. Under the condition of ensuring the time point, the queued waiting task is dynamically pre-allocated to the AGV of the normal execution task, and the algorithm does not have to be in the task transformation. After recalculation, the algorithm cycle is compressed. After the system releases a new task or completes the task, it determines the location and quantity of the material handling task, and assigns the material handling task again based on the improved particle algorithm.

6. References

[1] Hu, Guiyan and Sun, Weihua. A study of scheduling rules of flexible manufacturing systems: a simulation approach [J]. Advanced Materials Research, 2012: 627-641.

[2] Cheded L, Mulla. Design and operational issues in AGV-served manufacturing systems [C]. IEEE Trans, 2007.

[3] Yin, Y. Multicriteria meta-heuristics for AGV dispatching control based on computational intelligence [J]. International Journal of Advanced Manufacturing Technology. 2006:874-885.

[4] J.M.Apple. Dispatching multi-load AGVs in highly automated seaport container terminals [C].IEEE Control Systems Magazine. 2007.

[5] Obata M, Nishida T. Spring-based negotiation for conflict resolution in AGV scheduling [C].IEEJ Transactions on Electronics, Information and Systems, 2006.

[6] Fong T, Nourbakhsh I. Integrated scheduling of handling equipment at automated container terminals [J]. Robotics and Autonomous Systems, 2003: 1104-1130.

IOP Conf. Series: Materials Science and Engineering **719** (2020) 012082 doi:10.1088/1757-899X/719/1/012082

[7] Cui Y J, Ge S. Dynamic dispatching algorithm for scheduling machines and automated guided vehicles in a flexible manufacturing system [C]. IEEE Transaction on Robotics and Automation, 2013.
[8] Obata M, Nishida T. Spring-based negotiation for conflict resolution in AGV scheduling [C].IEEJ Transactions on Electronics, Information and Systems,2006.
[9] Fong T, Nourbakhsh I. Integrated scheduling of handling equipment at automated container terminals [J]. Robotics and Autonomous Systems, 2003: 1104-1130.

AUTHOR INDEX